Regulated Grammars and Automata

Alexander Meduna • Petr Zemek

Regulated Grammars and Automata

Alexander Meduna
Department of Information Systems
Faculty of Information Technology
Brno University of Technology
Brno, the Czech Republic

Petr Zemek
Department of Information Systems
Faculty of Information Technology
Brno University of Technology
Brno, the Czech Republic

ISBN 978-1-4939-4316-6 ISBN 978-1-4939-0369-6 (eBook)
DOI 10.1007/978-1-4939-0369-6
Springer New York Heidelberg Dordrecht London

Printed on acid-free paper

Springer is part of Springer Science+Business Media (www.springer.com)

for Daniela and Ivana

Preface

Motivation and Subject

Language processors have become an inseparable part of our daily life. For instance, all the sophisticated modern means of communication, such as Internet with its numerous information processing tools, are based upon them to some extent, and indisputably, literally billions of people use these means on a daily basis. It thus comes as no surprise that the scientific development and study of languages and their processors fulfill a more important role today than ever before. Naturally, we expect that this study produces concepts and results that are as reliable as possible. As a result, we tend to base this study upon mathematics as a systematized body of unshakable knowledge obtained by exact and infallible reasoning. In this respect, we pay our principal attention to *formal language theory* as a branch of mathematics that formalizes languages and devices that define them strictly rigorously.

This theory defines languages mathematically as sets of sequences consisting of symbols. This definition encompasses almost all languages as they are commonly understood. Indeed, natural languages, such as English, are included in this definition. Of course, all artificial languages introduced by various scientific disciplines can be viewed as formal languages as well; perhaps most illustratively, every programming language represents a formal language in terms of this definition. Consequently, formal language theory is important to all the scientific areas that make use of these languages to a certain extent.

The strictly mathematical approach to languages necessitates introducing *formal language models* that define them, and formal language theory has introduced a great variety of them over its history. Most of them are based upon rules by which they repeatedly rewrite sequences of symbols, called strings. Despite their diversity, they can be classified into two basic categories—generative and recognition language models. Generative models, better known as *grammars*, define strings of their language and so their rewriting process generates them from a special start symbol. On the other hand, recognition models, better known as *automata*, define strings of their language by rewriting process that starts from these strings and ends in a special set of strings, usually called *final configurations*.

Like any branch of mathematics, formal language theory has defined its language models generally. Unfortunately, from a practical viewpoint, this generality actually means that the models work in a completely non-deterministic way, and as such, they are hardly implementable and, therefore, applicable in practice. Being fully aware of this pragmatic difficulty, formal language theory has introduced fully deterministic versions of these models; sadly, their application-oriented perspectives are also doubtful. First and foremost, in an ever-changing environment in which real language processors work, it is utterly naive, if not absurd, that these deterministic versions might adequately reflect and simulate real language processors applied in such pragmatically oriented areas as various engineering techniques for language analysis. Second, in many case, this determinism decreases the power of their general counterparts—another highly undesirable feature of this strict determinism. Considering all these difficulties, formal language theory has introduced yet another version of language models, generally referred to as *regulated language models*, which formalize real language processors perhaps most adequately. In essence, these models are based upon their general versions extended by an additional mathematical mechanism that prescribes the use of rules during the generation of their languages. From a practical viewpoint, an important advantage of these models consists in controlling their language-defining process and, therefore, operating in a more deterministic way than general models, which perform their derivations in a quite unregulated way. Perhaps even more significantly, the regulated versions of language models are stronger than their unregulated versions. Considering these advantages, it comes as no surprise that formal language theory has paid an incredibly high attention to *regulated grammars and automata*, which represent the principal subject of the present book.

Purpose

Over the past quarter century, literally hundreds of studies were written about regulated grammars, and their investigation represents an exciting trend within formal language theory. Although this investigation has introduced a number of new regulated grammatical concepts and achieved many remarkable results, all these concepts and results are scattered in various conference and journal papers. The principal *theoretical purpose* of the present book is to select crucially important concepts of this kind and summarize key results about them in a compact, systematic, and uniform way.

From a more practical viewpoint, as already stated, the developers of current and future language processing technologies need a systematized body of mathematically precise knowledge upon which they can rely and build up their methods and techniques. The *practical purpose* of this book is to provide them with this knowledge.

Focus

The material concerning regulated grammars and automata is so huge that it is literally impossible to cover it completely. Considering the purpose of this book, we restrict our attention to four crucially important topics concerning these grammars and automata—their power, properties, reduction, and convertibility.

As obvious, the *power* of the regulated language models under consideration represents perhaps the most important information about them. Indeed, we always want to know the family of languages that these models define.

A special attention is paid to algorithms that arrange regulated grammars and automata and so they satisfy some prescribed *properties* while the generated languages remain unchanged because many language processors strictly require their satisfaction in practice. From a theoretical viewpoint, these properties frequently simplify proofs demonstrating results about these grammars and automata.

The *reduction* of regulated grammars and automata also represents an important investigation area of this book because their reduced versions define languages in a succinct and easy-to-follow way. As obvious, this reduction simplifies the development of language processing technologies, which then work economically and effectively.

Of course, the same languages can be defined by different language models. We obviously tend to define them by the most appropriate models under given circumstances. Therefore, whenever discussing different types of equally powerful language models, we also study their mutual *convertibility*. More specifically, given a language model of one type, we explain how to convert it to a language model of another equally powerful type and so both the original model and the model produced by this conversion define the same language.

We prove most of the results concerning the topics mentioned above *effectively*. That is, within proofs demonstrating them, we give algorithms that describe how to achieve these results. For instance, we often present conversions between equally powerful models as algorithms, whose correctness is then rigorously verified. In this way, apart from their theoretical value, we actually demonstrate how to implement them.

Organization

The text is divided into nine parts, each of which consists of several chapters. Every part starts with an abstract that summarizes its chapters. Altogether, the book contains twenty-two chapters.

Part I, consisting of Chaps. 1 through 3, gives an introduction to this monograph in order to express all its discussion clearly and, in addition, make it completely self-contained. It places all the coverage of the book into scientific context and reviews important mathematical concepts with a focus on formal language theory.

Part II, consisting of Chaps. 4 and 5, gives the fundamentals of regulated grammars. It distinguishes between context-based regulated grammars and rule-based regulated grammars. First, it gives an extensive and thorough coverage of regulated grammars that generate languages under various context-related restrictions. Then, it studies grammatical regulation underlain by restrictions placed on the use of rules.

Part III, consisting of Chaps. 6 through 9, covers special topics concerning grammatical regulation. First, it studies special cases of context-based regulated grammars. Then, it discusses problems concerning the erasure of symbols in strings generated by regulated grammars. Finally, this part presents an algebraic way of grammatical regulation.

Part IV, consisting of Chaps. 10 through 12, studies parallel versions of regulated grammars. First, it studies generalized parallel versions of context-free grammars, generally referred to as regulated ET0L grammars. Then, it studies how to perform the parallel generation of languages in a uniform way. Finally, it studies algebraically regulated parallel grammars.

Part V, consisting of Chaps. 13 and 14, studies sets of mutually communicating grammars working under regulating restrictions. First, it studies their regulation based upon a simultaneous generation of several strings composed together by some basic operation after the generation is completed. Then, it studies their regulated pure versions, which have only one type of symbols.

Part VI, consisting of Chaps. 15 and 16, presents the fundamentals of regulated automata. First, it studies self-regulating automata. Then, it covers the essentials concerning automata regulated by control languages.

Part VII, consisting of Chaps. 17 and 18, studies modified versions of classical automata closely related to regulated automata—namely, jumping finite automata and deep pushdown automata.

Part VIII, consisting of Chaps. 19 and 20, demonstrates applications of regulated language models. It narrows its attention to regulated grammars rather than automata. First, it describes these applications and their perspectives from a rather general viewpoint. Then, it adds several case studies to show quite specific real-world applications concerning computational linguistics, molecular biology, and compiler writing.

Part IX, consisting of Chaps. 21 and 22, closes the entire book by adding several remarks concerning its coverage. First, it sketches the entire development of regulated grammars and automata. Then, it points out many new investigation trends and long-time open problems. Finally, it briefly summarizes all the material covered in the text.

Approach

This book represents a theoretically oriented treatment of regulated grammars and automata. We introduce all formalisms concerning these grammars with enough rigor to make all results quite clear and valid. Every complicated mathematical passage is preceded by its intuitive explanation so that even the most complex parts of the book are easy to grasp. As most proofs of the achieved results contain many transformations of regulated grammars and automata, the present book also maintains an emphasis on algorithmic approach to regulated grammars and automata under discussion and, thereby, their use in practice. Several worked-out examples illustrate the theoretical notions and their applications.

Use

Primarily, this book is useful to all researchers, ranging from mathematicians through computer scientists up to linguists, who deal with language processors based upon regulated grammars or automata.

Secondarily, the entire book can be used as a text for a two-term course in regulated grammars and automata at a graduate level. The text allows the flexibility needed to select some of the discussed topics and, thereby, use it for a one-term course on this subject.

Tertiarily and finally, serious undergraduate students may find this book useful as an accompanying text for a course that deals with formal languages and their models.

WWW Support

Further backup materials, such as lectures about selected topics covered in the book, are available at

http://www.fit.vutbr.cz/~meduna/books/rga

Brno, the Czech Republic — Alexander Meduna
Brno, the Czech Republic — Petr Zemek

Acknowledgements

This book is based on many papers published by us as well as other authors over the last three decades or so. To some extent, we have also made use of our lecture notes for talks and lectures given at various universities throughout the world. Notes made at the Kyoto Sangyo University in Japan were particularly helpful.

This work was supported by several grants—namely, BUT FIT grant FIT-S-11-2, European Regional Development Fund in the IT4Innovations Centre of Excellence (MŠMT CZ1.1.00/02.0070), research plan CEZ MŠMT MSM0021630528, and Visual Computing Competence Center (TE01010415).

Our thanks go to many colleagues from our home university for fruitful discussions about regulated grammars and automata. We are grateful to Susan Lagerstrom-Fife and Courtney Clark at Springer for their invaluable assistance during the preparation of this book. Finally, we thank our families for their enthusiastic encouragement; most importantly, we deeply appreciate the great patience and constant support of Petr's girlfriend Daniela and Alexander's wife Ivana, to whom this book is dedicated.

Alexander Meduna
Petr Zemek

Contents

Part I Introduction and Terminology

1 Introduction ... 3
References ... 7

2 Mathematical Background ... 9
2.1 Sets and Sequences ... 10
2.2 Relations ... 11
2.3 Functions ... 12
2.4 Graphs ... 12
References ... 13

3 Rudiments of Formal Language Theory ... 15
3.1 Strings and Languages ... 16
3.2 Language Families ... 19
3.3 Grammars ... 20
3.4 Automata ... 31
References ... 35

Part II Regulated Grammars: Fundamentals

4 Context-Based Grammatical Regulation ... 39
4.1 Classical Grammars Viewed as Tight-Context Regulated Grammars ... 40
4.1.1 Normal Forms ... 40
4.1.2 Uniform Rewriting ... 47
4.2 Context-Conditional Grammars ... 56
4.2.1 Definitions ... 56
4.2.2 Generative Power ... 57
4.3 Random Context Grammars ... 63
4.3.1 Definitions and Examples ... 64
4.3.2 Generative Power ... 65

4.4 Generalized Forbidding Grammars 68
4.4.1 Definitions 69
4.4.2 Generative Power and Reduction 69
4.5 Semi-Conditional Grammars 84
4.5.1 Definitions and Examples 84
4.5.2 Generative Power 86
4.6 Simple Semi-Conditional Grammars 88
4.6.1 Definitions and Examples 88
4.6.2 Generative Power and Reduction 89
4.7 Scattered Context Grammars 119
4.7.1 Definitions and Examples 119
4.7.2 Generative Power 123
4.7.3 Normal Forms 124
4.7.4 Reduction 126
4.7.5 LL Scattered Context Grammars 149
References 153

5 Rule-Based Grammatical Regulation 155
5.1 Regular-Controlled Grammars 156
5.1.1 Definitions and Examples 156
5.1.2 Generative Power 160
5.2 Matrix Grammars 160
5.2.1 Definitions and Examples 161
5.2.2 Generative Power 163
5.3 Programmed Grammars 163
5.3.1 Definitions and Examples 163
5.3.2 Generative Power 165
5.3.3 Normal Forms 166
5.3.4 Restricted Non-Determinism 174
5.4 State Grammars 184
5.4.1 Definitions and Examples 184
5.4.2 Generative Power 186
References 187

Part III Regulated Grammars: Special Topics

6 One-Sided Versions of Random Context Grammars 191
6.1 Definitions and Examples 194
6.2 Generative Power 198
6.2.1 One-Sided Random Context Grammars 198
6.2.2 One-Sided Forbidding Grammars 203
6.2.3 One-Sided Permitting Grammars 213

6.3 Normal Forms 215
6.4 Reduction 223
6.4.1 Total Number of Nonterminals 224
6.4.2 Number of Left and Right Random Context Nonterminals 231
6.4.3 Number of Right Random Context Rules 237
6.5 Leftmost Derivations 244
6.5.1 Type-1 Leftmost Derivations 245
6.5.2 Type-2 Leftmost Derivations 248
6.5.3 Type-3 Leftmost Derivations 253
6.6 Generalized One-Sided Forbidding Grammars 256
6.6.1 Definitions and Examples 257
6.6.2 Generative Power 259
6.7 LL One-Sided Random Context Grammars 267
6.7.1 Definitions 268
6.7.2 A Motivational Example 270
6.7.3 Generative Power 271
References 277

7 On Erasing Rules and Their Elimination 281
7.1 Elimination of Erasing Rules from Context-Free Grammars 282
7.1.1 The Standard Algorithm 282
7.1.2 A New Algorithm 284
7.1.3 Can Erasing Rules Be Eliminated from Regulated Grammars? 291
7.2 Workspace Theorems for Regular-Controlled Grammars 293
7.3 Generalized Restricted Erasing in Scattered Context Grammars .. 310
References 328

8 Extension of Languages Resulting from Regulated Grammars 329
8.1 Regular-Controlled Generators 330
8.2 Coincidental Extension of Scattered Context Languages 346
References 348

9 Sequential Rewriting Over Word Monoids 351
9.1 Definitions 352
9.2 Generative Power 352
References 361

Part IV Regulated Grammars: Parallelism

10 Regulated ET0L Grammars 365
10.1 Context-Conditional ET0L Grammars 367
10.1.1 Definitions 367
10.1.2 Generative Power 368

10.2 Forbidding ET0L Grammars 375
10.2.1 Definitions and Examples 375
10.2.2 Generative Power and Reduction 377
10.3 Simple Semi-Conditional ET0L Grammars 396
10.3.1 Definitions 396
10.3.2 Generative Power and Reduction 397
10.4 Left Random Context ET0L Grammars 411
10.4.1 Definitions and Examples 411
10.4.2 Generative Power and Reduction 414
References 428

11 Uniform Regulated Rewriting in Parallel 431
11.1 Semi-Parallel Uniform Rewriting 432
11.2 Parallel Uniform Rewriting 439
References 444

12 Parallel Rewriting Over Word Monoids 445
12.1 Definitions 445
12.2 Generative Power 446
References 453

Part V Regulated Grammar Systems

13 Regulated Multigenerative Grammar Systems 457
13.1 Multigenerative Grammar Systems 459
13.2 Leftmost Multigenerative Grammar Systems 475
References 489

14 Controlled Pure Grammar Systems 491
14.1 Definitions and Examples 492
14.2 Generative Power 495
References 504

Part VI Regulated Automata

15 Self-Regulating Automata 509
15.1 Self-Regulating Finite Automata 510
15.1.1 Definitions and Examples 511
15.1.2 Accepting Power 513
15.2 Self-Regulating Pushdown Automata 526
15.2.1 Definitions 526
15.2.2 Accepting Power 527
References 530

16 Automata Regulated by Control Languages 531
16.1 Finite Automata Regulated by Control Languages 532
16.1.1 Definitions 533
16.1.2 Conversions 534
16.1.3 Regular-Controlled Finite Automata 536
16.1.4 Context-Free-Controlled Finite Automata 537
16.1.5 Program-Controlled Finite Automata 537
16.2 Pushdown Automata Regulated by Control Languages 547
16.2.1 Definitions 548
16.2.2 Regular-Controlled Pushdown Automata 549
16.2.3 Linear-Controlled Pushdown Automata 550
16.2.4 One-Turn Linear-Controlled Pushdown Automata 558
References 563

Part VII Related Unregulated Automata

17 Jumping Finite Automata 567
17.1 Definitions and Examples 569
17.2 Basic Properties 571
17.3 Relations with Well-Known Language Families 573
17.4 Closure Properties 574
17.5 Decidability 578
17.6 An Infinite Hierarchy of Language Families 579
17.7 Left and Right Jumps 580
17.8 A Variety of Start Configurations 582
References 585

18 Deep Pushdown Automata 587
18.1 Definitions and Examples 589
18.2 Accepting Power 591
References 601

Part VIII Applications

19 Applications: Overview 605
19.1 Current Applications 605
19.2 Perspectives 609
References 612

20 Case Studies 615
20.1 Linguistics 616
20.1.1 Syntax and Related Linguistic Terminology 617
20.1.2 Transformational Scattered Context Grammars 621
20.1.3 Scattered Context in English Syntax 624

20.2 Biology 634
20.2.1 Simulation of Biological Organisms 634
20.2.2 Implementation 642
20.3 Compilers 648
20.3.1 Underlying Formal Model 648
20.3.2 Implementation 649
References 650

Part IX Conclusion

21 Concluding Remarks 653
21.1 New Trends and Their Expected Investigation 654
21.2 Open Problem Areas 655
21.3 Bibliographical and Historical Remarks 657
References 660

22 Summary 669
References 678

Language Family Index 679

Subject Index 685

Part I
Introduction and Terminology

This part gives an introduction to the present monograph in order to express all its upcoming discussion clearly and precisely. First, it places all the material covered in the book into the scientific context. Then, it reviews all the necessary mathematical concepts so no other sources are needed to grasp all the topics covered in the book. Finally, it gives an overview of formal language theory in order to make the entire monograph completely self-contained. This part consists of three chapters.

Chapter 1 demonstrates that regulated grammars and automata represent a significant investigation area in theoretical computer science. Specifically, they fulfill a crucially important role in formal language theory and its applications.

Chapter 2 reviews the mathematical ideas, notions, concepts, and methods underlying some mathematical areas because they are needed for understanding this book. These areas include set theory, discrete mathematics, and graph theory.

Chapter 3 gives an overview of formal language theory. Apart from its classical rudiments, it covers several less-known areas of this theory, such as fundamentals concerning parallel grammars, because these areas are also needed to fully grasp some upcoming topics included in the book.

Readers having solid background in the topics covered in Chaps. 2 and 3 can only treat them as a reference for the terminology used throughout the rest of the book.

Chapter 1
Introduction

Abstract This chapter gives an introduction to the present monograph as a whole. It intuitively conceptualizes regulated grammars and automata, which represent the subject of this book, and places this subject into general scientific context. It demonstrates that the subject of this book represents a vivid investigation area of today's computer science. In particular, this investigation is central to the theory of formal languages and automata, and the present chapter gives an informal insight into all the upcoming material, including crucially important concepts and results, in terms of this theory. In addition, the chapter sketches significant applications of regulated grammars and automata in several scientific areas. Some of these application areas, such as compiler writing, belong to computer science while others, such as biology, lie out of it.

Keywords Introduction • Formulation of the book subject • Regulated grammars and automata • Their conceptualization and significance • Their study in formal language theory • Application areas

Formal Languages and Their Regulated Models

Formal languages, such as programming languages, are applied in a great number of scientific disciplines, ranging from biology through linguistics up to informatics (see [1]). As obvious, to use them properly, they have to be precisely specified in the first place. Most often, they are defined by mathematical models with finitely many rules by which the models rewrite sequences of symbols, called strings. Over its history, the theory of formal languages have introduced a great variety of these language-defining models. The present monograph deals with their regulated versions, which are extended by additional mathematical mechanisms by which these models regulate the string-rewriting process.

A. Meduna and P. Zemek, *Regulated Grammars and Automata*,
DOI 10.1007/978-1-4939-0369-6_1,

This book is primarily interested in establishing the *power* of regulated language models. Secondarily, it studies algorithmic *transformations* that modify the models so they satisfy some desired *properties* while the defined languages remain unchanged. It pays a principal attention to two kinds of transformations. First, it studies the *reduction* of these models because their small size makes, in effect, the definition of languages easy-to-follow, economical and succinct. Second, whenever dealing with different types of equally powerful language models, this book is obviously interested in the *conversion* between them so they can be turned to each other at will.

In principle, formal language theory classifies language models into two basic categories—generative and recognition language models. Generative models, better known as *grammars*, define strings of their language so their rewriting process generates them from a special start symbol. On the other hand, recognition models, better known as *automata*, define strings of their language by a rewriting process that starts from these strings and ends in a prescribed set of final strings. Accordingly, this book deals with regulated versions of these two types of language models, hence its title—*regulated grammars and automata*.

Regulated Grammars

Concerning grammars, the classical formal language theory has often classified all grammars into two fundamental categories—context-free grammars and non-context-free grammars. As their name suggests, context-free grammars are based upon context-free rules, by which these grammars rewrite symbols regardless of the context surrounding them. As opposed to them, non-context-free grammars rewrite symbols according to context-dependent rules, whose application usually depends on rather strict conditions placed upon the context surrounding the rewritten symbols, and this way of context-dependent rewriting often makes them clumsy and inapplicable in practice. From this point of view, we obviously always prefer using context-free grammars, but they have their drawbacks, too. Perhaps most importantly, context-free grammars are significantly less powerful than non-context-free grammars. Considering all these pros and cons, it comes as no surprise that modern formal language theory has intensively and systematically struggled to come with new types of grammars that are underlined by context-free rules, but they are more powerful than ordinary context-free grammars. Regulated versions of context-free grammars, briefly referred to as *regulated grammars* in this book, represent perhaps the most successful and significant achievement in this direction. They are based upon context-free grammars extended by additional regulating mechanisms by which they control the way the language generation is performed.

Although regulated grammars are based upon context-free rules just like ordinary context-free grammars, they are significantly stronger than the unregulated versions. In fact, many of them are as powerful as classical non-context-free grammars with context-dependent rules. In addition, regulated grammars control their rewriting process and, thereby, operate in a more deterministic way than unregulated context-free grammars, which work, in a general case, completely non-deterministically. Considering these indisputable advantages over classical grammars, it comes as no surprise that the theory of formal languages has paid a special attention to regulated grammars, and their investigation has introduced several important types of grammatical regulation and achieved significant results about them, most of which were published in various conference and journal papers. The present book systematically and compactly summarizes the knowledge about grammatical regulation while focusing its attention on the latest trends in the theory of regulated grammars.

However, the knowledge concerning regulated grammars is so huge that we cannot cover it all. Instead, we restrict our main attention to these two types of grammatical regulation—*context-based regulation* and *rule-based regulation*. As their names indicate, the former is based upon various context-related restrictions while the latter is underlain by restrictions placed on the use of rules during the language generation process.

Concerning context-based regulated grammars (see Chap. 4), this monograph primarily deal with two types of these grammars—conditional grammars and scattered context grammars, sketched next.

- In a *context-conditional grammar* G, a set of permitting strings and a set of forbidding strings are assigned to every rule. During a derivation step, a rule like this is applicable provided that all its permitting strings occur in the rewritten sentential form while all the forbidding strings do not occur there.
- A *scattered context grammar* G is based on finite sequences of context-free rules. By rules of this form, G simultaneously rewrites several nonterminals during a single derivation step.

Concerning rule-based regulated grammars (see Chap. 5), this book covers four crucially important types of these grammars—regular-controlled grammars, matrix grammars, programmed grammars, and state grammars. Next, we give their gist. Notice how they differ in their regulating mathematical mechanisms as well as in the way they apply these mechanisms.

- A *regular-controlled grammar* G is extended by a regular control language defined over the set of rules in G. A terminal string x is in the language generated by G if and only if the control set contains a control string according to which G generates x.
- A *matrix grammar* G is extended by a finite language defined over the set of rules in G, and the strings in this language are referred to as matrices. In essence, G makes a derivation so it selects a matrix, and after this selection, it applies all its rules one by one until it reaches the very last rule. Then, it completes its derivation, or it makes another selection of a matrix and continues the derivation in the same way.

- A *programmed grammar* G has a finite set of rules attached to each rule r. If G applies r during a derivation step, then during the next derivation step, it has to apply a rule from its attached set.
- A *state grammar* G is extended by an additional state mechanism that strongly resembles a finite-state control of finite automata. During every derivation step, G applies a rule in a left-fashion way, and in addition, it moves from a state to another state, which influences the choice of the rule to be applied in the next step.

Apart from these fundamentals of grammatical regulation, the text discusses several closely related topics, such as special cases of context-based regulation (see Chap. 6), the role of erasing rules in this regulation (see Chap. 7), generation of languages extended by extra symbols (see Chap. 8), grammatical regulation in parallel (see Chaps. 10 and 11), algebraically regulated grammars (see Chaps. 9 and 12), and sets of mutually communicating regulated grammars (see Chaps. 13 and 14).

Regulated Automata

Concerning automata, this book studies regulated versions of finite and pushdown automata, which are central to formal language theory.

This theory conceptualizes the general notion of a *finite automaton* M as a strictly finitary model—that is, all components of M are of a fixed size, and none of them can be extended during the course of the rewriting process. More precisely, M consists of an input tape, a read head, and a finite state control. The input tape is divided into squares. Each square contains one symbol of an input string w. The symbol under the read head is the current input symbol. The finite control is represented by a finite set of states together with a relation, specified as a set of computational rules in this book. On w, M works by making moves. Each move is made according to a computational rule that describes how the current state is changed and whether the current input symbol is read. If the symbol is read, the read head is shifted one square to the right; otherwise, the read head is kept stationary. The main task of M is to decide whether w is accepted. For this purpose, M has one state defined as the start state and some states designated as final states. If M can read w by making a sequence of moves from the start state to a final state, M accepts w; otherwise, M rejects w. The language of M consists of all strings that M accepts in this way.

A *pushdown automaton* represents a finite automaton extended by a potentially infinite pushdown list. During a move, according to one of its rules, it reads a symbol, changes the current state, and rewrites the symbol occurring on the pushdown top. If it reads the entire input string, empties the pushdown list and enters a final state, the automaton accepts the input string; the set of all strings accepted in this way is the language that the automaton accepts.

Concerning regulated finite and pushdown automata, this monograph covers the following two fundamental types.

- *Self-regulating automata* select a rule according to which the current move is made based upon the rule applied during the previous move (see Chap. 15). As obvious, this regulation strongly resembles the regulation of programmed grammars.
- *Language-controlled automata* regulate the application of rules by control languages by analogy with context-free grammars regulated by control languages (see Chap. 16). This book considers finite and pushdown automata regulated by control languages from various language families.

In addition, this monograph covers modified versions of finite and pushdown automata closely related to regulated automata, too. Specifically, it presents *jumping finite automata* that work just like classical finite automata except that they can, after reading a symbol, jump in either direction within their input tapes and continue making moves from there (see Chap. 17). Furthermore, it covers *deep pushdown automata* that can make expansions deeper in the pushdown, not just on the very pushdown top (see Chap. 18).

Applications

Although this book primarily deals with regulated language-defining models from a theoretical viewpoint, it sketches their applications, too. It narrows its attention to the applications of regulated grammars rather than regulated automata, and it restricts their coverage only to three application areas—computational linguistics, molecular biology, and compiler writing. It describes these applications and their perspectives from a general viewpoint (see Chap. 19). In addition, it presents many case studies to show quite specific real-world applications concerning the three scientific fields mentioned above (see Chap. 20).

References

1. Rozenberg, G., Salomaa, A. (eds.): Handbook of Formal Languages, vol. 1 through 3. Springer, New York (1997)

Chapter 2
Mathematical Background

Abstract To follow all the upcoming discussion of this book clearly and accurately, this chapter describes the mathematical language used throughout. That is, it reviews all the necessary mathematical concepts so that no other sources are needed to grasp all the topics covered in the book. Simply stated, the present chapter makes this book completely self-contained. The concepts covered in the present chapter, divided into four sections, primarily include fundamental areas of discrete mathematics. First, Sect. 2.1 reviews some of the most basic concepts from set theory. Then, Sect. 2.2 give the essentials concerning relations, after which Sect. 2.3 concentrates its attention on their crucially important special cases—functions. Finally, Sect. 2.4 reviews a number of concepts from graph theory, used throughout the remainder of the book.

Keywords Discrete mathematics • Sets • Relations • Functions • Graphs • Trees

Although this book is self-contained in the sense that no other sources are needed to grasp all the presented results, the reader is expected to have at least basic knowledge regarding mathematical background of formal language theory. Of course, a deeper knowledge of mathematics is more than welcome because this introductory chapter describes the necessary mathematical terminology rather briefly. For a good treatment of mathematical foundations underlying formal language theory, consult [2].

The present four-section chapter reviews rudimentary concepts concerning sets (Sect. 2.1), relations (Sect. 2.2), functions (Sect. 2.3), and graphs (Sect. 2.4). For readers familiar with these concepts, this chapter can be skimmed and treated as a reference for notation and definitions.

A. Meduna and P. Zemek, *Regulated Grammars and Automata*,
DOI 10.1007/978-1-4939-0369-6_2,

2.1 Sets and Sequences

A *set* Q is a collection of differentiable elements taken from some prescribed *universe* $\mathbb{U}$ without any structure other than membership. To indicate that x is an element (a *member*) of Q, we write $x \in Q$. The statement that x is not in Q is written as $x \notin Q$. If Q has a finite number of members, then Q is a *finite set*; otherwise, Q is an *infinite set*. The set that has no members is the *empty set*, denoted by $\emptyset$. The *cardinality* of Q, denoted by $\text{card}(Q)$, is, for finite sets, the number of members of Q. If Q is infinite, then we assume that $\text{card}(Q) > k$ for every integer k. Note that $\text{card}(\emptyset) = 0$.

Sets can specified by enclosing some description of their elements in curly brackets; for example, the set Q of three consecutive integers, $1, 2$ and 3, is denoted by

$$Q = \{1, 2, 3\}$$

Ellipses can be used whenever the meaning is clear. Thus, $\{a, b, \dots, z\}$ stands for all the lower-case letters of the English alphabet. When the need arises, we use a more explicit notation, in which a set Q is specified by a property σ so Q contains all elements satisfying σ. This specification has the following format

$$Q = \{x \mid \sigma(x)\}$$

Let $\mathbb{N}_0$ denote the set of all nonnegative integers. Then, for example, the set of all even nonnegative integers can be defined as $\mathbb{N}_0^{even} = \{i \mid i \in \mathbb{N}_0, i \text{ is even}\}$.

The usual set operations are *union* ($\cup$), *intersection* ($\cap$), and *difference* ($-$). Let Q_1 and Q_2 be two sets. Then,

$$\begin{aligned} Q_1 \cup Q_2 &= \{x \mid x \in Q_1 \text{ or } x \in Q_2\} \\ Q_1 \cap Q_2 &= \{x \mid x \in Q_1 \text{ and } x \in Q_2\} \\ Q_1 - Q_2 &= \{x \mid x \in Q_1 \text{ and } x \notin Q_2\} \end{aligned}$$

For n sets, $Q_1, Q_2, \dots, Q_n$, where $n \geq 1$, instead of $Q_1 \cup Q_2 \cup \dots \cup Q_n$ and $Q_1 \cap Q_2 \cap \dots \cap Q_n$, we usually write $\bigcup_{1 \leq i \leq n} Q_i$ and $\bigcap_{1 \leq i \leq n} Q_i$, respectively. If there are infinitely many sets, we omit the upper bound n.

Another basic operation is the *complementation* of Q, which is denoted by $\overline{Q}$ and defined as

$$\overline{Q} = \{x \mid x \in \mathbb{U} \text{ and } x \notin Q\}$$

A set P is said to be a *subset* of Q if every element of P is also an element of Q; we write this as $P \subseteq Q$. We also say that Q is a *superset* of P. If $P \subseteq Q$ and $Q - P \neq \emptyset$, we say that P is a *proper subset* of Q, written as $P \subset Q$. Similarly, we also say that Q is a *proper superset* of P.

Let Q_1 and Q_2 be two sets. If $Q_1 \cap Q_2 = \emptyset$, then Q_1 and Q_2 are *disjoint*. If $Q_1 \subseteq Q_2$ and $Q_2 \subseteq Q_1$, then Q_1 and Q_2 are *identical*, written as $Q_1 = Q_2$; otherwise, they are *non-identical*, written as $Q_1 \neq Q_2$.

Let $Q_1, Q_2, \ldots, Q_n$ be n sets, for some $n \geq 1$. If $Q_i \cap Q_j = \emptyset$ for all $i = 1, 2, \ldots, n$ and all $j = 1, 2, \ldots, n$ such that $i \neq j$, then we say that all Q_i are *pairwise disjoint*.

The *power set* of Q, denoted by 2^Q, is the set of all subsets of Q. In symbols,

$$2^Q = \{U \mid U \subseteq Q\}$$

Sets whose members are sets are customarily called *families of sets*, rather than sets of sets. For instance, 2^Q is a family of sets. As obvious, all notions and operations from sets apply to families of sets as well.

For a finite set $Q \subseteq \mathbb{N}_0$, let $\max(Q)$ denote the smallest integer m such that m is greater or equal to all members of Q. Similarly, $\min(Q)$ denotes the greatest integer n such that n is lesser or equal to all members of Q.

A *sequence* is a list of elements. Contrary to a set, a sequence can contain an element more than once and the elements appear in a certain order. Elements in sequences are usually separated by a comma. As sets, sequences can be either *finite* or *infinite*. Finite sequences are also called *tuples*. More specifically, sequences of two, three, four, five, six, and seven elements are called *pairs*, *triples*, *quadruples*, *quintuples*, *sextuples*, and *septuples*, respectively.

2.2 Relations

Let Q_1 and Q_2 be two sets. The *Cartesian product* of Q_1 and Q_2, denoted by $Q_1 \times Q_2$, is a set of pairs defined as

$$Q_1 \times Q_2 = \{(x_1, x_2) \mid x_1 \in Q_1 \text{ and } x_2 \in Q_2\}$$

A *binary relation* ρ from Q_1 to Q_2 is any subset of their Cartesian product. That is,

$$\rho \subseteq Q_1 \times Q_2$$

The *inverse* of ρ, denoted by ρ^{-1}, is defined as

$$\rho^{-1} = \{(y, x) \mid (x, y) \in \rho\}$$

Instead of $(x, y) \in \rho$, we often write $x\rho y$; in other words, $(x, y) \in \rho$ and $x\rho y$ are used interchangeably. If $Q_1 = Q_2$, then we say that ρ is a *relation on* Q_1 or *relation over* Q_1. As relations are sets, all common operations over sets apply to relations as well. If ρ is a finite set, then ρ is a *finite relation*; otherwise, ρ is an *infinite relation*.

For every $k \geq 0$, the *kth power* of ρ, denoted by ρ^k, is recursively defined as follows:

(1) $x\rho^0 y$ if and only if $x = y$
(2) for $k \geq 1, x\rho^k y$ if and only if $x\rho z$ and $z\rho^{k-1} y$, for some $z \in Q$

The *transitive closure* of ρ, denoted by ρ^+, is defined as $x\rho^+ y$ if and only if $x\rho^k y$, for some $k \geq 1$. The *reflexive-transitive closure* of ρ, denoted by ρ^*, is defined as $x\rho^* y$ if and only if $x\rho^k y$, for some $k \geq 0$.

2.3 Functions

A *function* ψ from Q_1 to Q_2 is a relation from Q_1 to Q_2 such that for every $x \in Q_1$,

$$\text{card}\left(\{y \mid y \in Q_2 \text{ and } (x, y) \in \psi\}\right) \leq 1$$

If for every $y \in Q_2, \text{card}(\{x \mid x \in Q_1 \text{ and } (x, y) \in \psi\}) \leq 1$, ψ is an *injection*. If for every $y \in Q_2, \text{card}(\{x \mid x \in Q_1 \text{ and } (x, y) \in \psi\}) \geq 1$, ψ is a *surjection*. If ψ is both an injection and a surjection, ψ represents a *bijection*.

The *domain* of ψ, denoted by domain(ψ), and the *range* of ψ, denoted by range(ψ), are defined as

$$\text{domain}(\psi) = \left\{x \mid x \in Q_1 \text{ and } (x, y) \in \psi, \text{ for some } y \in Q_2\right\}$$

and

$$\text{range}(\psi) = \left\{y \mid y \in Q_2 \text{ and } (x, y) \in \psi, \text{ for some } x \in Q_1\right\}$$

If domain(ψ) $= Q_1$, ψ is *total*; otherwise, ψ is *(strictly) partial*. Instead of $(x, y) \in \psi$, we usually write $\psi(x) = y$.

Let P and Q be two sets. A *unary operation* over P is a function f from P to Q. We say that P is *closed under a unary operation* f if for each $a \in P$, $f(a) \in P$.

A *binary operation* over P is a function $\odot$ from $(P \times P)$ to Q. We say that P is *closed under a binary operation* $\odot$ if for each $a, b \in P$, $a \odot b \in P$ (we use the infix notation $a \odot b$ instead of $\odot(a, b)$). A set that is closed under an operation is said to satisfy a *closure property*.

2.4 Graphs

In this section, we review the basics of graph theory. For more information, please consult [1, 3, 4].

A *directed graph* is a pair, $G = (V, \rho)$, where V is a finite set and ρ is a relation over V. For brevity, by a *graph*, we automatically mean a directed graph throughout the book. Members of V are called *nodes* and pairs in ρ are called *edges*. If $e = (a, b)$ and $e \in \rho$, then e *leaves* a and *enters* b; at this point, a is a *direct predecessor* of b and b is a *direct descendant* of a. A sequence of nodes, $a_0, a_1, \ldots, a_n$, where $n \geq 1$, forms a *walk* from a_0 to a_n if $(a_{i-1}, a_i) \in \rho$ for all $i = 1, \ldots, n$. If, in addition, $a_0 = a_n$, then $a_0, a_1, \ldots, a_n$ is a *cycle*. If there is no cycle in G, then G is an *acyclic graph*.

Let Q be a nonempty set. An *ordered labeled tree* or, briefly, a *tree* is an acyclic graph, $T = (V, \rho)$, satisfying the following four conditions:

(1) there is exactly one special node, called the *root* such that no edge enters it;
(2) for each node $a \in V$ other than the root, there is exactly one walk from the root to a;
(3) every node is labeled with a member of Q (there is a total function from V to Q);
(4) each node $a \in V$ has its direct descendants $b_1, b_2, \ldots, b_n$ ordered from the left to the right, so b_1 is the leftmost descendant of a and b_n is the rightmost descendant of a.

For brevity, we often refer to nodes by their labels.

Let $T = (V, \rho)$ be a tree and $a \in V$ be a node. If no edge leaves a, then a is a *leaf*. The *frontier* of T is the sequence of leaves of T, ordered from the left to the right. A tree, $T' = (V', \rho')$, is a *subtree* of T if it satisfies the following three conditions

(1) $\emptyset \subset V' \subseteq V$;
(2) $\rho' = (V' \times V') \cap \rho$;
(3) in T, no node in $V - V'$ is a direct descendant of a node in V'.

Let $r \in V'$ be the root of T'. Then, we say that T' is *rooted* at r.

A *multigraph* is a triple $G = (V, \rho, Q)$, where V and Q are two finite sets and $\rho \subseteq V \times Q \times V$ is a ternary relation. Informally, it is a graph where there can be more than one edge between two nodes, each being labeled with a symbol from Q.

References

1. Diestel, R.: Graph Theory, 3rd edn. Springer, New York (2005)
2. Gathen, J., Gerhard, J.: Modern Computer Algebra, 2nd edn. Cambridge University Press, New York (2003)
3. Gross, J.L., Yellen, J.: Graph Theory and Its Applications (Discrete Mathematics and Its Applications), 2nd edn. Chapman & Hall/CRC, London (2005)
4. McHugh, J.A.: Algorithmic Graph Theory. Prentice-Hall, New Jersey (1990)

Chapter 3
Rudiments of Formal Language Theory

Abstract The subject of this book represents a vivid investigation area of formal language theory, whose basics are given in the present chapter. Apart from the classical rudiments, however, the chapter covers several less known areas of this theory, such as parallel grammars, because these areas are also needed to fully grasp some upcoming topics covered in this book. This chapter consists of four sections. Section 3.1 introduces the very basics concerning strings, languages, and operations over them. Section 3.2 covers language families and closure properties. Section 3.3 overviews a variety of formal grammars, and Sect. 3.4 covers automata needed to follow the rest of this book. Readers familiar with the classical concepts used in formal languages theory should primarily concentrate their attention to non-classical concepts covered in this chapter, such as language-defining devices working in parallel.

Keywords Formal language theory • Strings • Languages • Language operations • Closure properties • Grammars • Automata • Parallelism

The present chapter briefly reviews formal language theory. It covers all the notions that are necessary to follow the rest of this book. Apart from well-known essentials of formal language theory, it includes less known notions, such as a variety of parallel grammars, which are also needed to establish several upcoming results, so the reader should pay a special attention to them, too.

This chapter consists of four sections. Section 3.1 covers strings, languages, and operations over them. Section 3.2 deals with language families and closure properties. Sections 3.3 and 3.4 concern grammars and automata, respectively.

A. Meduna and P. Zemek, *Regulated Grammars and Automata*,

DOI 10.1007/978-1-4939-0369-6_3,

3.1 Strings and Languages

An *alphabet* Σ is a finite, nonempty set of elements called *symbols*. If $\text{card}(\Sigma) = 1$, then Σ is a *unary alphabet*. A *string* or, synonymously, a *word* over Σ is any finite sequence of symbols from Σ. We omit all separating commas in strings; that is, for a string $a_1, a_2, \ldots, a_n$, for some $n \geq 1$, we write $a_1 a_2 \cdots a_n$ instead. The *empty string*, denoted by ε, is the string that is formed by no symbols, i.e. the empty sequence. By Σ^*, we denote the set of all strings over Σ (including ε). Set $\Sigma^+ = \Sigma^* - \{\varepsilon\}$.

Let x be a string over Σ, i.e. $x \in \Sigma^*$, and express x as $= a_1 a_2 \cdots a_n$, where $a_i \in \Sigma$, for all $i = 1 \ldots, n$, for some $n \geq 0$ (the case when $n = 0$ means that $x = \varepsilon$). The *length* of x, denoted by $|x|$, is defined as $|x| = n$. The *reversal* of x, denoted by $\text{rev}(x)$, is defined as $\text{rev}(x) = a_n a_{n-1} \cdots a_1$. The *alphabet* of x, denoted by $\text{alph}(x)$, is defined as $\text{alph}(x) = \{a_1, a_2, \ldots, a_n\}$; informally, it is the set of symbols appearing in x. For $U \subseteq \Sigma$, $\text{occur}(x, U)$ denotes the number of occurrences of symbols from U in x. If $U = \{a\}$, then instead of $\text{occur}(x, \{a\})$, we write just $\text{occur}(x, a)$. The *leftmost symbol* of x, denoted by $\text{lms}(x)$, is defined as $\text{lms}(x) = a_1$ if $n \geq 1$ and $\text{lms}(x) = \varepsilon$ otherwise. The *rightmost symbol* of x, denoted by $\text{rms}(x)$, is defined analogously. If $n \geq 1$, then for every $i = 1, \ldots, n$, let $\text{sym}(x, i)$ denote the ith symbol in x. Notice that $|\varepsilon| = 0$, $\text{rev}(\varepsilon) = \varepsilon$, and $\text{alph}(\varepsilon) = \emptyset$,

Let x and y be two strings over Σ. Then, xy is the *concatenation* of x and y. Note that $x\varepsilon = \varepsilon x = x$. If x can be written in the form $x = uv$, for some $u, v \in \Sigma^*$, then u is a *prefix* of x and v is a *suffix* of x. If $0 < |u| < |x|$, then u is a *proper prefix* of x; similarly, if $0 < |v| < |x|$, then v is a *proper suffix* of x. Define $\text{prefix}(x) = \{u \mid u \text{ is a prefix of } x\}$ and $\text{suffix}(x) = \{v \mid v \text{ is a suffix of } x\}$. For every $i \geq 0$, $\text{prefix}(x, i)$ is the prefix of x of length i if $|x| \geq i$, and $\text{prefix}(x, i) = x$ if $|x| < i$. If $x = uvw$, for some $u, v, w \in \Sigma^*$, then v is a *substring* of x. The set of all substrings of x is denoted by $\text{sub}(x)$. Moreover,

$$\text{sub}(y, k) = \{x \mid x \in \text{sub}(y), |x| \leq k\}$$

Let n be a nonnegative integer. Then, the *nth power* of x, denoted by x^n, is a string over Σ recursively defined as

$$\begin{aligned}&(1)\ x^0 = \varepsilon\\ &(2)\ x^n = xx^{n-1} \text{ for } n \geq 1\end{aligned}$$

Let $x = a_1 a_2 \cdots a_n$ be a string over Σ, for some $n \geq 0$. The set of all *permutations* of x, denoted by $\text{perm}(x)$, is defined as

$$\begin{aligned}\text{perm}(x) = \{b_1 b_2 \cdots b_n \mid\ & b_i \in \text{alph}(x), \text{ for all } i = 1, \ldots, n, \text{ and}\\ & (b_1, b_2, \ldots, b_n) \text{ is a permutation of } (a_1, a_2, \ldots, a_n)\}\end{aligned}$$

Note that $\text{perm}(\varepsilon) = \varepsilon$.

A *language* L over Σ is any set of strings over Σ, i.e. $L \subseteq \Sigma^*$. The set Σ^* is called the *universal language* because it consists of all strings over Σ. If L is a finite set, then it is a *finite language*; otherwise, it is an *infinite language*. The set of all finite languages over Σ is denoted by $\text{fin}(\Sigma)$. For $L \in \text{fin}(\Sigma)$, $\text{max-len}(L)$ denotes the length of the longest string in L. We set $\text{max-len}(\emptyset) = 0$. If $\text{card}(\Sigma) = 1$, then L is a *unary language*. The *empty language* is denoted by $\emptyset$.

The *alphabet* of L, denoted by $\text{alph}(L)$, is defined as

$$\text{alph}(L) = \bigcup_{x \in L} \text{alph}(x)$$

The *permutation* of L, denoted by $\text{perm}(L)$, is defined as

$$\text{perm}(L) = \{\text{perm}(x) \mid x \in L\}$$

The *reversal* of L, denoted by $\text{rev}(L)$, is defined as

$$\text{rev}(L) = \{\text{rev}(x) \mid x \in L\}$$

For every $L \subseteq \Sigma^*$, where $\{\varepsilon\} \subseteq L$, and every $x \in \Sigma^*$, $\text{max-prefix}(x, L)$ denotes the longest prefix of x that is in L; analogously, $\text{max-suffix}(x, L)$ denotes the longest suffix of x that is in L.

Let L_1 and L_2 be two languages over Σ. Throughout the book, we consider L_1 and L_2 to be *equal*, symbolically written as $L_1 = L_2$, if $L_1 \cup \{\varepsilon\}$ and $L_2 \cup \{\varepsilon\}$ are identical. Similarly, $L_1 \subseteq L_2$ if $L_1 \cup \{\varepsilon\}$ is a subset of $L_2 \cup \{\varepsilon\}$.

As all languages are sets, all common operations over sets can be applied to them. Therefore,

$$\begin{aligned} L_1 \cup L_2 &= \{x \mid x \in L_1 \text{ or } x \in L_2\} \\ L_1 \cap L_2 &= \{x \mid x \in L_1 \text{ and } x \in L_2\} \\ L_1 - L_2 &= \{x \mid x \in L_1 \text{ and } x \notin L_2\} \end{aligned}$$

The *complement* of L, denoted by $\overline{L}$, is defined as

$$\overline{L} = \{x \mid x \in \Sigma^*, x \notin L\}$$

There are also some special operations which apply only to languages. The *concatenation* of L_1 and L_2, denoted by $L_1 L_2$, is the set

$$L_1 L_2 = \{x_1 x_2 \mid x_1 \in L_1 \text{ and } x_2 \in L_2\}$$

Note that $L\{\varepsilon\} = \{\varepsilon\}L = L$. For $n \geq 0$, the *nth power* of L, denoted by L^n, is recursively defined as

$$(1)\ L^0 = \{\varepsilon\}$$
$$(2)\ L^n = L^{n-1}L$$

The *closure* (*Kleene star*) of a language L, denoted by L^*, is the set

$$L^* = \bigcup_{i \geq 0} L^i$$

The *positive closure* of a language L, denoted by L^+, is the set

$$L^+ = \bigcup_{i \geq 1} L^i$$

The *right quotient* of L_1 with respect to L_2, denoted by L_1/L_2, is defined as

$$L_1/L_2 = \{y \mid yx \in L_1, \text{ for some } x \in L_2\}$$

Similarly, the *left quotient* of L_1 with respect to L_2, denoted by $L_2 \backslash L_1$, is defined as

$$L_2 \backslash L_1 = \{y \mid xy \in L_1, \text{ for some } x \in L_2\}$$

We also use special types of the right and left quotients. The *exhaustive right quotient* of L_1 with respect to L_2, denoted by $L_1 \,/\!\!/\, L_2$, is defined as

$$L_1 \,/\!\!/\, L_2 = \{y \mid yx \in L_1, \text{ for some } x \in L_2, \text{ and no } x' \in L_2 \text{ such that } |x'| > |x| \text{ is a proper suffix of } yx\}$$

Similarly, the *exhaustive left quotient* of L_1 with respect to L_2, denoted by $L_2 \,\backslash\!\!\backslash\, L_1$, is defined as

$$L_2 \,\backslash\!\!\backslash\, L_1 = \{x \mid yx \in L_1, \text{ for some } y \in L_2, \text{ and no } y' \in L_2 \text{ such that } |y'| > |y| \text{ is a proper prefix of } yx\}$$

Let $L_2 = \{\$\}^*$, where \$ is a symbol. Then, $L_1 \,/\!\!/\, L_2$ is the *symbol-exhaustive right quotient* of L_1 with respect to \$, and $L_2 \,\backslash\!\!\backslash\, L_1$ is the *symbol-exhaustive left quotient* of L_1 with respect to \$.

Let Σ be an alphabet. For $x, y \in \Sigma^*$, the *shuffle* of x and y, denoted by shuffle(x, y), is defined as

$$\text{shuffle}(x, y) = \{x_1y_1x_2y_2 \cdots x_ny_n \mid x = x_1x_2 \ldots x_n, y = y_1y_2 \cdots y_n, x_i, y_i \in \Sigma^*, 1 \leq i \leq n, n \geq 1\}$$

We extend the shuffle operation on languages in the following way. For $K_1, K_2 \subseteq \Sigma^*$,

$$\text{shuffle}(K_1, K_2) = \{z \mid z \in \text{shuffle}(x, y), x \in K_1, y \in K_2\}$$

Let Σ and Γ be two alphabets. A total function σ from Σ^* to 2^{Γ^*} such that $\sigma(uv) = \sigma(u)\sigma(v)$, for every $u, v \in \Sigma^*$, is a *substitution*. A substitution is ε*-free* if it is defined from Σ^* to 2^{Γ^+}. If $\sigma(a)$ for every $a \in \Sigma$ is finite, then σ is said to be *finite*. By this definition, $\sigma(\varepsilon) = \{\varepsilon\}$ and $\sigma(a_1 a_2 \cdots a_n) = \sigma(a_1)\sigma(a_2)\cdots\sigma(a_n)$, where $n \geq 1$ and $a_i \in \Sigma$, for all $i = 1, 2, \ldots, n$, so σ is completely specified by defining $\sigma(a)$ for each $a \in \Sigma$. For $L \subseteq \Sigma^*$, we extend the definition of σ to

$$\sigma(L) = \bigcup_{w \in L} \sigma(w)$$

A total function φ from Σ^* to Γ^* such that $\varphi(uv) = \varphi(u)\varphi(v)$, for every $u, v \in \Sigma^*$, is a *homomorphism* or, synonymously, a *morphism*. As any homomorphism is a special case of finite substitution, we specify φ by analogy with the specification of σ. For $L \subseteq \Sigma^*$, we extend the definition of φ to

$$\varphi(L) = \{\varphi(w) \mid w \in L\}$$

By analogy with substitution, φ is ε*-free* if $\varphi(a) \neq \varepsilon$, for every $a \in \Sigma$. By φ^{-1}, we denote the *inverse homomorphism*, defined as

$$\varphi^{-1}(u) = \{w \mid \varphi(u) = w\}$$

A homomorphism ω from Σ^* represents an *almost identity* if there exists a symbol $\# \in \Sigma$ such that $\omega(a) = a$, for every $a \in \Sigma - \{\#\}$, and $\omega(\#) \in \{\#, \varepsilon\}$. A homomorphism τ from Σ^* to Γ^* is a *coding* if $\tau(a) \in \Gamma$, for every $a \in \Sigma$.

Let L be a language over Σ, and let k be a positive integer. A homomorphism λ over Σ^* is a k*-linear erasing* with respect to L if and only if for each $y \in L$, $|y| \leq k|\lambda(y)|$. Furthermore, if $L \subseteq (\Sigma\{\varepsilon, c, c^2, \ldots, c^k\})^*$, for some $c \notin \Sigma$ and $k \geq 1$, and λ is defined by $\lambda(c) = \varepsilon$ and $\lambda(a) = a$, for all $a \in \Sigma$, then we say that λ is k*-restricted* with respect to L. Clearly, each k-restricted homomorphism is a k-linear erasing.

3.2 Language Families

By analogy with set theory, sets whose members are languages are called *families of languages*. A family of languages $\mathscr{L}$ is ε*-free* if for every $L \in \mathscr{L}$, $\varepsilon \notin L$. The family of finite languages is denoted by **FIN**.

Just like for languages, we consider two language families, $\mathscr{L}_1$ and $\mathscr{L}_2$, *equal* if and only if

$$\bigcup_{L \in \mathscr{L}_1} L \cup \{\varepsilon\} = \bigcup_{K \in \mathscr{L}_2} K \cup \{\varepsilon\}$$

If $\mathscr{L}_1$ and $\mathscr{L}_2$ are equal, we write $\mathscr{L}_1 = \mathscr{L}_2$. We also say that these two families *coincide*. $\mathscr{L}_1$ is a *subset* of $\mathscr{L}_2$, written as $\mathscr{L}_1 \subseteq \mathscr{L}_2$, if and only if

$$\bigcup_{L \in \mathscr{L}_1} L \cup \{\varepsilon\} \subseteq \bigcup_{K \in \mathscr{L}_2} K \cup \{\varepsilon\}$$

The closure of a language family under an operation is defined by analogy with the definition of the closure of a set. Next, we define three closure properties, discussed later in this book.

Definition 3.2.1. A language family $\mathscr{L}$ is *closed under linear erasing* if and only if for all $L \in \mathscr{L}$, $\lambda(L)$ is also in $\mathscr{L}$, where λ is a k-linear erasing with respect to L, for some $k \geq 1$. □

Definition 3.2.2. A language family $\mathscr{L}$ is *closed under restricted homomorphism* if and only if for all $L \in \mathscr{L}$, $\lambda(L)$ is also in $\mathscr{L}$, where λ is a k-restricted homomorphism with respect to L, for some $k \geq 1$. □

Definition 3.2.3. Let $\mathscr{L}$ be a language family. We say that $\mathscr{L}$ is *closed under endmarking* if and only if for every $L \in \mathscr{L}$, where $L \subseteq \Sigma^*$, for some alphabet Σ, $\# \notin \Sigma$ implies that $L\{\#\} \in \mathscr{L}$. □

In Sects. 3.3 and 3.4, we introduce several language families resulting from language-defining models, such as various types of grammars and automata. If some language models define the same language family $\mathscr{L}$, we say that they are *equivalent* or, synonymously, *equally powerful*. Regarding $\mathscr{L}$, we say that these models *characterize* $\mathscr{L}$. For instance, in the next section, we introduce phrase-structure grammars that characterize the family of recursively enumerable languages.

3.3 Grammars

In this section, we define devices that generate languages. These devices are called grammars, and they play a major role in formal language theory.

Definition 3.3.1. A *phrase-structure grammar* is a quadruple

$$G = (N, T, P, S)$$

where

- N is an alphabet of *nonterminals*;
- T is an alphabet of *terminals* such that $N \cap T = \emptyset$;
- P is a finite relation from $(N \cup T)^* N (N \cup T)^*$ to $(N \cup T)^*$;
- $S \in N$ is the *start symbol*.

Pairs $(u, v) \in P$ are called *rewriting rules* (abbreviated *rules*) or *productions*, and are written as $u \to v$. The set $V = N \cup T$ is the *total alphabet* of G. A rewriting rule $u \to v \in P$ satisfying $v = \varepsilon$ is called an *erasing rule*. If there is no such rule in P, then we say that G is a *propagating* (or *ε-free*) grammar.

The G-based *direct derivation relation* over V^* is denoted by $\Rightarrow_G$ and defined as

$$x \Rightarrow_G y$$

if and only if $x = x_1 u x_2$, $y = y_1 v y_2$, and $u \to v \in P$, where $x_1, x_2, y_1, y_2 \in V^*$.

Since $\Rightarrow_G$ is a relation, $\Rightarrow_G^k$ is the kth power of $\Rightarrow_G$, for $k \geq 0$, $\Rightarrow_G^+$ is the transitive closure of $\Rightarrow_G$, and $\Rightarrow_G^*$ is the reflexive-transitive closure of $\Rightarrow_G$. Let $D\colon S \Rightarrow_G^* x$ be a derivation, for some $x \in V^*$. Then, x is a *sentential form*. If $x \in T^*$, then x is a *sentence*. If x is a sentence, then D is a *successful* (or *terminal*) *derivation*.

The *language* of G, denoted by $L(G)$, is the set of all sentences defined as

$$L(G) = \{w \in T^* \mid S \Rightarrow_G^* w\} \qquad \square$$

Next, for every phrase-structure grammar G, we define two sets, $F(G)$ and $\Delta(G)$. $F(G)$ contains all sentential forms of G. $\Delta(G)$ contains all sentential forms from which there is a derivation of a string in $L(G)$.

Definition 3.3.2. Let $G = (N, T, P, S)$ be a phrase-structure grammar, and let $V = N \cup T$. Set

$$F(G) = \{x \in V^* \mid S \Rightarrow_G^+ x\}$$

and

$$\Delta(G) = \{x \in F(G)^* \mid x \Rightarrow_G^* y,\ y \in T^*\} \qquad \square$$

For brevity, we often denote a rule $u \to v$ with a unique label r as $r\colon u \to v$, and instead of $u \to v \in P$, we simply write $r \in P$. The notion of rule labels is formalized in the following definition.

Definition 3.3.3. Let $G = (N, T, P, S)$ be a phrase-structure grammar. Let Ψ be a set of symbols called *rule labels* such that $\mathrm{card}(\Psi) = \mathrm{card}(P)$, and ψ be a bijection from P to Ψ. For simplicity and brevity, to express that ψ maps a rule, $u \to v \in P$, to r, where $r \in \Psi$, we write $r\colon u \to v \in P$; in other words, $r\colon u \to v$ means that $\psi(u \to v) = r$. For $r\colon u \to v \in P$, u and v represent the *left-hand side* of r, denoted by $\mathrm{lhs}(r)$, and the *right-hand side* of r, denoted by $\mathrm{rhs}(r)$, respectively. Let P^* and Ψ^* denote the set of all sequences of rules from P and the set of all sequences of rule labels from Ψ, respectively. Set $P^+ = P^* - \{\varepsilon\}$ and $\Psi^+ = \Psi^* - \{\varepsilon\}$. As with strings, we omit all separating commas in these sequences.

We extend ψ from P to P^* in the following way

(1) $\psi(\varepsilon) = \varepsilon$
(2) $\psi(r_1 r_2 \cdots r_n) = \psi(r_1)\psi(r_2)\cdots\psi(r_n)$

for any sequence of rules $r_1 r_2 \cdots r_n$, where $r_i \in P$, for all $i = 1, 2, \ldots, n$, for some $n \geq 1$.

Let $w_0, w_1, \ldots, w_n$ be a sequence of strings, where $w_i \in V^*$, for all $i = 0, 1, \ldots, n$, for some $n \geq 1$. If $w_{j-1} \Rightarrow_G w_j$ according to r_j, where $r_j \in P$, for all $j = 1, 2, \ldots, n$, then we write

$$w_0 \Rightarrow_G^n w_n \; [\psi(r_1 r_2 \cdots r_n)]$$

For any string w, we write

$$w \Rightarrow_G^0 w \; [\varepsilon]$$

For any two strings w and y, if $w \Rightarrow_G^n y \; [\rho]$ for $n \geq 0$ and $\rho \in \Psi^*$, then we write

$$w \Rightarrow_G^* y \; [\rho]$$

If $n \geq 1$, which means that $|\rho| \geq 1$, then we write

$$w \Rightarrow_G^+ y \; [\rho]$$

If $w = S$, then ρ is called the *sequence of rules (rule labels)* used in the derivation of y or, more briefly, the *parse*[1] of y. □

For any phrase-structure grammar G, we automatically assume that V, N, T, S, P, and Ψ denote its total alphabet, the alphabet of nonterminals, the alphabet of terminals, the start symbol, the set of rules, and the set of rule labels, respectively. Sometimes, we write $G = (N, T, \Psi, P, S)$ or $G = (V, T, P, S)$ instead of $G = (N, T, P, S)$ with V and Ψ having the above-defined meaning.

In the literature, a phrase-structure grammar is also often defined with rules of the form $x \rightarrow y$, where $x \in V^+$ and $y \in V^*$ (see, for instance, [20]). Both definitions are interchangeable in the sense that the grammars defined in these two ways generate the same family of languages—the family of recursively enumerable languages.

Definition 3.3.4. A *recursively enumerable language* is a language generated by a phrase-structure grammar. The family of recursively enumerable languages is denoted by **RE**. □

Throughout this book, in the proofs, we frequently make use of Church's thesis (see [3]), which we next state in terms of formal language theory. Before this, however, we need to explain how we understand the *intuitive notion of an effective procedure* or, briefly, a *procedure*, contained in this thesis. We surely agree that each

[1]Let us note that the notion of a parse represents a synonym of several other notions, including a *derivation word*, a *Szilard word*, and a *control word*.

procedure describes how to perform a task in an unambiguous and detailed way. We also agree that it consists of finitely many instructions, each of which can be executed mechanically in a fixed amount of time. When performed, a procedure reads input data, executes its instructions, and produces output data; of course, both the input data and the output data may be nil. We are now ready to state Church's thesis in terms of **RE**—that is, the family of recursively enumerable languages, defined by phrase-structure grammars (see Definition 3.3.1).

Church's Thesis. *Let L be a language. Then, $L \in$ **RE** if and only if there is a procedure that defines L by listing all its strings.*

All the grammars and automata discussed in this book obviously constitutes procedures in the above sense. Consequently, whenever grammars or automata of a new type are considered in this book, Church's thesis automatically implies that the language family they define is necessarily contained in **RE**, and we frequently make use of this implication in the sequel.

Observe that Church's thesis is indeed a thesis, not a theorem because it cannot be proved. Indeed, any proof of this kind would necessitate a formalization of our intuitive notion of a language-defining procedure so it can be rigorously compared with the notion of a phrase-structure grammar. At this point, however, there would be a problem whether this newly formalized notion is equivalent to the intuitive notion of a procedure, which would give rise to another thesis similar to Church's thesis. Therefore, any attempt to prove this thesis inescapably ends up with an infinite regression. However, the evidence supporting Church's thesis is hardly disputable because throughout its history, computer science has formalized the notion of a procedure in the intuitive sense by other language-defining models, such as Post systems (see [13]) and Markov algorithms (see [9]), and all of them have turned out to be equivalent with phrase-structure grammars. Even more importantly, nobody has ever come with a procedure that defines a language and demonstrated that the language cannot be generated by any phrase-structure grammar.

Originally, Church's thesis have been stated in terms of Turing machines in [19]. Indeed, Church and Turing hypothesized that any computational process which could be reasonably called as a procedure could be simulated by a Turing machine (see [14] for an in-depth discussion concerning to Church's thesis). In the present monograph, however, we do not need the notion of a Turing machine while we frequently make use of the notion of a phrase-structure grammar. Therefore, for the purposes of this book, we have reformulated Church's thesis in the above way. As phrase-structure grammars and Turing machines are equivalent (see [10]), this reformulation is obviously perfectly correct and legal from a mathematical viewpoint.

Any language models that characterize **RE** are said to be *computationally complete* because they are as strong as all possible procedures in terms of language-defining power according to Church's thesis. Apart from them, however, this book also discusses many *computationally incomplete* language models, which define proper subfamilies of **RE**. For instance, the following special versions of phrase-structure grammars are all computationally incomplete.

Definition 3.3.5. A *context-sensitive grammar* is a phrase-structure grammar

$$G = (N, T, P, S)$$

such that every $u \to v$ in P is of the form

$$u = x_1 A x_2, \ v = x_1 y x_2$$

where $x_1, x_2 \in V^*$, $A \in N$, and $y \in V^+$. A *context-sensitive language* is a language generated by a context-sensitive grammar. The family of context-sensitive languages is denoted by **CS**. □

Definition 3.3.6. A *context-free grammar* is a phrase-structure grammar

$$G = (N, T, P, S)$$

such that every rule in P is of the form

$$A \to x$$

where $A \in N$ and $x \in V^*$. A *context-free language* is a language generated by a context-free grammar. The family of context-free languages is denoted by **CF**. □

Definition 3.3.7. A *linear grammar* is a phrase-structure grammar

$$G = (N, T, P, S)$$

such that every rule in P is of the form

$$A \to xBy \text{ or } A \to x$$

where $A, B \in N$ and $x, y \in T^*$. A *linear language* is a language generated by a linear grammar. The family of linear languages is denoted by **LIN**. □

Definition 3.3.8. A *regular grammar* is a phrase-structure grammar

$$G = (N, T, P, S)$$

such that every rule in P is of the form

$$A \to aB \text{ or } A \to a$$

where $A, B \in N$ and $a \in T$. A *regular language* is a language generated by a regular grammar. The family of regular languages is denoted by **REG**. □

Alternatively, the family of regular languages is characterized by right-linear grammars, defined next.

Definition 3.3.9. A *right-linear grammar* is a phrase-structure grammar

$$G = (N, T, P, S)$$

such that every rule in P is of the form

$$A \to xB \text{ or } A \to x$$

where $A, B \in N$ and $x \in T^*$. A *right-linear language* is a language generated by a right-linear grammar. The family of right-linear languages is denoted by **RLIN**. □

Concerning the families of finite, regular, right-linear, context-free, context-sensitive, and recursively enumerable languages, the next important theorem holds true.

Theorem 3.3.10 (Chomsky Hierarchy, see [1,2]).

$$\mathbf{FIN} \subset \mathbf{REG} = \mathbf{RLIN} \subset \mathbf{LIN} \subset \mathbf{CF} \subset \mathbf{CS} \subset \mathbf{RE}$$

Next, we recall canonical derivations in context-free grammars.

Definition 3.3.11. Let $G = (N, T, \Psi, P, S)$ be a context-free grammar. The relation of a *direct leftmost derivation*, denoted by ${}_{\mathrm{lm}}\!\Rightarrow_G$, is defined as follows: if $u \in T^*$, $v \in V^*$, and $r\colon A \to x \in P$, then

$$uAv \;{}_{\mathrm{lm}}\!\Rightarrow_G uxv \; [r]$$

Let ${}_{\mathrm{lm}}\!\Rightarrow^n_G$, ${}_{\mathrm{lm}}\!\Rightarrow^*_G$, and ${}_{\mathrm{lm}}\!\Rightarrow^+_G$ denote the nth power of ${}_{\mathrm{lm}}\!\Rightarrow_G$, for some $n \geq 0$, the reflexive-transitive closure of ${}_{\mathrm{lm}}\!\Rightarrow_G$, and the transitive closure of ${}_{\mathrm{lm}}\!\Rightarrow_G$, respectively. The *language that G generates by using leftmost derivations* is denoted by $L(G, {}_{\mathrm{lm}}\!\Rightarrow)$ and defined as

$$L(G, {}_{\mathrm{lm}}\!\Rightarrow) = \{w \in T^* \mid S \;{}_{\mathrm{lm}}\!\Rightarrow^*_G w\}$$

If $S \;{}_{\mathrm{lm}}\!\Rightarrow^*_G w\; [\rho]$, where $w \in T^*$, then ρ is the *left parse* of w. □

By analogy with leftmost derivations and left parses, we define rightmost derivations and right parses.

Definition 3.3.12. Let $G = (N, T, \Psi, P, S)$ be a context-free grammar. The relation of a *direct rightmost derivation*, denoted by ${}_{\mathrm{rm}}\!\Rightarrow_G$, is defined as follows: if $u \in V^*$, $v \in T^*$, and $r\colon A \to x \in P$, then

$$uAv \;{}_{\mathrm{rm}}\!\Rightarrow_G uxv \; [r]$$

Let ${}_{\mathrm{rm}}\!\Rightarrow^n_G$, ${}_{\mathrm{rm}}\!\Rightarrow^*_G$, and ${}_{\mathrm{rm}}\!\Rightarrow^+_G$ denote the nth power of ${}_{\mathrm{rm}}\!\Rightarrow_G$, for some $n \geq 0$, the reflexive-transitive closure of ${}_{\mathrm{rm}}\!\Rightarrow_G$, and the transitive closure of ${}_{\mathrm{rm}}\!\Rightarrow_G$, respectively. The *language that G generates by using rightmost derivations* is denoted by $L(G, {}_{\mathrm{rm}}\Rightarrow_G)$ and defined as

$$L(G, {}_{\mathrm{rm}}\Rightarrow) = \{w \in T^* \mid S \ {}_{\mathrm{rm}}\Rightarrow_G^* w\}$$

If $S \ {}_{\mathrm{rm}}\Rightarrow_G^* w\ [\rho]$, where $w \in T^*$, then ρ is the *right parse* of w. □

Without any loss of generality, in context-free grammars, we may consider only canonical derivations, which is formally stated in the following theorem.

Theorem 3.3.13 (see [10]). *Let G be a context-free grammar. Then,*

$$L(G, {}_{\mathrm{lm}}\Rightarrow) = L(G, {}_{\mathrm{rm}}\Rightarrow) = L(G)$$

The following theorem gives a characterization of the family of recursively enumerable languages by context-free languages.

Theorem 3.3.14 (see [4]). *For every recursively enumerable language K, there exist two context-free languages, L_1 and L_2, and a homomorphism h such that*

$$K = h(L_1 \cap L_2)$$

The next theorem says that if a phrase-structure grammar generates each of its sentences by a derivation satisfying a length-limited condition, then the generated language is, in fact, context sensitive.

Theorem 3.3.15 (Workspace Theorem, see [18]). *Let $G = (N, T, P, S)$ be a phrase-structure grammar. If there is a positive integer k such that for every nonempty $y \in L(G)$, there exists a derivation*

$$D: S \Rightarrow_G x_1 \Rightarrow_G x_2 \Rightarrow_G \cdots \Rightarrow_G x_n = y$$

where $x_i \in V^*$ and $|x_i| \leq k|y|$, for all $i = 1, 2, \ldots, n$, for some $n \geq 1$, then $L(G) \in$ **CS**.

Queue Grammars

As their name indicates, queue grammars (see [7]) rewrite strings in a way that resemble the standard way of working with an abstract data type referred to as a queue. Indeed, these grammars work on strings based upon the well-known first-in-first-out principle—that is, the first symbol added to the string will be the first one to be removed. More specifically, during every derivation step, these grammars attach a string as a suffix to the current sentential form while eliminating the leftmost symbol of this form; as a result, all symbols that were attached prior to this step have to be removed before the newly attached suffix is removed. Next, we define these grammars rigorously.

Definition 3.3.16. A *queue grammar* is a sixtuple

$$Q = \big(V, T, W, F, R, g\big)$$

where V and W are two alphabets satisfying $V \cap W = \emptyset$, $T \subset V$, $F \subset W$, $g \in (V - T)(W - F)$, and

$$R \subseteq V \times (W - F) \times V^* \times W$$

is a finite relation such that for each $a \in V$, there exists an element $(a, b, x, c) \in R$. If $u = arb$, $v = rxc$, and $(a, b, x, c) \in R$, $r, x \in V^*$, where $a \in V$ and $b, c \in W$, then Q makes a *derivation step* from u to v according to (a, b, x, c), symbolically written as

$$u \Rightarrow_Q v \; [(a, b, x, c)]$$

or, simply, $u \Rightarrow_Q v$. We define $\Rightarrow_Q^n$ $(n \geq 0)$, $\Rightarrow_Q^+$, and $\Rightarrow_Q^*$ in the standard way. The *language* of Q, denoted by $L(Q)$, is defined as

$$L(Q) = \big\{x \in T^* \mid g \Rightarrow_Q^* xf, f \in F\big\}$$ □

Theorem 3.3.17 (see [7]). *For every recursively enumerable language K, there is a queue grammar Q such that $L(Q) = K$.*

Next, we slightly modify the definition of a queue grammar.

Definition 3.3.18 (see [8, 11, 12]). A *left-extended queue grammar* is a sixtuple

$$Q = \big(V, T, W, F, R, g\big)$$

where V, T, W, F, R, g have the same meaning as in a queue grammar; in addition, assume that $\# \notin V \cup W$. If $u, v \in V^*\{\#\}V^*W$ so $u = w\#arb$, $v = wa\#rzc$, $a \in V$, $r, z, w \in V^*$, $b, c \in W$, and $(a, b, z, c) \in R$, then

$$u \Rightarrow_Q v \; [(a, b, z, c)]$$

or, simply, $u \Rightarrow_Q v$. In the standard manner, extend $\Rightarrow_Q$ to $\Rightarrow_Q^n$, where $n \geq 0$. Based on $\Rightarrow_Q^n$, define $\Rightarrow_Q^+$ and $\Rightarrow_Q^*$. The *language* of Q, denoted by $L(Q)$, is defined as

$$L(Q) = \big\{v \in T^* \mid \#g \Rightarrow_Q^* w\#vf \text{ for some } w \in V^* \text{ and } f \in F\big\}$$ □

Less formally, during every step of a derivation, a left-extended queue grammar shifts the rewritten symbol over #; in this way, it records the derivation history, which represents a property fulfilling a crucial role in several proofs later in this book.

Selective Substitution Grammars

Selective substitution grammars (see [5, 6, 15]) use context-free-like rules that have a terminal or a nonterminal on their left-hand sides. By using extremely simple regular languages, referred to as *selectors*, they specify symbols where one of them is rewritten during a derivation step and, in addition, place some restrictions on the context appearing before and after the rewritten symbol. Otherwise, they work by analogy with context-free grammars.

Definition 3.3.19. A *selective substitution grammar* (an *s-grammar* for short) is a quintuple

$$G = \big(V, T, P, S, K\big)$$

where V is the *total alphabet*, $T \subseteq V$ is an alphabet of *terminals*,

$$P \subseteq V \times V^*$$

is a finite relation called the set of *rules*, $S \in V - T$ is the *start symbol*, and K is a finite set of *selectors* of the form

$$X^*\overline{Y}Z^*$$

where $X, Y, Z \subseteq V$; in words, the barred symbols are said to be *activated*. If $A \to x \in P$ implies that $x \neq \varepsilon$, then G is said to be *propagating*.

The *direct derivation relation* over V^*, symbolically denoted by $\Rightarrow_G$, is defined as follows:

$$uAv \Rightarrow_G uxv$$

if and only if $A \to x \in P$ and $X^*\overline{Y}Z^* \in K$ such that $u \in X^*$, $A \in Y$, and $v \in Z^*$. Let $\Rightarrow_G^n$ and $\Rightarrow_G^*$ denote the nth power of $\Rightarrow_G$, for some $n \geq 0$, and the reflexive-transitive closure of $\Rightarrow_G$, respectively.

The *language* of G is denoted by $L(G)$ and defined as

$$L(G) = \big\{w \in T^* \mid S \Rightarrow_G^* w\big\}$$ □

The families of languages generated by s-grammars and propagating s-grammars are denoted by **S** and $\mathbf{S}^{-\varepsilon}$, respectively.

ET0L Grammars and Their Variants

Next, we define ET0L grammars and their variants (see [16, 17]). Contrary to all the previously defined types of grammars, during every derivation step, these grammars rewrite all symbols appearing in the current sentential form in parallel.

Definition 3.3.20. An *ET0L grammar* is a $(t + 3)$-tuple

$$G = \big(V, T, P_1, \ldots, P_t, w\big)$$

where $t \geq 1$, and V, T, and w are the *total alphabet*, the *terminal alphabet* ($T \subseteq V$), and the *start string* ($w \in V^+$), respectively. Each P_i is a finite set of *rules* of the form

$$a \rightarrow x$$

where $a \in V$ and $x \in V^*$. If $a \rightarrow x \in P_i$ implies that $x \neq \varepsilon$ for all $i = 1, \ldots, t$, then G is said to be *propagating* (an *EPT0L grammar* for short).

Let $u, v \in V^*$, $u = a_1 a_2 \cdots a_q$, $v = v_1 v_2 \cdots v_q$, $q = |u|$, $a_j \in V$, $v_j \in V^*$, and $p_1, p_2, \ldots, p_q$ is a sequence of rules of the form $p_j = a_j \rightarrow v_j \in P_i$ for all $j = 1, \ldots, q$, for some $i \in \{1, \ldots, t\}$. Then, u *directly derives* v according to the rules p_1 through p_q, denoted by

$$u \Rightarrow_G v \; [p_1, p_2, \ldots, p_q]$$

If p_1 through p_q are immaterial, we write just $u \Rightarrow_G v$. In the standard manner, we define the relations $\Rightarrow_G^n$ ($n \geq 0$), $\Rightarrow_G^*$, and $\Rightarrow_G^+$.

The *language* of G, denoted by $L(G)$, is defined as

$$L(G) = \big\{y \in T^* \mid w \Rightarrow_G^* y\big\}$$ □

The families of languages generated by ET0L and EPT0L grammars are denoted by **ET0L** and **EPT0L**, respectively.

Definition 3.3.21. Let $G = (V, T, P_1, \ldots, P_t, w)$ be an ET0L grammar, for some $t \geq 1$. If $t = 1$, then G is called an *E0L grammar*. □

The families of languages generated by E0L and propagating E0L grammars (EP0L grammars for short) are denoted by **E0L** and **EP0L**, respectively.

Definition 3.3.22. An *0L grammar* is defined by analogy with an E0L grammar except that $V = T$. □

For simplicity, we specify an 0L grammar as a triple $G = (T, P, S)$ rather than a quadruple $G = (T, T, P, S)$. By **0L**, we denote the family of languages generated by 0L grammars.

Theorem 3.3.23 (see [16]). $\mathbf{CF} \subset \mathbf{E0L} = \mathbf{EP0L} \subset \mathbf{ET0L} = \mathbf{EPT0L} \subset \mathbf{CS}$

Finally, we define EIL grammars (see [16, 17]).

Definition 3.3.24. Given integers $m, n \geq 0$, an *E(m, n)L grammar* is defined as a quadruple

$$G = \big(V, T, P, w\big)$$

where V, T, and w are the *total alphabet*, the *terminal alphabet* ($T \subseteq V$), and the *start string* ($w \in V^+$), respectively. P is a finite set of rules of the form

$$(u, a, v) \to y$$

such that $a \in V$, $u, v, y \in V^*$, $0 \le |u| \le m$, and $0 \le |v| \le n$. Let $x, y \in V^*$. Then, x *directly derives* y in G, written as

$$x \Rightarrow_G y$$

provided that $x = a_1 a_2 \cdots a_k$, $y = y_1 y_2 \cdots y_k$, $k \ge 1$, and for all i, $1 \le i \le k$,

$$(a_{i-m} \cdots a_{i-1}, a_i, a_{i+1} \cdots a_{i+n}) \to y_i \in P$$

We assume that $a_j = \varepsilon$ for all $j \le 0$ or $j \ge k + 1$. In the standard way, $\Rightarrow_G^i$, $\Rightarrow_G^+$, and $\Rightarrow_G^*$ denote the ith power of $\Rightarrow_G$, $i \ge 0$, the transitive closure of $\Rightarrow_G$, and the transitive and reflexive closure of $\Rightarrow_G$, respectively. The *language* of G, denoted by $L(G)$, is defined as

$$L(G) = \{z \in T^* \mid w \Rightarrow_G^* z\}$$ □

Let $G = (V, T, P, s)$ be an E$(0, n)$L grammar, $n \ge 0$, and $p = (\varepsilon, A, v) \to y \in P$. We simplify the notation of p so that $p = (A, v) \to y$ throughout this book. By *EIL grammars*, we refer to E(m, n)L grammars, for all $m, n \ge 0$.

Derivation Trees

A derivation tree graphically represents the structure of a derivation in a context-free grammar and suppresses the order in which individual rules are used.

Definition 3.3.25. Let $G = (N, T, P, S)$ be a context-free grammar, and let $S \Rightarrow_G^* w$ be a derivation of the form

$$S = w_1 \Rightarrow_G w_2 \Rightarrow_G \cdots \Rightarrow_G w_n = w$$

where $w_i \in V^*$, for all $i = 1, 2, \ldots, n$, for some $n \ge 1$. A *derivation tree* corresponding to this derivation is denoted by $\Delta(S \Rightarrow_G^* w)$ and defined as a tree with the following three properties:

1. nodes of the derivation tree are labeled with members of $V \cup \{\varepsilon\}$;
2. the root of the derivation tree is labeled with S;
3. for a direct derivation $w_{i-1} \Rightarrow_G w_i$, for all $i = 1, 2, \ldots, n$, where

- $w_{i-1} = xAz$, $x, z \in V^*$, $A \in N$,
- $w_i = xyz$,
- $A \to y \in P$, where $y = Y_1Y_2 \cdots Y_k$, $Y_j \in V$, for all $j = 1, 2, \ldots, k$, for some $k \geq 0$ (if $k = 0$, then $y = \varepsilon$),

if $y \neq \varepsilon$, then there are exactly k edges, (A, Y_j), $1 \leq j \leq k$, leaving A, which are ordered from the left to the right in order (A, Y_1), (A, Y_2), …, (A, Y_k). If $y = \varepsilon$, then there is only one edge leaving A, (A, ε). □

Definition 3.3.26. Let $G = (N, T, P, S)$ be a context-free grammar, and let $S \Rightarrow_G^* w$ be a derivation of the form

$$S \Rightarrow_G^* uXv \Rightarrow_G^* w$$

where $u, v \in V^*$, $X \in V$, and $w \in L(G)$.

If $X \in N$ and the frontier of the subtree rooted at this occurrence of X in $\Delta(S \Rightarrow_G^* w)$ is ε, then G *erases* this occurrence of X in $S \Rightarrow_G^* w$, symbolically written as

$$^{\varepsilon}X$$

If either

(a) $X \in T$ or
(b) $X \in N$ and the frontier of the subtree rooted at this occurrence of X in $\Delta(S \Rightarrow_G^* w)$ differs from ε,

then G *does not erase* this occurrence of X in $S \Rightarrow_G^* w$, symbolically written as

$$^{\not\varepsilon}X$$

Furthermore, let $S \Rightarrow_G^* yxz \Rightarrow_G^* w$, where $x, y, z \in V^*$, and $w \in L(G)$. Let $x = X_1X_2 \cdots X_n$, where $X_i \in V$, for all $i = 1, 2, \ldots, n$, $n = |x|$. If $x = \varepsilon$ or $^{\varepsilon}X_i$, for all $i = 1, 2, \ldots, n$, then we write $^{\varepsilon}x$; informally, it means that x is completely erased in the rest of the derivation. If $^{\not\varepsilon}X_j$, for some $j = 1, 2, \ldots, n$, then we write $^{\not\varepsilon}x$; informally, it means that x is not completely erased in the rest of the derivation. □

3.4 Automata

In this section, basic devices for recognizing strings of a given (regular or context-free) language are defined—finite and pushdown automata (see [10]).

First, we define the notion of a general finite automaton.

Definition 3.4.1. A *general finite automaton* is a quintuple

$$M = \left(Q, \Sigma, R, s, F\right)$$

where

- Q is a finite set of *states*;
- Σ is an *input alphabet*;
- $R \subseteq Q \times \Sigma^* \times Q$ is a finite relation, called the set of *rules* (or *transitions*);
- $s \in Q$ is the *start state*;
- $F \subseteq Q$ is the set of *final states*.

Instead of $(p, y, q) \in R$, we write $py \to q \in R$. If $py \to q \in R$ implies that $y \neq \varepsilon$, then M is said to be *ε-free*.

A *configuration* of M is any string from $Q\Sigma^*$. The relation of a *move*, symbolically denoted by $\vdash_M$, is defined over $Q\Sigma^*$ as follows:

$$pyx \vdash_M qx$$

if and only if $pyx, qx \in Q\Sigma^*$ and $py \to q \in R$.

Let $\vdash_M^n$, $\vdash_M^*$, and $\vdash_M^+$ denote the nth power of $\vdash_M$, for some $n \geq 0$, the reflexive-transitive closure of $\vdash_M$, and the transitive closure of $\vdash_M$, respectively. The *language* of M is denoted by $L(M)$ and defined as

$$L(M) = \{w \in \Sigma^* \mid sw \vdash_M^* f, f \in F\}$$ □

Next, we define three special variants of general finite automata.

Definition 3.4.2. Let $M = (Q, \Sigma, R, s, F)$ be a general finite automaton. M is a *finite automaton* if and only if $py \to q \in R$ implies that $|y| \leq 1$. M is said to be *deterministic* if and only if $py \to q \in R$ implies that $|y| = 1$ and $py \to q_1, py \to q_2 \in R$ implies that $q_1 = q_2$, for all $p, q, q_1, q_2 \in Q$ and $y \in \Sigma^*$. M is said to be *complete* if and only if M is deterministic and for all $p \in Q$ and all $a \in \Sigma$, $pa \to q \in R$ for some $q \in Q$. □

To make several definitions and proofs concerning finite automata more concise, we sometimes denote a rule $pa \to q$ with a unique label r as $r\colon pa \to q$. This notion of rule labels is formalized in the following definition.

Definition 3.4.3. Let $M = (Q, \Sigma, R, s, F)$ be a finite automaton. Let Ψ be an alphabet of *rule labels* such that $\mathrm{card}(\Psi) = \mathrm{card}(R)$, and ψ be a bijection from R to Ψ. For simplicity, to express that ψ maps a rule, $pa \to q \in R$, to r, where $r \in \Psi$, we write $r\colon pa \to q \in R$; in other words, $r\colon pa \to q$ means $\psi(pa \to q) = r$.

For every $y \in \Sigma^*$ and $r\colon pa \to q \in R$, M makes a *move* from configuration pay to configuration qy according to r, written as

$$pay \vdash_M qy\ [r]$$

Let χ be any configuration of M. M makes *zero moves* from χ to χ according to ε, symbolically written as

$$\chi \vdash_M^0 \chi\ [\varepsilon]$$

Let there exist a sequence of configurations $\chi_0, \chi_1, \ldots, \chi_n$ for some $n \geq 1$ such that $\chi_{i-1} \vdash_M \chi_i \; [r_i]$, where $r_i \in \Psi$, for $i = 1, \ldots, n$, then M makes n *moves* from χ_0 to χ_n according to $r_1 \cdots r_n$, symbolically written as

$$\chi_0 \vdash_M^n \chi_n \; [r_1 \cdots r_n]$$

Define $\vdash_M^*$ and $\vdash_M^+$ in the standard manner. □

Sometimes, we specify a finite automaton as $M = (Q, \Sigma, \Psi, R, s, F)$, where Q, Σ, Ψ, R, s, and F are the set of states, the input alphabet, the alphabet of rule labels, the set of rules, the start state, and the set of final states, respectively.

Theorem 3.4.4 (see [20]). *For every general finite automaton M, there is a complete finite automaton M' such that $L(M') = L(M)$.*

Finite automata accept precisely the family of regular languages.

Theorem 3.4.5 (see [20]). *A language K is regular if and only if there is a complete finite automaton M such that $K = L(M)$.*

Pushdown automata represent finite automata extended by a potentially unbounded pushdown store. We first define their general version, customarily referred to as extended pushdown automata.

Definition 3.4.6. An *extended pushdown automaton* is a septuple

$$M = \big(Q, \Sigma, \Gamma, R, s, S, F\big)$$

where

- Q, Σ, s, and F are defined as in a finite automaton;
- Γ is a *pushdown alphabet*;
- $R \subseteq \Gamma^* \times Q \times (\Sigma \cup \{\varepsilon\}) \times \Gamma^* \times Q$ is a finite relation, called the set of *rules* (or *transitions*);
- S is the *initial pushdown symbol*.

Q and $(\Sigma \cup \Gamma)$ are always assumed to be disjoint. By analogy with finite automata, instead of $(\gamma, p, a, w, q) \in R$, we write $\gamma pa \rightarrow wq$.

A *configuration* of M is any string from $\Gamma^* Q \Sigma^*$. The relation of a *move*, symbolically denoted by $\vdash_M$, is defined over $\Gamma^* Q \Sigma^*$ as follows:

$$x\gamma pay \vdash_M xwqy$$

if and only if $x\gamma pay, xwqy \in \Gamma^* Q \Sigma^*$ and $\gamma pa \rightarrow wq \in R$.

Let $\vdash_M^k$, $\vdash_M^*$, and $\vdash_M^+$ denote the kth power of $\vdash_M$, for some $k \geq 0$, the reflexive-transitive closure of $\vdash_M$, and the transitive closure of $\vdash_M$, respectively.

□

For an extended pushdown automaton, there exist three ways of language acceptance: (1) by entering a final state, (2) by emptying its pushdown, and (3) by entering a final state and emptying its pushdown. All of them are defined next.

Definition 3.4.7. Let $M = (Q, \Sigma, \Gamma, R, s, S, F)$ be an extended pushdown automaton. The *language accepted by M by final state* is denoted by $L_f(M)$ and defined as

$$L_f(M) = \{w \in \Sigma^* \mid Ssw \vdash_M^* \gamma f, f \in F, \gamma \in \Gamma^*\}$$

The *language accepted by M by empty pushdown* is denoted by $L_e(M)$ and defined as

$$L_e(M) = \{w \in \Sigma^* \mid Ssw \vdash_M^* q, q \in Q\}$$

The *language accepted by M empty pushdown and final state*, denoted by $L_{ef}(M)$, is defined as

$$L_{ef}(M) = \{w \in \Sigma^* \mid Ssw \vdash_M^* f, f \in F\}$$ □

Let $\mathbf{EPDA}_f$, $\mathbf{EPDA}_e$, and $\mathbf{EPDA}_{ef}$ denote the language families accepted by extended pushdown automata accepting by final state, by empty pushdown, and by final state and empty pushdown, respectively.

All of the three ways of acceptance are equivalent:

Theorem 3.4.8 (see [10]). $\mathbf{EPDA}_f = \mathbf{EPDA}_e = \mathbf{EPDA}_{ef}$

If an extended pushdown automaton rewrites a single symbol on its pushdown top during every move, we obtain a pushdown automaton, defined next.

Definition 3.4.9. Let $M = (Q, \Sigma, \Gamma, R, s, S, F)$ be an extended pushdown automaton. Then, M is a *pushdown automaton* if and only if $\gamma pa \rightarrow wq \in R$ implies that $|\gamma| = 1$. □

To make definitions and proofs concerning pushdown automata more readable, we sometimes denote a rule $Apa \rightarrow wq$ with a unique label r as $r\colon Apa \rightarrow wq$, which is formalized in the following definition.

Definition 3.4.10. Let $M = (Q, \Sigma, \Gamma, R, s, S, F)$ be a pushdown automaton. Let Ψ be an alphabet of *rule labels* such that $\text{card}(\Psi) = \text{card}(R)$, and ψ be a bijection from R to Ψ. For simplicity, to express that ψ maps a rule, $Apa \rightarrow wq \in R$, to r, where $r \in \Psi$, we write $r\colon Apa \rightarrow wq \in R$; in other words, $r\colon Apa \rightarrow wq$ means $\psi(Apa \rightarrow wq) = r$.

For every $x \in \Gamma^*$, $y \in \Sigma^*$, and $r\colon Apa \rightarrow wq \in R$, M makes a *move* from configuration $xApay$ to configuration $xwqy$ according to r, written as

$$xApay \vdash_M xwqy \; [r]$$

Let χ be any configuration of M. M makes *zero moves* from χ to χ according to ε, symbolically written as

$$\chi \vdash_M^0 \chi \ [\varepsilon]$$

Let there exist a sequence of configurations $\chi_0, \chi_1, \ldots, \chi_n$ for some $n \geq 1$ such that $\chi_{i-1} \vdash_M \chi_i \ [r_i]$, where $r_i \in \Psi$, for $i = 1, \ldots, n$, then M makes n *moves* from χ_0 to χ_n according to $r_1 \cdots r_n$, symbolically written as

$$\chi_0 \vdash_M^n \chi_n \ [r_1 \cdots r_n]$$

Define $\vdash_M^*$ and $\vdash_M^+$ in the standard manner. □

Let $\mathbf{PDA}_f$, $\mathbf{PDA}_e$, and $\mathbf{PDA}_{ef}$ denote the language families accepted by pushdown automata accepting by final state, by empty pushdown, and by final state and empty pushdown, respectively.

Theorem 3.4.11 (see [10]). $\mathbf{PDA}_f = \mathbf{PDA}_e = \mathbf{PDA}_{ef}$

As the next theorem states, pushdown automata characterize the family of context-free languages.

Theorem 3.4.12 (see [10]).

$$\mathbf{CF} = \mathbf{EPDA}_f = \mathbf{EPDA}_e = \mathbf{EPDA}_{ef} = \mathbf{PDA}_f = \mathbf{PDA}_e = \mathbf{PDA}_{ef}$$

References

1. Chomsky, N.: Three models for the description of language. IRE Trans. Inform. Theory **2**(3), 113–124 (1956)
2. Chomsky, N.: On certain formal properties of grammars. Inform. Control **2**, 137–167 (1959)
3. Church, A.: An unsolvable problem of elementary number theory. Am. J. Math. **58**(2), 345–363 (1936)
4. Ginsburg, S., Greibach, S.A., Harrison, M.: One-way stack automata. J. ACM **14**(2), 389–418 (1967)
5. Kleijn, H.C.M.: Selective substitution grammars based on context-free productions. Ph.D. thesis, Leiden University, Netherlands (1983)
6. Kleijn, H.C.M.: Basic ideas of selective substitution grammars. In: Trends, Techniques, and Problems in Theoretical Computer Science. Lecture Notes in Computer Science, vol. 281, pp. 75–95. Springer, Berlin (1987)
7. Kleijn, H.C.M., Rozenberg, G.: On the generative power of regular pattern grammars. Acta Inform. **20**, 391–411 (1983)
8. Kolář, D., Meduna, A.: Regulated pushdown automata. Acta Cybern. **2000**(4), 653–664 (2000)
9. Markov, A.A.: The theory of algorithms. Am. Math. Soc. Transl. **15**(2), 1–14 (1960)
10. Meduna, A.: Automata and Languages: Theory and Applications. Springer, London (2000)
11. Meduna, A.: Simultaneously one-turn two-pushdown automata. Int. J. Comput. Math. **2003**(80), 679–687 (2003)

12. Meduna, A.: Two-way metalinear PC grammar systems and their descriptional complexity. Acta Cybern. **2004**(16), 385–397 (2004)
13. Post, E.: Formal reductions of the general combinatorial decision problem. Am. J. Math. **65**(2), 197–215 (1943)
14. Rogers, H.: Theory of Recursive Functions and Effective Computability. The MIT Press, Cambridge (1987)
15. Rozenberg, G.: Selective substitution grammars (towards a framework for rewriting systems), part 1: Definitions and examples. Elektronische Informationsverarbeitung und Kybernetik **13**(9), 455–463 (1977)
16. Rozenberg, G., Salomaa, A.: Mathematical Theory of L Systems. Academic, Orlando (1980)
17. Rozenberg, G., Salomaa, A.: The Book of L. Springer, New York (1986)
18. Salomaa, A.: Formal Languages. Academic, London (1973)
19. Turing, A.: On computable numbers, with an application to the entscheidungs problem. Proc. London Math. Soc. **42**(2), 230–265 (1936)
20. Wood, D.: Theory of Computation: A Primer. Addison-Wesley, Boston (1987)

Part II
Regulated Grammars: Fundamentals

This part, consisting of two comprehensive chapters, gives the fundamentals of regulated grammars. It classifies these grammars into two categories—context-based regulated grammars (Chap. 4) and rule-based regulated grammars (Chap. 5).

Chapter 4 gives an extensive and thorough coverage of regulated grammars that generate languages under various context-related restrictions. First, it views classical grammars as context-based regulated grammars. Then, it studies *context-conditional grammars* and their variants, including *random context grammars*, *generalized forbidding grammars*, *semi-conditional grammars*, and *simple semi-conditional grammars*. They all have their rules enriched by permitting and forbidding strings, referred to as permitting and forbidding conditions, respectively. These grammars regulate the language generation process so they require the presence of permitting conditions and, simultaneously, the absence of forbidding conditions in the rewritten sentential forms. Finally, this chapter covers *scattered context grammars*, which regulate their language generation so they simultaneously rewrite several prescribed nonterminals scattered throughout sentential forms.

Chapter 5 studies grammatical regulation underlain by restrictions placed on the use of rules. Four types of grammars regulated in this way are covered—namely, regular-controlled, matrix, programmed, and state grammars. *Regular-controlled grammars* control the use of rules by regular languages. *Matrix grammars* represent special cases of regular-control grammars whose control languages have the form of the iteration of finite languages. *Programmed grammars* base their regulation on binary relations over the sets of rules. Finally, *state grammars* regulate the use of rules by states in a way that strongly resembles the finite-state control of finite automata.

Chapter 4
Context-Based Grammatical Regulation

Abstract The generation of languages under various contextual restrictions represents perhaps the most natural and significant way of grammatical regulation. Therefore, this comprehensive over-100-page seven-section chapter gives an extensive and thorough coverage of grammars regulated in this way. First, Sect. 4.1 considers classical grammars, including phrase-structure and context-sensitive grammars, as contextually regulated grammars. It pays a special attention to their normal forms and uniform rewriting; thereby, it actually extends the material covered in Sect. 3.3. In a sense, Sects. 4.2 through 4.6 deal with a single language-generating regulated model—*context-conditional grammars*. Indeed, Sect. 4.2 introduces their general version and establishes key results concerning them. Sects. 4.3 through 4.6 discusses their special cases—namely, Sects. 4.3–4.6 cover *random context grammars*, *generalized forbidding grammars*, *semi-conditional grammars*, and *simple semi-conditional grammars*, respectively. Section 4.7 closes this chapter by discussing yet another type of contextually regulated grammars—*scattered context grammars*. These grammars regulate their language generation so they simultaneously rewrite several prescribed nonterminals scattered throughout sentential forms during derivation steps.

Keywords Context-based grammatical regulation • Context-sensitive grammars • Phrase-structure grammars • Context-conditional grammars • Random context grammars • Generalized forbidding grammars • Semi-conditional grammars • Scattered context grammars

The present seven-section chapter discusses context-regulated grammars, which regulate their derivations by placing context-related restrictions upon their rewritten sentential forms. Section 4.1 demonstrates that classical grammars can be viewed as grammars regulated in this way. It concentrates its attention on their normal forms and uniform rewriting in them. Section 4.2 introduces general versions of context-conditional grammars—the main grammatical model of the

A. Meduna and P. Zemek, *Regulated Grammars and Automata*,
DOI 10.1007/978-1-4939-0369-6_4,

present chapter as a whole—and establishes fundamental results about them. Then, Sects. 4.3 through 4.6 discusses four special cases of these general versions—namely, Sects. 4.3–4.6 study random context grammars, generalized forbidding grammars, semi-conditional grammars, and simple semi-conditional grammars, respectively. Section 4.7 closes this chapter by discussing another important type of context-regulated grammars—scattered context grammars.

4.1 Classical Grammars Viewed as Tight-Context Regulated Grammars

Classical grammars, such as context-sensitive and phrase-structure grammars (see Sect. 3.3), can be seen, in a quite natural way, as context-regulated grammars. Indeed, on the left-hand sides of their rules, they have strings—that is, sequences of symbols, not single symbols. In effect, they thus regulate their derivations by prescribing sequences of neighboring symbols that can be rewritten during a derivation step; this kind of regulation is generally referred to as tight-context regulation to distinct it from scattered-context regulation, in which the symbol-neighborhood requirement is dropped (see Sect. 4.7).

In general, tight-context regulated grammars, represented by context-sensitive and phrase-structure grammars in this section, may have rules of various forms, and they may generate a very broad variety of completely different sentential forms during the generation of their languages. As obvious, this inconsistency concerning the form of rules as well as rewritten strings represents an undesirable phenomenon in theory as well as in practice. From a theoretical viewpoint, the demonstration of properties concerning languages generated in this inconsistent way usually lead to unbearably tedious proofs. From a practical viewpoint, this kind of language generation is obviously difficult to apply and implement. Therefore, we pay a special attention to arranging these grammars so they generate their languages in a more uniform way.

The present section consists of Sects. 4.1.1 and 4.1.2. Section 4.1.1 modifies grammatical rules so they all satisfy some simple prescribed forms, generally referred to as normal forms. Then, Sect. 4.1.2 explains how to perform tight-context rewriting over strings that have a uniform form.

4.1.1 Normal Forms

In this section, we convert context-sensitive and phrase-structure grammars into several normal forms, including the Kuroda, Penttonen, and Geffert normal forms. We also reduce the number of context-free rules in these grammars. In addition, we describe the Greibach and Chomsky normal forms for context-free grammars.

Definition 4.1.1. Let $G = (N, T, P, S)$ be a phrase-structure grammar. G is in the *Kuroda normal form* (see [7]) if every rule in P is of one of the following four forms

$$\text{(i) } AB \to CD \qquad \text{(ii) } A \to BC \qquad \text{(iii) } A \to a \qquad \text{(iv) } A \to \varepsilon$$

where $A, B, C, D \in N$, and $a \in T$. □

Theorem 4.1.2 (See [7]). *For every phrase-structure grammar G, there is a phrase-structure grammar G' in the Kuroda normal form such that $L(G') = L(G)$.*

Definition 4.1.3. Let $G = (N, T, P, S)$ be a phrase-structure grammar. G is in the *Penttonen normal form* (see [17]) if every rule in P is of one of the following four forms

$$\text{(i) } AB \to AC \qquad \text{(ii) } A \to BC \qquad \text{(iii) } A \to a \qquad \text{(iv) } A \to \varepsilon$$

where $A, B, C \in N$, and $a \in T$. □

In other words, G is in the Penttonen normal form if G is in the Kuroda normal form an every $AB \to CD \in P$ satisfies that $A = C$.

Theorem 4.1.4 (See [17]). *For every phrase-structure grammar G, there is a phrase-structure grammar G' in the Penttonen normal form such that $L(G') = L(G)$.*

Theorem 4.1.5 (See [17]). *For every context-sensitive grammar G, there is a context-sensitive grammar G' in the Penttonen normal form such that $L(G') = L(G)$.*

Observe that if G is a context-sensitive grammar in the Pentonnen normal form, then none of its rules is of the form (iv), which is not context-sensitive.

Theorems 4.1.4 and 4.1.5 can be further modified so that for every context-sensitive rule of the form $AB \to AC \in P$, where $A, B, C \in N$, there exist no $B \to x$ or $BD \to BE$ in P for any $x \in V^*$, $D, E \in N$:

Theorem 4.1.6. *Every context-sensitive language can be generated by a context-sensitive grammar $G = (N_{CF} \cup N_{CS}, T, P, S)$, where N_{CF}, N_{CS}, and T are pairwise disjoint alphabets, and every rule in P is either of the form $AB \to AC$, where $B \in N_{CS}$, $A, C \in N_{CF}$, or of the form $A \to x$, where $A \in N_{CF}$ and $x \in N_{CS} \cup T \cup N_{CF}^2$.*

Proof. Let $G' = (N, T, P', S)$ be a context-sensitive grammar in the Penttonen normal form (see Theorem 4.1.5). Then, let

$$G = \big(N_{CF} \cup N_{CS}, T, P, S\big)$$

be the context-sensitive grammar defined as follows:

$$N_{CF} = N$$
$$N_{CS} = \{\tilde{B} \mid AB \to AC \in P',\ A, B, C \in N\}$$

$$P = \{A \to x \mid A \to x \in P',\ A \in N,\ x \in T \cup N^2\} \cup$$
$$\{B \to \tilde{B},\ A\tilde{B} \to AC \mid AB \to AC \in P',\ A, B, C \in N\}$$

Obviously, $L(G') = L(G)$ and G is of the required form, so the theorem holds. □

Theorem 4.1.7. *Every recursively enumerable language can be generated by a phrase-structure grammar* $G = (N_{CF} \cup N_{CS}, T, P, S)$, *where* N_{CF}, N_{CS}, *and* T *are pairwise disjoint alphabets, and every rule in* P *is either of the form* $AB \to AC$, *where* $B \in N_{CS}$, $A, C \in N_{CF}$, *or of the form* $A \to x$, *where* $A \in N_{CF}$ *and* $x \in N_{CS} \cup T \cup N_{CF}^2 \cup \{\varepsilon\}$.

Proof. The reader can prove this theorem by analogy with the proof of Theorem 4.1.6. □

The next two normal forms limit the number of nonterminals and non-context-sensitive rules in phrase-structure grammars.

Definition 4.1.8. Let G be a phrase-structure grammar. G is in the *first Geffert normal form* (see [4]) if it is of the form

$$G = \big(\{S, A, B, C, D\}, T, P \cup \{ABC \to \varepsilon\}, S\big)$$

where P contains context-free rules of the following three forms

(i) $S \to uSa$ (ii) $S \to uSv$ (iii) $S \to uv$

where $u \in \{A, AB\}^*$, $v \in \{BC, C\}^*$, and $a \in T$. □

Theorem 4.1.9 (See [4]). *For every recursively enumerable language* K, *there exists a phrase-structure grammar* G *in the first Geffert normal form such that* $L(G) = K$. *In addition, every successful derivation in* G *is of the form* $S \Rightarrow_G^* w_1 w_2 w$ *by rules from* P, *where* $w_1 \in \{A, AB\}^*$, $w_2 \in \{BC, C\}^*$, $w \in T^*$, *and* $w_1 w_2 w \Rightarrow_G^* w$ *is derived by* $ABC \to \varepsilon$.

Definition 4.1.10. Let G be a phrase-structure grammar. G is in the *second Geffert normal form* (see [4]) if it is of the form

$$G = \big(\{S, A, B, C, D\}, T, P \cup \{AB \to \varepsilon, CD \to \varepsilon\}, S\big)$$

where P contains context-free rules of the following three forms

(i) $S \to uSa$ (ii) $S \to uSv$ (iii) $S \to uv$

where $u \in \{A, C\}^*$, $v \in \{B, D\}^*$, and $a \in T$. □

Theorem 4.1.11 (See [4]). *For every recursively enumerable language* K, *there exists a phrase-structure grammar* G *in the second Geffert normal form such that* $L(G) = K$. *In addition, every successful derivation in* G *is of the form* $S \Rightarrow_G^*$

w_1w_2w by rules from P, where $w_1 \in \{A, C\}^*$, $w_2 \in \{B, D\}^*$, $w \in T^*$, and $w_1w_2w \Rightarrow_G^* w$ is derived by $AB \rightarrow \varepsilon$ and $CD \rightarrow \varepsilon$.

Next, we establish two new normal forms for phrase-structure grammars with a limited number of context-free rules in a prescribed form and, simultaneously, with non-context-free rules in a prescribed form. Specifically, we establish the following two normal forms of this kind.

(I) First, we explain how to turn any phrase-structure grammar to an equivalent phrase-structure grammar that has $2 + n$ context-free rules, where n is the number of terminals, and every context-free rule is of the form $A \rightarrow x$, where x is a terminal, a two-nonterminal string, or the empty string. In addition, every non-context-free rule is of the form $AB \rightarrow CD$, where A, B, C, D are nonterminals.

(II) In the second normal form, phrase-structure grammars have always only two context-free rules—that is, the number of context-free rules is reduced independently of the number of terminals as opposed to the first normal form. Specifically, we describe how to turn any phrase-structure grammar to an equivalent phrase-structure grammar that has two context-free rules of the forms $A \rightarrow \varepsilon$ and $A \rightarrow BC$, where A, B, C are nonterminals and ε denotes the empty string, and in addition, every non-context-free rule is of the form $AB \rightarrow CD$, where A, B, D are nonterminals and C is nonterminal or a terminal.

Theorem 4.1.12. *Let G be a phrase-structure grammar. Then, there is an equivalent phrase-structure grammar*

$$H = \left(N, T, P_1 \cup P_2 \cup P_3, S\right)$$

with

$$\begin{aligned} P_1 &= \{AB \rightarrow CD \mid A, B, C, D \in N\} \\ P_2 &= \{S \rightarrow S\#, \# \rightarrow \varepsilon\} \\ P_3 &= \{A \rightarrow a \mid A \in N, a \in T\} \end{aligned}$$

where $\# \in N$.

Proof. Let $G = (N, T, P, S)$ be a phrase-structure grammar. By Theorem 4.1.2, we assume that G is in the Kuroda normal form. Set $\bar{T} = \{\bar{a} \mid a \in T\}$. Without any loss of generality, we assume that N, T, $\bar{T}$, and $\{\#\}$ are pairwise disjoint. Construct the phrase-structure grammar

$$H = \left(N', T, P_1' \cup P_2' \cup P_3', S\right)$$

as follows. Initially, set $N' = N \cup \bar{T} \cup \{\#\}$, $P_1' = \emptyset$, $P_2' = \{S \rightarrow S\#, \# \rightarrow \varepsilon\}$, and $P_3' = \{\bar{a} \rightarrow a \mid a \in T\}$. Perform (1) through (5), given next.

(1) For each $AB \to CD \in P$, where $A, B, C, D \in N$, add $AB \to CD$ to P_1'.
(2) For each $A \to BC \in P$, where $A, B, C \in N$, add $A\# \to BC$ to P_1'.
(3) For each $A \to a \in P$, where $A \in N$ and $a \in T$, add $A\# \to \bar{a}\#$ to P_1'.
(4) For each $A \to \varepsilon \in P$, where $A \in N$, add $A\# \to \#\#$ to P_1'.
(5) For each $A \in N$, add $A\# \to \#A$ and $\#A \to A\#$ to P_1'.

Before proving that $L(H) = L(G)$, let us give an insight into the construction. We simulate G by H using the following sequences of derivation steps.

First, by repeatedly using $S \to S\#$, we generate a proper number of #s. Observe that if the number of #s is too low, the derivation can be blocked since rules from (2) consume # during their application. Furthermore, notice that only rules from (4) and the initial rule $S \to S\#$ increase the number of #s in sentential forms of H.

Next, we simulate each application of a rule in G by several derivation steps in H. By using rules from (5), we can move # in the current sentential form as needed. If we have # or B in a proper position next to A, we can apply a rule from (1) through (4). We can also apply $\# \to \varepsilon$ to remove any occurrence of # from a sentential form of H.

To conclude the simulation, we rewrite the current sentential form by rules of the form $\bar{a} \to a$ to generate a string of terminals. Observe that a premature application of a rule of this kind may block the derivation in H. Indeed, #s then cannot move freely through such a sentential form.

To establish $L(H) = L(G)$, we prove four claims. Claim 1 demonstrates that every $w \in L(H)$ can be generated in two stages; first, only nonterminals are generated, and then, all nonterminals are rewritten to terminals. Claim 2 shows that we can arbitrarily generate and move #s within sentential forms of H during the first stage. Claim 3 shows how derivations of G are simulated by H. Finally, Claim 4 shows how derivations of every $w \in L(H)$ in H are simulated by G.

Set $V = N \cup T$ and $V' = N' \cup T$. Define the homomorphism τ from V'^* to V^* as $\tau(X) = X$ for all $X \in V$, $\tau(\bar{a}) = a$ for all $a \in T$, and $\tau(\#) = \varepsilon$.

Claim 1. Let $w \in L(H)$. Then, there exists a derivation $S \Rightarrow_H^* x \Rightarrow^* w$, where $x \in N'^+$, and during $x \Rightarrow_H^* w$, only rules of the form $\bar{a} \to a$, where $a \in T$, are applied.

Proof. Let $w \in L(H)$. Since there are no rules in $P_1' \cup P_2' \cup P_3'$ with symbols from T on their left-hand sides, we can always rearrange all the applications of the rules occurring in $S \Rightarrow_H^* w$ so the claim holds. □

Claim 2. If $S \Rightarrow_H^* uv$, where $u, v \in V'^*$, then $S \Rightarrow_H^* u\#v$.

Proof. By an additional application of $S \to S\#$, we get $S \Rightarrow_H^* S\#^{n+1}$ instead of $S \Rightarrow_H^* S\#^n$ for some $n \geq 0$, so we derive one more # in the sentential form. From Claim 1, by applying rules from (5), # can freely migrate through the sentential form as needed, until a rule of the form $\bar{a} \to a$ is used. □

Claim 3. If $S \Rightarrow_G^k x$, where $x \in V^*$, for some $k \geq 0$, then $S \Rightarrow_H^* x'$, where $\tau(x') = x$.

Proof. This claim is established by induction on $k \geq 0$.

Basis. For $S \Rightarrow_G^0 S$, there is $S \Rightarrow_H^0 S$.

Induction Hypothesis. For some $k \geq 0$, $S \Rightarrow_G^k x$ implies that $S \Rightarrow_H^* x'$ such that $x = \tau(x')$.

Induction Step. Let $u, v \in N'^*$, $A, B, C, D \in N$, and $m \geq 0$. Assume that $S \Rightarrow_G^k y \Rightarrow_G x$. By the induction hypothesis, $S \Rightarrow_H^* y'$ with $y = \tau(y')$. Let us show the simulation of $y \Rightarrow_G x$ by an application of several derivation steps in H to get $y' \Rightarrow_H^+ x'$ with $\tau(x') = x$. This simulation is divided into the following four cases—(i) through (iv).

(i) *Simulation of* $AB \to CD$: $y' = uA\#^m Bv \Rightarrow_H^m u\#^m ABv \Rightarrow_H u\#^m CDv = x'$ using m derivation steps according to rules $A\# \to \#A$ from (5), and concluding the derivation by rule $AB \to CD$ from (1).

By the induction hypothesis and Claim 2, $y = \tau(u)A\tau(v)$ allows $y' = uA\#v$.

(ii) *Simulation of* $A \to BC$: $y' = uA\#v \Rightarrow_H uBCv = x'$ using rule $A\# \to BC$ from (2).

(iii) *simulation of* $A \to a$: $y' = uA\#v \Rightarrow_H u\bar{a}\#v = x'$ using rule $A\# \to \bar{a}\#$ from (3).

(iv) *Simulation of* $A \to \varepsilon$: $y' = uA\#v \Rightarrow_H u\#\#v = x'$ using rule $A\# \to \#\#$ from (4). □

Claim 4. If $S \Rightarrow_H^k x'$, *where* $x' \in N'^*$, *for some* $k \geq 0$, *then* $S \Rightarrow_G^* x$ *with* $x = \tau(x')$.

Proof. This claim is established by induction on $k \geq 0$.

Basis. For $S \Rightarrow_H^0 S$, there is $S \Rightarrow_G^0 S$.

Induction Hypothesis. For some $k \geq 0$, $S \Rightarrow_H^k x'$ implies that $S \Rightarrow_G^* x$ such that $x = \tau(x')$.

Induction Step. Let $u, v, w \in N'^*$ and $A, B, C, D \in N$. Assume that $S \Rightarrow_H^k y' \Rightarrow_H x'$. By the induction hypothesis, $S \Rightarrow_G^* y$ such that $y = \tau(y')$. Let us examine the following seven possibilities of $y' \Rightarrow_H x'$.

(i) $y' = uSv \Rightarrow_H uS\#v = x'$: Then,

$$\tau(y') = y = \tau(uSv) \Rightarrow_G^0 \tau(uS\#v) = \tau(uSv) = x = \tau(x')$$

(ii) $y' = uABv \Rightarrow_H uCDv = x'$: According to (1),

$$y = \tau(u)AB\tau(v) \Rightarrow_G \tau(u)CD\tau(v) = x$$

(iii) $y' = uA\#v \Rightarrow_H uBCv = x'$: According to the source rule in (2),

$$y = \tau(u)A\tau(\#v) \Rightarrow_G \tau(u)BC\tau(\#v) = \tau(u)BC\tau(v) = x$$

(iv) $y' = uA\#v \Rightarrow_H u\#\#v = x'$: By the corresponding rule $A \to \varepsilon$,

$$y = \tau(u)A\tau(v) \Rightarrow_G \tau(u\#\#v) = \tau(uv) = x$$

(v) $y' = uA\#v \Rightarrow_H u\#Av = x'$ or $y' = u\#Av \Rightarrow_H uA\#v = x'$: In G,

$$y = \tau(uA\#v) = \tau(u)A\tau(\#v) \Rightarrow_G^0 \tau(u\#)A\tau(v) = x$$

or

$$y = \tau(u\#Av) = \tau(u\#)A\tau(v) \Rightarrow_G^0 \tau(u)A\tau(\#v) = x$$

(vi) $y' = u\#v \Rightarrow_H uv = x'$: In G,

$$y = \tau(u\#v) \Rightarrow_G^0 \tau(uv) = x$$

(vii) $y' = u\bar{a}v \Rightarrow_H uav = x'$: In G,

$$y = \tau(u\bar{a}v) = \tau(u)a\tau(v) \Rightarrow_G^0 \tau(u)a\tau(v) = x$$ □

Next, we establish $L(H) = L(G)$. Consider Claim 3 with $x \in T^*$. Then, $S \Rightarrow_G^* x$ implies that $S \Rightarrow_H^* x$, so $L(G) \subseteq L(H)$. Let $w \in L(H)$. By Claim 1, $S \Rightarrow_H^* x \Rightarrow_H^* w$, where $x \in N'^+$, and during $x \Rightarrow_H^* w$, only rules of the form $\bar{a} \to a$, where $a \in T$, are applied. By Claim 4, $S \Rightarrow_G^* \tau(x) = w$, so $L(H) \subseteq L(G)$. Hence, $L(H) = L(G)$.

Since H is of the required form, the theorem holds. □

From the construction in the proof of Theorem 4.1.12, we obtain the following corollary concerning the number of nonterminals and rules in the resulting grammar.

Corollary 4.1.13. *Let $G = (N, T, P, S)$ be a phrase-structure grammar in the Kuroda normal form. Then, there is an equivalent phrase-structure grammar in the normal form from Theorem 4.1.12*

$$H = (N', T, P', S)$$

where

$$\text{card}(N') = \text{card}(N) + \text{card}(T) + 1$$

and

$$\text{card}(P') = \text{card}(P) + \text{card}(T) + 2(\text{card}(N) + 1)$$ □

If we drop the requirement on each symbol in the non-context-free rules to be a nonterminal, we can reduce the number of context-free rules even more.

Theorem 4.1.14. *Let G be a phrase-structure grammar. Then, there is an equivalent phrase-structure grammar*

$$H = (N, T, P_1 \cup P_2, S)$$

with

$$P_1 = \{AB \to XD \mid A, B, C \in N, X \in N \cup T\}$$
$$P_2 = \{S \to S\#, \# \to \varepsilon\}$$

where $\# \in N$.

Proof. Reconsider the proof of Theorem 4.1.12. Observe that we can obtain the new normal form by omitting the construction of P_3' and modifying step (3) in the following way

(3) For each $A \to a \in P$, where $A \in N$ and $a \in T$, add $A\# \to a\#$ to P_1'.

The rest of the proof is analogical to the proof of Theorem 4.1.12 and it is left to the reader. □

Finally, we define two normal forms for context-free grammars—the Chomsky and Greibach normal forms (see [1,5]).

Definition 4.1.15. Let G be a context-free grammar. G is in the *Chomsky normal form* if every $A \to x \in P$ satisfies that $x \in NN \cup T$. □

Theorem 4.1.16 (See [1]). *For every context-free grammar G, there is a context-free grammar G' in the Chomsky normal form such that $L(G') = L(G)$.*

Definition 4.1.17. Let G be a context-free grammar. G is in the *Greibach normal form* if every $A \to x \in P$ satisfies that $x \in TN^*$. □

Theorem 4.1.18 (See [5]). *For every context-free grammar G, there is a context-free grammar G' in the Greibach normal form such that $L(G') = L(G)$.*

4.1.2 Uniform Rewriting

Classical grammars can produce a very broad variety of quite different sentential forms during the generation of their languages. This inconsistent generation represents a highly undesirable grammatical phenomenon. In theory, the demonstration of properties concerning languages generated in this way lead to extremely tedious proofs. In practice, the inconsistent generation of languages is uneasy to apply and implement. Therefore, in this section, we explain how to reduce or even overcome this difficulty by making the language generation more uniform. Specifically, phrase-structure grammars are transformed so that they generate only strings that have a uniform permutation-based form.

More precisely, the present section demonstrates that for every phrase-structure grammar G, there exists an equivalent phrase-structure grammar

$$G' = (\{S, 0, 1\} \cup T, T, P, S)$$

so that every $x \in F(G')$ satisfies

$$x \in T^* \Pi(w)^*$$

where $w \in \{0, 1\}^*$ (recall that $F(G')$ is defined in Definition 3.3.2). Then, it makes this conversion so that for every $x \in F(G)$,

$$x \in \Pi(w)^* T^*$$

Let $G = (V, T, P, S)$ be a phrase-structure grammar. Notice that $\text{alph}(L(G)) \subseteq T$. If $a \in T - \text{alph}(L(G))$, then a actually acts as a pseudoterminal because it appears in no string of $L(G)$. Every transformation described in this section assumes that its input grammar contains no pseudoterminals of this kind, and does not contain any useless nonterminals either.

Let j be a natural number. Set

$$\begin{aligned}\mathbf{PS}[.j] = \{L \mid L = L(G),\ & \text{where } G = (V, T, P, S) \text{ is a phrase-structure} \\ & \text{grammar such that } \text{card}(\text{alph}(F(G)) - T) = j \text{ and} \\ & F(G) \subseteq T^* \Pi(w)^*, \text{ where } w \in (V - T)^*\}\end{aligned}$$

Analogously, set

$$\begin{aligned}\mathbf{PS}[j.] = \{L \mid L = L(G),\ & \text{where } G = (V, T, P, S) \text{ is a phrase-structure} \\ & \text{grammar such that } \text{card}(\text{alph}(F(G)) - T) = j \text{ and} \\ & F(G) \subseteq \Pi(w)^* T^*, \text{ where } w \in (V - T)^*\}\end{aligned}$$

Lemma 4.1.19. *Let G be a phrase-structure grammar. Then, there exists a phrase-structure grammar, $G' = (\{S, 0, 1\} \cup T, T, P, S)$, satisfying $L(G') = L(G)$ and $F(G') \subseteq T^* \Pi(1^{n-2}00)^*$.*

Proof. Let $G = (V, T, Q, \$)$ be a phrase-structure grammar, where V is the total alphabet of G, T is the terminal alphabet of G, Q is the set of rules of G, and $\$$ is the start symbol of G. Without any loss of generality, assume that $V \cap \{0, 1\} = \emptyset$. The following construction produces an equivalent phrase-structure grammar

$$G' = (\{S, 0, 1\} \cup T, T, P, S)$$

such that $F(G') \subseteq T^* \Pi(1^{n-2}00)^*$, for some natural number n.

For some integers m, n such that $m \geq 3$ and $2m = n$, introduce an injective homomorphism β from V to

$$(\{1\}^m \{1\}^* \{0\} \{1\}^* \{0\} \cap \{0, 1\}^n) - \{1^{n-2}00\}$$

Extend the domain of β to V^*. Define the phrase-structure grammar

$$G' = \big(\{S, 0, 1\} \cup T, T, P, S\big)$$

with

$$\begin{aligned} P = \{&S \rightarrow 1^{n-1}00\beta(\$)1^{n-1}00\} \cup \\ \{&\beta(x) \rightarrow \beta(y) \mid x \rightarrow y \in Q\} \cup \\ \{&1^{n-2}00\beta(a) \rightarrow a1^{n-2}00 \mid a \in T\} \cup \\ \{&1^{n-2}001^{n-2}00 \rightarrow \varepsilon\} \end{aligned}$$

Claim 1. Let $S \Rightarrow_{G'}^{h} w$, where $w \in V^*$ and $h \geq 1$. Then,

$$w \in T^*(\{\varepsilon\} \cup \{1^{n-2}00\}(\beta(V))^*\{1^{n-2}00\})$$

Proof. The claim is proved by induction on $h \geq 1$.

Basis. Let $h = 1$. That is,

$$S \Rightarrow_{G'} 1^{n-1}00\beta(\$)1^{n-1}00 \; [\$ \rightarrow 1^{n-1}00\beta(\$)1^{n-1}00]$$

As

$$1^{n-2}00\beta(S)1^{n-2}00 \in T^*(\{1^{n-2}00\}(\beta(V))^*\{1^{n-2}00\} \cup \{\varepsilon\})$$

the basis holds.

Induction Hypothesis. Suppose that for some $k \geq 0$, if $S \Rightarrow_{G'}^{i} w$, where $i = 1, \ldots, k$ and $w \in V^*$, then $w \in T^*(\{1^{n-2}00\}(\beta(V))^*\{1^{n-2}00\} \cup \{\varepsilon\})$.

Induction Step. Consider any derivation of the form

$$S \Rightarrow_{G'}^{k+1} w$$

where $w \in V^* - T^*$. Express $S \Rightarrow_{G'}^{k+1} w$ as

$$\begin{aligned} S &\Rightarrow_{G'}^{k} u\,\mathrm{lhs}(p)v \\ &\Rightarrow_{G'} u\,\mathrm{rhs}(p)v \; [p] \end{aligned}$$

where $p \in P$ and $w = u\,\mathrm{rhs}(p)v$. Less formally, after k steps, G' derives $u\,\mathrm{lhs}(p)v$. Then, by using p, G' replaces $\mathrm{lhs}(p)$ with $\mathrm{rhs}(p)$ in $u\,\mathrm{lhs}(p)v$, so it obtains $u\,\mathrm{rhs}(p)v$. By the induction hypothesis,

$$u\,\mathrm{lhs}(p)v \in T^*(\{1^{n-1}00\}(\beta(V))^*\{1^{n-2}00\} \cup \{\varepsilon\})$$

As $\text{lhs}(p) \notin T^*$, $u\,\text{lhs}(p)v \notin T^*$. Therefore,

$$u\,\text{lhs}(p)v \in T^*\{1^{n-2}00\}(\beta(V))^*\{1^{n-2}00\}$$

Let

$$u\,\text{lhs}(p)v \in T^*\{1^{n-2}00\}(\beta(V))^j\{1^{n-2}00\}$$

in G', for some $j \geq 1$. By the definition of P, p satisfies one of the following three properties.

(i) Let $\text{lhs}(p) = \beta(x)$ and $\text{rhs}(p) = \beta(y)$, where $x \to y \in Q$, At this point,

$$u \in T^*\{1^{n-2}00\}\{\beta(V)\}^r$$

for some $r \geq 0$, and

$$v \in \{\beta(V)\}^{(j-|\text{lhs}(p)|-r)}\{1^{n-2}00\}$$

Distinguish between these two cases: $|x| \leq |y|$ and $|x| > |y|$.

(i.i) Let $|x| \leq |y|$. Set $s = |y| - |x|$. Observe that

$$u\,\text{rhs}(p)v \in T^*\{1^{n-2}00\}(\beta(V))^{(j+s)}\{1^{n-2}00\}$$

As $w = u\,\text{rhs}(p)v$,

$$w \in T^*(\{1^{n-2}00\}(\beta(V))^*\{1^{n-2}00\} \cup \{\varepsilon\})$$

(i.ii) Let $|x| > |y|$. By analogy with (a), prove that

$$w \in T^*(\{1^{n-2}00\}(\beta(V))^*\{1^{n-2}00\} \cup \{\varepsilon\})$$

(ii) Assume that $\text{lhs}(p) = 1^{n-1}00\beta(a)$ and $\text{rhs}(p) = a1^{n-2}00$, for some $a \in T$. Notice that

$$u\,\text{lhs}(p)v \in T^*\{1^{n-2}00\}(\beta(V))^j\{1^{n-2}00\}$$

implies that $u \in T^*$ and

$$v \in (\beta(V))^{(j-1)}\{1^{n-2}00\}$$

Then,

$$u\,\text{rhs}(p)v \in T^*\{a\}\{1^{n-2}00\}(\beta(V))^{(j-1)}\{1^{n-2}00\}$$

As $w = u\,\text{rhs}(p)v$,

$$w \in T^*(\{1^{n-2}00\}(\beta(V))^*\{1^{n-2}00\} \cup \{\varepsilon\})$$

(iii) Assume that $\text{lhs}(p) = 1^{n-2}001^{n-2}00$ and $\text{rhs}(p) = \varepsilon$. Then, $j = 0$ in

$$T^*\{1^{n-2}00\}(\beta(V))^j\{1^{n-2}00\}$$

so

$$u\,\text{lhs}(p)v \in T^*\{1^{n-2}00\}\{1^{n-2}00\}$$

and $u\,\text{rhs}(p)v \in T^*$. As $w = u\,\text{rhs}(p)v$,

$$w \in T^*(\{1^{n-2}00\}(\beta(V))^*\{1^{n-2}00\} \cup \{\varepsilon\})$$ □

Claim 2. Let $S \Rightarrow_{G'}^+ u \Rightarrow_{G'}^* z$, where $z \in T^*$. Then, $u \in T^*\Pi(1^{n-2}00)^*$.

Proof. Let $S \Rightarrow_{G'}^+ u \Rightarrow_{G'}^* z$, where $z \in T^*$. By Claim 1,

$$u \in T^*(\{1^{n-2}00\}(\beta(V))^*\{1^{n-2}00\} \cup \{\varepsilon\})$$

and by the definition of β, $u \in T^*\Pi(1^{n-2}00)^*$. □

Claim 3. Let $\$ \Rightarrow_G^m w$, for some $m \geq 0$. Then, $S \Rightarrow_{G'}^+ 1^{n-2}00\beta(w)1^{n-2}00$.

Proof. The claim is proved by induction on $m \geq 0$.

Basis. Let $m = 0$. That is, $\$ \Rightarrow_G^0 \$$. As

$$S \Rightarrow_{G'} 1^{n-1}00\beta(\$)1^{n-1}00\ [S \rightarrow 1^{n-1}00\beta(\$)1^{n-1}00]$$

the basis holds.

Induction Hypothesis. Suppose that for some $j \geq 1$, if $\$ \Rightarrow_G^i w$, where $i = 1, \ldots, j$ and $w \in V^*$, then $S \Rightarrow_{G'}^* \beta(w)$.

Induction Step. Let $\$ \Rightarrow_G^{j+1} w$. Express $\$ \Rightarrow_G^{j+1} w$ as

$$\$ \Rightarrow_G^j uxv \Rightarrow_G uyv\ [x \rightarrow y]$$

where $x \rightarrow y \in Q$ and $w = uyv$. By the induction hypothesis,

$$S \Rightarrow_{G'}^+ 1^{n-2}00\beta(uxv)1^{n-2}00$$

Express $\beta(uxv)$ as $\beta(uxv) = \beta(u)\beta(x)\beta(v)$. As $x \rightarrow y \in P$, $\beta(x) \rightarrow \beta(y) \in P$. Therefore,

$$\begin{aligned} S &\Rightarrow^{+}_{G'} 1^{n-2}00\beta(u)\beta(x)\beta(v)1^{n-2}00 \\ &\Rightarrow_{G'} 1^{n-2}00\beta(u)\beta(y)\beta(v)1^{n-2}00 \ [\beta(x) \to \beta(y)] \end{aligned}$$

Because $w = uyv$, $\beta(w) = \beta(u)\beta(y)\beta(v)$, so

$$S \Rightarrow^{+}_{G'} 1^{n-2}00\beta(w)1^{n-2}00$$

□

Claim 4. $L(G) \subseteq L(G')$

Proof. Let $w \in L(G)$. Thus, $\$ \Rightarrow^{*}_{G} w$ with $w \in T^*$. By Claim 3,

$$S \Rightarrow^{+}_{G'} 1^{n-2}00\beta(w)1^{n-2}00$$

Distinguish between these two cases: $w = \varepsilon$ and $w \neq \varepsilon$.

(i) If $w = \varepsilon$, $1^{n-2}00\beta(w)1^{n-2}00 = 1^{n-2}001^{n-2}00$. As $1^{n-2}001^{n-2}00 \to \varepsilon \in P$,

$$\begin{aligned} S &\Rightarrow^{*}_{G'} 1^{n-2}001^{n-2}00 \\ &\Rightarrow_{G'} \varepsilon \ [1^{n-2}001^{n-2}00 \to \varepsilon] \end{aligned}$$

Thus, $w \in L(G')$.

(ii) Assume that $w \neq \varepsilon$. Express w as $w = a_1a_2 \cdots a_{n-1}a_n$ with $a_i \in T$ for $i = 1, \dots, n$, $n \geq 0$. Because

$$(\{1^{n-2}00\beta(a) \to a1^{n-2}00 \mid a \in T\} \cup \{1^{n-2}001^{n-2}00 \to \varepsilon\}) \subseteq P$$

there exists

$$\begin{aligned} S &\Rightarrow^{*}_{G'} 1^{n-2}00\beta(a_1)\beta(a_2)\cdots\beta(a_{n-1})\beta(a_n)1^{n-2}00 \\ &\Rightarrow_{G'} a_1 1^{n-2}00\beta(a_2)\cdots\beta(a_{n-1})\beta(a_n)1^{n-2}00 \\ &\qquad [1^{n-2}00\beta(a_1) \to a_1 1^{n-2}00] \\ &\Rightarrow_{G'} a_1a_2 1^{n-2}00\beta(a_3)\cdots\beta(a_{n-1})\beta(a_n)1^{n-2}00 \\ &\qquad [1^{n-2}00\beta(a_2) \to a_2 1^{n-2}00] \\ &\ \ \vdots \\ &\Rightarrow_{G'} a_1a_2\cdots a_{n-2} 1^{n-2}00\beta(a_{n-1})\beta(a_n)1^{n-2}00 \\ &\qquad [1^{n-2}00\beta(a_{n-2}) \to a_{n-2} 1^{n-2}00] \\ &\Rightarrow_{G'} a_1a_2\cdots a_{n-2}a_{n-1} 1^{n-2}00\beta(a_n)1^{n-2}00 \\ &\qquad [1^{n-2}00\beta(a_{n-1}) \to a_{n-1} 1^{n-2}00] \\ &\Rightarrow_{G'} a_1a_2\cdots a_{n-2}a_{n-1}a_n 1^{n-2}001^{n-2}00 \\ &\qquad [1^{n-2}00\beta(a_n) \to a_n 1^{n-2}00] \\ &\Rightarrow_{G'} a_1a_2\cdots a_{n-2}a_{n-1}a_n \\ &\qquad [1^{n-2}001^{n-2}00 \to \varepsilon] \end{aligned}$$

Therefore, $w \in L(G')$.

□

Claim 5. Let $S \Rightarrow_{G'}^{m} 1^{n-2}00w1^{n-2}00$, *where* $w \in \{0, 1\}^*$, *for some* $m \geq 1$. *Then,* $\$ \Rightarrow_G^* \beta^{-1}(w)$.

Proof. This claim is proved by induction on $m \geq 1$.

Basis. Let $m = 1$. That is,

$$S \Rightarrow_{G'} 1^{n-2}00w1^{n-2}00$$

where $w \in \{0, 1\}^*$. Then, $w = \beta(\$)$. As $\$ \Rightarrow_G^0 \$$, the basis holds.

Induction Hypothesis. Suppose that for some $j \geq 1$, if $S \Rightarrow_{G'}^{i} 1^{n-2}00w1^{n-2}00$, where $i = 1, \ldots, j$ and $w \in \{0, 1\}^*$, then $\$ \Rightarrow_G^+ \beta^{-1}(w)$.

Induction Step. Let

$$S \Rightarrow_{G'}^{j+1} 1^{n-2}00w1^{n-2}00$$

where $w \in \{0, 1\}^*$. As $w \in \{0, 1\}^*$,

$$S \Rightarrow_{G'}^{j+1} 1^{n-2}00w1^{n-2}00$$

can be expressed as

$$\begin{aligned} S &\Rightarrow_{G'}^{j} 1^{n-2}00u\beta(x)v1^{n-2}00 \\ &\Rightarrow_{G'} 1^{n-2}00u\beta(y)v1^{n02}00 \ [\beta(x) \to \beta(y)] \end{aligned}$$

where $x, y \in V^*$, $x \to y \in Q$, and $w = u\beta(y)v$. By the induction hypothesis,

$$S \Rightarrow_{G'}^{+} 1^{n-2}00\beta^{-1}(u\beta(x)v)1^{n-2}00$$

Express $\beta^{-1}(u\beta(x)v)$ as

$$\beta^{-1}(u\beta(x)v) = \beta^{-1}(u)x\beta^{-1}(v)$$

Since $x \to y \in Q$,

$$\begin{aligned} \$ &\Rightarrow_G^+ \beta^{-1}(u)x\beta^{-1}(v) \\ &\Rightarrow_G \beta^{-1}(u)y\beta^{-1}(v) \ [x \to y] \end{aligned}$$

Because $w = u\beta(y)v$, $\beta^{-1}(w) = \beta^{-1}(u)y\beta^{-1}(v)$, so

$$\$ \Rightarrow_G^+ \beta^{-1}(w)$$

□

Claim 6. $L(G') \subseteq L(G)$

Proof. Let $w \in L(G')$. Distinguish between $w = \varepsilon$ and $w \neq \varepsilon$.

(i) Let $w = \varepsilon$. Observe that G' derives ε as

$$\begin{aligned} S &\Rightarrow_{G'}^{*} 1^{n-2}001^{n-2}00 \\ &\Rightarrow_{G'} \varepsilon\ [1^{n-2}001^{n-2}00 \rightarrow \varepsilon] \end{aligned}$$

Because

$$S \Rightarrow_{G'}^{*} 1^{n-2}001^{n-2}00$$

Claim 5 implies that $\$ \Rightarrow_{G}^{*} \varepsilon$. Therefore, $w \in L(G)$.

(ii) Assume that $w \neq \varepsilon$. Let $w = a_1a_2\cdots a_{n-1}a_n$ with $a_i \in T$ for $i = 1,\dots,n$, where $n \geq 1$. Examine P to see that in G', there exists this derivation

$$\begin{aligned} S &\Rightarrow_{G'}^{*} 1^{n-2}00\beta(a_1)\beta(a_2)\cdots\beta(a_{n-1})\beta(a_n)1^{n-2}00 \\ &\Rightarrow_{G'} a_1 1^{n-2}00\beta(a_2)\cdots\beta(a_{n-1})\beta(a_n)1^{n-2}00 \\ &\qquad [1^{n-2}00\beta(a_1) \rightarrow a_1 1^{n-2}00] \\ &\Rightarrow_{G'} a_1a_2 1^{n-2}00\beta(a_3)\cdots\beta(a_{n-1})\beta(a_n)1^{n-2}00 \\ &\qquad [1^{n-2}00\beta(a_2) \rightarrow a_2 1^{n-2}00] \\ &\ \vdots \\ &\Rightarrow_{G'} a_1a_2\cdots a_{n-2} 1^{n-2}00\beta(a_{n-1})\beta(a_n)1^{n-2}00 \\ &\qquad [1^{n-2}00\beta(a_{n-2}) \rightarrow a_{n-2} 1^{n-2}00] \\ &\Rightarrow_{G'} a_1a_2\cdots a_{n-2}a_{n-1} 1^{n-2}00\beta(a_n)1^{n-2}00 \\ &\qquad [1^{n-2}00\beta(a_{n-1}) \rightarrow a_{n-1} 1^{n-2}00] \\ &\Rightarrow_{G'} a_1a_2\cdots a_{n-2}a_{n-1}a_n 1^{n-2}001^{n-2}00 \\ &\qquad [1^{n-2}00\beta(a_n) \rightarrow a_n 1^{n-2}00] \\ &\Rightarrow_{G'} a_1a_2\cdots a_{n-2}a_{n-1}a_n \\ &\qquad [1^{n-2}001^{n-2}00 \rightarrow \varepsilon] \end{aligned}$$

Because

$$S \Rightarrow_{G'}^{*} 1^{n-2}00\beta(a_1)\beta(a_2)\cdots\beta(a_{n-1})\beta(a_n)1^{n-2}00$$

Claim 5 implies that

$$\$ \Rightarrow_{G}^{*} a_1a_2\cdots a_{n-1}a_n$$

Hence, $w \in L(G)$. □

By Claims 4 and 6, $L(G) = L(G')$. By Claim 2, $F(G') \subseteq T^{*}\Pi(1^{n-2}00)^{*}$. Thus, Lemma 4.1.19 holds. □

Theorem 4.1.20. **PS**[.2] = **RE**

Proof. The inclusion **PS**[.2] $\subseteq$ **RE** follows from Church's thesis (see p. 23). By Lemma 4.1.19, **RE** $\subseteq$ **PS**[.2]. Therefore, this theorem holds. □

Lemma 4.1.21. *Let G be a phrase-structure grammar. Then, there exists a phrase-structure grammar $G' = (\{S, 0, 1\} \cup T, T, P, S)$ satisfying $L(G) = L(G')$ and $F(G') \subseteq \Pi(1^{n-2}00)^* T^*$, for some $n \geq 1$.*

Proof. Let $G = (V, T, Q, \$)$ be a phrase-structure grammar, where V is the total alphabet of G, T is the terminal alphabet of G, Q is the set of rules of G, and \$ is the start symbol of G. Without any loss of generality, assume that $V \cap \{0, 1\} = \emptyset$. The following construction produces an equivalent phrase-structure grammar

$$G' = \left(\{S, 0, 1\} \cup T, T, P, S\right)$$

such that $F(G') \subseteq \Pi(1^{n-2}00)^* T^*$, for some $n \geq 1$.

For some $m \geq 3$ and n such that $2m = n$, introduce an injective homomorphism β from V to

$$\left(\{1\}^m \{1\}^* \{0\} \{1\}^* \cap \{0, 1\}^n\right) - \{1^{n-2}00\}$$

Extend the domain of β to V^*. Define the phrase-structure grammar

$$G' = \left(T \cup \{S, 0, 1\}, P, S, T\right)$$

with

$$\begin{aligned} P = {} & \{S \to 1^{n-1}00\beta(\$)1^{n-1}00\} \cup \\ & \{\beta(x) \to \beta(y) \mid x \to y \in Q\} \cup \\ & \{\beta(a)1^{n-2}00 \to 1^{n-2}00a \mid a \in T\} \cup \\ & \{1^{n-2}001^{n-2}00 \to \varepsilon\} \end{aligned}$$

Complete this proof by analogy with the proof of Lemma 4.1.19. □

Theorem 4.1.22. **PS**[2.] = **RE**

Proof. Clearly, **PS**[2.] $\subseteq$ **RE**. By Lemma 4.1.21, **RE** $\subseteq$ **PS**[2.]. Therefore, this theorem holds. □

Corollary 4.1.23. **PS**[.2] = **PS**[2.] = **RE** □

There is an open problem area related to the results above.

Open Problem 4.1.24. Recall that in this section we converted any phrase-structure grammar G to an equivalent phrase-structure grammar, $G' = (V, T, P, S)$, so that for every $x \in F(G')$, $x \in T^* \Pi(w)^*$, where w is a string over $V - T$. Then, we made this conversion so that for every $x \in F(G')$, $x \in \Pi(w)^* T^*$. Take into account the length of w. More precisely, for $j, k \geq 1$ set

$$\mathbf{PS}[.j,k] = \{L \mid L = L(G), \text{ where } G = (V,T,P,S) \text{ is a phrase-structure grammar such that } \mathrm{card}(\mathrm{alph}(F(G)) - T) = j \text{ and } F(G) \subseteq T^*\Pi(w)^*, \text{ where } w \in (V-T)^* \text{ and } |w| = k\}$$

Analogously, set

$$\mathbf{PS}[j,k.] = \{L \mid L = L(G), \text{ where } G = (V,T,P,S) \text{ is a phrase-structure grammar such that } \mathrm{card}(\mathrm{alph}(F(G)) - T) = j \text{ and } F(G) \subseteq \Pi(w)^*T^*, \text{ where } w \in (V-T)^* \text{ and } |w| = k\}$$

Reconsider this section in terms of these families of languages. □

4.2 Context-Conditional Grammars

Context-conditional grammars are based on context-free rules, each of which may be extended by finitely many *permitting* and *forbidding strings*. A rule like this can rewrite a sentential form on the condition that all its permitting strings occur in the current sentential form while all its forbidding strings do not occur there.

This section consists of Sects. 4.2.1 and 4.2.2. The former defines context-conditional grammars, and the latter establishes their power.

4.2.1 Definitions

Without further ado, we define the basic versions of context-regulated grammars.

Definition 4.2.1. A *context-conditional grammar* is a quadruple

$$G = (V,T,P,S)$$

where V, T, and S are the *total alphabet*, the *terminal alphabet* ($T \subset V$), and the *start symbol* ($S \in V - T$), respectively. P is a finite set of *rules* of the form

$$(A \to x, Per, For)$$

where $A \in V - T$, $x \in V^*$, and $Per, For \subseteq V^+$ are two finite sets. If $Per \neq \emptyset$ or $For \neq \emptyset$, the rule is said to be *conditional*; otherwise, it is called *context-free*. G has *degree* (r,s), where r and s are natural numbers, if for every $(A \to x, Per, For) \in P$, max-len$(Per) \leq r$ and max-len$(For) \leq s$. If $(A \to x, Per, For) \in P$ implies that $x \neq \varepsilon$, G is said to be *propagating*. Let $u, v \in V^*$ and $(A \to x, Per, For) \in P$. Then, *u directly derives v* according to $(A \to x, Per, For)$ in G, denoted by

$$u \Rightarrow_G v\ [(A \to x, Per, For)]$$

provided that for some $u_1, u_2 \in V^*$, the following conditions hold:

(a) $u = u_1 A u_2$,
(b) $v = u_1 x u_2$,
(c) $Per \subseteq \mathrm{sub}(u)$,
(d) $For \cap \mathrm{sub}(u) = \emptyset$.

When no confusion exists, we simply write $u \Rightarrow_G v$ instead of $u \Rightarrow_G v$ $[(A \rightarrow x, Per, For)]$. By analogy with context-free grammars, we extend $\Rightarrow_G$ to $\Rightarrow_G^k$ (where $k \geq 0$), $\Rightarrow_G^+$, and $\Rightarrow_G^*$. The *language* of G, denoted by $L(G)$, is defined as

$$L(G) = \{w \in T^* \mid S \Rightarrow_G^* w\}$$ □

The families of languages generated by context-conditional grammars and propagating context-conditional grammars of degree (r, s) are denoted by $\mathbf{CG}(r, s)$ and $\mathbf{CG}^{-\varepsilon}(r, s)$, respectively. Furthermore, set

$$\mathbf{CG} = \bigcup_{r=0}^{\infty} \bigcup_{s=0}^{\infty} \mathbf{CG}(r, s)$$

and

$$\mathbf{CG}^{-\varepsilon} = \bigcup_{r=0}^{\infty} \bigcup_{s=0}^{\infty} \mathbf{CG}^{-\varepsilon}(r, s)$$

4.2.2 Generative Power

Next, we prove several theorems concerning the generative power of the general versions of context-conditional grammars. Let us point out, however, that Sects. 4.3 through 4.6 establish many more results about special cases of these grammars.

Theorem 4.2.2. $\mathbf{CG}^{-\varepsilon}(0, 0) = \mathbf{CG}(0, 0) = \mathbf{CF}$

Proof. This theorem follows immediately from the definition. Clearly, context-conditional grammars of degree $(0, 0)$ are ordinary context-free grammars. □

Lemma 4.2.3. $\mathbf{CG}^{-\varepsilon} \subseteq \mathbf{CS}$

Proof. Let $r = s = 0$. Then, $\mathbf{CG}^{-\varepsilon}(0, 0) = \mathbf{CF} \subset \mathbf{CS}$. The rest of the proof establishes the inclusion for degrees (r, s) such that $r + s > 0$.

Consider a propagating context-conditional grammar

$$G = (V, T, P, S)$$

of degree (r, s), where $r + s > 0$, for some $r, s \geq 0$. Let k be the greater number of r and s. Set

$$M = \{x \in V^+ \mid |x| \leq k\}$$

Next, define

$$\text{cf-rules}(P) = \{A \to x \mid (A \to x, Per, For) \in P,\ A \in (V - T),\ x \in V^+\}$$

Then, set

$$\begin{aligned} N_F &= \{\lfloor X, x \rfloor \mid X \subseteq M,\ x \in M \cup \{\varepsilon\}\} \\ N_T &= \{\langle X \rangle \mid X \subseteq M\} \\ N_B &= \{\lceil p \rceil \mid p \in \text{cf-rules}(P)\} \cup \{\lceil \emptyset \rceil\} \\ V' &= V \cup N_F \cup N_T \cup N_B \cup \{\rhd, \lhd, \$, S', \#\} \\ T' &= T \cup \{\#\} \end{aligned}$$

Construct the context-sensitive grammar

$$G' = (V', T', P', S')$$

with the finite set of rules P' defined as follows:

(1) Add $S' \to \rhd \lfloor \emptyset, \varepsilon \rfloor S \lhd$ to P'
(2) For all $X \subseteq M, x \in (V^k \cup \{\varepsilon\})$ and $y \in V^k$, extend P' by adding

$$\lfloor X, x \rfloor y \to y \lfloor X \cup \text{sub}(xy, k), y \rfloor$$

(3) For all $X \subseteq M, x \in (V^k \cup \{\varepsilon\})$ and $y \in V^+$, $|y| \leq k$, extend P' by adding

$$\lfloor X, x \rfloor y \lhd \to y \langle X \cup \text{sub}(xy, k) \rangle \lhd$$

(4) For all $X \subseteq M$ and every $p = A \to x \in \text{cf-rules}(P)$ such that there exists $(A \to x, Per, For) \in P$ satisfying $Per \subseteq X$ and $For \cap X = \emptyset$, extend P' by adding

$$\langle X \rangle \lhd \to \lceil p \rceil \lhd$$

(5) For every $p \in \text{cf-rules}(P)$ and $a \in V$, extend P' by adding

$$a \lceil p \rceil \to \lceil p \rceil a$$

(6) For every $p = A \to x \in \text{cf-rules}(P)$, $A \in (V - T)$, $x \in V^+$, extend P' by adding

$$A \lceil p \rceil \to \lceil \emptyset \rceil x$$

(7) For every $a \in V$, extend P' by adding

$$a\lceil\emptyset\rceil \rightarrow \lceil\emptyset\rceil a$$

(8) Add $\rhd\lceil\emptyset\rceil \rightarrow \rhd\lfloor\emptyset,\varepsilon\rfloor$ to P'
(9) Add $\rhd\lfloor\emptyset,\varepsilon\rfloor \rightarrow \#\$$, $\$\lhd \rightarrow \#\#$, and $\$a \rightarrow a\$$, for all $a \in T$, to P'

Claim 1. Every successful derivation in G' has the form

$$\begin{aligned} S' &\Rightarrow_{G'} \rhd\lfloor\emptyset,\varepsilon\rfloor S\lhd \\ &\Rightarrow_{G'}^{+} \rhd\lfloor\emptyset,\varepsilon\rfloor x\lhd \\ &\Rightarrow_{G'} \#\$x\lhd \\ &\Rightarrow_{G'}^{|x|} \#x\$\lhd \\ &\Rightarrow_{G'} \#x\#\# \end{aligned}$$

such that $x \in T^+$, and during

$$\rhd\lfloor\emptyset,\varepsilon\rfloor S\lhd \Rightarrow_{G'}^{+} \rhd\lfloor\emptyset,\varepsilon\rfloor x\lhd$$

every sentential form w satisfies $w \in \{\rhd\}H^+\{\lhd\}$, where $H \subseteq V'-\{\rhd, \lhd, \#, \$, S'\}$.

Proof. Observe that the only rule that rewrites S' is $S' \rightarrow \rhd\lfloor\emptyset,\varepsilon\rfloor S\lhd$; thus,

$$S' \Rightarrow_{G'} \rhd\lfloor\emptyset,\varepsilon\rfloor S\lhd$$

After that, every sentential form that occurs in

$$\rhd\lfloor\emptyset,\varepsilon\rfloor S\lhd \Rightarrow_{G'}^{+} \rhd\lfloor\emptyset,\varepsilon\rfloor x\lhd$$

can be rewritten by using any of the rules (2) through (8) from the construction of P'. By the inspection of these rules, it is obvious that the delimiting symbols $\rhd$ and $\lhd$ remain unchanged and no other occurrences of them appear inside the sentential form. Moreover, there is no rule generating a symbol from $\{\#, \$, S'\}$. Therefore, all these sentential forms belong to $\{\rhd\}H^+\{\lhd\}$.

Next, let us explain how G' generates a string from $L(G')$. Only $\rhd\lfloor\emptyset,\varepsilon\rfloor \rightarrow \#\$$ can rewrite $\rhd$ to a symbol from T (see (9) in the definition of P'). According to the left-hand side of this rule, we obtain

$$S' \Rightarrow_{G'} \rhd\lfloor\emptyset,\varepsilon\rfloor S\lhd \Rightarrow_{G'}^{*} \rhd\lfloor\emptyset,\varepsilon\rfloor x\lhd \Rightarrow_{G'} \#\$x\lhd$$

where $x \in H^+$. To rewrite $\lhd$, G' uses $\$\lhd \rightarrow \#\#$. Thus, G' needs $\$$ as the left neighbor of $\lhd$. Suppose that $x = a_1a_2\cdots a_q$, where $q = |x|$ and $a_i \in T$, for all $i \in \{1,\ldots,q\}$. Since for every $a \in T$ there is $\$a \rightarrow a\$ \in P'$ [see (9)], we can construct

$$\begin{aligned}\#\$a_1a_2\cdots a_n\lhd &\Rightarrow_{G'} \#a_1\$a_2\cdots a_n\lhd\\ &\Rightarrow_{G'} \#a_1a_2\$\cdots a_n\lhd\\ &\Rightarrow_{G'}^{|x|-2} \#a_1a_2\cdots a_n\$\lhd\end{aligned}$$

Notice that this derivation can be constructed only for x that belong to T^+. Then, $\$\lhd$ is rewritten to ##. As a result,

$$S' \Rightarrow_{G'} \rhd\lfloor\emptyset,\varepsilon\rfloor S\lhd \Rightarrow_{G'}^{+} \rhd\lfloor\emptyset,\varepsilon\rfloor x\lhd \Rightarrow_{G'} \#\$x\lhd \Rightarrow_{G'}^{|x|} \#x\$\lhd \Rightarrow_{G'} \#x\#\#$$

with the required properties. Thus, the claim holds. □

The following claim demonstrates how G' simulates a direct derivation from G—the heart of the construction.

Let $x \Rightarrow_{G'}^{\oplus} y$ denote the derivation $x \Rightarrow_{G'}^{+} y$ such that $x = \rhd\lfloor\emptyset,\varepsilon\rfloor u\lhd$, $y = \rhd\lfloor\emptyset,\varepsilon\rfloor v\lhd$, $u, v \in V^+$, and during $x \Rightarrow_{G'}^{+} y$, there is no other occurrence of a string of the form $\rhd\lfloor\emptyset,\varepsilon\rfloor z\lhd$, $z \in V^*$.

Claim 2. For every $u, v \in V^*$, it holds that

$$\rhd\lfloor\emptyset,\varepsilon\rfloor u\lhd \Rightarrow_{G'}^{\oplus} \rhd\lfloor\emptyset,\varepsilon\rfloor v\lhd \quad \textit{if and only if} \quad u \Rightarrow_G v$$

Proof. The proof is divided into the only-if part and the if part.

Only If. Let us show how G' rewrites $\rhd\lfloor\emptyset,\varepsilon\rfloor u\lhd$ to $\rhd\lfloor\emptyset,\varepsilon\rfloor v\lhd$. The simulation consists of two phases.

During the first, forward phase, G' scans u to get all nonempty substrings of length k or less. By repeatedly using rules $\lfloor X,x\rfloor y \rightarrow y\lfloor X \cup \mathrm{sub}(xy,k), y\rfloor$, $X \subseteq M$, $x \in (V^k \cup \{\varepsilon\})$, $y \in V^k$ (see (2) in the definition of P'), the occurrence of a symbol of the form $\lfloor X,x\rfloor$ is moved toward the end of the sentential form. Simultaneously, the substrings of u are recorded in X. The forward phase is finished by applying $\lfloor X,x\rfloor y\lhd \rightarrow y\langle X \cup \mathrm{sub}(xy,k)\rangle\lhd$, $x \in (V^k \cup \{\varepsilon\})$, $y \in V^+$, $|y| \le k$ [see (3)]; this rule reaches the end of u and completes $X = \mathrm{sub}(u,k)$. Formally,

$$\rhd\lfloor\emptyset,\varepsilon\rfloor u\lhd \Rightarrow_{G'}^{+} \rhd u\langle X\rangle\lhd$$

with $X = \mathrm{sub}(u,k)$.

The second, backward phase simulates the application of a conditional rule. Assume that $u = u_1Au_2$, $u_1, u_2 \in V^*$, $A \in (V - T)$, and there exists a rule $A \rightarrow x \in$ cf-rules(P) such that $(A \rightarrow x, Per, For) \in P$ for some $Per, For \subseteq M$, where $Per \subseteq X$, $For \cap X = \emptyset$. Let $u_1xu_2 = v$. Then, G' derives

$$\rhd u\langle X\rangle\lhd \Rightarrow_{G'}^{+} \rhd\lfloor\emptyset,\varepsilon\rfloor v\lhd$$

by performing the following five steps

(i) $\langle X \rangle$ is changed to $\lceil p \rceil$, where $p = A \rightarrow x$ satisfies the conditions above (see (4) in the definition of P');
(ii) $\rhd u_1 A u_2 \lceil p \rceil \lhd$ is rewritten to $\rhd u_1 A \lceil p \rceil u_2 \lhd$ by using the rules of the form $a\lceil p \rceil \rightarrow \lceil p \rceil a$, $a \in V$ [see (5)];
(iii) $\rhd u_1 A \lceil p \rceil u_2 \lhd$ is rewritten to $\rhd u_1 \lceil \emptyset \rceil x u_2 \lhd$ by using $A\lceil p \rceil \rightarrow \lceil \emptyset \rceil x$ [see (6)];
(iv) $\rhd u_1 \lceil \emptyset \rceil x u_2 \lhd$ is rewritten to $\rhd \lceil \emptyset \rceil u_1 x u_2 \lhd$ by using the rules of the form $a\lceil \emptyset \rceil \rightarrow \lceil \emptyset \rceil a$, $a \in V$ [see (7)];
(v) Finally, $\rhd \lceil \emptyset \rceil$ is rewritten to $\rhd \lfloor \emptyset, \varepsilon \rfloor$ by $\rhd \lceil \emptyset \rceil \rightarrow \rhd \lfloor \emptyset, \varepsilon \rfloor$.

As a result, we obtain

$$\begin{aligned} \rhd \lfloor \emptyset, \varepsilon \rfloor u \lhd &\Rightarrow_{G'}^{+} \rhd u \langle X \rangle \lhd \Rightarrow_{G'} \rhd u \lceil p \rceil \lhd \\ &\Rightarrow_{G'}^{|u|} \rhd \lceil \emptyset \rceil v \lhd \Rightarrow_{G'} \rhd \lfloor \emptyset, \varepsilon \rfloor v \lhd \end{aligned}$$

Observe that this is the only way of deriving

$$\rhd \lfloor \emptyset, \varepsilon \rfloor u \lhd \Rightarrow_{G'}^{\oplus} \rhd \lfloor \emptyset, \varepsilon \rfloor v \lhd$$

Let us show that $u \Rightarrow_G v$. Indeed, the application of $A\lceil p \rceil \rightarrow \lceil \emptyset \rceil x$ implies that there exists $(A \rightarrow x, Per, For) \in P$, where $Per \subseteq \text{sub}(u, k)$ and $For \cap \text{sub}(u, k) = \emptyset$. Hence, there exists a derivation of the form

$$u \Rightarrow_G v\ [p]$$

where $u = u_1 A u_2$, $v = u_1 x u_2$ and $p = (A \rightarrow x, Per, For) \in P$.

If. The converse implication is similar to the only-if part, so we leave it to the reader. □

Claim 3. $S' \Rightarrow_{G'}^{+} \rhd \lfloor \emptyset, \varepsilon \rfloor x \lhd$ *if and only if* $S \Rightarrow_G^{*} x$*, for all* $x \in V^+$.

Proof. The proof is divided into the only-if part and the if part.

Only If. The only-if part is proved by induction on the ith occurrence of the sentential form w satisfying $w = \rhd \lfloor \emptyset, \varepsilon \rfloor u \lhd$, $u \in V^+$ during the derivation in G'.

Basis. Let $i = 1$. Then, $S' \Rightarrow_{G'} \rhd \lfloor \emptyset, \varepsilon \rfloor S \lhd$ and $S \Rightarrow_G^0 S$.

Induction Hypothesis. Suppose that the claim holds for all $i \leq h$, for some $h \geq 1$.

Induction Step. Let $i = h + 1$. Since $h + 1 \geq 2$, we can express

$$S' \Rightarrow_{G'}^{+} \rhd \lfloor \emptyset, \varepsilon \rfloor x_i \lhd$$

as

$$S' \Rightarrow_{G'}^{+} \rhd \lfloor \emptyset, \varepsilon \rfloor x_{i-1} \lhd \Rightarrow_{G'}^{\oplus} \rhd \lfloor \emptyset, \varepsilon \rfloor x_i \lhd$$

where $x_{i-1}, x_i \in V^+$. By the induction hypothesis,

$$S \Rightarrow_G^* x_{i-1}$$

Claim 2 says that

$$\triangleright \lfloor \emptyset, \varepsilon \rfloor x_{i-1} \triangleleft \Rightarrow_{G'}^{\oplus} \triangleright \lfloor \emptyset, \varepsilon \rfloor x_i \triangleleft \quad \text{if and only if} \quad x_{i-1} \Rightarrow_G x_i$$

Hence,

$$S \Rightarrow_G^* x_{i-1} \Rightarrow_G x_i$$

and the only-if part holds.

If. By induction on n, we prove that

$$S \Rightarrow_G^n x \text{ implies that } S' \Rightarrow_{G'}^+ \triangleright \lfloor \emptyset, \varepsilon \rfloor x \triangleleft$$

for all $n \geq 0$, $x \in V^+$.

Basis. For $n = 0$, $S \Rightarrow_G^0 S$ and $S' \Rightarrow_{G'} \triangleright \lfloor \emptyset, \varepsilon \rfloor S \triangleleft$.

Induction Hypothesis. Assume that the claim holds for all n or less, for some $n \geq 0$.

Induction Step. Let

$$S \Rightarrow_G^{n+1} x$$

with $x \in V^+$. Because $n + 1 \geq 1$, there exists $y \in V^+$ such that

$$S \Rightarrow_G^n y \Rightarrow_G x$$

By the induction hypothesis, there is also a derivation

$$S' \Rightarrow_{G'}^+ \triangleright \lfloor \emptyset, \varepsilon \rfloor y \triangleleft$$

From Claim 2 it follows that

$$\triangleright \lfloor \emptyset, \varepsilon \rfloor y \triangleleft \Rightarrow_{G'}^{\oplus} \triangleright \lfloor \emptyset, \varepsilon \rfloor x \triangleleft$$

Therefore,

$$S' \Rightarrow_{G'}^+ \triangleright \lfloor \emptyset, \varepsilon \rfloor x \triangleleft$$

and the converse implication holds as well. □

From Claims 1 and 3, we see that any successful derivation in G' is of the form

$$S' \Rightarrow_{G'}^{+} \rhd \lfloor \emptyset, \varepsilon \rfloor x \lhd \Rightarrow_{G'}^{+} \#x\#\#$$

such that

$$S \Rightarrow_{G}^{*} x, \ x \in T^{+}$$

Therefore, for each $x \in T^{+}$,

$$S' \Rightarrow_{G'}^{+} \#x\#\# \quad \text{if and only if} \quad S \Rightarrow_{G}^{*} x$$

Define the homomorphism h over $(T \cup \{\#\})^{*}$ as $h(\#) = \varepsilon$ and $h(a) = a$ for all $a \in T$. Observe that h is 4-linear erasing with respect to $L(G')$. Furthermore, notice that $h(L(G')) = L(G)$. Because **CS** is closed under linear erasing (see Theorem 10.4 on p. 98 in [21]), $L \in$ **CS**. Thus, Lemma 4.2.3 holds. □

Theorem 4.2.4. $\mathbf{CG}^{-\varepsilon} = \mathbf{CS}$

Proof. By Lemma 4.2.3, we have $\mathbf{CG}^{-\varepsilon} \subseteq \mathbf{CS}$. $\mathbf{CS} \subseteq \mathbf{CG}^{-\varepsilon}$ holds as well. In fact, later in this book, we introduce several special cases of propagating context-conditional grammars and prove that even these grammars generate **CS** (see Theorems 4.6.4 and 4.6.9). As a result, $\mathbf{CG}^{-\varepsilon} = \mathbf{CS}$. □

Lemma 4.2.5. $\mathbf{CG} \subseteq \mathbf{RE}$

Proof. This lemma follows from Church's thesis (see p. 23). To obtain an algorithm converting any context-conditional grammar to an equivalent phrase-structure grammar, use the technique presented in Lemma 4.2.3. □

Theorem 4.2.6. $\mathbf{CG} = \mathbf{RE}$

Proof. By Lemma 4.2.5, $\mathbf{CG} \subseteq \mathbf{RE}$. Later on, we define some special cases of context-conditional grammars and demonstrate that they characterize **RE** (see Theorems 4.4.4, 4.6.6, and 4.6.11). Thus, $\mathbf{RE} \subseteq \mathbf{CG}$. □

4.3 Random Context Grammars

This section discusses three special cases of context-conditional grammars whose conditions are nonterminal symbols, so their degree is not greater than $(1, 1)$. Specifically, *permitting grammars* are of degree $(1, 0)$. *Forbidding grammars* are of degree $(0, 1)$. Finally, *random context grammars* are of degree $(1, 1)$.

The present section consists of Sects. 4.3.1 and 4.3.2. The former defines and illustrates all the grammars under discussion. The latter establishes their generative power.

4.3.1 Definitions and Examples

We open this section by defining random context grammars and their two important special cases—permitting and forbidding grammars. Later in this section, we illustrate them.

Definition 4.3.1. Let $G = (V, T, P, S)$ be a context-conditional grammar. G is called a *random context grammar* provided that every $(A \to x, Per, For) \in P$ satisfies $Per \subseteq N$ and $For \subseteq N$. $\square$

Definition 4.3.2. Let $G = (V, T, P, S)$ be a random context grammar. G is called a *permitting grammar* provided that every $(A \to x, Per, For) \in P$ satisfies $For = \emptyset$. $\square$

Definition 4.3.3. Let $G = (V, T, P, S)$ be a random context grammar. G is called a *forbidding grammar* provided that every $(A \to x, Per, For) \in P$ satisfies $Per = \emptyset$. $\square$

The following conventions simplify rules in permitting and forbidding grammars.

Let $G = (V, T, P, S)$ be a permitting grammar, and let $p = (A \to x, Per, For) \in P$. Since $For = \emptyset$, we usually omit the empty set of forbidding conditions. That is, we write $(A \to x, Per)$ when no confusion arises.

Let $G = (V, T, P, S)$ be a forbidding grammar, and let $p = (A \to x, Per, For) \in P$. We write $(A \to x, For)$ instead of $(A \to x, Per, For)$ because $Per = \emptyset$ for all $p \in P$.

The families of languages defined by permitting grammars, forbidding grammars, and random context grammars are denoted by **Per**, **For**, and **RC**, respectively. To indicate that only propagating grammars are considered, we use the upper index $-\varepsilon$. That is, $\mathbf{Per}^{-\varepsilon}$, $\mathbf{For}^{-\varepsilon}$, and $\mathbf{RC}^{-\varepsilon}$ denote the families of languages defined by propagating permitting grammars, propagating forbidding grammars, and propagating random context grammars, respectively.

Example 4.3.4 (See [3]). Let

$$G = \big(\{S, A, B, C, D, A', B', C', a, b, c\}, \{a, b, c\}, P, S\big)$$

be a permitting grammar, where P is defined as follows:

$$\begin{aligned} P = \big\{&(S \to ABC, \emptyset),\\ &(A \to aA', \{B\}),\\ &(B \to bB', \{C\}),\\ &(C \to cC', \{A'\}),\\ &(A' \to A, \{B'\}),\\ &(B' \to B, \{C'\}),\\ &(C' \to C, \{A\}),\\ &(A \to a, \{B\}),\\ &(B \to b, \{C\}),\\ &(C \to c, \emptyset)\big\} \end{aligned}$$

Consider the string $aabbcc$. G generates this string in the following way

$$\begin{aligned} S \Rightarrow\ & ABC \Rightarrow aA'BC \Rightarrow aA'bB'C \Rightarrow aA'bB'cC' \Rightarrow \\ & aAbB'cC' \Rightarrow aAbBcC' \Rightarrow aAbBcC \Rightarrow \\ & aabBcC \Rightarrow aabbcC \Rightarrow aabbcc \end{aligned}$$

Observe that G is propagating and

$$L(G) = \{a^n b^n c^n \mid n \geq 1\}$$

which is a non-context-free language. □

Example 4.3.5 (See [3]). Let

$$G = \big(\{S, A, B, D, a\}, \{a\}, P, S\big)$$

be a random context grammar. The set of rules P is defined as follows:

$$\begin{aligned} P = \big\{ & (S \rightarrow AA, \emptyset, \{B, D\}), \\ & (A \rightarrow B, \emptyset, \{S, D\}), \\ & (B \rightarrow S, \emptyset, \{A, D\}), \\ & (A \rightarrow D, \emptyset, \{S, B\}), \\ & (D \rightarrow a, \emptyset, \{S, A, B\})\big\} \end{aligned}$$

Notice that G is a propagating forbidding grammar. For $aaaaaaaa$, G makes the following derivation

$$\begin{aligned} S \Rightarrow\ & AA \Rightarrow AB \Rightarrow BB \Rightarrow BS \Rightarrow SS \Rightarrow AAS \Rightarrow AAAA \Rightarrow BAAA \Rightarrow \\ & BABA \Rightarrow BBBA \Rightarrow BBBB \Rightarrow SBBB \Rightarrow SSBB \Rightarrow SSSB \Rightarrow \\ & SSSS \Rightarrow AASSS \Rightarrow^3 AAAAAAAA \Rightarrow^8 DDDDDDDD \Rightarrow^8 aaaaaaaa \end{aligned}$$

Clearly, G generates this non-context-free language

$$L(G) = \left\{a^{2^n} \mid n \geq 1\right\}$$

□

4.3.2 Generative Power

We next establish several theorems concerning the generative power of the grammars defined in the previous section.

Theorem 4.3.6. $\mathbf{CF} \subset \mathbf{Per}^{-\varepsilon} \subset \mathbf{RC}^{-\varepsilon} \subset \mathbf{CS}$

Proof. $\mathbf{CF} \subset \mathbf{Per}^{-\varepsilon}$ follows from Example 4.3.4. $\mathbf{Per}^{-\varepsilon} \subset \mathbf{RC}^{-\varepsilon}$ follows from Theorem 2.7 in Chap. 3 in [20]. Finally, $\mathbf{RC}^{-\varepsilon} \subset \mathbf{CS}$ follows from Theorems 1.2.4 and 1.4.5 in [3]. □

Theorem 4.3.7. $\mathbf{Per}^{-\varepsilon} = \mathbf{Per} \subset \mathbf{RC} = \mathbf{RE}$

Proof. $\mathbf{Per}^{-\varepsilon} = \mathbf{Per}$ follows from Theorem 1 in [23]. By Theorem 1.2.5 in [3], $\mathbf{RC} = \mathbf{RE}$. Furthermore, from Theorem 2.7 in Chap. 3 in [20], it follows that $\mathbf{Per} \subset \mathbf{RC}$; thus, the theorem holds. □

Lemma 4.3.8. $\mathbf{ET0L} \subset \mathbf{For}^{-\varepsilon}$.

Proof (See [18]). Let $L \in \mathbf{ET0L}$, $L = L(G)$ for some ET0L grammar,

$$G = (V, T, P_1, \ldots, P_t, S)$$

Without loss of generality, we assume that G is propagating (see Theorem 3.3.23). We introduce the alphabets

$$\begin{aligned} V^{(i)} &= \{a^{(i)} \mid a \in V\},\ 1 \le i \le t \\ V' &= \{a' \mid a \in V\} \\ V'' &= \{a'' \mid a \in V\} \\ \bar{V} &= \{\bar{a} \mid a \in T\} \end{aligned}$$

For $w \in V^*$, by $w^{(i)}$, w', w'', and $\bar{w}$, we denote the strings obtained from w by replacing each occurrence of a symbol $a \in V$ by $a^{(i)}$, a', a'', and $\bar{a}$, respectively. Let P' be the set of all random context rules defined as follows:

(1) For every $a \in V$, add $(a' \to a'', \emptyset, \bar{V} \cup V^{(1)} \cup V^{(2)} \cup \cdots \cup V^{(t)})$ to P';
(2) For every $a \in V$ for all $1 \le i \le t$, add

$$(a'' \to a^{(i)}, \emptyset, \bar{V} \cup V' \cup V^{(1)} \cup V^{(2)} \cup \cdots \cup V^{(i-1)} \cup V^{(i+1)} \cup \cdots \cup V^{(t)})$$

to P';
(3) For all $i \in \{1, \ldots, t\}$ for every $a \to u \in P_i$, add $(a^{(i)} \to u', \emptyset, V'' \cup \bar{V})$ to P';
(4) For all $a \in T$, add $(a' \to \bar{a}, \emptyset, V'' \cup V^{(1)} \cup V^{(2)} \cup \cdots \cup V^{(t)})$ to P';
(5) For all $a \in T$, add $(\bar{a} \to a, \emptyset, V' \cup V'' \cup V^{(1)} \cup V^{(2)} \cup \cdots \cup V^{(t)})$ to P'.

Then, define the random context grammar

$$G' = (V' \cup V'' \cup \bar{V} \cup V^{(1)} \cup V^{(2)} \cup \cdots \cup V^{(t)}, T, P', S)$$

that has only forbidding context conditions.

Let x' be a string over V'. To x', we can apply only rules whose left-hand side is in V'.

(i) We use $a' \to a''$ for some $a' \in V'$. The obtained sentential form contains symbols of V' and V''. Hence, we can use only rules of type (1). Continuing in this way, we get $x' \Rightarrow_{G'}^{*} x''$. By analogous arguments, we now have to rewrite all symbols of x'' by rules of (2) with the same index (i). Thus, we obtain $x^{(i)}$. To each symbol $a^{(i)}$ in $x^{(i)}$, we apply a rule $a^{(i)} \to u'$, where $a \to u \in P_i$. Since again all symbols in $x^{(i)}$ have to be replaced before starting with rules of another type, we simulate a derivation step in G and get z', where $x \Rightarrow_G z$ in G. Therefore, starting with a rule of (1), we simulate a derivation step in G, and conversely, each derivation step in G can be simulated in this way.

(ii) We apply a rule $a' \to \bar{a}$ to x'. Next, each a' of T' occurring in x' has to be substituted by $\bar{a}$ and then by a by using the rules constructed in (5). Therefore, we obtain a terminal string only if $x' \in T'^{*}$.

By these considerations, any successful derivation in G' is of the form

$$\begin{aligned} S' &\Rightarrow_{G'} S'' \Rightarrow_{G'} S^{(i_0)} \\ &\Rightarrow_{G'} z_1' \Rightarrow_{G'}^{*} z_1'' \Rightarrow_{G'}^{*} z_1^{(i_1)} \\ &\vdots \\ &\Rightarrow_{G'}^{*} z_n' \Rightarrow_{G'}^{*} z_n'' \Rightarrow_{G'}^{*} z_n^{(i_n)} \\ &\Rightarrow_{G'}^{*} z_{n+1} \Rightarrow_{G'}^{*} \bar{z}_{n+1} \Rightarrow_{G'}^{*} z_{n+1} \end{aligned}$$

and such a derivation exists if and only if

$$S \Rightarrow_G z_1 \Rightarrow_G z_2 \Rightarrow_G \cdots \Rightarrow_G z_n \Rightarrow_G z_{n+1}$$

is a successful derivation in G. Thus, $L(G) = L(G')$.

In order to finish the proof, it is sufficient to find a language that is not in **ET0L** and can be generated by a forbidding grammar. A language of this kind is

$$L = \{b(ba^m)^n \mid m \geq n \geq 0\}$$

which can be generated by the grammar

$$G = (\{S, A, A', B, B', B'', C, D, E\}, \{a, b\}, P, s)$$

with P consisting of the following rules

$$(S \to SA, \emptyset, \emptyset)$$
$$(S \to C, \emptyset, \emptyset)$$
$$(C \to D, \emptyset, \{S, A', B', B'', D, E\})$$
$$(B \to B'a, \emptyset, \{S, C, E\})$$
$$(A \to B''a, \emptyset, \{S, C, E, B''\})$$
$$(A \to A'a, \emptyset, \{S, C, E\})$$

$$(D \to C, \emptyset, \{A, B\})$$
$$(B' \to B, \emptyset, \{D\})$$
$$(B'' \to B, \emptyset, \{D\})$$
$$(A' \to A, \emptyset, \{D\})$$
$$(D \to E, \emptyset, \{S, A, A', B', B'', C, E\})$$
$$(B \to b, \emptyset, \{S, A, A', B', B'', C, D\})$$
$$(E \to b, \emptyset, \{S, A, A', B, B', B'', C, D\})$$

First, we have the derivation

$$S \Rightarrow_G^* SA^n \Rightarrow_G CA^n \Rightarrow_G DA^n$$

Then, we have to replace all occurrences of A. If we want to replace an occurrence of A by a terminal string in some steps, it is necessary to use $A \to B''a$. However, this can be done at most once in a phase that replaces all As. Therefore, $m \geq n$. □

Theorem 4.3.9. $\mathbf{CF} \subset \mathbf{ET0L} \subset \mathbf{For}^{-\varepsilon} \subseteq \mathbf{For} \subset \mathbf{CS}$

Proof. According to Example 4.3.5, we already have $\mathbf{CF} \subset \mathbf{For}^{-\varepsilon}$. By [19] and Lemma 4.3.8, $\mathbf{CF} \subset \mathbf{ET0L} \subset \mathbf{For}^{-\varepsilon}$. Moreover, in [18], it has been proved that $\mathbf{For}^{-\varepsilon} \subseteq \mathbf{For} \subset \mathbf{CS}$. Therefore, the theorem holds. □

The following corollary summarizes the relations of language families generated by random context grammars.

Corollary 4.3.10.

$$\mathbf{CF} \subset \mathbf{Per}^{-\varepsilon} \subset \mathbf{RC}^{-\varepsilon} \subset \mathbf{CS}$$
$$\mathbf{Per}^{-\varepsilon} = \mathbf{Per} \subset \mathbf{RC} = \mathbf{RE}$$
$$\mathbf{CF} \subset \mathbf{ET0L} \subset \mathbf{For}^{-\varepsilon} \subseteq \mathbf{For} \subset \mathbf{CS}$$

Proof. This corollary follows from Theorems 4.3.6, 4.3.7, and 4.3.9. □

Open Problem 4.3.11. Are $\mathbf{For}^{-\varepsilon}$ and **For** identical? □

4.4 Generalized Forbidding Grammars

Generalized forbidding grammars represent a generalized variant of forbidding grammars (see Sect. 4.3) in which forbidding context conditions are formed by finite languages.

This section consists of Sects. 4.4.1 and 4.4.2. The former defines generalized forbidding grammars, and the latter establishes their power.

4.4.1 Definitions

Next, we define generalized forbidding grammars.

Definition 4.4.1. Let $G = (V, T, P, S)$ be a context-conditional grammar. If every $(A \to x, Per, For)$ satisfies $Per = \emptyset$, then G is said to be a *generalized forbidding grammar* (a *gf-grammar* for short). □

The following convention simplifies the notation of gf-grammars. Let $G = (V, T, P, S)$ be a gf-grammar of degree (r, s). Since every $(A \to x, Per, For) \in P$ implies that $Per = \emptyset$, we omit the empty set of permitting conditions. That is, we write $(A \to x, For)$ instead of $(A \to x, Per, For)$. For simplicity, we also say that the degree of G is s instead of (r, s).

The families generated by gf-grammars and propagating gf-grammars of degree s are denoted by $\mathbf{GF}(s)$ and $\mathbf{GF}^{-\varepsilon}(s)$, respectively. Furthermore, set

$$\mathbf{GF} = \bigcup_{s=0}^{\infty} \mathbf{GF}(s)$$

and

$$\mathbf{GF}^{-\varepsilon} = \bigcup_{s=0}^{\infty} \mathbf{GF}^{-\varepsilon}(s)$$

4.4.2 Generative Power and Reduction

In the present section, we establish the generative power of generalized forbidding grammars, defined in the previous section. In fact, apart from establishing this power, we also give several related results concerning the reduction of these grammars. Indeed, we reduce these grammars with respect to the number of nonterminals, the number of forbidding rules, and the length of forbidding strings.

By analogy with Theorem 4.2.2, it is easy to see that gf-grammars of degree 0 are ordinary context-free grammars.

Theorem 4.4.2. $\mathbf{GF}^{-\varepsilon}(0) = \mathbf{GF}(0) = \mathbf{CF}$ □

Furthermore, gf-grammars of degree 1 are as powerful as forbidding grammars.

Theorem 4.4.3. $\mathbf{GF}(1) = \mathbf{For}$

Proof. This simple proof is left to the reader. □

Theorem 4.4.4. $\mathbf{GF}(2) = \mathbf{RE}$

Proof. It is straightforward to prove that $\mathbf{GF}(2) \subseteq \mathbf{RE}$; hence it is sufficient to prove the converse inclusion.

Let L be a recursively enumerable language. Without any loss of generality we assume that L is generated by a phrase-structure grammar

$$G = \big(V, T, P, S\big)$$

in the Penttonen normal form (see Theorem 4.1.4). Set $N = V - T$.

Let @, \$, S' be new symbols and m be the cardinality of $V \cup \{@\}$. Clearly, $m \geq 1$. Furthermore, let f be an arbitrary bijection from $V \cup \{@\}$ onto $\{1, \dots, m\}$ and f^{-1} be the inverse of f.

The gf-grammar

$$G' = \big(V' \cup \{@, \$, S'\}, T, P', S'\big)$$

of degree 2 is defined as follows:

$$V' = W \cup V, \text{ where}$$
$$W = \{[AB \to AC, j] \mid AB \to AC \in P,\ A, B, C \in N, 1 \leq j \leq m + 1\}$$

We assume that W, $\{@, \$, S'\}$, and V are pairwise disjoint alphabets. The set of rules P' is defined in the following way

(1) Add $(S' \to @S, \emptyset)$ to P';
(2) If $A \to x \in P$, $A \in N$, $x \in \{\varepsilon\} \cup T \cup N^2$, then add $(A \to x, \{\$\})$ to P';
(3) If $AB \to AC \in P$, $A, B, C \in N$, then

 (a) Add $(B \to \$[AB \to AC, 1], \{\$\})$ to P';
 (b) For all $j = 1, \dots, m$, $f(A) \neq j$, extend P' by adding

$$\big([AB \to AC, j] \to [AB \to AC, j + 1], \{f^{-1}(j)\$\}\big);$$

 (c) Add $([AB \to AC, f(A)] \to [AB \to AC, f(A) + 1], \emptyset)$ and $([AB \to AC, m + 1] \to C, \emptyset)$ to P';

(4) Add$(@ \to \varepsilon, N \cup W \cup \{\$\})$ and $(\$ \to \varepsilon, W)$ to P'.

Basically, the application of $AB \to AC$ in G is simulated in G' in the following way. An occurrence of B is rewritten with $\$[AB \to AC, 1]$. Then, the left adjoining symbol of \$ is checked not to be any symbol from $(V \cup \{@\})$ except A. After this, the right adjoining symbol of \$ is $[AB \to AC, m + 1]$. This symbol is rewritten with C. A formal proof is given below.

Immediately from the definition of P' it follows that

$$S' \Rightarrow_{G'}^{+} x$$

where $x \in (V' \cup \{@, S'\})^*$, implies that

(I) $S' \notin \text{sub}(x)$;
(II) $\text{occur}(x, \text{sub}(\{\$\}W) - \{\varepsilon\}) \leq 1$ such that if $\text{occur}(x, W) = 1$, then we have $\text{occur}(x, \{\$\}W) = 1$;
(III) If $x \notin T^*$, then the leftmost symbol of x is $@$.

Next, we define a finite substitution g from V^* into V'^* such that for all $B \in V$,

$$g(B) = \{B\} \cup \{[AB \to AC, j] \in W \mid AB \to AC \in P,\ A, C \in N,\ j = 1, \ldots, m+1\}$$

Let g^{-1} be the inverse of g.

To show that $L(G) = L(G')$, we first prove that

$$S \Rightarrow_G^n x \quad \text{if and only if} \quad S \Rightarrow_{G'}^{n'} x'$$

where $x' = @v'Xw'$, $X \in \{\$, \varepsilon\}$, $v'w' \in g(x)$, $x \in V^*$, for some $n \geq 0$, $n' \geq 1$.

Only If. This is established by induction on $n \geq 0$. That is, we have to demonstrate that $S \Rightarrow_G^n x$, $x \in V^*$, $n \geq 0$, implies that $S \Rightarrow_{G'}^{+} x'$ for some x' such that $x' = @v'Xw'$, $X \in \{\$, \varepsilon\}$, $v'w' \in g(x)$.

Basis. Let $n = 0$. The only x is S because $S \Rightarrow_G^0 S$. Clearly, $S' \Rightarrow_{G'} @S$ and $S \in g(S)$.

Induction Hypothesis. Suppose that the claim holds for all derivations of length n or less, for some $n \geq 0$.

Induction Step. Let us consider any derivation of the form

$$S \Rightarrow_G^{n+1} x$$

with $x \in V^*$. Since $n + 1 \geq 1$, there is some $y \in V^+$ and $p \in P$ such that

$$S \Rightarrow_G^n y \Rightarrow_G x\ [p]$$

and by the induction hypothesis, there is also a derivation of the form

$$S \Rightarrow_{G'}^{n'} y'$$

for some $n' \geq 1$, such that $y' = @r'Ys'$, $Y \in \{\$, \varepsilon\}$, and $r's' \in g(y)$.

(i) Let us assume that $p = D \to y_2 \in P$, $D \in N$, $y_2 \in \{\varepsilon\} \cup T \cup N^2$, $y = y_1 D y_3$, $y_1, y_3 \in V^*$, and $x = y_1 y_2 y_3$. From (2) it is clear that $(D \to y_2, \{\$\}) \in P'$.

(i.a) Let $\$ \notin \text{alph}(y')$. Then, we have $y' = @r's' = @y_1Dy_3$,

$$S' \Rightarrow_{G'}^{n'} @y_1Dy_3 \Rightarrow_{G'} @y_1y_2y_3 \ [(D \to y_2, \{\$\})]$$

and $y_1y_2y_3 \in g(y_1y_2y_3) = g(x)$.

(i.b) Let $Y = \$ \in \text{sub}(y')$ and $W \cap \text{sub}(y') = \emptyset$. Then, there is the following derivation in G'

$$S' \Rightarrow_{G'}^{n'} @r'\$s' \Rightarrow_{G'} @r's' \ [(\$ \to \varepsilon, W)]$$

By analogy with (a) above, we have $@r's' = @y_1Dy_2$, so

$$S' \Rightarrow_{G'}^{n'+1} @y_1Dy_3 \Rightarrow_{G'} @y_1y_2y_3 \ [(D \to y_2, \{\$\})]$$

where $y_1y_2y_3 \in g(x)$.

(i.c) Let $\$[AB \to AC, i] \in \text{sub}(y')$ for some $i \in \{1, \ldots, m+1\}$, $AB \to AC \in P$, $A, B, C \in N$. Thus, $y' = @r'\$[AB \to AC, i]t'$, where $s' = [AB \to AC, i]t'$. By the inspection of the rules [see (3)] it can be seen (and the reader should be able to produce a formal proof) that we can express the derivation

$$S' \Rightarrow_{G'}^{*} y'$$

in the following form

$$\begin{aligned} S' &\Rightarrow_{G'}^{*} @r'Bt' \\ &\Rightarrow_{G'} @r'\$[AB \to AC, 1]t' \ [(B \to \$[AB \to AC, 1], \{\$\})] \\ &\Rightarrow_{G'}^{i-1} @r'\$[AB \to AC, i]t' \end{aligned}$$

Clearly, $r'Bt' \in g(y)$ and $\$ \notin \text{sub}(r'Bt')$. Thus, $r'Bt' = y_1Dy_3$, and there is a derivation

$$S' \Rightarrow_{G'}^{*} @y_1Dy_3 \Rightarrow_{G'} @y_1y_2y_3 \ [(D \to y_2, \{\$\})]$$

and $y_1y_2y_3 \in g(x)$.

(ii) Let $p = AB \to AC \in P$, $A, B, C \in N$, $y = y_1ABy_2$, $y_1, y_2 \in V^*$, and $x = y_1ACy_2$.

(ii.a) Let $\$ \notin \text{sub}(y')$. Thus, $r's' = y_1ABy_2$. By the inspection of the rules introduced in (3) (technical details are left to the reader), there is the following derivation in G'

$$
\begin{aligned}
S' &\Rightarrow_{G'}^{n'} y_1 A B y_2 \\
&\Rightarrow_{G'} @ y_1 A\$[AB \to AC, 1] y_2 \\
&\qquad [(B \to \$[AB \to AC, 1], \{\$\})] \\
&\Rightarrow_{G'} @ y_1 A\$[AB \to AC, 2] y_2 \\
&\qquad [([AB \to AC, 1] \to [AB \to AC, 2], \{f^{-1}(1)\$\})] \\
&\ \vdots \\
&\Rightarrow_{G'} @ y_1 A\$[AB \to AC, f(A)] y_2 \\
&\qquad [([AB \to AC, f(A) - 1] \to [AB \to AC, f(A)], \\
&\qquad\quad \{f^{-1}(f(A) - 1)\$\})] \\
&\Rightarrow_{G'} @ y_1 A\$[AB \to AC, f(A) + 1] y_2 \\
&\qquad [([AB \to AC, f(A)] \to [AB \to AC, f(A) + 1], \emptyset)] \\
&\ \vdots \\
&\Rightarrow_{G'} @ y_1 A\$[AB \to AC, m + 1] y_2 \\
&\qquad [([AB \to AC, m] \to [AB \to AC, m + 1], \{f^{-1}(m)\$\})] \\
&\Rightarrow_{G'} @ y_1 A\$C y_2 \\
&\qquad [([AB \to AC, m + 1] \to C, \emptyset)]
\end{aligned}
$$

such that $y_1 AC y_2 \in g(y_1 AC y_2) = g(x)$.

(ii.b) Let $\$ \in \mathrm{sub}(y')$, $\mathrm{sub}(y') \cap W = \emptyset$. By analogy with (i.b), the derivation

$$S' \Rightarrow_{G'}^{*} @r's'$$

with $@r's' = @y_1 AB y_2$, can be constructed in G'. Then, by analogy with (ii.a), one can construct the derivation

$$S' \Rightarrow_{G'}^{*} @y_1 AB y_2 \Rightarrow_{G'}^{*} @y_1 A\$C y_2$$

such that $y_1 AC y_2 \in g(x)$.

(ii.c) Let $\mathrm{occur}(y', \{\$\}W - \{\varepsilon\}) = 1$. By analogy with (i.c), one can construct the derivation

$$S' \Rightarrow_{G'}^{*} @y_1 AB y_2$$

Next, by using an analogue from (ii.a), the derivation

$$S' \Rightarrow_{G'}^{*} @y_1 AB y_2 \Rightarrow_{G'}^{*} @y_1 A\$C y_2$$

can be constructed in G' so $y_1 AC y_2 \in g(x)$.

In (i) and (ii) above we have considered all possible forms of p. In cases (a), (b), (c) of (i) and (ii), we have considered all possible forms of y'. In any of these cases, we have constructed the desired derivation of the form

$$S' \Rightarrow_{G'}^{+} x'$$

such that $x' = @r'Xs'$, $X \in \{\$, \varepsilon\}$, $r's' \in g(x)$. Hence, we have established the only-if part of our claim by the principle of induction.

If. This is also demonstrated by induction on $n' \geq 1$. We have to demonstrate that if $S' \Rightarrow_{G'}^{n'} x'$, $x' = @r'Xs'$, $X \in \{\$, \varepsilon\}$, $r's' \in g(x)$, $x \in V^*$, for some $n' \geq 1$, then $S \Rightarrow_G^* x$.

Basis. For $n' = 1$ the only x' is $@S$ since $S' \Rightarrow_{G'} @S$. Because $S \in g(S)$, we have $x = S$. Clearly, $S \Rightarrow_G^0 S$.

Induction Hypothesis. Assume that the claim holds for all derivations of length at most n' for some $n' \geq 1$. Let us show that it also holds for $n' + 1$.

Induction Step. Consider any derivation of the form

$$S' \Rightarrow_{G'}^{n'+1} x'$$

with $x' = @r'Xs'$, $X \in \{\$, \varepsilon\}$, $r's' \in g(x)$, $x \in V^*$. Since $n' + 1 \geq 2$, we have

$$S' \Rightarrow_{G'}^{n'} y' \Rightarrow_{G'} x' \; [p']$$

for some $p' = (Z' \to w', For) \in P'$, $y' = @q'Yt'$, $Y \in \{\$, \varepsilon\}$, $q't' \in g(y)$, $y \in V^*$, and by the induction hypothesis,

$$S \Rightarrow_G^* y$$

Suppose:

(i) $Z' \in N$, $w' \in \{\varepsilon\} \cup T \cup N^2$. By inspecting P' [see (2)], we have $For = \{\$\}$ and $Z' \to w' \in P$. Thus, $\$ \notin \mathrm{sub}(y')$ and so $q't' = y$. Hence, there is the following derivation

$$S \Rightarrow_G^* y \Rightarrow_G x \; [Z' \to w']$$

(ii) $g^{-1}(Z') = g^{-1}(w')$. But then $y = x$, and by the induction hypothesis, we have the derivation

$$S \Rightarrow_G^* y$$

(iii) $p' = (B \to \$[AB \to AC, 1], \{\$\})$; that is, $Z' = B$, $w' = \$[AB \to AC, 1]$, $For = \{\$\}$ and so $w' \in \{\$\}g(Z')$, $Y = \varepsilon$, $X = \$$. By analogy with (ii), we have

$$S \Rightarrow_G^* y$$

and $y = x$.

(iv) $Z' = Y = \$$; that is, $p' = (\$ \to \varepsilon, W)$. Then, $X = \varepsilon$, $r's' = q't' \in g(y)$, and

$$S \Rightarrow_G^* y$$

(v) $p' = ([AB \to AC, m+1] \to C, \emptyset)$; that is, $Z' = [AB \to AC, m+1]$, $w' = C$, $For = \emptyset$. From (3), it follows that there is a rule of the form $AB \to AC \in P$. Moreover, by inspecting (3), it is not too difficult to see (the technical details are left to the reader) that $Y = \$$, $r' = q'$, $t' = [AB \to AC, m+1]o'$, $s' = Co'$, and the derivation

$$S' \Rightarrow_{G'}^{n'} y' \Rightarrow_{G'} x' \; [p']$$

can be expressed as

$$\begin{aligned} S' &\Rightarrow_{G'}^* @q'Bo' \\ &\Rightarrow_{G'} @q'\$[AB \to AC, 1]o' && [(B \to \$[AB \to AC, 1], \{\$\})] \\ &\Rightarrow_{G'}^{m+1} @q'\$[AB \to AC, m+1]o' && [h] \\ &\Rightarrow_{G'} @q'\$Co' && [([AB \to AC, m+1] \to C, \emptyset)] \end{aligned}$$

where

$$\begin{aligned} h &= h_1([AB \to AC, f(A)] \to [AB \to AC, f(A)+1], \emptyset)h_2, \\ h_1 &= ([AB \to AC, 1] \to [AB \to AC, 2], \{f^{-1}(1)\$\}) \\ &\quad ([AB \to AC, 2] \to [AB \to AC, 3], \{f^{-1}(2)\$\}) \\ &\quad \vdots \\ &\quad ([AB \to AC, f(A)-1] \to [AB \to AC, f(A)], \{f^{-1}(f(A)-1)\$\}) \end{aligned}$$

in which $f(A) = 1$ implies that $h_1 = \varepsilon$,

$$\begin{aligned} h_2 = {} & ([AB \to AC, f(A)+1] \to [AB \to AC, f(A)+2], \{f^{-1}(f(A)+1)\$\}) \\ & \vdots \\ & ([AB \to AC, m] \to [AB \to AC, m+1], \{f^{-1}(m)\$\}) \end{aligned}$$

in which $f(A) = m$ implies that $h_2 = \varepsilon$; that is, the rightmost symbol of $q' = r'$ must be A.

Since $q't' \in g(y)$, we have $y = q'Bo'$. Because the rightmost symbol of q' is A and $AB \to AC \in P$, we have

$$S \Rightarrow_G^* q'Bo' \Rightarrow_G q'Co' \; [AB \to AC]$$

where $q'Co' = x$.

By inspecting P', we see that (i) through (v) cover all possible derivations of the form

$$S' \Rightarrow_{G'}^{n'} y' \Rightarrow_{G'} x'$$

and thus we have established that

$$S \Rightarrow_G^* x \quad \text{if and only if} \quad S' \Rightarrow_{G'}^+ x'$$

where $x' = @r'Xs', r's' \in g(x), X \in \{\$, \varepsilon\}, x \in V^*$, by the principle of induction.

A proof of the equivalence of G and G' can easily be derived from above. By the definition of g, we have $g(a) = \{a\}$ for all $a \in T$. Thus, we have for any $x \in T^*$,

$$S \Rightarrow_G^* x \quad \text{if and only if} \quad S' \Rightarrow_{G'}^* @rXs$$

where $X \in \{\$, \varepsilon\}, rs = x$. If $X = \varepsilon$, then

$$@x \Rightarrow_{G'} x \; [(@ \rightarrow \varepsilon, N \cup W \cup \{\$\})]$$

If $X = \$$, then

$$@r\$s \Rightarrow_{G'} @x \; [(\$ \rightarrow \varepsilon, W)] \Rightarrow_{G'} x \; [(@ \rightarrow \varepsilon, N \cup W \cup \{\$\})]$$

Hence,

$$S \Rightarrow_G^+ x \quad \text{if and only if} \quad S' \Rightarrow_{G'}^+ x$$

for all $x \in T^*$, and so $L(G) = L(G')$. Thus, **RE** = **GF**(2). □

Theorem 4.4.5. GF(2) = GF = RE

Proof. This theorem follows immediately from the definitions and Theorem 4.4.4. □

Examine the rules in G' in the proof of Theorem 4.4.4 to establish the following normal form.

Corollary 4.4.6. *Every recursively enumerable language L over some alphabet T can be generated by a gf-grammar $G = (V, T, P \cup \{p_1, p_2\}, S)$ of degree 2 such that*

(i) $(A \rightarrow x, For) \in P$ implies that $|x| = 2$ and the cardinality of For is at most 1;
(ii) $p_i = (A_i \rightarrow \varepsilon, For_i)$, $i = 1, 2$, where $For_i \subseteq V$; that is, max-len$(For_i) \leq 1$. □

In fact, the corollary above represents one of the reduced forms of gf-grammars of degree 2. Perhaps most importantly, it reduces the cardinality of the sets of forbidding conditions so that if a rule contains a condition of length two, this condition is the only context condition attached to the rule. Next, we study another reduced form of gf-grammars of degree 2. We show that we can simultaneously reduce the number of conditional rules and the number of nonterminals in gf-grammars of degree 2 without any decrease of their generative power.

Theorem 4.4.7. *Every recursively enumerable language can be defined by a gf-grammar of degree 2 with no more than 13 forbidding rules and 15 nonterminals.*

Proof. Let $L \in \mathbf{RE}$. By Theorem 4.1.11, without any loss of generality, we assume that L is generated by a phrase-structure grammar G of the form

$$G = \big(V, T, P \cup \{AB \to \varepsilon, CD \to \varepsilon\}, S\big)$$

such that P contains only context-free rules and

$$V - T = \big\{S, A, B, C, D\big\}$$

We construct a gf-grammar of degree 2

$$G' = \big(V', T, P', S'\big)$$

where

$$\begin{aligned} V' &= V \cup W \\ W &= \{S', @, \tilde{A}, \tilde{B}, \langle\varepsilon_A\rangle, \$, \tilde{C}, \tilde{D}, \langle\varepsilon_C\rangle, \#\},\ V \cap W = \emptyset \end{aligned}$$

in the following way. Let

$$N' = (V' - T) - \{S', @\}$$

Informally, N' denotes the set of all nonterminals in G' except S' and $@$. Then, the set of rules P' is constructed by performing (1) through (4), given next.

(1) If $H \to y \in P$, $H \in V - T$, $y \in V^*$, then add $(H \to y, \emptyset)$ to P';
(2) Add $(S' \to @S@, \emptyset)$ and $(@ \to \varepsilon, N')$ to P';
(3) Extend P' by adding

$$\begin{aligned} &(A \to \tilde{A}, \{\tilde{A}\}) \\ &(B \to \tilde{B}, \{\tilde{B}\}) \\ &(\tilde{A} \to \langle\varepsilon_A\rangle, \{\tilde{A}a \mid a \in V' - \{\tilde{B}\}\}) \\ &(\tilde{B} \to \$, \{a\tilde{B} \mid a \in V' - \{\langle\varepsilon_A\rangle\}\}) \\ &(\langle\varepsilon_A\rangle \to \varepsilon, \{\tilde{B}\}) \\ &(\$ \to \varepsilon, \{\langle\varepsilon_A\rangle\}) \end{aligned}$$

(4) Extend P' by adding

$$(C \to \tilde{C}, \{\tilde{C}\})$$
$$(D \to \tilde{D}, \{\tilde{D}\})$$
$$(\tilde{C} \to \langle\varepsilon_C\rangle, \{\tilde{C}a \mid a \in V' - \{\tilde{D}\}\})$$
$$(\tilde{D} \to \#, \{a\tilde{D} \mid a \in V' - \{\langle\varepsilon_C\rangle\}\})$$
$$(\langle\varepsilon_C\rangle \to \varepsilon, \{\tilde{D}\})$$
$$(\# \to \varepsilon, \{\langle\varepsilon_C\rangle\})$$

Next, we prove that $L(G') = L(G)$.

Notice that G' has degree 2 and contains only 13 forbidding rules and 15 nonterminals. The rules of (3) simulate the application of $AB \to \varepsilon$ in G' and the rules of (4) simulate the application of $CD \to \varepsilon$ in G'.

Let us describe the simulation of $AB \to \varepsilon$. First, one occurrence of A and one occurrence of B are rewritten with $\tilde{A}$ and $\tilde{B}$, respectively (no sentential form contains more than one occurrence of $\tilde{A}$ or $\tilde{B}$). The right neighbor of $\tilde{A}$ is checked to be $\tilde{B}$ and $\tilde{A}$ is rewritten with $\langle\varepsilon_A\rangle$. Then, analogously, the left neighbor of $\tilde{B}$ is checked to be $\langle\varepsilon_A\rangle$ and $\tilde{B}$ is rewritten with \$. Finally, $\langle\varepsilon_A\rangle$ and \$ are erased. The simulation of $CD \to \varepsilon$ is analogical.

To establish $L(G) = L(G')$, we first prove several claims.

Claim 1. *$S' \Rightarrow_{G'}^{+} w'$ implies that w' has one of the following two forms*

(I) $w' = @x'@$, $x' \in (N' \cup T)^*$, $\mathrm{alph}(x') \cap N' \neq \emptyset$;
(II) $w' = Xx'Y$, $x' \in T^*$, $X, Y \in \{@, \varepsilon\}$.

Proof. The start symbol S' is always rewritten with $@S@$. After this initial step, @ can be erased in a sentential form provided that any nonterminal occurring in the sentential form belongs to $\{@, S'\}$ (see N' and (2) in the definition of P'). In addition, notice that only rules of (2) contain @ and S'. Thus, any sentential form containing some nonterminals from N' is of form (I).

Case (II) covers sentential forms containing no nonterminal from N'. At this point, @ can be erased, and we obtain a string from $L(G')$. □

Claim 2. *$S' \Rightarrow_{G'}^{*} w'$ implies that* $\mathrm{occur}(w', \tilde{X}) \leq 1$ *for all $\tilde{X} \in \{\tilde{A}, \tilde{B}, \tilde{C}, \tilde{D}\}$ and some $w' \in V'^*$.*

Proof. By the inspection of rules in P', the only rule that can generate $\tilde{X}$ is of the form $(X \to \tilde{X}, \{\tilde{X}\})$. This rule can be applied only when no $\tilde{X}$ occurs in the rewritten sentential form. Thus, it is impossible to derive w' from S' such that $\mathrm{occur}(w', \tilde{X}) \geq 2$. □

Informally, next claim says that every occurrence of $\langle\varepsilon_A\rangle$ in derivations from S' is always followed either by $\tilde{B}$ or \$, and every occurrence of $\langle\varepsilon_C\rangle$ is always followed either by $\tilde{D}$ or #.

Claim 3. *The following two statements hold true.*

(I) *$S' \Rightarrow_{G'}^{*} y_1'\langle\varepsilon_A\rangle y_2'$ implies that $y_2' \in V'^{+}$ and* $\mathrm{first}(y_2') \in \{\tilde{B}, \$\}$ *for any $y_1' \in V'^*$.*

(II) $S' \Rightarrow^*_{G'} y_1'\langle\varepsilon_C\rangle y_2'$ *implies that* $y_2' \in V'^+$ *and* $\text{first}(y_2') \in \{\tilde{D}, \#\}$ *for any* $y_1' \in V'^*$.

Proof. We establish this claim by examination of all possible forms of derivations that may occur when deriving a sentential form containing $\langle\varepsilon_A\rangle$ or $\langle\varepsilon_C\rangle$.

(I) By the definition of P', the only rule that can generate $\langle\varepsilon_A\rangle$ is $p = (\tilde{A} \to \langle\varepsilon_A\rangle, \{\tilde{A}a \mid a \in V' - \{\tilde{B}\}\})$. The rule can be applied provided that $\tilde{A}$ occurs in a sentential form. It also holds that $\tilde{A}$ has always a right neighbor (as follows from Claim 1), and according to the set of forbidding conditions in p, $\tilde{B}$ is the only allowed right neighbor of $\tilde{A}$. Furthermore, by Claim 2, no other occurrence of $\tilde{A}$ or $\tilde{B}$ can appear in the given sentential form. Consequently, we obtain a derivation

$$S' \Rightarrow^*_{G'} u_1'\tilde{A}\tilde{B}u_2' \Rightarrow_{G'} u_1'\langle\varepsilon_A\rangle\tilde{B}u_2' \ [p]$$

for some $u_1', u_2' \in V'^*$, $\tilde{A}, \tilde{B} \notin \text{sub}(u_1'u_2')$. Obviously, $\langle\varepsilon_A\rangle$ is always followed by $\tilde{B}$ in $u_1'\langle\varepsilon_A\rangle\tilde{B}u_2'$.

Next, we discuss how G' can rewrite the substring $\langle\varepsilon_A\rangle\tilde{B}$ in $u_1'\langle\varepsilon_A\rangle\tilde{B}u_2'$. There are only two rules having the nonterminals $\langle\varepsilon_A\rangle$ or $\tilde{B}$ on their left-hand side, $p_1 = (\tilde{B} \to \$, \{a\tilde{B} \mid a \in V' - \{\langle\varepsilon_A\rangle\}\})$ and $p_2 = (\langle\varepsilon_A\rangle \to \varepsilon, \{\tilde{B}\})$. G' cannot use p_2 to erase $\langle\varepsilon_A\rangle$ in $u_1'\langle\varepsilon_A\rangle\tilde{B}u_2'$ because p_2 forbids an occurrence of $\tilde{B}$ in the rewritten string. However, we can rewrite $\tilde{B}$ to \$ by using p_1 because its set of forbidding conditions defines that the left neighbor of $\tilde{B}$ must be just $\langle\varepsilon_A\rangle$. Hence, we obtain a derivation of the form

$$\begin{aligned} S' &\Rightarrow^*_{G'} u_1'\tilde{A}\tilde{B}u_2' &&\Rightarrow_{G'} u_1'\langle\varepsilon_A\rangle\tilde{B}u_2' \ [p] \\ &\Rightarrow^*_{G'} v_1'\langle\varepsilon_A\rangle\tilde{B}v_2' &&\Rightarrow_{G'} v_1'\langle\varepsilon_A\rangle\$v_2' \ [p_1] \end{aligned}$$

Notice that during this derivation, G' may rewrite u_1' and u_2' with some v_1' and v_2', respectively ($v_1', v_2' \in V'^*$); however, $\langle\varepsilon_A\rangle\tilde{B}$ remains unchanged after this rewriting.

In this derivation we obtained the second symbol \$, which can appear as the right neighbor of $\langle\varepsilon_A\rangle$. It is sufficient to show that there is no other symbol that can appear immediately after $\langle\varepsilon_A\rangle$. By the inspection of P', only $(\$ \to \varepsilon, \{\langle\varepsilon_A\rangle\})$ can rewrite \$. However, this rule cannot be applied when $\langle\varepsilon_A\rangle$ occurs in the given sentential form. In other words, the occurrence of \$ in the substring $\langle\varepsilon_A\rangle\$$ cannot be rewritten before $\langle\varepsilon_A\rangle$ is erased by p_2. Hence, $\langle\varepsilon_A\rangle$ is always followed either by $\tilde{B}$ or \$, so the first part of Claim 3 holds.

(II) By the inspection of rules simulating $AB \to \varepsilon$ and $CD \to \varepsilon$ in G' (see (3) and (4) in the definition of P'), these two sets of rules work analogously. Thus, part (II) of Claim 3 can be proved by analogy with part (I). □

Let us return to the main part of the proof. Let g be a finite substitution from $(N' \cup T)^*$ to V^* defined as follows:

(a) For all $X \in V$, $g(X) = \{X\}$;
(b) $g(\tilde{A}) = \{A\}$, $g(\tilde{B}) = \{B\}$, $g(\langle \varepsilon_A \rangle) = \{A\}$, $g(\$) = \{B, AB\}$;
(c) $g(\tilde{C}) = \{C\}$, $g(\tilde{D}) = \{D\}$, $g(\langle \varepsilon_C \rangle) = \{C\}$, $g(\#) = \{C, CD\}$.

Having this substitution, we can now prove the following claim.

Claim 4. $S \Rightarrow_G^* x$ *if and only if* $S' \Rightarrow_{G'}^+ @x'@$ *for some* $x \in g(x')$, $x \in V^*$, $x' \in (N' \cup T)^*$.

Proof. The claim is proved by induction on the length of derivations.

Only If. We show that

$$S \Rightarrow_G^m x \quad \text{implies} \quad S' \Rightarrow_{G'}^+ @x@$$

where $m \geq 0$, $x \in V^*$; clearly $x \in g(x)$. This is established by induction on $m \geq 0$.

Basis. Let $m = 0$. That is, $S \Rightarrow_G^0 S$. Clearly, $S' \Rightarrow_{G'} @S@$.

Induction Hypothesis. Suppose that the claim holds for all derivations of length m or less, for some $m \geq 0$.

Induction Step. Let us consider any derivation of the form

$$S \Rightarrow_G^{m+1} x,\ x \in V^*$$

Since $m + 1 \geq 1$, there is some $y \in V^+$ and $p \in P \cup \{AB \to \varepsilon, CD \to \varepsilon\}$ such that

$$S \Rightarrow_G^m y \Rightarrow_G x\ [p]$$

By the induction hypothesis, there is a derivation

$$S' \Rightarrow_{G'}^+ @y@$$

There are the following three cases that cover all possible forms of p.

(i) Let $p = H \to y_2 \in P$, $H \in V - T$, $y_2 \in V^*$. Then, $y = y_1 H y_3$ and $x = y_1 y_2 y_3$, $y_1, y_3 \in V^*$. Because we have $(H \to y_2, \emptyset) \in P'$,

$$S' \Rightarrow_{G'}^+ @y_1 H y_3@ \Rightarrow_{G'} @y_1 y_2 y_3@\ [(H \to y_2, \emptyset)]$$

and $y_1 y_2 y_3 = x$.

(ii) Let $p = AB \to \varepsilon$. Then, $y = y_1 AB y_3$ and $x = y_1 y_3$, $y_1, y_3 \in V^*$. In this case, there is the following derivation

$$\begin{aligned}
S' &\Rightarrow^+_{G'} @y_1ABy_3@ \\
&\Rightarrow_{G'} @y_1\tilde{A}By_3@ \quad [(A \to \tilde{A}, \{\tilde{A}\})] \\
&\Rightarrow_{G'} @y_1\tilde{A}\tilde{B}y_3@ \quad [(B \to \tilde{B}, \{\tilde{B}\})] \\
&\Rightarrow_{G'} @y_1\langle\varepsilon_A\rangle\tilde{B}y_3@ \quad [(\tilde{A} \to \langle\varepsilon_A\rangle, \{\tilde{A}a \mid a \in V' - \{\tilde{B}\}\})] \\
&\Rightarrow_{G'} @y_1\langle\varepsilon_A\rangle\$y_3@ \quad [(\tilde{B} \to \$, \{a\tilde{B} \mid a \in V' - \{\langle\varepsilon_A\rangle\}\})] \\
&\Rightarrow_{G'} @y_1\$y_3@ \quad [(\langle\varepsilon_A\rangle \to \varepsilon, \{\tilde{B}\})] \\
&\Rightarrow_{G'} @y_1y_3@ \quad [(\$ \to \varepsilon, \{\langle\varepsilon_A\rangle\})]
\end{aligned}$$

(iii) Let $p = CD \to \varepsilon$. Then, $y = y_1CDy_3$ and $x = y_1y_3$, $y_1, y_3 \in V^*$. In this case, there exists the following derivation

$$\begin{aligned}
S' &\Rightarrow^+_{G'} @y_1CDy_3@ \\
&\Rightarrow_{G'} @y_1\tilde{C}Dy_3@ \quad [(C \to \tilde{C}, \{\tilde{C}\})] \\
&\Rightarrow_{G'} @y_1\tilde{C}\tilde{D}y_3@ \quad [(D \to \tilde{D}, \{\tilde{D}\})] \\
&\Rightarrow_{G'} @y_1\langle\varepsilon_C\rangle\tilde{D}y_3@ \quad [(\tilde{C} \to \langle\varepsilon_C\rangle, \{\tilde{C}a \mid a \in V' - \{\tilde{D}\}\})] \\
&\Rightarrow_{G'} @y_1\langle\varepsilon_C\rangle\#y_3@ \quad [(\tilde{D} \to \#, \{a\tilde{D} \mid a \in V' - \{\langle\varepsilon_C\rangle\}\})] \\
&\Rightarrow_{G'} @y_1\#y_3@ \quad [(\langle\varepsilon_C\rangle \to \varepsilon, \{\tilde{D}\})] \\
&\Rightarrow_{G'} @y_1y_3@ \quad [(\# \to \varepsilon, \{\langle\varepsilon_C\rangle\})]
\end{aligned}$$

If. By induction on the length n of derivations in G', we prove that

$$S' \Rightarrow^n_{G'} @x'@ \quad \text{implies} \quad S \Rightarrow^*_G x$$

for some $x \in g(x')$, $x \in V^*$, $x' \in (N' \cup T)^*$, $n \geq 1$.

Basis. Let $n = 1$. According to the definition of P', the only rule rewriting S' is $(S' \to @S@, \emptyset)$, so $S' \Rightarrow_{G'} @S@$. It is obvious that $S \Rightarrow^0_G S$ and $S \in g(S)$.

Induction Hypothesis. Assume that the claim holds for all derivations of length n or less, for some $n \geq 1$.

Induction Step. Consider any derivation of the form

$$S' \Rightarrow^{n+1}_{G'} @x'@, \ x' \in (N' \cup T)^*$$

Since $n + 1 \geq 2$, there is some $y' \in (N' \cup T)^+$ and $p' \in P'$ such that

$$S' \Rightarrow^n_{G'} @y'@ \Rightarrow_{G'} @x'@ \ [p']$$

and by the induction hypothesis, there is also a derivation

$$S \Rightarrow^*_G y$$

such that $y \in g(y')$.

By the inspection of P', the following cases (i) through (xiii) cover all possible forms of p'.

(i) Let $p' = (H \rightarrow y_2, \emptyset) \in P'$, $H \in V - T$, $y_2 \in V^*$. Then, $y' = y'_1 H y'_3$, $x' = y'_1 y_2 y'_3$, $y'_1, y'_3 \in (N' \cup T)^*$, and y has the form $y = y_1 Z y_3$, where $y_1 \in g(y'_1)$, $y_3 \in g(y'_3)$, and $Z \in g(H)$. Because for all $X \in V - T$: $g(X) = \{X\}$, the only Z is H; thus, $y = y_1 H y_3$. By the definition of P' [see (1)], there exists a rule $p = H \rightarrow y_2$ in P, and we can construct the derivation

$$S \Rightarrow^*_G y_1 H y_3 \Rightarrow_G y_1 y_2 y_3 \; [p]$$

such that $y_1 y_2 y_3 = x$, $x \in g(x')$.

(ii) Let $p' = (A \rightarrow \tilde{A}, \{\tilde{A}\})$. Then, $y' = y'_1 A y'_3$, $x' = y'_1 \tilde{A} y'_3$, $y'_1, y'_3 \in (N' \cup T)^*$ and $y = y_1 Z y_3$, where $y_1 \in g(y'_1)$, $y_3 \in g(y'_3)$ and $Z \in g(A)$. Because $g(A) = \{A\}$, the only Z is A, so we can express $y = y_1 A y_3$. Having the derivation $S \Rightarrow^*_G y$ such that $y \in g(y')$, it is easy to see that also $y \in g(x')$ because $A \in g(\tilde{A})$.

(iii) Let $p' = (B \rightarrow \tilde{B}, \{\tilde{B}\})$. By analogy with (ii), $y' = y'_1 B y'_3$, $x' = y'_1 \tilde{B} y'_3$, $y = y_1 B y_3$, where $y'_1, y'_3 \in (N' \cup T)^*$, $y_1 \in g(y'_1)$, $y_3 \in g(y'_3)$; thus, $y \in g(x')$ because $B \in g(\tilde{B})$.

(iv) Let $p' = (\tilde{A} \rightarrow \langle \varepsilon_A \rangle, \{\tilde{A}a \mid a \in V' - \{\tilde{B}\}\})$. In this case, it holds that

   (iv.i) Application of p' implies that $\tilde{A} \in \text{alph}(y')$, and moreover, by Claim 2, we have $\text{occur}(y', \tilde{A}) \leq 1$;
   (iv.ii) $\tilde{A}$ has always a right neighbor in $@y'@$;
   (iv.iii) According to the set of forbidding conditions in p', the only allowed right neighbor of $\tilde{A}$ is $\tilde{B}$.

Hence, y' must be of the form $y' = y'_1 \tilde{A}\tilde{B} y'_3$, where $y'_1, y'_3 \in (N' \cup T)^*$ and $\tilde{A} \notin \text{sub}(y'_1 y'_3)$. Then, $x' = y'_1 \langle \varepsilon_A \rangle \tilde{B} y'_3$ and y is of the form $y = y_1 Z y_3$, where $y_1 \in g(y'_1)$, $y_3 \in g(y'_3)$ and $Z \in g(\tilde{A}\tilde{B})$. Because $g(\tilde{A}\tilde{B}) = \{AB\}$, the only Z is AB; thus, we obtain $y = y_1 A B y_3$. By the induction hypothesis, we have a derivation $S \Rightarrow^*_G y$ such that $y \in g(y')$. According to the definition of g, $y \in g(x')$ as well because $A \in g(\langle \varepsilon_A \rangle)$ and $B \in g(\tilde{B})$.

(v) Let $p' = (\tilde{B} \rightarrow \$, \{a\tilde{B} \mid a \in V' - \{\langle \varepsilon_A \rangle\}\})$. Then, it holds that

   (v.i) $\tilde{B} \in \text{alph}(y')$ and, by Claim 2, $\text{occur}(y', \tilde{B}) \leq 1$;
   (v.ii) $\tilde{B}$ has always a left neighbor in $@y'@$;
   (v.iii) By the set of forbidding conditions in p', the only allowed left neighbor of $\tilde{B}$ is $\langle \varepsilon_A \rangle$.

Therefore, we can express $y' = y'_1 \langle \varepsilon_A \rangle \tilde{B} y'_3$, where $y'_1, y'_3 \in (N' \cup T)^*$ and $\tilde{B} \notin \text{sub}(y'_1 y'_3)$. Then, $x' = y'_1 \langle \varepsilon_A \rangle \$ y'_3$ and $y = y_1 Z y_3$, where $y_1 \in g(y'_1)$, $y_3 \in g(y'_3)$, and $Z \in g(\langle \varepsilon_A \rangle \tilde{B})$. By the definition of g, $g(\langle \varepsilon_A \rangle \tilde{B}) = \{AB\}$, so $Z = AB$ and $y = y_1 A B y_3$. By the induction hypothesis, we have a derivation $S \Rightarrow_G^* y$ such that $y \in g(y')$. Because $A \in g(\langle \varepsilon_A \rangle)$ and $B \in g(\$)$, $y \in g(x')$ as well.

(vi) Let $p' = (\langle \varepsilon_A \rangle \to \varepsilon, \{\tilde{B}\})$. An application of $(\langle \varepsilon_A \rangle \to \varepsilon, \{\tilde{B}\})$ implies that $\langle \varepsilon_A \rangle$ occurs in y'. Claim 3 says that $\langle \varepsilon_A \rangle$ has either $\tilde{B}$ or \$ as its right neighbor. Since the forbidding condition of p' forbids an occurrence of $\tilde{B}$ in y', the right neighbor of $\langle \varepsilon_A \rangle$ must be \$. As a result, we obtain $y' = y_1' \langle \varepsilon_A \rangle \$ y_3'$, where $y_1', y_3' \in (N' \cup T)^*$. Then, $x' = y_1' \$ y_3'$, and y is of the form $y = y_1 Z y_3$, where $y_1 \in g(y_1')$, $y_3 \in g(y_3')$, and $Z \in g(\langle \varepsilon_A \rangle \$)$. By the definition of g, $g(\langle \varepsilon_A \rangle \$) = \{AB, AAB\}$. If $Z = AB$, $y = y_1 AB y_3$. Having the derivation $S \Rightarrow_G^* y$, it holds that $y \in g(x')$ because $AB \in g(\$)$.

(vii) Let $p' = (\$ \to \varepsilon, \{\langle \varepsilon_A \rangle\})$. Then, $y' = y_1' \$ y_3'$ and $x' = y_1' y_3'$, where $y_1', y_3' \in (N' \cup T)^*$. Express $y = y_1 Z y_3$ so that $y_1 \in g(y_1')$, $y_3 \in g(y_3')$, and $Z \in g(\$)$, where $g(\$) = \{B, AB\}$. Let $Z = AB$. Then, $y = y_1 AB y_3$, and there exists the derivation

$$S \Rightarrow_G^* y_1 AB y_3 \Rightarrow_G y_1 y_3 \ [AB \to \varepsilon]$$

where $y_1 y_3 = x$, $x \in g(x')$.

In cases (ii) through (vii), we discussed all six rules simulating the application of $AB \to \varepsilon$ in G' (see (3) in the definition of P'). Cases (viii) through (xiii) should cover the rules simulating the application of $CD \to \varepsilon$ in G' [see (4)]. However, by the inspection of these two sets of rules, it is easy to see that they work analogously. Therefore, we leave this part of the proof to the reader.

We have completed the proof and established Claim 4 by the principle of induction. □

Observe that $L(G) = L(G')$ can be easily derived from the above claim. According to the definition of g, we have $g(a) = \{a\}$ for all $a \in T$. Thus, from Claim 4, we have for any $x \in T^*$

$$S \Rightarrow_G^* x \quad \text{if and only if} \quad S' \Rightarrow_{G'}^+ @x@$$

Since

$$@x@ \Rightarrow_{G'}^2 x \ [(@ \to \varepsilon, N')(@ \to \varepsilon, N')]$$

we obtain for any $x \in T^*$:

$$S \Rightarrow_G^* x \quad \text{if and only if} \quad S' \Rightarrow_{G'}^+ x$$

Consequently, $L(G) = L(G')$, and the theorem holds. □

4.5 Semi-Conditional Grammars

The notion of a semi-conditional grammar, discussed in this section, is defined as a context-conditional grammar in which the cardinality of any context-conditional set is no more than 1.

The present section consists of two subsections. Section 4.5.1 defines and illustrates semi-conditional grammars. Section 4.5.2 studies their generative power and reduction.

4.5.1 Definitions and Examples

The definition of a semi-conditional grammar opens this section.

Definition 4.5.1. Let $G = (V, T, P, S)$ be a context-conditional grammar. G is called a *semi-conditional grammar* (an *sc-grammar* for short) provided that every $(A \rightarrow x, Per, For) \in P$ satisfies $\text{card}(Per) \leq 1$ and $\text{card}(For) \leq 1$. □

Let $G = (V, T, P, S)$ be an sc-grammar, and let $(A \rightarrow x, Per, For) \in P$. For brevity, we omit braces in each $(A \rightarrow x, Per, For) \in P$, and instead of $\emptyset$, we write 0. For instance, we write $(A \rightarrow x, BC, 0)$ instead of $(A \rightarrow x, \{BC\}, \emptyset)$.

The families of languages generated by sc-grammars and propagating sc-grammars of degree (r, s) are denoted by $\mathbf{SC}(r, s)$ and $\mathbf{SC}^{-\varepsilon}(r, s)$, respectively. The families of languages generated by sc-grammars and propagating sc-grammars of any degree are defined as

$$\mathbf{SC} = \bigcup_{r=0}^{\infty} \bigcup_{s=0}^{\infty} \mathbf{SC}(r, s)$$

and

$$\mathbf{SC}^{-\varepsilon} = \bigcup_{r=0}^{\infty} \bigcup_{s=0}^{\infty} \mathbf{SC}^{-\varepsilon}(r, s)$$

First, we give examples of sc-grammars with degrees $(1, 0)$, $(0, 1)$, and $(1, 1)$.

Example 4.5.2 (See [16]). Let us consider an sc-grammar

$$G = \big(\{S, A, B, A', B', a, b\}, \{a, b\}, P, S\big)$$

where

$$P = \big\{(S \rightarrow AB, 0, 0), (A \rightarrow A'A', B, 0),\\ (B \rightarrow bB', 0, 0), (A' \rightarrow A, B', 0),\\ (B' \rightarrow B, 0, 0), (B \rightarrow b, 0, 0),\\ (A' \rightarrow a, 0, 0), (A \rightarrow a, 0, 0)\big\}$$

Observe that A can be replaced by $A'A'$ only if B occurs in the rewritten string, and A' can be replaced by A only if B' occurs in the rewritten string. If there is an occurrence of B, the number of occurrences of A and A' can be doubled. However, the application of $(B \rightarrow bB', 0, 0)$ implies an introduction of one occurrence of b. As a result,

$$L(G) = \{a^n b^m \mid m \geq 1,\ 1 \leq n \leq 2^m\}$$

which is a non-context-free language. □

Example 4.5.3 (See [16]). Let

$$G = \left(\{S, A, B, A', A'', B', a, b, c\}, \{a, b, c\}, P, S\right)$$

be an sc-grammar, where

$$\begin{aligned} P = \{&(S \rightarrow AB, 0, 0), (A \rightarrow A', 0, B'), \\ &(A' \rightarrow A''A'', 0, c), (A'' \rightarrow A, 0, B), \\ &(B \rightarrow bB', 0, 0), (B' \rightarrow B, 0, 0), \\ &(B \rightarrow c, 0, 0), (A \rightarrow a, 0, 0), \\ &(A'' \rightarrow a, 0, 0)\} \end{aligned}$$

In this case, we get the non-context-free language

$$L(G) = \{a^n b^m c \mid m \geq 0,\ 1 \leq n \leq 2^{m+1}\}$$ □

Example 4.5.4. Let

$$G = \left(\{S, P, Q, R, X, Y, Z, a, b, c, d, e, f\}, \{a, b, c, d, e, f\}, P, S\right)$$

be an sc-grammar, where

$$\begin{aligned} P = \{&(S \rightarrow PQR, 0, 0), \\ &(P \rightarrow aXb, Q, Z), \\ &(Q \rightarrow cYd, X, Z), \\ &(R \rightarrow eZf, X, Q), \\ &(X \rightarrow P, Z, Q), \\ &(Y \rightarrow Q, P, R), \\ &(Z \rightarrow R, P, Y), \\ &(P \rightarrow \varepsilon, Q, Z), \\ &(Q \rightarrow \varepsilon, R, P), \\ &(R \rightarrow \varepsilon, 0, Y)\} \end{aligned}$$

Note that this grammar is an sc-grammar of degree $(1, 1)$. Consider $aabbccddeeff$. For this string, G makes the following derivation

$$S \Rightarrow PQR \Rightarrow aXbQR \Rightarrow aXbcYdR \Rightarrow aXbcYdeZf \Rightarrow \\ aPbcYdeZf \Rightarrow aPbcQdeZf \Rightarrow aPbcQdeRf \Rightarrow \\ aaXbbcQdeRf \Rightarrow aaXbbccYddeRf \Rightarrow aaXbbccYddeeZff \Rightarrow \\ aaPbbccYddeeZff \Rightarrow aaPbbccQddeeZff \Rightarrow aaPbbccQddeeRff \Rightarrow \\ aabbccQddeeRff \Rightarrow aabbccddeeRff \Rightarrow aabbccddeeff$$

Clearly, G generates the following language

$$L(G) = \{a^n b^n c^n d^n e^n f^n \mid n \geq 0\}$$

As is obvious, this language is non-context-free. □

4.5.2 *Generative Power*

The present section establishes the generative power of sc-grammars.

Theorem 4.5.5. $\mathbf{SC}^{-\varepsilon}(0, 0) = \mathbf{SC}(0, 0) = \mathbf{CF}$

Proof. Follows directly from the definitions. □

Theorem 4.5.6. $\mathbf{CF} \subset \mathbf{SC}^{-\varepsilon}(1, 0)$, $\mathbf{CF} \subset \mathbf{SC}^{-\varepsilon}(0, 1)$

Proof. In Examples 4.5.2 and 4.5.3, we show propagating sc-grammars of degrees $(1, 0)$ and $(0, 1)$ that generate non-context-free languages. Therefore, the theorem holds. □

Theorem 4.5.7. $\mathbf{SC}^{-\varepsilon}(1, 1) \subset \mathbf{CS}$

Proof. Consider a propagating sc-grammar of degree $(1, 1)$

$$G = (V, T, P, S)$$

If $(A \to x, A, \beta) \in P$, then the permitting condition A does not impose any restriction. Hence, we can replace this rule by $(A \to x, 0, \beta)$. If $(A \to x, \alpha, A) \in P$, then this rule cannot ever by applied; thus, we can remove it from P. Let $T' = \{a' \mid a \in T\}$ and $V' = V \cup T' \cup \{S', X, Y\}$. Define a homomorphism τ from V^* to $((V-T) \cup (T'))^*$ as $\tau(a) = a'$ for all $a \in T$ and $\tau(A) = A$ for every $A \in V-T$. Furthermore, introduce a function g from $V \cup \{0\}$ to $2^{((V-T) \cup T')}$ as $g(0) = \emptyset$, $g(a) = \{a'\}$ for all $a \in T$, and $g(A) = \{A\}$ for all $A \in V - T$. Next, construct the propagating random context grammar

$$G' = (V', T \cup \{c\}, P', S')$$

where

$$\begin{aligned} P' = \{&(S' \to SX, \emptyset, \emptyset), (X \to Y, 0, 0), (Y \to c, 0, 0)\} \cup \\ &\{(A \to \tau(x), g(\alpha) \cup \{X\}, g(\beta)) \mid (A \to x, \alpha, \beta) \in P\} \cup \\ &\{(a' \to a, \{Y\}, \emptyset) \mid a \in T\} \end{aligned}$$

It is obvious that $L(G') = L(G)\{c\}$. Therefore, $L(G)\{c\} \in \mathbf{RC}^{-\varepsilon}$. Recall that $\mathbf{RC}^{-\varepsilon}$ is closed under restricted homomorphisms (see p. 48 in [3]), and by Theorem 4.3.6, it holds that $\mathbf{RC}^{-\varepsilon} \subset \mathbf{CS}$. Thus, we obtain $\mathbf{SC}^{-\varepsilon}(1,1) \subset \mathbf{CS}$. □

The following corollary summarizes the generative power of propagating sc-grammars of degrees $(1,0)$, $(0,1)$, and $(1,1)$—that is, the propagating sc-grammars containing only symbols as their context conditions.

Corollary 4.5.8.

$$\mathbf{CF} \subset \mathbf{SC}^{-\varepsilon}(0,1) \subseteq \mathbf{SC}^{-\varepsilon}(1,1)$$
$$\mathbf{CF} \subset \mathbf{SC}^{-\varepsilon}(1,0) \subseteq \mathbf{SC}^{-\varepsilon}(1,1)$$
$$\mathbf{SC}^{-\varepsilon}(1,1) \subseteq \mathbf{RC}^{-\varepsilon} \subset \mathbf{CS}$$

Proof. This corollary follows from Theorems 4.5.5, 4.5.6, and 4.5.7. □

The next theorem says that propagating sc-grammars of degrees $(1,2)$, $(2,1)$ and propagating sc-grammars of any degree generate exactly the family of context-sensitive languages. Furthermore, if we allow erasing rules, these grammars generate the family of recursively enumerable languages.

Theorem 4.5.9.

$$\begin{gathered} \mathbf{CF} \\ \subset \\ \mathbf{SC}^{-\varepsilon}(2,1) = \mathbf{SC}^{-\varepsilon}(1,2) = \mathbf{SC}^{-\varepsilon} = \mathbf{CS} \\ \subset \\ \mathbf{SC}(2,1) = \mathbf{SC}(1,2) = \mathbf{SC} = \mathbf{RE} \end{gathered}$$

Proof. In the next section, we prove a stronger result in terms of a special variant of sc-grammars—simple semi-conditional grammars (see Theorems 4.6.9 and 4.6.11). Therefore, we omit the proof here. □

In [15], the following theorem is proved. It shows that **RE** can be characterized even by sc-grammars of degree $(2,1)$ with a reduced number of nonterminals and conditional rules.

Theorem 4.5.10 (See Theorem 1 in [15]). *Every recursively enumerable language can be generated by an sc-grammar of degree* $(2,1)$ *having no more than 9 conditional rules and 10 nonterminals.*

4.6 Simple Semi-Conditional Grammars

The notion of a simple semi-conditional grammar—that is, the subject of this section—is defined as an sc-grammar in which every rule has no more than 1 condition.

The present section consists of two subsections. Section 4.6.1 defines simple semi-conditional grammars, and Sect. 4.6.2 discusses their generative power and reduction.

4.6.1 Definitions and Examples

First, we define simple semi-conditional grammars. Then, we illustrate them.

Definition 4.6.1. Let $G = (V, T, P, S)$ be a semi-conditional grammar. G is a *simple semi-conditional grammar* (an *ssc-grammar* for short) if $(A \rightarrow x, \alpha, \beta) \in P$ implies that $0 \in \{\alpha, \beta\}$. □

The families of languages generated by ssc-grammars and propagating ssc-grammars of degree (r, s) are denoted by $\mathbf{SSC}(r, s)$ and $\mathbf{SSC}^{-\varepsilon}(r, s)$, respectively. Furthermore, set

$$\mathbf{SSC} = \bigcup_{r=0}^{\infty} \bigcup_{s=0}^{\infty} \mathbf{SSC}(r, s)$$

and

$$\mathbf{SSC}^{-\varepsilon} = \bigcup_{r=0}^{\infty} \bigcup_{s=0}^{\infty} \mathbf{SSC}^{-\varepsilon}(r, s)$$

The following proposition provides an alternative definition based on context-conditional grammars. Let $G = (V, T, P, S)$ be a context-conditional grammar. G is an ssc-grammar if and only if every $(A \rightarrow x, Per, For) \in P$ satisfies $\mathrm{card}(Per) + \mathrm{card}(For) \leq 1$.

Example 4.6.2. Let

$$G = \big(\{S, A, X, C, Y, a, b\}, \{a, b\}, P, S\big)$$

be an ssc-grammar, where

$$\begin{aligned} P = \big\{&(S \rightarrow AC, 0, 0),\\ &(A \rightarrow aXb, Y, 0),\\ &(C \rightarrow Y, A, 0),\\ &(Y \rightarrow Cc, 0, A),\\ &(A \rightarrow ab, Y, 0),\\ &(Y \rightarrow c, 0, A),\\ &(X \rightarrow A, C, 0)\big\} \end{aligned}$$

Notice that G is propagating, and it has degree $(1, 1)$. Consider $aabbcc$. G derives this string as follows:

$$S \Rightarrow AC \Rightarrow AY \Rightarrow aXbY \Rightarrow aXbCc \Rightarrow \\ aAbCc \Rightarrow aAbYc \Rightarrow aabbYc \Rightarrow aabbcc$$

Obviously,

$$L(G) = \{a^n b^n c^n \mid n \geq 1\}$$ □

Example 4.6.3. Let

$$G = (\{S, A, B, X, Y, a\}, \{a\}, P, S)$$

be an ssc-grammar, where

$$\begin{aligned} P = \{&(S \to a, 0, 0), \\ &(S \to X, 0, 0), \\ &(X \to YB, 0, A), \\ &(X \to aB, 0, A), \\ &(Y \to XA, 0, B), \\ &(Y \to aA, 0, B), \\ &(A \to BB, XA, 0), \\ &(B \to AA, YB, 0), \\ &(B \to a, a, 0)\} \end{aligned}$$

G is a propagating ssc-grammar of degree $(2, 1)$. Consider the string $aaaaaaaa$. G derives this string as follows:

$$S \Rightarrow X \Rightarrow YB \Rightarrow YAA \Rightarrow XAAA \Rightarrow XBBAA \Rightarrow XBBABB \Rightarrow \\ XBBBBBB \Rightarrow aBBBBBBB \Rightarrow aBBaBBBB \Rightarrow^6 aaaaaaaa$$

Observe that G generates the following non-context-free language

$$L(G) = \left\{a^{2^n} \mid n \geq 0\right\}$$ □

4.6.2 Generative Power and Reduction

The power and reduction of ssc-grammars represent the central topic discussed in this section.

Theorem 4.6.4. $\mathbf{SSC}^{-\varepsilon}(2, 1) = \mathbf{CS}$

Proof. Because $\mathbf{SSC}^{-\varepsilon}(2,1) \subseteq \mathbf{CG}^{-\varepsilon}$ and Lemma 4.2.3 implies that $\mathbf{CG}^{-\varepsilon} \subseteq \mathbf{CS}$, it is sufficient to prove the converse inclusion.

Let $G = (V, T, P, S)$ be a context-sensitive grammar in the Penttonen normal form (see Theorem 4.1.5). We construct an ssc-grammar

$$G' = \big(V \cup W, T, P', S\big)$$

that generates $L(G)$. Let

$$W = \big\{\tilde{B} \mid AB \to AC \in P,\ A, B, C \in V - T\big\}$$

Define P' in the following way

(1) If $A \to x \in P$, $A \in V - T$, $x \in T \cup (V - T)^2$, then add $(A \to x, 0, 0)$ to P';
(2) If $AB \to AC \in P$, $A, B, C \in V - T$, then add $(B \to \tilde{B}, 0, \tilde{B})$, $(\tilde{B} \to C, A\tilde{B}, 0)$, $(\tilde{B} \to B, 0, 0)$ to P'.

Notice that G' is a propagating ssc-grammar of degree $(2, 1)$. Moreover, from (2), we have for any $\tilde{B} \in W$,

$$S \Rightarrow_{G'}^{*} w \quad \text{implies} \quad \text{occur}(w, \tilde{B}) \leq 1$$

for all $w \in V'^{*}$ because the only rule that can generate $\tilde{B}$ is of the form $(B \to \tilde{B}, 0, \tilde{B})$.

Let g be a finite substitution from V^* into $(V \cup W)^*$ defined as follows: for all $D \in V$,

(1) If $\tilde{D} \in W$, then $g(D) = \{D, \tilde{D}\}$;
(2) If $\tilde{D} \notin W$, then $g(D) = \{D\}$.

Claim 1. For any $x \in V^+$, $m, n \geq 0$, $S \Rightarrow_G^m x$ if and only if $S \Rightarrow_{G'}^n x'$ with $x' \in g(x)$.

Proof. The proof is divided into the only-if part and the if part.

Only If. This is proved by induction on $m \geq 0$.

Basis. Let $m = 0$. The only x is S as $S \Rightarrow_G^0 S$. Clearly, $S \Rightarrow_{G'}^n S$ for $n = 0$ and $S \in g(S)$.

Induction Hypothesis. Assume that the claim holds for all derivations of length m or less, for some $m \geq 0$.

Induction Step. Consider any derivation of the form

$$S \Rightarrow_G^{m+1} x$$

where $x \in V^+$. Because $m + 1 \geq 1$, there is some $y \in V^*$ and $p \in P$ such that

$$S \Rightarrow_G^m y \Rightarrow_{G'} x \; [p]$$

By the induction hypothesis,

$$S \Rightarrow_{G'}^n y'$$

for some $y' \in g(y)$ and $n \geq 0$. Next, we distinguish between two cases: case (i) considers p with one nonterminal on its left-hand side, and case (ii) considers p with two nonterminals on its left-hand side.

(i) Let $p = D \to y_2 \in P$, $D \in V - T$, $y_2 \in T \cup (V - T)^2$, $y = y_1 D y_3$, $y_1, y_3 \in V^*$, $x = y_1 y_2 y_3$, $y' = y_1' X y_3'$, $y_1' \in g(y_1)$, $y_3' \in g(y_3)$, and $X \in g(D)$. By (1) in the definition of P', $(D \to y_2, 0, 0) \in P$. If $X = D$, then

$$S \Rightarrow_{G'}^n y_1' D y_3' \Rightarrow_{G'} y_1' y_2 y_3' \; [(D \to y_2, 0, 0)]$$

Because $y_1' \in g(y_1)$, $y_3' \in g(y_3)$, and $y_2 \in g(y_2)$, we obtain $y_1' y_2 y_3' \in g(y_1 y_2 y_3) = g(x)$. If $X = \tilde{D}$, we have $(X \to D, 0, 0)$ in P', so

$$S \Rightarrow_{G'}^n y_1' X y_3' \Rightarrow_{G'} y_1' D y_3' \Rightarrow_{G'} y_1' y_2 y_3' \; [(X \to D, 0, 0)(D \to y_2, 0, 0)]$$

and $y_1' y_2 y_3' \in g(x)$.

(ii) Let $p = AB \to AC \in P$, $A, B, C \in V - T$, $y = y_1 AB y_2$, $y_1, y_2 \in V^*$, $x = y_1 AC y_2$, $y' = y_1' XY y_2'$, $y_1' \in g(y_1)$, $y_2' \in g(y_2)$, $X \in g(A)$, and $Y \in g(B)$. Recall that for any $\tilde{B}$, occur$(y', \tilde{B}) \leq 1$ and $(\tilde{B} \to B, 0, 0) \in P'$. Then,

$$y' \Rightarrow_{G'}^i y_1' AB y_2'$$

for some $i \in \{0, 1, 2\}$. At this point, we have

$$\begin{aligned} S &\Rightarrow_{G'}^* y_1' AB y_2' \\ &\Rightarrow_{G'} y_1' A\tilde{B} y_2' \; [(B \to \tilde{B}, 0, \tilde{B})] \\ &\Rightarrow_{G'} y_1' AC y_2' \; [(\tilde{B} \to C, A\tilde{B}, 0)] \end{aligned}$$

where $y_1' AC y_2' \in g(x)$.

If. This is established by induction on $n \geq 0$; in other words, we demonstrate that if $S \Rightarrow_{G'}^n x'$ with $x' \in g(x)$ for some $x \in V^+$, then $S \Rightarrow_G^* x$.

Basis. For $n = 0$, x' surely equals S as $S \Rightarrow_{G'}^0 S$. Because $S \in g(S)$, we have $x = S$. Clearly, $S \Rightarrow_G^0 S$.

Induction Hypothesis. Assume that the claim holds for all derivations of length n of less, for some $n \geq 0$.

Induction Step. Consider any derivation of the form

$$S \Rightarrow_{G'}^{n+1} x'$$

with $x' \in g(x)$, $x \in V^+$. As $n + 1 \geq 1$, there exists some $y \in V^+$ such that

$$S \Rightarrow_{G'}^{n} y' \Rightarrow_{G'} x' \ [p]$$

where $y' \in g(y)$. By the induction hypothesis,

$$S \Rightarrow_G^* y$$

Let $y' = y_1' B' y_2'$, $y = y_1 B y_2$, $y_1' \in g(y_1)$, $y_2' \in g(y_2)$, $y_1, y_2 \in V^*$, $B' \in g(B)$, $B \in V - T$, $x' = y_1' z' y_2'$, and $p = (B' \to z', \alpha, \beta) \in P'$. The following three cases cover all possible forms of the derivation step $y' \Rightarrow_{G'} x'$ $[p]$.

(i) Let $z' \in g(B)$. Then,

$$S \Rightarrow_G^* y_1 B y_2$$

where $y_1' z' y_2' \in g(y_1 B y_2)$; that is, $x' \in g(y_1 B y_2)$.

(ii) Let $B' = B \in V - T$, $z' \in T \cup (V - T)^2$, $\alpha = \beta = 0$. Then, there exists a rule, $B \to z' \in P$, so

$$S \Rightarrow_G^* y_1 B y_2 \Rightarrow_G y_1 z' y_2 \ [B \to z']$$

Since $z' \in g(z')$, we have $x = y_1 z' y_2$ such that $x' \in g(x)$.

(iii) Let $B' = \tilde{B}$, $z' = C$, $\alpha = A\tilde{B}$, $\beta = 0$, $A, B, C \in V - T$. Then, there exists a rule of the form $AB \to AC \in P$. Since $\text{occur}(y', Z) \leq 1$, $Z = \tilde{B}$, and $A\tilde{B} \in \text{sub}(y')$, we have $y_1' = u'A$, $y_1 = uA$, $u' \in g(u)$ for some $u \in V^*$. Thus,

$$S \Rightarrow_G^* uABy_2 \Rightarrow_G uACy_2 \ [AB \to AC]$$

where $uACy_2 = y_1 C y_2$. Because $C \in g(C)$, we get $x = y_1 C y_2$ such that $x' \in g(x)$.

As cases (i) through (iii) cover all possible forms of a derivation step in G', we have completed the induction step and established Claim 1 by the principle of induction. □

The statement of Theorem 4.6.4 follows immediately from Claim 1. Because for all $a \in T$, $g(a) = \{a\}$, we have for every $w \in T^+$,

$$S \Rightarrow_G^* w \quad \text{if and only if} \quad S \Rightarrow_{G'}^* w$$

Therefore, $L(G) = L(G')$, so the theorem holds. □

Corollary 4.6.5. $\mathbf{SSC}^{-\varepsilon}(2,1) = \mathbf{SSC}^{-\varepsilon} = \mathbf{SC}^{-\varepsilon}(2,1) = \mathbf{SC}^{-\varepsilon} = \mathbf{CS}$

Proof. This corollary follows from Theorem 4.6.4 and the definitions of propagating ssc-grammars. □

Next, we turn our investigation to ssc-grammars of degree $(2, 1)$ with erasing rules. We prove that these grammars generate precisely the family of recursively enumerable languages.

Theorem 4.6.6. $\mathbf{SSC}(2,1) = \mathbf{RE}$

Proof. Clearly, $\mathbf{SSC}(2,1) \subseteq \mathbf{RE}$; hence, it is sufficient to show that $\mathbf{RE} \subseteq \mathbf{SSC}(2,1)$. Every recursively enumerable language $L \in \mathbf{RE}$ can be generated by a phrase-structure grammar G in the Penttonen normal form (see Theorem 4.1.4). That is, the rules of G are of the form $AB \rightarrow AC$ or $A \rightarrow x$, where $A, B, C \in V - T$, $x \in \{\varepsilon\} \cup T \cup (V-T)^2$. Thus, the inclusion $\mathbf{RE} \subseteq \mathbf{SSC}(2,1)$ can be proved by analogy with the proof of Theorem 4.6.4. The details are left to the reader. □

Corollary 4.6.7. $\mathbf{SSC}(2,1) = \mathbf{SSC} = \mathbf{SC}(2,1) = \mathbf{SC} = \mathbf{RE}$ □

To demonstrate that propagating ssc-grammars of degree $(1, 2)$ characterize **CS**, we first establish a normal form for context-sensitive grammars.

Lemma 4.6.8. *Every $L \in \mathbf{CS}$ can be generated by a context-sensitive grammar*

$$G = \big(\{S\} \cup N_{CF} \cup N_{CS} \cup T, T, P, S\big)$$

where $\{S\}$, N_{CF}, N_{CS}, and T are pairwise disjoint alphabets, and every rule in P is either of the form $S \rightarrow aD$ or $AB \rightarrow AC$ or $A \rightarrow x$, where $a \in T$, $D \in N_{CF} \cup \{\varepsilon\}$, $B \in N_{CS}$, $A, C \in N_{CF}$, $x \in N_{CS} \cup T \cup (\bigcup_{i=1}^{2} N_{CF}^{i})$.

Proof. Let L be a context-sensitive language over an alphabet, T. Without any loss of generality, we can express L as $L = L_1 \cup L_2$, where $L_1 \subseteq T$ and $L_2 \subseteq TT^+$. Thus, by analogy with the proofs of Theorems 1 and 2 in [16], L_2 can be represented as $L_2 = \bigcup_{a \in T} aL_a$, where each L_a is a context-sensitive language. Let L_a be generated by a context-sensitive grammar

$$G_a = \big(N_{CF_a} \cup N_{CS_a} \cup T, T, P_a, S_a\big)$$

of the form of Theorem 4.1.6. Clearly, we assume that for all as, the nonterminal alphabets N_{CF_a} and N_{CS_a} are pairwise disjoint. Let S be a new start symbol. Consider the context-sensitive grammar

$$G = \big(\{S\} \cup N_{CF} \cup N_{CS} \cup T, T, P, S\big)$$

where

$$\begin{aligned} N_{CF} &= \bigcup_{a \in T} N_{CF_a} \\ N_{CS} &= \bigcup_{a \in T} N_{CS_a} \\ P &= \bigcup_{a \in T} P_a \cup \{S \to aS_a \mid a \in T\} \cup \{S \to a \mid a \in L_1\} \end{aligned}$$

Obviously, G satisfies the required form, and we have

$$L(G) = L_1 \cup \left(\bigcup_{a \in T} aL(G_a)\right) = L_1 \cup \left(\bigcup_{a \in T} aL_a\right) = L_1 \cup L_2 = L$$

Consequently, the lemma holds. □

We are now ready to characterize **CS** by propagating ssc-grammars of degree $(1, 2)$.

Theorem 4.6.9. $\mathbf{CS} = \mathbf{SSC}^{-\varepsilon}(1, 2)$

Proof. By Lemma 4.2.3, $\mathbf{SSC}^{-\varepsilon}(1, 2) \subseteq \mathbf{CG}^{-\varepsilon} \subseteq \mathbf{CS}$; thus, it is sufficient to prove the converse inclusion.

Let L be a context-sensitive language. Without any loss of generality, we assume that L is generated by a context-sensitive grammar

$$G = \left(\{S\} \cup N_{CF} \cup N_{CS} \cup T, T, P, S\right)$$

of the form of Lemma 4.6.8. Set

$$V = \{S\} \cup N_{CF} \cup N_{CS} \cup T$$

Let q be the cardinality of V; $q \geq 1$. Furthermore, let f be an arbitrary bijection from V onto $\{1, \dots, q\}$, and let f^{-1} be the inverse of f. Let

$$\tilde{G} = (\tilde{V}, T, \tilde{P}, S)$$

be a propagating ssc-grammar of degree $(1, 2)$, in which

$$\tilde{V} = \left(\bigcup_{i=1}^{4} W_i\right) \cup V$$

where

$$\begin{aligned} W_1 &= \{\langle a, AB \to AC, j\rangle \mid a \in T,\ AB \to AC \in P,\ 1 \leq j \leq 5\}, \\ W_2 &= \{[a, AB \to AC, j] \mid a \in T,\ AB \to AC \in P,\ 1 \leq j \leq q + 3\} \\ W_3 &= \{\hat{B}, B', B'' \mid B \in N_{CS}\} \\ W_4 &= \{\bar{a} \mid a \in T\} \end{aligned}$$

$\tilde{P}$ is defined as follows:

(1) If $S \to aA \in P, a \in T, A \in (N_{CF} \cup \{\varepsilon\})$, then add $(S \to \bar{a}A, 0, 0)$ to $\tilde{P}$;
(2) If $a \in T, A \to x \in P, A \in N_{CF}, x \in (V - \{S\}) \cup (N_{CF})^2$, then add $(A \to x, \bar{a}, 0)$ to $\tilde{P}$;
(3) If $a \in T, AB \to AC \in P, A, C \in N_{CF}, B \in N_{CS}$, then add the following rules to P' (an informal explanation of these rules can be found below):

(3.1) $(\bar{a} \to \langle a, AB \to AC, 1\rangle, 0, 0)$
(3.2) $(B \to B', \langle a, AB \to AC, 1\rangle, 0)$
(3.3) $(B \to \hat{B}, \langle a, AB \to AC, 1\rangle, 0)$
(3.4) $(\langle a, AB \to AC, 1\rangle \to \langle a, AB \to AC, 2\rangle, 0, B)$
(3.5) $(\hat{B} \to B'', 0, B'')$
(3.6) $(\langle a, AB \to AC, 2\rangle \to \langle a, AB \to AC, 3\rangle, 0, \hat{B})$
(3.7) $(B'' \to [a, AB \to AC, 1], \langle a, AB \to AC, 3\rangle, 0)$
(3.8) $([a, AB \to AC, j] \to [a, AB \to AC, j+1], 0, f^{-1}(j)[a, AB \to AC, j])$, for all $j = 1, \dots, q$, $f(A) \neq j$
(3.9) $([a, AB \to AC, f(A)] \to [a, AB \to AC, f(A)+1], 0, 0)$
(3.10) $([a, AB \to AC, q+1] \to [a, AB \to AC, q+2], 0, B'[a, AB \to AC, q+1])$
(3.11) $([a, AB \to AC, q+2] \to [a, AB \to AC, q+3], 0, \langle a, AB \to AC, 3\rangle[a, AB \to AC, q+2])$
(3.12) $(\langle a, AB \to AC, 3\rangle \to \langle a, AB \to AC, 4\rangle, [a, AB \to AC, q+3], 0)$
(3.13) $(B' \to B, \langle a, AB \to AC, 4\rangle, 0)$
(3.14) $(\langle a, AB \to AC, 4\rangle \to \langle a, AB \to AC, 5\rangle, 0, B')$
(3.15) $([a, AB \to AC, q+3] \to C, \langle a, AB \to AC, 5\rangle, 0)$
(3.16) $(\langle a, AB \to AC, 5\rangle \to \bar{a}, 0, [a, AB \to AC, q+3])$

(4) If $a \in T$, then add $(\bar{a} \to a, 0, 0)$ to $\tilde{P}$.

Let us informally explain the basic idea behind (3)—the heart of the construction. The rules introduced in (3) simulate the application of rules of the form $AB \to AC$ in G as follows: an occurrence of B is chosen, and its left neighbor is checked not to belong to $\tilde{V} - \{A\}$. At this point, the left neighbor necessarily equals A, so B is rewritten with C.

Formally, we define a finite substitution g from V^* into $\tilde{V}^*$ as follows:

(a) If $D \in V$, then add D to $g(D)$;
(b) If $\langle a, AB \to AC, j\rangle \in W_1, a \in T, AB \to AC \in P, B \in N_{CS}, A, C \in N_{CF}, j \in \{1, \dots, 5\}$, then add $\langle a, AB \to AC, j\rangle$ to $g(a)$;
(c) If $[a, AB \to AC, j] \in W_2, a \in T, AB \to AC \in P, B \in N_{CS}, A, C \in N_{CF}, j \in \{1, \dots, q+3\}$, then add $[a, AB \to AC, j]$ to $g(B)$;
(d) If $\{\hat{B}, B', B''\} \subseteq W_3, B \in N_{CS}$, then include $\{\hat{B}, B', B''\}$ into $g(B)$;
(e) If $\bar{a} \in W_4, a \in T$, then add $\bar{a}$ to $g(a)$.

Let g^{-1} be the inverse of g. To show that $L(G) = L(\tilde{G})$, we first prove three claims.

Claim 1. $S \Rightarrow_G^+ x$, $x \in V^*$, *implies that* $x \in T(V - \{S\})^*$.

Proof. Observe that the start symbol S does not appear on the right side of any rule and that $S \rightarrow x \in P$ implies that $x \in T \cup T(V - \{S\})$. Hence, the claim holds. □

Claim 2. If $S \Rightarrow_{\tilde{G}}^+ x$, $x \in \tilde{V}^*$, *then* x *has one of the following seven forms*

(i) $x = ay$, *where* $a \in T$, $y \in (V - \{S\})^*$;
(ii) $x = \bar{a}y$, *where* $\bar{a} \in W_4$, $y \in (V - \{S\})^*$;
(iii) $x = \langle a, AB \rightarrow AC, 1\rangle y$, *where* $\langle a, AB \rightarrow AC, 1\rangle \in W_1$, $y \in ((V - \{S\}) \cup \{B', \hat{B}, B''\})^*$, $\text{occur}(y, B'') \leq 1$;
(iv) $x = \langle a, AB \rightarrow AC, 2\rangle y$, *where* $\langle a, AB \rightarrow AC, 2\rangle \in W_1$, $y \in ((V - \{S, B\}) \cup \{B', \hat{B}, B''\})^*$, $\text{occur}(y, B') \leq 1$;
(v) $x = \langle a, AB \rightarrow AC, 3\rangle y$, *where* $\langle a, AB \rightarrow AC, 3\rangle \in W_1$, $y \in ((V - \{S, B\}) \cup \{B'\})^*(\{[a, AB \rightarrow AC, j] \mid 1 \leq j \leq q+3\} \cup \{\varepsilon, B''\})((V - \{S, B\}) \cup \{B'\})^*$;
(vi) $x = \langle a, AB \rightarrow AC, 4\rangle y$, *where* $\langle a, AB \rightarrow AC, 4\rangle \in W_1$, $y \in ((V - \{S\}) \cup \{B'\})^*[a, AB \rightarrow AC, q + 3]((V - \{S\}) \cup \{B'\})^*$;
(vii) $x = \langle a, AB \rightarrow AC, 5\rangle y$, *where* $\langle a, AB \rightarrow AC, 5\rangle \in W_1$, $y \in (V - \{S\})^*\{[a, AB \rightarrow AC, q + 3], \varepsilon\}(V - \{S\})^*$.

Proof. The claim is proved by induction on the length of derivations.

Basis. Consider $S \Rightarrow_{\tilde{G}} x$, $x \in \tilde{V}^*$. By the inspection of the rules, we have

$$S \Rightarrow_{\tilde{G}} \bar{a}A \; [(S \rightarrow \bar{a}A, 0, 0)]$$

for some $\bar{a} \in W_4$, $A \in (\{\varepsilon\} \cup N_{CF})$. Therefore, $x = \bar{a}$ or $x = \bar{a}A$; in either case, x is a string of the required form.

Induction Hypothesis. Assume that the claim holds for all derivations of length n or less, for some $n \geq 1$.

Induction Step. Consider any derivation of the form

$$S \Rightarrow_{\tilde{G}}^{n+1} x$$

where $x \in \tilde{V}^*$. Since $n \geq 1$, we have $n + 1 \geq 2$. Thus, there is some z of the required form, $z \in \tilde{V}^*$, such that

$$S \Rightarrow_{\tilde{G}}^{n} z \Rightarrow_{\tilde{G}} x \; [p]$$

for some $p \in \tilde{P}$.

Let us first prove by contradiction that the first symbol of z does not belong to T. Assume that the first symbol of z belongs to T. As z is of the required form, we have $z = ay$ for some $a \in (V - \{S\})^*$. By the inspection of $\tilde{P}$, there is no $p \in \tilde{P}$ such that $ay \Rightarrow_{\tilde{G}} x \; [p]$, where $x \in \tilde{V}^*$. We have thus obtained a contradiction, so the first symbol of z is not in T.

Because the first symbol of z does not belong to T, z cannot have form (i); as a result, z has one of forms (ii) through (vii). The following cases (I) through (VI) demonstrate that if z has one of these six forms, then x has one of the required forms, too.

(I) Assume that z is of form (ii); that is, $z = \bar{a}y$, $\bar{a} \in W_4$, and $y \in (V - \{S\})^*$. By the inspection of the rules in $\tilde{P}$, we see that p has one of the following forms (a), (b), and (c)

(a) $p = (A \to u, \bar{a}, 0)$, where $A \in N_{CF}$ and $u \in (V - \{S\}) \cup N_{CF}^2$;
(b) $p = (\bar{a} \to \langle a, AB \to AC, 1\rangle, 0, 0)$, where $\langle a, AB \to AC, 1\rangle \in W_1$;
(c) $p = (\bar{a} \to a, 0, 0)$, where $a \in T$.

Note that rules of forms (a), (b), and (c) are introduced in construction steps (2), (3), and (4), respectively. If p has form (a), then x has form (ii). If p has form (b), then x has form (iii). Finally, if p has form (c), then x has form (i). In any of these three cases, we obtain x that has one of the required forms.

(II) Assume that z has form (iii); that is, $z = \langle a, AB \to AC, 1\rangle y$ for some $\langle a, AB \to AC, 1\rangle \in W_1$, $y \in ((V - \{S\}) \cup \{B', \hat{B}, B''\})^*$, and $\mathrm{occur}(y, B'') \leq 1$. By the inspection of $\tilde{P}$, we see that z can be rewritten by rules of these four forms

(a) $(B \to B', \langle a, AB \to AC, 1\rangle, 0)$.
(b) $(B \to \hat{B}, \langle a, AB \to AC, 1\rangle, 0)$.
(c) $(\hat{B} \to B'', 0, B'')$ if $B'' \notin \mathrm{alph}(y)$; that is, $\mathrm{occur}(y, B'') = 0$.
(d) $(\langle a, AB \to AC, 1\rangle \to \langle a, AB \to AC, 2\rangle, 0, B)$ if $B \notin \mathrm{alph}(y)$; that is, $\mathrm{occur}(y, B) = 0$.

Clearly, in cases (a) and (b), we obtain x of form (iii). If $z \Rightarrow_{\tilde{G}} x\ [p]$, where p is of form (c), then $\mathrm{occur}(x, B'') = 1$, so we get x of form (iii). Finally, if we use the rule of form (d), then we obtain x of form (iv) because $\mathrm{occur}(z, B) = 0$.

(III) Assume that z is of form (iv); that is, $z = \langle a, AB \to AC, 2\rangle y$, where $\langle a, AB \to AC, 2\rangle \in W_1$, $y \in ((V - \{S, B\}) \cup \{B', \hat{B}, B''\})^*$, and $\mathrm{occur}(y, B'') \leq 1$. By the inspection of $\tilde{P}$, we see that the following two rules can be used to rewrite z

(a) $(\hat{B} \to B'', 0, B'')$ if $B'' \notin \mathrm{alph}(y)$.
(b) $(\langle a, AB \to AC, 2\rangle \to \langle a, AB \to AC, 3\rangle, 0, \hat{B})$ if $\hat{B} \notin \mathrm{alph}(y)$.

In case (a), we get x of form (iv). In case (b), we have $\mathrm{occur}(y, \hat{B}) = 0$, so $\mathrm{occur}(x, \hat{B}) = 0$. Moreover, notice that $\mathrm{occur}(x, B'') \leq 1$ in this case. Indeed, the symbol B'' can be generated only if there is no occurrence of B'' in a given rewritten string, so no more than 1 occurrence of B'' appears in any sentential form. As a result, we have $\mathrm{occur}(\langle a, AB \to AC, 3\rangle y, B'') \leq 1$; that is, $\mathrm{occur}(x, B'') \leq 1$. In other words, we get x of form (v).

(IV) Assume that z is of form (v); that is, $z = \langle a, AB \to AC, 3\rangle y$ for some $\langle a, AB \to AC, 3\rangle \in W_1$, $y \in ((V - \{S, B\}) \cup \{B'\})^*(\{[a, AB \to AC, j] \mid 1 \leq j \leq q + 3\} \cup \{B'', \varepsilon\})((V - \{S, B\}) \cup \{B'\})^*$. Assume that $y = y_1 Y y_2$

with $y_1, y_2 \in ((V - \{S, B\}) \cup \{B'\})^*$. If $Y = \varepsilon$, then we can use no rule from $\tilde{P}$ to rewrite z. Because $z \Rightarrow_{\tilde{G}} x$, we have $Y \neq \varepsilon$. The following cases (a) through (f) cover all possible forms of Y.

(a) Assume $Y = B''$. By the inspection of $\tilde{P}$, we see that the only rule that can rewrite z has the form

$$(B'' \rightarrow [a, AB \rightarrow AC, 1], \langle a, AB \rightarrow AC, 3\rangle, 0)$$

In this case, we get x of form (v).

(b) Assume $Y = [a, AB \rightarrow AC, j]w$, $j \in \{1, \dots, q\}$, and $f(A) \neq j$. Then, z can be rewritten only according to the rule

$$([a, AB \rightarrow AC, j] \rightarrow [a, AB \rightarrow AC, j+1]0, f^{-1}(j)[a, AB \rightarrow AC, j])$$

which can be used if the rightmost symbol of $\langle a, AB \rightarrow AC, 3\rangle y_1$ differs from $f^{-1}(j)$. Clearly, in this case, we again get x of form (v).

(c) Assume $Y = [a, AB \rightarrow AC, j]$, $j \in \{1, \dots, q\}$, $f(A) = j$. This case forms an analogy to case (b) except that the rule of the form

$$([a, AB \rightarrow AC, f(A)] \rightarrow [a, AB \rightarrow AC, f(A) + 1], 0, 0)$$

is now used.

(d) Assume $Y = [a, AB \rightarrow AC, q + 1]$. This case forms an analogy to case (b); the only change is the application of the rule

$$([a, AB \rightarrow AC, q+1] \rightarrow [a, AB \rightarrow AC, q+2], 0, B'[a, AB \rightarrow AC, q+1]).$$

(e) Assume $Y = [a, AB \rightarrow AC, q + 2]$. This case forms an analogy to case (b) except that the rule

$$([a, AB \rightarrow AC, q + 2] \rightarrow [a, AB \rightarrow AC, q + 3], 0,$$
$$\langle a, AB \rightarrow AC, 3\rangle[a, AB \rightarrow AC, q + 2])$$

is used.

(f) Assume $Y = [a, AB \rightarrow AC, q + 3]$. By the inspection of $\tilde{P}$, we see that the only rule that can rewrite z is

$$(\langle a, AB \rightarrow AC, 3\rangle \rightarrow \langle a, AB \rightarrow AC, 4\rangle, [a, AB \rightarrow AC, q + 3], 0)$$

If this rule is used, we get x of form (vi).

(V) Assume that z is of form (vi); that is, $z = \langle a, AB \rightarrow AC, 4\rangle y$, where $\langle a, AB \rightarrow AC, 4\rangle \in W_1$ and $y \in ((V - \{S\}) \cup \{B'\})^*[a, AB \rightarrow AC, q + 3]((V - \{S\}) \cup \{B'\})^*$. By the inspection of $\tilde{P}$, these two rules can rewrite z

(a) $(B' \rightarrow B, \langle a, AB \rightarrow AC, 4\rangle, 0)$;
(b) $(\langle a, AB \rightarrow AC, 4\rangle \rightarrow \langle a, AB \rightarrow AC, 5\rangle, 0, B')$ if $B' \notin \mathrm{alph}(y)$.

Clearly, in case (a), we get x of form (vi). In case (b), we get x of form (vii) because $\mathrm{occur}(y, B') = 0$, so $y \in (V - \{S\})^*\{[a, AB \rightarrow AC, q+3], \varepsilon\}(V - \{S\})^*$.

(VI) Assume that z is of form (vii); that is, $z = \langle a, AB \rightarrow AC, 5\rangle y$, where $\langle a, AB \rightarrow AC, 5\rangle \in W_1$ and $y \in (V - \{S\})^*\{[a, AB \rightarrow AC, q+3], \varepsilon\}(V - \{S\})^*$. By the inspection of $\tilde{P}$, one of the following two rules can be used to rewrite z

(a) $([a, AB \rightarrow AC, q+3] \rightarrow C, \langle a, AB \rightarrow AC, 5\rangle, 0)$.
(b) $(\langle a, AB \rightarrow AC, 5\rangle \rightarrow \bar{a}, 0, [a, AB \rightarrow AC, q+3])$ if $[a, AB \rightarrow AC, q+3] \notin \mathrm{alph}(z)$.

In case (a), we get x of form (vii). Case (b) implies that $\mathrm{occur}(y, [a, AB \rightarrow AC, q+3]) = 0$; thus, x is of form (ii).

This completes the induction step and establishes Claim 2. □

Claim 3. It holds that

$$S \Rightarrow_G^m w \quad \textit{if and only if} \quad S \Rightarrow_{\tilde{G}}^n v$$

where $v \in g(w)$ *and* $w \in V^+$, *for some* $m, n \geq 0$.

Proof. The proof is divided into the only-if part and the if part.

Only If. The only-if part is established by induction on m; that is, we have to demonstrate that

$$S \Rightarrow_G^m w \quad \text{implies} \quad S \Rightarrow_{\tilde{G}}^* v$$

for some $v \in g(w)$ and $w \in V^+$.

Basis. Let $m = 0$. The only w is S because $S \Rightarrow_G^0 S$. Clearly, $S \Rightarrow_{\tilde{G}}^0 S$, and $S \in g(S)$.

Induction Hypothesis. Suppose that the claim holds form all derivations of length m or less, for some $m \geq 0$.

Induction Step. Let us consider any derivation of the form

$$S \Rightarrow_G^{m+1} x$$

where $x \in V^+$. Because $m+1 \geq 1$, there are $y \in V^+$ and $p \in P$ such that

$$S \Rightarrow_G^m y \Rightarrow_G x\ [p]$$

and by the induction hypothesis, there is also a derivation

$$S \Rightarrow^n_{\tilde{G}} \tilde{y}$$

for some $\tilde{y} \in g(y)$. The following cases (i) through (iii) cover all possible forms of p.

(i) Let $p = S \to aA \in P$ for some $a \in T$, $A \in N_{CF} \cup \{\varepsilon\}$. Then, by Claim 1, $m = 0$, so $y = S$ and $x = aA$. By (1) in the construction of $\tilde{G}$, $(S \to \bar{a}A, 0, 0) \in \tilde{P}$. Hence,

$$S \Rightarrow_{\tilde{G}} \bar{a}A$$

where $\bar{a}A \in g(aA)$.

(ii) Let us assume that $p = D \to y_2 \in P$, $D \in N_{CF}$, $y_2 \in (V - \{S\}) \cup N^2_{CF}$, $y = y_1 D y_3$, $y_1, y_3 \in V^*$, and $x = y_1 y_2 y_3$. From the definition of g, it is clear that $g(Z) = \{Z\}$ for all $Z \in N_{CF}$; therefore, we can express $\tilde{y} = z_1 D z_3$, where $z_1 \in g(y_1)$ and $z_3 \in g(y_3)$. Without any loss of generality, we can also assume that $y_1 = au$, $a \in T$, $u \in (V - \{S\})^*$ (see Claim 1), so $z_1 = a''u''$, $a'' \in g(a)$, and $u'' \in g(u)$. Moreover, by (2) in the construction, we have $(D \to y_2, \bar{a}, 0) \in \tilde{P}$. The following cases (a) through (e) cover all possible forms of a''.

(ii.a) Let $a'' = \bar{a}$ (see (ii) in Claim 2). Then, we have

$$S \Rightarrow^n_{\tilde{G}} \bar{a}u''Dz_3 \Rightarrow_{\tilde{G}} \bar{a}u''y_2z_3 \; [(D \to y_2, \bar{a}, 0)]$$

and $\bar{a}u''y_2z_3 = z_1y_2z_3 \in g(y_1y_2y_3) = g(x)$.

(ii.b) Let $a'' = a$ (see (i) in Claim 2). By (4) in the construction of $\tilde{G}$, we can express the derivation

$$S \Rightarrow^n_{\tilde{G}} au''Dz_3$$

as

$$S \Rightarrow^{n-1}_{\tilde{G}} \bar{a}u''Dz_3 \Rightarrow_{\tilde{G}} au''Dz_3 \; [(\bar{a} \to a, 0, 0)]$$

Thus, there exists the derivation

$$S \Rightarrow^{n-1}_{\tilde{G}} \bar{a}u''Dz_3 \Rightarrow_{\tilde{G}} \bar{a}u''y_2z_3 \; [(D \to y_2, \bar{a}, 0)]$$

with $\bar{a}u''y_2z_3 \in g(x)$.

(ii.c) Let $a'' = \langle a, AB \to AC, 5\rangle$ for some $AB \to AC \in P$ (see (vii) in Claim 2), and let $u''Dz_3 \in (V - \{S\})^*$; that is, $[a, AB \to AC, q+3] \notin \text{alph}(u''Dz_3)$. Then, there exists the derivation

$$
\begin{aligned}
S &\Rightarrow^{n}_{\tilde{G}} \langle a, AB \rightarrow AC, 5\rangle u'' D z_3 \\
&\Rightarrow_{\tilde{G}} \bar{a} u'' D z_3 \; [(\langle a, AB \rightarrow AC, 5\rangle \rightarrow \bar{a}, 0, [a, AB \rightarrow AC, q+3])] \\
&\Rightarrow_{\tilde{G}} \bar{a} u'' y_2 z_3 \; [(D \rightarrow y_2, \bar{a}, 0)]
\end{aligned}
$$

and $\bar{a}u''y_2z_3 \in g(x)$.

(ii.d) Let $a'' = \langle a, AB \rightarrow AC, 5\rangle$ (see (vii) in Claim 2). Let $[a, AB \rightarrow AC, q+3] \in \mathrm{alph}(u''Dz_3)$. Without any loss of generality, we can assume that $\tilde{y} = \langle a, AB \rightarrow AC, 5\rangle u''Do''[a, AB \rightarrow AC, q+3]t''$, where $o''[a, AB \rightarrow AC, q+3]t'' = z_3$, $oBt = y_3$, $o'' \in g(t)$, $o, t \in (V - \{S\})^*$. By the inspection of $\tilde{P}$ (see (3) in the construction of $\tilde{G}$), we can express the derivation

$$S \Rightarrow^{n}_{\tilde{G}} \tilde{y}$$

as

$$
\begin{aligned}
S \Rightarrow^{*}_{\tilde{G}} \quad & \bar{a} u'' D o'' B t'' \\
\Rightarrow_{\tilde{G}} \quad & \langle a, AB \rightarrow AC, 1\rangle u'' D o'' B t'' \\
& \quad [(\bar{a} \rightarrow \langle a, AB \rightarrow AC, 1\rangle, 0, 0)] \\
\Rightarrow^{1+|m_1m_2|}_{\tilde{G}} \quad & \langle a, AB \rightarrow AC, 1\rangle u' D o' \hat{B} t' \\
& \quad [m_1(B \rightarrow \hat{B}, \langle a, AB \rightarrow AC, 1\rangle, 0) m_2] \\
\Rightarrow_{\tilde{G}} \quad & \langle a, AB \rightarrow AC, 2\rangle u' D o' \hat{B} t' \\
& \quad [(\langle a, AB \rightarrow AC, 1\rangle \rightarrow \langle a, AB \rightarrow AC, 2\rangle, 0, B)] \\
\Rightarrow_{\tilde{G}} \quad & \langle a, AB \rightarrow AC, 2\rangle u' D o' B'' t' \\
& \quad [\hat{B} \rightarrow B'', 0, B''] \\
\Rightarrow_{\tilde{G}} \quad & \langle a, AB \rightarrow AC, 3\rangle u' D o' B'' t' \\
& \quad [(\langle a, AB \rightarrow AC, 2\rangle \rightarrow \langle a, AB \rightarrow AC, 3\rangle, 0, \hat{B})] \\
\Rightarrow_{\tilde{G}} \quad & \langle a, AB \rightarrow AC, 3\rangle u' D o' [a, AB \rightarrow AC, 1] t' \\
& \quad [(B'' \rightarrow [a, AB \rightarrow AC, 1], \langle a, AB \rightarrow AC, 3\rangle, 0)] \\
\Rightarrow^{q+2}_{\tilde{G}} \quad & \langle a, AB \rightarrow AC, 3\rangle u' D o' [a, AB \rightarrow AC, q+3] t' \\
& \quad [\omega] \\
\Rightarrow_{\tilde{G}} \quad & \langle a, AB \rightarrow AC, 4\rangle u' D o' [a, AB \rightarrow AC, q+3] t' \\
& \quad [(\langle a, AB \rightarrow AC, 3\rangle \rightarrow \langle a, AB \rightarrow AC, 4\rangle,
\end{aligned}
$$

$$[a, AB \to AC, q+3], 0)]$$

$$\Rightarrow_{\tilde{G}}^{|m_3|} \quad \langle a, AB \to AC, 4\rangle u''Do''[a, AB \to AC, q+3]t''$$

$$[m_3]$$

$$\Rightarrow_{\tilde{G}} \quad \langle a, AB \to AC, 5\rangle u''Do''[a, AB \to AC, q+3]t''$$

$$[(\langle a, AB \to AC, 4\rangle \to \langle a, AB \to AC, 5\rangle, 0, B')]$$

where $m_1, m_2 \in \{(B \to B', \langle a, AB \to AC, 1\rangle, 0)\}^*$, $m_3 \in \{(B' \to B, \langle a, AB \to AC, 4\rangle, 0)\}^*$, $|m_3| = |m_1 m_2|$,

$$\begin{aligned}
\omega = {} & ([a, AB \to AC, 1] \to [a, AB \to AC, 2], 0, \\
& \quad f^{-1}(1)[a, AB \to AC, 1]) \cdots \\
& ([a, AB \to AC, f(A)-1] \to [a, AB \to AC, f(A)], 0, \\
& \quad f^{-1}(f(A)-1)[a, AB \to AC, f(A)-1]) \\
& ([a, AB \to AC, f(A)] \to [a, AB \to AC, f(A)+1], 0, 0) \\
& ([a, AB \to AC, f(A)+1] \to [a, AB \to AC, f(A)+2], 0, \\
& \quad f^{-1}(f(A)+1)[a, AB \to AC, f(A)+1]) \cdots \\
& ([a, AB \to AC, q] \to [a, AB \to AC, q+1], 0, \\
& \quad f^{-1}(q)[a, AB \to AC, q]) \\
& ([a, AB \to AC, q+1] \to [a, AB \to AC, q+2], 0, \\
& \quad B'[a, AB \to AC, q+1]) \\
& ([a, AB \to AC, q+2] \to [a, AB \to AC, q+3]), 0, \\
& \quad \langle a, AB \to AC, 3\rangle[a, AB \to AC, q+2])
\end{aligned}$$

$u' \in ((\text{alph}(u'') - \{B\}) \cup \{B'\})^*$, $g^{-1}(u') = u$, $o' \in ((\text{alph}(o'') - \{B\}) \cup \{B''\})^*$, $g^{-1}(o') = g^{-1}(o'') = o$, $t' \in ((\text{alph}(t'') - \{B\}) \cup \{B'\})^*$, $g^{-1}(t') = g^{-1}(t'') = t$.

Clearly, $\bar{a}u''Do''Bt'' \in g(auDoBt) = g(auDy_3) = g(y)$. Thus, there exists the derivation

$$S \Rightarrow_{\tilde{G}}^* \bar{a}u''Do''Bt'' \Rightarrow_{\tilde{G}} \bar{a}u''y_2o''Bt'' \ [(D \to y_2, \bar{a}, 0)]$$

where $z_1 y_2 z_3 = \bar{a}u''y_2o''Bt'' \in g(auy_2oBt) = g(y_1y_2y_3) = g(x)$.

(ii.e) Let $a'' = \langle a, AB \to AC, i\rangle$ for some $AB \to AC \in P$ and $i \in \{1, \dots, 4\}$ (see (iii)–(vi) in Claim 2). By analogy with (ii.d), we can construct the derivation

$$S \Rightarrow_{\tilde{G}}^* \bar{a}u''Do''Bt'' \Rightarrow_{\tilde{G}} \bar{a}u''y_2o''Bt'' \ [(D \to y_2, \bar{a}, 0)]$$

such that $\bar{a}u''y_2o''Bt'' \in g(y_1y_2y_3) = g(x)$. The details are left to the reader.

(iii) Let $p = AB \to AC \in P$, $A, C \in N_{CF}$, $B \in N_{CS}$, $y = y_1 ABy_3$, $y_1, y_3 \in V^*$, $x = y_1 ACy_3$, $\tilde{y} = z_1 AYz_3$, $Y \in g(B)$, $z_i \in g(y_i)$ where $i \in \{1, 3\}$. Moreover, let $y_1 = au$ (see Claim 1), $z_1 = a''u''$, $a'' \in g(a)$, and $u'' \in g(u)$. The following cases (a) through (e) cover all possible forms of a''.

(iii.a) Let $a'' = \bar{a}$. Then, by Claim 2, $Y = B$. By (3) in the construction of $\tilde{G}$, there exists the following derivation

$$
\begin{aligned}
S \Rightarrow_{\tilde{G}}^{n} \;& \bar{a}u''ABz_3 \\
\Rightarrow_{\tilde{G}} \;& \langle a, AB \to AC, 1\rangle u''ABz_3 \\
& [(\bar{a} \to \langle a, AB \to AC, 1\rangle, 0, 0)] \\
\Rightarrow_{\tilde{G}}^{1+|m_1|} \;& \langle a, AB \to AC, 1\rangle u'A\hat{B}u_3 \\
& [m_1(B \to \hat{B}, \langle a, AB \to AC, 1\rangle, 0)] \\
\Rightarrow_{\tilde{G}} \;& \langle a, AB \to AC, 2\rangle u'A\hat{B}u_3 \\
& [(\langle a, AB \to AC, 1\rangle \to \langle a, AB \to AC, 2\rangle, 0, B)] \\
\Rightarrow_{\tilde{G}} \;& \langle a, AB \to AC, 2\rangle u'AB''u_3 \\
& [(\hat{B} \to B'', 0, B'')] \\
\Rightarrow_{\tilde{G}} \;& \langle a, AB \to AC, 3\rangle u'AB''u_3 \\
& [(\langle a, AB \to AC, 2\rangle \to \langle a, AB \to AC, 3\rangle, 0, \hat{B})] \\
\Rightarrow_{\tilde{G}} \;& \langle a, AB \to AC, 3\rangle u'A[a, AB \to AC, 1]u_3 \\
& [(B'' \to [a, AB \to AC, 1], \langle a, AB \to AC, 3\rangle, 0)] \\
\Rightarrow_{\tilde{G}}^{q+2} \;& \langle a, AB \to AC, 3\rangle u'A[a, AB \to AC, q+3]u_3 \\
& [\omega] \\
\Rightarrow_{\tilde{G}} \;& \langle a, AB \to AC, 4\rangle u'A[a, AB \to AC, q+3]u_3 \\
& [(\langle a, AB \to AC, 3\rangle \to \langle a, AB \to AC, 4\rangle, \\
& \quad [a, AB \to AC, q+3], 0)] \\
\Rightarrow_{\tilde{G}}^{|m_2|} \;& \langle a, AB \to AC, 4\rangle u''A[a, AB \to AC, q+3]z_3 \\
& [m_2] \\
\Rightarrow_{\tilde{G}} \;& \langle a, AB \to AC, 5\rangle u''A[a, AB \to AC, q+3]z_3 \\
& [(\langle a, AB \to AC, 4\rangle \to \langle a, AB \to AC, 5\rangle, 0, B')] \\
\Rightarrow_{\tilde{G}} \;& \langle a, AB \to AC, 5\rangle u''ACz_3 \\
& [([a, AB \to AC, q+3] \to C, \langle a, AB \to AC, 5\rangle, 0)]
\end{aligned}
$$

where $m_1 \in \{(B \to B', \langle a, AB \to AC, 1\rangle, 0)\}^*$, $m_2 \in \{(B' \to B, \langle a, AB \to AC, 4\rangle, 0)\}^*$, $|m_1| = |m_2|$,

$$\begin{aligned}\omega = {} & ([a, AB \to AC, 1] \to [a, AB \to AC, 2], 0, \\ & \quad f^{-1}(1)[a, AB \to AC, 1]) \cdots \\ & ([a, AB \to AC, f(A) - 1] \to [a, AB \to AC, f(A)], 0, \\ & \quad f^{-1}(f(A) - 1)[a, AB \to AC, f(A) - 1]) \\ & ([a, AB \to AC, f(A)] \to [a, AB \to AC, f(A) + 1], 0, 0) \\ & ([a, AB \to AC, f(A) + 1] \to [a, AB \to AC, f(A) + 2], 0, \\ & \quad f^{-1}(f(A) + 1)[a, AB \to AC, f(A) + 1]) \cdots \\ & ([a, AB \to AC, q] \to [a, AB \to AC, q + 1], 0, \\ & \quad f^{-1}(q)[a, AB \to AC, q]) \\ & ([a, AB \to AC, q + 1] \to [a, AB \to AC, q + 2], 0, \\ & \quad B'[a, AB \to AC, q + 1]) \\ & ([a, AB \to AC, q + 2] \to [a, AB \to AC, q + 3]), 0, \\ & \quad \langle a, AB \to AC, 3\rangle[a, AB \to AC, q + 2])\end{aligned}$$

$u_3 \in ((\mathrm{alph}(z_3) - \{B\}) \cup \{B'\})^*$, $g^{-1}(u_3) = g^{-1}(z_3) = y_3$, $u' \in ((\mathrm{alph}(u'') - \{B\}) \cup \{B'\})^*$, $g^{-1}(u') = g^{-1}(u'') = u$. It is clear that $\langle a, AB \to AC, 5\rangle \in g(a)$; thus, $\langle a, AB \to AC, 5\rangle u'' AC z_3 \in g(auACy_3) = g(x)$.

(iii.b) Let $a'' = a$. Then, by Claim 2, $Y = B$. By analogy with (ii.b) and (iii.a) in the proof of this claim (see above), we obtain

$$S \Rightarrow_{\tilde{G}}^{n-1} \bar{a}u''ABz_3 \Rightarrow_{\tilde{G}}^{*} \langle a, AB \to AC, 5\rangle u'' AC z_3$$

so $\langle a, AB \to AC, 5\rangle u'' AC z_3 \in g(x)$.

(iii.c) Let $a'' = \langle a, AB \to AC, 5\rangle$ for some $AB \to AC \in P$ (see (vii) in Claim 2), and let $u''AYz_3 \in (V - \{S\})^*$. At this point, $Y = B$. By analogy with (ii.c) and (iii.a) in the proof of this claim (see above), we can construct

$$S \Rightarrow_{\tilde{G}}^{n+1} \bar{a}u''ABz_3 \Rightarrow_{\tilde{G}}^{*} \langle a, AB \to AC, 5\rangle u'' AC z_3$$

so $\langle a, AB \to AC, 5\rangle u'' AC z_3 \in g(x)$.

(iii.d) Let $a'' = \langle a, AB \to AC, 5\rangle$ for some $AB \to AC \in P$ (see (vii) in Claim 2), and let $[a, AB \to AC, q + 3] \in \mathrm{alph}(u''AYz_3)$. By analogy with (ii.d) and (iii.a) in the proof of this claim (see above), we can construct

$$S \Rightarrow_{\tilde{G}}^{*} \bar{a}u''ABz_3$$

and then

$$S \Rightarrow_{\tilde{G}}^{*} \bar{a}u''ABz_3 \Rightarrow_{\tilde{G}}^{*} \langle a, AB \to AC, 5\rangle u'' AC z_3$$

so that $\langle a, AB \to AC, 5\rangle u'' AC z_3 \in g(auACy_3) = g(x)$.

(iii.e) Let $a'' = \langle a, AB \to AC, i \rangle$ for some $AB \to AC \in P$, $i \in \{1, \dots, 4\}$, see (III)–(IV) in Claim 2. By analogy with (ii.e) and (iii.d) in the proof of this claim, we can construct

$$S \Rightarrow_{\tilde{G}}^{*} \bar{a}u''ACz_3$$

where $\bar{a}u''ACz_3 \in g(x)$.

If. By induction on n, we next prove that if $S \Rightarrow_{\tilde{G}}^{n} v$ with $v \in g(w)$ and $w \in V^*$, for some $n \geq 0$, then $S \Rightarrow_{G}^{*} w$.

Basis. For $n = 0$, the only v is S as $S \Rightarrow_{\tilde{G}}^{0} S$. Because $\{S\} = g(S)$, we have $w = S$. Clearly, $S \Rightarrow_{G}^{0} S$.

Induction Hypothesis. Assume that the claim holds for all derivations of length n or less, for some $n \geq 0$. Let us show that it also holds true for $n + 1$.

Induction Step. For $n + 1 = 1$, there only exists a direct derivation of the form

$$S \Rightarrow_{\tilde{G}} \bar{a}A \; [(S \to \bar{a}A, 0, 0)]$$

where $A \in N_{CF} \cup \{\varepsilon\}$, $a \in T$, and $\bar{a}A \in g(aA)$. By (1), we have in P a rule of the form $S \to aA$ and, thus, a direct derivation $S \Rightarrow_G aA$.

Suppose that $n + 1 \geq 2$ (i.e. $n \geq 1$). Consider any derivation of the form

$$S \Rightarrow_{G}^{n+1} x'$$

where $x' \in g(x)$, $x \in V^*$. Because $n + 1 \geq 2$, there exist $\bar{a} \in W_4$, $A \in N_{CF}$, and $y \in V^+$ such that

$$S \Rightarrow_{\tilde{G}} \bar{a}A \Rightarrow_{\tilde{G}}^{n-1} y' \Rightarrow_{\tilde{G}} x' \; [p]$$

where $p \in \tilde{P}$, $y' \in g(y)$, and by the induction hypothesis,

$$S \Rightarrow_{G}^{*} y$$

Let us assume that $y' = z_1 Z z_2$, $y = y_1 D y_2$, $z_j \in g(y_j)$, $y_j \in (V - \{S\})^*$, $j = 1, 2$, $Z \in g(D)$, $D \in V - \{S\}$, $p = (Z \to u', \alpha, \beta) \in P'$, $\alpha = 0$ or $\beta = 0$, $x' = z_1 u' z_2$, $u' \in g(u)$ for some $u \in V^*$; that is, $x' \in g(y_1 u y_2)$. The following cases (i) through (iii) cover all possible forms of

$$y' \Rightarrow_{\tilde{G}} x' \; [p]$$

(i) Let $Z \in N_{CF}$. By the inspection of $\tilde{P}$, we see that $Z = D$, $p = (D \to u', \bar{a}, 0) \in \tilde{P}$, $D \to u \in P$ and $u = u'$. Thus,

$$S \Rightarrow_{G}^{*} y_1 B y_2 \Rightarrow_G y_1 u y_2 \; [B \to u].$$

(ii) Let $u = D$. Then, by the induction hypothesis, we have the derivation

$$S \Rightarrow_G^* y_1 D y_2$$

and $y_1 D y_2 = y_1 u y_2$ in G.

(iii) Let $p = ([a, AB \rightarrow AC, q+3] \rightarrow C, \langle a, AB \rightarrow AC, 5\rangle, 0)$, $Z = [a, AB \rightarrow AC, q+3]$. Thus, $u' = C$ and $D = B \in N_{CS}$. By case (VI) in Claim 2 and the form of p, we have $z_1 = \langle a, AB \rightarrow AC, 5\rangle t$ and $y_1 = ao$, where $t \in g(o)$, $\langle a, AB \rightarrow AC, 5\rangle \in g(a)$, $o \in (V - \{S\})^*$, and $a \in T$. From (3) in the construction of $\tilde{G}$, it follows that there exists a rule of the form $AB \rightarrow AC \in P$. Moreover, (3) and Claim 2 imply that the derivation

$$S \Rightarrow_{\tilde{G}} \bar{a}A \Rightarrow_{\tilde{G}}^{n-1} y' \Rightarrow_{\tilde{G}} x' \; [p]$$

can be expressed in the form

$$
\begin{aligned}
S \Rightarrow_{\tilde{G}} \; & \bar{a}A \\
\Rightarrow_{\tilde{G}}^* \; & \bar{a}tBz_2 \\
\Rightarrow_{\tilde{G}} \; & \langle a, AB \rightarrow AC, 1\rangle vtBz_2 \\
& \quad [(\bar{a} \rightarrow \langle a, AB \rightarrow AC, 1\rangle, 0, 0)] \\
\Rightarrow_{\tilde{G}}^{|\omega'|} \; & \langle a, AB \rightarrow AC, 1\rangle v\hat{B}w_2 \\
& \quad [\omega'] \\
\Rightarrow_{\tilde{G}} \; & \langle a, AB \rightarrow AC, 1\rangle vB''w_2 \\
& \quad [(\hat{B} \rightarrow B'', 0, B'')] \\
\Rightarrow_{\tilde{G}} \; & \langle a, AB \rightarrow AC, 2\rangle vB''w_2 \\
& \quad [(\langle a, AB \rightarrow AC, 1\rangle \rightarrow \langle a, AB \rightarrow AC, 2\rangle, 0, B)] \\
\Rightarrow_{\tilde{G}} \; & \langle a, AB \rightarrow AC, 3\rangle vB''w_2 \\
& \quad [(\langle a, AB \rightarrow AC, 2\rangle \rightarrow \langle a, AB \rightarrow AC, 3\rangle, 0, \hat{B})] \\
\Rightarrow_{\tilde{G}} \; & \langle a, AB \rightarrow AC, 3\rangle v[a, AB \rightarrow AC, 1]w_2 \\
& \quad [(B'' \rightarrow [a, AB \rightarrow AC, 1], \langle a, AB \rightarrow AC, 3\rangle, 0)] \\
\Rightarrow_{\tilde{G}}^{|\omega|} \; & \langle a, AB \rightarrow AC, 3\rangle v[a, AB \rightarrow AC, q+3]w_2 \\
& \quad [\omega] \\
\Rightarrow_{\tilde{G}} \; & \langle a, AB \rightarrow AC, 4\rangle v[a, AB \rightarrow AC, q+3]w_2 \\
& \quad [(\langle a, AB \rightarrow AC, 3\rangle \rightarrow \langle a, AB \rightarrow AC, 4\rangle, \\
& \quad\quad [a, AB \rightarrow AC, q+3], 0)] \\
\Rightarrow_{\tilde{G}}^{|\omega'|-1} \; & \langle a, AB \rightarrow AC, 4\rangle t[a, AB \rightarrow AC, q+3]z_2 \\
& \quad [\omega''] \\
\Rightarrow_{\tilde{G}} \; & \langle a, AB \rightarrow AC, 5\rangle t[a, AB \rightarrow AC, q+3]z_2 \\
& \quad [(\langle a, AB \rightarrow AC, 4\rangle \rightarrow \langle a, AB \rightarrow AC, 5\rangle, 0, B')]
\end{aligned}
$$

$$
\begin{aligned}
\Rightarrow_{\tilde{G}} \; & \langle a, AB \rightarrow AC, 5\rangle tCz_2 \\
& \quad [([a, AB \rightarrow AC, q+3] \rightarrow C, \langle a, AB \rightarrow AC, 5\rangle, 0)]
\end{aligned}
$$

where

$$\begin{aligned}\omega' \in \ & \{(B \to B', \langle a, AB \to AC, 1\rangle, 0)\}^* \\ & \{(B \to \hat{B}, \langle a, AB \to AC, 1\rangle, 0)\} \\ & \{(B \to B', \langle a, AB \to AC, 1\rangle, 0)\}^*\end{aligned}$$

$g(B) \cap \text{alph}(vw_2) \subseteq \{B'\}$, $g^{-1}(v) = g^{-1}(t)$, $g^{-1}(w_2) = g^{-1}(z_2)$,

$$\begin{aligned}\omega = \ & \omega_1 \\ & ([a, AB \to AC, f(A)] \to [a, AB \to AC, f(A)+1], 0, 0)\omega_2 \\ & ([a, AB \to AC, q+1] \to [a, AB \to AC, q+2], 0 \\ & \quad B'[a, AB \to AC, q+1]) \\ & ([a, AB \to AC, q+2] \to [a, AB \to AC, q+3], 0, \\ & \quad \langle a, AB \to AC, 3\rangle[a, AB \to AC, q+2]) \\ \omega_1 = \ & ([a, AB \to AC, 1] \to [a, AB \to AC, 2], 0, \\ & \quad f^{-1}(1)[a, AB \to AC, 1]) \cdots \\ & ([a, AB \to AC, f(A)-1] \to [a, AB \to AC, f(A)], 0, \\ & \quad f^{-1}(f(A)-1)[a, AB \to AC, f(A)-1])\end{aligned}$$

where $f(A)$ implies that $q_1 = \varepsilon$, $\omega_2 = ([a, AB \to AC, f(A)+1] \to [a, AB \to AC, f(A)+2], 0, f^{-1}(f(A)+1)[a, AB \to AC, f(A)+1]) \cdots ([a, AB \to AC, q] \to [a, AB \to AC, q+1], 0, f^{-1}(q)[a, AB \to AC, q])$, where $f(A) = q$ implies that $q_2 = \varepsilon$, $\omega'' \in \{(B' \to B, \langle a, AB \to AC, 4\rangle, 0)\}^*$.

The derivation above implies that the rightmost symbol of t must be A. As $t \in g(o)$, the rightmost symbol of o must be A as well. That is, $t = s'A$, $o = sA$ and $s' \in g(s)$, for some $s \in (V - \{S\})^*$. By the induction hypothesis, there exists a derivation

$$S \Rightarrow_G^* asABy_2$$

Because $AB \to AC \in P$, we get

$$S \Rightarrow_G^* asABy_2 \Rightarrow_G asACy_2 \ [AB \to AC]$$

where $asACy_2 = y_1uy_2$.

By (i), (ii), and (iii) and the inspection of $\tilde{P}$, we see that we have considered all possible derivations of the form

$$S \Rightarrow_{\tilde{G}}^{n+1} x'$$

so we have established Claim 3 by the principle of induction. □

The equivalence of G and $\tilde{G}$ can be easily derived from Claim 3. By the definition of g, we have $g(a) = \{a\}$ for all $a \in T$. Thus, by Claim 3, we have for all $x \in T^*$,

$$S \Rightarrow_G^* x \quad \text{if and only if} \quad S \Rightarrow_{\tilde{G}}^* x$$

Consequently, $L(G) = L(\tilde{G})$, and the theorem holds. □

Corollary 4.6.10. $\mathbf{SSC}^{-\varepsilon}(1,2) = \mathbf{SSC}^{-\varepsilon} = \mathbf{SC}^{-\varepsilon}(1,2) = \mathbf{SC}^{-\varepsilon} = \mathbf{CS}$ □

We now turn to the investigation of ssc-grammars of degree $(1, 2)$ with erasing rules.

Theorem 4.6.11. $\mathbf{SSC}(1,2) = \mathbf{RE}$

Proof. Clearly, we have $\mathbf{SSC}(1,2) \subseteq \mathbf{RE}$. Thus, we only need to show that $\mathbf{RE} \subseteq \mathbf{SSC}(1,2)$. Every language $L \in \mathbf{RE}$ can be generated by a phrase-structure grammar $G = (V, T, P, S)$ in which each rule is of the form $AB \to AC$ or $A \to x$, where $A, B, C \in V - T$, $x \in \{\varepsilon\} \cup T \cup (V - T)^2$ (see Theorem 4.1.7). Thus, the inclusion can be established by analogy with the proof of Theorem 4.6.9. The details are left to the reader. □

Corollary 4.6.12. $\mathbf{SSC}(1,2) = \mathbf{SSC} = \mathbf{SC}(1,2) = \mathbf{SC} = \mathbf{RE}$ □

The following corollary summarizes the relations of language families generated by ssc-grammars.

Corollary 4.6.13.

$$\begin{array}{c}
\mathbf{CF} \\
\subset \\
\mathbf{SSC}^{-\varepsilon} = \mathbf{SSC}^{-\varepsilon}(2,1) = \mathbf{SSC}^{-\varepsilon}(1,2) = \\
= \mathbf{SC}^{-\varepsilon} = \mathbf{SC}^{-\varepsilon}(2,1) = \mathbf{SC}^{-\varepsilon}(1,2) = \mathbf{CS} \\
\subset \\
\mathbf{SSC} = \mathbf{SSC}(2,1) = \mathbf{SSC}(1,2) = \mathbf{SC} = \mathbf{SC}(2,1) = \mathbf{SC}(1,2) = \mathbf{RE}
\end{array}$$

Proof. This corollary follows from Corollaries 4.6.5, 4.6.7, 4.6.10, and 4.6.12. □

Next, we turn our attention to reduced versions of ssc-grammars. More specifically, we demonstrate that there exist several normal forms of ssc-grammars with a limited number of conditional rules and nonterminals.

Theorem 4.6.14. *Every recursively enumerable language can be defined by an ssc-grammar of degree* $(2, 1)$ *with no more than 12 conditional rules and 13 nonterminals.*

Proof. Let L be a recursively enumerable language. By Theorem 4.1.11, we assume that L is generated by a grammar G of the form

$$G = \bigl(V, T, P \cup \{AB \to \varepsilon, CD \to \varepsilon\}, S\bigr)$$

such that P contains only context-free rules and

$$V - T = \{S, A, B, C, D\}$$

Construct an ssc-grammar G' of degree (2, 1),

$$G' = (V', T, P', S)$$

where

$$V' = V \cup W$$
$$W = \{\tilde{A}, \tilde{B}, \langle\varepsilon_A\rangle, \$, \tilde{C}, \tilde{D}, \langle\varepsilon_C\rangle, \#\}, \ V \cap W = \emptyset$$

The set of rules P' is defined in the following way

(1) If $H \rightarrow y \in P$, $H \in V - T$, $y \in V^*$, then add $(H \rightarrow y, 0, 0)$ to P'
(2) Add the following six rules to P'

$$(A \rightarrow \tilde{A}, 0, \tilde{A})$$
$$(B \rightarrow \tilde{B}, 0, \tilde{B})$$
$$(\tilde{A} \rightarrow \langle\varepsilon_A\rangle, \tilde{A}\tilde{B}, 0)$$
$$(\tilde{B} \rightarrow \$, \langle\varepsilon_A\rangle\tilde{B}, 0)$$
$$(\langle\varepsilon_A\rangle \rightarrow \varepsilon, 0, \tilde{B})$$
$$(\$ \rightarrow \varepsilon, 0, \langle\varepsilon_A\rangle)$$

(3) Add the following six rules to P'

$$(C \rightarrow \tilde{C}, 0, \tilde{C})$$
$$(D \rightarrow \tilde{D}, 0, \tilde{D})$$
$$(\tilde{C} \rightarrow \langle\varepsilon_C\rangle, \tilde{C}\tilde{D}, 0)$$
$$(\tilde{D} \rightarrow \#, \langle\varepsilon_C\rangle\tilde{D}, 0)$$
$$(\langle\varepsilon_C\rangle \rightarrow \varepsilon, 0, \tilde{D})$$
$$(\# \rightarrow \varepsilon, 0, \langle\varepsilon_C\rangle)$$

Notice that G' has degree (2,1) and contains only 12 conditional rules and 13 nonterminals. The rules of (2) simulate the application of $AB \rightarrow \varepsilon$ in G' and the rules of (3) simulate the application of $CD \rightarrow \varepsilon$ in G'.

Let us describe the simulation of $AB \rightarrow \varepsilon$. First, one occurrence of A and one occurrence of B are rewritten to $\tilde{A}$ and $\tilde{B}$, respectively (no more than 1 $\tilde{A}$ and one $\tilde{B}$ appear in any sentential form). The right neighbor of $\tilde{A}$ is checked to be $\tilde{B}$ and $\tilde{A}$ is rewritten to $\langle\varepsilon_A\rangle$. Then, analogously, the left neighbor of $\tilde{B}$ is checked to be $\langle\varepsilon_A\rangle$ and $\tilde{B}$ is rewritten to \$. Finally, $\langle\varepsilon_A\rangle$ and \$ are erased. The simulation of $CD \rightarrow \varepsilon$ is analogous.

To establish $L(G) = L(G')$, we first prove two claims.

Claim 1. $S \Rightarrow^*_{G'} x'$ *implies that* $\text{occur}(x', \tilde{X}) \leq 1$ *for all* $\tilde{X} \in \{\tilde{A}, \tilde{B}, \tilde{C}, \tilde{D}\}$ *and* $x' \in V'^*$.

Proof. By the inspection of rules in P', the only rule that can generate $\tilde{X}$ is of the form $(X \rightarrow \tilde{X}, 0, \tilde{X})$. This rule can be applied only when no $\tilde{X}$ occurs in the rewritten sentential form. Thus, it is not possible to derive x' from S such that $\text{occur}(x', \tilde{X}) \geq 2$. □

Informally, the next claim says that every occurrence of $\langle \varepsilon_A \rangle$ in derivations from S is always followed by either $\tilde{B}$ or \$, and every occurrence of $\langle \varepsilon_C \rangle$ is always followed by either $\tilde{D}$ or #.

Claim 2. It holds that

(I) $S \Rightarrow^*_{G'} y_1' \langle \varepsilon_A \rangle y_2'$ *implies* $y_2' \in V'^+$ *and* $\text{first}(y_2') \in \{\tilde{B}, \$\}$ *for any* $y_1' \in V'^*$;
(II) $S \Rightarrow^*_{G'} y_1' \langle \varepsilon_C \rangle y_2'$ *implies* $y_2' \in V'^+$ *and* $\text{first}(y_2') \in \{\tilde{D}, \#\}$ *for any* $y_1' \in V'^*$.

Proof. We base this proof on the examination of all possible forms of derivations that may occur during a derivation of a sentential form containing $\langle \varepsilon_A \rangle$ or $\langle \varepsilon_C \rangle$.

(I) By the definition of P', the only rule that can generate $\langle \varepsilon_A \rangle$ is $p = (\tilde{A} \rightarrow \langle \varepsilon_A \rangle, \tilde{A}\tilde{B}, 0)$. This rule has the permitting condition $\tilde{A}\tilde{B}$, so it can be used provided that $\tilde{A}\tilde{B}$ occurs in a sentential form. Furthermore, by Claim 1, no other occurrence of $\tilde{A}$ or $\tilde{B}$ can appear in the given sentential form. Consequently, we obtain a derivation

$$S \Rightarrow^*_{G'} u_1' \tilde{A}\tilde{B} u_2' \Rightarrow_{G'} u_1' \langle \varepsilon_A \rangle \tilde{B} u_2' \; [p]$$

for some $u_1', u_2' \in V'^*$, $\tilde{A}, \tilde{B} \notin \text{sub}(u_1' u_2')$, which represents the only way of getting $\langle \varepsilon_A \rangle$. Obviously, $\langle \varepsilon_A \rangle$ is always followed by $\tilde{B}$ in $u_1' \langle \varepsilon_A \rangle \tilde{B} u_2'$.

Next, we discuss how G' can rewrite the substring $\langle \varepsilon_A \rangle \tilde{B}$ in $u_1' \langle \varepsilon_A \rangle \tilde{B} u_2'$. There are only two rules having the nonterminals $\langle \varepsilon_A \rangle$ or $\tilde{B}$ on their left-hand side, $p_1 = (\tilde{B} \rightarrow \$, \langle \varepsilon_A \rangle \tilde{B}, 0)$ and $p_2 = (\langle \varepsilon_A \rangle \rightarrow \varepsilon, 0, \tilde{B})$. G' cannot use p_2 to erase $\langle \varepsilon_A \rangle$ in $u_1' \langle \varepsilon_A \rangle \tilde{B} u_2'$ because p_2 forbids an occurrence of $\tilde{B}$ in the rewritten string. Rule p_1 has also a context condition, but $\langle \varepsilon_A \rangle \tilde{B} \in \text{sub}(u_1' \langle \varepsilon_A \rangle \tilde{B} u_2')$, and thus p_1 can be used to rewrite $\tilde{B}$ with \$. Hence, we obtain a derivation of the form

$$\begin{aligned} S \Rightarrow^*_{G'} u_1' \tilde{A}\tilde{B} u_2' \quad & \Rightarrow_{G'} u_1' \langle \varepsilon_A \rangle \tilde{B} u_2' \; [p] \\ \Rightarrow^*_{G'} v_1' \langle \varepsilon_A \rangle \tilde{B} v_2' & \Rightarrow_{G'} v_1' \langle \varepsilon_A \rangle \$ v_2' \; [p_1] \end{aligned}$$

Notice that during this derivation, G' may rewrite u_1' and u_2' to some v_1' and v_2', respectively, where $v_1', v_2' \in V'^*$; however, $\langle \varepsilon_A \rangle \tilde{B}$ remains unchanged after this rewriting.

In this derivation, we obtained the second symbol \$ that can appear as the right neighbor of $\langle \varepsilon_A \rangle$. It is sufficient to show that there is no other symbol that can appear immediately after $\langle \varepsilon_A \rangle$. By the inspection of P', only

($\$ \rightarrow \varepsilon, 0, \langle\varepsilon_A\rangle$) can rewrite \$. However, this rule cannot be applied when $\langle\varepsilon_A\rangle$ occurs in the given sentential form. In other words, the occurrence of \$ in the substring $\langle\varepsilon_A\rangle\$$ cannot be rewritten before $\langle\varepsilon_A\rangle$ is erased by rule p_2. Hence, $\langle\varepsilon_A\rangle$ is always followed by either $\tilde{B}$ or \$, and thus, the first part of Claim 2 holds.

(II) By the inspection of rules simulating $AB \rightarrow \varepsilon$ and $CD \rightarrow \varepsilon$ in G' (see (2) and (3) in the definition of P'), these two sets of rules work analogously. Thus, part (II) of Claim 2 can be proved by analogy with part (I). □

Let us return to the main part of the proof. Let g be a finite substitution from V'^* to V^* defined as follows:

(1) For all $X \in V$, $g(X) = \{X\}$.
(2) $g(\tilde{A}) = \{A\}$, $g(\tilde{B}) = \{B\}$, $g(\langle\varepsilon_A\rangle) = \{A\}$, $g(\$) = \{B, AB\}$.
(3) $g(\tilde{C}) = \{C\}$, $g(\tilde{D}) = \{D\}$, $g(\langle\varepsilon_C\rangle) = \{C\}$, $g(\#) = \{C, CD\}$.

Having this substitution, we can prove the following claim.

Claim 3. *$S \Rightarrow_G^* x$ if and only if $S \Rightarrow_{G'}^* x'$ for some $x \in g(x')$, $x \in V^*$, $x' \in V'^*$.*

Proof. The claim is proved by induction on the length of derivations.

Only If. We show that

$$S \Rightarrow_G^m x \quad \text{implies} \quad S \Rightarrow_{G'}^* x$$

where $m \geq 0$, $x \in V^*$; clearly $x \in g(x)$. This is established by induction on $m \geq 0$.

Basis. Let $m = 0$. That is, $S \Rightarrow_G^0 S$. Clearly, $S \Rightarrow_{G'}^0 S$.

Induction Hypothesis. Suppose that the claim holds for all derivations of length m or less, for some $m \geq 0$.

Induction Step. Consider any derivation of the form

$$S \Rightarrow_G^{m+1} x,\ x \in V^*$$

Since $m + 1 \geq 1$, there is some $y \in V^+$ and $p \in P \cup \{AB \rightarrow \varepsilon, CD \rightarrow \varepsilon\}$ such that

$$S \Rightarrow_G^m y \Rightarrow_G x\ [p]$$

By the induction hypothesis, there is a derivation

$$S \Rightarrow_{G'}^* y$$

The following three cases cover all possible forms of p.

(i) Let $p = H \to y_2 \in P$, $H \in V - T$, $y_2 \in V^*$. Then, $y = y_1Hy_3$ and $x = y_1y_2y_3$, $y_1, y_3 \in V^*$. Because we have $(H \to y_2, 0, 0) \in P'$,

$$S \Rightarrow_{G'}^{*} y_1Hy_3 \Rightarrow_{G'} y_1y_2y_3\ [(H \to y_2, 0, 0)]$$

and $y_1y_2y_3 = x$.

(ii) Let $p = AB \to \varepsilon$. Then, $y = y_1ABy_3$ and $x = y_1y_3$, $y_1, y_3 \in V^*$. In this case, there is the derivation

$$\begin{aligned} S &\Rightarrow_{G'}^{*} y_1ABy_3 \\ &\Rightarrow_{G'} y_1\tilde{A}By_3 && [(A \to \tilde{A}, 0, \tilde{A})] \\ &\Rightarrow_{G'} y_1\tilde{A}\tilde{B}y_3 && [(B \to \tilde{B}, 0, \tilde{B})] \\ &\Rightarrow_{G'} y_1\langle\varepsilon_A\rangle\tilde{B}y_3 && [(\tilde{A} \to \langle\varepsilon_A\rangle, \tilde{A}\tilde{B}, 0)] \\ &\Rightarrow_{G'} y_1\langle\varepsilon_A\rangle\$y_3 && [(\tilde{B} \to \$, \langle\varepsilon_A\rangle\tilde{B}, 0)] \\ &\Rightarrow_{G'} y_1\$y_3 && [(\langle\varepsilon_A\rangle \to \varepsilon, 0, \tilde{B})] \\ &\Rightarrow_{G'} y_1y_3 && [(\$ \to \varepsilon, 0, \langle\varepsilon_A\rangle)] \end{aligned}$$

(iii) Let $p = CD \to \varepsilon$. Then, $y = y_1CDy_3$ and $x = y_1y_3$, $y_1, y_3 \in V^*$. By analogy with (ii), there exists the derivation

$$\begin{aligned} S &\Rightarrow_{G'}^{*} y_1CDy_3 \\ &\Rightarrow_{G'} y_1\tilde{C}Dy_3 && [(C \to \tilde{C}, 0, \tilde{C})] \\ &\Rightarrow_{G'} y_1\tilde{C}\tilde{D}y_3 && [(D \to \tilde{D}, 0, \tilde{D})] \\ &\Rightarrow_{G'} y_1\langle\varepsilon_C\rangle\tilde{D}y_3 && [(\tilde{C} \to \langle\varepsilon_C\rangle, \tilde{C}\tilde{D}, 0)] \\ &\Rightarrow_{G'} y_1\langle\varepsilon_C\rangle\#y_3 && [(\tilde{D} \to \#, \langle\varepsilon_C\rangle\tilde{D}, 0)] \\ &\Rightarrow_{G'} y_1\#y_3 && [(\langle\varepsilon_C\rangle \to \varepsilon, 0, \tilde{D})] \\ &\Rightarrow_{G'} y_1y_3 && [(\# \to \varepsilon, 0, \langle\varepsilon_C\rangle)] \end{aligned}$$

If. By induction on the length n of derivations in G', we prove that

$$S \Rightarrow_{G'}^{n} x' \quad \text{implies} \quad S \Rightarrow_{G}^{*} x$$

for some $x \in g(x')$, $x \in V^*$, $x' \in V'^*$.

Basis. Let $n = 0$. That is, $S \Rightarrow_{G'}^{0} S$. It is obvious that $S \Rightarrow_{G}^{0} S$ and $S \in g(S)$.

Induction Hypothesis. Assume that the claim holds for all derivations of length n or less, for some $n \geq 0$.

Induction Step. Consider any derivation of the form

$$S \Rightarrow_{G'}^{n+1} x',\ x' \in V'^*$$

Since $n + 1 \geq 1$, there is some $y' \in V'^+$ and $p' \in P'$ such that

$$S \Rightarrow_{G'}^{n} y' \Rightarrow_{G'} x'\ [p']$$

and by the induction hypothesis, there is also a derivation

$$S \Rightarrow_G^* y$$

such that $y \in g(y')$.

By the inspection of P', the following cases (i) through (xiii) cover all possible forms of p'.

(i) Let $p' = (H \to y_2, 0, 0) \in P'$, $H \in V - T$, $y_2 \in V^*$. Then, $y' = y_1' H y_3'$, $x' = y_1' y_2 y_3'$, $y_1', y_3' \in V'^*$ and y has the form $y = y_1 Z y_3$, where $y_1 \in g(y_1')$, $y_3 \in g(y_3')$ and $Z \in g(H)$. Because $g(X) = \{X\}$ for all $X \in V - T$, the only Z is H, and thus $y = y_1 H y_3$. By the definition of P' (see (1)), there exists a rule $p = H \to y_2$ in P, and we can construct the derivation

$$S \Rightarrow_G^* y_1 H y_3 \Rightarrow_G y_1 y_2 y_3 \ [p]$$

such that $y_1 y_2 y_3 = x$, $x \in g(x')$.

(ii) Let $p' = (A \to \tilde{A}, 0, \tilde{A})$. Then, $y' = y_1' A y_3'$, $x' = y_1' \tilde{A} y_3'$, $y_1', y_3' \in V'^*$, and $y = y_1 Z y_3$, where $y_1 \in g(y_1')$, $y_3 \in g(y_3')$ and $Z \in g(A)$. Because $g(A) = \{A\}$, the only Z is A, so we can express $y = y_1 A y_3$. Having the derivation $S \Rightarrow_G^* y$ such that $y \in g(y')$, it is easy to see that also $y \in g(x')$ because $A \in g(\tilde{A})$.

(iii) Let $p' = (B \to \tilde{B}, 0, \tilde{B})$. By analogy with (ii), $y' = y_1' B y_3'$, $x' = y_1' \tilde{B} y_3'$, $y = y_1 B y_3$, where $y_1', y_3' \in V'^*$, $y_1 \in g(y_1')$, $y_3 \in g(y_3')$, and thus $y \in g(x')$ because $B \in g(\tilde{B})$.

(iv) Let $p' = (\tilde{A} \to \langle \varepsilon_A \rangle, \tilde{A}\tilde{B}, 0)$. By the permitting condition of this rule, $\tilde{A}\tilde{B}$ surely occurs in y'. By Claim 1, no more than 1 $\tilde{A}$ can occur in y'. Therefore, y' must be of the form $y' = y_1' \tilde{A}\tilde{B} y_3'$, where $y_1', y_3' \in V'^*$ and $\tilde{A} \notin \mathrm{sub}(y_1' y_3')$. Then, $x' = y_1' \langle \varepsilon_A \rangle \tilde{B} y_3'$ and y is of the form $y = y_1 Z y_3$, where $y_1 \in g(y_1')$, $y_3 \in g(y_3')$ and $Z \in g(\tilde{A}\tilde{B})$. Because $g(\tilde{A}\tilde{B}) = \{AB\}$, the only Z is AB; thus, we obtain $y = y_1 A B y_3$. By the induction hypothesis, we have a derivation $S \Rightarrow_G^* y$ such that $y \in g(y')$. According to the definition of g, $y \in g(x')$ as well because $A \in g(\langle \varepsilon_A \rangle)$ and $B \in g(\tilde{B})$.

(v) Let $p' = (\tilde{B} \to \$, \langle \varepsilon_A \rangle \tilde{B}, 0)$. This rule can be applied provided that $\langle \varepsilon_A \rangle \tilde{B} \in \mathrm{sub}(y')$. Moreover, by Claim 1, $\mathrm{occur}(y', \tilde{B}) \leq 1$. Hence, we can express $y' = y_1' \langle \varepsilon_A \rangle \tilde{B} y_3'$, where $y_1', y_3' \in V'^*$ and $\tilde{B} \notin \mathrm{sub}(y_1' y_3')$. Then, $x' = y_1' \langle \varepsilon_A \rangle \$ y_3'$ and $y = y_1 Z y_3$, where $y_1 \in g(y_1')$, $y_3 \in g(y_3')$ and $Z \in g(\langle \varepsilon_A \rangle \tilde{B})$. By the definition of g, $g(\langle \varepsilon_A \rangle \tilde{B}) = \{AB\}$, so $Z = AB$ and $y = y_1 A B y_3$. By the induction hypothesis, we have a derivation $S \Rightarrow_G^* y$ such that $y \in g(y')$. Because $A \in g(\langle \varepsilon_A \rangle)$ and $B \in g(\$)$, $y \in g(x')$ as well.

(vi) Let $p' = (\langle \varepsilon_A \rangle \to \varepsilon, 0, \tilde{B})$. Application of $(\langle \varepsilon_A \rangle \to \varepsilon, 0, \tilde{B})$ implies that $\langle \varepsilon_A \rangle$ occurs in y'. Claim 2 says that $\langle \varepsilon_A \rangle$ has either $\tilde{B}$ or \$ as its right neighbor. Since the forbidding condition of p' forbids an occurrence of $\tilde{B}$ in y', the right neighbor of $\langle \varepsilon_A \rangle$ must be \$. As a result, we obtain $y' = y_1' \langle \varepsilon_A \rangle \$ y_3'$ where $y_1', y_3' \in V'^*$. Then, $x' = y_1' \$ y_3'$ and y is of the form $y = y_1 Z y_3$, where

$y_1 \in g(y'_1)$, $y_3 \in g(y'_3)$ and $Z \in g((\langle\varepsilon_A\rangle\$)$. By the definition of g, $g((\langle\varepsilon_A\rangle\$) = \{AB, AAB\}$. If $Z = AB$, $y = y_1 AB y_3$. Having the derivation $S \Rightarrow_G^* y$, it holds that $y \in g(x')$ because $AB \in g(\$)$.

(vii) Let $p' = (\$ \rightarrow \varepsilon, 0, \langle\varepsilon_A\rangle)$. Then, $y' = y'_1 \$ y'_3$ and $x' = y'_1 y'_3$, where $y'_1, y'_3 \in V'^*$. Express $y = y_1 Z y_3$ so that $y_1 \in g(y'_1)$, $y_3 \in g(y'_3)$ and $Z \in g(\$)$, where $g(\$) = \{B, AB\}$. Let $Z = AB$. Then, $y = y_1 AB y_3$, and there exists the derivation

$$S \Rightarrow_G^* y_1 AB y_3 \Rightarrow_G y_1 y_3 \; [AB \rightarrow \varepsilon]$$

where $y_1 y_3 = x$, $x \in g(x')$.

In cases (ii) through (vii), we discussed all six rules simulating the application of $AB \rightarrow \varepsilon$ in G' (see (2) in the definition of P'). Cases (viii) through (xiii) should cover rules simulating the application of $CD \rightarrow \varepsilon$ in G' [see (3)]. However, by the inspection of these two sets of rules, it is easy to see that they work analogously. Therefore, we leave this part of the proof to the reader.

We have completed the proof and established Claim 3 by the principle of induction. □

Observe that $L(G) = L(G')$ follows from Claim 3. Indeed, according to the definition of g, we have $g(a) = \{a\}$ for all $a \in T$. Thus, from Claim 3, we have for any $x \in T^*$

$$S \Rightarrow_G^* x \quad \text{if and only if} \quad S \Rightarrow_{G'}^* x$$

Consequently, $L(G) = L(G')$, and the theorem holds. □

Let us note that in [8], Theorem 4.6.14 has been improved by demonstrating that even nine conditional rules and ten nonterminals are enough to generate every recursively enumerable language.

Theorem 4.6.15 (See [8]). *Every recursively enumerable language can be generated by an ssc-grammar of degree* $(2, 1)$ *having no more than 9 conditional rules and 10 nonterminals.*

Continuing with the investigation of reduced ssc-grammars, we point out that Vaszil in [22] proved that if we allow permitting conditions of length three—that is, ssc-grammars of degree $(3, 1)$, then the number of conditional rules and nonterminals can be further decreased.

Theorem 4.6.16. *Every recursively enumerable language can be generated by an ssc-grammar of degree* $(3, 1)$ *with no more than 8 conditional rules and 11 nonterminals.*

Proof (See [22]). Let L be a recursively enumerable language. Without any loss of generality, we assume that L is generated by a phrase-structure grammar

$$G = \big(V, T, P \cup \{ABC \to \varepsilon\}, S\big)$$

where

$$V - T = \{S, S', A, B, C\}$$

and P contains only context-free rules of the forms $S \to zSx$, $z \in \{A, B\}^*$, $x \in T$, $S \to S'$, $S' \to uS'v$, $u \in \{A, B\}^*$, $v \in \{B, C\}^*$, $S' \to \varepsilon$ (see Theorem 4.1.9). Every successful derivation in G consists of the following two phases.

(1) $S \Rightarrow_G^* z_n \cdots z_1 S x_1 \cdots x_n \Rightarrow_G z_n \cdots z_1 S' x_1 \cdots x_n$; $z_i \in \{A, B\}^*$, $1 \le i \le n$.
(2) $z_n \cdots z_1 S' x_1 \cdots x_n \Rightarrow_G^* z_n \cdots z_1 u_m \cdots u_1 S' v_1 \cdots v_m x_1 \cdots x_n \Rightarrow_G z_n \cdots z_1 u_m \cdots u_1 v_1 \cdots v_m x_1 \cdots x_n$, where $u_j \in \{A, B\}^*$, $v_j \in \{B, C\}^*$, $1 \le j \le m$, and the terminal string $x_1 \cdots x_n$ is generated by G if and only if by using the erasing rule $ABC \to \varepsilon$, the substring $z_n \cdots z_1 u_m \cdots u_1 v_1 \cdots v_m$ can be deleted.

Next, we introduce the ssc-grammar

$$G' = \big(V', T, P', S\big)$$

of degree (3, 1), where

$$V' = \big\{S, S', A, A', A'', B, B', B'', C, C', C''\big\} \cup T$$

and P' is constructed as follows:

(1) For every $H \to y \in P$, add $(H \to y, 0, 0)$ to P';
(2) For every $X \in \{A, B, C\}$, add $(X \to X', 0, X')$ to P';
(3) Add the following six rules to P'

$$\begin{aligned}
&(C' \to C'', A'B'C', 0)\\
&(A' \to A'', A'B'C'', 0)\\
&(B' \to B'', A''B'C'', 0)\\
&(A'' \to \varepsilon, 0, C'')\\
&(C'' \to \varepsilon, 0, B')\\
&(B'' \to \varepsilon, 0, 0).
\end{aligned}$$

Observe that G' satisfies all the requirements of this theorem—that is, it contains only 8 conditional rules and 11 nonterminals. G' reproduces the first two phases of generating a terminal string in G by using the rules of the form $(H \to y, 0, 0) \in P'$. The third phase, during which $ABC \to \varepsilon$ is applied, is simulated by the additional rules. Examine these rules to see that all strings generated by G can also be generated by G'. Indeed, for every derivation step

$$y_1ABCy_2 \Rightarrow_G y_1y_2\ [ABC \to \varepsilon]$$

in G, $y_1, y_2 \in V^*$, there exists the following derivation in G'

$$\begin{aligned}
y_1ABCy_2 &\Rightarrow_{G'} y_1A'BCy_2 && [(A \to A', 0, A')]\\
&\Rightarrow_{G'} y_1A'B'Cy_2 && [(B \to B', 0, B')]\\
&\Rightarrow_{G'} y_1A'B'C'y_2 && [(C \to C', 0, C')]\\
&\Rightarrow_{G'} y_1A'B'C''y_2 && [(C' \to C'', A'B'C', 0)]\\
&\Rightarrow_{G'} y_1A''B'C''y_2 && [(A' \to A'', A'B'C'', 0)]\\
&\Rightarrow_{G'} y_1A''B''C''y_2 && [(B' \to B'', A''B'C'', 0)]\\
&\Rightarrow_{G'} y_1A''B''y_2 && [(C'' \to \varepsilon, 0, B')]\\
&\Rightarrow_{G'} y_1B''y_2 && [(A'' \to \varepsilon, 0, C'')]\\
&\Rightarrow_{G'} y_1y_2 && [(B'' \to \varepsilon, 0, 0)]
\end{aligned}$$

As a result, $L(G) \subseteq L(G')$. In the following, we show that G' does not generate strings that cannot be generated by G; thus, $L(G') - L(G) = \emptyset$, so $L(G') = L(G)$.

Let us study how G' can generate a terminal string. All derivations start from S. While the sentential form contains S or S', its form is zSw or $zuS'vw$, $z, u, v \in \{A, B, C, A', B', C'\}^*$, $w \in T^*$, where if $g(X') = X$ for $X \in \{A, B, C\}$ and $g(X) = X$ for all other symbols of V, then $g(zSw)$ or $g(zuS'vw)$ are valid sentential forms of G. Furthermore, zu contains at most one occurrence of A', v contains at most one occurrence of C', and zuv contains at most one occurrence of B' (see (2) in the construction of P'). After $(S' \to \varepsilon, 0, 0)$ is used, we get a sentential form $zuvw$ with z, u, v, and w as above such that

$$S \Rightarrow_G^* g(zuvw)$$

Next, we demonstrate that

$$zuv \Rightarrow_{G'}^* \varepsilon \quad \text{implies} \quad g(zuv) \Rightarrow_G^* \varepsilon$$

More specifically, we investigate all possible derivations rewriting a sentential form containing a single occurrence of each of the letters A', B', and C'.

Consider a sentential form of the form $zuvw$, where $z, u, v \in \{A, B, C, A', B', C'\}^*$, $w \in T^*$, and $\text{occur}(zu, A') = \text{occur}(zuv, B') = \text{occur}(v, C') = 1$. By the definition of rules rewriting A', B', and C' (see (3) in the construction of P'), we see that these three symbols must form a substring $A'B'C'$; otherwise, no next derivation step can be made. That is, $zuvw = z\bar{u}A'B'C'\bar{v}w$ for some $\bar{u}, \bar{v} \in \{A, B, C\}^*$. Next, observe that the only applicable rule is $(C' \to C'', A'B'C', 0)$. Thus, we get

$$z\bar{u}A'B'C'\bar{v}w \Rightarrow_{G'} z\bar{u}A'B'C''\bar{v}w$$

This sentential form can be rewritten in two ways. First, we can rewrite A' to A'' by $(A' \to A'', A'B'C'', 0)$. Second, we can replace another occurrence of C with C'. Let us investigate the derivation

$$z\bar{u}A'B'C''\bar{v}w \Rightarrow_{G'} z\bar{u}A''B'C''\bar{v}w \ [(A' \to A'', A'B'C'', 0)]$$

As before, we can either rewrite another occurrence of A to A' or rewrite an occurrence of C to C' or rewrite B' to B'' by using $(B' \to B'', A''B'C'', 0)$. Taking into account all possible combinations of the above-described steps, we see that after the first application of $(B' \to B'', A''B'C'', 0)$, the whole derivation is of the form

$$z\bar{u}A'B'C'\bar{v}w \Rightarrow^{+}_{G'} zu_1Xu_2A''B''C''v_1Yv_2w$$

where $X \in \{A', \varepsilon\}$, $Y \in \{C', \varepsilon\}$, $u_1g(X)u_2 = \bar{u}$, and $v_1g(Y)v_2 = \bar{v}$. Let $zu_1Xu_2 = x$ and $v_1Yv_2 = y$. The next derivation step can be made in four ways. By an application of $(B \to B', 0, B')$, we can rewrite an occurrence of B in x or y. In both cases, this derivation is blocked in the next step. The remaining two derivations are

$$xA''B''C''yw \Rightarrow_{G'} xA''C''yw \ [(B'' \to \varepsilon, 0, 0)]$$

and

$$xA''B''C''yw \Rightarrow_{G'} xA''B''yw \ [(C'' \to \varepsilon, 0, B')]$$

Let us examine how G' can rewrite $xA''C''yw$. The following three cases cover all possible steps.

(i) If $xA''C''yw \Rightarrow_{G'} x_1B'x_2A''C''yw \ [(B \to B', 0, B')]$, where $x_1Bx_2 = x$, then the derivation is blocked.
(ii) If $xA''C''yw \Rightarrow_{G'} xA''C''y_1B'y_2w \ [(B \to B', 0, B')]$, where $y_1By_2 = y$, then no next derivation step can be made.
(iii) Let $xA''C''yw \Rightarrow_{G'} xA''yw \ [(C'' \to \varepsilon, 0, B')]$. Then, all the following derivations

$$xA''yw \Rightarrow_{G'} xyw$$

and

$$xA''yw \Rightarrow_{G'} x_1B'x_2A''yw \Rightarrow_{G'} x_1B'x_2yw$$

where $x_1Bx_2 = x$, and

$$xA''yw \Rightarrow_{G'} xA''y_1B'y_2w \Rightarrow_{G'} xy_1B'y_2w$$

where $y_1 B y_2 = y$, produce a sentential form in which the substring $A''B''C''$ is erased. This sentential form contains at most one occurrence of A', B', and C'.

Return to

$$xA''B''C''yw \Rightarrow_{G'} xA''B''yw$$

Observe that by analogy with case (iii), any rewriting of $xA''B''yw$ removes the substring $A''B''$ and produces a sentential form containing at most one occurrence of A', B', and C'.

To summarize the considerations above, the reader can see that as long as there exists an occurrence of A'', B'', or C'' in the sentential form, only the erasing rules or $(B \to B', 0, B')$ can be applied. The derivation either enters a sentential form that blocks the derivation or the substring $A'B'C'$ is completely erased, after which new occurrences of A, B, and C can be changed to A', B', and C'. That is,

$$z\bar{u}A'B'C'\bar{v}w \Rightarrow_{G'}^{+} xyw \quad \text{implies} \quad g(z\bar{u}A'B'C'\bar{v}w) \Rightarrow_G g(xyw)$$

where $z, \bar{u}, \bar{v} \in \{A, B, C\}^*$, $x, y \in \{A, B, C, A', B', C'\}^*$, $w \in T^*$, and $z\bar{u} = g(x)$, $\bar{v}w = g(yw)$. In other words, the rules constructed in (2) and (3) correctly simulate the application of the only non-context-free rule $ABC \to \varepsilon$. Recall that $g(a) = a$, for all $a \in T$. Hence, $g(xyw) = g(xy)w$. Thus, $L(G') - L(G) = \emptyset$.

Having $L(G) \subseteq L(G')$ and $L(G') - L(G) = \emptyset$, we get $L(G) = L(G')$, and the theorem holds. □

Theorem 4.6.16 was further slightly improved in [15], where the following result was proved (the number of nonterminals was reduced from 11 to 9).

Theorem 4.6.17 (See [15]). *Every recursively enumerable language can be generated by an ssc-grammar of degree* (3, 1) *with no more than 8 conditional rules and 9 nonterminals.*

Let us close this section by stating several open problems.

Open Problem 4.6.18. In Theorems 4.6.4, 4.6.6, 4.6.9, and 4.6.11, we proved that ssc-grammars of degrees (1, 2) and (2, 1) generate the family of recursively enumerable languages, and propagating ssc-grammars of degrees (1, 2) and (2, 1) generate the family of context-sensitive languages. However, we discussed no ssc-grammars of degree (1, 1). According to Penttonen (see Theorem 4.5.7), propagating sc-grammars of degree (1, 1) generate a proper subfamily of context-sensitive languages. That is, $\mathbf{SSC}^{-\varepsilon}(1, 1) \subseteq \mathbf{SC}^{-\varepsilon}(1, 1) \subset \mathbf{CS}$. Are propagating ssc-grammars of degree (1, 1) as powerful as propagating sc-grammars of degree (1, 1)? Furthermore, consider ssc-grammars of degree (1, 1) with erasing rules. Are they more powerful than propagating ssc-grammars of degree (1, 1)? Do they generate the family of all context-sensitive languages or, even more, the family of recursively enumerable languages? □

Open Problem 4.6.19. In Theorems 4.6.14 through 4.6.17, several reduced normal forms of these grammars were presented. These normal forms give rise to the following questions. Can any of the results be further improved with respect to the number of conditional rules or nonterminals? Are there analogical reduced forms of ssc-grammars with degrees $(1, 2)$ and $(1, 3)$? Moreover, reconsider these results in terms of propagating ssc-grammars. Is it possible to achieve analogical results if we disallow erasing rules? □

4.7 Scattered Context Grammars

Up until now, in the context-regulated grammars discussed in this chapter, only a single rule was applied during every derivation step. As obvious, this one-rule application can be quite naturally generalized to the application of several rules during a single step. This generalization underlies scattered context grammars the subject of this section. More specifically, the notion of a scattered context grammar G is based on sequences of context-free rules, according to which G can simultaneously rewrite several nonterminals during a single derivation step.

Section 4.7 consists of five subsections. Section 4.7.1 define and illustrate scattered context grammars, whose power is established in Sect. 4.7.2. Section 4.7.3 gives a normal form of these grammars, and Sect. 4.7.4 reduces these grammars. Finally, Sect. 4.7.5 studies LL versions of scattered context grammars, which are important in terms of parsing-related applications of scattered context grammars.

4.7.1 *Definitions and Examples*

In this section, we define scattered context grammars and illustrate them by examples.

Definition 4.7.1. A *scattered context grammar* is a quadruple

$$G = \big(V, T, P, S\big)$$

where

- V is a *total alphabet*;
- $T \subset V$ an alphabet of *terminals*;
- P is a finite set of *rules* of the form

$$(A_1, \ldots, A_n) \to (x_1, \ldots, x_n)$$

  where $n \geq 1$, $A_i \in V - T$, and $x_i \in V^*$, for all i, $1 \leq i \leq n$ (each rule may have different n);

- $S \in V - T$ is the *start symbol*.

If

$$u = u_1 A_1 \dots u_n A_n u_{n+1}$$
$$v = u_1 x_1 \dots u_n x_n u_{n+1}$$

and $p = (A_1, \dots, A_n) \to (x_1, \dots, x_n) \in P$, where $u_i \in V^*$, for all i, $1 \leq i \leq n + 1$, then G makes a *derivation step* from u to v according to p, symbolically written as

$$u \Rightarrow_G v \; [p]$$

or, simply, $u \Rightarrow_G v$. Set

$$\text{lhs}(p) = A_1 \dots A_n$$
$$\text{rhs}(p) = x_1 \dots x_n$$

and

$$\text{len}(p) = n$$

If $\text{len}(p) \geq 2$, p is said to be a *context-sensitive rule* while for $\text{len}(p) = 1$, p is said to be *context-free*. Define $\Rightarrow_G^k$, $\Rightarrow_G^*$, and $\Rightarrow_G^+$ in the standard way. The *language* of G is denoted by $L(G)$ and defined as

$$L(G) = \{w \in T^* \mid S \Rightarrow_G^* w\}$$

A language L is a *scattered context language* if there exists a scattered context grammar G such that $L = L(G)$. □

Definition 4.7.2. A *propagating scattered context grammar* is a scattered context grammar

$$G = (V, T, P, S)$$

in which every $(A_1, \dots, A_n) \to (x_1, \dots, x_n) \in P$ satisfies $|x_i| \geq 1$, for all i, $1 \leq i \leq n$. A *propagating scattered context language* is a language generated by a propagating scattered context grammar. □

Example 4.7.3. Consider the non-context-free language $L = \{a^n b^n c^n \mid n \geq 1\}$. This language can be generated by the scattered context grammar

$$G = (\{S, A, a, b, c\}, \{a, b, c\}, P, S)$$

where

$$P = \{(S) \to (aAbAcA),\\ (A, A, A) \to (aA, bA, cA),\\ (A, A, A) \to (\varepsilon, \varepsilon, \varepsilon)\}$$

For example, the sentence $aabbcc$ is generated by G as follows:

$$S \Rightarrow_G aAbAcA \Rightarrow_G aaAbbAccA \Rightarrow_G aabbcc$$

Notice, however, that L can be also generated by the propagating scattered context grammar

$$G' = (\{S, A, a, b, c\}, \{a, b, c\}, P', S)$$

where

$$P' = \{(S) \to (AAA),\\ (A, A, A) \to (aA, bA, cA),\\ (A, A, A) \to (a, b, c)\}$$

□

For brevity, we often label rules of scattered context grammars with labels (just like we do in other grammars), as illustrated in the next example.

Example 4.7.4. Consider the non-context-free language

$$L = \{(ab^n)^m \mid m \geq n \geq 2\}$$

This language is generated by the propagating scattered context grammar

$$G = (\{S, S_1, S_2, B, M, X, Y, Z, a\}, \{a\}, P, S)$$

with P containing the following rules

$$\begin{aligned}
&1 : (S) \to (MS)\\
&2 : (S) \to (S_1S_2)\\
&3 : (S_1, S_2) \to (MS_1, BS_2)\\
&4 : (S_1, S_2) \to (MX, BY)\\
&5 : (X, B, Y) \to (BX, Y, b)\\
&6 : (M, X, Y) \to (X, Y, ab)\\
&7 : (M, X, Y) \to (Z, Y, ab)\\
&8 : (Z, B, Y) \to (Z, b, Y)\\
&9 : (Z, Y) \to (a, b)
\end{aligned}$$

Clearly, by applying rules 1 through 4, G generates a string from

$$\{M\}^+\{X\}\{B\}^+\{Y\}$$

In what follows, we demonstrate that the string is of the form $M^{m-1}XB^{n-1}Y$, where $m, n \geq 2$. Rule 1 allows G to add Ms to the beginning of the sentential form, so $m \geq n$ holds true. Observe that each of the rules 5 through 8 either shifts the last nonterminal Y left or keeps its position unchanged. As a result, always the rightmost nonterminal preceding Y has to be replaced with Y by rules 5 through 7; otherwise, the skipped nonterminals cannot be rewritten during the rest of the derivation. For the same reason, the rightmost nonterminal M preceding X has to be rewritten by the rule 6. Rules 5 and 6 are applied in a cycle consisting of $n - 1$ applications of 5 and one application of 6:

$$\begin{aligned} M^{m-1}XB^{n-1}Y &\Rightarrow_G^{n-1} M^{m-1}B^{n-1}XYb^{n-1} \; [5^{n-1}] \\ &\Rightarrow_G \; M^{m-2}XB^{n-1}Yab^n \quad [6] \end{aligned}$$

At this point, the substring preceding Y differs from the original string only in the number of Ms decremented by 1, and the cycle can be repeated again. After repeating this cycle $m - 2$ times, we obtain $MXB^{n-1}Y(ab^n)^{m-2}$. The derivation is completed as follows:

$$\begin{aligned} MXB^{n-1}Y(ab^n)^{m-2} &\Rightarrow_G^{n-1} MB^{n-1}XYb^{n-1}(ab^n)^{m-2} && [5^{n-1}] \\ &\Rightarrow_G \; ZB^{n-1}Y(ab^n)^{m-1} && [7] \\ &\Rightarrow_G^{n-1} Zb^{n-1}Y(ab^n)^{m-1} && [8^{n-1}] \\ &\Rightarrow_G \; (ab^n)^m && [9] \end{aligned}$$

□

Example 4.7.5 (See [9, 10]). Consider the non-context-free language

$$L = \left\{a^{2^n} \mid n \geq 0\right\}$$

This language is generated by the propagating scattered context grammar

$$G = \left(\{S, W, X, Y, Z, A, a\}, \{a\}, P, S\right)$$

with P containing these rules

$$\begin{aligned} &1 : (S) \to (a) \\ &2 : (S) \to (aa) \\ &3 : (S) \to (WAXY) \\ &4 : (W, A, X, Y) \to (a, W, X, AAY) \\ &5 : (W, X, Y) \to (a, W, AXY) \\ &6 : (W, X, Y) \to (Z, Z, a) \\ &7 : (Z, A, Z) \to (Z, a, Z) \\ &8 : (Z, Z) \to (a, a) \end{aligned}$$

In what follows, we demonstrate that $L(G) = L$. Rules 1 and 2 generate a and aa, respectively. Rule 3 starts off the derivation of longer strings in L. Consider the following derivation of $a^{16} \in L(G)$

$$\begin{aligned} S &\Rightarrow_G WAXY && [3] \\ &\Rightarrow_G aWXA^2Y && [4] \\ &\Rightarrow_G a^2WA^3XY && [5] \\ &\Rightarrow_G a^3WA^2XA^2Y && [4] \\ &\Rightarrow_G a^4WAXA^4Y && [4] \\ &\Rightarrow_G a^5WXA^6Y && [4] \\ &\Rightarrow_G a^6WA^7XY && [5] \\ &\Rightarrow_G a^6ZA^7Za && [6] \\ &\Rightarrow_G^7 a^{13}ZZa && [7^7] \\ &\Rightarrow_G a^{16} && [8] \end{aligned}$$

Observe that in any successful derivation, rules 4 and 5 are applied in a cycle, and after the required number of As is obtained, the derivation is finished by rules 6, 7, and 8. In a greater detail, observe that the rule $(W, A, X, Y) \rightarrow (a, W, X, AAY)$ removes one A between W and X, and inserts two As between X and Y. In a successful derivation, this rule has to rewrite the leftmost nonterminal A. After all As are removed between W and X, the rule $(W, X, Y) \rightarrow (a, W, AXY)$ can be used to bring all As occurring between X and Y back between W and X, and the cycle can be repeated again. Alternatively, rule 6 can be used, which initializes the final phase of the derivation in which all As are replaced with as by rules 7 and 8.

By adding one more stage, the above grammar can be extended so that it generates the language

$$\left\{a^{2^{2^n}} \mid n \geq 0\right\}$$

The first stage, similar to the above grammar, generates 2^n identical symbols that serve as a counter for the second stage. In the second stage, a string consisting of identical symbols, which are different from those generated during the first stage, is doubled 2^n times, thus obtaining 2^{2^n} identical symbols. This doubling starts from a string consisting of a single symbol. See [9] for the details. □

The families of languages generated by scattered context grammars and propagating scattered context grammars are denoted by **SCAT** and $\mathbf{SCAT}^{-\varepsilon}$, respectively.

4.7.2 Generative Power

This brief section establishes the power of scattered context grammars. In addition, it points out a crucially important open problem, referred to as the *PSCAT = CS problem*, which asks whether $\mathbf{SCAT}^{-\varepsilon}$ and **CS** coincide.

Theorem 4.7.6 (See [14]). **CF** $\subset$ **SCAT**$^{-\varepsilon}$ $\subseteq$ **CS** $\subset$ **SCAT** $=$ **RE**

Open Problem 4.7.7. Is the inclusion **SCAT**$^{-\varepsilon}$ $\subseteq$ **CS**, in fact, an identity? □

4.7.3 Normal Forms

This section demonstrates how to transform any propagating scattered context grammar to an equivalent 2*-limited propagating scattered context grammar*, which represent an important normal form of propagating scattered context grammars. More specifically, in a 2-limited propagating scattered context grammar, each rule consist of no more than 2 context-free rules, either of which has on their right-hand side no more than 2 symbols.

Definition 4.7.8. A 2*-limited propagating scattered context grammar* is a propagating scattered context grammar, $G = (V, T, P, S)$, such that

- $(A_1, \dots, A_n) \rightarrow (w_1, \dots, w_n) \in P$ implies that $n \leq 2$, and for every i, $1 \leq i \leq n$, $1 \leq |w_i| \leq 2$, and $w_i \in \left(V - \{S\}\right)^*$;
- $(A) \rightarrow (w) \in P$ implies that $A = S$. □

The proof of the transformation is divided into two lemmas.

Lemma 4.7.9. *If $L \subseteq T^*$ is a language generated by a propagating scattered context grammar, $G = (V, T, P, S)$, and if c is a symbol such that $c \notin T$, then there is a 2-limited propagating scattered context grammar, $\bar{G}$, such that $L(\bar{G}) = L\{c\}$.*

Proof. Let $\bar{n}$ be the number of the rules in P. Number the rules of P from 1 to $\bar{n}$. Let $(A_{i1}, \dots, A_{in_i}) \rightarrow (w_{i1}, \dots, w_{in_i})$ be the ith rule. Let C and $\bar{S}$ be new symbols,

$$\begin{aligned} W &= \left\{\langle i, j\rangle \mid 1 \leq i \leq \bar{n}, 1 \leq j \leq n_i\right\} \\ \bar{V} &= V \cup \{C, \bar{S}\} \cup W \cup \left\{\langle C, i\rangle \mid 1 \leq i \leq \bar{n}\right\} \end{aligned}$$

Let $G' = \left(\bar{V}, T \cup \{c\}, P', \bar{S}\right)$ be a propagating scattered context grammar, where P' is defined as follows:

(1) For each $1 \leq i \leq \bar{n}$, add
  $(\bar{S}) \rightarrow (S\langle C, i\rangle)$ to P';
(2) For each i such that $n_i = 1$ and $1 \leq k \leq \bar{n}$, add
  $(A_{i1}, \langle C, i\rangle) \rightarrow (w_{i1}, \langle C, k\rangle)$ to P';
(3) For each i such that $n_i > 1$, $1 \leq j \leq n_i - 1$, $1 \leq k \leq \bar{n}$, add

  (3.1) $(A_{i1}, \langle C, i\rangle) \rightarrow (\langle i, 1\rangle, C)$,
  (3.2) $(\langle i, j\rangle, A_{i(j+1)}) \rightarrow (w_{ij}, \langle i, j+1\rangle)$, and
  (3.3) $(\langle i, n_i\rangle, C) \rightarrow (w_{in_i}, \langle C, k\rangle)$ to P';

(4)

(4.1) For each i such that $n_i = 1$, add
$(A_{i1}, \langle C, i \rangle) \to (w_{i1}, c)$ to P';
(4.2) For each i such that $n_i > 1$, add
$(\langle i, n_i \rangle, C) \to (w_{in_i}, c)$ to P'.

Clearly, $L(G') = L\{c\}$. Since for some i and j, w_{ij} may satisfy $|w_{ij}| > 2$, G' may not be a 2-limited propagating scattered context grammar. However, by making use of standard techniques, one can obtain a 2-limited propagating scattered context grammar $\bar{G}$ from G' such that $L(\bar{G}) = L(G')$. □

Lemma 4.7.10. *If $L \subseteq T^+$, c is a symbol such that $c \notin T$, and $G = (V, T \cup \{c\}, P, S)$ is a 2-limited propagating scattered context grammar satisfying $L(G) = L\{c\}$, then there is a 2-limited propagating scattered context grammar $\bar{G}$ such that $L(\bar{G}) = L$.*

Proof. For each $a \in T \cup \{S\}$, let $\bar{a}$ be a new symbol. Let

$$L_1 = \{A_1A_2A_3 \mid S \Rightarrow_G^* A_1A_2A_3, A_i \in V, \text{ for all } i = 1, 2, 3\}$$
$$L_2 = \{A_1A_2A_3A_4 \mid S \Rightarrow_G^* A_1A_2A_3A_4, A_i \in V, \text{ for all } i = 1, 2, 3, 4\}$$

Let h be the homomorphism from V^* to $(\{\bar{a} \mid a \in T\} \cup (V - T))^*$ defined as $h(a) = \bar{a}$, for each $a \in T$, and $h(A) = A$, for each $A \in V - T$. Let

$$V' = h(V) \cup T \cup \{S'\} \cup \{\langle a, b \rangle \mid a, b \in V\}$$

Let $G' = (V', T, P', S')$, where for all $a, b \in T$, $A_1, \ldots, A_6 \in V$, $A \in h(V)$, P' is defined as follows:

(1)

(1.1) For each $a \in T \cap L$, add
$(S') \to (a)$ to P';
(1.2) For each $A_1A_2A_3 \in L_1$, add
$(S') \to (h(A_1)\langle A_2, A_3 \rangle)$ to P';
(1.3) For each $A_1A_2A_3A_4 \in L_2$, add
$(S') \to (h(A_1A_2)\langle A_3, A_4 \rangle)$ to P';

(2)

(2.1) For each $(A_1, A_2) \to (w_1, w_2) \in P$, add
$(A_1, A_2) \to (h(w_1), h(w_2))$ to P';
(2.2) For each $(A_1, A_2) \to (w_1, w_2) \in P$,

(i) where $|w_2| = 1$, add

A. $(A_1, \langle A_2, A_3 \rangle) \to (h(w_1), \langle w_2, A_3 \rangle)$, and
B. $(A_1, \langle A_3, A_2 \rangle) \to (h(w_1), \langle A_3, w_2 \rangle)$ to P';

(ii) where $w_2 = A_4 A_5$, add

A. $(A_1, \langle A_2, A_3 \rangle) \rightarrow (h(w_1), h(A_4)\langle A_5, A_3 \rangle)$, and
B. $(A_1, \langle A_3, A_2 \rangle) \rightarrow (h(w_1), h(A_3)\langle A_4, A_5 \rangle)$ to P'.

(2.3) For each $(A_1, A_2) \rightarrow (w_1, w_2) \in P$,

(i) where $|w_1| = |w_2| = 1$, add
$(A, \langle A_1, A_2 \rangle) \rightarrow (A, \langle w_1, w_2 \rangle)$ to P';
(ii) where $w_1 w_2 = A_3 A_4 A_5$, add
$(A, \langle A_1, A_2 \rangle) \rightarrow (A, h(A_3)\langle A_4, A_5 \rangle)$ to P';
(iii) where $w_1 w_2 = A_3 A_4 A_5 A_6$, add
$(A, \langle A_1, A_2 \rangle) \rightarrow (A, h(A_3 A_4)\langle A_5, A_6 \rangle)$ to P';

(3) For each $a, b \in T$, add

(3.1) $(\bar{a}, \langle b, c \rangle) \rightarrow (a, \langle b, c \rangle)$, and
(3.2) $(\bar{a}, \langle b, c \rangle) \rightarrow (a, b)$ to P'.

Note that the construction simply combines the symbol c with the symbol to its left. The reason for introducing a new symbol $\bar{a}$, for each $a \in T$, is to guarantee that there always exists a nonterminal A whenever a rule from (2.3) is to be applied, and a nonterminal $\bar{a}$ that enables $\langle b, c \rangle$ to be converted to b by a rule from (3). Clearly, $L(G') = L$. G' may not be a 2-limited propagating scattered context grammar since in (2.3), $|h(A_3 A_4)\langle A_5, A_6 \rangle| = 3$. Once again, by standard techniques, we can obtain a 2-limited propagating scattered context grammar $\bar{G}$ from G' such that $L(\bar{G}) = L(G')$. □

By Lemmas 4.7.9 and 4.7.10, any propagating scattered context grammar can be converted to an equivalent 2-limited propagating scattered context grammar as stated in the following theorem.

Theorem 4.7.11. *If G is a propagating scattered context grammar, then there exists a 2-limited propagating scattered context grammar $\bar{G}$ such that $L(\bar{G}) = L(G)$.* □

4.7.4 Reduction

The present section discusses the reduction of scattered context grammars. Perhaps most importantly, it studies how to reduce the size of their components, such as the number of nonterminals or the number of context-sensitive rules, without any decrease of their generative power. Indeed, any reduction like this is highly appreciated in both theory and practice because it makes scattered context rewriting more succinct and economical while preserving its power.

Definition 4.7.12. Let $G = (V, T, P, S)$ be a scattered context grammar. Then, its *degree of context sensitivity*, symbolically written as $\text{dcs}(G)$, is defined as

$$\operatorname{dcs}(G) = \operatorname{card}\left(\{p \mid p \in P, \operatorname{lhs}(p) \geq 2\}\right)$$

The *maximum context sensitivity* of G, denoted by $\operatorname{mcs}(G)$, is defined as

$$\operatorname{mcs}(G) = \max\left(\{\operatorname{len}(p) - 1 \mid p \in P\}\right)$$

The *overall context sensitivity* of G, denoted by $\operatorname{ocs}(G)$, is defined as

$$\operatorname{ocs}(G) = \operatorname{len}(p_1) + \cdots + \operatorname{len}(p_n) - n$$

where $P = \{p_1, \ldots, p_n\}$. □

We present several results that reduce one of these measures while completely ignoring the other measures. Frequently, however, results of this kind are achieved at the cost of an enormous increase of the other measures. Therefore, we also undertake a finer approach to this descriptional complexity by simultaneously reducing several of these measures while keeping the generative power unchanged.

We start by pointing out a result regarding scattered context grammars with a single nonterminal.

Theorem 4.7.13 (See Theorem 5 in [12]). *One-nonterminal scattered context grammars cannot generate all recursively enumerable languages.*

For scattered context grammars containing only one context-sensitive rule (see Definition 4.7.1), the following theorem holds.

Theorem 4.7.14. *There exists a scattered context grammar G such that G defines a non-context-free language, and*

$$\operatorname{dcs}(G) = \operatorname{mcs}(G) = \operatorname{ocs}(G) = 1$$

Proof. Consider the scattered context grammar

$$G = \left(\{S, A, B, C, D, a, b, c\}, \{a, b, c\}, P, S\right)$$

where the set of rules P is defined as

$$\begin{aligned} P = \{&(S) \to (AC), \\ &(A) \to (aAbB), \\ &(A) \to (\varepsilon), \\ &(C) \to (cCD), \\ &(C) \to (\varepsilon), \\ &(B, D) \to (\varepsilon, \varepsilon)\} \end{aligned}$$

It is easy to verify that $L(G) = \{a^n b^n c^n \mid n \geq 0\}$ and $\operatorname{dcs}(G) = \operatorname{mcs}(G) = \operatorname{ocs}(G) = 1$. □

Next, we concentrate our attention on reducing the number of nonterminals in scattered context grammars. We first demonstrate how the number of nonterminals can be reduced to 3. To do this, we will need the following normal form of a queue grammar.

Lemma 4.7.15. *Let Q' be a queue grammar. Then, there exists a queue grammar*

$$Q = \big(V, T, W' \cup \{1, f\}, \{f\}, R, g\big)$$

such that $L(Q') = L(Q)$, where $W' \cap \{1, f\} = \emptyset$, each $(a, b, x, c) \in R$ satisfies $a \in V - T$ and either

$$b \in W', x \in (V - T)^*, c \in W' \cup \{1, f\}$$

or

$$b = 1, x \in T, c \in \{1, f\}$$

Proof. Let $Q' = (V', T, W', F', R', g')$ be any queue grammar. Set $\Phi = \{\bar{a} \mid a \in T\}$. Define the homomorphism α from V'^* to $((V'-T) \cup \Phi)^*$ as $\alpha(a) = \bar{a}$, for each $a \in T$ and $\alpha(A) = A$, for each $A \in V' - T$. Set $V = V' \cup \Phi$, $W = W' \cup \{1, f\}$, $F = \{f\}$, and $g = \alpha(a_0)q_0$ for $g' = a_0 q_0$. Define the queue grammar $Q = (V, T, W, F, R, g)$, with R constructed in the following way.

(1) For each $(a, b, x, c) \in R'$, where $c \in W' - F'$, add $(\alpha(a), b, \alpha(x), c)$ to R.
(2)

(2.1) For each $(a, b, x, c) \in R'$, where $c \in F'$, add $(\alpha(a), b, \alpha(x), 1)$ to R.
(2.2) For each $(a, b, \varepsilon, c) \in R'$, where $c \in F'$, add $(\alpha(a), b, \varepsilon, f)$ to R.

(3) For each $a \in T$,

(3.1) Add $(\bar{a}, 1, a, 1)$ to R;
(3.2) Add $(\bar{a}, 1, a, f)$ to R.

Clearly, each $(a, b, x, c) \in R$ satisfies $a \in V - T$ and either $b \in W'$, $x \in (V - T)^*$, $c \in W' \cup \{1, f\}$ or $b = 1$, $x \in T$, $c \in \{1, f\}$.

To see that $L(Q') \subseteq L(Q)$, consider any $v \in L(Q')$. As $v \in L(Q')$, $g' \Rightarrow^*_{Q'} vt$, where $v \in T^*$ and $t \in F'$. Express $g' \Rightarrow^*_{Q'} vt$ as

$$g' \Rightarrow^*_{Q'} axc \Rightarrow_{Q'} vt\ [(a, c, y, t)]$$

where $a \in V'$, $x, y \in T^*$, $xy = v$, and $c \in W' - F'$. This derivation is simulated by Q as follows. First, Q uses rules from (1) to simulate $g' \Rightarrow^*_{Q'} axc$. Then, it uses a rule from (2) to simulate $axc \Rightarrow_{Q'} vt$. For $x = \varepsilon$, a rule from (2.2) can be used to generate $\varepsilon \in L(Q)$ in the case of $\varepsilon \in L(Q')$; otherwise, a rule from (2.1) is used. This part of simulation can be expressed as

$$g \Rightarrow^*_Q \alpha(ax)c \Rightarrow_Q \alpha(v)1$$

At this point, $\alpha(v)$ satisfies $\alpha(v) = \bar{a}_1 \cdots \bar{a}_n$, where $a_i \in T$ for all i, $1 \leq i \leq n$, for some $n \geq 1$. The rules from (3) of the form $(\bar{a}, 1, a, 1)$, where $a \in T$, replace every $\bar{a}_j$ with a_j, where $1 \leq j \leq n-1$, and, finally, $(\bar{a}, 1, a, f)$, where $a \in T$, replaces $\alpha(a_n)$ with a_n. As a result, we obtain the sentence vf, so $L(Q') \subseteq L(Q)$.

To establish $L(Q) \subseteq L(Q')$, observe that the use of a rule from (2.2) in Q before the sentential form is of the form $\alpha(ax)c$, where $a \in V'$, $x \in T^*$, $c \in W' - F'$, leads to an unsuccessful derivation. Similarly, the use of (2.2) if $x \neq \varepsilon$ leads to an unsuccessful derivation as well. The details are left to the reader. As a result, $L(Q) \subseteq L(Q')$.

As $L(Q') \subseteq L(Q)$ and $L(Q) \subseteq L(Q')$, we obtain $L(Q) = L(Q')$. □

Consider the queue grammar $Q = (V, T, W, F, R, g)$ from Lemma 4.7.15. Its properties imply that Q generates every string in $L(Q) - \{\varepsilon\}$ so it passes through 1. Before it enters 1, it generates only strings from $(V - T)^*$; after entering 1, it generates only strings from T^*. The following corollary express this property formally.

Corollary 4.7.16. *Let Q be a queue grammar that satisfies the properties given in Lemma 4.7.15. Then, Q generates every $y \in L(Q) - \{\varepsilon\}$ in this way*

$$
\begin{array}{lll}
a_0 q_0 & \Rightarrow_Q x_0 q_1 & [(a_0, q_0, z_0, q_1)] \\
 & \vdots & \\
 & \Rightarrow_Q x_{k-1} q_k & [(a_{k-1}, q_{k-1}, z_{k-1}, q_k)] \\
 & \Rightarrow_Q x_k 1 & [(a_k, q_k, z_k, 1)] \\
 & \Rightarrow_Q x_{k+1} b_1 1 & [(a_{k+1}, 1, b_1, 1)] \\
 & \vdots & \\
 & \Rightarrow_Q x_{k+m-1} b_1 \cdots b_{m-1} 1 & [(a_{k+m-1}, 1, b_{m-1}, 1)] \\
 & \Rightarrow_Q b_1 \cdots b_m f & [(a_{k+m}, 1, b_m, f)]
\end{array}
$$

where $k, m \geq 1$, $g = a_0 q_0$, $a_1, \ldots, a_{k+m} \in V - T$, $b_1, \ldots, b_m \in T$, $z_0, \ldots, z_k \in (V - T)^*$, $q_0, \ldots, q_k, 1 \in W - F$, $f \in F$, $x_0, \ldots, x_{k+m-1} \in (V - T)^+$, and $y = b_1 \cdots b_m$. □

Theorem 4.7.17. *For every recursively enumerable language L, there is a scattered context grammar $G = (V, T, P, S)$ such that $L(G) = L$, and*

$$\operatorname{card}\big(V - T\big) = 3$$

Proof. Let L be a recursively enumerable language. By Theorem 3.3.17, there exists a queue grammar $Q = (\bar{V}, T, W, F, R, g)$ such that $L = L(Q)$. Without any loss of generality, assume that Q satisfies the properties described in Lemma 4.7.15. Set $n = \operatorname{card}(\bar{V} \cup W)$. Introduce a bijective homomorphism β from $\bar{V} \cup W$ to $\{B\}^*\{A\}\{B\}^* \cap \{A, B\}^n$. Without any loss of generality, assume that $(\bar{V} \cup W) \cap \{A, B, S\} = \emptyset$. Define the scattered context grammar

$$G = \big(T \cup \{A, B, S\}, T, P, S\big)$$

where P is constructed in the following way

(1) For $g = ab$, where $a \in \bar{V} - T$ and $b \in W - F$, add $(S) \to \big(\beta(b)SS\beta(a)SA\big)$ to P;
(2) For each $a \in \{A, B\}$, add $(S, S, a, S) \to (S, \varepsilon, aS, S)$ to P;
(3) For each $(a, b, x, c) \in R$, where $a \in \bar{V} - T$, $x \in (\bar{V} - T)^*$, and b, $c \in W - F - \{1\}$, extend P by adding

$$(b_1, \dots, b_n, S, a_1, \dots, a_n, S, S)$$
$$\to \big(c_1, \dots, c_n, \varepsilon, e_1, \dots, e_n, SS, \beta(x)S\big)$$

where $b_1 \cdots b_n = \beta(b)$, $a_1 \cdots a_n = \beta(a)$, $c_1 \cdots c_n = \beta(c)$, and $e_1 \cdots e_n = \varepsilon$;
(4) For each $(a, b, x, c) \in R$, where $a \in \bar{V} - T$, $b \in W - F - \{1\}$, $x \in (\bar{V} - T)^*$, and

(4.1) $c = 1$, extend P by adding

$$(b_1, \dots, b_n, S, a_1, \dots, a_n, S, S)$$
$$\to \big(c_1, \dots, c_n, \varepsilon, e_1, \dots, e_n, SS, \beta(x)S\big)$$

(4.2) $c \in F$ and $x = \varepsilon$, extend P by adding

$$(b_1, \dots, b_n, S, a_1, \dots, a_n, S, S, A)$$
$$\to (e_1, \dots, e_n, \varepsilon, e_{n+1}, \dots, e_{2n}, \varepsilon, \varepsilon, \varepsilon)$$

where $b_1 \cdots b_n = \beta(b)$, $a_1 \cdots a_n = \beta(a)$, $c_1 \cdots c_n = \beta(c)$, and $e_1 \cdots e_{2n} = \varepsilon$;
(5) For each $(a, 1, x, c) \in R$, where $a \in \bar{V} - T$, $x \in T$, and

(5.1) $c = 1$, extend P by adding

$$(b_1, \dots, b_n, S, a_1, \dots, a_n, S, S)$$
$$\to (c_1, \dots, c_n, \varepsilon, e_1, \dots, e_n, SS, xS)$$

(5.2) $c \in F$, extend P by adding

$$(b_1, \dots, b_n, S, a_1, \dots, a_n, S, S, A)$$
$$\to (e_1, \dots, e_n, \varepsilon, e_{n+1}, \dots, e_{2n}, \varepsilon, \varepsilon, x)$$

where $b_1 \cdots b_n = \beta(1)$, $a_1 \cdots a_n = \beta(a)$, $c_1 \cdots c_n = \beta(c)$, and $e_1 \cdots e_{2n} = \varepsilon$.

The constructed scattered context grammar G simulates the queue grammar Q that satisfies the properties described in Lemma 4.7.15. The rule from (1), applied only once, initializes the derivation. One of the rules from (4.2) and (5.2) terminates the derivation. In a greater detail, a rule from (4.2) is used in the derivation of $\varepsilon \in L(Q)$; in a derivation of every other string, a rule from (5.2) is used in the last step of the derivation.

Every sentential form of G can be divided into two parts. The first n nonterminals encode the state of Q. The second part represents the queue, where the first symbol S always occurs at the beginning of the queue and the third S always occurs at the end of the queue, followed by the ultimate nonterminal A.

During any successful derivation of G, a rule introduced in (2) is always applied after the application of a rule introduced in (1), (3), (4.1), and (5.1). More precisely, to go on performing the successful derivation, after applying rules from (1), (3), (4.1), and (5.1), G shifts the second occurrence of S right in the current sentential form. G makes this shift by using rules introduced in (2) to obtain a sentential form having precisely n occurrences of $d \in \{A, B\}$ between the first occurrence of S and the second occurrence of S.

The following claims demonstrate that the rule from (1) can be used only once during a successful derivation.

Claim 1. Let $S \Rightarrow_G^* x$ *be a derivation during which* G *uses the rules introduced in (1)* i *times, for some* $i \geq 1$. *Then* $\text{occur}(x, \{S\}) = 1 + 2i - 3j$, $\text{occur}(x, \{B\}) = (n-1)k$, *and* $\text{occur}(x, \{A\}) = k + i - j$, *where* k *is a non-negative integer and* j *is the number of applications of rules introduced in (4.2) and (5.2) such that* $j \geq 1$ *and* $1 + 2i \geq 3j$.

Proof. Notice that the rules introduced in (2), (3), (4.1), and (5.1) preserve the number of As, Bs, and Ss present in the sentential form. Next, observe that every application of the rule from (1) adds 2 Ss to the sentential form and every application of a rule from (4.2) or (5.2) removes 3 Ss from the sentential form. Finally, notice the last A on the right-hand side of the rule from (1) and on the left-hand sides of the rules from (4.2) and (5.2). Based on these observations, it is easy to see that Claim 1 holds. □

Claim 2. Let $S \Rightarrow_G^* x$ *be a derivation during which* G *applies the rule introduced in (1) two or more times. Then,* $x \notin T^*$.

Proof. Let $S \Rightarrow_G^* x$, where $x \in T^*$. Because $x \in T^*$, $\text{occur}(x, \{S\}) = \text{occur}(x, \{B\}) = \text{occur}(x, \{A\}) = 0$. As a result, we get $k = 0$, and $i = j = 1$ from the equations introduced in Claim 1. Thus, for $i \geq 2$, $x \notin T^*$. □

Next, we demonstrate that rules from (4.2) and (5.2) can only be used during the last derivation step of a successful derivation.

Claim 3. G *generates every* $w \in L(G)$ *as follows:*

$$S \Rightarrow_G u \; [p] \Rightarrow_G^* v \Rightarrow_G w \; [q]$$

where p is the rule introduced in (1), q is a rule introduced in (4.2) or (5.2), and during $u \Rightarrow_G^* v$, G makes every derivation step by a rule introduced in (2), (3), (4.1), or (5.1).

Proof. Let $w \in L(G)$. By Claim 2, as $w \in T^*$, G uses the rule introduced in (1) only once. Because $S \Rightarrow_G^* w$ begins from S, we can express $S \Rightarrow_G^* w$ as

$$S \Rightarrow_G u \; [p] \Rightarrow_G^* w$$

where p is the rule introduced in (1), and G never uses this rule during $u \Rightarrow_G^* w$. Observe that every rule r introduced in (2), (3), (4.1), and (5.1) satisfies $\text{occur}(\text{lhs}(r), \{S\}) = 3$ and $\text{occur}(\text{rhs}(r), \{S\}) = 3$. Furthermore, notice that every rule q introduced in (4.2) and (5.2) satisfies $\text{occur}(\text{lhs}(q), \{S\}) = 3$ and $\text{occur}(\text{rhs}(q), \{S\}) = 0$. These observations imply

$$S \Rightarrow_G u \; [p] \Rightarrow_G^* v \Rightarrow_G w \; [q]$$

where p is the rule introduced in (1), q is a rule introduced in (4.2) or (5.2), and during $u \Rightarrow_G^* v$, G makes every step by a rule introduced in (2), (3), (4.1), or (5.1). □

In what follows, we demonstrate that in order to apply a rule from (3) through (5), there have to be exactly n nonterminals between the first and the second occurrence of S. This can be accomplished by one or more applications of a rule from (2).

Claim 4. If $x \Rightarrow_G y \; [p]$ is a derivation step in a successful derivation of G, where p is a rule from (3) through (5), then $x = x_1 S x_2 S x_3 S A$, where x_1, x_2, $x_3 \in (T \cup \{A, B\})^+$, $\text{occur}(x_1, \{A, B\}) = k$, $\text{occur}(x_2, \{A, B\}) = m$, and $k = m = n$.

Proof. If $k < n$ or $m < n$, no rule introduced in (3) through (5) can be used. Therefore, $k \geq n$ and $m \geq n$.

Assume that $k > n$. The only rules that remove the symbols from $\{A, B\}$ in front of the first symbol S are those introduced in (4.2) and (5.2), and these rules remove precisely n nonterminals preceding the first symbol S. For $k > n$, $k - n$ nonterminals remain in the sentential form after the last derivation step so the derivation is unsuccessful. Therefore, $k = n$.

Assume that $m > n$. Then, after the application of a rule introduced in (3) through (5), m symbols from $\{A, B\}$ appear in front of the first S. Therefore, the number of nonterminals appearing in front of the first occurrence of S is greater than n, which contradicts the argument given in the previous paragraph. As a result, $m = n$. □

Based on Claims 1 through 4 and the properties of Q, we can express every successful derivation of G as

- either $S \Rightarrow_G \text{rhs}(p_1)\ [p_1] \Rightarrow_G^* u\ [\Xi] \Rightarrow_G v\ [p_{4a}] \Rightarrow_G^* w\ [\Psi] \Rightarrow_G z\ [p_{5b}]$ for $z \neq \varepsilon$;
- or $S \Rightarrow_G \text{rhs}(p_1)\ [p_1] \Rightarrow_G^* u\ [\Xi] \Rightarrow_G \varepsilon\ [p_{4b}]$;

where p_1, p_{4a}, p_{4b}, and p_{5b}, are rules introduced in (1), (4.1), (4.2), and (5.2), respectively, Ξ is a sequence of rules from (2) and (3), Ψ is a sequence of rules from (2) and (5.1), and the derivation satisfies the following properties.

- Every derivation step in $\text{rhs}(p_1) \Rightarrow_G^* u\ [\Xi]$ has one of these forms:

$$\beta(b_1)Sa_1'Sa_1''d_1y_1'SA \Rightarrow_G \beta(b_1)Sa_1'a_1''d_1Sy_1'SA\ [p_2], \text{ or}$$
$$\beta(b_1)S\beta(a_1)S\beta(y_1)SA \Rightarrow_G \beta(c_1)SS\beta(y_1x_1)SA\ [p_3]$$

  where $a_1', a_1'', y_1' \in \{A, B\}^*$, $d_1 \in \{A, B\}$, $(a_1, b_1, x_1, c_1) \in R$, $b_1 \neq 1$, $c_1 \neq 1$, $y_1 \in (\bar{V} - T)^*$, and p_2, p_3 are rules introduced in (2), (3), respectively.
- The derivation step $u \Rightarrow_G v\ [p_{4a}]$ has this form

$$\beta(b_2)S\beta(a_2)S\beta(y_2)SA \Rightarrow_G \beta(1)SS\beta(y_2x_2)SA\ [p_{4a}]$$

  where $(a_2, b_2, x_2, 1) \in R$, $b_2 \neq 1$, and $y_2 \in (\bar{V} - T)^+$. Observe that if $y_2x_2 = \varepsilon$, no rule is applicable after this step and the derivation is blocked.
- The derivation step $u \Rightarrow_G \varepsilon\ [p_{4b}]$ has this form

$$\beta(b_3)S\beta(a_3)S\beta(y_3)SA \Rightarrow_G \varepsilon\ [p_{4b}]$$

  where $(a_3, b_3, \varepsilon, c_3) \in R$, $b_3 \neq 1$, $c_3 \in F$, and $y_3 = \varepsilon$. As no rule can be applied after a rule from (4.2) is used, if $y_3 \neq \varepsilon$, there remain some nonterminals in the sentential form so the derivation is unsuccessful.
- Every derivation step in $v \Rightarrow_G^* w\ [\Psi]$ has one of these forms

$$\beta(1)Sa_4'Sa_4''d_4y_4't_4SA \Rightarrow_G \beta(1)Sa_4'a_4''dSy_4t_4SA\ [p_2], \text{ or}$$
$$\beta(1)S\beta(a_4)S\beta(y_4)t_4SA \Rightarrow_G \beta(1)SS\beta(y_4)t_4x_4SA\ [p_{5a}]$$

  where $a_4', a_4'', y_4' \in \{A, B\}^*$, $d_4 \in \{A, B\}$, $(a_4, 1, x_4, 1) \in R$, $y_4 \in (\bar{V} - T)^*$, $t_4 \in T^*$, and p_2, p_{5a} are rules introduced in (2), (5.1), respectively.
- The derivation step $w \Rightarrow_G z\ [p_{5b}]$ has this form

$$\beta(1)S\beta(a_5)St_5SA \Rightarrow_G t_5x_5\ [p_{5b}]$$

  where $(a_5, 1, x_5, c_5) \in R$, $c_5 \in F$, and $t_5 \in T^*$.

  Observe that

$$S \Rightarrow_G \text{rhs}(p_1)\ [p_1] \Rightarrow_G^* u\ [\Xi] \Rightarrow_G v\ [p_{4a}] \Rightarrow_G^* w\ [\Psi] \Rightarrow_G z\ [p_{5b}], \text{ for } z \neq \varepsilon$$

if and only if

$$\begin{aligned} g \Rightarrow_Q^* a_2 y_2 b_2 &\Rightarrow_Q y_2 x_2 1 \; [(a_2, b_2, x_2, 1)] \\ \Rightarrow_Q^* a_5 t_5 1 &\Rightarrow_Q z c_5 \quad\; [(a_5, 1, x_5, c_5)] \end{aligned}$$

or

$$S \Rightarrow_G \text{rhs}(p_1)\; [p_1] \Rightarrow_G^* u\; [\Xi] \Rightarrow_G \varepsilon\; [p_{4b}]$$

if and only if

$$g \Rightarrow_Q^* a_3 y_3 b_3 \Rightarrow_Q c_3 \; [(a_3, b_3, \varepsilon, c_3)]$$

As a result, $L(Q) = L(G)$, so the theorem holds. □

Recall that one-nonterminal scattered context grammars are incapable of generating all recursively enumerable languages (see Theorem 4.7.13). By Theorem 4.7.17, three-nonterminal scattered context grammars characterize **RE**. As stated in the following theorem, the optimal bound for the needed number of nonterminals is, in fact, two. This very recent result is proved in [2].

Theorem 4.7.18 (See [2]). *For every recursively enumerable language L, there is a scattered context grammar $G = (V, T, P, S)$ such that $L(G) = L$, and*

$$\text{card}\left(V - T\right) = 2$$

Up until now, we have reduced only one measure of descriptional complexity regardless of all the other measures. We next reconsider this topic in a finer way by simultaneously reducing several measures. It turns out that this simultaneous reduction results in an increase of all the measures involved. In addition, reducing the number of nonterminals necessarily leads to an increase of the number of context-sensitive rules and vice versa.

Theorem 4.7.19. *For every recursively enumerable language L, there is a scattered context grammar $G = (V, T, P, S)$ such that $L(G) = L$, and*

$$\begin{aligned} \text{card}(V - T) &= 5 \\ \text{dcs}(G) &= 2 \\ \text{mcs}(G) &= 3 \\ \text{ocs}(G) &= 6 \end{aligned}$$

Proof (See [22]). Let

$$G' = \left(\{S', A, B, C, D\} \cup T, T, P' \cup \{AB \to \varepsilon, CD \to \varepsilon\}, S'\right)$$

be a phrase-structure grammar in the Geffert normal form, where P' is a set of context-free rules, and $L(G') = L$ (see Theorem 4.1.11). Define the homomorphism h from $\{A, B, C, D\}^*$ to $\{0, 1\}^*$ so that $h(A) = h(B) = 00$, $h(C) = 10$, and $h(D) = 01$. Define the scattered context grammar

$$G = \big(\{S, \bar{S}, 0, 1, \$\} \cup T, T, P, S\big)$$

with P constructed as follows:

(1) For each $S' \to zS'a \in P'$, where $z \in \{A, C\}^*$, $a \in T$, extend P by adding

$$(S) \to \big(h(z)Sa\big)$$

(2) Add $(S) \to (\bar{S})$ to P
(3) For each $S' \to uS'v \in P'$, where $u \in \{A, C\}^*$, $v \in \{B, D\}^*$, extend P by adding

$$(\bar{S}) \to \big(h(u)\bar{S}h(v)\big)$$

(4) Extend P by adding

(4.1) $(\bar{S}) \to (\$\$)$,
(4.2) $(0, \$, \$, 0) \to (\$, \varepsilon, \varepsilon, \$)$,
(4.3) $(1, \$, \$, 1) \to (\$, \varepsilon, \varepsilon, \$)$,
(4.4) $(\$) \to (\varepsilon)$.

Observe that G' generates every $a_1 \cdots a_k \in L(G')$ in the following way

$$\begin{aligned}
S' &\Rightarrow_{G'} z_{a_k} S' a_k \\
&\Rightarrow_{G'} z_{a_k} z_{a_{k-1}} S' a_{k-1} a_k \\
&\ \ \vdots \\
&\Rightarrow_{G'} z_{a_k} \cdots z_{a_2} S' a_2 \cdots a_k \\
&\Rightarrow_{G'} z_{a_k} \cdots z_{a_2} z_{a_1} S' a_1 a_2 \cdots a_k \\
&\Rightarrow_{G'} z_{a_k} \cdots z_{a_2} z_{a_1} u_l S' v_l a_1 a_2 \cdots a_k \\
&\ \ \vdots \\
&\Rightarrow_{G'} z_{a_k} \cdots z_{a_1} u_l \cdots u_2 S' v_2 \cdots v_l a_1 \cdots a_k \\
&\Rightarrow_{G'} z_{a_k} \cdots z_{a_1} u_l \cdots u_2 u_1 v_1 v_2 \cdots v_l a_1 \cdots a_k \\
&= \quad d_m \cdots d_2 d_1 e_1 e_2 \cdots e_n a_1 \cdots a_k \\
&\Rightarrow_{G'} d_m \cdots d_2 e_2 \cdots e_n a_1 \cdots a_k \\
&\ \ \vdots \\
&\Rightarrow_{G'} d_m e_n a_1 \cdots a_k \\
&\Rightarrow_{G'} a_1 \cdots a_k
\end{aligned}$$

where $a_1, \ldots, a_k \in T$, $z_{a_1}, \ldots, z_{a_k}$, $u_1, \ldots, u_l \in \{A, C\}^*$, $v_1, \ldots, v_l \in \{B, D\}^*$, $d_1, \ldots, d_m \in \{A, C\}$, and $e_1, \ldots, e_n \in \{B, D\}$. After erasing S' from the sentential form, G' verifies that the generated strings $z_{a_k} \cdots z_{a_1} u_l \cdots u_1$ and $v_1 \cdots v_l$ are identical. If $m \neq n$, or $d_i e_i \notin \{AB, CD\}$, for some $i \geq 1$, the generated strings do not coincide, and the derivation is blocked, so $a_1 \cdots a_k$ does not belong to the generated language.

The above derivation can be straightforwardly simulated by G as follows:

$$
\begin{aligned}
S &\Rightarrow_G h(z_{a_k})Sa_k \\
&\Rightarrow_G h(z_{a_k})h(z_{a_{k-1}})Sa_{k-1}a_k \\
&\quad\vdots \\
&\Rightarrow_G h(z_{a_k})\cdots h(z_{a_2})Sa_2\cdots a_k \\
&\Rightarrow_G h(z_{a_k})\cdots h(z_{a_2})h(z_{a_1})Sa_1a_2\cdots a_k \\
&\Rightarrow_G h(z_{a_k}\cdots z_{a_2}z_{a_1})\bar{S}a_1a_2\cdots a_k && [p_2] \\
&\Rightarrow_G h(z_{a_k}\cdots z_{a_2}z_{a_1})h(u_l)\bar{S}h(v_l)a_1a_2\cdots a_k \\
&\quad\vdots \\
&\Rightarrow_G h(z_{a_k}\cdots z_{a_1})h(u_l)\cdots h(u_2)\bar{S}h(v_2)\cdots h(v_l)a_1\cdots a_k \\
&\Rightarrow_G h(z_{a_k}\cdots z_{a_1})h(u_l)\cdots h(u_2)h(u_1)\bar{S}h(v_1)h(v_2)\cdots h(v_l)a_1\cdots a_k \\
&\Rightarrow_G h(z_{a_k}\cdots z_{a_1})h(u_l\cdots u_2u_1)\$\$h(v_1v_2\cdots v_l)a_1\cdots a_k && [p_{4a}] \\
&= f_r\cdots f_2f_1\$\$g_1g_2\cdots g_sa_1\cdots a_k \\
&\Rightarrow_G f_r\cdots f_2\$\$g_2\cdots g_sa_1\cdots a_k \\
&\quad\vdots \\
&\Rightarrow_G f_r\$\$g_sa_1\cdots a_k \\
&\Rightarrow_G \$\$a_1\cdots a_k \\
&\Rightarrow_G \$a_1\cdots a_k && [p_{4d}] \\
&\Rightarrow_G a_1\cdots a_k && [p_{4d}]
\end{aligned}
$$

where $f_1, \ldots, f_r, g_1, \ldots, g_s \in \{0, 1\}$, and p_2, p_{4a}, and p_{4d} are rules introduced in (2), (4.1), and (4.4), respectively. In this derivation, the context-free rules of G' are simulated by the rules introduced in (1) through (3), and the context-sensitive rules of G' are simulated by the rules introduced in (4.2) and (4.3). There are the following differences between the derivations in G' and G.

- Instead of verifying the identity of $z_{a_k} \cdots z_{a_1} u_l \cdots u_1$ and $v_1 \cdots v_l$, G verifies that $h(z_{a_k} \cdots z_{a_1} u_l \cdots u_1)$ and $h(v_1 \cdots v_l)$ coincide. This means that instead of comparing strings over $\{A, B, C, D\}$, G compares the strings $f_r \cdots f_1$ and $g_1 \cdots g_s$ over $\{0, 1\}$.
- The rule introduced in (2) guarantees that no rule from (1) can be used after its application. Similarly, the rule introduced in (4.1) prevents the rules of (1) through (3) from being applied.

- When applying the rules from (4.2) and (4.3), some symbols f_i and g_j, where i, $j \geq 1$, can be skipped. However, if some 0s and 1s that do not directly neighbor with the \$s are rewritten, the form of these rules guarantees that the skipped nonterminals can never be rewritten later in the derivation, so the derivation is necessarily unsuccessful in this case.
- The rule from (4.4) can be used anytime the symbol \$ appears in the sentential form. However, when this rule is used and some nonterminals from $\{0, 1\}$ occur in the sentential form, these nonterminals can never be removed from the sentential form, so the derivation is blocked. As a result, the rule from (4.4) has to be applied at the very end of the derivation.

These observations imply that $L = L(G) = L(G')$. As obvious, $\text{card}(V - T) = 5$, $\text{dcs}(G) = 2$, $\text{mcs}(G) = 3$, $\text{ocs}(G) = 6$. Thus, the theorem holds. □

Theorem 4.7.20. *For every recursively enumerable language L, there is a scattered context grammar* $\bar{G} = (V, T, \bar{P}, S)$ *such that* $L(\bar{G}) = L$, *and*

$$\begin{aligned} \text{card}(V - T) &= 8 \\ \text{dcs}(\bar{G}) &= 6 \\ \text{mcs}(\bar{G}) &= 1 \\ \text{ocs}(\bar{G}) &= 6 \end{aligned}$$

Proof. We slightly modify the construction given in the proof of Theorem 4.7.19. Define the scattered context grammar

$$\bar{G} = \big(\{S, \bar{S}, 0, 1, \$_L, \$_R, \$_0, \$_1\} \cup T, T, \bar{P}, S\big)$$

and initialize $\bar{P}$ with the set of all rules introduced in steps (1) through (3) of the construction given in the proof of Theorem 4.7.19. Then, add the following rules to $\bar{P}$

(4)

(4.1) $(\bar{S}) \rightarrow (\$_L\$_R)$,
(4.2) $(0, \$_L) \rightarrow (\$_0, \varepsilon)$, $(\$_R, 0) \rightarrow (\varepsilon, \$_0)$, $(\$_0, \$_0) \rightarrow (\$_L, \$_R)$,
(4.3) $(1, \$_L) \rightarrow (\$_1, \varepsilon)$, $(\$_R, 1) \rightarrow (\varepsilon, \$_1)$, $(\$_1, \$_1) \rightarrow (\$_L, \$_R)$,
(4.4) $(\$_L) \rightarrow (\varepsilon)$, and $(\$_R) \rightarrow (\varepsilon)$.

Observe that a single derivation step made by a rule introduced in step (4.2) or (4.3) of the construction of G is simulated in $\bar{G}$ by the above rules from (4.2) or (4.3) in three derivation steps. In a greater detail, a derivation of the form

$$x0\$\$0yz \Rightarrow_G x\$\$yz \; [(0, \$, \$, 0) \rightarrow (\$, \varepsilon, \varepsilon, \$)]$$

is simulated by $\bar{G}$ as follows:

$$\begin{aligned} x0\$_L\$_R0yz &\Rightarrow_{\bar{G}} x\$_0\$_R0yz \; [(0, \$_L) \to (\$_0, \varepsilon)] \\ &\Rightarrow_{\bar{G}} x\$_0\$_0yz \; [(\$_R, 0) \to (\varepsilon, \$_0)] \\ &\Rightarrow_{\bar{G}} x\$_L\$_Ryz \; [(\$_0, \$_0) \to (\$_L, \$_R)] \end{aligned}$$

where $x, y \in \{0, 1\}^*$, and $z \in T^*$. The rest of the proof resembles the proof of Theorem 4.7.19 and is, therefore, left to the reader. □

Theorem 4.7.21. *For every recursively enumerable language L, there is a scattered context grammar $G = (V, T, P, S)$ such that $L(G) = L$, and*

$$\begin{aligned} \text{card}(V - T) &= 4 \\ \text{dcs}(G) &= 4 \\ \text{mcs}(G) &= 5 \\ \text{ocs}(G) &= 20 \end{aligned}$$

Proof. Let

$$G' = \big(\{S', A, B, C, D\} \cup T, T, P' \cup \{AB \to \varepsilon, CD \to \varepsilon\}, S'\big)$$

be a phrase-structure grammar in the Geffert normal form, where P' is a set of context-free rules, and $L(G') = L$ (see Theorem 4.1.11). Define the homomorphism h from $\{A, B, C, D\}^*$ to $\{0, 1\}^*$ so that $h(A) = h(B) = 00$, $h(C) = 10$, and $h(D) = 01$. Define the scattered context grammar

$$G = \big(\{S, 0, 1, \$\} \cup T, T, P, S\big)$$

with P constructed as follows:

(1) Add $(S) \to (11S11)$ to P;
(2) For each $S' \to zS'a \in P'$, add $(S) \to (h(z)S1a1)$ to P;
(3) For each $S' \to uS'v \in P'$, add $(S) \to (h(u)Sh(v))$ to P;
(4) For each $S' \to uv \in P'$, add $(S) \to (h(u)\$\$h(v))$ to P;
(5) Add

(5.1) $(0, 0, \$, \$, 0, 0) \to (\$, \varepsilon, \varepsilon, \varepsilon, \varepsilon, \$)$,
(5.2) $(1, 0, \$, \$, 0, 1) \to (\$, \varepsilon, \varepsilon, \varepsilon, \varepsilon, \$)$,
(5.3) $(1, 1, \$, \$, 1, 1) \to (11\$, \varepsilon, \varepsilon, \varepsilon, \varepsilon, \$)$, and
(5.4) $(1, 1, \$, \$, 1, 1) \to (\varepsilon, \varepsilon, \varepsilon, \varepsilon, \varepsilon, \varepsilon)$ to P.

Every successful derivation starts by an application of the rule introduced in (1), and this rule is not used during the rest of the derivation. Rules from (2) through (4) simulate the context-free rules of G'. After the rule from (4) is used, only rules from (5) are applicable. The rules from (5.1) and (5.2) verify that the strings over $\{0, 1\}$, generated by the rules from (2) through (4), coincide. The rule from (5.3) removes the 1s between the terminal symbols and, in addition, makes sure that rules from (2) can never be used in a successful derivation after a rule from (3) is applied. Finally, the rule from (5.4) completes the derivation.

The proof of the theorem is based on five claims, established next.

Claim 1. Every successful derivation in G can be expressed as

$$\begin{aligned} S &\Rightarrow_G^* v \ [\Xi] \\ &\Rightarrow_G w \ [p_4] \\ &\Rightarrow_G^* y \ [\Psi] \\ &\Rightarrow_G z \ [p_{5d}] \end{aligned}$$

where

$$\begin{aligned} v &\in \{0,1\}^*\{S\}(\{0,1\} \cup T)^* \\ w &\in \{0,1\}^+\{\$\}\{\$\}(\{0,1\} \cup T)^* \\ y &\in \{1\}\{1\}\{\$\}T^*\{\$\}T^*\{1\}T^*\{1\}T^* \end{aligned}$$

$z \in T^*$, p_4 and p_{5d} are rules introduced in (4) and (5.4), respectively, and Ξ and Ψ are sequences of rules introduced in (1) through (3) and (5.1) through (5.3), respectively.

Proof. As S appears on the left-hand side of every rule introduced in (1) through (4), all of them are applicable while S occurs in the sentential form. On the other hand, no rule from (5) can be used at this point. After p_4 is used, it replaces S with \$\$, so rules from (1) through (4) are not applicable and only rules from (5) can be used. Therefore, the beginning of the derivation can be expressed as

$$\begin{aligned} S &\Rightarrow_G^* v \ [\Xi] \\ &\Rightarrow_G w \ [p_4] \end{aligned}$$

Because all rules, except for p_{5d}, contain nonterminals on their right-hand sides, p_{5d} has to be applied in the last derivation step and no other rule can be applied after its use. Applications of rules from (2) through (4) may introduce some nonterminals 0 and 1 to the sentential form, so in this case, the rules from (5.1) and (5.2) are applied to remove them. As a result,

$$\begin{aligned} w &\Rightarrow_G^* y \ [\Psi] \\ &\Rightarrow_G z \ [p_{5d}] \end{aligned}$$

and the sentential forms satisfy the conditions given in the claim. □

Claim 2. In $w \Rightarrow_G^+ z$ from Claim 1, every sentential form s satisfies

$$s \in \{0,1\}^*\{\$\}T^*\{\$\}(\{0,1\} \cup T)^*$$

Proof. The form of the rules introduced in (5) implies that whenever a nonterminal appears between the two occurrences of \$, it can never be removed during the rest of the derivation. Therefore, the claim holds. □

Claim 3. In $w \Rightarrow_G^+ z$ from Claim 1, every sentential form s satisfies

$$s \in \{1\}\{1\}(\{1\}\{0\} \cup \{0\}\{0\})^* \{\$\} T^* \{\$\} (\{0, 1\} \cup T)^*$$

Proof. Claim 2 implies that whenever rules from (5) are used, each of these rules is applied to the nonterminals from $\{0, 1\}$ immediately preceding the first occurrence of \$ and immediately following the second occurrence of \$; otherwise, the derivation is unsuccessful. As a result, the only rule that removes the substring 11 preceding the first occurrence of \$ is (5.4). However, by Claim 1, (5.4) is used during the very last derivation step, so the substring 11 has to appear at the beginning of the sentential form in order to generate a string over T. □

Claim 4. The derivation

$$S \Rightarrow_G^* v \; [\Xi]$$

from Claim 1 can be expressed, in a greater detail, as

$$\begin{aligned} S &\Rightarrow_G 11S11 \; [p_1] \\ &\Rightarrow_G^* v \end{aligned}$$

where p_1 is the rule introduced in (1), and this rule is not used during the rest of the derivation.

Proof. The rule introduced in (1) is the only rule that introduces the substring 11 in front of the first occurrence of \$. By Claim 3, in front of the first \$, this substring appears only at the beginning of every sentential form in $w \Rightarrow_G^+ z$, so p_1 has to be applied at the beginning of the derivation and cannot be used later in the derivation. □

Claim 5. The derivation

$$\begin{aligned} w &\Rightarrow_G^* y \; [\Psi] \\ &\Rightarrow_G z \; [p_{5d}] \end{aligned}$$

from Claim 1 can be expressed, in a greater detail, as

$$\begin{aligned} w &\Rightarrow_G^* x \; [\Psi_1] \\ &\Rightarrow_G^* y \; [\Psi_2] \\ &\Rightarrow_G z \; [p_{5d}] \end{aligned}$$

where

$$x \in \{1\}\{1\}(\{1\}\{0\} \cup \{0\}\{0\})^*\{\$\}T^*\{\$\}(\{0, 1\} \cup T)^*$$

Ψ_1 is a sequence of rules introduced in (5.1) and (5.2), and Ψ_2 is a sequence of rules introduced in (5.3).

Proof. The proof of this claim follows immediately from Claims 2 and 3. □

Claim 6. The derivation

$$11S11 \Rightarrow_G^* v$$

from Claims 1 and 4 can be expressed in a greater detail as

$$\begin{aligned} 11S11 &\Rightarrow_G^* u\ [\Xi_1] \\ &\Rightarrow_G^* v\ [\Xi_2] \end{aligned}$$

where

$$u \in \{1\}\{1\}(\{1\}\{0\} \cup \{0\}\{0\})^* S(\{0\}\{1\} \cup \{0\}\{0\})^*(\{1\}T\{1\})^*\{1\}\{1\}$$

and Ξ_1, Ξ_2 are sequences of rules introduced in (2), (3), respectively.

Proof. By Claim 4, every derivation starts by an application of the rule from (1). Therefore, u ends with 11. Next, notice that the two nonterminals 1 surrounding a, where $a \in T$, introduced by every application of a rule from (2) can only be removed by the rule from (5.3). Indeed, by Claim 2, any other rule leaves a nonterminal between the two symbols \$, so the derivation is unsuccessful. By Claim 5, rules from (5.1) and (5.2) cannot be applied after the rule from (5.3) is used. As a result, the generation of the strings over $\{0, 1\}$ by rules from (2) and (3) has to correspond to their removal by (5.1), (5.2), and (5.3). This implies that rules from (2) have to be applied before rules from (3). □

Based upon Claims 1 through 6, we see that every successful derivation is of this form

$$\begin{aligned} S &\Rightarrow_G 11S11 && [p_1] \\ &\Rightarrow_G^* u && [\Xi_1] \\ &\Rightarrow_G^* v && [\Xi_2] \\ &\Rightarrow_G w && [p_4] \\ &\Rightarrow_G^* x && [\Psi_1] \\ &\Rightarrow_G^* y && [\Psi_2] \\ &\Rightarrow_G z && [p_{5d}] \end{aligned}$$

As the rest of this proof can be made by analogy with the proof of Theorem 4.7.19, we leave it to the reader. □

Economical Transformations

The generation of languages is frequently performed in a specifically required way and based upon a prescribed set of grammatical components, such as a certain collection of nonterminals or rules. On the other hand, if these requirements are met, the generation can be based upon grammars of various types. For this purpose, we often make use of transformations that convert grammars of some type to equivalent grammars of another type so the transformed grammars strongly resemble the original grammars regarding the way they work as well as the components they consist of. In other words, we want the output grammars resulting from these transformations to work similarly to the way the given original grammars work and, perhaps even more importantly, to contain the same set of grammatical components possibly extended by very few additional components. Transformations that produce scattered context grammars in this economical way are discussed throughout the rest of this section. Because phrase-structure grammars represent one of the very basic grammatical models in formal language theory (see Sect. 3.3), this section pays a special attention to the economical transformations that convert these fundamental grammars to equivalent scattered context grammars.

To compare the measures of scattered context and phrase-structure grammars, we first define the degree of context-sensitivity of phrase-structure grammars analogously to the degree of context sensitivity of scattered context grammars (see Definition 4.7.12).

Definition 4.7.22. Let $G = (N, T, P, S)$ be a phrase-structure grammar. Its *degree of context sensitivity*, symbolically written as $\mathrm{dcs}(G)$, is defined as

$$\mathrm{dcs}(G) = \mathrm{card}\left(\{x \to y \mid x \to y \in P, |x| \geq 2\}\right) \qquad \square$$

Theorem 4.7.23. *For every phrase-structure grammar $G = (N, T, P, S)$ in the Kuroda normal form, there is a scattered context grammar $\bar{G} = (\bar{V}, T, \bar{P}, \bar{S})$ such that $L(\bar{G}) = L(G)$, and*

$$\begin{aligned}\mathrm{card}(\bar{V}) &= \mathrm{card}(V) + 5\\ \mathrm{card}(\bar{P}) &= \mathrm{card}(P) + 4\\ \mathrm{dcs}(\bar{G}) &= \mathrm{dcs}(G) + 2\end{aligned}$$

where $V = N \cup T$.

Proof. Let $G = (N, T, P, S)$ be a phrase-structure grammar in the Kuroda normal form. Without any loss of generality, assume that $V \cap \{\bar{S}, F, 0, 1, \$\} = \emptyset$, where $V = N \cup T$. Set $\bar{V} = V \cup \{\bar{S}, F, 0, 1, \$\}$. Define the scattered context grammar

$$\bar{G} = \left(\bar{V}, T, \bar{P}, \bar{S}\right)$$

where $\bar{P}$ is constructed as follows:

(1) Add $(\bar{S}) \rightarrow (FFFS)$ to $\bar{P}$;
(2) For each $AB \rightarrow CD \in P$, add $(A, B) \rightarrow (C0, 1D)$ to $\bar{P}$;
(3) For each $A \rightarrow BC \in P$, add $(A) \rightarrow (BC)$ to $\bar{P}$;
(4) For each $A \rightarrow a \in P$, where $a \in T \cup \{\varepsilon\}$, add $(A) \rightarrow (\$a)$ to $\bar{P}$;
(5) Add

(5.1) $(F, 0, 1, F, F) \rightarrow (\varepsilon, F, F, \varepsilon, F)$,
(5.2) $(F, F, F, \$) \rightarrow (\varepsilon, \varepsilon, F, FF)$,
(5.3) $(F) \rightarrow (\varepsilon)$ to $\bar{P}$.

The rule from (1) starts a derivation and introduces three occurrences of the nonterminal F, which are present in every sentential form until three applications of the rule from (5.3) complete the derivation. Rules from (2), (3), and (4) simulate the corresponding rules of the Kuroda normal form behind the last occurrence of F. The rules from (5.1) and (5.2) guarantee that before (5.3) is applied for the first time, every sentential form in a successful derivation belongs to

$$T^*\{F\}(T \cup \{\varepsilon\})\{0^i 1^i \mid i \geq 0\}\{F\}\{F\}(V \cup \{0, 1, \$\})^*$$

and, thereby, the simulation of every derivation of G is performed properly. Notice that there are only terminals in front of the first nonterminal F. Moreover, the only nonterminals appearing between the first occurrence and the second occurrence of F are from $\{0, 1\}$, and there is no symbol between the second and the third occurrence of F in a successful derivation.

Next, we establish several claims to demonstrate that $L(G) = L(\bar{G})$ in a rigorous way.

Claim 1. Every successful derivation of $\bar{G}$ can be expressed as

$$\begin{aligned} \bar{S} &\Rightarrow_{\bar{G}} FFFS && [p_1] \\ &\Rightarrow_{\bar{G}}^* uFvFxFy && [\Psi] \\ &\Rightarrow_{\bar{G}}^* w \\ &\Rightarrow_{\bar{G}}^3 z && [p_{5c}p_{5c}p_{5c}] \end{aligned}$$

where $u, z \in T^*$, $v, x, y \in (\bar{V} - \{\bar{S}, F\})^*$, $w \in (\bar{V} - \{\bar{S}\})^*$, p_1 and p_{5c} are rules introduced in (1) and (5.3), respectively, and Ψ is a sequence of rules introduced in (2) through (5.2).

Proof. The only rule with $\bar{S}$ on its left-hand side is the rule introduced in (1), and because no rule contains $\bar{S}$ on its right-hand side, this rule is not used during the rest of the derivation process. As a result,

$$\bar{S} \Rightarrow_{\bar{G}} FFFS \ [p_1]$$

Observe that no rule from (2) through (4) contains the nonterminal F and rules from (5.1) and (5.2) contain three nonterminals F on their left-hand sides as well as their right-hand sides. The rule from (5.3), which is the only rule with its right-hand side over T, removes F from the sentential form, so no rule from (5.1) and (5.2) can be used once it is applied. Notice that rules from (4) simulate $A \to a$, where $A \in V - T$, $a \in T \cup \{\varepsilon\}$, and these rules introduce \$ to the sentential form. In addition, observe that only the rule from (5.2) rewrites \$. Consequently, to generate a string over T, rules from (2) through (4) cannot be used after the rule from (5.3) is applied. Therefore,

$$w \Rightarrow_{\bar{G}}^{3} z\ [p_{5c}\, p_{5c}\, p_{5c}]$$

Notice that rules from (5.1) and (5.2) cannot rewrite any symbol in u. If $\mathrm{alph}(u) \cap (\bar{V} - T) \neq \emptyset$, then a nonterminal from $\{0, 1, \$\}$ remains in front of the first F because rules from (2) through (4) cannot rewrite u to a string over T, so the derivation would be unsuccessful in this case. Therefore, $u \in T^*$, and the claim holds. □

Claim 2. Let

$$\bar{S} \Rightarrow_{\bar{G}}^{+} uFvFxFy \Rightarrow_{\bar{G}}^{*} w \Rightarrow_{\bar{G}}^{3} z$$

where $u, z \in T^*$, $v, x, y \in \left(\bar{V} - \{\bar{S}, F\}\right)^*$, *and* $w \in \left(\bar{V} - \{\bar{S}\}\right)^*$. *Then,* $x \in T^*$.

Proof. First, notice that if $(\bar{V} - T) \cap \mathrm{alph}(x) \neq \emptyset$, x cannot be rewritten to a string over T by using only rules from (2) through (4). Next, examine the rules from (5.1) and (5.2) to see that these rules cannot rewrite any symbol from x, and the rule from (5.2) moves x in front of the first occurrence of F. However, by Claim 1, no nonterminal can appear in front of the first F. As a result, $(\bar{V} - T) \cap \mathrm{alph}(x) = \emptyset$, so $x \in T^*$. □

Claim 3. Let

$$\bar{S} \Rightarrow_{\bar{G}}^{+} uFvFxFy \Rightarrow_{\bar{G}}^{*} w \Rightarrow_{\bar{G}}^{3} z$$

where $u, z \in T^*$, $v, x, y \in \left(\bar{V} - \{\bar{S}, F\}\right)^*$, *and* $w \in \left(\bar{V} - \{\bar{S}\}\right)^*$. *Then,* $v = v'v''$, *where* $v' \in \left(\{0\} \cup T\right)^*$, $v'' \in \left(\{1\} \cup T\right)^*$, *and* $\mathrm{occur}(v', \{0\}) = \mathrm{occur}(v'', \{1\})$.

Proof. First, notice that if $(\bar{V} - T) \cap \mathrm{alph}(v) \neq \emptyset$, v cannot be rewritten to a string over T by using only rules from (2) through (4). Next, examine the rules from (5.1) and (5.2).

First, observe that the rule from (5.2) can only be applied if $v \in T^*$. Indeed, (5.2) moves v in front of the first F, and if $(\bar{V} - T) \cap \mathrm{alph}(v) \neq \emptyset$, then Claim 1 implies that the derivation is unsuccessful. Therefore, $(\bar{V} - T) \cap \mathrm{alph}(v) = \emptyset$ before the rule from (5.2) is applied. Second, observe that because the rule from (5.1) rewrites only nonterminals over $\{0, 1\}$ in v, $\left((V - T) \cup \{\$\}\right) \cap \mathrm{alph}(v) = \emptyset$. Finally, observe that the rule from (5.1) has to be applied so that the first 0 following the first F and the first 1 preceding the second F is rewritten by (5.1). If this property is not satisfied, the form of (5.1) implies that 0 appears in front of the first F or 1 appears in between the second F and the third F. However, by Claims 1 and 2, this results into an unsuccessful derivation.

Based on these observations, we see that in order to generate $z \in T^*$, v has to satisfy $v = v'v''$, where $v' \in (\{0\} \cup T)^*$, $v'' \in (\{1\} \cup T)^*$, and occur(v', $\{0\}$) = occur(v'', $\{1\}$). □

Claim 4. Let

$$\bar{S} \Rightarrow_{\bar{G}}^{+} uFvFxFy \Rightarrow_{\bar{G}}^{*} w \Rightarrow_{\bar{G}}^{3} z$$

where $u, z \in T^*$, $v, x, y \in (\bar{V} - \{\bar{S}, F\})^*$, *and* $w \in (\bar{V} - \{\bar{S}\})^*$. *Then,*

$$y \in (T \cup \{\varepsilon\})(\{0^i 1^i \mid i \geq 0\}K)^*$$

with $K = (V - T) \cup \{\$\}(T \cup \{\varepsilon\})$, $v \in (T \cup \{\varepsilon\})\{0^i 1^i \mid i \geq 0\}$, *and* $x = \varepsilon$.

Proof. First, consider the rule introduced in (5.2). This rule rewrites \$ to FF in its last component. Because the nonterminal \$ is introduced by rules from (4), \$ may be followed by $a \in T$. Therefore, after (5.2) is applied, the last nonterminal F may be followed by a. As a result, the prefix of y is always over $T \cup \{\varepsilon\}$.

Second, notice that when the rule (5.2) is used, the first nonterminal \$ following the third nonterminal F has to be rewritten. In addition, the substring appearing between these symbols has to be in $\{0^i 1^i \mid i \geq 0\}$. The form of the rule introduced in (5.2) implies that after its application, this substring is moved in between the first occurrence of F and the second occurrence of F, so the conditions given by Claim 3 are satisfied. Therefore,

$$v \in (T \cup \{\varepsilon\})\{0^i 1^i \mid i \geq 0\}$$

and because no terminal appears in the suffix of v, the proof of Claim 3 implies that $x = \varepsilon$. By induction, prove that

$$y \in (T \cup \{\varepsilon\})(\{0^i 1^i \mid i \geq 0\}K)^*$$

The induction part is left to the reader. □

Next, we define the homomorphism α from $\bar{V}^*$ to V^* as $\alpha(\bar{A}) = \varepsilon$, for all $\bar{A} \in \bar{V} - V$, and $\alpha(A) = A$, for all $A \in V$, and use this homomorphism in the following claims.

Claim 5. Let $\bar{S} \Rightarrow_{\bar{G}}^{m} w \Rightarrow_{\bar{G}}^{*} z$, *where* $m \geq 1$, $z \in T^*$, *and* $w \in \bar{V}^*$. *Then,* $S \Rightarrow_{G}^{*} \alpha(w)$.

Proof. This claim is established by induction on $m \geq 1$.

Basis. Let $m = 1$. Then, $\bar{S} \Rightarrow_{\bar{G}} FFFS$. Because $\alpha(FFFS) = S$, $S \Rightarrow_{G}^{0} S$, so the basis holds.

Induction Hypothesis. Suppose that the claim holds for every $m \leq j$, for some $j \geq 1$.

Induction Step. Let $\bar{S} \Rightarrow_{\bar{G}}^{j+1} w \Rightarrow_{\bar{G}}^{*} z$, where $z \in T^*$ and $w \in \bar{V}^*$. Based on Claims 1 and 4, express this derivation as

$$\begin{aligned}\bar{S} &\Rightarrow_{\bar{G}}^{j} uFvFxFy \\ &\Rightarrow_{\bar{G}} w\ [p] \\ &\Rightarrow_{\bar{G}}^{*} z\end{aligned}$$

where $u \in T^*$, $x = \varepsilon$,

$$y \in (T \cup \{\varepsilon\})\big(\{0^i 1^i \mid i \geq 0\}K\big)^*$$

with $K = (V - T) \cup \{\$\}(T \cup \{\varepsilon\})$, and

$$v \in (T \cup \{\varepsilon\})\{0^i 1^i \mid i \geq 0\}$$

By the induction hypothesis, $S \Rightarrow_G^* \alpha(uFvFxFy)$. Next, this proof considers all possible forms of p.

- Assume that $p = (A, B) \rightarrow (C0, 1D) \in \bar{P}$, where $A, B, C, D \in V - T$. Claim 4 and its proof imply $y = y'Ay''By'''$, where $y'' \in \{0^i 1^i \mid i \geq 0\}$, and

  $$w = uFvFxFy'C0y''1Dy'''$$

  As $(A, B) \rightarrow (C0, 1D) \in \bar{P}$, $AB \rightarrow CD \in P$ holds true. Because $\alpha(y'') = \varepsilon$,

  $$\alpha(uFvFxFy'Ay''By''') \Rightarrow_G \alpha(uFvFxFy'C0y''1Dy''')$$

  Therefore, $S \Rightarrow_G^* \alpha(w)$.
- Assume that $p = (A) \rightarrow (BC) \in \bar{P}$, where $A, B, C \in V - T$. Claim 4 implies that $y = y'Ay''$, and

  $$w = uFvFxFy'BCy''$$

  As $(A) \rightarrow (BC) \in \bar{P}$, $A \rightarrow BC \in P$ holds true. Notice that

  $$\alpha(uFvFxFy'Ay'') \Rightarrow_G \alpha(uFvFxFy'BCy'')$$

  Therefore, $S \Rightarrow_G^* \alpha(w)$.
- Assume that $p = (A) \rightarrow (\$a) \in \bar{P}$, where $A \in V - T$ and $a \in T \cup \{\varepsilon\}$. Claim 4 implies that $y = y'Ay''$, and

  $$w = uFvFxFy'\$ay''$$

As $(A) \rightarrow (\$a) \in \bar{P}$, $A \rightarrow a \in P$ holds true. Notice that

$$\alpha(uFvFxFy'Ay'') \Rightarrow_G \alpha(uFvFxFy'\$ay'')$$

Therefore, $S \Rightarrow_G^* \alpha(w)$.

- Assume that p is a rule from (5). Notice that these rules rewrite only nonterminals over $\{0, 1, F, \$\}$. Therefore, $\alpha(w) = \alpha(uFvFxFy)$, so $S \Rightarrow_G^* \alpha(w)$.

Based on the arguments above, $\bar{S} \Rightarrow_{\bar{G}}^j uFvFxFy \Rightarrow_{\bar{G}} w\ [p]$, for any $p \in \bar{P}$, implies that $S \Rightarrow_G^* \alpha(w)$. Thus, the claim holds. □

Claim 6. $L(\bar{G}) \subseteq L(G)$

Proof. By Claim 5, if $\bar{S} \Rightarrow_{\bar{G}}^+ z$ with $z \in T^*$, then $S \Rightarrow_G^* z$. Therefore, the claim holds. □

Claim 7. Let $S \Rightarrow_G^m w \Rightarrow_G^* z$, *where* $m \geq 0$, $w \in V^*$, *and* $z \in T^*$. *Then,* $\bar{S} \Rightarrow_{\bar{G}}^+ uFvFxFy$, *where* $u \in T^*$,

$$y \in \big(T \cup \{\varepsilon\}\big)\big(\{0^i 1^i \mid i \geq 0\}K\big)^*$$

with $K = (V - T) \cup \{\$\}\big(T \cup \{\varepsilon\}\big)$,

$$v \in \big(T \cup \{\varepsilon\}\big)\{0^i 1^i \mid i \geq 0\}$$

and $x = \varepsilon$, *so that* $w = \alpha(uFvFxFy)$.

Proof. This claim is established by induction on $m \geq 0$.

Basis. Let $m = 0$. Then, $S \Rightarrow_G^0 S \Rightarrow_G^* z$. Notice that $\bar{S} \Rightarrow_{\bar{G}} FFFS$ by using the rule introduced in (1), and $S = \alpha(FFFS)$. Thus, the basis holds.

Induction Hypothesis. Suppose that the claim holds for every $m \leq j$, where $j \geq 1$.

Induction Step. Let $S \Rightarrow_G^{j+1} w \Rightarrow_G^* z$, where $w \in V^*$, and $z \in T^*$. Express this derivation as

$$\begin{aligned} S &\Rightarrow_G^j t \\ &\Rightarrow_G w\ [p] \\ &\Rightarrow_G^* z \end{aligned}$$

where $w \in V^*$ and $p \in P$. By the induction hypothesis, $\bar{S} \Rightarrow_{\bar{G}}^+ uFvFxFy$, where $u \in T^*$,

$$y \in \big(T \cup \{\varepsilon\}\big)\big(\{0^i 1^i \mid i \geq 0\}K\big)^*$$

with $K = (V - T) \cup \{\$\}\big(T \cup \{\varepsilon\}\big)$,

$$v \in \big(T \cup \{\varepsilon\}\big)\{0^i 1^i \mid i \geq 0\}$$

and $x = \varepsilon$ so that $t = \alpha(uFvFxFy)$. Next, this proof considers all possible forms of p:

- Assume that $p = AB \to CD \in P$, where $A, B, C, D \in V - T$. Express $t \Rightarrow_G w$ as $t'ABt'' \Rightarrow_G t'CDt''$, where $t'ABt'' = t$ and $t'CDt'' = w$. Claim 4 implies that $y = y'A0^k1^kBy''$, where $k \geq 0$, $\alpha(uFvFxFy') = t'$, and $\alpha(y'') = t''$. As $AB \to CD \in P$, $(A, B) \to (C0, 1D) \in \bar{P}$ holds true. Then,

$$uFvFxFy'A0^k1^kBy'' \Rightarrow_{\bar{G}} uFvFxFy'C0^{k+1}1^{k+1}Dy''$$

  Therefore,

$$\bar{S} \Rightarrow_{\bar{G}}^{+} uFvFxFy'C0^{k+1}1^{k+1}Dy''$$

  and $w = \alpha(uFvFxFy'C0^{k+1}1^{k+1}Dy'')$.
- Assume that $p = A \to BC \in P$, where $A, B, C \in V - T$. Express $t \Rightarrow_G w$ as $t'At'' \Rightarrow_G t'BCt''$, where $t'At'' = t$ and $t'BCt'' = w$. Claim 4 implies that $y = y'Ay''$, where $\alpha(uFvFxFy') = t'$ and $\alpha(y'') = t''$. As $A \to BC \in P$, $(A) \to (BC) \in \bar{P}$ holds true. Then,

$$uFvFxFy'Ay'' \Rightarrow_{\bar{G}} uFvFxFy'BCy''$$

  Therefore,

$$\bar{S} \Rightarrow_{\bar{G}}^{+} uFvFxFy'BCy''$$

  and $w = \alpha(uFvFxFy'BCy'')$.
- Assume that $p = A \to a \in P$, where $A \in V - T$ and $a \in T \cup \{\varepsilon\}$. Express $t \Rightarrow_G w$ as $t'At'' \Rightarrow_G t'at''$, where $t'At'' = t$ and $t'at'' = w$. Claim 4 implies that $y = y'Ay''$, where $\alpha(uFvFxFy') = t'$ and $\alpha(y'') = t''$. As $A \to a \in P$, $(A) \to (\$a) \in \bar{P}$ holds true. Then,

$$uFvFxFy'Ay'' \Rightarrow_{\bar{G}} uFvFxFy'\$ay''$$

  Therefore,

$$\bar{S} \Rightarrow_{\bar{G}}^{+} uFvFxFy'\$ay''$$

  and $w = \alpha(uFvFxFy'\$ay'')$.

Consider the arguments above to see that $S \Rightarrow_G^j t \Rightarrow_G w\ [p]$, for any $p \in P$, implies that $\bar{S} \Rightarrow_{\bar{G}}^{+} s$, where $w = \alpha(s)$. Thus, the claim holds. □

Claim 8. $L(G) \subseteq L(\bar{G})$

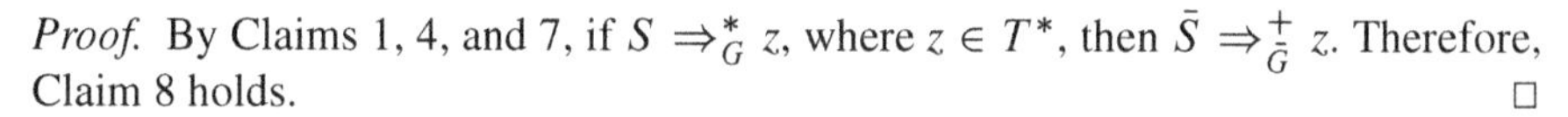

Proof. By Claims 1, 4, and 7, if $S \Rightarrow_G^* z$, where $z \in T^*$, then $\bar{S} \Rightarrow_{\bar{G}}^+ z$. Therefore, Claim 8 holds. □

By Claims 6 and 8, $L(\bar{G}) = L(G)$. Observe that $\text{card}(\bar{V}) = \text{card}(V) + 5$, $\text{card}(\bar{P}) = \text{card}(P) + 4$, and $\text{dcs}(\bar{G}) = \text{dcs}(G) + 2$. Thus, the theorem holds. □

In the conclusion of this section, we point out several open problem areas.

Open Problem 4.7.24. By Theorem 4.7.19, scattered context grammars with two context-sensitive rules characterize **RE**. What is the generative power of scattered context grammars with one context-sensitive rule? □

Open Problem 4.7.25. Revert the transformation under discussion and study economical transformations of scattered context grammars to phrase-structure grammars. □

Open Problem 4.7.26. From a much broader perspective, apart from the transformations between scattered context grammars and phrase-structure grammars, study economical transformations between other types of grammars. □

4.7.5 LL Scattered Context Grammars

In this section, we introduce, illustrate and study propagating liner-rule LL(k) versions of scattered context grammars. We demonstrate an infinite hierarchy of language families resulting from them.

We modify the notion of a scattered context grammar in the following way.

(1) Every compound rule can be composed only of *linear rules*—that is, every x_i can contain at most one nonterminal. The only exception is the rule of the form $(S) \rightarrow (x)$, where S is the start symbol and x is a nonempty string of at most k nonterminals, where k is a positive integer for every grammar. Furthermore, S cannot appear on the right-hand side of any rule, hence the term k-*linear* (see p. 72 in [21]; also, cf. Sect. 5.5 of [14]).
(2) If we take the first components of all rules, e.g. from $(A, B, C) \rightarrow (a, b, c)$, we obtain $A \rightarrow a$, the resulting context-free grammar has to be an *LL grammar* (see, for instance, [13]; also, cf. [6]), hence the term *LL*.
(3) Every rule has to be applied in a *leftmost* way, e.g. $ABCABC$ can be rewritten by $(A, B, C) \rightarrow (a, b, c)$ only to $abcABC$, and $AABBCC$ cannot be rewritten by this rule at all. Note that our meaning of a leftmost application of a rule differs from the one used in Sect. 5.2 of [14] or in [11].

Before defining LL k-linear scattered context grammars, we introduce two auxiliary notions.

Definition 4.7.27. Let $G = (N, T, P, S)$ be a context-free grammar. For every $x \in (N \cup T)^*$, set

$$\text{first}(G, x) = \{a \in T \mid x \Rightarrow_G^* ay \text{ in } G, y \in (N \cup T)^*\}$$ □

Definition 4.7.28. Let $G = (N, T, P, S)$ be a scattered context grammar. The *first-component core grammar underlying* G is denoted by fcore(G) and defined as the context-free grammar

$$\text{fcore}(G) = \big(N, T, P', S\big)$$

with

$$P' = \big\{A_1 \to x_1 \mid (A_1, A_2, \ldots, A_m) \to (x_1, x_2, \ldots, x_m) \in P\big\} \qquad \square$$

Next, we define an LL k-linear scattered context grammar and illustrate the definition by examples.

Definition 4.7.29. Let $G = (N, T, P, S)$ be a scattered context grammar and k be a positive integer. G is referred to as an *LL k-linear scattered context grammar* if it satisfies the following three conditions.

(1) *k-Linear condition.* Every rule in P is either of the form

$$(S) \to (x)$$

where $x \in (N - \{S\})^+$, $|x| \leq k$, or

$$(A_1, A_2, \ldots, A_m) \to (x_1, x_2, \ldots, x_m)$$

where $A_i \in (N - \{S\})$, $x_i \in T^*(N - \{S\})T^* \cup T^+$, for all i, $1 \leq i \leq m$, for some $m \geq 1$.

(2) *LL condition, part 1.* The relation $\Rightarrow_G$ is defined as follows:

$$u \Rightarrow_G v$$

if and only if

$$u = u_1 A_1 u_2 A_2 \cdots u_m A_m y$$

$$v = u_1 x_1 u_2 x_2 \cdots u_m x_m y$$

and

$$(A_1, A_2, \ldots, A_m) \to (x_1, x_2, \ldots, x_m) \in P$$

where $y \in V^*$, $u_i \in T^*$, for all i, $1 \leq i \leq m$. $\square$

(3) *LL condition, part 2.* For every two distinct rules

$$(A_1, A_2, \ldots, A_m) \to (x_1, x_2, \ldots, x_m) \in P$$

and

$$(B_1, B_2, \ldots, B_n) \to (y_1, y_2, \ldots, y_n) \in P$$

satisfying $A_1 = B_1$, it holds that

$$\text{first}\left(\text{fcore}(G), x_1\right) \cap \text{first}\left(\text{fcore}(G), y_1\right) = \emptyset$$ □

Example 4.7.30. Consider the LL 2-linear scattered context grammar

$$G = \left(\{S, A, B\}, \{a, b, c, d\}, P, S\right)$$

where

$$P = \{ (S) \to (AB),$$
$$(A, B) \to (aAb, Bc),$$
$$(A, B) \to (d, d) \}$$

Observe that $|AB| = 2$, S does not appear on the right-hand side of any rule, and

$$\text{first}\left(\text{fcore}(G), aAb\right) \cap \text{first}\left(\text{fcore}(G), d\right) = \emptyset$$

Clearly,

$$L(G) = \left\{a^n db^n dc^n \mid n \geq 0\right\}$$

which is a non-context free language. For example, $aadbbdcc$ is generated by

$$S \Rightarrow_G aAbBc \Rightarrow_G aaAbbBcc \Rightarrow_G aadbbdcc$$ □

Example 4.7.31. Consider the LL 2-linear scattered context grammar

$$I = \left(\{S, X\}, \{0, 1, \#\}, P, S\right)$$

where

$$P = \{ (S) \to (XX),$$
$$(X, X) \to (0X, 0X),$$
$$(X, X) \to (1X, 1X),$$
$$(X, X) \to (\#, \#) \}$$

Clearly,

$$L(I) = \left\{w\#w\# \mid w \in \{0, 1\}^*\right\}$$

which is also a non-context free language. □

Example 4.7.32. Consider the scattered context grammar

$$J = \big(\{S, A\}, \{a\}, P, S\big)$$

where

$$\begin{aligned} P = \{ \ &(S) \rightarrow (A), \\ &(A) \rightarrow (aA), \\ &(A) \rightarrow (a) \ \} \end{aligned}$$

Observe that J is not an LL k-linear scattered context grammar, for any $k \geq 0$, because

$$\text{first}\big(\text{fcore}(J), Aa\big) \cap \text{first}\big(\text{fcore}(J), a\big) = \{a\}$$ □

Next, we discusses the generative power of LL k-linear scattered context grammars. We show that even though they can generate some non-context-free languages, they cannot generate some very basic languages. Furthermore, we prove that they form an infinite hierarchy of language families based on $k \geq 1$.

Theorem 4.7.33. *The finite language $K = \{a, aa\}$ cannot be generated by any LL k-linear scattered context grammar, for any $k \geq 1$.*

Proof. By contradiction. Assume that the LL k-linear scattered context grammar

$$G = \big(N, \{a\}, P, S\big)$$

generates K, for some $k \geq 1$. Without any loss of generality, we assume that P contains only rules that can be used in some derivation of a or aa. As there is only one terminal symbol, there can be only one rule of the form $(S) \rightarrow (x)$. Furthermore, $|x| = 1$ because derivations are non-decreasing, so $k = 1$. As there is only one terminal symbol, there can be only one rule for each $A \in N$ to satisfy the LL condition from Definition 4.7.29. Thus, G can generate only one string, which is a contradiction. Hence, the theorem holds. □

Lemma 4.7.34. *The language $L_k = \{x^k \mid x = a^m b c^m, m \geq 0\}$ cannot be generated by any LL $(k-1)$-linear scattered context grammar, for all $k \geq 2$.*

Proof. We only give a basic idea of this proof. Since strings from L_k contain k concatenations of some $a^m b c^m$, $m \geq 0$, and only linear rules can be used, we cannot generate this language with any LL $(k-1)$-linear scattered context grammar. A rigorous proof of this lemma is left to the reader. □

Define

$$\textbf{LL - }_k\textbf{SCAT} = \big\{L(G) \mid G \text{ is an LL } k\text{-linear scattered context grammar}\big\}$$

The following theorem says that LL k-linear scattered context grammars are more powerful than LL $(k-1)$-linear scattered context grammars.

Theorem 4.7.35. $\mathbf{LL}\text{-}_{k-1}\mathbf{SCAT} \subset \mathbf{LL}\text{-}_{k}\mathbf{SCAT}$, *for all* $k \geq 2$.

Proof. Clearly, $\mathbf{LL}\text{-}_{k-1}\mathbf{SCAT} \subseteq \mathbf{LL}\text{-}_{k}\mathbf{SCAT}$, for all $k \geq 2$. The strictness of this inclusion follows from Lemma 4.7.34 and from the fact that L_k is generated by the LL k-linear scattered context grammar

$$G = \big(\{S, A_1, \dots, A_k\}, \{a, b, c\}, P, S\big)$$

with

$$\begin{aligned} P = \big\{ & (S) \to (A_1 \cdots A_k), \\ & (A_1, \dots, A_k) \to (aA_1c, \dots, aA_kc), \\ & (A_1, \dots, A_k) \to (b, \dots, b) \big\} \end{aligned}$$

Hence, the theorem holds. □

We close this section by present an open problem area.

Open Problem 4.7.36. Define an *LL k-linear scattered context grammar with erasing rules* by analogy with the definition of an LL k-linear scattered context grammar. Of course, the LL condition has to be changed accordingly by introducing other sets, like follow(A), for all $A \in N$; see, for instance, [13]. What is the generative power of LL k-linear scattered context grammar with erasing rules? □

References

1. Chomsky, N.: On certain formal properties of grammars. Inform. Control **2**, 137–167 (1959)
2. Csuhaj-Varjú, E., Vaszil, G.: Scattered context grammars generate any recursively enumerable language with two nonterminals. Inform. Process. Lett. **110**(20), 902–907 (2010)
3. Dassow, J., Păun, G.: Regulated Rewriting in Formal Language Theory. Springer, New York (1989)
4. Geffert, V.: Normal forms for phrase-structure grammars. Theor. Informat. Appl. **25**(5), 473–496 (1991)
5. Greibach, S.A.: A new normal-form theorem for context-free phrase structure grammars. J. ACM **12**(1), 42–52 (1965)
6. Jirák, O.: Table-driven parsing of scattered context grammar. In: Proceedings of the 16th Conference and Competition EEICT 2010, pp. 171–175. Brno University of Technology, Brno (2010)
7. Kuroda, S.Y.: Classes of languages and linear-bounded automata. Inform. Control **7**(2), 207–223 (1964)
8. Masopust, T.: An improvement of the descriptional complexity of grammars regulated by context conditions. In: 2nd Doctoral Workshop on Mathematical and Engineering Methods in Computer Science, pp. 105–112. Faculty of Information Technology BUT, Brno (2006)
9. Masopust, T.: Formal models: regulation and reduction. Ph.D. thesis, Faculty of Information Technology, Brno University of Technology (2007)

10. Masopust, T.: Scattered context grammars can generate the powers of 2. In: Proceedings of the 13th Conference and Competition EEICT 2007, vol. 4, pp. 401–404. Brno University of Technology, Brno (2007)
11. Masopust, T., Techet, J.: Leftmost derivations of propagating scattered context grammars: a new proof. Discrete Math. Theor. Comput. Sci. **10**(2), 39–46 (2008)
12. Meduna, A.: Terminating left-hand sides of scattered context grammars. Theor. Comput. Sci. **2000**(237), 424–427 (2000)
13. Meduna, A.: Elements of Compiler Design. Auerbach Publications, Boston (2007)
14. Meduna, A., Techet, J.: Scattered Context Grammars and Their Applications. WIT Press, Southampton (2010)
15. Okubo, F.: A note on the descriptional complexity of semi-conditional grammars. Inform. Process. Lett. **110**(1), 36–40 (2009)
16. Păun, G.: A variant of random context grammars: semi-conditional grammars. Theor. Comput. Sci. **41**(1), 1–17 (1985)
17. Penttonen, M.: One-sided and two-sided context in formal grammars. Inform. Control **25**(4), 371–392 (1974)
18. Penttonen, M.: ET0L-grammars and N-grammars. Inform. Process. Lett. **4**(1), 11–13 (1975)
19. Rozenberg, G., Salomaa, A.: Mathematical Theory of L Systems. Academic Press, Orlando (1980)
20. Rozenberg, G., Salomaa, A. (eds.): Handbook of Formal Languages, vol. 2. Linear Modeling: Background and Application. Springer, New York (1997)
21. Salomaa, A.: Formal Languages. Academic Press, London (1973)
22. Vaszil, G.: On the descriptional complexity of some rewriting mechanisms regulated by context conditions. Theor. Comput. Sci. **330**(2), 361–373 (2005)
23. Zetzsche, G.: On erasing productions in random context grammars. In: ICALP'10: Proceedings of the 37th International Colloquium on Automata, Languages and Programming, pp. 175–186. Springer, New York (2010)

Chapter 5
Rule-Based Grammatical Regulation

Abstract The present chapter explores grammatical regulation underlain by restrictions placed on the use of rules. Four types of regulated grammars of this kind are covered–namely, regular-controlled, matrix, programmed, and state grammars, each of which is dedicated one section of this chapter. Section 5.1 discusses *regular-control grammars* that control the use of rules by regular languages over rule labels. Section 5.2 studies *matrix grammars* as special cases of regular-control grammars whose control languages have the form of the iteration of finite languages. Section 5.3 deals with *programmed grammars* that regulate the use of their rules by relations over rule labels. Finally, Sect. 5.4 outlines *state grammars* that regulate the use of rules by states in a way that strongly resembles the finite-state control of finite automata.

Keywords Rule-based grammatical regulation • Regular-control grammars • Matrix grammars • Programmed grammars • State grammars • Generative power

While the previous chapter has discussed context-based grammatical regulation, the present chapter covers grammatical regulation based on restrictions placed on the use of rules. It presents four types of grammars whose regulation is underlain by restrictions of this kind. Namely, they include regular-controlled, matrix, programmed, and state grammars.

As their name indicates, *regular-control grammars* control the use of rules by regular languages over labels attached to the rules (Sect. 5.1). *Matrix grammars* can be seen as special cases of regular-control grammars whose regular control languages have always the form of the iteration of finite languages, whose members are called *matrices* (Sect. 5.2). *Programmed grammars* regulate the use of their rules by relations over rule labels (Sect. 5.3). In an intuitive way, these relations actually act as programs that prescribe how the grammars work, which explains their name. Finally, *state grammars* regulate the use of rules by states in a way that strongly resembles the finite-state control of finite automata (Sect. 5.4).

A. Meduna and P. Zemek, *Regulated Grammars and Automata*,
DOI 10.1007/978-1-4939-0369-6_5,

5.1 Regular-Controlled Grammars

In essence, a *regular-controlled grammar* H is a context-free grammar G extended by a regular *control language* Ξ defined over the set of rules of G. Thus, each control string in Ξ represents, in effect, a sequence of rules in G. A terminal string w is in the language generated by H if and only if Ξ contains a control string according to which G generates w.

The present section is divided into two subsections—Sects. 5.1.1 and 5.1.2. The former defines and illustrates regular-controlled grammars. The latter states their generative power.

5.1.1 Definitions and Examples

In this section, we define the notion of a regular-controlled grammar and illustrate it by examples. Before reading this definition, recall the notion of rule labels, formalized in Definition 3.3.3.

Definition 5.1.1. A *regular-controlled (context-free) grammar* (see [7]) is a pair

$$H = (G, \Xi)$$

where

- $G = (N, T, \Psi, P, S)$ is a context-free grammar, called *core grammar*;
- $\Xi \subseteq \Psi^*$ is a regular language, called *control language*.

The *language* of H, denoted by $L(H)$, is defined as

$$L(H) = \{w \in T^* \mid S \Rightarrow_G^* w\ [\alpha] \text{ with } \alpha \in \Xi\} \qquad \square$$

In other words, $L(H)$ in the above definition consists of all strings $w \in T^*$ such that there is a derivation in G,

$$S \Rightarrow_G w_1\ [r_1] \Rightarrow_G w_2\ [r_2] \Rightarrow_G \cdots \Rightarrow_G w_n\ [r_n]$$

where

$$w = w_n \text{ and } r_1 r_2 \cdots r_n \in \Xi \text{ for some } n \geq 1$$

In what follows, instead of $x \Rightarrow_G y$, we sometimes write $x \Rightarrow_H y$—that is, we use $\Rightarrow_G$ and $\Rightarrow_H$ interchangeably.

Note that if $\Xi = \Psi^*$, then there is no regulation, and thus $L(H) = L(G)$ in this case.

Example 5.1.2. Let $H = (G, \Xi)$ be a regular-controlled grammar, where

$$G = \big(\{S, A, B, C\}, \{a, b, c\}, \Psi, P, S\big)$$

is a context-free grammar with P consisting of the following seven rules

$$\begin{array}{lll} r_1: S \rightarrow ABC & r_2: A \rightarrow aA & r_5: A \rightarrow \varepsilon \\ & r_3: B \rightarrow bB & r_6: B \rightarrow \varepsilon \\ & r_4: C \rightarrow cC & r_7: C \rightarrow \varepsilon \end{array}$$

and $\Xi = \{r_1\}\{r_2r_3r_4\}^*\{r_5r_6r_7\}$.

First, r_1 has to be applied. Then, r_2, r_3, and r_4 can be consecutively applied any number of times. The derivation is finished by applying r_5, r_6, and r_7. As a result, this grammar generates the non-context-free language

$$L(H) = \big\{a^n b^n c^n \mid n \geq 0\big\}$$

For example, the sentence $aabbcc$ is obtained by the following derivation

$$\begin{array}{rll} S & \Rightarrow_H ABC & [r_1] \\ & \Rightarrow_H aABC & [r_2] \\ & \Rightarrow_H aAbBC & [r_3] \\ & \Rightarrow_H aAbBcC & [r_4] \\ & \Rightarrow_H aaAbBcC & [r_2] \\ & \Rightarrow_H aaAbbBcC & [r_3] \\ & \Rightarrow_H aaAbbBccC & [r_4] \\ & \Rightarrow_H aabbBccC & [r_5] \\ & \Rightarrow_H aabbccC & [r_6] \\ & \Rightarrow_H aabbcc & [r_7] \end{array}$$

As another example, the empty string is derived in this way

$$S \Rightarrow_H ABC\ [r_1] \Rightarrow_H BC\ [r5] \Rightarrow_H C\ [r6] \Rightarrow_H \varepsilon\ [r7]$$ □

Next, we introduce the concept of appearance checking. Informally, it allows us to skip the application of certain rules if they are not applicable to the current sentential form.

Definition 5.1.3. A *regular-controlled grammar with appearance checking* (see [7]) is a triple

$$H = \big(G, \Xi, W\big)$$

where

- G and Ξ are defined as in a regular-controlled grammar;
- $W \subseteq \Psi$ is the *appearance checking set*.

We say that $x \in V^+$ directly derives $y \in V^*$ in G in the *appearance checking mode* W by application of $r: A \to w \in P$, symbolically written as

$$x \Rightarrow_{(G,W)} y\ [r]$$

if either

$$x = x_1 A x_2 \text{ and } y = x_1 w x_2$$

or

$$A \notin \text{alph}(x), r \in W, \text{ and } x = y$$

Define $\Rightarrow^k_{(G,W)}$ for $k \geq 0$, $\Rightarrow^+_{(G,W)}$, and $\Rightarrow^*_{(G,W)}$ in the standard way. The *language* of H, denoted by $L(H)$, is defined as

$$L(H) = \{w \in T^* \mid S \Rightarrow^*_{(G,W)} w\ [\alpha] \text{ with } \alpha \in \Xi\}$$ □

According to Definition 5.1.1, in a regular-controlled grammar without appearance checking, once a control string has been started by G, all its rules have to be applied. G with an appearance checking set somewhat relaxes this necessity, however. Indeed, if the left-hand side of a rule is absent in the sentential form under scan and, simultaneously, this rule is in the appearance checking set, then G skips its application and moves on to the next rule in the control string.

Observe that the only difference between a regular-controlled grammar with and without appearance checking is the derivation mode ($\Rightarrow_{(G,W)}$ instead of $\Rightarrow_G$). Furthermore, note that when $W = \emptyset$, these two modes coincides, so any regular-controlled grammar represents a special case of a regular-controlled grammar with appearance checking.

Example 5.1.4 (from Chap. 3 of [9]). Let $H = (G, \Xi, W)$ be a regular-controlled grammar with appearance checking, where

$$G = (\{S, A, X\}, \{a\}, \Psi, P, S)$$

is a context-free grammar with P consisting of the following rules

$r_1: S \to AA$
$r_2: S \to X$
$r_3: A \to S$
$r_4: A \to X$
$r_5: S \to a$

and $\Xi = (\{r_1\}^*\{r_2\}\{r_3\}^*\{r_4\})^*\{r_5\}^*$, $W = \{r_2, r_4\}$.

Assume that we have the sentential form

$$S^{2^m}$$

for some $m \geq 0$, obtained by using a sequence of rules from $(\{r_1\}^*\{r_2\}\{r_3\}^*\{r_4\})^*$. This holds for the start symbol ($m = 0$). We can either repeat this sequence or finish the derivation by using r_5 until we have

$$a^{2^m}$$

In the former case, we might apply r_1 as many times as we wish. However, if we apply it only k many times, where $k < m$, then we have to use r_2, which blocks the derivation. Indeed, there is no rule with X on its left hand side. Thus, this rule guarantees that every S is eventually rewritten to AA. Notice that $r_2 \in W$. As a result, if no S occurs in the sentential form, we can skip it (it is not applicable), so we get

$$S^{2^m} \Rightarrow^*_{(G,W)} (AA)^{2^m} = A^{2^{m+1}}$$

Then, by analogy, we have to rewrite each A to S, so we get

$$A^{2^{m+1}} \Rightarrow^*_{(G,W)} S^{2^{m+1}}$$

which is of the same form as the sentential form from which we started the derivation. Therefore, this grammar generates the non-context-free language

$$L(H) = \left\{a^{2^n} \mid n \geq 0\right\}$$

For example, the sentence $aaaa$ is obtained by the following derivation

$$\begin{aligned}
S &\Rightarrow_{(G,W)} AA && [r_1] \\
&\Rightarrow_{(G,W)} AS && [r_3] \\
&\Rightarrow_{(G,W)} SS && [r_3] \\
&\Rightarrow_{(G,W)} AAS && [r_1] \\
&\Rightarrow_{(G,W)} AAAA && [r_1] \\
&\Rightarrow_{(G,W)} AASA && [r_3] \\
&\Rightarrow_{(G,W)} AASS && [r_3] \\
&\Rightarrow_{(G,W)} SASS && [r_3] \\
&\Rightarrow_{(G,W)} SSSS && [r_3] \\
&\Rightarrow_{(G,W)} SSSa && [r_5] \\
&\Rightarrow_{(G,W)} aSSa && [r_5] \\
&\Rightarrow_{(G,W)} aaSa && [r_5] \\
&\Rightarrow_{(G,W)} aaaa && [r_5]
\end{aligned}$$

As another example, a single a is generated by

$$S \Rightarrow_{(G,W)} a\ [r5]$$ □

We can disallow erasing rules in the underlying core grammar. This is formalized in the following definition.

Definition 5.1.5. Let $H = (G, \Xi)$ $(H = (G, \Xi, W))$ be a regular-controlled grammar (with appearance checking). If G is propagating, then H is said to be a *propagating regular-controlled grammar (with appearance checking).* □

By $\mathbf{rC}_{ac}$, $\mathbf{rC}_{ac}^{-\varepsilon}$, $\mathbf{rC}$, and $\mathbf{rC}^{-\varepsilon}$, we denote the families of languages generated by regular-controlled grammars with appearance checking, propagating regular-controlled grammars with appearance checking, regular-controlled grammars, and propagating regular-controlled grammars, respectively.

5.1.2 Generative Power

The present section concerns the generative power of regular-controlled grammars. More specifically, the next theorem summarizes the relations between the language families defined in the conclusion of the previous section.

Theorem 5.1.6 (see Theorem 1 in [7]).

(i) *All languages in* $\mathbf{rC}$ *over a unary alphabet are regular.*
(ii) $\mathbf{CF} \subset \mathbf{rC}^{-\varepsilon} \subset \mathbf{rC}_{ac}^{-\varepsilon} \subset \mathbf{CS}$
(iii) $\mathbf{CF} \subset \mathbf{rC}^{-\varepsilon} \subseteq \mathbf{rC} \subset \mathbf{rC}_{ac} = \mathbf{RE}$

Open Problem 5.1.7. Is $\mathbf{rC} - \mathbf{rC}^{-\varepsilon}$ empty? Put in other words, can any regular-controlled grammar be converted to an equivalent propagating regular-controlled grammar? □

5.2 Matrix Grammars

As already pointed out in the beginning of this chapter, in essence, any matrix grammar can be viewed as a special regular-control grammar with a control language that has the form of the iteration of a finite language. More precisely, a *matrix grammar* H is a context-free grammar G extended by a finite set of sequences of its rules, referred to as *matrices*. In essence, H makes a derivation so it selects a matrix, and after this selection, it applies all its rules one by one until it reaches the very last rule. Then, it either completes its derivation, or it makes another selection of a matrix and continues the derivation in the same way.

Structurally, the present section parallels the previous section. That is, this section is also divided into two subsections. The first subsection defines and illustrates matrix grammars rigorous. The other states their power.

5.2.1 *Definitions and Examples*

We open this section by giving the rigorous definition of matrix grammars. Then, we illustrate this definition by an example.

Definition 5.2.1. A *matrix grammar with appearance checking* (see [3]) is a triple

$$H = (G, M, W)$$

where

- $G = (N, T, \Psi, P, S)$ is a context-free grammar, called *core grammar*;
- $M \subseteq \Psi^+$ is a finite language whose elements are called *matrices*;
- $W \subseteq \Psi$ is the *appearance checking set*.

The *direct derivation relation*, symbolically denoted by $\Rightarrow_H$, is defined over V^* as follows: for $r_1 r_2 \cdots r_n \in M$, for some $n \geq 1$, and $x, y \in V^*$,

$$x \Rightarrow_H y$$

if and only if

$$x = x_0 \Rightarrow_{(G,W)} x_1 \; [r_1] \Rightarrow_{(G,W)} x_2 \; [r_2] \Rightarrow_{(G,W)} \cdots \Rightarrow_{(G,W)} x_n = y \; [r_n]$$

where $x_i \in V^*$, for all i, $1 \leq i \leq n - 1$, and the application of rules in the appearance checking mode is defined as in Definition 5.1.3.

Define $\Rightarrow_H^k$ for $k \geq 0$, $\Rightarrow_H^+$, and $\Rightarrow_H^*$ in the standard way. The *language* of H, denoted by $L(H)$, is defined as

$$L(H) = \{w \in T^* \mid S \Rightarrow_H^* w\}$$

□

Note that if $M = \Psi$, then there is no regulation, and thus $L(H) = L(G)$ in this case. Sometimes, especially in Chap. 13, for brevity, we use rules and rule labels interchangeably.

Definition 5.2.2. Let $H = (G, M, W)$ be a matrix grammar with appearance checking. If $W = \emptyset$, then we say that H is a *matrix grammar without appearance checking* or, simply, a *matrix grammar* and we just write $H = (G, M)$. □

Without appearance checking, once a matrix has been started, H has to apply all its rules. However, with an appearance checking set of the rules in the matrices, H may sometimes skip the application of a rule within a matrix. More precisely, if the left-hand side of a rule is absent in the current sentential form while the corresponding rule of the applied matrix occurs in the appearance checking set, then H moves on to the next rule in the matrix.

Example 5.2.3 (from [7]). Let $H = (G, M)$ be a matrix grammar, where $G = (N, T, \Psi, P, S)$ is a context-free grammar with $N = \{S, A, B\}$, $T = \{a, b\}$, P consists of the following rules

$$
\begin{array}{lll}
r_1 \colon S \to AB & r_4 \colon A \to bA & r_7 \colon B \to a \\
r_2 \colon A \to aA & r_5 \colon B \to bB & r_8 \colon A \to b \\
r_3 \colon B \to aB & r_6 \colon A \to a & r_9 \colon B \to b
\end{array}
$$

and $M = \{r_1, r_2r_3, r_4r_5, r_6r_7, r_8r_9\}$.

We start with the only applicable matrix r_1 and we get AB. Next, we can either

- Terminate the derivation by using the matrix r_6r_7 and obtain aa,
- Terminate the derivation by using the matrix r_8r_9 and obtain bb,
- Rewrite AB to $aAaB$ by using the matrix r_2r_3, or
- Rewrite AB to $bAbB$ by using the matrix r_4r_5.

If the derivation is not terminated, we can continue analogously. For example, the sentence $aabaab$ is obtained by the following derivation

$$
\begin{aligned}
S &\Rightarrow_H AB \\
&\Rightarrow_H aAaB \\
&\Rightarrow_H aaAaaB \\
&\Rightarrow_H aabaab
\end{aligned}
$$

Clearly, this grammar generates the non-context-free language

$$L(H) = \{ww \mid w \in \{a, b\}^+\}$$

□

As with regular-controlled grammars, we can disallow erasing rules in the underlying core grammar.

Definition 5.2.4. Let $H = (G, M, W)$ be a matrix grammar (with appearance checking). If G is propagating, then H is a *propagating matrix grammar (with appearance checking)*. □

The families of languages generated by matrix grammars with appearance checking, propagating matrix grammars with appearance checking, matrix grammars, and propagating matrix grammars are denoted by $\mathbf{M}_{ac}$, $\mathbf{M}_{ac}^{-\varepsilon}$, $\mathbf{M}$, and $\mathbf{M}^{-\varepsilon}$, respectively.

5.2.2 Generative Power

This section states the relations between the language families defined in the conclusion of the previous section.

Theorem 5.2.5 (see Theorem 2 in [7]).

(i) $\mathbf{M}_{ac} = \mathbf{rC}_{ac}$
(ii) $\mathbf{M}_{ac}^{-\varepsilon} = \mathbf{rC}_{ac}^{-\varepsilon}$
(iii) $\mathbf{M} = \mathbf{rC}$
(iv) $\mathbf{M}^{-\varepsilon} = \mathbf{rC}^{-\varepsilon}$

Notice that the relations between the language families generated by matrix grammars are analogical to the relations between language families generated by regular-controlled grammars (see Theorem 5.1.6).

5.3 Programmed Grammars

In essence, the regulation of a programmed grammar is based upon two binary relations, represented by two sets attached to the grammatical rules. More precisely, a *programmed grammar* G is a context-free grammar, in which two sets, σ_r and φ_r, are attached to each rule r, where σ_r and φ_r are subsets of the entire set of rules in G. G can apply r in the following two ways.

(1) If the left-hand side of r occurs in the sentential form under scan, G rewrites the left-hand side of r to its right-hand side, and during the next derivation step, it has to apply a rule from σ_r.
(2) If the left-hand side of r is absent in the sentential form under scan, then G skips the application of r, and during the next derivation step, it has to apply a rule from φ_r.

This section is divided into four subsections. Section 5.3.1 defines and illustrates programmed grammars. Section 5.3.2 describes their generative power. Section 5.3.3 establishes three normal forms of these grammars. Finally, Sect. 5.3.4 demonstrates an infinite hierarchy of language families resulting from programmed grammars with a limitation placed on their non-determinism.

5.3.1 Definitions and Examples

In this section, we define programmed grammars and illustrate them by an example.

Definition 5.3.1. A *programmed grammar with appearance checking* (see [3]) is a quintuple

$$G = (N, T, \Psi, P, S)$$

where

- N, T, Ψ, and S are defined as in a context-free grammar;
- $P \subseteq \Psi \times N \times (N \cup T)^* \times 2^\Psi \times 2^\Psi$ is a finite relation, called the set of *rules*, such that $\text{card}(\Psi) = \text{card}(P)$ and if $(r, A, x, \sigma_r, \varphi_r), (s, A, x, \sigma_s, \varphi_s) \in P$, then $(r, A, x, \sigma_r, \varphi_r) = (s, A, x, \sigma_s, \varphi_s)$.

Instead of $(r, A, x, \sigma_r, \varphi_r) \in P$, we write $(r: A \to x, \sigma_r, \varphi_r) \in P$. For $(r: A \to x, \sigma_r, \varphi_r) \in P$, A is referred to as the *left-hand side* of r, and x is referred to as the *right-hand side* of r.

Let $V = N \cup T$ be the *total alphabet*. The *direct derivation relation*, symbolically denoted by $\Rightarrow_G$, is defined over $V^* \times \Psi$ as follows: for $(x_1, r), (x_2, s) \in V^* \times \Psi$,

$$(x_1, r) \Rightarrow_G (x_2, s)$$

if and only if either

$$x_1 = yAz, x_2 = ywz, (r: A \to w, \sigma_r, \varphi_r) \in P, \text{ and } s \in \sigma_r$$

or

$$x_1 = x_2, (r: A \to w, \sigma_r, \varphi_r) \in P, A \notin \text{alph}(x_1), \text{ and } s \in \varphi_r$$

Let $(r: A \to w, \sigma_r, \varphi_r) \in P$. Then, σ_r and φ_r are called the *success field* of r and the *failure field* of r, respectively. Observe that due to our definition of the relation of a direct derivation, if $\sigma_r \cup \varphi_r = \emptyset$, then r is never applicable. Therefore, we assume that $\sigma_r \cup \varphi_r \neq \emptyset$, for all $(r: A \to w, \sigma_r, \varphi_r) \in P$. Define $\Rightarrow_G^k$ for $k \geq 0$, $\Rightarrow_G^*$, and $\Rightarrow_G^+$ in the standard way. Let $(S, r) \Rightarrow_G^* (w, s)$, where $r, s \in \Psi$ and $w \in V^*$. Then, (w, s) is called a *configuration*. The *language* of G is denoted by $L(G)$ and defined as

$$L(G) = \{w \in T^* \mid (S, r) \Rightarrow_G^* (w, s), \text{ for some } r, s \in \Psi\}$$

□

Definition 5.3.2. Let $G = (N, T, \Psi, P, S)$ be a programmed grammar with appearance checking. G is *propagating* if every $(r: A \to x, \sigma_r, \varphi_r) \in P$ satisfies that $|x| \geq 1$. Rules of the form $(r: A \to \varepsilon, \sigma_r, \varphi_r)$ are called *erasing rules*. If every $(r: A \to x, \sigma_r, \varphi_r) \in P$ satisfies that $\varphi_r = \emptyset$, then G is a *programmed grammar without appearance checking* or, simply, a *programmed grammar*. □

Example 5.3.3 (from [3]). Consider the programmed grammar with appearance checking

$$G = (\{S, A\}, \{a\}, \{r_1, r_2, r_3\}, P, S)$$

where P consists of the three rules

$(r_1: S \to AA, \{r_1\}, \{r_2, r_3\})$
$(r_2: A \to S, \{r_2\}, \{r_1\})$
$(r_3: A \to a, \{r_3\}, \emptyset)$

Since the success field of r_i is $\{r_i\}$, for each $i \in \{1, 2, 3\}$, the rules r_1, r_2, and r_3 have to be used as many times as possible. Therefore, starting from S^n, for some $n \geq 1$, the successful derivation has to pass to A^{2n} and then, by using r_2, to S^{2n}, or, by using r_3, to a^{2n}. A cycle like this, consisting of the repeated use of r_1 and r_2, doubles the number of symbols. In conclusion, we obtain the non-context-free language

$$L(G) = \left\{a^{2^n} \mid n \geq 1\right\}$$

For example, the sentence $aaaa$ is obtained by the following derivation

$$\begin{aligned}
(S, r_1) &\Rightarrow_G (AA, r_2) \\
&\Rightarrow_G (AS, r_2) \\
&\Rightarrow_G (SS, r_1) \\
&\Rightarrow_G (AAS, r_1) \\
&\Rightarrow_G (AAAA, r_2) \\
&\Rightarrow_G (AASA, r_2) \\
&\Rightarrow_G (AASS, r_2) \\
&\Rightarrow_G (SASS, r_2) \\
&\Rightarrow_G (SSSS, r_3) \\
&\Rightarrow_G (SSSa, r_3) \\
&\Rightarrow_G (aSSa, r_3) \\
&\Rightarrow_G (aaSa, r_3) \\
&\Rightarrow_G (aaaa, r_3)
\end{aligned}$$

Notice the similarity between G from this example and H from Example 5.1.4. □

By $\mathbf{P}_{ac}$, $\mathbf{P}_{ac}^{-\varepsilon}$, $\mathbf{P}$, and $\mathbf{P}^{-\varepsilon}$, we denote the families of languages generated by programmed grammars with appearance checking, propagating programmed grammars with appearance checking, programmed grammars, and propagating programmed grammars, respectively.

5.3.2 Generative Power

The next theorem states the power of programmed grammars.

Theorem 5.3.4 (see Theorem 5 in [7]).

(i) $\mathbf{P}_{ac} = \mathbf{M}_{ac}$
(ii) $\mathbf{P}_{ac}^{-\varepsilon} = \mathbf{M}_{ac}^{-\varepsilon}$
(iii) $\mathbf{P} = \mathbf{M}$
(iv) $\mathbf{P}^{-\varepsilon} = \mathbf{M}^{-\varepsilon}$

Observe that programmed grammars, matrix grammars, and regular-controlled grammars are equally powerful (see Theorems 5.2.5 and 5.3.4).

5.3.3 Normal Forms

In this section, we establish the following three normal forms of programmed grammars.

1. In the first form, each rule has a nonempty success field and a nonempty failure field.
2. In the second form, the set of rules is divided into two disjoint subsets, P_1 and P_2. In P_1, the rules have precisely one rule in its success field and precisely one rule in its failure field. In P_2, each rule has no rule in its success field and no more than two rules in its failure field.
3. Finally, in the third form, the right-hand side of each rule consists either of the empty string, a single terminal, a single nonterminal, or a two-nonterminal string.

Algorithm 5.3.5. *Conversion of a programmed grammar with appearance checking to an equivalent programmed grammar with appearance checking in which every rule contains nonempty success and failure fields.*

Input: A programmed grammar with appearance checking, $G = (N, T, \Psi, P, S)$.
Output: A programmed grammar with appearance checking, $G' = (N', T, \Psi', P', S)$, such that $L(G') = L(G)$ and $(r\colon A \to x, \sigma_r, \varphi_r) \in P'$ implies that $\mathrm{card}(\sigma_r) > 0$ and $\mathrm{card}(\varphi_r) > 0$.
Note: Without any loss of generality, we assume that $\$ \notin N \cup T \cup \Psi$.
Method: Initially, set

$$\begin{aligned} N' &= N \cup \{\$\} \\ \Psi' &= \Psi \cup \{\$\} \\ P &= \{(r\colon A \to x, \sigma_r, \varphi_r) \mid (r\colon A \to x, \sigma_r, \varphi_r) \in P, \mathrm{card}(\sigma_r) > 0, \\ &\qquad \mathrm{card}(\varphi_r) > 0\} \cup \{(\$\colon \$ \to \$, \{\$\}, \{\$\})\} \end{aligned}$$

Perform (1) and (2), given next.

(1) For each $(r\colon A \to x, \sigma_r, \emptyset) \in P$, add $(r\colon A \to x, \sigma_r, \{\$\})$ to P'.
(2) For each $(r\colon A \to x, \emptyset, \varphi_r) \in P$, add $(r\colon A \to x\$, \{\$\}, \varphi_r)$ to P'. □

Lemma 5.3.6. *Algorithm 5.3.5 is correct.*

Proof. Clearly, the algorithm always halts and all the rules in G' satisfy the required form. To see that $L(G') = L(G)$, we make the following observations. Rules from P satisfying the desired form are placed directly into P'. In (1), every $(r: A \rightarrow x, \sigma_r, \emptyset) \in P$ is replaced with $(r: A \rightarrow x, \sigma_r, \{\$\})$. Notice that the original rule cannot be applied in the appearance checking mode, while the new rule can. However, this does not affect the generated language in any way. Indeed, the new rule $(\$: \$ \rightarrow \$, \{\$\}, \{\$\})$ cannot be used to derive a string of terminals if the current sentential form was not already formed only by terminals.

In (2), every $(r: A \rightarrow x, \emptyset, \varphi_r) \in P$ is replaced with $(r: A \rightarrow x\$, \{\$\}, \varphi_r)$. Notice that the original rule has to be used only in the appearance checking mode. We allow the new rule to rewrite A, but this rewriting always introduces a new occurrence of \$. Therefore, if the new rule is not used in the appearance checking mode, then the generation of a terminal string is ruled out.

Based on these observations, we see that $L(G') = L(G)$. □

Theorem 5.3.7. *For every programmed grammar with appearance checking, $G = (N, T, \Psi, P, S)$, there exists a programmed grammar with appearance checking, $G' = (N', T, \Psi', P', S)$, such that $L(G') = L(G)$ and $(r: A \rightarrow x, \sigma_r, \varphi_r) \in P'$ implies that* $\mathrm{card}(\sigma_r) > 0$ *and* $\mathrm{card}(\varphi_r) > 0$. *Furthermore, if G is propagating, then so is G'.*

Proof. This theorem follows from Algorithm 5.3.5 and Lemma 5.3.6 (observe that Algorithm 5.3.5 does not introduce any erasing rules). □

Next, we establish a normal form of programmed grammars with appearance checking where every rule satisfies a certain form of its success and failure field. More precisely, a programmed grammar with appearance checking is in this normal form if its set of rules is divided into two disjoint subsets, P_1 and P_2. In P_1, the rules have precisely one rule in its success field and precisely one rule in its failure field. In P_2, each rule has no rule in its success field and no more than two rules in its failure field.

Algorithm 5.3.8. *Conversion of a programmed grammar with appearance checking to an equivalent programmed grammar with appearance checking in the above-described normal form.*

Input: A programmed grammar with appearance checking, $G = (N, T, \Psi, P, S)$, where $(r: A \rightarrow y, \sigma_r, \varphi_r) \in P$ implies that $\mathrm{card}(\sigma_r) > 0$ and $\mathrm{card}(\varphi_r) > 0$.

Output: A programmed grammar with appearance checking, $G' = (N', T, \Psi', P', S)$, such that $L(G') = L(G)$ and $P' = P'_1 \cup P'_2$, $P'_1 \cap P'_2 = \emptyset$, where

- $(r: A \rightarrow y, \sigma_r, \varphi_r) \in P'_1$ implies that $\mathrm{card}(\sigma_r) = \mathrm{card}(\varphi_r) = 1$;
- $(r: A \rightarrow y, \sigma_r, \varphi_r) \in P'_2$ implies that $\sigma_r = \emptyset$ and $1 \leq \mathrm{card}(\varphi_r) \leq 2$.

Note: Without any loss of generality, we assume that $\# \notin N$.

Method: Initially, set $N' = N \cup \{\#\}$, $\Psi' = \Psi$, and $P' = \emptyset$. Next, for each $(r: A \rightarrow y, \sigma_r, \varphi_r) \in P$, let $\sigma_r = \{s_1, s_2, \dots, s_h\}$, $\varphi_r = \{t_1, t_2, \dots, t_k\}$, and

(1) For each $p \in \sigma_r \cup \varphi_r$, add $\langle r, p\rangle$ to N' and to Ψ';
(2) Add $(r: A \rightarrow y, \{\langle r, s_1\rangle\}, \{\langle r, t_1\rangle\})$ to P';
(3) Extend $P'$by adding the following $h + k$ rules

$$
\begin{array}{l}
(\langle r, s_1\rangle: \# \rightarrow \langle r, s_1\rangle, \emptyset, \{s_1, \langle r, s_2\rangle\}) \\
(\langle r, s_2\rangle: \# \rightarrow \langle r, s_2\rangle, \emptyset, \{s_2, \langle r, s_3\rangle\}) \\
\quad\vdots \\
(\langle r, s_h\rangle: \# \rightarrow \langle r, s_h\rangle, \emptyset, \{s_h\}) \\
\\
(\langle r, t_1\rangle: \# \rightarrow \langle r, t_1\rangle, \emptyset, \{t_1, \langle r, t_2\rangle\}) \\
(\langle r, t_2\rangle: \# \rightarrow \langle r, t_2\rangle, \emptyset, \{t_2, \langle r, t_3\rangle\}) \\
\quad\vdots \\
(\langle r, t_k\rangle: \# \rightarrow \langle r, t_k\rangle, \emptyset, \{t_k\})
\end{array}
$$

□

Lemma 5.3.9. *Algorithm 5.3.8 is correct.*

Proof. Clearly, the algorithm always halts and rules of G' are in the required form. To see that $L(G') = L(G)$, we make the following observations. Recall that $(r: A \rightarrow y, \sigma_r, \varphi_r) \in P$ implies that $\mathrm{card}(\sigma_r) > 0$ and $\mathrm{card}(\varphi_r) > 0$ (see the form of the input of the algorithm).

A single rule, $(r: A \rightarrow y, \sigma_r, \varphi_r) \in P$, where $\sigma_r = \{s_1, s_2, \dots, s_h\}$ and $\varphi_r = \{t_1, t_2, \dots, t_k\}$, is simulated by G' as follows. First, $(r: A \rightarrow y, \{\langle r, s_1\rangle\}, \{\langle r, t_1\rangle\})$, introduced in (2), is used. It either rewrites A to y and passes to $\langle r, s_1\rangle$, or it passes directly to $\langle r, t_1\rangle$ if A does not appear in the current sentential form. Then, depending on whether A was rewritten, rules from (3) are used to pass to the next rule label. Observe that all of these rules can be applied only in the appearance checking mode. Since there is no rule with # on its right-hand side, # never appears in any sentential form. Therefore, rules introduced in (3) are always applicable in the appearance checking mode.

Based on these observations, we see that $L(G') = L(G)$. □

Theorem 5.3.10. *For any programmed grammar with appearance checking, $G = (N, T, \Psi, P, S)$, there exists a programmed grammar with appearance checking, $G' = (N', T, \Psi', P', S)$, such that $L(G') = L(G)$ and $P' = P'_1 \cup P'_2$, $P'_1 \cap P'_2 = \emptyset$, where*

- $(r: A \rightarrow y, \sigma_r, \varphi_r) \in P'_1$ *implies that* $\mathrm{card}(\sigma_r) = \mathrm{card}(\varphi_r) = 1$;
- $(r: A \rightarrow y, \sigma_r, \varphi_r) \in P'_2$ *implies that* $\sigma_r = \emptyset$ *and* $1 \leq \mathrm{card}(\varphi_r) \leq 2$.

Furthermore, if G is propagating, then so is G'.

Proof. Let $G = (N, T, \Psi, P, S)$ be a programmed grammar with appearance checking. If G does not satisfy the form required by Algorithm 5.3.8, we can convert it to such a form by Theorem 5.3.7. Therefore, this theorem follows from Algorithm 5.3.8 and Lemma 5.3.9 (observe that Algorithm 5.3.8 does not introduce any erasing rules). □

Definition 5.3.11. Let $G = (N, T, \Psi, P, S)$ be a programmed grammar with appearance checking. G is in the *binary normal form* if every $(r: A \to x, \sigma_r, \varphi_r) \in P$ satisfies that $x \in NN \cup N \cup T \cup \{\varepsilon\}$. □

The following algorithm converts any programmed grammar with appearance checking G to an equivalent programmed grammar with appearance checking G' in the binary normal form. To give an insight into this conversion, we first explain the fundamental idea underlying this algorithm. It is based on the well-known transformation of context-free grammars to the Chomsky normal form (see, for instance, page 348 in [8]). First, we create a new set of rules, where every occurrence of a terminal a is replaced with its barred version $\bar{a}$. These barred terminals are rewritten to terminals at the end of every successful derivation. From this new arisen set of rules, we directly use rules which are of the required form. The other rules—that is, the rules with more than two nonterminals on their right-hand sides—are simulated by a sequence of rules having the desired form.

Algorithm 5.3.12. *Conversion of a programmed grammar with appearance checking to an equivalent programmed grammar with appearance checking in the binary normal form.*

Input: A programmed grammar with appearance checking, $G = (N, T, \Psi, P, S)$.
Output: A programmed grammar with appearance checking in the binary normal form, $G' = (N', T, \Psi', P', S)$, such that $L(G') = L(G)$.
Note: Without any loss of generality, we assume that N, T, Ψ, and $\{\bar{a} \mid a \in T\}$ are pairwise disjoint.
Method: Let $V = N \cup T$. Set $\bar{T} = \{\bar{a} \mid a \in T\}$. Define the homomorphism τ from V^* to $(N \cup \bar{T})^*$ as $\tau(A) = A$, for all $A \in N$, and $\tau(a) = \bar{a}$, for all $a \in T$. Define the function δ from 2^{Ψ} to $2^{\Psi \cup \bar{T}}$ as $\delta(\emptyset) = \emptyset$ and $\delta(U) = U \cup \bar{T}$ if $U \neq \emptyset$. Set $\bar{P} = \{(r: A \to \tau(y), \sigma_r, \varphi_r) \mid (r: A \to y, \sigma_r, \varphi_r) \in P\}$. Initially, set

$$\begin{aligned} N' &= N \cup \bar{T} \\ \Psi' &= \Psi \cup \bar{T} \\ P' &= \{(r: A \to y, \delta(\sigma_r), \delta(\varphi_r)) \mid (r: A \to y, \sigma_r, \varphi_r) \in \bar{P}, |y| \leq 2\} \\ &\quad \cup \{(\bar{a}: \bar{a} \to a, \bar{T}, \emptyset) \mid a \in T\} \end{aligned}$$

To complete the construction, for every $(r: A \to Y_1 Y_2 \cdots Y_k, \sigma_r, \varphi_r) \in \bar{P}$, where $Y_i \in (N \cup \bar{T})$, for all i, $1 \leq i \leq k$, for some $k > 2$, extend P' by adding the following k rules

$$\begin{aligned} &(r: A \to Y_1 C_1, \{s_1\}, \delta(\varphi_r)) \\ &(s_1: C_1 \to Y_2 C_2, \{s_2\}, \emptyset) \\ &\quad \vdots \\ &(s_{k-1}: C_{k-1} \to Y_k, \delta(\sigma_r), \emptyset) \end{aligned}$$

where each s_j is a new unique label in Ψ' and each C_j is a new unique nonterminal in N', for all j, $1 \leq j \leq k-1$. □

Lemma 5.3.13. *Algorithm 5.3.12 is correct.*

Proof. Clearly, the algorithm always halts and G' is in the binary normal form. To establish $L(G') = L(G)$, we first prove two claims. The first claim shows how derivations of G are simulated by G'. The second claim demonstrates the converse—that is, it shows how derivations of G' are simulated by G. These two claims are used to prove that $L(G') = L(G)$ later in this proof.

Claim 1. Let $(S, p) \Rightarrow_G^m (x, r)$, *where* $p, r \in \Psi$, $x \in V^*$, *for some* $m \geq 0$. *Then,* $(S, p) \Rightarrow_{G'}^* (\tau(x), r')$, *where* $r' = r$ *if* $\mathrm{alph}(x) \cap N \neq \emptyset$ *and* r' *is any member of* $\bar{T}$ *if* $\mathrm{alph}(x) \cap N = \emptyset$.

Proof. This claim is established by induction on $m \geq 0$.

Basis. Let $m = 0$. Then, for $(S, p) \Rightarrow_G^0 (S, p)$, where $p \in \Psi$, there is $(S, p) \Rightarrow_{G'}^0 (S, p)$. Since $\mathrm{alph}(S) \cap N = \{S\}$, the basis holds.

Induction Hypothesis. Suppose that the claim holds for all derivations of length ℓ or less, where $0 \leq \ell \leq m$, for some $m \geq 0$.

Induction Step. Consider any derivation of the form

$$(S, p) \Rightarrow_G^{m+1} (w, q)$$

where $w \in V^*$, $p, q \in \Psi$. Since $m + 1 \geq 1$, this derivation can be expressed as

$$(S, p) \Rightarrow_G^m (x, r) \Rightarrow_G (w, q)$$

where $x \in V^*$, $r \in \Psi$. Next, we consider all possible forms of $(x, r) \Rightarrow_G (w, q)$, covered by the next two cases—(i) and (ii).

(i) Let $(r: A \rightarrow y, \sigma_r, \varphi_r) \in P$, where $|y| \leq 2$. From the initialization part of the algorithm, we see that $(r: A \rightarrow \tau(y), \delta(\sigma_r), \delta(\varphi_r)) \in P'$. There are two subcases, (i.a) and (i.b), depending on whether $A \in \mathrm{alph}(x)$ or not.

  (i.a) Assume that $A \notin \mathrm{alph}(x)$. Then, $(x, r) \Rightarrow_G (x, q)$, for some $q \in \varphi_r$. By the induction hypothesis,

$$(S, p) \Rightarrow_{G'}^* (\tau(x), r')$$

  where $r' = r$ if $\mathrm{alph}(x) \cap N \neq \emptyset$ and r' is any member of $\bar{T}$ if $\mathrm{alph}(x) \cap N = \emptyset$. Since $A \notin \mathrm{alph}(x)$, $A \notin \mathrm{alph}(\tau(x))$. If $\mathrm{alph}(x) \cap N = \emptyset$, then

$$(x, r') \Rightarrow_{G'}^0 (x, r')$$

  If $\mathrm{alph}(x) \cap N \neq \emptyset$, then $r' = r$, so

$$(x, r) \Rightarrow_{G'} (x, q)$$

  which completes the induction step for (i.a).

(i.b) Assume that $A \in \text{alph}(x)$. Therefore, $x = x_1 A x_2$, for some $x_1, x_2 \in V^*$. Then, $(x_1 A x_2, r) \Rightarrow_G (x_1 y x_2, q)$, for some $q \in \sigma_r$. By the induction hypothesis,

$$(S, p) \Rightarrow_{G'}^* (\tau(x_1 A x_2), r)$$

(recall that $\text{alph}(x) \cap N \neq \emptyset$). Since $\tau(x_1 A x_2) = \tau(x_1) A \tau(x_2)$,

$$(\tau(x_1) A \tau(x_2), r) \Rightarrow_{G'} (\tau(x_1)\tau(y)\tau(x_2), q')$$

where $q' = q$ if $\text{alph}(w) \cap N \neq \emptyset$ and q' is any member of $\bar{T}$ if $\text{alph}(w) \cap N = \emptyset$. Since $\tau(x_1)\tau(y)\tau(x_2) = \tau(w)$, the induction step for (i.b) is completed.

(ii) Let $(r: A \to y, \sigma_r, \varphi_r) \in P$, where $|y| > 2$. Let $y = Y_1 Y_2 \cdots Y_k$, where $k = |y|$. From the algorithm, P' contains

$$\begin{aligned}
&(r: A \to \tau(Y_1)C_1, \{s_1\}, \delta(\varphi_r))\\
&(s_1: C_1 \to \tau(Y_2)C_2, \{s_2\}, \emptyset)\\
&\quad\vdots\\
&(s_{k-1}: C_{k-1} \to \tau(Y_k), \delta(\sigma_r), \emptyset)
\end{aligned}$$

where $C_j \in N'$, for all j, $1 \leq j \leq k-1$. If $A \notin \text{alph}(x)$, then we establish the induction step by analogy with (i.a). Therefore, assume that $A \in \text{alph}(x)$. Then, $x = x_1 A x_2$, for some $x_1, x_2 \in V^*$, and $(x_1 A x_2, r) \Rightarrow_G (x_1 y x_2, q)$, for some $q \in \sigma_r$. By the induction hypothesis,

$$(S, p) \Rightarrow_{G'}^* (\tau(x_1 A x_2), r)$$

(recall that $\text{alph}(x) \cap N \neq \emptyset$). Since $\tau(x_1 A x_2) = \tau(x_1) A \tau(x_2)$,

$$\begin{aligned}
(\tau(x_1) A \tau(x_2), r) &\Rightarrow_{G'} (\tau(x_1)\tau(Y_1)C_1\tau(x_2), s_1)\\
&\Rightarrow_{G'} (\tau(x_1)\tau(Y_1)\tau(Y_2)C_2\tau(x_2), s_2)\\
&\quad\vdots\\
&\Rightarrow_{G'} (\tau(x_1)\tau(Y_1)\tau(Y_2)\cdots\tau(Y_{k-1})C_{k-1}\tau(x_2), s_{k-1})\\
&\Rightarrow_{G'} (\tau(x_1)\tau(Y_1)\tau(Y_2)\cdots\tau(Y_{k-1})\tau(Y_k)\tau(x_2), q')
\end{aligned}$$

where $q' = q$ if $\text{alph}(w) \cap N \neq \emptyset$ and q' is any member of $\bar{T}$ if $\text{alph}(w) \cap N = \emptyset$. Since $\tau(x_1)\tau(Y_1)\tau(Y_2)\cdots\tau(Y_k)\tau(x_2) = \tau(x_1)\tau(y)\tau(x_2) = \tau(w)$, the induction step for (ii) is completed.

Observe that cases (i) and (ii) cover all possible forms of $(x, r) \Rightarrow_G (w, q)$. Thus, the claim holds. □

Let $V' = N' \cup T$. Define the homomorphism π from V'^* to V^* as $\pi(A) = A$, for all $A \in N$, and $\pi(a) = \pi(\bar{a}) = a$, for all $a \in T$.

Claim 2. Let $(S, p) \Rightarrow_{G'}^{m} (x, r)$, *where* $x \in V'^*$, $p \in \Psi$, $r \in \Psi'$, *for some* $m \geq 0$. *Then,* $x \in (V \cup \bar{T})^*$, $r \in \Psi \cup \bar{T}$, *and* $(S, p) \Rightarrow_G^* (\pi(x), r')$, *where* $r' = r$ *if* $r \in \Psi$ *and* $r' \in \Psi$ *if* $r \notin \Psi$.

Proof. This claim is established by induction on $m \geq 0$.

Basis. Let $m = 0$. Then, for $(S, p) \Rightarrow_{G'}^{0} (S, p)$, where $p \in \Psi$, there is $(S, p) \Rightarrow_G^0 (S, p)$, so the basis holds.

Induction Hypothesis. Suppose that the claim holds for all derivations of length ℓ or less, where $0 \leq \ell \leq m$, for some $m \geq 0$.

Induction Step. Consider any derivation of the form

$$(S, p) \Rightarrow_{G'}^{m+1} (w, q)$$

where $w \in V'^*$ and $q \in \Psi'$. Since $m + 1 \geq 1$, this derivation can be expressed as

$$(S, p) \Rightarrow_{G'}^{m} (x, r) \Rightarrow_{G'} (w, q)$$

where $x \in V'^*$, $r \in \Psi'$. By the induction hypothesis, $x \in (V \cup \bar{T})^*$, $r \in \Psi \cup \bar{T}$, and

$$(S, p) \Rightarrow_G^* (\pi(x), r')$$

where $r' = r$ if $r \in \Psi$ and $r' \in \Psi$ if $r \notin \Psi$.

Next, we consider all possible forms of $(x, r) \Rightarrow_{G'} (w, q)$, covered by the following three cases—(i) through (iii).

(i) Let $(r: A \to y, \delta(\sigma_r), \delta(\varphi_r)) \in P'$ be constructed from $(r: A \to \pi(y), \sigma_r, \varphi_r) \in P$ in the initialization part of the algorithm. Therefore, $r' = r$. There are two subcases, (i.a) and (i.b), depending on whether $A \in \text{alph}(x)$ or not.

(i.a) Assume that $A \notin \text{alph}(x)$. Then, $(x, r) \Rightarrow_{G'} (x, q)$, for some $q \in \delta(\varphi_r)$. Since $A \notin \text{alph}(x)$, $A \notin \text{alph}(\pi(x))$. Recall that $r' = r$. If $q \in \Psi$, then

$$(\pi(x), r) \Rightarrow_G (\pi(x), q)$$

If $q \notin \Psi$, then

$$(\pi(x), r) \Rightarrow_G^0 (\pi(x), r)$$

Since $x \in (V \cup \bar{T})^*$ by the induction hypothesis and $q \in \Psi \cup \bar{T}$, the induction step for (i.a) is completed.

(i.b) Assume that $A \in \text{alph}(x)$. Therefore, $x = x_1 A x_2$ and $(x_1 A x_2, r) \Rightarrow_{G'} (x_1 y x_2, q)$, for some $q \in \delta(\sigma_r)$. Since $A \in \text{alph}(x)$, $A \in \text{alph}(\pi(x))$. Recall that $r' = r$. Since $\pi(x_1 A x_2) = \pi(x_1) A \pi(x_2)$,

$$(\pi(x_1) A \pi(x_2), r) \Rightarrow_G (\pi(x_1)\pi(y)\pi(x_2), q')$$

where $q' = q$ if $q \in \Psi$ and q is an arbitrary member of φ_r if $q \notin \Psi$ (notice that $\varphi_r = \emptyset$). Since $x \in (V \cup \bar{T})^*$ by the induction hypothesis and $y \in (V \cup \bar{T})^*$, $x_1 y x_2 \in (V \cup \bar{T})^*$. Furthermore, $q \in \Psi \cup \bar{T}$. Therefore, the induction step for (i.b) is completed.

(ii) Let $(r\colon A \to \tau(Y_1)C_1, \{s_1\}, \delta(\varphi_r)) \in P'$ be a rule that is constructed from $(r\colon A \to Y_1 Y_2 \cdots Y_k, \sigma_r, \varphi_r) \in P$ in the algorithm, where $Y_i \in V$, for all i, $1 \leq i \leq k$, for some $k > 2$. If $A \notin \text{alph}(x)$, then complete this by analogy with (i.a). Therefore, assume that $A \in \text{alph}(x)$, so $x = x_1 A x_2$. From Algorithm 5.3.12, P' also contains

$$\begin{gathered}(s_1\colon C_1 \to \tau(Y_2)C_2, \{s_2\}, \emptyset)\\ \vdots\\ (s_{k-1}\colon C_{k-1} \to \tau(Y_k), \delta(\sigma_r), \emptyset)\end{gathered}$$

where $C_j \in N'$, for all j, $1 \leq j \leq k-1$. Observe that all of these rules have to be consecutively applied; otherwise, the derivation is blocked. Hence,

$$\begin{aligned}(x_1 A x_2, r) &\Rightarrow_{G'} (x_1 \tau(Y_1) C_1 x_2, s_1)\\ &\Rightarrow_{G'} (x_1 \tau(Y_1)\tau(Y_2) C_2 x_2, s_2)\\ &\ \ \vdots\\ &\Rightarrow_{G'} (x_1 \tau(Y_1)\tau(Y_2)\cdots\tau(Y_{k-1}) C_{k-1} x_2, s_{k-1})\\ &\Rightarrow_{G'} (x_1 \tau(Y_1)\tau(Y_2)\cdots\tau(Y_{k-1})\tau(Y_k) x_2, h)\end{aligned}$$

for some $h \in \delta(\sigma_r)$. Let $y = Y_1 Y_2 \cdots Y_k$. Since $x \in (V \cup \bar{T})^*$ by the induction hypothesis and $y \in (V \cup \bar{T})^*$, $x_1 y x_2 \in (V \cup \bar{T})^*$. Since $\pi(x_1 A x_2) = \pi(x_1) A \pi(x_2)$,

$$(\pi(x_1) A \pi(x_2), r) \Rightarrow_G (\pi(x_1)\pi(y)\pi(x_2), h')$$

where $h' = h$ if $h \in \Psi$ and h' is an arbitrary member of σ_r if $h \notin \Psi$ (notice that $\sigma_r = \emptyset$). Since $\pi(x_1)\pi(y)\pi(x_2) = \pi(x_1 y x_2)$, the induction step for (ii) is completed.

(iii) Let $r = \bar{a}$, $(\bar{a}\colon \bar{a} \to a, \bar{T}, \emptyset) \in P$, for some $a \in T$. Then,

$$(\pi(x), r') \Rightarrow_G^0 (\pi(x), r')$$

Since $\pi(w) = \pi(x)$, the induction step for (iii) is completed.

Observe that cases (i) through (iii) cover all possible forms of $(x, r) \Rightarrow_{G'} (w, q)$. Thus, the claim holds. □

Next, we prove that $L(G') = L(G)$. Consider Claim 1 with $x \in T^*$. Then, $(S, p) \Rightarrow_G^* (x, r)$ implies that $(S, p) \Rightarrow_{G'}^* (\tau(x), r')$, where r' is any member of $\bar{T}$ if $\text{alph}(x) \cap N = \emptyset$. Let $x = a_1 a_2 \cdots a_k$, where $k = |x|$, so $\tau(x) = \bar{a}_1 \bar{a}_2 \cdots \bar{a}_k$. By the initialization part of Algorithm 5.3.12, there are $(\bar{a}_i : \bar{a}_i \to a_i, \bar{T}, \emptyset) \in P'$, for all i, $1 \leq i \leq k$. Let $r' = \bar{a}_1$. Then,

$$
\begin{aligned}
(\bar{a}_1 \bar{a}_2 \cdots \bar{a}_k, \bar{a}_1) &\Rightarrow_{G'} (a_1 \bar{a}_2 \cdots \bar{a}_k, \bar{a}_2) \\
&\Rightarrow_{G'} (a_1 a_2 \cdots \bar{a}_k, \bar{a}_3) \\
&\;\;\vdots \\
&\Rightarrow_{G'} (a_1 a_2 \cdots a_k, \bar{a}_1)
\end{aligned}
$$

Hence, $L(G) \subseteq L(G')$. Consider Claim 2 with $x \in T^*$. Then, $(S, p) \Rightarrow_{G'}^* (x, r)$ implies that $(S, p) \Rightarrow_{G'}^* (\pi(x), r')$, where $r' = r$ if $r \in \Psi$ and $r' \in \Psi$ if $r \in \bar{T}$. Since $\pi(x) = x$, $L(G') \subseteq L(G)$. As $L(G) \subseteq L(G')$ and $L(G') \subseteq L(G)$, $L(G') = L(G)$, so the lemma holds. □

Theorem 5.3.14. *For any programmed grammar with appearance checking G, there is a programmed grammar with appearance checking in the binary normal form G' such that $L(G') = L(G)$. Furthermore, if G is propagating, then so is G', and if G is without appearance checking, then so is G'.*

Proof. Consider Algorithm 5.3.12. Observe that if G is propagating, then so is G'. Furthermore, observe that if G is without appearance checking, then so is G'. Therefore, this theorem follows from Algorithm 5.3.12 and Lemma 5.3.13. □

5.3.4 Restricted Non-Determinism

Sections 5.3.1 through 5.3.3 have discussed programmed grammars in a general and, therefore, non-deterministic form. As a result, they have not studied the role of determinism in programmed grammars at all although this study is obviously significant from a practical point of view. Therefore, to fill this gap to some extent, the present section pays a special attention to this pragmatically important open problem area.

In [1] and [2], two important results concerning this open problem area have been achieved. First, if we require every rule in a programmed grammar to have at most one successor, then we can generate only finite languages. Second, any programmed grammar can be converted to an equivalent programmed grammar with every rule having at most two successors. Considering the importance of these results, it comes as no surprise that some other determinism-related measures of programmed as well as other regulated grammars have been discussed (see Sect. 4.3 of [3] and [4, 5]). To illustrate, it has been demonstrated an infinite hierarchy of language families resulting from the measure consisting in the number of matrices in matrix grammars; simply put, the more matrices we have, the greater power we obtain (see page 192 in [3]). To give an example in terms of programmed grammars, in [4, 5], the necessary number of nonterminals to keep the power of programmed grammars is discussed. Continuing with this vivid investigation trend, the present section considers rules with two or more successors and discusses the following two topics concerning the generative power of programmed grammars

(i) A reduction of the number of these rules without decreasing the generative power;
(ii) The impact of the number of successors in these rules to the generative power of programmed grammars.

Regarding (i), we reduce this number to one. Regarding (ii), we propose a new measure, called the *overall non-determinism* of a programmed grammar, as the sum of all successors of these rules . Then, we show that this measure gives rise to an infinite hierarchy of language families. More precisely, we prove that programmed grammars with $n + 1$ successors are stronger than programmed grammars with n successors, for all $n \geq 1$.

First, we define a new normal form and measure for programmed grammars. Then, we illustrate them by an example.

Definition 5.3.15. Let $G = (N, T, \Psi, P, S)$ be a programmed grammar. G is in the *one-ND rule normal form* (ND stands for *n*on*d*eterministic) if at most one $(r\colon A \to x, \sigma_r, \emptyset) \in P$ satisfies that $\mathrm{card}(\sigma_r) \geq 1$ and every other $(s\colon B \to y, \sigma_s, \emptyset) \in P$ satisfies that $\mathrm{card}(\sigma_s) \leq 1$. □

Definition 5.3.16. Let $G = (N, T, \Psi, P, S)$ be a programmed grammar. For each $(r\colon A \to x, \sigma_r, \emptyset) \in P$, let $\zeta(r)$ be defined as

$$\zeta(r) = \begin{cases} \mathrm{card}(\sigma_r) & \text{if } \mathrm{card}(\sigma_r) \geq 2 \\ 0 & \text{otherwise} \end{cases}$$

The *overall non-determinism* of G is denoted by $\mathrm{ond}(G)$ and defined as

$$\mathrm{ond}(G) = \sum_{r \in \Psi} \zeta(r)$$ □

By ${}_1\mathbf{P}$, we denote the family of languages generated by programmed grammars in the one-ND rule normal form. Recall that $\mathbf{P}$ denotes the family of languages generated by programmed grammars (see Sect. 5.3.1).

Definition 5.3.17. Let $X \in \{\mathbf{P}, {}_1\mathbf{P}\}$ and $L \in X$. Then, we define

- $\mathrm{ond}(X, L) = \min(\{\mathrm{ond}(G) \mid G \text{ is in the form generating } X, L = L(G)\})$,
- $\mathbf{OND}(X, n) = \{M \in X \mid \mathrm{ond}(X, M) \leq n\}$. □

By "G is in the form generating X" in terms of Definition 5.3.17, we mean that if $X = \mathbf{P}$, then G is a programmed grammar, and if $X = {}_1\mathbf{P}$, then G is a programmed grammar in the one-ND rule normal form.

Example 5.3.18. Consider the language $K = \{a^n b^n c^n \mid n \geq 1\}$. This non-context-free language is generated by the programmed grammar

$$G = \left(N, T, \Psi, P, S\right)$$

where $N = \{S, A, B, C\}$, $T = \{a, b, c\}$, and P contains the seven rules

$$
\begin{array}{ll}
(r_1: S \to ABC, \{r_2, r_5\}, \emptyset) & (r_5: A \to a, \{r_6\}, \emptyset) \\
(r_2: A \to aA, \{r_3\}, \emptyset) & (r_6: B \to b, \{r_7\}, \emptyset) \\
(r_3: B \to bB, \{r_4\}, \emptyset) & (r_7: C \to c, \{r_7\}, \emptyset) \\
(r_4: C \to cC, \{r_2, r_5\}, \emptyset) &
\end{array}
$$

For example, $aabbcc$ is generated by

$$
\begin{aligned}
(S, r_1) &\Rightarrow_G (ABC, r_2) \\
&\Rightarrow_G (aABC, r_3) \\
&\Rightarrow_G (aAbBC, r_4) \\
&\Rightarrow_G (aAbBcC, r_5) \\
&\Rightarrow_G (aabBcC, r_6) \\
&\Rightarrow_G (aabbcC, r_7) \\
&\Rightarrow_G (aabbcc, r_7)
\end{aligned}
$$

Since the only non-deterministic rules are r_1 and r_2, each containing two rules in their successor sets, the overall non-determinism of G is $\mathrm{ond}(G) = 4$. Therefore, $K \in \mathbf{OND}(\mathbf{P}, 4)$.

Examine r_1 and $_2$ to see that G is not in the one-ND rule normal form. However, consider the programmed grammar

$$H = \left(N, T, \Psi', P', S\right)$$

where P' contains the eight rules

$$
\begin{array}{ll}
(r_0: S \to abc, \{r_0\}, \emptyset) & (r_4: C \to cC, \{r_2, r_5\}, \emptyset) \\
(r_1: S \to ABC, \{r_2\}, \emptyset) & (r_5: A \to a, \{r_6\}, \emptyset) \\
(r_2: A \to aA, \{r_3\}, \emptyset) & (r_6: B \to b, \{r_7\}, \emptyset) \\
(r_3: B \to bB, \{r_4\}, \emptyset) & (r_7: C \to c, \{r_7\}, \emptyset)
\end{array}
$$

Clearly, $L(H) = K$, and H is in the one-ND rule normal form. Observe that $\mathrm{ond}(H) = 2$, so $K \in \mathbf{OND}(\mathbf{P}, 2)$. Finally, since H is in the one-ND rule normal form, $K \in \mathbf{OND}({}_1\mathbf{P}, 2)$. In what follows, we show that $\mathbf{OND}(\mathbf{P}, n) = \mathbf{OND}({}_1\mathbf{P}, n)$. □

The following algorithm converts any programmed grammar G to an equivalent programmed grammar G' in the one-ND rule normal form. To give an insight into this conversion, we first explain the underlying idea behind it. First, we introduce the only non-deterministic rule of G', $(X\colon \# \to \varepsilon, \sigma_X, \emptyset)$. Obviously, each non-deterministic choice of some rule $(r\colon A \to x, \{s_1, s_2, \dots, s_n\}, \emptyset)$ of G has to be simulated by using X. To ensure the proper simulation, we have to guarantee that (i) one of s_i is applied after r, and (ii) no other rules can be applied after r.

To satisfy both of these requirements, we introduce a special nonterminal symbol $\langle r \rangle$ for each rule r of G. These symbols are used to encode the information about the last applied rule in a derivation. Then, for each successor of r, s_i, we introduce the following sequence of rules

- $(r\colon A \to \langle r \rangle\#, \{X\}, \emptyset)$ to preserve the information that r is the currently simulated rule,
- X to make a non-deterministic choice of the successor of r, and
- $(\langle r \rhd s_i \rangle\colon \langle r \rangle \to x, \{s_i\}, \emptyset)$ to simulate r and continue with s_i.

Note that if X chooses some $\langle p \rhd q \rangle$ with $p \neq r$ instead, the derivation gets blocked because $\langle p \rangle$ is not present in the current sentential form.

Algorithm 5.3.19. *Conversion of a programmed grammar to an equivalent programmed grammar in the one-ND rule normal form.*

Input: A programmed grammar, $G = (N, T, \Psi, P, S)$.
Output: A programmed grammar in the one-ND rule normal form, $G' = (N', T, \Psi', P', S')$, such that $L(G') = L(G)$.
Method: Initially, set

$$
\begin{aligned}
N' &= N \cup \{\#\} \cup \{\langle r \rangle \mid r \in \Psi\} \\
\Psi' &= \Psi \cup \{X\} \text{ with } X \text{ being a new unique symbol} \\
P' &= \{(r\colon A \to x, \sigma_r, \emptyset) \mid (r\colon A \to x, \sigma_r, \emptyset) \in P, \mathrm{card}(\sigma_r) = 1\} \\
&\quad \cup \{(X\colon \# \to \varepsilon, \sigma_X, \emptyset)\}
\end{aligned}
$$

with $\sigma_X = \emptyset$. For each $(r\colon A \to \omega, \sigma_r, \emptyset) \in P$ satisfying that $\mathrm{card}(\sigma_r) > 1$, apply the following two steps

(1) Add $(r\colon A \to \langle r \rangle\#, \{X\}, \emptyset)$ to P';
(2) For each $q \in \sigma_r$, add $(\langle r \rhd q \rangle\colon \langle r \rangle \to \omega, \{q\}, \emptyset)$ to P', $\langle r \rhd q \rangle$ to Ψ', and $\langle r \rhd q \rangle$ to σ_X. □

Lemma 5.3.20. *Algorithm 5.3.19 is correct.*

Proof. Clearly, the algorithm always halts and G' is in the one-ND rule normal form. To establish $L(G) = L(G')$, we first prove that $L(G) \subseteq L(G')$ by showing how derivations of G are simulated by G', and then, we prove that $L(G') \subseteq L(G)$ by showing how every $w \in L(G')$ can be generated by G.

First, we introduce some notions used later in the proof. Set $V = N \cup T$, $V' = N' \cup T$, and $\bar{N} = \{\langle r\rangle \mid r \in \Psi\}$. For $(p\colon A \to x, \sigma_p, \emptyset) \in P'$, let $\mathrm{lhs}(p) = A$ and $\mathrm{rhs}(p) = x$.

For two rule labels q and r, we say that q *follows* r in a derivation if the derivation can be expressed as

$$(u_1, p_1) \Rightarrow^* (v, r) \Rightarrow (w, q) \Rightarrow^* (u_2, p_2)$$

Next, we establish four claims, which we use later in the proof of $L(G') \subseteq L(G)$.

Claim 1. Let $(S, s) \Rightarrow_G^m (w, q)$*, where* $s, q \in \Psi$*,* $w \in V^*$*, for some* $m \geq 0$*. Then,* $(S, s) \Rightarrow_{G'}^* (w, q)$*.*

Proof. This claim is established by induction on $m \geq 0$.

Basis. Let $m = 0$. Then, for $(S, s) \Rightarrow_G^0 (S, s)$, where $s \in \Psi$, there is $(S, s) \Rightarrow_{G'}^0 (S, s)$, so the basis holds.

Induction Hypothesis. Suppose that the claim holds for all derivations of length ℓ or less, where $0 \leq \ell \leq m$, for some $m \geq 0$.

Induction Step. Consider any derivation of the form

$$(S, s) \Rightarrow_G^{m+1} (w, q)$$

where $w \in V^*$ and $s, q \in \Psi$. Since $m + 1 \geq 1$, this derivation can be expressed as

$$(S, s) \Rightarrow_G^m (u, p) \Rightarrow_G (w, q)$$

where $u \in V^*$, $p \in \Psi$. By the induction hypothesis, $(S, s) \Rightarrow_{G'}^* (u, p)$.

Next, we consider two possible cases, (i) and (ii), based on whether $\mathrm{card}(\sigma_p) = 1$ or $\mathrm{card}(\sigma_p) > 1$ in $(p\colon A \to x, \sigma_p, \emptyset) \in P$.

(i) Let $\mathrm{card}(\sigma_p) = 1$. Then, $(p\colon A \to x, \sigma_p, \emptyset) \in P'$ by the initialization part of the algorithm, so the induction step is completed for (i).
(ii) Let $\mathrm{card}(\sigma_p) > 1$. Then, P' contains the following three rules

- $(p\colon A \to \langle p\rangle\#, \{X\}, \emptyset)$, constructed in (1),
- $(X\colon\# \to \varepsilon, \sigma_X, \emptyset)$, constructed in the initialization part of the algorithm, and
- $(\langle p \rhd q\rangle\colon \langle p\rangle \to x, \{q\}, \emptyset)$, constructed in (2) from q, such that $\langle p \rhd q\rangle \in \sigma_X$.

Based on these rules, $(u, p) \Rightarrow_{G'}^3 (w, q)$, which completes the induction step for (ii). □

Claim 2. Let $(S, s) \Rightarrow_{G'}^* (w, q)$*, where* $s, q \in \Psi'$ *and* $w \in V'^*$*. Then, either* $w \in V^*$*, or* w *can be expressed as* $u\langle r\rangle\#v$ *or* $u\langle r\rangle v$*, where* $u, v \in V^*$ *and* $\langle r\rangle \in \bar{N}$*.*

Proof. Observe that only the rules constructed in (1) have symbols from $N' - N$ on their right-hand side. These rules have to be followed by X, which have to be followed by some rule constructed in (2). As X erases # and rules constructed in (2) are rewriting symbols from $\bar{N}$ to some $\omega \in V^*$, each string derived from S can be expressed as u, $u\langle r\rangle\#v$, or $u\langle r\rangle v$, where $u, v \in V^*$ and $\langle r\rangle \in \bar{N}$. □

Claim 3. Let $(S, s) \Rightarrow^*_{G'} (w, q)$, *where* $s, q \in \Psi'$, *and* $w \in V'^*$. *If* $w \in V^*$, *then* $q \in \Psi$.

Proof. Let $w \in V^*$ and assume that $q \in \Psi' - \Psi$ for the sake of contradiction. Then, q is either X, or it is constructed in (2). As X is only in the success field of some $(r\colon A \rightarrow \langle r\rangle\#, \{X\}, \emptyset) \in P'$, constructed in (1), and labels constructed in (2) are only in the success field of X, the derivation of (w, q) can be expressed in one of the following two forms

- $(S, s) \Rightarrow^*_{G'} (w_1 A w_2, r) \Rightarrow_{G'} (w_1\langle r\rangle\# w_2, X)$, or
- $(S, s) \Rightarrow^*_{G'} (w_1 A w_2, r) \Rightarrow_{G'} (w_1\langle r\rangle\# w_2, X) \Rightarrow_{G'} (w_1\langle r\rangle w_2, \langle r \rhd q\rangle)$,

where $w_1, w_2 \in V^*$, and $\langle r \rhd q\rangle \in \sigma_X$. Note that in both forms, w would have to contain some $\langle r\rangle \in \bar{N}$, thus contradicting $w \in V^*$. Thus, $q' \in \Psi$, so the claim holds. □

The following claim shows that for each $w \in V^*$ derived in G', there is a derivation of w in G.

Claim 4. Let $(S, s) \Rightarrow^m_{G'} (w, q)$, *where* $s, q \in \Psi$, $w \in V^*$, *for some* $m \geq 0$. *Then,* $(S, s) \Rightarrow^*_G (w, q)$.

Proof. This claim is established by induction on $m \geq 0$.

Basis. Let $m = 0$. Then, for $(S, s) \Rightarrow^0_{G'} (S, s)$, where $s \in \Psi$, there is $(S, s) \Rightarrow^0_G (S, s)$, so the basis holds.

Induction Hypothesis. Suppose that the claim holds for all derivations of length ℓ or less, where $0 \leq \ell \leq m$, for some $m \geq 0$.

Induction Step. Consider any derivation of the form

$$(S, s) \Rightarrow^{m+1}_{G'} (w, q)$$

where $w \in V^*$ and $s, q \in \Psi$. Since $m + 1 \geq 1$, this derivation can be expressed as

$$(S, s) \Rightarrow^m_{G'} (u, p) \Rightarrow_{G'} (w, q)$$

where $u \in V'^*$, $p \in \Psi'$. Now, we consider all possible forms of $(u, p) \Rightarrow_{G'} (w, q)$, covered by the next two cases.

(i) Let $u \in V^*$. Then, by Claim 3, $p \in \Psi$. Therefore, $(p\colon A \rightarrow \omega, \{q\}, \emptyset) \in P'$ is one of the rules constructed in the initialization part of the algorithm, so it is also in P. Thus, $(u, p) \Rightarrow_G (w, q)$. By the induction hypothesis, $(S, s) \Rightarrow^*_G (u, p)$, so the induction step is completed for (i).

(ii) Let $u = u_1\langle r\rangle u_2$, where $u_1, u_2 \in V^*$ and $\langle r\rangle \in \bar{N}$ for some $r \in \Psi$. Observe that $(r\colon A \to \langle r\rangle\#, \{X\}, \emptyset) \in P'$, constructed in (1), is the only rule with $\langle r\rangle$ on its right-hand side. Furthermore, observe that X has to follow r in the derivation. As rhs(r) contains # and only the rule labeled with X erases #, X has to be used after the last occurrence of r in the derivation.

Consider all the labels in σ_X. These labels belong to the rules constructed in (2). Only these rules rewrite the symbols from $\bar{N}$ to some $\omega \in V^*$, so they have to be used after (u, p) to satisfy $(u, p) \Rightarrow_{G'} (w, q)$. As p is followed by q, $p = \langle r \rhd q\rangle$, so the derivation can be expressed as

$$\begin{aligned}(S, s) &\Rightarrow_{G'}^{m-3} (u_1 A u_2, r)\\ &\Rightarrow_{G'} (u_1\langle r\rangle\# u_2, X)\\ &\Rightarrow_{G'} (u_1\langle r\rangle u_2, \langle r \rhd q\rangle)\\ &\Rightarrow_{G'} (w, q)\end{aligned}$$

Observe that $(r\colon A \to \langle r\rangle\#, \{X\}, \emptyset) \in P'$ is constructed in (1) from some $(r\colon A \to \omega, \sigma_r, \emptyset) \in P$ with $q \in \sigma_r$. Thus, $(u_1 A u_2, r) \Rightarrow_G (w, q)$. Clearly, $u_1 A u_2 \in V^*$. Therefore, by the induction hypothesis, $(S, s) \Rightarrow_G^* (u_1 A u_2, r)$, so the induction step is completed for (ii).

These cases cover only two of the three possible forms of u (see Claim 2). However, if $u = u_1\langle r\rangle\# u_2$, then $p = X$. As σ_X contains only rules constructed in (2), q would have to be in $\Psi' - \Psi$, contradicting $q \in \Psi$. Thus, cases (i) and (ii) cover all possible forms of $(u, p) \Rightarrow_{G'} (w, q)$, so the claim holds. □

To establish $L(G) = L(G')$, it is sufficient to show the following two statements

- By Claim 1, for each $(S, s) \Rightarrow_G^* (w, q)$, where $s, q \in \Psi$ and $w \in T^*$, there is $(S, s) \Rightarrow_{G'}^* (w, q)$, so $L(G) \subseteq L(G')$;
- Let $(S, s) \Rightarrow_{G'}^* (w, q)$, where $s, q \in \Psi'$ and $w \in T^*$. As $w \neq S$, s is used in the derivation, and so lhs$(s) = S$. Observe that only rules with labels from Ψ have S on their left-hand side, so $s \in \Psi$. As $w \in T^*$, $q \in \Psi$ by Claim 3. Then, by Claim 4, $(S, s) \Rightarrow_G^* (w, q)$, so $L(G') \subseteq L(G)$.

As $L(G) \subseteq L(G')$ and $L(G') \subseteq L(G)$, $L(G) = L(G')$, so the lemma holds. □

The following theorem represents the first main achievement of this section.

Theorem 5.3.21. *For any programmed grammar G, there is a programmed grammar in the one-ND rule normal form G' such that $L(G') = L(G)$.*

Proof. This theorem follows from Algorithm 5.3.19 and Lemma 5.3.20. □

Next, we study the impact of the overall number of successors in rules with two or more successors to the generative power of programmed grammars. First, however, we introduce some terminology. Let $G = (N, T, \Psi, P, S)$be a programmed grammar in the one-ND rule normal form. Set $V = N \cup T$, and for each $(r: A \to x, \sigma_r, \emptyset) \in P$, let $\sigma(r) = \sigma_r$. Recall that for $u \in V^*$ and $W \subseteq V$, occur(u, W) denotes the number of occurrences of symbols from W in u.

Let $s = (r_1, r_2, \ldots, r_k)$ be a sequence of labels, where $r_i \in \Psi$, for all $i = 1, 2, \ldots, k$. We say that s is *deterministic* if $\sigma(r_{i-1}) = \{r_i\}$, for $i = 2, 3, \ldots, k$. Each sequence of labels of the form $(r_1, r_2, \ldots, r_j)$, where $j \leq k$, is called a *prefix* of s.

We say that s *generates* a, where $a \in V$, if some $(r_i: A \to x, \sigma, \emptyset) \in P$ satisfies that $x = uav$, where $u, v \in V^*$, for some i, $1 \leq i \leq k$. We say that a derivation *contains* s if it can be expressed as

$$(u, p) \Rightarrow^* (w_1, r_1) \Rightarrow (w_2, r_2) \Rightarrow \cdots \Rightarrow (w_k, r_k) \Rightarrow^* (v, q)$$

where $u, v, w_i \in V^*$, $p, q \in \Psi$, for all $i = 1, 2, \ldots, k$. Let $\Rightarrow_{[s]}$ be a binary relation defined over V^* as follows: for $u, v \in V^*$, $u \Rightarrow_{[s]} v$ if and only if there is a derivation

$$(w_1, r_1) \Rightarrow (w_2, r_2) \Rightarrow \cdots \Rightarrow (w_k, r_k) \Rightarrow (v, p)$$

such that $w_1 = u$.

Let Q_r be the set of all deterministic sequences beginning with $r \in P$, and let $\leq$ be a binary relation over Q_r defined as $s \leq t$ if and only if s is a prefix of t. Observe that $\leq$ is reflexive, antisymmetric, transitive, and total (as all the sequences are deterministic and starting with the same rule). As there is a least element for every nonempty subset of Q_r, $(Q_r, \leq)$ is a well-ordered set.

We say that Q_r *generates* $a \in V$ if and only if there is some $s \in Q_r$ such that s generates a. We say that Q_r *reduces the number of nonterminals* if there is some $s \in Q_r$ such that for each $t \geq s$ and for each $u, v \in V^*$, if $u \Rightarrow_{[t]} v$, then occur$(u, N) >$ occur(v, N).

Lemma 5.3.22. $\mathbf{OND(P}, n) = \mathbf{OND(}_1\mathbf{P}, n)$

Proof. Let $G = (N, T, \Psi, P, S)$ be a programmed grammar. Then, by Algorithm 5.3.19 and Lemma 5.3.20, we can construct a programmed grammar in the one-ND rule form

$$G' = \left(N', T, \Psi', S', P'\right)$$

such that $L(G) = L(G')$ and

$$(r: A \to x, \sigma_r, \emptyset) \in P' - \left\{(X: \# \to \varepsilon, \sigma_X), \emptyset\right\}$$

satisfies that card$(\sigma_r) \leq 1$. Observe that for each $(r: A \to x, \sigma_r, \emptyset) \in P$, where card$(\sigma_r) > 1$, there are card$(\sigma_r)$ labels in σ_X constructed in (2) of Algorithm 5.3.19. As these are the only labels in σ_X, and all other rules in P' have at most one label in their success field, by Definition 5.3.16, $\text{ond}(G') = \text{ond}(G)$. Thus, $\mathbf{OND}(\mathbf{P}, n) \subseteq \mathbf{OND}({}_1\mathbf{P}, n)$. Obviously, $\mathbf{OND}({}_1\mathbf{P}, n) \subseteq \mathbf{OND}(\mathbf{P}, n)$, so the lemma holds. □

Lemma 5.3.23. *Let* $G = (N, T, \Psi, P, S)$ *be a programmed grammar in the one-ND rule normal form such that* $L(G)$ *is infinite. Then, there is exactly one* $r \in \Psi$ *such that* $\text{card}(\sigma(r)) > 1$.

Proof. This lemma follows from Definition 5.3.15 in this section and from Lemma 8 in [2], which says that programmed grammars with every rule having at most one successor generate only finite languages. □

Lemma 5.3.24. *Let* $G = (N, T, \Psi, P, S)$ *be a programmed grammar in the one-ND rule normal form such that* $L(G)$ *is infinite, and let* r_x *denote the only rule satisfying that* $\text{card}(\sigma(r_x)) > 1$. *Then, there are* $p, q \in \sigma(r_x)$ *such that* Q_p *reduces the number of nonterminals and* Q_q *does not reduce the number of nonterminals.*

Proof. Observe that there is a finite number of $u \in V^*$ that can be derived without r_x being used in their derivation. Furthermore, to derive a string of an arbitrary length, we cannot limit the number of uses of r_x (see Lemma 5.3.23).

We prove by contradiction that there is at least one $r \in \sigma(r_x)$ such that Q_r reduces the number of nonterminals. For the sake of contradiction, assume that each Q_r, where $r \in \sigma(r_x)$, does not reduce the number of nonterminals. Then, for sufficiently large k, $(u, r_x) \Rightarrow_G^k (w, q)$ implies that $w \notin T^*$. Therefore, $L(G)$ would be finite, which leads to a contradiction. Thus, there is at least one $r \in \sigma(r_x)$ such that Q_r reduces the number of nonterminals.

Next, we prove by contradiction that there is at least one $r \in \sigma(r_x)$ such that Q_r does not reduce the number of nonterminals. For the sake of contradiction, assume that each Q_r, where $r \in \sigma(r_x)$, reduces the number of nonterminals. Then, there exists some $k \geq 0$ such that each $(u, r_x) \Rightarrow_G^+ (w, q)$, where $w \in T^*$, implies that $|w| \leq k|u|$. As there is a finite number of such u that can be derived from S without using r_x, $L(G)$ is finite, which leads to a contradiction. Thus, the lemma holds. □

Lemma 5.3.25. $\mathbf{OND}({}_1\mathbf{P}, n) \subset \mathbf{OND}({}_1\mathbf{P}, n + 1)$

Proof. Let L_n be a language over $\Sigma = \{a_1, a_2, \dots a_n\}$, defined as

$$L_n = \bigcup_{i=1}^{n} \{a_i\}^+$$

We show that $\text{ond}({}_1\mathbf{P}, L_n) = n + 1$.

First, observe that L_n is generated by the propagating programmed grammar

$$G = \big(\{S\}, \{a_1, a_2, \dots, a_n\}, S, \{r_S, r_1, r_2, \dots, r_n\}, P'\big)$$

with

$$P' = \big\{(r_S \colon S \to SS, \{r_S, r_1, r_2, \dots, r_n\}, \emptyset)\big\} \cup \big\{(r_i \colon S \to a_i, \{r_i\}, \emptyset) \mid 1 \leq i \leq n\big\}$$

As the cardinality of the success field of r_S is $n + 1$, $\text{ond}({}_1\mathbf{P}, L_n) \leq n + 1$.

We show that every programmed grammar in the one-ND rule normal form generating L_n requires at least one rule with $n + 1$ labels in its success field. Let $G' = (N, T, \Psi, P, S)$, where $T = \{a_1, a_2, \ldots, a_n\}$, be a programmed grammar in the one-ND rule normal form such that $L(G') = L_n$. As L_n is infinite, by Lemma 5.3.23, there is exactly one $(r: A \rightarrow x, \sigma_r, \emptyset) \in P$ satisfying $\text{card}(\sigma_r) > 1$. Let r_x denote this rule.

First, we prove that there is at least one $r_a \in \sigma(r_x)$ for each $a \in T$. Then, we show that there is at least one additional rule in $\sigma(r_x)$ to generate all the strings in L_n.

Claim 1. For each $a \in T$, there is $r \in \sigma(r_x)$ such that Q_r generates a and Q_r does not generate any $b \in T$, $b \neq a$.

Proof. For the sake of contradiction, assume that there is $a \in T$ such that each Q_r generating a generates also some $b \in T$, $b \neq a$. Let s denote the shortest sequence in Q_r generating both a and b. Observe that there is no string in L_n for which there is a derivation in G' containing s, or some $t \geq s$ (such a string would have to contain both a and b). As all the sequences in Q_r are deterministic, any prefix of s could be contained at most once in any successful derivation. As there is a limited number of such prefixes, it would be impossible to derive a^m for arbitrary m, contradicting $L(G') = L_n$, so the claim holds. □

By Claim 1, there is $r \in \sigma(r_x)$ for each $a \in T$ such that Q_r generates only a. Let $Q(a)$ denote such Q_r. We show that there is at least one additional rule in $\sigma(r_x)$. Consider the following two cases, based on whether each $Q(a)$ reduces the number of nonterminals or not.

(i) Each $Q(a)$ does not reduce the number of nonterminals. Since L_n is infinite, by Lemma 5.3.24, there is at least one additional $p \in \sigma(r_x)$ such that Q_p reduces the number of nonterminals.

(ii) At least one $Q(a)$ reduces the number of nonterminals. For the sake of contradiction, assume that $\text{card}(\sigma(r_x)) = n$. As only $Q(a)$ generates a, and it also reduces the number of nonterminals, there is some $k \geq 0$ such that each $(u, r_x) \Rightarrow_G^+ (w, q)$, where $q \in \Psi$, $u \in V^*$, and $w \in \{a\}^*$, implies that $|w| \leq k|u|$. As there is a limited number of such u that can be derived from S without using r_x, a^m cannot be derived for arbitrary m, which leads to a contradiction. Thus, there is an additional rule in $\sigma(r_x)$.

Observe that these two cases cover all possible $Q(a)$ for each $a \in T$. Therefore, $\text{card}(\sigma(r_x)) \geq n + 1$, which implies that $\text{ond}({}_1\mathbf{P}, L_n) \geq n + 1$. Hence, $L_n \notin \mathbf{OND}({}_1\mathbf{P}, n)$. Since $\text{ond}({}_1\mathbf{P}, L_n) \leq n + 1$ implies that $L_n \in \mathbf{OND}({}_1\mathbf{P}, n + 1)$, the lemma holds. □

The following theorem represents the second main achievement of this section.

Theorem 5.3.26. $\mathbf{OND}(\mathbf{P}, n) \subset \mathbf{OND}(\mathbf{P}, n + 1)$

Proof. This theorem follows Lemma 5.3.22 and Lemma 5.3.25. □

We close this section by formulating two important open problem areas.

Open Problem 5.3.27. Do the achieved results also hold in terms of programmed grammars with appearance checking? For a preliminary result, see [10]. □

Open Problem 5.3.28. Reconsider Algorithm 5.3.19. Observe that it introduces erasing rules to G', even if the input grammar G is propagating. Can we modify this algorithm so that when G is propagating, then so is G'? Furthermore, do Theorems 5.3.21 and 5.3.26 hold in terms of propagating programmed grammars? Observe that the argument Lemma 5.3.25 is based on holds in terms of propagating programmed grammars, too. □

5.4 State Grammars

A *state grammar* G is a context-free grammar extended by an additional state mechanism that strongly resembles a finite-state control of finite automata. During every derivation step, G rewrites the leftmost occurrence of a nonterminal that can be rewritten under the current state; in addition, it moves from a state to another state, which influences the choice of the rule to be applied in the next step. If the application of a rule always takes place within the first n occurrences of nonterminals, G is referred to as *n-limited*.

The present section consists of Sects. 5.4.1 and 5.4.2. The former defines and illustrates state grammars. The latter describes their generative power.

5.4.1 Definitions and Examples

In this section, we define state grammars and illustrate them by an example.

Definition 5.4.1. A *state grammar* (see [6]) is a quintuple

$$G = \big(V, W, T, P, S\big)$$

where

- V is a *total alphabet*;
- W is a finite set of *states*;
- $T \subset V$ is an alphabet of *terminals*;
- $S \in V - T$ is the *start symbol*;
- $P \subseteq (W \times (V - T)) \times (W \times V^+)$ is a finite relation.

Instead of $(q, A, p, v) \in P$, we write $(q, A) \to (p, v) \in P$. For every $z \in V^*$, define

$$\text{states}_G(z) = \big\{q \in W \mid (q, A) \to (p, v) \in P, A \in \text{alph}(z)\big\}$$

If $(q, A) \to (p, v) \in P$, $x, y \in V^*$, and $\text{states}_G(x) = \emptyset$, then G makes a *derivation step* from (q, xAy) to (p, xvy), symbolically written as

$$(q, xAy) \Rightarrow (p, xvy)\ [(q, A) \to (p, v)]$$

In addition, if n is a positive integer satisfying that $\text{occur}(xA, V - T) \leq n$, we say that $(q, xAy) \Rightarrow (p, xvy)\ [(q, A) \to (p, v)]$ is *n-limited*, symbolically written as

$$(q, xAy)\ {}_n\!\Rightarrow (p, xvy)\ [(q, A) \to (p, v)]$$

Whenever there is no danger of confusion, we simplify $(q, xAy) \Rightarrow (p, xvy)$ $[(q, A) \to (p, v)]$ and $(q, xAy)\ {}_n\!\Rightarrow (p, xvy)\ [(q, A) \to (p, v)]$ to

$$(q, xAy) \Rightarrow (p, xvy)$$

and

$$(q, xAy)\ {}_n\!\Rightarrow (p, xvy)$$

respectively. In the standard manner, we extend $\Rightarrow$ *to* $\Rightarrow^m$, where $m \geq 0$; then, based on $\Rightarrow^m$, we define $\Rightarrow^+$ *and* $\Rightarrow^*$.

Let n be a positive integer, and let $\nu, \omega \in W \times V^+$. To express that every derivation step in $\nu \Rightarrow^m \omega$, $\nu \Rightarrow^+ \omega$, and $\nu \Rightarrow^* \omega$ is n-limited, we write $\nu\ {}_n\!\Rightarrow^m \omega$, $\nu\ {}_n\!\Rightarrow^+ \omega$, and $\nu\ {}_n\!\Rightarrow^* \omega$ instead of $\nu \Rightarrow^m \omega$, $\nu \Rightarrow^+ \omega$, and $\nu \Rightarrow^* \omega$, respectively.

By $\text{strings}(\nu\ {}_n\!\Rightarrow^* \omega)$, we denote the set of all strings occurring in the derivation $\nu\ {}_n\!\Rightarrow^* \omega$. The *language* of G, denoted by $L(G)$, is defined as

$$L(G) = \{w \in T^* \mid (q, S) \Rightarrow^* (p, w), q, p \in W\}$$

Furthermore, for every $n \geq 1$, define

$$L(G, n) = \{w \in T^* \mid (q, S)\ {}_n\!\Rightarrow^* (p, w), q, p \in W\}$$

A derivation of the form $(q, S)\ {}_n\!\Rightarrow^* (p, w)$, where $q, p \in W$ and $w \in T^*$, represents a *successful n-limited generation* of w in G. □

Next, we illustrate the previous definition by an example.

Example 5.4.2. Consider the state grammar

$$G = (\{S, X, Y, a, b\}, \{p_0, p_1, p_2, p_3, p_4, p_5\}, \{a, b\}, P, S)$$

with the following nine rules in P

$$
\begin{array}{ll}
(p_0, S) \to (p_0, XY) & (p_0, X) \to (p_3, a) \\
(p_0, X) \to (p_1, aX) & (p_3, Y) \to (p_0, a) \\
(p_1, Y) \to (p_0, aY) & (p_0, X) \to (p_4, b) \\
(p_0, X) \to (p_2, bX) & (p_4, Y) \to (p_0, b) \\
(p_2, Y) \to (p_0, bY) &
\end{array}
$$

Observe that G generates the non-context-free language

$$L(G) = \{ww \mid w \in \{a, b\}^+\}$$

Indeed, first, S is rewritten to XY. Then, by using its states, G ensures that whenever X is rewritten to aX, the current state is changed to force the rewrite of Y to aY. Similarly, whenever X is rewritten to bX, the current state is changed to force the rewrite of Y to bY. Every successful derivation is finished by rewriting X to a or b and then Y to a or b, respectively.

For example, $abab$ is produced by the following derivation

$$
\begin{array}{rll}
(p_0, S) & \Rightarrow (p_0, XY) & [(p_0, S) \to (p_0, XY)] \\
& \Rightarrow (p_1, aXY) & [(p_0, X) \to (p_1, aX)] \\
& \Rightarrow (p_0, aXaY) & [(p_1, Y) \to (p_0, aY)] \\
& \Rightarrow (p_4, abaY) & [(p_0, X) \to (p_4, b)] \\
& \Rightarrow (p_0, abab) & [(p_4, Y) \to (p_0, b)]
\end{array}
$$

□

By **ST**, we denote the family of languages generated by state grammars. For every $n \geq 1$, $\mathbf{ST}_n$ denotes the family of languages generated by n-limited state grammars. Set

$$\mathbf{ST}_\infty = \bigcup_{n \geq 1} \mathbf{ST}_n$$

5.4.2 *Generative Power*

In this section, which closes the chapter, we give the key result concerning state grammars, originally established in [6].

Theorem 5.4.3. $\mathbf{CF} = \mathbf{ST}_1 \subset \mathbf{ST}_2 \subset \cdots \subset \mathbf{ST}_\infty \subset \mathbf{ST} = \mathbf{CS}$ □

References

1. Barbaiani, M., Bibire, C., Dassow, J., Delaney, A., Fazekas, S., Ionescu, M., Liu, G., Lodhi, A., Nagy, B.: The power of programmed grammars with graphs from various classes. J. Appl. Math. Comput. **22**(1–2), 21–38 (2006)
2. Bordihn, H., Holzer, M.: Programmed grammars and their relation to the LBA problem. Acta Informatica **43**(4), 223–242 (2006)
3. Dassow, J., Păun, G.: Regulated Rewriting in Formal Language Theory. Springer, New York (1989)
4. Fernau, H.: Nonterminal complexity of programmed grammars. Theor. Comp. Sci. **296**(2), 225–251 (2003)
5. Fernau, H., Freund, R., Oswald, M., Reinhardt, K.: Refining the nonterminal complexity of graph-controlled, programmed, and matrix grammars. J. Autom. Lang. Combin. **12**(1–2), 117–138 (2007)
6. Kasai, T.: An hierarchy between context-free and context-sensitive languages. J. Comput. Syst. Sci. **4**, 492–508 (1970)
7. Martín-Vide, C., Mitrana, V., Păun, G. (eds.): Formal Languages and Applications, chap. 13, pp. 249–274. Springer, Berlin (2004)
8. Meduna, A.: Automata and Languages: Theory and Applications. Springer, London (2000)
9. Rozenberg, G., Salomaa, A. (eds.): Handbook of Formal Languages, Volume 2: Linear Modeling: Background and Application. Springer, New York (1997)
10. Vrábel, L.: A new normal form for programmed grammars with appearance checking. In: Proceedings of the 18th Conference STUDENT EEICT 2012, vol. 3, pp. 420–425. Brno University of Technology, Brno, CZ (2012)

Part III
Regulated Grammars: Special Topics

This part covers several special, but vivid and important topics concerning grammatical regulation. It consists of Chaps. 6 through 9.

Chapter 6 studies *one-sided random context grammars* as random context grammars that verify their context conditions only in one direction from the rewritten nonterminal within sentential forms. Chapter 7 discusses a vivid investigation area concerning the elimination of erasing rules, which have the empty string on their right-hand sides, from regulated grammars. Chapter 8 studies regulated grammars that generate their languages extended by some additional strings, which may represent useful information closely related to this generation. Chapter 9 defines the relation of a direct derivation over free monoids generated by finitely many strings. It explains that this definition represents, in effect, an algebraic way of grammatical regulation, whose power is also studied therein.

Chapter 6
One-Sided Versions of Random Context Grammars

Abstract Consider random context grammars. The present chapter discusses their special cases referred to as *one-sided random context grammars*. In every one-sided random context grammar, the set of rules is divided into the set of *left random context rules* and the set of *right random context rules*. When applying a left random context rule, the grammar checks the existence and absence of its permitting and forbidding symbols, respectively, only in the prefix to the left of the rewritten nonterminal. Analogously, when applying a right random context rule, it checks the existence and absence of its permitting and forbidding symbols, respectively, only in the suffix to the right of the rewritten nonterminal. Otherwise, it works just like any ordinary random context grammar. In Sects. 6.1 and 6.2, the present chapter demonstrates that propagating versions of one-sided random context grammars characterize the family of context-sensitive languages, and with erasing rules, they characterize the family of recursively enumerable languages. Furthermore, it discusses the generative power of special cases of one-sided random context grammars. Specifically, it proves that *one-sided permitting grammars*, which have only permitting rules, are more powerful than context-free grammars; on the other hand, they are no more powerful than scattered context grammars. *One-sided forbidding grammars*, which have only forbidding rules, are equivalent to selective substitution grammars. Finally, *left forbidding grammars*, which have only left-sided forbidding rules, are only as powerful as context-free grammars. Section 6.3 establishes four normal forms of one-sided random context grammars. After that, Sect. 6.4 studies reduction of one-sided random context grammars with respect to the number of nonterminals and rules. Section 6.5 places various leftmost restrictions on derivations in one-sided random context grammars and investigate how they affect the generative power of these grammars. Section 6.6 discusses a generalized version of one-sided forbidding grammars. Finally, Sect. 6.7 investigates parsing-related variants of one-sided forbidding grammars in order to sketch their future application-related perspectives.

A. Meduna and P. Zemek, *Regulated Grammars and Automata*,
DOI 10.1007/978-1-4939-0369-6_6,

Keywords Special cases of random context grammars • One-sided random context grammars • One-sided permitting grammars • One-sided forbidding grammars • Left forbidding grammars • Generative power • Normal forms • Reduction • Leftmost derivation restrictions • Generalization • Parsing

The present chapter returns to the discussion of random context grammars, whose fundamental properties were established in Chap. 4. Recall that in these grammars, two finite sets of nonterminals are attached to every context-free rule. One set contains *permitting symbols* while the other has *forbidding symbols*. A rule like this can rewrite a nonterminal in an ordinary context-free way provided that each of its permitting symbols occurs in the current sentential form while each of its forbidding symbols does not occur there. There exist two important special cases of random context grammars—*permitting grammars* and *forbidding grammars*. If no rule has any forbidding symbols, the grammar is called a permitting grammar. If no rule has any permitting symbols, the grammar is called a forbidding grammar.

As is demonstrated in Chap. 4, random context grammars are significantly stronger than ordinary context-free grammars (see Theorem 4.3.7). In fact, they characterize the family of recursively enumerable languages, and this computational completeness obviously represents their indisputable advantage.

From a pragmatical standpoint, however, random context grammars have a drawback consisting in the necessity of scanning the current sentential form in its entirety during every single derivation step. From this viewpoint, it is highly desirable to modify these grammars so they scan only a part of the sentential form, yet they keep their computational completeness. *One-sided random context grammars*, discussed in this chapter, represent a modification like this.

Specifically, in every one-sided random context grammar (see Sect. 6.1), the set of rules is divided into the set of *left random context rules* and the set of *right random context rules*. When applying a left random context rule, the grammar checks the existence and absence of its permitting and forbidding symbols, respectively, only in the prefix to the left of the rewritten nonterminal in the current sentential form. Analogously, when applying a right random context rule, it checks the existence and absence of its permitting and forbidding symbols, respectively, only in the suffix to the right of the rewritten nonterminal. Otherwise, it works just like any ordinary random context grammar.

As the main result of this chapter, we demonstrate that propagating versions of one-sided random context grammars, which possess no erasing rules, characterize the family of context-sensitive languages (see Theorem 6.2.3), and with erasing rules, they characterize the family of recursively enumerable languages (see Theorem 6.2.4).

Furthermore, we discuss the generative power of several special cases of one-sided random context grammars. Specifically, we prove that *one-sided permitting grammars*, which have only permitting rules, are more powerful than context-free grammars; on the other hand, they are no more powerful than scattered context grammars (see Theorem 6.2.21). *One-sided forbidding grammars*, which have only forbidding rules, are equivalent to selective substitution grammars (see Theorem 6.2.7). Finally, *left forbidding grammars*, which have only left-sided forbidding rules, are only as powerful as context-free grammars (see Theorem 6.2.13).

Taking into account the definition of one-sided random context grammars and all the results sketched above, we see that these grammars may fulfill an important role in the language theory and its applications for the following four reasons.

(I) From a practical viewpoint, one-sided random context grammars examine the existence of permitting and the absence of forbidding symbols only within a portion of the current sentential form while ordinary random context grammars examine the entire current sentential form. As a result, the one-sided versions of these grammars work in a more economical and, therefore, efficient way than the ordinary versions.

(II) The one-sided versions of propagating random context grammars are stronger than ordinary propagating random context grammars. Indeed, the language family defined by propagating random context grammars is properly included in the family of context-sensitive languages (see Theorem 4.3.6). One-sided random context grammars are as powerful as ordinary random context grammars. These results come as a surprise because one-sided random context grammars examine only parts of sentential forms as pointed out in (I).

(III) Left forbidding grammars were introduced in [22], which also demonstrated that these grammars only define the family of context-free languages (see Theorem 1 in [22]). It is more than natural to generalize left forbidding grammars to one-sided forbidding grammars, which are stronger than left forbidding grammars (see Corollary 6.2.24). As a matter of fact, even *propagating left permitting grammars*, introduced in [8], are stronger than left forbidding grammars because they define a proper superfamily of the family of context-free languages (see Corollary 6.2.22).

(IV) In the future, one might find results achieved in this chapter useful when attempting to solve some well-known open problems. Specifically, recall that every propagating scattered context grammar can be turned to an equivalent context-sensitive grammar (see Theorem 4.7.6), but it is a longstanding open problem whether these two kinds of grammars are actually equivalent—the PSCAT = CS problem (see Sect. 4.7.2). If in the future one proves that propagating one-sided permitting grammars and propagating one-sided random context grammars are equivalent, then so are propagating scattered context grammars and context-sensitive grammars (see Theorem 6.2.21), so the PSCAT = CS problem would be solved.

This chapter is organized as follows. First, Sect. 6.1 defines one-sided random context grammars and their variants, and illustrates them by examples. Then, Sect. 6.2 establishes their generative power. Section 6.3 presents four normal forms of these grammars. After that, Sect. 6.4 studies reduction of one-sided random context grammars with respect to the number of nonterminals and rules. In Sect. 6.5, we place three leftmost derivation restrictions on one-sided random context grammars and investigate their generative power. Section 6.6 introduces a generalized version of one-sided forbidding grammars. Finally, Sect. 6.7 concludes the present chapter by investigating one-sided random context grammars from a more practical viewpoint by investigating their parsing-related variants.

6.1 Definitions and Examples

Next, we formally define one-sided random context grammars and their variants. In addition, we illustrate them by examples.

Definition 6.1.1. A *one-sided random context grammar* is a quintuple

$$G = (N, T, P_L, P_R, S)$$

where N and T are two disjoint alphabets, $S \in N$, and

$$P_L, P_R \subseteq N \times (N \cup T)^* \times 2^N \times 2^N$$

are two finite relations. Set $V = N \cup T$. The components V, N, T, P_L, P_R, and S are called the *total alphabet*, the alphabet of *nonterminals*, the alphabet of *terminals*, the set of *left random context rules*, the set of *right random context rules*, and the *start symbol*, respectively. Each $(A, x, U, W) \in P_L \cup P_R$ is written as

$$(A \to x, U, W)$$

throughout this chapter. For $(A \to x, U, W) \in P_L$, U and W are called the *left permitting context* and the *left forbidding context*, respectively. For $(A \to x, U, W) \in P_R$, U and W are called the *right permitting context* and the *right forbidding context*, respectively. □

When applying a left random context rule, the grammar checks the existence and absence of its permitting and forbidding symbols, respectively, only in the prefix to the left of the rewritten nonterminal in the current sentential form. Analogously, when applying a right random context rule, it checks the existence and absence of its permitting and forbidding symbols, respectively, only in the suffix to the right of the rewritten nonterminal. The following definition states this formally.

Definition 6.1.2. Let $G = (N, T, P_L, P_R, S)$ be a one-sided random context grammar. The *direct derivation relation* over V^* is denoted by $\Rightarrow_G$ and defined as follows. Let $u, v \in V^*$ and $(A \to x, U, W) \in P_L \cup P_R$. Then,

$$uAv \Rightarrow_G uxv$$

if and only if

$$(A \to x, U, W) \in P_L, U \subseteq \text{alph}(u), \text{ and } W \cap \text{alph}(u) = \emptyset$$

or

$$(A \to x, U, W) \in P_R, U \subseteq \text{alph}(v), \text{ and } W \cap \text{alph}(v) = \emptyset$$

Let $\Rightarrow_G^n$, $\Rightarrow_G^+$, and $\Rightarrow_G^*$ denote the nth power of $\Rightarrow_G$, for some $n \geq 0$, the transitive closure of $\Rightarrow_G$, and the reflexive-transitive closure of $\Rightarrow_G$, respectively. □

The language generated by a one-sided random context grammar is defined as usual—that is, it consists of strings over the terminal alphabet that can be generated from the start symbol.

Definition 6.1.3. Let $G = (N, T, P_L, P_R, S)$ be a one-sided random context grammar. The *language* of G is denoted by $L(G)$ and defined as

$$L(G) = \{w \in T^* \mid S \Rightarrow_G^* w\}$$ □

Next, we define several special variants of one-sided random context grammars.

Definition 6.1.4. Let $G = (N, T, P_L, P_R, S)$ be a one-sided random context grammar. If $(A \to x, U, W) \in P_L \cup P_R$ implies that $|x| \geq 1$, then G is a *propagating one-sided random context grammar*. If $(A \to x, U, W) \in P_L \cup P_R$ implies that $W = \emptyset$, then G is a *one-sided permitting grammar*. If $(A \to x, U, W) \in P_L \cup P_R$ implies that $U = \emptyset$, then G is a *one-sided forbidding grammar*. By analogy with propagating one-sided random context grammars, we define a *propagating one-sided permitting grammar* and a *propagating one-sided forbidding grammar*, respectively. □

Definition 6.1.5. Let $G = (N, T, P_L, P_R, S)$ be a one-sided random context grammar. If $P_R = \emptyset$, then G is a *left random context grammar*. By analogy with one-sided permitting and forbidding grammars, we define a *left permitting grammar* and a *left forbidding grammar* (see [8, 22]). Their propagating versions are defined analogously as well. □

Next, we illustrate the above definitions by three examples.

Example 6.1.6. Consider the one-sided random context grammar

$$G = (\{S, A, B, \bar{A}, \bar{B}\}, \{a, b, c\}, P_L, P_R, S)$$

where P_L contains the following four rules

$$(S \to AB, \emptyset, \emptyset) \qquad (\bar{B} \to B, \{A\}, \emptyset)$$
$$(B \to b\bar{B}c, \{\bar{A}\}, \emptyset) \qquad (B \to \varepsilon, \emptyset, \{A, \bar{A}\})$$

and P_R contains the following three rules

$$(A \to a\bar{A}, \{B\}, \emptyset) \qquad (\bar{A} \to A, \{\bar{B}\}, \emptyset) \qquad (A \to \varepsilon, \{B\}, \emptyset)$$

It is rather easy to see that every derivation that generates a nonempty string of $L(G)$ is of the form

$$\begin{aligned} S &\Rightarrow_G AB \\ &\Rightarrow_G a\bar{A}B \\ &\Rightarrow_G a\bar{A}b\bar{B}c \\ &\Rightarrow_G aAb\bar{B}c \\ &\Rightarrow_G aAbBc \\ &\Rightarrow_G^* a^nAb^nBc^n \\ &\Rightarrow_G a^nb^nBc^n \\ &\Rightarrow_G a^nb^nc^n \end{aligned}$$

where $n \geq 1$. The empty string is generated by

$$S \Rightarrow_G AB \Rightarrow_G B \Rightarrow_G \varepsilon$$

Based on the previous observations, we see that G generates the non-context-free language $\{a^nb^nc^n \mid n \geq 0\}$. □

Example 6.1.7. Consider $K = \{a^nb^mc^m \mid 1 \leq m \leq n\}$. This non-context-free language is generated by the one-sided permitting grammar

$$G = \big(\{S, A, B, X, Y\}, \{a, b, c\}, P_L, \emptyset, S\big)$$

with P_L containing the following seven rules

$$\begin{array}{lll} (S \to AX, \emptyset, \emptyset) & (A \to a, \emptyset, \emptyset) & (X \to bc, \emptyset, \emptyset) \\ & (A \to aB, \emptyset, \emptyset) & (X \to bYc, \{B\}, \emptyset) \\ & (B \to A, \emptyset, \emptyset) & (Y \to X, \{A\}, \emptyset) \end{array}$$

Notice that G is, in fact, a propagating left permitting grammar. Observe that $(X \to bYc, \{B\}, \emptyset)$ is applicable if B, produced by $(A \to aB, \emptyset, \emptyset)$, occurs to the left of X in the current sentential form. Similarly, $(Y \to X, \{A\}, \emptyset)$ is applicable if A, produced by $(B \to A, \emptyset, \emptyset)$, occurs to the left of Y in the current sentential form. Consequently, it is rather easy to see that every derivation that generates $w \in L(G)$ is of the form[1]

[1] Notice that after X is rewritten to bc by $(X \to bc, \emptyset, \emptyset)$, more as can be generated by $(A \to aB, \emptyset, \emptyset)$. However, observe that this does not affect the generated language.

$$
\begin{aligned}
S &\Rightarrow_G AX \\
&\Rightarrow_G^* a^u AX \\
&\Rightarrow_G a^{u+1} BX \\
&\Rightarrow_G a^{u+1} BbYc \\
&\Rightarrow_G a^{u+1} AbYc \\
&\Rightarrow_G^* a^{u+1+v} AbYc \\
&\Rightarrow_G a^{u+1+v} AbXc \\
&\ \vdots \\
&\Rightarrow_G^* a^{n-1} Ab^{m-1} Xc^{m-1} \\
&\Rightarrow_G^2 a^n b^m c^m = w
\end{aligned}
$$

where $u, v \geq 0$, $1 \leq m \leq n$. Hence, $L(G) = K$. □

Example 6.1.8. Consider the one-sided forbidding grammar

$$G = \big(\{S, A, B, A', B', \bar{A}, \bar{B}\}, \{a, b, c\}, P_L, P_R, S\big)$$

where P_L contains the following five rules

$$
\begin{array}{lll}
(S \to AB, \emptyset, \emptyset) & (B \to bB'c, \emptyset, \{A, \bar{A}\}) & (B' \to B, \emptyset, \{A'\}) \\
 & (B \to \bar{B}, \emptyset, \{A, A'\}) & (\bar{B} \to \varepsilon, \emptyset, \{\bar{A}\})
\end{array}
$$

and P_R contains the following four rules

$$
\begin{array}{ll}
(A \to aA', \emptyset, \{B'\}) & (A' \to A, \emptyset, \{B\}) \\
(A \to \bar{A}, \emptyset, \{B'\}) & (\bar{A} \to \varepsilon, \emptyset, \{B\})
\end{array}
$$

Notice that every derivation that generates a nonempty string of $L(G)$ is of the form

$$
\begin{aligned}
S &\Rightarrow_G AB \\
&\Rightarrow_G aA'B \\
&\Rightarrow_G aA'bB'c \\
&\Rightarrow_G aAbB'c \\
&\Rightarrow_G aAbBc \\
&\Rightarrow_G^* a^n Ab^n Bc^n \\
&\Rightarrow_G a^n \bar{A}b^n Bc^n \\
&\Rightarrow_G a^n \bar{A}b^n \bar{B}c^n \\
&\Rightarrow_G a^n b^n \bar{B}c^n \\
&\Rightarrow_G a^n b^n c^n
\end{aligned}
$$

where $n \geq 1$. The empty string is generated by

$$S \Rightarrow_G AB \Rightarrow_G \bar{A}B \Rightarrow_G \bar{A}\bar{B} \Rightarrow_G \bar{B} \Rightarrow_G \varepsilon$$

Based on the previous observations, we see that G generates the non-context-free language $\{a^n b^n c^n \mid n \geq 0\}$. □

Denotation of Language Families

Throughout the rest of this chapter, the language families under discussion are denoted in the following way. **ORC**, **OPer**, and **OFor** denote the language families generated by one-sided random context grammars, one-sided permitting grammars, and one-sided forbidding grammars, respectively. **LRC**, **LPer**, and **LFor** denote the language families generated by left random context grammars, left permitting grammars, and left forbidding grammars, respectively.

The notation with the upper index $-\varepsilon$ stands for the corresponding propagating family. For example, $\mathbf{ORC}^{-\varepsilon}$ denotes the family of languages generated by propagating one-sided random context grammars.

6.2 Generative Power

In this section, we establish relations between the above-mentioned language families and some well-known language families. Most importantly, we show that

(I) $\mathbf{ORC}^{-\varepsilon} = \mathbf{CS}$ (Theorem 6.2.3);
(II) $\mathbf{ORC} = \mathbf{RE}$ (Theorem 6.2.4);
(III) $\mathbf{OFor} = \mathbf{S}$ (Theorem 6.2.7);
(IV) $\mathbf{OFor}^{-\varepsilon} = \mathbf{S}^{-\varepsilon}$ (Theorem 6.2.8);
(V) $\mathbf{LFor}^{-\varepsilon} = \mathbf{LFor} = \mathbf{CF}$ (Theorem 6.2.13);
(VI) $\mathbf{CF} \subset \mathbf{OPer}^{-\varepsilon} \subseteq \mathbf{SCAT}^{-\varepsilon}$ (Theorem 6.2.21).

The present section consists of three subsections. First, Sect. 6.2.1 studies the generative power of one-sided random context grammars. Then, Sect. 6.2.2 investigates the power of one-sided forbidding grammars. Finally, Sect. 6.2.3 discusses one-sided permitting grammars and their generative power.

6.2.1 One-Sided Random Context Grammars

First, we consider one-sided random context grammars and their propagating versions. We prove that $\mathbf{ORC}^{-\varepsilon} = \mathbf{CS}$ and $\mathbf{ORC} = \mathbf{RE}$.

Lemma 6.2.1. $\mathbf{CS} \subseteq \mathbf{ORC}^{-\varepsilon}$

Proof. Let $G = (N, T, P, S)$ be a context-sensitive grammar. Without any loss of generality, making use of Theorem 4.1.5, we assume that G is in the Penttonen normal form. We next construct a propagating one-sided random context grammar H such that $L(H) = L(G)$. Set $\bar{N} = \{\bar{A} \mid A \in N\}$, $\hat{N} = \{\hat{A} \mid A \in N\}$, and $N' = N \cup \bar{N} \cup \hat{N}$. Define H as

$$H = \left(N', T, P_L, P_R, S\right)$$

with P_L and P_R constructed as follows:

(1) for each $A \to a \in P$, where $A \in N$ and $a \in T$, add $(A \to a, \emptyset, N')$ to P_L;
(2) for each $A \to BC \in P$, where $A, B, C \in N$, add $(A \to BC, \emptyset, \bar{N} \cup \hat{N})$ to P_L;
(3) for each $AB \to AC \in P$, where $A, B, C \in N$, add $(B \to C, \{\hat{A}\}, N)$ to P_L;
(4) for each $A \in N$, add $(A \to \bar{A}, \emptyset, N \cup \hat{N})$ and $(A \to \hat{A}, \emptyset, N \cup \hat{N})$ to P_L;
(5) for each $A \in N$, add $(\bar{A} \to A, \emptyset, \bar{N} \cup \hat{N})$ and $(\hat{A} \to A, \emptyset, \bar{N} \cup \hat{N})$ to P_R.

Before proving that $L(H) = L(G)$, we give an insight into the construction. The simulation of context-free rules of the form $A \to BC$, where $A, B, C \in N$, is performed directly by rules introduced in (2). H simulates context-sensitive rules—that is, rules of the form $AB \to AC$, where $A, B, C \in N$—as follows. H first rewrites all nonterminals to the left of an occurrence of A to their barred versions by rules from (4), starting from the leftmost nonterminal of the current sentential form. Then, it rewrites A to $\hat{A}$ by $(A \to \hat{A}, \emptyset, N \cup \hat{N})$ from (4). After this, it rewrites B to C by $(B \to C, \{\hat{A}\}, N)$ from (3). Finally, H rewrites $\hat{A}$ back to A and all barred nonterminals back to their corresponding original versions by rules from (5) in the right-to-left way.

To prevent $AAB \Rightarrow_H AaB \Rightarrow_H \hat{A}aB \Rightarrow_H AaC$, rules simulating $A \to a$, where $A \in A$ and $a \in T$, introduced in (1), can be used only if there are no nonterminals to the left of A. Therefore, a terminal can never appear between two nonterminals. Consequently, every sentential form generated by H is of the form x_1x_2, where $x_1 \in T^*$ and $x_2 \in N'^*$.

To prove that $L(H) = L(G)$, we first prove three claims. The first claim shows that every $y \in L(G)$ can be generated by G in two stages; first, only nonterminals are generated, and then, all nonterminals are rewritten to terminals. To prove that $L(G) \subseteq L(H)$, it then suffices to show how H simulates these derivations of G.

Claim 1. Let $y \in L(G)$. Then, there exists a derivation $S \Rightarrow_G^* x \Rightarrow_G^* y$, where $x \in N^+$, and during $x \Rightarrow_G^* y$, G applies only rules of the form $A \to a$, $A \in N$, where $a \in T$.

Proof. Let $y \in L(G)$. Since there are no rules in P with symbols from T on their left-hand sides, we can always rearrange all the applications of the rules occurring in $S \Rightarrow_G^* y$ so the claim holds. □

The second claim shows how certain derivations of G are simulated by H. Together with the previous claim, it is used to demonstrate that $L(G) \subseteq L(H)$ later in the proof.

Claim 2. If $S \Rightarrow_G^n x$, *where* $x \in N^+$, *for some* $n \geq 0$, *then* $S \Rightarrow_H^* x$.

Proof. This claim is established by induction on $n \geq 0$.

Basis. Let $n = 0$. Then, for $S \Rightarrow_G^0 S$, there is $S \Rightarrow_H^0 S$, so the basis holds.

Induction Hypothesis. Suppose that there exists $m \geq 0$ such that the claim holds for all derivations of length ℓ, where $0 \leq \ell \leq m$.

Induction Step. Consider any derivation of the form

$$S \Rightarrow_G^{n+1} w$$

where $w \in N^+$. Since $n + 1 \geq 1$, this derivation can be expressed as

$$S \Rightarrow_G^n x \Rightarrow_G w$$

for some $x \in N^+$. By the induction hypothesis, $S \Rightarrow_H^* x$.

Next, we consider all possible forms of $x \Rightarrow_G w$, covered by the following two cases—(i) and (ii).

(i) Let $A \rightarrow BC \in P$ and $x = x_1 A x_2$, where $A, B, C \in N$, and $x_1, x_2 \in N^*$. Then,

$$x_1 A x_2 \Rightarrow_G x_1 BC x_2$$

By (2), $(A \rightarrow BC, \emptyset, \bar{N} \cup \hat{N}) \in P_L$, so

$$x_1 A x_2 \Rightarrow_H x_1 BC x_2$$

which completes the induction step for (i).

(ii) Let $AB \rightarrow AC \in P$ and $x = x_1 AB x_2$, where $A, B, C \in N$, and $x_1, x_2 \in N^*$. Then,

$$x_1 AB x_2 \Rightarrow_G x_1 BC x_2$$

Let $x_1 = X_1 X_2 \cdots X_k$, where $X_i \in N$, for all i, $1 \leq i \leq k$, $k = |x_1|$. By (4), $(X_i \rightarrow \bar{X}_i, \emptyset, N \cup \hat{N}) \in P_L$, for all i, $1 \leq i \leq k$, so

$$\begin{aligned} X_1 X_2 \cdots X_k AB x_2 &\Rightarrow_H \bar{X}_1 X_2 \cdots X_k AB x_2 \\ &\Rightarrow_H \bar{X}_1 \bar{X}_2 \cdots X_k AB x_2 \\ &\vdots \\ &\Rightarrow_H \bar{X}_1 \bar{X}_2 \cdots \bar{X}_k AB x_2 \end{aligned}$$

Let $\bar{x}_1 = \bar{X}_1\bar{X}_2\cdots\bar{X}_k$. By (4), $(A \rightarrow \hat{A}, \emptyset, N \cup \hat{N}) \in P_L$, so

$$\bar{x}_1 ABx_2 \Rightarrow_H \bar{x}_1 \hat{A}Bx_2$$

By (3), $(B \rightarrow C, \{\hat{A}\}, N) \in P_L$, so

$$\bar{x}_1 \hat{A}Bx_2 \Rightarrow_H \bar{x}_1 \hat{A}Cx_2$$

Finally, by (5), $(\hat{A} \rightarrow A, \emptyset, \bar{N} \cup \hat{N}), (\bar{X}_i \rightarrow X_i, \emptyset, \bar{N} \cup \hat{N}) \in P_R$, for all i, $1 \leq i \leq k$, so

$$\begin{aligned}\bar{X}_1\bar{X}_2\cdots\bar{X}_k\hat{A}Cx_2 &\Rightarrow_H \bar{X}_1\bar{X}_2\cdots\bar{X}_kACx_2\\ &\Rightarrow_H \bar{X}_1\bar{X}_2\cdots X_kACx_2\\ &\;\;\vdots\\ &\Rightarrow_H \bar{X}_1X_2\cdots X_kACx_2\\ &\Rightarrow_H X_1X_2\cdots X_kACx_2\end{aligned}$$

which completes the induction step for (ii).

Observe that cases (i) and (ii) cover all possible forms of $x \Rightarrow_G w$. Thus, Claim 2 holds. □

Next, we prove how G simulates derivations of H. The following claim is used to prove that $L(H) \subseteq L(G)$ later in the proof. Set $V = N \cup T$ and $V' = N' \cup T$. Define the homomorphism τ from V'^* to V^* as $\tau(\bar{A}) = A$, $\tau(\hat{A}) = A$, and $\tau(A) = A$, for all $A \in N$, and $\tau(a) = a$, for all $a \in T$.

Claim 3. If $S \Rightarrow_H^n x$, where $x \in V'^+$, for some $n \geq 0$, then $S \Rightarrow_G^* \tau(x)$ and x is of the form $x'X_1X_2\cdots X_h$, where $x' \in T^*$ and $X_i \in N'$, for all i, $1 \leq i \leq h$, for some $h \geq 0$. Furthermore, if $X_j \in \hat{N}$, for some j, $1 \leq j \leq h$, then $X_k \in \bar{N}$, for all k, $1 \leq k < j$, and $X_l \in N$, for all l, $j < l \leq h$.

Proof. This claim is established by induction on $n \geq 0$.

Basis. Let $n = 0$. Then, for $S \Rightarrow_H^0 S$, there is $S \Rightarrow_G^0 S$, so the basis holds.

Induction Hypothesis. Suppose that there exists $m \geq 0$ such that the claim holds for all derivations of length ℓ, where $0 \leq \ell \leq m$.

Induction Step. Consider any derivation of the form

$$S \Rightarrow_H^{n+1} w$$

where $w \in V'^+$. Since $n + 1 \geq 1$, this derivation can be expressed as

$$S \Rightarrow_H^n x \Rightarrow_H w$$

for some $x \in V'^{+}$. By the induction hypothesis, $S \Rightarrow_G^* \tau(x)$ and x is of the form $x'X_1X_2\cdots X_h$, where $x' \in T^*$ and $X_i \in N'$, for all i, $1 \leq i \leq h$, for some $h \geq 0$. Furthermore, if $X_j \in \hat{N}$, for some j, $1 \leq j \leq h$, then $X_k \in \bar{N}$, for all k, $1 \leq k < j$, and $X_l \in N$, for all l, $j < l \leq h$.

Next, we consider all possible forms of $x \Rightarrow_H w$, covered by the following four cases—(i) through (iv). That is, we show how rules introduced to P_L and P_R in (1) through (5) are simulated by G. The last case covers the simulation of rules from both (4) and (5) because these rules are simulated in the same way.

(i) Let $x'X_1X_2\cdots X_h \Rightarrow_H x'aX_2\cdots X_h$ by $(X_1 \rightarrow a, \emptyset, N') \in P_L$, introduced in (1) from $X_1 \rightarrow a \in P$, where $a \in T$. Then,

$$x'\tau(X_1X_2\cdots X_h) \Rightarrow_G x'\tau(aX_2\cdots X_h)$$

which completes the induction step for (i).

(ii) Let $x'X_1X_2\cdots X_{j-1}X_jX_{j+1}\cdots X_h \Rightarrow_H x'X_1X_2\cdots X_{j-1}BCX_{j+1}\cdots X_h$ by $(X_j \rightarrow BC, \emptyset, \bar{N} \cup \hat{N}) \in P_L$, introduced in (2) from $X_j \rightarrow BC \in P$, for some j, $1 \leq j \leq h$, and $B, C \in N$. Then,

$$x'\tau(X_1X_2\cdots X_{j-1}X_jX_{j+1}\cdots X_h) \Rightarrow_G x'\tau(X_1X_2\cdots X_{j-1}BCX_{j+1}\cdots X_h)$$

which completes the induction step for (ii).

(iii) Let $x'X_1X_2\cdots X_{j-1}X_jX_{j+1}\cdots X_h \Rightarrow_H x'X_1X_2\cdots X_{j-1}CX_{j+1}\cdots X_h$ by $(X_j \rightarrow C, \{\hat{A}\}, N) \in P_L$, introduced in (3) from $AX_j \rightarrow AC \in P$, for some j, $2 \leq j \leq h$, and $A, C \in N$. By the induction hypothesis, $X_{j-1} = \hat{A}$. Therefore,

$$x'\tau(X_1X_2\cdots X_{j-1}X_jX_{j+1}\cdots X_h) \Rightarrow_G x'\tau(X_1X_2\cdots X_{j-1}CX_{j+1}\cdots X_h)$$

which completes the induction step for (iii).

(iv) Let $x'X_1X_2\cdots X_{j-1}X_jX_{j+1}\cdots X_h \Rightarrow_H x'X_1X_2\cdots X_{j-1}X_j'X_{j+1}\cdots X_h$ by $(X_j \rightarrow X_j', \emptyset, W) \in P_L \cup P_R$, introduced in (4) or (5), for some j, $1 \leq j \leq h$, where X_j' depends on the particular rule that was used. Then,

$$x'\tau(X_1X_2\cdots X_h) \Rightarrow_G^0 x'\tau(X_1X_2\cdots X_h)$$

which completes the induction step for (iv).

Observe that cases (i) through (iv) cover all possible forms of $x \Rightarrow_H w$. Thus, Claim 3 holds. □

We next prove that $L(H) = L(G)$. Let $y \in L(G)$. Then, by Claim 1, there is $S \Rightarrow_G^* x \Rightarrow_G^* y$ such that $x \in N^+$ and during $x \Rightarrow_G^* y$, G uses only rules of the form $A \rightarrow a$, where $A \in N$, $a \in T$. By Claim 2, $S \Rightarrow_H^* x$. Let $x = A_1A_2\cdots A_k$ and $y = a_1a_2\cdots a_k$, where $A_i \in N$, $A_i \rightarrow a_i \in P$, $a_i \in T$, for all i, $1 \leq i \leq k$, $k = |x|$. By (1), $(A_i \rightarrow a_i, \emptyset, N') \in P_L$, for all i, $1 \leq i \leq k$, so

$$\begin{aligned} A_1A_2\cdots A_k &\Rightarrow_H a_1A_2\cdots A_k \\ &\Rightarrow_H a_1a_2\cdots A_k \\ &\vdots \\ &\Rightarrow_H a_1a_2\cdots a_k \end{aligned}$$

Consequently, $y \in L(G)$ implies that $y \in L(H)$, so $L(G) \subseteq L(H)$.

Consider Claim 3 for $x \in T^+$. Then, $x \in L(H)$ implies that $\tau(x) = x \in L(G)$, so $L(H) \subseteq L(G)$. Since $L(G) \subseteq L(H)$ and $L(H) \subseteq L(G)$, $L(H) = L(G)$, so Lemma 6.2.1 holds. □

Lemma 6.2.2. $\mathbf{ORC}^{-\varepsilon} \subseteq \mathbf{CS}$

Proof. Let $G = (N, T, P_L, P_R, S)$ be a propagating one-sided random context grammar. From G, we can construct a phrase-structure grammar, $H = (N', T, P', S')$, such that $L(G) = L(H)$ and if $S' \Rightarrow_H^* x \Rightarrow_H^* w$, where $x \in (N' \cup T)^+$ and $w \in T^+$, then $|x| \leq 4|w|$. Consequently, by the workspace theorem (see Theorem 3.3.15), $L(H) \in \mathbf{CS}$. Since $L(G) = L(H)$, $L(G) \in \mathbf{CS}$, so the lemma holds. □

Theorem 6.2.3. $\mathbf{ORC}^{-\varepsilon} = \mathbf{CS}$

Proof. This theorem follows from Lemmas 6.2.1 and 6.2.2. □

Theorem 6.2.4. $\mathbf{ORC} = \mathbf{RE}$

Proof. The inclusion $\mathbf{ORC} \subseteq \mathbf{RE}$ follows from Church's thesis. $\mathbf{RE} \subseteq \mathbf{ORC}$ can be proved by analogy with the proof of Lemma 6.2.1. Observe that by Theorem 4.1.4, G can additionally contain rules of the form $A \rightarrow \varepsilon$, where $A \in N$. We can simulate these context-free rules in the same way we simulate $A \rightarrow BC$, where $A, B, C \in N$—that is, for each $A \rightarrow \varepsilon \in P$, we introduce $(A \rightarrow \varepsilon, \emptyset, \bar{N} \cup \hat{N})$ to P_L. □

6.2.2 *One-Sided Forbidding Grammars*

Next, we consider one-sided forbidding grammars. First, we prove that $\mathbf{OFor} = \mathbf{S}$ and $\mathbf{OFor}^{-\varepsilon} = \mathbf{S}^{-\varepsilon}$. Then, we show that one-sided forbidding grammars with the set of left forbidding rules coinciding with the set of right forbidding rules characterize only the family of context-free languages. This characterization also holds in terms of left forbidding grammars. Indeed, we prove that $\mathbf{LFor}^{-\varepsilon} = \mathbf{LFor} = \mathbf{CF}$.

Lemma 6.2.5. *For every s-grammar G, there is a one-sided forbidding grammar H such that $L(H) = L(G)$.*

Proof. Let $G = (V, T, P, S, K)$ be an s-grammar. We next construct a one-sided forbidding grammar H such that $L(H) = L(G)$. Set $N = V - T$, $\hat{T} = \{\hat{a} \mid a \in T\}$, $T_1 = \{\langle a, 1\rangle \mid a \in T\}$, $T_2 = \{\langle a, 2\rangle \mid a \in T\}$, $T_{12} = T_1 \cup T_2$, and

$$M_{12} = \{\langle r, s, i \rangle \mid r \in P, s = (X^* \overline{Y} Z^*) \in K, i = 1, 2\}$$

Without any loss of generality, we assume that N, T, $\hat{T}$, T_1, T_2, and M_{12} are pairwise disjoint. Construct

$$H = (N', T, P_L, P_R, S)$$

as follows. Initially, set

$$\begin{aligned} N' &= N \cup \hat{T} \cup T_{12} \cup M_{12} \\ P_L &= \emptyset \\ P_R &= \emptyset \end{aligned}$$

Define the homomorphism τ from V^* to N'^* as $\tau(A) = A$, for all $A \in N$, and $\tau(a) = \hat{a}$, for all $a \in T$. Define the function $\mathscr{T}$ from 2^V to $2^{N'}$ as $\mathscr{T}(\emptyset) = \emptyset$ and

$$\mathscr{T}(\{A_1, \dots, A_n\}) = \{\tau(A_1), \dots, \tau(A_n)\}$$

Perform (1) and (2), given next.

(1) For each $s = (X^* \overline{Y} Z^*) \in K$ and each $A \in Y$ such that $r = (A \to y) \in P$,

(1.1) add $(\tau(A) \to \langle r, s, 1 \rangle, \emptyset, N' - \mathscr{T}(X))$ to P_L;
(1.2) add $(\langle r, s, 1 \rangle \to \langle r, s, 2 \rangle, \emptyset, N' - \mathscr{T}(Z))$ to P_R;
(1.3) add $(\langle r, s, 2 \rangle \to \tau(y), \emptyset, T_{12} \cup M_{12})$ to P_L.

(2) For each $a \in T$,

(2.1) add $(\hat{a} \to \langle a, 1 \rangle, \emptyset, N' - \hat{T})$ to P_L;
(2.2) add $(\langle a, 1 \rangle \to \langle a, 2 \rangle, \emptyset, N' - T_2)$ to P_R;
(2.3) add $(\langle a, 2 \rangle \to a, \emptyset, N')$ to P_L.

Before proving that $L(H) = L(G)$, let us informally describe (1) and (2). Let $X^* \overline{Y} Z^* \in K$ and $x_1 A x_2 \in V^*$, where $x_1, x_2 \in V^*$ and $A \in Y$. Observe that $x_1 \in X^*$ if and only if $(V - X) \cap \text{alph}(x_1) = \emptyset$, and $x_2 \in Z^*$ if and only if $(V - Z) \cap \text{alph}(x_2) = \emptyset$. Therefore, to simulate the application of $A \to y \in P$ in H, we first check the absence of all symbols from $V - X$ to the left of A, and then, we check the absence of all symbols from $V - Z$ to the right of A.

We need to guarantee the satisfaction of the following two conditions. First, we need to make sure that only a single rule is simulated at a time. For this purpose, we have the three-part construction of rules in (1). Examine it to see that whenever H tries to simultaneously simulate more than one rule of G, the derivation is blocked. Second, as opposed to s-grammars, one-sided forbidding grammars can neither rewrite terminals nor forbid their occurrence. To circumvent this restriction, rules introduced in (1) rewrite and generate hatted terminals which act as nonterminals. For example, $a \to bDc \in P$, where $a, b, c \in T$ and $D \in N$, is simulated by $\hat{a} \to \hat{b} D \hat{c}$ in H.

Hatted terminals can be rewritten to terminals by rules introduced in (2). Observe that this can be done only if there are no symbols from $N' - \hat{T}$ present in the current sentential form; otherwise, the derivation is blocked. Furthermore, observe that after a rule from (2.1) is applied, no rule of G can be simulated anymore. Based on these observations, we see that every successful derivation of $a_1 a_2 \cdots a_h$ in H is of the form

$$S \Rightarrow_H^* \hat{a}_1 \hat{a}_2 \cdots \hat{a}_h \Rightarrow_H^* a_1 a_2 \cdots a_h$$

and during $S \Rightarrow_H^* \hat{a}_1 \hat{a}_2 \cdots \hat{a}_h$, no sentential form contains any symbols from T.

To establish $L(H) = L(G)$, we prove two claims. First, Claim 1 shows how derivations of G are simulated by H. Then, Claim 2 demonstrates the converse—that is, it shows how derivations of H are simulated by G.

Claim 1. If $S \Rightarrow_G^n w \Rightarrow_G^* z$, where $w \in V^*$ and $z \in T^*$, for some $n \geq 0$, then $S \Rightarrow_H^* \tau(w)$.

Proof. This claim is established by induction on $n \geq 0$.

Basis. For $n = 0$, this claim obviously holds.

Induction Hypothesis. Suppose that there exists $n \geq 0$ such that the claim holds for all derivations of length ℓ, where $0 \leq \ell \leq n$.

Induction Step. Consider any derivation of the form

$$S \Rightarrow_G^{n+1} w \Rightarrow_G^* z$$

where $w \in V^*$ and $z \in T^*$. Since $n + 1 \geq 1$, this derivation can be expressed as

$$S \Rightarrow_G^n x \Rightarrow_G w \Rightarrow_G^* z$$

for some $x \in V^+$. Let $x = x_1 A x_2$ and $w = x_1 y x_2$ so $s = (X^* \overline{Y} Z^*) \in K$ such that $x_1 \in X^*$, $A \in Y$, $x_2 \in Z^*$, and $r = (A \rightarrow y) \in P$.

By the induction hypothesis, $S \Rightarrow_H^* \tau(x)$. By (1),

$$\begin{aligned}
&(\tau(A) \rightarrow \langle r, s, 1\rangle, \emptyset, N' - \mathscr{T}(X)) \in P_L \\
&(\langle r, s, 1\rangle \rightarrow \langle r, s, 2\rangle, \emptyset, N' - \mathscr{T}(Z)) \in P_R \\
&(\langle r, s, 2\rangle \rightarrow \tau(y), \emptyset, T_{12} \cup M_{12}) \in P_L
\end{aligned}$$

By the induction hypothesis and by $x_1 A x_2 \Rightarrow_G x_1 y x_2$, $\tau(x) = \tau(x_1)\tau(A)\tau(x_2)$, $(N' - \mathscr{T}(X)) \cap \mathrm{alph}(\tau(x_1)) = \emptyset$, and $(N' - \mathscr{T}(Z)) \cap \mathrm{alph}(\tau(x_2)) = \emptyset$, so

$$\begin{aligned}
\tau(x_1)\tau(A)\tau(x_2) &\Rightarrow_H \tau(x_1)\langle r, s, 1\rangle\tau(x_2) \\
&\Rightarrow_H \tau(x_1)\langle r, s, 2\rangle\tau(x_2) \\
&\Rightarrow_H \tau(x_1)\tau(y)\tau(x_2)
\end{aligned}$$

Since $\tau(w) = \tau(x_1)\tau(y)\tau(x_2)$, the induction step is completed. □

Set $V' = N' \cup T$. Define the homomorphism ψ from V'^* to V^* as $\psi(A) = A$, for all $A \in N$, $\psi(\langle r, s, 1\rangle) = \psi(\langle r, s, 2\rangle) = A$, for all $r = (A \to x) \in P$ and all $s = (X^*\overline{Y}Z^*) \in K$, and $\psi(\langle a, 1\rangle) = \psi(\langle a, 2\rangle) = \psi(\hat{a}) = a$, for all $a \in T$.

Claim 2. If $S \Rightarrow_H^n w \Rightarrow_H^* z$, where $w \in V'^*$ and $z \in T^*$, for some $n \geq 0$, then $S \Rightarrow_G^* \psi(w)$.

Proof. This claim is established by induction on $n \geq 0$.

Basis. For $n = 0$, this claim obviously holds.

Induction Hypothesis. Suppose that there exists $n \geq 0$ such that the claim holds for all derivations of length ℓ, where $0 \leq \ell \leq n$.

Induction Step. Consider any derivation of the form

$$S \Rightarrow_H^{n+1} w \Rightarrow_H^* z$$

where $w \in V'^*$ and $z \in T^*$. Since $n + 1 \geq 1$, this derivation can be expressed as

$$S \Rightarrow_H^n x \Rightarrow_H w \Rightarrow_H^* z$$

for some $x \in V'^+$. By the induction hypothesis, $S \Rightarrow_G^* \psi(x)$. Observe that if $x \Rightarrow_H w$ is derived by a rule introduced in (1.1), (1.2), or in (2), then the induction step follows directly from the induction hypothesis. Therefore, assume that $x = x_1\langle r, s, 2\rangle x_2$, $w = x_1\tau(y)x_2$, and $(\langle r, s, 2\rangle \to \tau(y), \emptyset, T_{12} \cup M_{12}) \in P_L$, introduced in (1.3) from $s = (X^*\overline{Y}Z^*) \in K$ such that $A \in Y$ and $r = (A \to y) \in P$. Recall that $(T_{12} \cup M_{12}) \cap \mathrm{alph}(x_1) = \emptyset$ has to hold; otherwise, the rule is not applicable. Next, we argue that $\psi(x) \Rightarrow_G \psi(w)$.

Observe that the two other rules from (1.1) and (1.2) have to be applied before $(\langle r, s, 2\rangle \to \tau(y), \emptyset, T_{12} \cup M_{12})$ is applicable. Therefore, $S \Rightarrow_H^* x$ has to be of the form

$$S \Rightarrow_H^* v_1\tau(A)v_2 \Rightarrow_H v_1\langle r, s, 1\rangle v_2 \Rightarrow_H^* x_1\langle r, s, 2\rangle x_2 = x$$

where $v_1, v_2 \in V'^*$ and $x_1\langle r, s, 2\rangle x_2$ is the first sentential form in $S \Rightarrow_H^* x$, where an occurrence of $\langle r, s, 1\rangle$ is rewritten to $\langle r, s, 2\rangle$. We next argue, by contradiction, that (i) $v_1 = x_1$ and (ii) $v_2 = x_2$. We then use (i) and (ii) to show that $\psi(x) \Rightarrow_G \psi(w)$.

(i) Assume that $v_1 \neq x_1$. The only possible way this case could happen is that after

$$S \Rightarrow_H^* v_1\tau(A)v_2 \Rightarrow_H v_1\langle r, s, 1\rangle v_2$$

v_1 is rewritten by a rule t. Since $\langle r, s, 1\rangle$ occurs to the right of v_1 and since it is generated by $(\tau(A) \to \langle r, s, 1\rangle, \emptyset, N' - \mathscr{T}(X)) \in P_L$ from (1.1), t is necessarily a rule from (1.1) or (2.1). However, in either case, we cannot rewrite the compound nonterminal generated by t because $\langle r, s, 1\rangle$ is still present to the right of v_1. Although we can rewrite $\langle r, s, 1\rangle$ to $\langle r, s, 2\rangle$ by $(\langle r, s, 1\rangle \rightarrow \langle r, s, 2\rangle, \emptyset, N' - \mathscr{T}(Z)) \in P_R$, we cannot rewrite $\langle r, s, 2\rangle$ because of the generated nonterminal. This contradicts the assumption that $v_1 \neq x_1$. Therefore, $v_1 = x_1$.

(ii) Assume that $v_2 \neq x_2$. The only possible way this case could happen is that after

$$S \Rightarrow_H^* v_1\tau(A)v_2 \Rightarrow_H v_1\langle r, s, 1\rangle v_2$$

v_2 is rewritten by a rule t. Since $\langle r, s, 1\rangle$ occurs to the left of v_2, t is necessarily a rule from (1.2) or (2.2). However, in either case, we cannot rewrite the compound nonterminal generated by t because $\langle r, s, 1\rangle$ is still present to the left of v_2. Furthermore, we cannot rewrite $\langle r, s, 1\rangle$ by $(\langle r, s, 1\rangle \rightarrow \langle r, s, 2\rangle, \emptyset, N' - \mathscr{T}(Z)) \in P_R$ from (1.2) because of the generated nonterminal. This contradicts the assumption that $v_2 \neq x_2$. Therefore, $v_2 = x_2$.

In a similar way, $\mathrm{alph}(x) \cap T = \emptyset$ can be demonstrated. Consequently, $(V' - \mathscr{T}(X)) \cap \mathrm{alph}(x_1) = \emptyset$ and $(V' - \mathscr{T}(Z)) \cap \mathrm{alph}(x_2) = \emptyset$. Recall that $X^*\overline{Y}Z^* \in K$, $A \in Y$, and $A \rightarrow y \in P$. Since $\psi(x) = \psi(x_1\langle r, s, 2\rangle x_2) = \psi(x_1)A\psi(x_2)$, $\mathrm{alph}(\psi(x_1)) \subseteq X$, $A \in Y$, and $\mathrm{alph}(\psi(x_2)) \subseteq Z$, we see that $\psi(x) \Rightarrow_G \psi(w)$, which completes the induction step. □

Next, we prove that $L(H) = L(G)$. Consider Claim 1 for $w \in T^*$. Then, $S \Rightarrow_G^* w$ implies that $S \Rightarrow_H^* \tau(w)$. Let $\tau(w) = \hat{a}_1\hat{a}_2 \cdots \hat{a}_h$, where $h = |w|$ (the case when $h = 0$ means $w = \varepsilon$). By (2),

$$\begin{aligned}
&(\hat{a}_i \rightarrow \langle a_i, 1\rangle, \emptyset, N' - \hat{T}) \in P_L \\
&(\langle a_i, 1\rangle \rightarrow \langle a_i, 2\rangle, \emptyset, N' - T_2) \in P_R \\
&(\langle a_i, 2\rangle \rightarrow a_i, \emptyset, N') \in P_L
\end{aligned}$$

for all i, $1 \leq i \leq h$. Therefore,

$$\begin{aligned}
\hat{a}_1 \cdots \hat{a}_{h-1}\hat{a}_h &\Rightarrow_H \hat{a}_1 \cdots \hat{a}_{h-1}\langle a_h, 1\rangle \\
&\Rightarrow_H \hat{a}_1 \cdots \langle a_{h-1}, 1\rangle\langle a_h, 1\rangle \\
&\ \ \vdots \\
&\Rightarrow_H \langle a_1, 1\rangle \cdots \langle a_{h-1}, 1\rangle\langle a_h, 1\rangle \\
&\Rightarrow_H \langle a_1, 1\rangle \cdots \langle a_{h-1}, 1\rangle\langle a_h, 2\rangle \\
&\Rightarrow_H \langle a_1, 1\rangle \cdots \langle a_{h-1}, 2\rangle\langle a_h, 2\rangle \\
&\ \ \vdots \\
&\Rightarrow_H \langle a_1, 2\rangle\langle a_2, 2\rangle \cdots \langle a_h, 2\rangle \\
&\Rightarrow_H a_1\langle a_2, 2\rangle \cdots \langle a_h, 2\rangle \\
&\Rightarrow_H a_1a_2 \cdots \langle a_h, 2\rangle \\
&\ \ \vdots \\
&\Rightarrow_H a_1a_2 \cdots a_h
\end{aligned}$$

Hence, $L(G) \subseteq L(H)$. Consider Claim 2 for $w \in T^*$. Then, $S \Rightarrow_H^* w$ implies that $S \Rightarrow_G^* \psi(w) = w$, so $L(H) \subseteq L(G)$. Consequently, $L(H) = L(G)$, and the lemma holds. □

Lemma 6.2.6. *For every one-sided forbidding grammar G, there is an s-grammar H such that $L(H) = L(G)$.*

Proof. Let $G = (N, T, P_L, P_R, S)$ be a one-sided forbidding grammar. We next construct an s-grammar H such that $L(H) = L(G)$. Set $V = N \cup T$ and

$$M = \{\langle r, L\rangle \mid r = (A \to y, \emptyset, F) \in P_L\} \cup \{\langle r, R\rangle \mid r = (A \to y, \emptyset, F) \in P_R\}$$

Without any loss of generality, we assume that $V \cap M = \emptyset$. Construct

$$H = (V', T, P', S, K)$$

as follows. Initially, set $V' = V \cup M$, $P' = \emptyset$, and

$$K = \{V^*\overline{\{A\}}V^* \mid A \in N\}$$

Perform (1) and (2), given next.

(1) For each $r = (A \to y, \emptyset, F) \in P_L$,

(1.1) add $A \to \langle r, L\rangle$ and $\langle r, L\rangle \to y$ to P';
(1.2) add $(V - F)^*\overline{\{\langle r, L\rangle\}}V^*$ to K.

(2) For each $r = (A \to y, \emptyset, F) \in P_R$,

(2.1) add $A \to \langle r, R\rangle$ and $\langle r, R\rangle \to y$ to P';
(2.2) add $V^*\overline{\{\langle r, R\rangle\}}(V - F)^*$ to K.

Before proving that $L(H) = L(G)$, let us informally explain (1) and (2). Since a rule $r = (A \to y, \emptyset, F)$ can be in both P_L and P_R and since there can be several rules with A on their left-hand sides, we simulate the application of a single rule of G in two steps. First, depending on whether $r \in P_L$ or $r \in P_r$, we rewrite an occurrence of A to a special compound nonterminal $\langle r, s\rangle$, which encodes the simulated rule r and the side on which we check the absence of forbidding symbols ($s = L$ or $s = R$). Then, we introduce a selector which checks the absence of all symbols from F to the proper side of $\langle r, s\rangle$, depending on whether $s = L$ or $s = R$.

To establish $L(H) = L(G)$, we prove two claims. First, Claim 1 shows how derivations of G are simulated by H. Then, Claim 2 demonstrates the converse—that is, it shows how derivations of H are simulated by G.

Claim 1. If $S \Rightarrow_G^n w \Rightarrow_G^* z$, where $w \in V^*$ and $z \in T^*$, for some $n \geq 0$, then $S \Rightarrow_H^* w$.

Proof. This claim is established by induction on $n \geq 0$.

Basis. For $n = 0$, this claim obviously holds.

Induction Hypothesis. Suppose that there exists $n \geq 0$ such that the claim holds for all derivations of length ℓ, where $0 \leq \ell \leq n$.

Induction Step. Consider any derivation of the form

$$S \Rightarrow_G^{n+1} w \Rightarrow_G^* z$$

where $w \in V^*$ and $z \in T^*$. Since $n + 1 \geq 1$, this derivation can be expressed as

$$S \Rightarrow_G^n x \Rightarrow_G w \Rightarrow_G^* z$$

for some $x \in V^+$. By the induction hypothesis, $S \Rightarrow_H^* x$.

Next, we consider all possible forms of $x \Rightarrow_G w$, covered by the following two cases—(i) and (ii).

(i) *Application of* $(A \rightarrow y, \emptyset, F) \in P_L$. Let $x = x_1 A x_2$, $w = x_1 y x_2$, and $r = (A \rightarrow y, \emptyset, F) \in P_L$, so $x \Rightarrow_G w$ by r. This implies that $\mathrm{alph}(x_1) \cap F = \emptyset$. By the initialization part of the construction, $V^*\overline{\{A\}}V^* \in K$, and by (1.1), $A \rightarrow \langle r, L\rangle \in P'$. Since $\mathrm{alph}(x_1) \subseteq V$ and $\mathrm{alph}(x_2) \subseteq V$,

$$x_1 A x_2 \Rightarrow_H x_1 \langle r, L\rangle x_2$$

By (1.2), $(V - F)^*\overline{\{\langle r, L\rangle\}}V^* \in K$, and by (1.1), $\langle r, L\rangle \rightarrow y \in P'$. Since $\mathrm{alph}(x_1) \cap F = \emptyset$,

$$x_1 \langle r, L\rangle x_2 \Rightarrow_H x_1 y x_2$$

which completes the induction step for (i).

(ii) *Application of* $(A \rightarrow y, \emptyset, F) \in P_R$. Proceed by analogy with (i), but use rules from (2) instead of rules from (1).

Observe that cases (i) and (ii) cover all possible forms of $x \Rightarrow_G w$. Thus, the claim holds. □

Define the homomorphism φ from V'^* to V^* as $\varphi(A) = A$, for all $A \in N$, $\varphi(\langle r, L\rangle) = A$, for all $r = (A \rightarrow x, \emptyset, F) \in P_L$, $\varphi(\langle r, R\rangle) = A$, for all $r = (A \rightarrow x, \emptyset, F) \in P_R$, and $\varphi(a) = a$, for all $a \in T$.

Claim 2. If $S \Rightarrow_H^n w \Rightarrow_H^* z$, where $w \in V'^*$ and $z \in T^*$, for some $n \geq 0$, then $S \Rightarrow_G^* \varphi(w)$.

Proof. This claim is established by induction on $n \geq 0$.

Basis. For $n = 0$, this claim obviously holds.

Induction Hypothesis. Suppose that there exists $n \geq 0$ such that the claim holds for all derivations of length ℓ, where $0 \leq \ell \leq n$.

Induction Step. Consider any derivation of the form

$$S \Rightarrow_H^{n+1} w \Rightarrow_H^* z$$

where $w \in V'^*$ and $z \in T^*$. Since $n + 1 \geq 1$, this derivation can be expressed as

$$S \Rightarrow_H^n x \Rightarrow_H w \Rightarrow_H^* z$$

for some $x \in V'^+$. By the induction hypothesis, $S \Rightarrow_G^* \varphi(x)$. Next, we consider all possible forms of $x \Rightarrow_H w$, covered by the following four cases—(i) through (iv).

(i) *Application of* $A \to \langle r, L \rangle \in P'$, *introduced in (1.1).* Let $x = x_1 A x_2$, $w = x_1 \langle r, L \rangle x_2$, so $x \Rightarrow_H w$ by $A \to \langle r, L \rangle \in P'$, introduced in (1.1) from $r = (A \to y, \emptyset, F) \in P_L$. Since $\varphi(\langle r, L \rangle) = A$, the induction step for (i) follows directly from the induction hypothesis.

(ii) *Application of* $A \to \langle r, R \rangle \in P'$, *introduced in (2.1).* Proceed by analogy with (i).

(iii) *Application of* $\langle r, L \rangle \to y \in P'$, *introduced in (1.1).* Let $x = x_1 \langle r, L \rangle x_2$, $w = x_1 y x_2$, $(V - F)^* \overline{\{\langle r, L \rangle\}} V^* \in K$ such that $x_1 \in (V - F)^*$ and $x_2 \in V^*$, so $x \Rightarrow_H w$ by $\langle r, L \rangle \to y \in P'$, introduced in (1.1) from $r = (A \to y, \emptyset, F) \in P_L$. Observe that $x_1 \in (V - F)^*$ implies that $\varphi(x_1) = x_1$ and $\text{alph}(x_1) \cap F = \emptyset$. Since $\varphi(x) = \varphi(x_1 \langle r, L \rangle x_2) = x_1 A \varphi(x_2)$,

$$x_1 A \varphi(x_2) \Rightarrow_G x_1 y \varphi(x_2) \text{ by } r$$

As $\varphi(w) = \varphi(x_1 y x_2) = x_1 y \varphi(x_2)$, the induction step is completed for (iii).

(iv) *Application of* $\langle r, R \rangle \to y \in P'$, *introduced in (2.1).* Proceed by analogy with (iii).

Observe that cases (i) through (iv) cover all possible forms of $x \Rightarrow_H w$. Thus, the claim holds. □

Next, we prove that $L(H) = L(G)$. Consider Claim 1 for $w \in T^*$. Then, $S \Rightarrow_G^* w$ implies that $S \Rightarrow_H^* w$. Hence, $L(G) \subseteq L(H)$. Consider Claim 2 for $w \in T^*$. Then, $S \Rightarrow_H^* w$ implies that $S \Rightarrow_G^* \varphi(w) = w$, so $L(H) \subseteq L(G)$. Consequently, $L(H) = L(G)$, and the theorem holds. □

Theorem 6.2.7. $\mathbf{OFor} = \mathbf{S}$

Proof. This theorem follows from Lemmas 6.2.5 and 6.2.6. □

Theorem 6.2.8. $\mathbf{OFor}^{-\varepsilon} = \mathbf{S}^{-\varepsilon}$

Proof. Reconsider the proof of Lemma 6.2.5. Observe that if G is propagating, so is H. Hence, $\mathbf{S}^{-\varepsilon} \subseteq \mathbf{OFor}^{-\varepsilon}$. Reconsider the proof of Lemma 6.2.6. Observe that if G is propagating, so is H. Hence, $\mathbf{OFor}^{-\varepsilon} \subseteq \mathbf{S}^{-\varepsilon}$, and the theorem holds. □

We next turn our attention to one-sided forbidding grammars with the set of left forbidding rules coinciding with the set of right forbidding rules. We prove that they characterize the family of context-free languages.

Lemma 6.2.9. *Let K be a context-free language. Then, there exists a one-sided forbidding grammar, $G = (N, T, P_L, P_R, S)$, satisfying $P_L = P_R$ and $L(G) = K$.*

Proof. Let K be a context-free language. Then, there exists a context-free grammar, $H = (N, T, P, S)$, such that $L(H) = K$. Define the one-sided forbidding grammar

$$G = \left(N, T, P', P', S\right)$$

where

$$P' = \left\{(A \to x, \emptyset, \emptyset) \mid A \to x \in P\right\}$$

Clearly, $L(G) = L(H) = K$, so the lemma holds. □

Lemma 6.2.10. *Let $G = (N, T, P_L, P_R, S)$ be a one-sided forbidding grammar satisfying $P_L = P_R$. Then, $L(G)$ is context-free.*

Proof. Let $G = (N, T, P_L, P_R, S)$ be a one-sided forbidding grammar satisfying $P_L = P_R$. Define the context free grammar $H = (N, T, P', S)$ with

$$P' = \left\{A \to x \mid (A \to x, \emptyset, F) \in P_L\right\}$$

Observe that since $P_L = P_R$, in the construction of P' above, it is sufficient to consider just the rules from P_L. As any successful derivation in G is also a successful derivation in H, the inclusion $L(G) \subseteq L(H)$ holds. On the other hand, let $w \in L(H)$ be a string successfully generated by H. Then, it is well-known that there exists a successful leftmost derivation of w in H (see Theorem 3.3.13). Observe that such a leftmost derivation is also possible in G because the leftmost nonterminal can always be rewritten. Indeed, P' contains only rules originating from the rules in P_L and all rules in P_L are applicable to the leftmost nonterminal. Thus, the other inclusion $L(H) \subseteq L(G)$ holds as well, which completes the proof. □

Theorem 6.2.11. *A language K is context-free if and only if there is a one-sided forbidding grammar, $G = (N, T, P_L, P_R, S)$, satisfying $K = L(G)$ and $P_L = P_R$.*

Proof. This theorem follows from Lemmas 6.2.9 and 6.2.10. □

Since erasing rules can be eliminated from any context-free grammar (see Theorem 7.1.8), we obtain the following corollary.

Corollary 6.2.12. *Let $G = (N, T, P_L, P_R, S)$ be a one-sided forbidding grammar satisfying $P_L = P_R$. Then, there is a propagating one-sided forbidding grammar H such that $L(H) = L(G)$.* □

The family of context-free languages is also characterized by left forbidding grammars.

Theorem 6.2.13. $\mathbf{LFor}^{-\varepsilon} = \mathbf{LFor} = \mathbf{CF}$

Proof. Since erasing rules can be eliminated from any context-free grammar (see Sect. 7.1), it is sufficient to prove that $\mathbf{LFor} = \mathbf{CF}$. As any context-free grammar is also a left forbidding grammar with empty sets attached to each of its rules, the inclusion $\mathbf{CF} \subseteq \mathbf{LFor}$ holds. To establish the other inclusion, $\mathbf{LFor} \subseteq \mathbf{CF}$, let $G = (N, T, P_L, \emptyset, S)$ be any left forbidding grammar, and let

$$H = \left(N, T, P', S\right)$$

be a context-free grammar, where

$$P' = \left\{A \to x \mid (A \to x, \emptyset, W) \in P_L\right\}$$

As any successful derivation of G is also a successful derivation of H, the inclusion $L(G) \subseteq L(H)$ holds. On the other hand, let $w \in L(H)$ be a string successfully generated by the context-free grammar H. Then, by Theorem 3.3.13, there exists a successful leftmost derivation of w in H. Such a leftmost derivation is, however, also possible in G because the leftmost nonterminal can always be rewritten. Thus, the other inclusion $L(H) \subseteq L(G)$ holds as well, which completes the proof. □

We conclude this section by several remarks closely related to the previous results. Recall that we have established an equivalence between one-sided forbidding grammars and s-grammars. In [25], it is proved that special versions of s-grammars, referred to as *symmetric s-grammars*, are equivalent to forbidding grammars. Recall that in a symmetric s-grammar, each selector is of the form $X^*\overline{Y}X^*$, where X and Y are alphabets. In a forbidding grammar, the absence of symbols is checked in the entire sentential form. Based on the achieved results, we see that one-sided forbidding grammars form a counterpart to s-grammars just like forbidding grammars form a counterpart to symmetric s-grammars. As symmetric s-grammars are just special versions of s-grammars, we see that one-sided forbidding grammars are at least as powerful as forbidding grammars. This result can be also proved directly, as demonstrated next.

Theorem 6.2.14. $\mathbf{For} \subseteq \mathbf{OFor}$

Proof. Let $G = (N, T, P, S)$ be a forbidding grammar. Without any loss of generality, we assume that $(A \to w, \emptyset, W) \in P$ implies that $A \notin W$ (otherwise, such a rule would not be applicable in G). We next construct a one-sided forbidding grammar H such that $L(H) = L(G)$. Set $R = \{\langle r, 1\rangle, \langle r, 2\rangle \mid r \in P\}$ and define H as

$$H = \left(N \cup R, T, P_L, P_R, S\right)$$

where P_L and P_R are constructed in the following way. Initially, set $P_L = \emptyset$ and $P_R = \emptyset$. To complete the construction, apply the following three steps for each $r = (A \to x, \emptyset, W) \in P$

(1) add $(A \to \langle r, 1\rangle, \emptyset, W \cup R)$ to P_L;
(2) add $(\langle r, 1\rangle \to \langle r, 2\rangle, \emptyset, W \cup R)$ to P_R;
(3) add $(\langle r, 2\rangle \to x, \emptyset, R)$ to P_L.

The simulation of every $r = (A \to x, \emptyset, W) \in P$ is done in three steps. First, we check the absence of all forbidding symbols from W to the left of A by a rule from (1). Then, we check the absence of all forbidding symbols from W to the right of A by a rule from (2). By our assumption, $A \notin W$, so we do not have to check the absence of A in the current sentential form. Finally, we rewrite $\langle r, 2\rangle$ to x by a rule from (3). In all these steps, we also check the absence of all nonterminals from R. In this way, we guarantee that only a single rule is simulated at a time (if this is not the case, then the derivation is blocked). Based on these observations, we see that $L(H) = L(G)$. □

Observe that the construction in the proof of Theorem 6.2.14 does not introduce any erasing rules. Hence, whenever G is propagating, so is H. This implies the following result.

Theorem 6.2.15. $\mathbf{For}^{-\varepsilon} \subseteq \mathbf{OFor}^{-\varepsilon}$ □

In [25], the question whether $\mathbf{S}^{-\varepsilon} = \mathbf{CS}$ is explicitly formulated (see open problem (5) in [25]). As $\mathbf{S}^{-\varepsilon} = \mathbf{OFor}^{-\varepsilon}$ (see Theorem 6.2.8), we obtain a reformulation of this longstanding open question.

Corollary 6.2.16. $\mathbf{S}^{-\varepsilon} = \mathbf{CS}$ *if and only if* $\mathbf{OFor}^{-\varepsilon} = \mathbf{CS}$. □

It is worth pointing out that it is not known whether s-grammars or one-sided forbidding grammars characterize **RE** either. From Theorem 6.2.7, we obtain the following corollary.

Corollary 6.2.17. $\mathbf{S} = \mathbf{RE}$ *if and only if* $\mathbf{OFor} = \mathbf{RE}$. □

Open Problem 6.2.18. What is the generative power of one-sided forbidding grammars and s-grammars? Do they characterize **RE**? □

6.2.3 One-Sided Permitting Grammars

Finally, we consider one-sided permitting grammars and their generative power. We prove that $\mathbf{CF} \subset \mathbf{OPer}^{-\varepsilon} \subseteq \mathbf{SCAT}^{-\varepsilon}$.

Lemma 6.2.19. $\mathbf{CF} \subset \mathbf{OPer}^{-\varepsilon} \subseteq \mathbf{OPer}$

Proof. Clearly, $\mathbf{CF} \subseteq \mathbf{OPer}^{-\varepsilon} \subseteq \mathbf{OPer}$. The strictness of the first inclusion follows from Example 6.1.7. □

Lemma 6.2.20. $\mathbf{OPer}^{-\varepsilon} \subseteq \mathbf{SCAT}^{-\varepsilon}$

Proof. Let $G = (N, T, P_L, P_R, S)$ be a propagating one-sided permitting grammar. We next construct a propagating scattered context grammar H such that $L(H) = L(G)$. Define H as

$$H = (N, T, P', S)$$

with P' constructed as follows:

(1) for each $(A \to x, \emptyset, \emptyset) \in P_L \cup P_R$, add $(A) \to (x)$ to P';
(2) for each $(A \to x, \{X_1, X_2, \dots, X_n\}, \emptyset) \in P_L$ and every permutation $(i_1, i_2, \dots, i_n)$ of $(1, 2, \dots, n)$, where $n \geq 1$, extend P' by adding

$$(X_{i_1}, X_{i_2}, \dots, X_{i_n}, A) \to (X_{i_1}, X_{i_2}, \dots, X_{i_n}, x)$$

(3) for each $(A \to x, \{X_1, X_2, \dots, X_n\}, \emptyset) \in P_R$ and every permutation $(i_1, i_2, \dots, i_n)$ of $(1, 2, \dots, n)$, where $n \geq 1$, extend P' by adding

$$(A, X_{i_1}, X_{i_2}, \dots, X_{i_n}) \to (x, X_{i_1}, X_{i_2}, \dots, X_{i_n})$$

Rules with no permitting symbols are simulated by ordinary context-free-like rules, introduced in (1). The presence of permitting symbols is checked by scattered context rules, introduced in (2) and (3), which have every permitting symbol to the left and to the right of the rewritten symbol, respectively. Because the exact order of permitting symbols in a sentential form is irrelevant in one-sided permitting grammars, we introduce every permutation of the all permitting symbols. Based on these observations, we see that $L(H) = L(G)$. □

Theorem 6.2.21. $\mathbf{CF} \subset \mathbf{OPer}^{-\varepsilon} \subseteq \mathbf{SCAT}^{-\varepsilon} \subseteq \mathbf{CS} = \mathbf{ORC}^{-\varepsilon}$

Proof. By Lemma 6.2.19, $\mathbf{CF} \subset \mathbf{OPer}^{-\varepsilon}$. By Lemma 6.2.20, $\mathbf{OPer}^{-\varepsilon} \subseteq \mathbf{SCAT}^{-\varepsilon}$. $\mathbf{SCAT}^{-\varepsilon} \subseteq \mathbf{CS}$ follows from Theorem 4.7.6. Finally, $\mathbf{CS} = \mathbf{ORC}^{-\varepsilon}$ follows from Theorem 6.2.3. □

Recall that the one-sided random context grammar from Example 6.1.7 is, in fact, a propagating left permitting grammar. Since every left permitting grammar is a special case of a one-sided permitting grammar, we obtain the following corollary of Theorem 6.2.21.

Corollary 6.2.22. $\mathbf{CF} \subset \mathbf{LPer}^{-\varepsilon} \subseteq \mathbf{SCAT}^{-\varepsilon} \subseteq \mathbf{CS} = \mathbf{ORC}^{-\varepsilon}$ □

In the conclusion of this section, we point out some consequences implied by the results achieved above. Then, we formulate some open problem areas.

Corollary 6.2.23. $\mathbf{RC}^{-\varepsilon} \subset \mathbf{ORC}^{-\varepsilon} \subset \mathbf{RC} = \mathbf{ORC}$.

Proof. These inclusions follow from Theorems 6.2.3 and 6.2.4 and from Corollary 4.3.10. □

Corollary 6.2.24. $\mathbf{LFor}^{-\varepsilon} = \mathbf{LFor} \subset \mathbf{For}^{-\varepsilon} \subseteq \mathbf{OFor}^{-\varepsilon} \subseteq \mathbf{OFor}$.

Proof. These inclusions follow from Theorems 6.2.14 and 6.2.13 and from Corollary 4.3.10. □

The previous results give rise to the following four open problem areas that are related to the achieved results.

Open Problem 6.2.25. Establish the relations between families $\mathbf{Per}^{-\varepsilon}$, $\mathbf{LPer}^{-\varepsilon}$, and $\mathbf{OPer}^{-\varepsilon}$. What is the generative power of left random context grammars? □

Open Problem 6.2.26. Recall that $\mathbf{Per}^{-\varepsilon} = \mathbf{Per}$ (see Theorem 4.3.7). Is it true that $\mathbf{OPer}^{-\varepsilon} = \mathbf{OPer}$? □

Open Problem 6.2.27. Theorem 6.2.13 implies that $\mathbf{LFor}^{-\varepsilon} = \mathbf{LFor}$. Is it also true that $\mathbf{OFor}^{-\varepsilon} = \mathbf{OFor}$? □

Open Problem 6.2.28. Does $\mathbf{OPer}^{-\varepsilon} = \mathbf{ORC}^{-\varepsilon}$ hold? If so, then Theorem 6.2.21 would imply $\mathbf{SCAT}^{-\varepsilon} = \mathbf{CS}$ and, thereby, solve a longstanding open question. □

6.3 Normal Forms

Formal language theory has always struggled to turn grammars into *normal forms*, in which grammatical rules satisfy some prescribed properties or format because they are easier to handle from a theoretical as well as practical standpoint. Concerning context-free grammars, there exist two famous normal forms—the Chomsky and Greibach normal forms (see Sect. 3.3). In the former, every grammatical rule has on its right-hand side either a terminal or two nonterminals. In the latter, every grammatical rule has on its right-hand side a terminal followed by zero or more nonterminals. Similarly, there exist normal forms for general grammars, such as the Kuroda, Penttonen, and Geffert normal forms (see Sect. 3.3).

The present section establishes four normal forms for one-sided random context grammars. The first of them has the set of left random context rules coinciding with the set of right random context rules. The second normal form, in effect, consists in demonstrating how to turn any one-sided random context grammar to an equivalent one-sided random context grammar with the sets of left and right random context rules being disjoint. The third normal form resembles the Chomsky normal form for context-free grammars, mentioned above. In the fourth normal form, each rule has its permitting or forbidding context empty.

In the first normal form, the set of left random context rules coincides with the set of right random context rules.

Theorem 6.3.1. *Let $G = (N, T, P_L, P_R, S)$ be a one-sided random context grammar. Then, there is a one-sided random context grammar, $H = (N', T, P'_L, P'_R, S)$, such that $L(H) = L(G)$ and $P'_L = P'_R$.*

Proof. Let $G = (N, T, P_L, P_R, S)$ be a one-sided random context grammar, and let $S', \#, \$$ be three new symbols not in $N \cup T$. Define the one-sided random context grammar

$$H = \big(N \cup \{S', \#, \$\}, T, P', P', S'\big)$$

with P' constructed in the following way. Initially, set

$$P' = \big\{(S' \rightarrow \#S\$, \emptyset, \emptyset), (\$ \rightarrow \varepsilon, \{\#\}, N), (\# \rightarrow \varepsilon, \emptyset, \emptyset)\big\}$$

Then, complete the construction by applying the following two steps

(1) for each $(A \rightarrow w, U, W) \in P_L$, add $(A \rightarrow w, U, W \cup \{\$\})$ to P';
(2) for each $(A \rightarrow w, U, W) \in P_R$, add $(A \rightarrow w, U, W \cup \{\#\})$ to P'.

In H, the side on which the rules check the presence and absence of symbols is not explicitly prescribed by their membership to a certain set of rules. Instead, the two new symbols, \$ and #, are used to force rules to check for their permitting and forbidding symbols on a proper side. These two new symbols are introduced by $(S' \rightarrow \#S\$, \emptyset, \emptyset)$, which is used at the very beginning of every derivation. Therefore, every sentential form of H has its symbols placed between these two end markers. If we want a rule to look to the left, we guarantee the absence of \$; otherwise, we guarantee the absence of #. At the end of a derivation, these two new symbols are erased by $(\$ \rightarrow \varepsilon, \{\#\}, N)$ and $(\# \rightarrow \varepsilon, \emptyset, \emptyset)$. The former rule checks whether, disregarding #, the only present symbols in the current sentential form are terminals. Observe that if $(\# \rightarrow \varepsilon, \emptyset, \emptyset)$ is used prematurely, H cannot derive a sentence because the presence of # is needed to erase \$ by $(\$ \rightarrow \varepsilon, \{\#\}, N)$. Based on these observations, we see that $L(H) = L(G)$. Since H has effectively only a single set of rules, the theorem holds. □

Next, we show that Theorem 6.3.1 also holds if we restrict ourselves only to propagating one-sided random context grammars.

Theorem 6.3.2. *Let $G = (N, T, P_L, P_R, S)$ be a propagating one-sided random context grammar. Then, there is a propagating one-sided random context grammar, $H = (N', T, P'_L, P'_R, S)$, such that $L(H) = L(G)$ and $P'_L = P'_R$.*

Proof. We prove this theorem by analogy with the proof of Theorem 6.3.1, but we give the present proof in a greater detail. Since H has to be propagating, instead of # and \$ as end markers, we use boundary symbols appearing in sentential forms. To this end, we keep the leftmost symbol marked by $\grave{}$ and the rightmost symbol marked by $\acute{}$. If there is only a single symbol in the current sentential form, we mark it by $\check{}$. At the end of every successful derivation, only terminals and two boundary marked terminals are present. To produce a string of terminals, we unmark these two marked terminals.

Let $G = (N, T, P_L, P_R, S)$ be a propagating one-sided random context grammar. Set $V = N \cup T$, $\grave{V} = \{\grave{X} \mid X \in V\}$, $\acute{V} = \{\acute{X} \mid X \in V\}$, and $\check{N} = \{\check{A} \mid A \in N\}$. Construct the propagating one-sided random context grammar

$$H = \big(N', T, P', P', \check{S}\big)$$

as follows. Initially, set $N' = N \cup \grave{V} \cup \acute{V} \cup \check{N}$ and $P' = \emptyset$. To keep the rest of the construction as readable as possible, we introduce several functions. Define the function $\grave{\pi}$ from 2^N to $2^{2^{N'}}$ as $\grave{\pi}(\emptyset) = \{\emptyset\}$ and

$$\begin{aligned}\grave{\pi}(\{A_1, A_2, \ldots, A_n\}) = \; & \{\{A_1, A_2, \ldots, A_n\}\} \cup \\ & \{\{\grave{A}_1, A_2, \ldots, A_n\}\} \cup \\ & \{\{A_1, \grave{A}_2, \ldots, A_n\}\} \cup \\ & \vdots \\ & \{\{A_1, A_2, \ldots, \grave{A}_n\}\}\end{aligned}$$

Define the function $\acute{\pi}$ from 2^N to $2^{2^{N'}}$ as $\acute{\pi}(\emptyset) = \{\emptyset\}$ and

$$\begin{aligned}\acute{\pi}(\{A_1, A_2, \ldots, A_n\}) = \; & \{\{A_1, A_2, \ldots, A_n\}\} \cup \\ & \{\{\acute{A}_1, A_2, \ldots, A_n\}\} \cup \\ & \{\{A_1, \acute{A}_2, \ldots, A_n\}\} \cup \\ & \vdots \\ & \{\{A_1, A_2, \ldots, \acute{A}_n\}\}\end{aligned}$$

Define the function $\grave{\sigma}$ from 2^N to $2^{N'}$ as $\grave{\sigma}(W) = W \cup \{\grave{Y} \mid Y \in W\} \cup \acute{V}$. Define the function $\acute{\sigma}$ from 2^N to $2^{N'}$ as $\acute{\sigma}(W) = W \cup \{\acute{Y} \mid Y \in W\} \cup \grave{V}$.

To complete the construction, apply the following ten steps.

(1) *Simulation of unit rules when there is only a single symbol present in the current sentential form.*
For each $(A \to B, \emptyset, W) \in P_L \cup P_R$, where $B \in N$, add $(\check{A} \to \check{B}, \emptyset, \emptyset)$ to P'.

(2) *Simulation of rules generating a single terminal when there is only a single symbol present in the current sentential form.*
For each $(A \to a, \emptyset, W) \in P_L \cup P_R$, where $a \in T$, add $(\check{A} \to a, \emptyset, \emptyset)$ to P'.

(3) *Simulation of rules forking the only symbol in the current sentential form into two or more symbols.*
For each $(A \to XwY, \emptyset, W) \in P_L \cup P_R$, where $X, Y \in V$, $w \in V^*$, add $(\check{A} \to \grave{X}w\acute{Y}, \emptyset, \emptyset)$ to P'.

(4) *Simulation of rules from P_L rewriting the leftmost nonterminal.*
For each $(A \to Xw, \emptyset, W) \in P_L$, where $X \in V$, $w \in V^*$, add $(\grave{A} \to \grave{X}w, \emptyset, \emptyset)$ to P'.

(5) *Simulation of rules from P_R rewriting the rightmost nonterminal.*
For each $(A \to wX, \emptyset, W) \in P_R$, where $X \in V$, $w \in V^*$, add $(\acute{A} \to w\acute{X}, \emptyset, \emptyset)$ to P'.

(6) *Simulation of rules from P_L rewriting the rightmost nonterminal.*
For each $(A \rightarrow wX, U, W) \in P_L$, where $X \in V$, $w \in V^*$, and every $U' \in \grave{\pi}(U)$, add $(\acute{A} \rightarrow w\acute{X}, U', \grave{\sigma}(W))$ to P'.

(7) *Simulation of rules from P_L rewriting a non-marked nonterminal.*
For each $(A \rightarrow w, U, W) \in P_L$, where $w \in V^*$, and every $U' \in \grave{\pi}(U)$, add $(A \rightarrow w, U', \grave{\sigma}(W))$ to P'.

(8) *Simulation of rules from P_R rewriting the leftmost nonterminal.*
For each $(A \rightarrow Xw, U, W) \in P_R$, where $X \in V$, $w \in V^*$, and every $U' \in \acute{\pi}(U)$, add $(\grave{A} \rightarrow \grave{X}w, U', \acute{\sigma}(W))$ to P'.

(9) *Simulation of rules from P_R rewriting a non-marked nonterminal.*
For each $(A \rightarrow w, U, W) \in P_R$, where $w \in V^*$, and every $U' \in \acute{\pi}(U)$, add $(A \rightarrow w, U', \acute{\sigma}(W)\})$ to P'.

(10) *Unmark both boundary terminals if only terminals and marked terminals are present.*
For each $a, c \in T$, add $(\grave{a} \rightarrow a, \emptyset, \emptyset)$ and $(\acute{c} \rightarrow c, \{\grave{a}\}, N)$ to P'.

Observe (i) through (vi), given next.

(i) Let $S = X_1 \Rightarrow_G X_2 \Rightarrow_G \cdots \Rightarrow_G X_n \Rightarrow_G a$ be a derivation, where $a \in T$, $X_i \in N$, for all i, $1 \leq i \leq n$, for some $n \geq 1$. Notice that every applied rule in such a derivation in G has to have an empty permitting context; otherwise, it would not be applicable. Then, there is

$$\check{S} = \check{X}_1 \Rightarrow_H \check{X}_2 \Rightarrow_H \cdots \Rightarrow_H \check{X}_n \Rightarrow_H a$$

by rules introduced in (1) and (2). Conversely, for every derivation in H by rules from (1) and (2), there is a corresponding derivation in G.

(ii) Rules from P_L used to rewrite the leftmost nonterminal of a sentential form and rules from P_R used to rewrite the rightmost nonterminal of a sentential form have to have empty permitting contexts; otherwise, they would not be applicable. Therefore, the assumption of empty permitting contexts in rules from P_L and P_R in (1) through (5) is without any loss of generality. Also, for the same reason, the resulting rules, introduced to P', have empty forbidding contexts.

(iii) Excluding case (i), every sentential form of H that has one or more nonterminals is bounded by marked symbols. If the leftmost marked symbol is unmarked prematurely by a rule of the form $(\grave{a} \rightarrow a, \emptyset, \emptyset)$, introduced in (10), no sentence can be obtained because the presence of a symbol marked by $\grave{}$ is needed to unmark the rightmost symbol marked by $\acute{}$ by a rule of the form $(\acute{c} \rightarrow c, \{\grave{a}\}, N)$, introduced in (10).

(iv) The simulation of a rewrite of a nonterminal that is not the leftmost nor the rightmost symbol in the current sentential form by a rule from P_L is done by rules from (6) and (7). To force the check to the left, the absence of all symbols marked by $\acute{}$ is required. Analogously, by (8) and (9), the forced check to the right is done by requiring the absence of all symbols marked by $\grave{}$. Because of the previous observation, this simulation is correct.

(v) Let $S \Rightarrow_G^* w$ be a derivation in G, where $w \in T^+$ such that $|w| \geq 2$. Using the corresponding rules introduced in steps (1) and (3) through (9) and then using two rules from (10), it is possible to derive w in H.

(vi) Every derivation in H leading to a sentence containing more than one terminal is of the form

$$\begin{aligned} \check{S} &\Rightarrow_H^* \check{A} && \text{(by rules from (1))} \\ &\Rightarrow_H \grave{X} w \acute{Y} && \text{(by a rule from (3))} \\ &\Rightarrow_H^* \grave{a}_1 a_2 \cdots a_{n-1} \acute{a}_n && \text{(by rules from (4) through (9))} \\ &\Rightarrow_H \grave{a}_1 a_2 \cdots a_{n-1} a_n && \text{(by } (\acute{a}_n \rightarrow a_n, \{\grave{a}_1\}, N) \text{ from (10))} \\ &\Rightarrow_H a_1 a_2 \cdots a_{n-1} a_n && \text{(by } (\grave{a}_1 \rightarrow a_1, \emptyset, \emptyset) \text{ from (10))} \end{aligned}$$

where $A \in N$, $X, Y \in V$, $w \in V^*$, $a_i \in T$, for all i, $1 \leq i \leq n$, for some $n \geq 2$. Such a derivation is also possible in G (of course, without marked symbols and the last two applied rules).

Based on these observations, we see that $L(H) = L(G)$. Since H has effectively only a single set of rules, the theorem holds. □

The second normal form represents a dual normal form to that in Theorems 6.3.1 and 6.3.2. Indeed, we show that every one-sided random context grammar can be turned into an equivalent one-sided random context grammar with the sets of left and right random context rules being disjoint.

Theorem 6.3.3. *Let $G = (N, T, P_L, P_R, S)$ be a one-sided random context grammar. Then, there is a one-sided random context grammar, $H = (N', T, P'_L, P'_R, S)$, such that $L(H) = L(G)$ and $P'_L \cap P'_R = \emptyset$. Furthermore, if G is propagating, so is H.*

Proof. Let $G = (N, T, P_L, P_R, S)$ be a one-sided random context grammar. Construct

$$H = (N', T, P'_L, P'_R, S)$$

where

$$\begin{aligned} N' &= N \cup \{L, R\} \\ P'_L &= \{(A \rightarrow x, U, W \cup \{L\}) \mid (A \rightarrow x, U, W) \in P_L\} \\ P'_R &= \{(A \rightarrow x, U, W \cup \{R\}) \mid (A \rightarrow x, U, W) \in P_R\} \end{aligned}$$

Without any loss of generality, we assume that $\{L, R\} \cap (N \cup T) = \emptyset$. Observe that the new nonterminals L and R cannot appear in any sentential form. Therefore, it is easy to see that $L(H) = L(G)$. Furthermore, observe that if G is propagating, so is H. Since $P'_L \cap P'_R = \emptyset$, the theorem holds. □

The third normal form represents an analogy of the well-known Chomsky normal form for context-free grammars (see Definition 4.1.15). However, since one-sided random context grammars with erasing rules are more powerful than their propagating versions, we allow the presence of erasing rules in the transformed grammar.

Theorem 6.3.4. *Let $G = (N, T, P_L, P_R, S)$ be a one-sided random context grammar. Then, there is a one-sided random context grammar, $H = (N', T, P'_L, P'_R, S)$, such that $L(H) = L(G)$ and $(A \to x, U, W) \in P'_L \cup P'_R$ implies that $x \in N'N' \cup T \cup \{\varepsilon\}$. Furthermore, if G is propagating, so is H.*

Proof. Let $G = (N, T, P_L, P_R, S)$ be a one-sided random context grammar. Set $V = N \cup T$ and $\bar{T} = \{\bar{a} \mid a \in T\}$. Define the homomorphism τ from V^* to $(N \cup \bar{T})^*$ as $\tau(A) = A$ for each $A \in N$, and $\tau(a) = \bar{a}$ for each $a \in T$. Let ℓ be the length of the longest right-hand side of a rule from $P_L \cup P_R$. Set

$$M = \{\langle y \rangle \mid y \in V^+, 2 \leq |y| \leq \ell - 1\}$$

Without any loss of generality, we assume that V, $\bar{T}$, and M are pairwise disjoint. Construct

$$H = (N', T, P'_L, P'_R, S)$$

as follows. Initially, set

$$\begin{aligned}
N' &= N \cup \bar{T} \cup M \\
P'_L &= \{(A \to x, U, W \cup M) \mid (A \to x, U, W) \in P_L, x \in T \cup \{\varepsilon\}\} \cup \\
&\quad\ \{(A \to \tau(x), U, W \cup M) \mid (A \to x, U, W) \in P_L, x \in VV\} \cup \\
&\quad\ \{(\bar{a} \to a, \emptyset, \emptyset) \mid a \in T\} \\
P'_R &= \{(A \to x, U, W \cup M) \mid (A \to x, U, W) \in P_R, x \in T \cup \{\varepsilon\}\} \cup \\
&\quad\ \{(A \to \tau(x), U, W \cup M) \mid (A \to x, U, W) \in P_R, x \in VV\}
\end{aligned}$$

Perform (1) and (2), given next.

(1) For each $(A \to X_1X_2 \cdots X_n, U, W) \in P_L$, where $X_i \in V$ for $i = 1, 2, \dots, n$, for some $n \geq 3$,

- add $(A \to \langle X_1X_2 \cdots X_{n-1} \rangle \tau(X_n), U, W \cup M)$ to P'_L;
- add $(\langle X_1X_2 \cdots X_{n-1} \rangle \to \langle X_1X_2 \cdots X_{n-2} \rangle \tau(X_{n-1}), \emptyset, M)$ to P'_L;

  ⋮

- add $(\langle X_1X_2 \rangle \to \tau(X_1X_2), \emptyset, M)$ to P'_L.

(2) For each $(A \to X_1X_2 \cdots X_n, U, W) \in P_R$, where $X_i \in V$ for $i = 1, 2, \dots, n$, for some $n \geq 3$,

- add $(A \to \langle X_1X_2 \cdots X_{n-1} \rangle \tau(X_n), U, W \cup M)$ to P'_R;
- add $(\langle X_1X_2 \cdots X_{n-1} \rangle \to \langle X_1X_2 \cdots X_{n-2} \rangle \tau(X_{n-1}), \emptyset, M)$ to P'_R;

  ⋮

- add $(\langle X_1X_2 \rangle \to \tau(X_1X_2), \emptyset, M)$ to P'_R.

To give an insight into the construction, notice that rules whose right-hand side is either a terminal or the empty string are directly added to P'_L and P'_R in the initialization part of the construction. When the right-hand side of a rule has two symbols, their homomorphic image (with respect to τ) is used, which results in the new right-hand side being formed by two nonterminals, even if the original right-hand side contained terminals. Barred nonterminals are rewritten to their corresponding terminals by rules of the form $(\bar{a} \rightarrow a, \emptyset, \emptyset)$, introduced in the initialization part of the construction. Notice that their permitting and forbidding contexts can be empty.

Rules with more than two symbols on their right-hand side are simulated in a several-step way by rules from (1) and (2). Compound nonterminals of the form $\langle X_1 X_2 \cdots X_n \rangle$, where each X_i is a symbol, are used to satisfy the required form of every rule in $P'_L \cup P'_R$. Each rule from (1) and (2) forbids the presence of these compound symbols to the left (or right) of the rewritten nonterminal to ensure a proper simulation.

Based on these observations, we see that $L(H) = L(G)$. Moreover, observe that if G is propagating, so is H. Since H is of the required form, the theorem holds. □

In the fourth normal form, every rule has its permitting or forbidding context empty.

Theorem 6.3.5. *Let $G = (N, T, P_L, P_R, S)$ be a one-sided random context grammar. Then, there is a one-sided random context grammar, $H = (N', T, P'_L, P'_R, S)$, such that $L(H) = L(G)$ and $(A \rightarrow x, U, W) \in P'_L \cup P'_R$ implies that $U = \emptyset$ or $W = \emptyset$. Furthermore, if G is propagating, so is H.*

Proof. Let $G = (N, T, P_L, P_R, S)$ be a one-sided random context grammar. Set $V = N \cup T$ and

$$F = \{\langle r, d, i \rangle \mid r = (A \rightarrow x, U, W) \in P_d, d \in \{L, R\}, i \in \{1, 2\}\}$$

Without any loss of generality, we assume that $F \cap V = \emptyset$. Construct

$$H = (N', T, P'_L, P'_R, S)$$

as follows. Initially, set $N' = N \cup F$, $P'_L = \emptyset$, and $P'_R = \emptyset$. Perform (1) and (2), given next.

(1) For each $r = (A \rightarrow x, U, W) \in P_L$,

  (1.1) add $(A \rightarrow \langle r, L, 1 \rangle, \emptyset, F)$ to P'_R;
  (1.2) add $(\langle r, L, 1 \rangle \rightarrow \langle r, L, 2 \rangle, \emptyset, W \cup F)$ to P'_L;
  (1.3) add $(\langle r, L, 2 \rangle \rightarrow x, U, \emptyset)$ to P'_L.

(2) For each $r = (A \rightarrow x, U, W) \in P_R$,

  (2.1) add $(A \rightarrow \langle r, R, 1 \rangle, \emptyset, F)$ to P'_L;
  (2.2) add $(\langle r, R, 1 \rangle \rightarrow \langle r, R, 2 \rangle, \emptyset, W \cup F)$ to P'_R;
  (2.3) add $(\langle r, R, 2 \rangle \rightarrow x, U, \emptyset)$ to P'_R.

To give an insight into the construction, notice that a single rule from P_L and P_R is simulated in three steps by rules introduced in (1) and (2), respectively. As we cannot check both the presence and absence of symbols in a single step, we split this check into two consecutive steps. Clearly, $L(G) \subseteq L(H)$, so we only prove that $L(H) \subseteq L(G)$.

Observe that if we apply the three rules from (1) in H, then we can apply the original rule in G. A similar application can be reformulated in terms of (2). Therefore, it remains to be shown that H cannot generate improper sentences by invalid intermixed simulations of more than one rule of G at a time. In what follows, we consider only simulations of rules from P_L; rules from P_R are simulated analogously.

Let us consider a simulation of some $r = (A \rightarrow x, U, W) \in P_L$. Observe that the only situation where a improper simulation may occur is that after a rule from (1.2) is applied, another simulation takes places which transforms a nonterminal to the left of $\langle r, L, 2\rangle$ that is not in U into a nonterminal that is in U. To investigate this possibility, set $V' = N' \cup T$ and consider any successful derivation in H, $S \Rightarrow_H^* z$, where $z \in L(H)$. This derivation can be written in the form

$$S \Rightarrow_H^* w \Rightarrow_H y \Rightarrow_H^* z$$

where $w = w_1\langle r, L, 1\rangle w_2$, $y = w_1\langle r, L, 2\rangle w_2$, and $w_1, w_2 \in V'^*$. Since $w \Rightarrow_H y$ by $(\langle r, L, 1\rangle \rightarrow \langle r, L, 2\rangle, \emptyset, W \cup F)$, introduced to P'_L in (1.2) from r,

$$\text{alph}(w_1) \cap (W \cup F) = \emptyset$$

From the presence of $\langle r, L, 2\rangle$, no rule from (1) is now applicable to w_1. Let $w_1 = w'_1 B w''_1$ and $(B \rightarrow \langle s, R, 1\rangle, \emptyset, F) \in P'_L$, introduced in (2.1) from some $s = (B \rightarrow v, X, Y) \in P_R$ such that $B \notin U$ and

$$\text{alph}(v) \cap \big(U - \text{alph}(w_1)\big) \neq \emptyset$$

This last requirement implies that by successfully simulating s prior to r, we necessarily end up with an invalid simulation of r. Then,

$$w'_1 B w''_1 \langle r, L, 2\rangle w_2 \Rightarrow_H w'_1 \langle s, R, 1\rangle w''_1 \langle r, L, 2\rangle w_2$$

Since $\langle s, R, 1\rangle$ cannot be rewritten to $\langle s, R, 2\rangle$ by a rule from (2.2) because $\langle r, L, 2\rangle$ occurs to the right of $\langle s, R, 1\rangle$, we can either

(a) correctly finish the simulation of r by rewriting $\langle r, L, 2\rangle$ to x (recall that $B \notin U$) or
(b) rewrite some nonterminal in w'_1 or w''_1.

However, observe that in (b), we end up in the same situation as we are now.

Based on these observations, we see that no invalid intermixed simulations of more than one rule of G at a time are possible in H. Hence, $L(H) \subseteq L(G)$, so

$L(H) = L(G)$. Clearly, $(A \to x, U, W) \in P'_L \cup P'_R$ implies that $U = \emptyset$ or $W = \emptyset$. Furthermore, observe that if G is propagating, so is H. Thus, the theorem holds. □

We conclude this section by suggesting an open problem.

Open Problem 6.3.6. Let $G = (N, T, P_L, P_R, S)$ be a one-sided random context grammar, and consider the following four normal forms

(I) either $P_L = \emptyset$ or $P_R = \emptyset$;
(II) $(A \to x, U, W) \in P_L \cup P_R$ implies that $\text{card}(U) + \text{card}(W) \leq 1$;
(III) $P_L = \emptyset$ and $(A \to x, U, W) \in P_R$ implies that $W = \emptyset$;
(IV) $P_R = \emptyset$ and $(A \to x, U, W) \in P_L$ implies that $W = \emptyset$.

Can we turn G into an equivalent one-sided random context grammar in any of the above-mentioned forms? □

6.4 Reduction

Recall that one-sided random context grammars characterize the family of recursively enumerable languages (see Theorem 6.2.4). Of course, it is more than natural to ask whether the family of recursively enumerable languages is characterized by one-sided random context grammars with a limited number of nonterminals or rules. The present section, consisting of three subsections, gives an affirmative answer to this question.

More specifically, in Sect. 6.4.1, we prove that every recursively enumerable language can be generated by a one-sided random context grammar with no more than ten nonterminals. In addition, we show that an analogous result holds for 13 nonterminals in terms of these grammars with the set of left random context rules coinciding with the set of right random context rules.

Then, in Sect. 6.4.2, we approach the discussion concerning the reduction of these grammars with respect to the number of nonterminals in a finer way. Indeed, we introduce the notion of a *right random context nonterminal*, defined as a nonterminal that appears on the left-hand side of a right random context rule, and demonstrate how to convert any one-sided random context grammar G to an equivalent one-sided random context grammar H with two right random context nonterminals. We also explain how to achieve an analogous conversion in terms of propagating versions of these grammars (recall that they characterize the family of context-sensitive languages, see Theorem 6.2.3). Similarly, we introduce the notion of a *left random context nonterminal* and demonstrate how to convert any one-sided random context grammar G to an equivalent one-sided random context grammar H with two left random context nonterminals. We explain how to achieve an analogous conversion in terms of propagating versions of these grammars, too.

Apart from reducing the number of nonterminals, we reduce the number of rules. More specifically, in Sect. 6.4.3, we prove that any recursively enumerable language can be generated by a one-sided random context grammar having no more than two right random context rules. As a motivation behind limiting the number of right random context rules in these grammars, consider left random context grammars, which are one-sided random context grammars with no right random context rules (see Sect. 6.1). Recall that it is an open question whether these grammars are equally powerful to one-sided random context grammars (see Open Problem 6.2.28). To give an affirmative answer to this question, it is sufficient to show that in one-sided random context grammars, no right random context rules are needed. From this viewpoint, the above-mentioned result may fulfill a useful role during the solution of this problem in the future.

The results sketched above can be also seen as a contribution to the investigation concerning the *descriptional complexity* of formal models, which represents an important trend in today's formal language theory as demonstrated by several recent studies (see [7, 10, 14–17, 23, 28–32, 43, 49]). As an important part, this trend discusses the *nonterminal complexity* of grammars—an investigation area that is primarily interested in reducing the number of nonterminals in grammars without affecting their power, so the results mentioned above actually represent new knowledge concerning the nonterminal complexity of one-sided random context grammars.

6.4.1 Total Number of Nonterminals

In this section, we prove that every recursively enumerable language can be generated by a one-sided random context grammar H that satisfies one of conditions (I) and (II), given next.

(I) H has ten nonterminals (Theorem 6.4.1).
(II) The set of left random context rules of H coincides with the set of right random context rules, and H has 13 nonterminals (Corollary 6.4.2).

Theorem 6.4.1. *Let K be a recursively enumerable language. Then, there is a one-sided random context grammar, $H = (N, T, P_L, P_R, S)$, such that $L(H) = K$ and* $\mathrm{card}(N) = 10$.

Proof. Let K be a recursively enumerable language. Then, by Theorem 4.1.8, there is a phrase-structure grammar in the Geffert normal form

$$G = \big(\{S, A, B, C\}, T, P \cup \{ABC \rightarrow \varepsilon\}, S\big)$$

satisfying $L(G) = K$. We next construct a one-sided random context grammar H such that $L(H) = L(G)$. Set $N = \{S, A, B, C\}$, $V = N \cup T$, and $N' = \{S, A, B, C, \bar{A}, \hat{A}, \bar{B}, \hat{B}, \tilde{C}, \#\}$. Without any loss of generality, we assume that $(\{\bar{A}, \hat{A}, \bar{B}, \hat{B}, \tilde{C}, \#\}) \cap V = \emptyset$. Construct

$$H = \big(N', T, P_L, P_R, S\big)$$

in the following way. Initially, set $P_L = \emptyset$ and $P_R = \emptyset$. Perform the following seven steps

(1) for each $S \to uSa \in P$, where $u \in \{A, AB\}^*$ and $a \in T$,
add $(S \to uS\#a, \emptyset, \{\bar{A}, \bar{B}, \hat{A}, \hat{B}, \tilde{C}, \#\})$ to P_L;
(2) for each $S \to uSv \in P$, where $u \in \{A, AB\}^*$ and $v \in \{BC, C\}^*$,
add $(S \to uSv, \emptyset, \{\bar{A}, \bar{B}, \hat{A}, \hat{B}, \tilde{C}, \#\})$ to P_L;
(3) for each $S \to uv \in P$, where $u \in \{A, AB\}^*$ and $v \in \{BC, C\}^*$,
add $(S \to uv, \emptyset, \{\bar{A}, \bar{B}, \hat{A}, \hat{B}, \tilde{C}, \#\})$ to P_L;
(4) add $(A \to \bar{A}, \emptyset, N \cup \{\hat{A}, \hat{B}, \tilde{C}, \#\})$ to P_L;
add $(B \to \bar{B}, \emptyset, N \cup \{\hat{A}, \hat{B}, \tilde{C}, \#\})$ to P_L;
add $(A \to \hat{A}, \emptyset, N \cup \{\hat{A}, \hat{B}, \tilde{C}, \#\})$ to P_L;
add $(B \to \hat{B}, \{\hat{A}\}, N \cup \{\hat{B}, \tilde{C}, \#\})$ to P_L;
add $(C \to \tilde{C}, \{\hat{A}, \hat{B}\}, N \cup \{\tilde{C}, \#\})$ to P_L;
(5) add $(\hat{B} \to \varepsilon, \{\tilde{C}\}, \{S, \bar{A}, \bar{B}, \hat{A}, \hat{B}\})$ to P_R;
add $(\hat{A} \to \varepsilon, \{\tilde{C}\}, \{S, \bar{A}, \bar{B}, \hat{A}, \hat{B}\})$ to P_R;
add $(\tilde{C} \to \varepsilon, \emptyset, N \cup \{\hat{A}, \hat{B}, \tilde{C}, \#\})$ to P_L;
(6) add $(\bar{A} \to A, \emptyset, \{S, \bar{A}, \bar{B}, \hat{A}, \hat{B}, \tilde{C}\})$ to P_R;
add $(\bar{B} \to B, \emptyset, \{S, \bar{A}, \bar{B}, \hat{A}, \hat{B}, \tilde{C}\})$ to P_R;
(7) add $(\# \to \varepsilon, \emptyset, N')$ to P_L.

Before proving that $L(H) = L(G)$, let us informally describe the purpose of rules introduced in (1) through (7). H simulates the derivations of G that satisfy the form described in Theorem 4.1.9. The context-free rules in P are simulated by rules from (1) through (3). The context-sensitive rule $ABC \to \varepsilon$ is simulated in a several-step way. First, rules introduced in (4) are used to prepare the erasure of ABC. These rules rewrite nonterminals from the left to the right. In this way, it is guaranteed that whenever $\hat{A}$, $\hat{B}$, and $\tilde{C}$ appear in a sentential form, then they form a substring of the form $\hat{A}\hat{B}\tilde{C}$. Then, rules from (5) sequentially erase $\hat{B}$, $\hat{A}$, and $\tilde{C}$. Finally, rules from (6) convert barred nonterminals back to their non-barred versions to prepare another simulation of $ABC \to \varepsilon$; this conversion is done from the right to the left. For example, $AABCBCab \Rightarrow_G ABCab$ is simulated by H as follows:

$$
\begin{aligned}
AABCBC\#a\#b &\Rightarrow_H \bar{A}ABCBC\#a\#b \\
&\Rightarrow_H \bar{A}\hat{A}BCBC\#a\#b \\
&\Rightarrow_H \bar{A}\hat{A}\hat{B}CBC\#a\#b \\
&\Rightarrow_H \bar{A}\hat{A}\hat{B}\tilde{C}BC\#a\#b \\
&\Rightarrow_H \bar{A}\hat{A}\tilde{C}BC\#a\#b \\
&\Rightarrow_H \bar{A}\tilde{C}BC\#a\#b \\
&\Rightarrow_H \bar{A}BC\#a\#b \\
&\Rightarrow_H ABC\#a\#b
\end{aligned}
$$

Symbol # is used to ensure that every sentential form of H is of the form w_1w_2, where $w_1 \in (N' - \{\#\})^*$ and $w_2 \in (T \cup \{\#\})^*$. Since permitting and forbidding contexts cannot contain terminals, a mixture of symbols from T and N in H could produce a terminal string out of $L(G)$. For example, observe that $AaBC \Rightarrow_H^* a$ by rules from (4) and (5), but such a derivation does not exist in G. #s can be eliminated by an application of rules from (7), provided that no nonterminals occur to the left of # in the current sentential form. Consequently, all #s are erased at the end of every successful derivation.

To establish $L(H) = L(G)$, we prove two claims. Claim 1 shows how derivations of G are simulated by H. This claim is then used to prove that $L(G) \subseteq L(H)$. Set $V' = N' \cup T$. Define the homomorphism φ from V^* to V'^* as $\varphi(X) = X$, for all $X \in N$, and $\varphi(a) = \#a$, for all $a \in T$.

Claim 1. If $S \Rightarrow_G^n x \Rightarrow_G^* z$, where $x \in V^*$ and $z \in T^*$, for some $n \geq 0$, then $S \Rightarrow_H^* \varphi(x)$.

Proof. This claim is established by induction on $n \geq 0$.

Basis. For $n = 0$, this claim is clear.

Induction Hypothesis. Suppose that there exists $n \geq 0$ such that the claim holds for all derivations of length ℓ, where $0 \leq \ell \leq n$.

Induction Step. Consider any derivation of the form

$$S \Rightarrow_G^{n+1} w \Rightarrow_G^* z$$

where $w \in V^*$ and $z \in T^*$. Since $n + 1 \geq 1$, this derivation can be expressed as

$$S \Rightarrow_G^n x \Rightarrow_G w \Rightarrow_G^* z$$

for some $x \in V^+$. Without any loss of generality, we assume that x is of the form $x = x_1x_2x_3x_4$, where $x_1 \in \{A, AB\}^*$, $x_2 \in \{S, \varepsilon\}$, $x_3 \in \{BC, C\}^*$, and $x_4 \in T^*$ (see Theorem 4.1.9 and [20]).

Next, we consider all possible forms of $x \Rightarrow_G w$, covered by the following four cases—(i) through (iv).

(i) *Application of $S \to uSa \in P$.* Let $x = x_1Sx_3x_4$, $w = x_1uSax_3x_4$, and $S \to uSa \in P$, where $x_1, u \in \{A, AB\}^*$, $x_3, v \in \{BC, C\}^*$, $x_4 \in T^*$, and $a \in T$. Then, by the induction hypothesis,

$$S \Rightarrow_H^* \varphi(x_1Sx_3x_4)$$

By (1), $(S \to uS\#a, \emptyset, \{\bar{A}, \bar{B}, \hat{A}, \hat{B}, \tilde{C}, \#\}) \in P_L$. Since $\varphi(x_1Sx_3x_4) = x_1S\varphi(x_3x_4)$ and $\text{alph}(x_1) \cap \{\bar{A}, \bar{B}, \hat{A}, \hat{B}, \tilde{C}, \#\} = \emptyset$,

$$x_1S\varphi(x_3x_4) \Rightarrow_H x_1uS\#a\varphi(x_3x_4)$$

As $\varphi(x_1uSax_3x_4) = x_1uS\#a\varphi(x_3x_4)$, the induction step is completed for (i).

(ii) *Application of* $S \to uSv \in P$. Let $x = x_1Sx_3x_4$, $w = x_1uSvx_3x_4$, and $S \to uSv \in P$, where $x_1, u \in \{A, AB\}^*$, $x_3, v \in \{BC, C\}^*$, and $x_4 \in T^*$. To complete the induction step for (ii), proceed by analogy with (i), but use a rule from (2) instead of a rule from (1).

(iii) *Application of* $S \to uv \in P$. Let $x = x_1Sx_3x_4$, $w = x_1uvx_3x_4$, and $S \to uv \in P$, where $x_1, u \in \{A, AB\}^*$, $x_3, v \in \{BC, C\}^*$, and $x_4 \in T^*$. To complete the induction step for (iii), proceed by analogy with (i), but use a rule from (3) instead of a rule from (1).

(iv) *Application of* $ABC \to \varepsilon$. Let $x = x_1ABCx_3x_4$, $w = x_1x_3x_4$, where $x_1 \in \{A, AB\}^*$, $x_3 \in \{BC, C\}^*$, and $x_4 \in T^*$, so $x \Rightarrow_G w$ by $ABC \to \varepsilon$. Then, by the induction hypothesis,

$$S \Rightarrow_H^* \varphi(x_1ABCx_3x_4)$$

Let $x_1 = X_1X_2 \cdots X_k$, where $k = |x_1|$ (the case when $k = 0$ means that $x_1 = \varepsilon$). Since $\varphi(x_1ABCx_3x_4) = x_1ABC\varphi(x_3x_4)$ and $\text{alph}(x_1) \subseteq N$, by rules introduced in (4),

$$\begin{aligned} X_1X_2 \cdots X_kABC\varphi(x_3x_4) &\Rightarrow_H \bar{X}_1X_2 \cdots X_kABC\varphi(x_3x_4) \\ &\Rightarrow_H \bar{X}_1\bar{X}_2 \cdots X_kABC\varphi(x_3x_4) \\ &\vdots \\ &\Rightarrow_H \bar{X}_1\bar{X}_2 \cdots \bar{X}_kABC\varphi(x_3x_4) \\ &\Rightarrow_H \bar{X}_1\bar{X}_2 \cdots \bar{X}_k\hat{A}BC\varphi(x_3x_4) \\ &\Rightarrow_H \bar{X}_1\bar{X}_2 \cdots \bar{X}_k\hat{A}\hat{B}C\varphi(x_3x_4) \\ &\Rightarrow_H \bar{X}_1\bar{X}_2 \cdots \bar{X}_k\hat{A}\hat{B}\tilde{C}\varphi(x_3x_4) \end{aligned}$$

Let $\bar{x}_1 = \bar{X}_1\bar{X}_2 \cdots \bar{X}_k$. Since $\text{alph}(\varphi(x_3x_4)) \cap \{S, \bar{A}, \bar{B}, \hat{A}, \hat{B}\} = \emptyset$, by rules introduced in (5),

$$\begin{aligned} \bar{x}_1\hat{A}\hat{B}\tilde{C}\varphi(x_3x_4) &\Rightarrow_H \bar{x}_1\hat{A}\tilde{C}\varphi(x_3x_4) \\ &\Rightarrow_H \bar{x}_1\tilde{C}\varphi(x_3x_4) \\ &\Rightarrow_H \bar{x}_1\varphi(x_3x_4) \end{aligned}$$

Finally, by rules from (6),

$$\begin{aligned} \bar{x}_1\varphi(x_3x_4) &\Rightarrow_H \bar{X}_1 \cdots \bar{X}_{k-1}X_k\varphi(x_3x_4) \\ &\Rightarrow_H \bar{X}_1 \cdots X_{k-1}X_k\varphi(x_3x_4) \\ &\vdots \\ &\Rightarrow_H X_1 \cdots X_{k-1}X_k\varphi(x_3x_4) \end{aligned}$$

Recall that $x_1 = X_1 \cdots X_{k-1}X_k$. Since $\varphi(x_1x_3x_4) = x_1\varphi(x_3x_4)$, the induction step is completed for (iv).

Observe that cases (i) through (iv) cover all possible forms of $x \Rightarrow_G w$. Thus, the claim holds. □

Claim 2 demonstrates how G simulates derivations of H. It is then used to prove that $L(H) \subseteq L(G)$. Define the homomorphism π from V'^* to V^* as $\pi(X) = X$, for all $X \in N$, $\pi(\bar{A}) = \pi(\hat{A}) = A$, $\pi(\bar{B}) = \pi(\hat{B}) = B$, $\pi(\tilde{C}) = C$, $\pi(a) = a$, for all $a \in T$, and $\pi(\#) = \varepsilon$. Define the homomorphism τ from V'^* to V^* as $\tau(X) = \pi(X)$, for all $X \in V' - \{\hat{A}, \hat{B}, \tilde{C}\}$, and $\tau(\hat{A}) = \tau(\hat{B}) = \tau(\tilde{C}) = \varepsilon$.

Claim 2. Let $S \Rightarrow_H^n x \Rightarrow_H^* z$, *where* $x \in V'^*$ *and* $z \in T^*$, *for some* $n \geq 0$. *Then,* $x = x_1x_2x_3x_4x_5$, *where* $x_1 \in \{\bar{A}, \bar{B}\}^*$, $x_2 \in \{A, B\}^*$, $x_3 \in \{S, \hat{A}BC, \hat{A}\hat{B}C, \hat{A}\hat{B}\tilde{C}, \hat{A}\tilde{C}, \tilde{C}, \varepsilon\}$, $x_4 \in \{B, C\}^*$, *and* $x_5 \in (T \cup \{\#\})^*$. *Furthermore,*

(a) if $x_3 \in \{S, \varepsilon\}$, *then* $S \Rightarrow_G^* \pi(x)$;
(b) if $x_3 \in \{\hat{A}BC, \hat{A}\hat{B}C, \hat{A}\hat{B}\tilde{C}\}$, *then* $x_2 = \varepsilon$ *and* $S \Rightarrow_G^* \pi(x)$;
(c) if $x_3 \in \{\hat{A}\tilde{C}, \tilde{C}\}$, *then* $x_2 = \varepsilon$ *and* $S \Rightarrow_G^* \tau(x)$.

Proof. This claim is established by induction on $n \geq 0$.

Basis. For $n = 0$, this claim is clear.

Induction Hypothesis. Suppose that there exists $n \geq 0$ such that the claim holds for all derivations of length ℓ, where $0 \leq \ell \leq n$.

Induction Step. Consider any derivation of the form

$$S \Rightarrow_H^{n+1} w \Rightarrow_H^* z$$

where $w \in V'^*$ and $z \in T^*$. Since $n + 1 \geq 1$, this derivation can be expressed as

$$S \Rightarrow_H^n x \Rightarrow_H w \Rightarrow_H^* z$$

for some $x \in V'^+$. By the induction hypothesis, $x = x_1x_2x_3x_4x_5$, where $x_1 \in \{\bar{A}, \bar{B}\}^*$, $x_2 \in \{A, B\}^*$, $x_3 \in \{S, \hat{A}BC, \hat{A}\hat{B}C, \hat{A}\hat{B}\tilde{C}, \hat{A}\tilde{C}, \tilde{C}, \varepsilon\}$, $x_4 \in \{B, C\}^*$, and $x_5 \in (T \cup \{\#\})^*$. Furthermore, (a) through (c), stated in the claim, hold.

Next, we consider all possible forms of $x \Rightarrow_H w$, covered by the following five cases—(i) through (v).

(i) *Application of a rule from (1).* Let $x_3 = S$, $x_1 = x_4 = \varepsilon$, and $(S \rightarrow uS\#a, \emptyset, \{\bar{A}, \bar{B}, \hat{A}, \hat{B}, \tilde{C}, \#\}) \in P_L$, introduced in (1), where $u \in \{A, AB\}^*$ and $a \in T$, so

$$x_2Sx_5 \Rightarrow_H x_2uS\#ax_5$$

Observe that if $x_4 \neq \varepsilon$, then $w \Rightarrow_H^* z$ does not hold. Indeed, if $x_4 \neq \varepsilon$, then to erase the nonterminals in x_4, there have to be As in x_2. However, the # symbol, introduced between x_2 and x_4, blocks the applicability of $(C \rightarrow \tilde{C}, \{\hat{A}, \hat{B}\}, N \cup \{\tilde{C}, \#\}) \in P_L$, introduced in (4), which is needed to erase the nonterminals in x_4. Since $(\# \to \varepsilon, \emptyset, N') \in P_L$, introduced in (7), requires that there are no nonterminals to the left of #, the derivation cannot be successfully finished. Hence, $x_4 = \varepsilon$. Since $u \in \{A, B\}^*$ and $\#a \in (T \cup \{\#\})^*$, $x_2uS\#ax_5$ is of the required form. As $x_3 = S$, $S \Rightarrow_G^* \pi(x)$. Observe that $\pi(x) = \pi(x_2)S\pi(x_5)$. By (1), $S \to uSa \in P$, so

$$\pi(x_2)S\pi(x_5) \Rightarrow_G \pi(x_2)uSa\pi(x_5)$$

Since $\pi(x_2)uSa\pi(x_5) = \pi(x_2uS\#ax_5)$ and both $\hat{A}\tilde{C}$ and $\tilde{C}$ are not substrings of $x_2uS\#ax_5$, the induction step is completed for (i).

(ii) *Application of a rule from (2).* Let $x_3 = S$, $x_1 = \varepsilon$, and $(S \to uSv, \emptyset, \{\bar{A}, \bar{B}, \hat{A}, \hat{B}, \tilde{C}, \#\}) \in P_L$, introduced in (2), where $u \in \{A, AB\}^*$ and $v \in \{BC, C\}^*$, so

$$x_2Sx_4x_5 \Rightarrow_H x_2uSvx_4x_5$$

To complete the induction step for (ii), proceed by analogy with (i), but use $S \to uSv \in P$ instead of $S \to uSa \in P$. Observe that x_4 may be nonempty in this case.

(iii) *Application of a rule from (3).* Let $x_3 = S$, $x_1 = \varepsilon$, and $(S \to uv, \emptyset, \{\bar{A}, \bar{B}, \hat{A}, \hat{B}, \tilde{C}, \#\}) \in P_L$, introduced in (3), where $u \in \{A, AB\}^*$ and $v \in \{BC, C\}^*$, so

$$x_2Sx_4x_5 \Rightarrow_H x_2uvx_4x_5$$

To complete the induction step for (iii), proceed by analogy with (i), but use $S \to uv \in P$ instead of $S \to uSa \in P$. Observe that x_4 may be nonempty in this case.

(iv) *Application of a rule from (5).* Let $x_3 \in \{\hat{A}\hat{B}\tilde{C}, \hat{A}\tilde{C}, \tilde{C}\}$. By the induction hypothesis [more specifically, by (b) and (c)], $x_2 = \varepsilon$. Then, there are three subcases, depending on what x_3 actually is.

(iv.i) Let $x_3 = \hat{A}\hat{B}\tilde{C}$. Then, $x_1\hat{A}\hat{B}\tilde{C}x_4x_5 \Rightarrow_H x_1\hat{A}\tilde{C}x_4x_5$ by $(\hat{B} \to \varepsilon, \{\tilde{C}\}, \{S, \bar{A}, \bar{B}, \hat{A}, \hat{B}\}) \in P_R$, introduced in (5). Observe that this is the only applicable rule from (5). By the induction hypothesis, $S \Rightarrow_G^* \pi(x)$. Since $\pi(x) = \pi(x_1)ABC\pi(x_4x_5)$,

$$\pi(x_1)ABC\pi(x_4x_5) \Rightarrow_G \pi(x_1)\pi(x_4x_5)$$

by $ABC \to \varepsilon$. As $w = x_1\hat{A}\tilde{C}x_4x_5$ is of the required form and $\pi(x_1)\pi(x_4x_5) = \tau(w)$, the induction step is completed for (iv.i).

(iv.ii) Let $x_3 = \hat{A}\tilde{C}$. Then, $x_1\hat{A}\tilde{C}x_4x_5 \Rightarrow_H x_1\tilde{C}x_4x_5$ by $(\hat{A} \to \varepsilon, \{\tilde{C}\}, \{S, \bar{A}, \bar{B}, \hat{A}, \hat{B}\}) \in P_R$, introduced in (5). Observe that this is the only applicable rule from (5). By the induction hypothesis, $S \Rightarrow_G^* \tau(x)$. As

$w = x_1\tilde{C}x_4x_5$ is of the required form and $\tau(x) = \tau(w)$, the induction step is completed for (iv.ii).

(iv.iii) Let $x_3 = \tilde{C}$. Then, $x_1\tilde{C}x_4x_5 \Rightarrow_H x_1x_4x_5$ by $(\tilde{C} \to \varepsilon, \emptyset, N \cup \{\hat{A}, \hat{B}, \tilde{C}, \#\}) \in P_L$, introduced in (5). Observe that this is the only applicable rule from (5). By the induction hypothesis, $S \Rightarrow_G^* \tau(x)$. As $w = x_1x_4x_5$ is of the required form and $\tau(x) = \tau(w)$, the induction step is completed for (iv.iii).

(v) *Application of a rule from (4), (6), or (7).* Let $x \Rightarrow_H w$ by a rule from (4), (6), or (7). Observe that $x_3 \notin \{\hat{A}\tilde{C}, \tilde{C}\}$ has to hold; otherwise, none of these rules is applicable. Indeed, if $x_3 \in \{\hat{A}\tilde{C}, \tilde{C}\}$, then $x_2 = \varepsilon$ by the induction hypothesis [more specifically, by (c)], which implies that no rule from (4) is applicable. Also, $x_3 \in \{\hat{A}\tilde{C}, \tilde{C}\}$ would imply that no rule from (6) and (7) is applicable. Therefore, $S \Rightarrow_G^* \pi(w)$ follows directly from the induction hypothesis (obviously, $\pi(w) = \pi(x)$, and since $x_3 \notin \{\hat{A}\tilde{C}, \tilde{C}\}$, $S \Rightarrow_G^* \pi(x)$ by the induction hypothesis). As w is clearly of the required form, the induction step is completed for (v).

Observe that cases (i) through (v) cover all possible forms of $x \Rightarrow_H w$. Thus, the claim holds. □

We next prove that $L(H) = L(G)$. Consider Claim 1 for $x \in T^*$. Then, $S \Rightarrow_H^* \varphi(x)$. Let $x = a_1a_2 \cdots a_k$, where $k = |x|$ (the case when $k = 0$ means that $x = \varepsilon$), so $\varphi(x) = \#a_1\#a_2 \cdots \#a_k$. By (7), $(\# \to \varepsilon, \emptyset, N') \in P_L$. Therefore,

$$
\begin{aligned}
\#a_1\#a_2 \cdots \#a_k &\Rightarrow_H a_1\#a_2 \cdots \#a_k \\
&\Rightarrow_H a_1a_2 \cdots \#a_k \\
&\vdots \\
&\Rightarrow_H a_2a_2 \cdots a_k
\end{aligned}
$$

Hence, $x \in L(G)$ implies that $x \in L(H)$, so $L(G) \subseteq L(H)$.

Consider Claim 2 for $x \in T^*$. Then, $S \Rightarrow_G^* \pi(x)$. Since $x \in T^*$, $\pi(x) = x$. Hence, $x \in L(H)$ implies that $x \in L(G)$, so $L(H) \subseteq L(G)$.

The two inclusions, $L(G) \subseteq L(H)$ and $L(H) \subseteq L(G)$, imply that $L(H) = L(G)$. As $\mathrm{card}(N') = 10$, the theorem holds. □

Let $G = (N, T, P_L, P_R, S)$ be a one-sided random context grammar. Recall that in the proof of Theorem 6.3.1, a construction of a one-sided random context grammar, $H = (N', T, P'_L, P'_R, S')$, satisfying $L(H) = L(G)$ and $P'_L = P'_R$, is given. Observe that this construction introduces three new nonterminals—that is, $\mathrm{card}(N') = \mathrm{card}(N) + 3$. Therefore, we obtain the following corollary.

Corollary 6.4.2. *Let K be a recursively enumerable language. Then, there is a one-sided random context grammar, $H = (N, T, P_L, P_R, S)$, such that $L(H) = K$, $P_L = P_R$, and* $\mathrm{card}(N) = 13$. □

6.4.2 Number of Left and Right Random Context Nonterminals

In this section, we approach the discussion concerning the reduction of one-sided random context grammars with respect to the number of nonterminals in a finer way. Indeed, we introduce the notion of a *right random context nonterminal*, defined as a nonterminal that appears on the left-hand side of a right random context rule, and demonstrate how to convert any one-sided random context grammar G to an equivalent one-sided random context grammar H with two right random context nonterminals. We also explain how to achieve an analogous conversion in terms of propagating versions of these grammars (recall that they characterize the family of context-sensitive languages, see Theorem 6.2.3). Similarly, we introduce the notion of a *left random context nonterminal* and demonstrate how to convert any one-sided random context grammar G to an equivalent one-sided random context grammar H with two left random context nonterminals. We explain how to achieve an analogous conversion in terms of propagating versions of these grammars, too.

First, we define these two new measures formally.

Definition 6.4.3. Let $G = (N, T, P_L, P_R, S)$ be a one-sided random context grammar. If $(A \rightarrow x, U, W) \in P_R$, then A is a *right random context nonterminal*. The *number of right random context nonterminals* of G is denoted by $\text{nrrcn}(G)$ and defined as

$$\text{nrrcn}(G) = \text{card}\big(\{A \mid (A \rightarrow x, U, W) \in P_R\}\big) \qquad \square$$

Left random context nonterminals and their number in a one-sided random context grammar are defined analogously.

Definition 6.4.4. Let $G = (N, T, P_L, P_R, S)$ be a one-sided random context grammar. If $(A \rightarrow x, U, W) \in P_L$, then A is a *left random context nonterminal*. The *number of left random context nonterminals* of G is denoted by $\text{nlrcn}(G)$ and defined as

$$\text{nlrcn}(G) = \text{card}\big(\{A \mid (A \rightarrow x, U, W) \in P_L\}\big) \qquad \square$$

Next, we prove that every recursively enumerable language can be generated by a one-sided random context grammar H that satisfies one of conditions (I) through (III), given next.

(I) H has four right random context nonterminals and six left random context nonterminals (Corollary 6.4.5).
(II) H has two right random context nonterminals (Theorem 6.4.7).
(III) H has two left random context nonterminals (Theorem 6.4.8).

In addition, we demonstrate that every context-sensitive language can be generated by a propagating one-sided random context grammar H with either two right random context nonterminals (Theorem 6.4.10), or with two left random context nonterminals (Theorem 6.4.11).

Observe that the construction in the proof of Theorem 6.4.1 implies the following result concerning the number of left and right random context nonterminals.

Corollary 6.4.5. *Let K be a recursively enumerable language. Then, there is a one-sided random context grammar, $H = (N, T, P_L, P_R, S)$, such that $L(H) = K$,* $\text{nrrcn}(H) = 4$, *and* $\text{nlrcn}(H) = 6$. □

Considering only the number of right random context nonterminals, we can improve the previous corollary as described in the following lemma.

Lemma 6.4.6. *Let G be a one-sided random context grammar. Then, there is a one-sided random context grammar H such that $L(H) = L(G)$ and* $\text{nrrcn}(H) = 2$.

Proof. Let $G = (N, T, P_L, P_R, S)$ be a one-sided random context grammar. We next construct a one-sided random context grammar H such that $L(H) = L(G)$ and $\text{nrrcn}(H) = 2$. Set $V = N \cup T$,

$$R = \{\langle r, i\rangle \mid r \in P_R, i = 1, 2\}$$

and

$$_{\$}R = \{\langle \$, r, i\rangle \mid r \in P_R, i = 1, 2\}$$

Without any loss of generality, we assume that R, $_{\$}R$, $\{S', \#_1, \#_2, \$\}$, and V are pairwise disjoint. Construct

$$H = (N', T, P'_L, P'_R, S')$$

as follows. Initially, set $N' = N \cup R \cup {}_{\$}R \cup \{\#_1, \#_2, \$, S'\}$, $P'_L = \emptyset$, and $P'_R = \emptyset$. Furthermore, set $\bar{N} = N' - N$. Perform (1) through (3), given next.

(1) Add $(S' \rightarrow S\$, \emptyset, \emptyset)$ and $(\$ \rightarrow \varepsilon, \emptyset, N')$ to P'_L.
(2) For each $(A \rightarrow y, U, W) \in P_L$, add $(A \rightarrow y, U, W \cup \bar{N})$ to P'_L.
(3) For each $r = (A \rightarrow y, U, W) \in P_R$,

  (3.1) add $(A \rightarrow \langle r, 1\rangle, \emptyset, \bar{N})$ to P'_L;
  (3.2) add $(\$ \rightarrow \langle \$, r, 1\rangle, \{\langle r, 1\rangle\}, \bar{N} - \{\langle r, 1\rangle\})$ to P'_L;
  (3.3) add $(\langle r, 1\rangle \rightarrow \#_1, \emptyset, \bar{N})$ to P'_L;
  (3.4) add $(\#_1 \rightarrow \langle r, 2\rangle, \{\langle \$, r, 1\rangle\}, \bar{N} - \{\langle \$, r, 1\rangle\})$ to P'_R;
  (3.5) add $(\langle \$, r, 1\rangle \rightarrow \langle \$, r, 2\rangle, \{\langle r, 2\rangle\}, \bar{N} - \{\langle r, 2\rangle\})$ to P'_L;
  (3.6) add $(\langle r, 2\rangle \rightarrow \#_2, \emptyset, \bar{N})$ to P'_L;
  (3.7) add $(\#_2 \rightarrow y, U \cup \{\langle \$, r, 2\rangle\}, W \cup (\bar{N} - \{\langle \$, r, 2\rangle\}))$ to P'_R;
  (3.8) add $(\langle \$, r, 2\rangle \rightarrow \$, \emptyset, \bar{N})$ to P'_L.

Before proving that $L(H) = L(G)$, let us informally describe the purpose of rules introduced in (1) through (3). The two rules from (1) are used to start and finish every derivation in H. As we want to reduce the number of right random context nonterminals, rules from P_L are simulated directly by rules from (2). An application of a single rule of G from P_R, $r = (A \to y, U, W) \in P_R$, is simulated by rules introduced in (3) in an eight-step way.

During the simulation of applying r, the very last symbol in sentential forms of H always encodes r for the following two reasons. First, as G can contain more than two right random context nonterminals, whenever a nonterminal is rewritten to $\#_1$ or $\#_2$, we keep track regarding the rule that is simulated. Second, it rules out intermixed simulations of two different rules from P_R, $r = (A \to y, U, W) \in P_R$ and $r' = (A' \to y', U', W') \in P_R$, where $r \neq r'$.

The only purpose of two versions of every compound nonterminal in angular brackets and the symbols $\#_1$, $\#_2$ is to enforce rewriting $\langle \$, r, i\rangle$ back to \$ before another rule from P_R is simulated. In this way, we guarantee that no terminal string out of $L(G)$ is generated.

To establish $L(H) = L(G)$, we prove four claims. Claim 1 shows how H simulates the application of rules from P_L, and how G simulates the application of rules constructed in (2).

Claim 1. In G, $x_1 A x_2 \Rightarrow_G x_1 y x_2$ by $(A \to y, U, W) \in P_L$, where $x_1, x_2 \in V^*$, if and only if in H, $x_1 A x_2 \$ \Rightarrow_H x_1 y x_2 \$$ by $(A \to y, U, W \cup \bar{N}) \in P'_L$, introduced in (2).

Proof. Notice that as $x_1 \in V^*$, $\mathrm{alph}(x_1) \cap \bar{N} = \emptyset$. Thus, this claim holds. □

Claim 2 shows how H simulates the application of rules from P_R, and how G simulates the application of rules constructed in (3).

Claim 2. In G, $x_1 A x_2 \Rightarrow_G x_1 y x_2$ by $r = (A \to y, U, W) \in P_R$, where $x_1, x_2 \in V^*$, if and only if in H, $x_1 A x_2 \$ \Rightarrow_H^8 x_1 y x_2 \$$ by the eight rules introduced in (3) from r.

Proof. The proof is divided into the only-if part and the if part.

Only If. Let $x_1 A x_2 \Rightarrow_G x_1 y x_2$ by $r = (A \to y, U, W) \in P_R$, where $x_1, x_2 \in V^*$. By (3.1), $(A \to \langle r, 1\rangle, \emptyset, \bar{N}) \in P'_L$. As $\mathrm{alph}(x_1) \cap \bar{N} = \emptyset$,

$$x_1 A x_2 \$ \Rightarrow_H x_1 \langle r, 1\rangle x_2 \$$$

By (3.2), $(\$ \to \langle \$, r, 1\rangle, \{\langle r, 1\rangle\}, \bar{N} - \{\langle r, 1\rangle\}) \in P'_L$. As $\langle r, 1\rangle \in \mathrm{alph}(x_1 \langle r, 1\rangle x_2)$ and $\mathrm{alph}(x_1 \langle r, 1\rangle x_2) \cap (\bar{N} - \{\langle r, 1\rangle\}) = \emptyset$,

$$x_1 \langle r, 1\rangle x_2 \$ \Rightarrow_H x_1 \langle r, 1\rangle x_2 \langle \$, r, 1\rangle$$

By (3.3), $(\langle r, 1\rangle \to \#_1, \emptyset, \bar{N}) \in P'_L$. As $\mathrm{alph}(x_1) \cap \bar{N} = \emptyset$,

$$x_1 \langle r, 1\rangle x_2 \langle \$, r, 1\rangle \Rightarrow_H x_1 \#_1 x_2 \langle \$, r, 1\rangle$$

By (3.4), $(\#_1 \to \langle r, 2\rangle, \{\langle \$, r, 1\rangle\}, \bar{N} - \{\langle \$, r, 1\rangle\}) \in P'_R$. As $\langle \$, r, 1\rangle \in \mathrm{alph}(x_2 \langle \$, r, 1\rangle)$ and $\mathrm{alph}(x_2 \langle \$, r, 1\rangle) \cap (\bar{N} - \{\langle \$, r, 1\rangle\}) = \emptyset$,

$$x_1 \#_1 x_2 \langle \$, r, 1\rangle \Rightarrow_H x_1 \langle r, 2\rangle x_2 \langle \$, r, 1\rangle$$

By (3.5), $(\langle \$, r, 1\rangle \rightarrow \langle \$, r, 2\rangle, \{\langle r, 2\rangle\}, \bar{N} - \{\langle r, 2\rangle\}) \in P'_L$. As $\{\langle r, 2\rangle\} \in \text{alph}(x_1\langle r, 2\rangle x_2)$ and $\text{alph}(x_1\langle r, 2\rangle x_2) \cap (\bar{N} - \{\langle r, 2\rangle\}) = \emptyset$,

$$x_1\langle r, 2\rangle x_2\langle \$, r, 1\rangle \Rightarrow_H x_1\langle r, 2\rangle x_2\langle \$, r, 2\rangle$$

By (3.6), $(\langle r, 2\rangle \rightarrow \#_2, \emptyset, \bar{N}) \in P'_L$. As $\text{alph}(x_1) \cap \bar{N} = \emptyset$,

$$x_1\langle r, 2\rangle x_2\langle \$, r, 2\rangle \Rightarrow_H x_1\#_2 x_2\langle \$, r, 2\rangle$$

By (3.7), $(\#_2 \rightarrow y, U \cup \{\langle \$, r, 2\rangle\}, W \cup (\bar{N} - \{\langle \$, r, 2\rangle\})) \in P'_R$. As $(U \cup \{\langle \$, r, 2\rangle\}) \subseteq \text{alph}(x_2\langle \$, r, 2\rangle)$ and $\text{alph}(x_2\langle \$, r, 2\rangle) \cap (W \cup (\bar{N} - \{\langle \$, r, 2\rangle\})) = \emptyset$,

$$x_1\#_2 x_2\langle \$, r, 2\rangle \Rightarrow_H x_1 y x_2\langle \$, r, 2\rangle$$

Finally, by (3.8), $(\langle \$, r, 2\rangle \rightarrow \$, \emptyset, \bar{N}) \in P'_L$. As $\text{alph}(x_1 y x_2) \cap \bar{N} = \emptyset$,

$$x_1 y x_2\langle \$, r, 2\rangle \Rightarrow_H x_1 y x_2 \$$$

Hence, the only-if part of the claim holds.

If. Let $x_1 A x_2 \$ \Rightarrow_H^8 x_1 y x_2 \$$ by the eight rules introduced in (3) from some $r = (A \rightarrow y, U, W) \in P_R$. Observe that this eight-step derivation is of the following form

$$\begin{aligned} x_1 A x_2 &\Rightarrow_H x_1\langle r, 1\rangle x_2 \\ &\Rightarrow_H x_1\langle r, 1\rangle x_2\langle r, 1\rangle \\ &\Rightarrow_H x_1\#_1 x_2\langle r, 1\rangle \\ &\Rightarrow_H x_1\langle r, 2\rangle x_2\langle r, 1\rangle \\ &\Rightarrow_H x_1\langle r, 2\rangle x_2\langle r, 2\rangle \\ &\Rightarrow_H x_1\#_2 x_2\langle r, 2\rangle \\ &\Rightarrow_H x_1 y x_2\langle r, 2\rangle \\ &\Rightarrow_H x_1 y x_2 \end{aligned}$$

As $x_1\#_2 x_2\langle \$, r, 2\rangle \Rightarrow_H x_1 y x_2\langle \$, r, 2\rangle$ by $(\#_2 \rightarrow y, U \cup \{\langle \$, r, 2\rangle\}, W \cup (\bar{N} - \{\langle \$, r, 2\rangle\})) \in P'_R$, introduced in (3.7) from r, $U \subseteq \text{alph}(x_2)$ and $W \cap \text{alph}(x_2) = \emptyset$. Therefore,

$$x_1 A x_2 \Rightarrow_G x_1 y x_2 \text{ by } r$$

Hence, the if part of the claim holds. □

Claim 3 shows that every $x \in L(H)$ can be derived in H by a derivation satisfying properties (i) through (iii), stated next. Set $V' = N' \cup T$.

Claim 3. Let $x \in V'^*$. Then, $x \in L(H)$ if and only if $S' \Rightarrow_H S\$ \Rightarrow_H^* x\$ \Rightarrow_H x$ so that during $S\$ \Rightarrow_H^* x\$$, (i) through (iii) hold:

- *(i) no rules from (1) are used;*
- *(ii) every application of a rule from (3.1) is followed by applying the remaining seven rules from (3.2) through (3.8) before another rule from (3.1) is applied;*
- *(iii) whenever some $A \in N$ is rewritten to $\langle r, 1\rangle$ by a rule from (3.1), constructed from $r = (A \to y, U, W) \in P_R$, $\langle r, 1\rangle$ cannot be rewritten to z with $z \neq y$.*

Proof. The if part of the claim is trivial; therefore, we only prove the only-if part. Let $x \in L(H)$. We argue that (i) through (iii) hold.

- (i) Observe that $(S' \to S\$, \emptyset, \emptyset) \in P_L$, introduced in (1), is the only rule with S' on its left-hand side. Therefore, this rule is used at the beginning of every derivation. Furthermore, observe that no rule has S' on its right-hand side. By (1), $(\$ \to \varepsilon, \emptyset, N') \in P_L'$. Notice that this rule is applicable if and only if the current sentential form is of the form $x\$$, where $x \in T^*$. Therefore, (i) holds.
- (ii) Let $x_1\langle r, 1\rangle x_2\$$ be the first sentential form in $S\$ \Rightarrow_H^* x\$$ after a rule from (3.1) is applied. Clearly, $x_1, x_2 \in V^*$. Observe that after this application, the remaining seven rules from (3.2) through (3.8) are applied before another rule from (3.1) is applied. As rules from (3.1) rule out the presence of symbols from $\bar{N}$, a rule from (3.1) can be only applied to some nonterminal in x_1. Then, however, no terminal string can be obtained anymore. Furthermore, observe that during the application of these seven rules, rules introduced in (2) can be applied to x_1 at any time without affecting the applicability of the seven rules to $\langle r, 1\rangle x_2\$$. Based on these observations, we see that (ii) holds.
- (iii) By (ii), once a rule from (3.1) is used, the remaining seven rules from (3.2) through (3.8) are applied. Observe that when rules from (3.3) and (3.6) are applied, the last nonterminal of the current sentential form encodes the currently simulated rule. Therefore, we cannot combine simulations of two different rules from P_R. Based on this observation, we see that (iii) holds.

Hence, the only-if part of the claim holds, so the claim holds. □

Claim 4. In G, $S \Rightarrow_G^* x$ if and only if in H, $S' \Rightarrow_H^* x$, where $x \in T^*$.

Proof. This claim follows from Claims 1, 2, and 3. □

Claim 4 implies that $L(G) = L(H)$. Clearly, $\#_1$ and $\#_2$ are the only right random context nonterminals in H, so $\mathrm{nrrcn}(H) = 2$. Hence, the lemma holds. □

Theorem 6.4.7. *For every recursively enumerable language K, there exists a one-sided random context grammar H such that $L(H) = K$ and* $\mathrm{nrrcn}(H) = 2$.

Proof. This theorem follows from Theorem 6.2.4 and Lemma 6.4.6. □

Theorem 6.4.8. *For every recursively enumerable language K, there exists a one-sided random context grammar H such that $L(H) = K$ and* $\mathrm{nlrcn}(H) = 2$.

Proof. This theorem can be proved by analogy with the proofs of Theorem 6.4.7 and Lemma 6.4.6. □

Next, we turn our attention to propagating one-sided random context grammars and their nonterminal complexity.

Lemma 6.4.9. *Let G be a propagating one-sided random context grammar. Then, there is a propagating one-sided random context grammar H such that $L(H) = L(G)$ and* $\mathrm{nrrcn}(H) = 2$.

Proof. Let $G = (N, T, P_L, P_R, S)$ be a propagating one-sided random context grammar. We prove this lemma by analogy with the proof of Lemma 6.4.6. However, since H has to be propagating, instead of using a special symbol \$, which is erased at the end of a successful derivation, we use the rightmost symbol of a sentential form for this purpose. Therefore, if X is the rightmost symbol of the current sentential form in G, we use $\langle X \rangle$ in H. By analogy with the construction given in Lemma 6.4.6, we introduce new nonterminals, $\langle X, r, 1\rangle$ and $\langle X, r, 2\rangle$, for every $r \in P_R$ and every $X \in N \cup T$, to keep track of the currently simulated rule r. At the end of a derivation, X has to be a terminal, so instead of erasing $\langle X \rangle$, we rewrite it to X, thus finishing the derivation.

We next construct a propagating one-sided random context grammar H such that $L(H) = L(G)$ and $\mathrm{nrrcn}(H) = 2$. Set $V = N \cup T$ and

$$\begin{aligned}\hat{V} &= \{\langle X \rangle \mid X \in V\}\\ R &= \{\langle r, i\rangle \mid r \in P_R, i = 1, 2\}\\ {}_{\$}R &= \{\langle X, r, i\rangle \mid X \in V, r \in P_R, i = 1, 2\}\end{aligned}$$

Without any loss of generality, we assume that $\hat{V}$, R, ${}_{\$}R$, $\{\#_1, \#_2\}$, and V are pairwise disjoint. Construct

$$H = \left(N', T, P'_L, P'_R, \langle S \rangle\right)$$

as follows. Initially, set $N' = N \cup \hat{V} \cup R \cup {}_{\$}R \cup \{\#_1, \#_2\}$, $P'_L = \emptyset$, and $P'_R = \emptyset$. Furthermore, set $\bar{N} = N' - N$. Perform the following five steps

(1) for each $a \in T$, add $(\langle a \rangle \to a, \emptyset, N')$ to P'_L;
(2) for each $(A \to y, U, W) \in P_L$, add $(A \to y, U, W \cup \bar{N})$ to P'_L;
(3) for each $r = (A \to y, U, W) \in P_R$ and each $X \in V$,

  (3.1) add $(A \to \langle r, 1\rangle, \emptyset, \bar{N})$ to P'_L;
  (3.2) add $(\langle X \rangle \to \langle X, r, 1\rangle, \{\langle r, 1\rangle\}, \bar{N} - \{\langle r, 1\rangle\})$ to P'_L;
  (3.3) add $(\langle r, 1\rangle \to \#_1, \emptyset, \bar{N})$ to P'_L;
  (3.4) add $(\#_1 \to \langle r, 2\rangle, \{\langle X, r, 1\rangle\}, \bar{N} - \{\langle X, r, 1\rangle\})$ to P'_R;
  (3.5) add $(\langle X, r, 1\rangle \to \langle X, r, 2\rangle, \{\langle r, 2\rangle\}, \bar{N} - \{\langle r, 2\rangle\})$ to P'_L;
  (3.6) add $(\langle r, 2\rangle \to \#_2, \emptyset, \bar{N})$ to P'_L;

(3.7) add $(\#_2 \rightarrow y, U \cup \{\langle X, r, 2\rangle\}, W \cup (\bar{N} - \{\langle X, r, 2\rangle\}))$ to P'_R;
(3.8) add $(\langle X, r, 2\rangle \rightarrow \langle X\rangle, \emptyset, \bar{N})$ to P'_L;

(4) for each $(A \rightarrow yY, U, W) \in P_L$, where $y \in V^*$ and $Y \in V$,
add $(\langle A\rangle \rightarrow y\langle Y\rangle, U, W \cup \bar{N})$ to P'_L;
(5) for each $(A \rightarrow yY, \emptyset, W) \in P_R$, where $y \in V^*$ and $Y \in V$,
add $(\langle A\rangle \rightarrow y\langle Y\rangle, \emptyset, \bar{N})$ to P'_L.

Steps (1) through (3) are similar to the corresponding three steps in the construction given in the proof of Lemma 6.4.6. Rules from (4) and (5) take care of rewriting the rightmost nonterminal. Note that every simulated rule from P_R rewriting this nonterminal has to have its permitting context empty; otherwise, it is not applicable to the rightmost nonterminal. Furthermore, observe that we can simulate such a right random context rule by a left random context rule. As obvious, there are no nonterminals to the right of the rightmost symbol.

The identity $L(H) = L(G)$ can be proved by analogy with proving Lemma 6.4.6, and we leave this proof to the reader. Clearly, $\#_1$ and $\#_2$ are the only right random context nonterminals in H, so $\text{nrrcn}(H) = 2$. Hence, the lemma holds. □

Theorem 6.4.10. *For every context-sensitive language K, there exists a propagating one-sided random context grammar H such that $L(H) = K$ and* $\text{nrrcn}(H) = 2$.

Proof. This theorem follows from Theorem 6.2.3 and Lemma 6.4.9. □

Theorem 6.4.11. *For every context-sensitive language K, there exists a propagating one-sided random context grammar H such that $L(H) = K$ and* $\text{nlrcn}(H) = 2$.

Proof. This theorem can be proved by analogy with the proofs of Theorem 6.4.10 and Lemma 6.4.9. □

6.4.3 Number of Right Random Context Rules

In this section, we prove that any recursively enumerable language can be generated by a one-sided random context grammar having no more than two right random context rules.

Theorem 6.4.12. *Let K be a recursively enumerable language. Then, there is a one-sided random context grammar, $H = (N, T, P_L, P_R, S)$, such that $L(H) = K$ and* $\text{card}(P_R) = 2$.

Proof. Let K be a recursively enumerable language. Then, by Theorem 4.1.9, there is a phrase-structure grammar in the Geffert normal form

$$G = \big(\{S, A, B, C\}, T, P \cup \{ABC \rightarrow \varepsilon\}, S\big)$$

satisfying $L(G) = K$. We next construct a one-sided random context grammar H such that $L(H) = L(G)$. Set $N = \{S, A, B, C\}$, $V = N \cup T$, and $N' = N \cup \{S', \$, \bar{\$}, \#, \bar{A}, \bar{B}, \hat{A}, \hat{B}, \hat{C}\}$. Without any loss of generality, we assume that $V \cap \{S', \$, \bar{\$}, \#, \bar{A}, \bar{B}, \hat{A}, \hat{B}, \hat{C}\} = \emptyset$. Construct

$$H = \left(N', T, P_L, P_R, S'\right)$$

as follows. Initially, set $P_L = \emptyset$ and $P_R = \emptyset$. Perform the following eleven steps

(1) add $(S' \rightarrow \$S, \emptyset, \emptyset)$ to P_L;
(2) for each $S \rightarrow uSa \in P$, where $u \in \{A, AB\}^*$ and $a \in T$,
 add $(S \rightarrow uS\#a, \emptyset, \{\bar{A}, \bar{B}, \hat{A}, \hat{B}, \hat{C}, \#\})$ to P_L;
(3) for each $S \rightarrow uSv \in P$, where $u \in \{A, AB\}^*$ and $v \in \{BC, C\}^*$,
 add $(S \rightarrow uSv, \emptyset, \{\bar{A}, \bar{B}, \hat{A}, \hat{B}, \hat{C}, \#\})$ to P_L;
(4) for each $S \rightarrow uv \in P$, where $u \in \{A, AB\}^*$ and $v \in \{BC, C\}^*$,
 add $(S \rightarrow uv, \emptyset, \{\bar{A}, \bar{B}, \hat{A}, \hat{B}, \hat{C}, \#\})$ to P_L;
(5) add $(A \rightarrow \bar{A}, \emptyset, N \cup \{\hat{\$}, \hat{A}, \hat{B}, \hat{C}, \#\})$ to P_L;
 add $(B \rightarrow \bar{B}, \emptyset, N \cup \{\hat{\$}, \hat{A}, \hat{B}, \hat{C}, \#\})$ to P_L;
 add $(A \rightarrow \hat{A}, \emptyset, N \cup \{\hat{\$}, \hat{A}, \hat{B}, \hat{C}, \#\})$ to P_L;
 add $(B \rightarrow \hat{B}, \{\hat{A}\}, N \cup \{\hat{\$}, \hat{B}, \hat{C}, \#\})$ to P_L;
 add $(C \rightarrow \hat{C}, \{\hat{A}, \hat{B}\}, N \cup \{\hat{\$}, \hat{C}, \#\})$ to P_L;
(6) add $(\$ \rightarrow \hat{\$}, \{\hat{A}, \hat{B}, \hat{C}\}, \emptyset)$ to P_R;
(7) add $(\hat{A} \rightarrow \varepsilon, \{\hat{\$}\}, N \cup \{\hat{A}, \hat{B}, \hat{C}, \#\})$ to P_L;
 add $(\hat{B} \rightarrow \varepsilon, \{\hat{\$}\}, N \cup \{\hat{A}, \hat{B}, \hat{C}, \#\})$ to P_L;
 add $(\hat{C} \rightarrow \varepsilon, \{\hat{\$}\}, N \cup \{\hat{A}, \hat{B}, \hat{C}, \#\})$ to P_L;
(8) add $(\bar{A} \rightarrow A, \{\hat{\$}\}, N \cup \{\hat{A}, \hat{B}, \hat{C}, \#\})$ to P_L;
 add $(\bar{B} \rightarrow B, \{\hat{\$}\}, N \cup \{\hat{A}, \hat{B}, \hat{C}, \#\})$ to P_L;
(9) add $(\hat{\$} \rightarrow \$, \emptyset, \{\bar{A}, \bar{B}, \hat{A}, \hat{B}, \hat{C}\})$ to P_R;
(10) add $(\$ \rightarrow \varepsilon, \emptyset, \emptyset)$ to P_L;
(11) add $(\# \rightarrow \varepsilon, \emptyset, N')$ to P_L.

Before proving that $L(H) = L(G)$, let us informally describe the meaning of rules introduced in (1) through (11). The rule from (1) starts every derivation of H. The leftmost symbol of every sentential form having at least one nonterminal is either $\$$ or $\hat{\$}$. The role of these two symbols is explained later. H simulates the derivations of G that satisfy the form described in Theorem 4.1.9. The context-free rules in P are simulated by rules from (2) through (4). The context-sensitive rule $ABC \rightarrow \varepsilon$ is simulated in a several-step way. First, rules introduced in (5) are used to prepare the erasure of ABC. These rules rewrite nonterminals from the left to the right. In this way, it is guaranteed that whenever $\hat{A}$, $\hat{B}$, and $\hat{C}$ appear in a sentential form, then they form a substring of the form $\hat{A}\hat{B}\hat{C}$. Then, $\$$ is changed to $\hat{\$}$ by using the rule from (6). After that, the rules from (7) erase $\hat{A}$, $\hat{B}$, and $\hat{C}$, one by one. Finally, rules from (8) convert the barred versions of nonterminals back to their non-barred versions to prepare another simulation of $ABC \to \varepsilon$; this conversion is done in a left-to-right way. After this conversion, $\hat{\$}$ is reverted back to \$ by the rule from (9). For example, $AABCBCab \Rightarrow_G ABCab$ is simulated by H as follows:

$$
\begin{aligned}
\$AABCBC\#a\#b &\Rightarrow_H \$\bar{A}ABCBC\#a\#b \\
&\Rightarrow_H \$\bar{A}\hat{A}BCBC\#a\#b \\
&\Rightarrow_H \$\bar{A}\hat{A}\hat{B}CBC\#a\#b \\
&\Rightarrow_H \$\bar{A}\hat{A}\hat{B}\hat{C}\,BC\#a\#b \\
&\Rightarrow_H \hat{\$}\bar{A}\hat{A}\hat{B}\hat{C}\,BC\#a\#b \\
&\Rightarrow_H \hat{\$}\bar{A}\hat{B}\hat{C}\,BC\#a\#b \\
&\Rightarrow_H \hat{\$}\bar{A}\hat{C}\,BC\#a\#b \\
&\Rightarrow_H \hat{\$}\bar{A}BC\#a\#b \\
&\Rightarrow_H \hat{\$}ABC\#a\#b \\
&\Rightarrow_H \$ABC\#a\#b
\end{aligned}
$$

Symbol # is used to ensure that every sentential form of H is of the form w_1w_2, where $w_1 \in (N' - \{\#\})^*$ and $w_2 \in (T \cup \{\#\})^*$. Since permitting and forbidding contexts cannot contain terminals, a mixture of symbols from T and N in H could produce a terminal string out of $L(G)$. For example, observe that $\$AaBC \Rightarrow_H^* \$a$ by rules from (5) through (9), but such a derivation does not exist in G. #s can be eliminated by an application of rules from (11) provided that no nonterminals occur to the left of # in the current sentential form. Consequently, all #s are erased at the end of every successful derivation.

The leftmost symbol \$ and its hatted version $\hat{\$}$ encode the current phase. When \$ is present, we use rules from (2) through (5). When $\hat{\$}$ is present, we use rules from (7) and (8). When none of these two symbols is present, which happens after the rule from (10) is applied, no substring ABC can be erased anymore so we have to finish the derivation by removing all #s. Notice that when \$ is erased prematurely, no terminal string can be derived.

To establish $L(H) = L(G)$, we prove two claims. Claim 1 shows how derivations of G are simulated by H. Then, Claim 2 demonstrates the converse simulation—that is, it shows how derivations of H are simulated by G.

Set $V' = N' \cup T$. Define the homomorphism φ from V^* to V'^* as $\varphi(X) = X$ for $X \in N$, and $\varphi(a) = \#a$ for $a \in T$.

Claim 1. If $S \Rightarrow_G^n x \Rightarrow_G^* z$, where $x \in V^*$ and $z \in T^*$, for some $n \geq 0$, then $S' \Rightarrow_H^* \$\varphi(x)$.

Proof. This claim is established by induction on $n \geq 0$.

Basis. For $n = 0$, this claim is clear [in H, we use $(S' \to \$S, \emptyset, \emptyset)$ from (1)].

Induction Hypothesis. Suppose that there exists $n \geq 0$ such that the claim holds for all derivations of length ℓ, where $0 \leq \ell \leq n$.

Induction Step. Consider any derivation of the form

$$S \Rightarrow_G^{n+1} w \Rightarrow_G^* z$$

where $w \in V^*$ and $z \in T^*$. Since $n + 1 \geq 1$, this derivation can be expressed as

$$S \Rightarrow_G^n x \Rightarrow_G w \Rightarrow_G^* z$$

for some $x \in V^+$. Without any loss of generality, we assume that x is of the form $x = x_1x_2x_3x_4$, where $x_1 \in \{A, AB\}^*$, $x_2 \in \{S, \varepsilon\}$, $x_3 \in \{BC, C\}^*$, and $x_4 \in T^*$ (see Theorem 4.1.9 and [20]).

Next, we consider all possible forms of $x \Rightarrow_G w$, covered by the following four cases—(i) through (iv).

(i) *Application of* $S \rightarrow uSa \in P$. Let $x = x_1Sx_3x_4$, $w = x_1uSax_3x_4$, and $S \rightarrow uSa \in P$, where $x_1, u \in \{A, AB\}^*$, $x_3 \in \{BC, C\}^*$, $x_4 \in T^*$, and $a \in T$. Then, by the induction hypothesis,

$$S' \Rightarrow_H^* \$\varphi(x_1Sx_3x_4)$$

By (2), $(S \rightarrow uS\#a, \emptyset, \{\bar{A}, \bar{B}, \hat{A}, \hat{B}, \hat{C}, \#\}) \in P_L$. Since $\$\varphi(x_1Sx_3x_4) = \$x_1S\varphi(x_3x_4)$ and $\mathrm{alph}(\$x_1) \cap \{\bar{A}, \bar{B}, \hat{A}, \hat{B}, \hat{C}, \#\} = \emptyset$,

$$\$x_1S\varphi(x_3x_4) \Rightarrow_H \$x_1uS\#a\varphi(x_3x_4)$$

As $x_1uS\#a\varphi(x_3x_4) = \varphi(x_1uSax_3x_4)$, the induction step is completed for (i).

(ii) *Application of* $S \rightarrow uSv \in P$. Let $x = x_1Sx_3x_4$, $w = x_1uSvx_3x_4$, and $S \rightarrow uSv \in P$, where $x_1, u \in \{A, AB\}^*$, $x_3, v \in \{BC, C\}^*$, and $x_4 \in T^*$. To complete the induction step for (ii), proceed by analogy with (i), but use a rule from (3) instead of a rule from (1).

(iii) *Application of* $S \rightarrow uv \in P$. Let $x = x_1Sx_3x_4$, $w = x_1uvx_3x_4$, and $S \rightarrow uv \in P$, where $x_1, u \in \{A, AB\}^*$, $x_3, v \in \{BC, C\}^*$, and $x_4 \in T^*$. To complete the induction step for (iii), proceed by analogy with (i), but use a rule from (4) instead of a rule from (1).

(iv) *Application of* $ABC \rightarrow \varepsilon$. Let $x = x_1ABCx_3x_4$ and $w = x_1x_3x_4$, where $x_1 \in \{A, AB\}^*$, $x_3 \in \{BC, C\}^*$, and $x_4 \in T^*$, so $x \Rightarrow_G w$ by $ABC \rightarrow \varepsilon$. Then, by the induction hypothesis,

$$S' \Rightarrow_H^* \$\varphi(x_1ABCx_3x_4)$$

Let $x_1 = X_1X_2 \cdots X_k$, where $k = |x_1|$ (the case when $k = 0$ means that $x_1 = \varepsilon$). Since $\$\varphi(x_1ABCx_3x_4) = \$x_1ABC\varphi(x_3x_4)$ and $\mathrm{alph}(x_1) \subseteq N$, by rules introduced in (5),

$$
\begin{aligned}
\$X_1X_2\cdots X_kABC\varphi(x_3x_4) &\Rightarrow_H \$\bar{X}_1X_2\cdots X_kABC\varphi(x_3x_4)\\
&\Rightarrow_H \$\bar{X}_1\bar{X}_2\cdots X_kABC\varphi(x_3x_4)\\
&\ \vdots\\
&\Rightarrow_H \$\bar{X}_1\bar{X}_2\cdots \bar{X}_kABC\varphi(x_3x_4)\\
&\Rightarrow_H \$\bar{X}_1\bar{X}_2\cdots \bar{X}_k\hat{A}BC\varphi(x_3x_4)\\
&\Rightarrow_H \$\bar{X}_1\bar{X}_2\cdots \bar{X}_k\hat{A}\hat{B}C\varphi(x_3x_4)\\
&\Rightarrow_H \$\bar{X}_1\bar{X}_2\cdots \bar{X}_k\hat{A}\hat{B}\hat{C}\varphi(x_3x_4)
\end{aligned}
$$

Let $\bar{x}_1 = \bar{X}_1\bar{X}_2\cdots\bar{X}_k$. By the rule introduced in (6),

$$\$\bar{x}_1\hat{A}\hat{B}\hat{C}\varphi(x_3x_4) \Rightarrow_H \hat{\$}\bar{x}_1\hat{A}\hat{B}\hat{C}\varphi(x_3x_4)$$

By the rules introduced in (7),

$$
\begin{aligned}
\hat{\$}\bar{x}_1\hat{A}\hat{B}\hat{C}\varphi(x_3x_4) &\Rightarrow_H \hat{\$}\bar{x}_1\hat{B}\hat{C}\varphi(x_3x_4)\\
&\Rightarrow_H \hat{\$}\bar{x}_1\hat{C}\varphi(x_3x_4)\\
&\Rightarrow_H \hat{\$}\bar{x}_1\varphi(x_3x_4)
\end{aligned}
$$

By rules from (8),

$$
\begin{aligned}
\hat{\$}\bar{x}_1\varphi(x_3x_4) &\Rightarrow_H \hat{\$}\bar{X}_1\cdots\bar{X}_{k-1}X_k\varphi(x_3x_4)\\
&\Rightarrow_H \hat{\$}\bar{X}_1\cdots X_{k-1}X_k\varphi(x_3x_4)\\
&\ \vdots\\
&\Rightarrow_H \hat{\$}X_1\cdots X_{k-1}X_k\varphi(x_3x_4)
\end{aligned}
$$

Recall that $X_1\cdots X_{k-1}X_k = x_1$. Finally, by the rule from (9),

$$\hat{\$}x_1\varphi(x_3x_4) \Rightarrow_H \$x_1\varphi(x_3x_4)$$

Since $x_1\varphi(x_3x_4) = \varphi(x_1x_3x_4)$, the induction step is completed for (iv).

Observe that cases (i) through (iv) cover all possible forms of $x \Rightarrow_G w$. Thus, the claim holds. □

Define the homomorphism π from $(V' - \{S'\})^*$ to V^* as $\pi(X) = X$ for $X \in N$, $\pi(\bar{A}) = \pi(\hat{A}) = A$, $\pi(\bar{B}) = \pi(\hat{B}) = B$, $\pi(\hat{C}) = C$, $\pi(a) = a$ for $a \in T$, and $\pi(\#) = \pi(\$) = \pi(\hat{\$}) = \varepsilon$. Define the homomorphism τ from $(V' - \{S'\})^*$ to V^* as $\tau(X) = \pi(X)$ for $X \in V' - \{S', \hat{A}, \hat{B}, \hat{C}\}$, and $\tau(\hat{A}) = \tau(\hat{B}) = \tau(\hat{C}) = \varepsilon$.

Claim 2. Let $S' \Rightarrow_H^n x \Rightarrow_H^* z$, where $x \in V'^*$ and $z \in T^*$, for some $n \geq 1$. Then, $x = x_0x_1x_2x_3x_4x_5$, where $x_0 \in \{\varepsilon, \$, \hat{\$}\}$, $x_1 \in \{\bar{A}, \bar{B}\}^*$, $x_2 \in \{A, B\}^*$, $x_3 \in \{S, \hat{A}BC, \hat{A}\hat{B}C, \hat{A}\hat{B}\hat{C}, \hat{B}\hat{C}, \hat{C}, \varepsilon\}$, $x_4 \in \{B, C\}^*$, and $x_5 \in (T \cup \{\#\})^*$. Furthermore,

(a) if $x_3 \in \{S, \varepsilon\}$, *then* $S \Rightarrow_G^* \pi(x)$;
(b) if $x_3 \in \{\hat{A}BC, \hat{A}\hat{B}C, \hat{A}\hat{B}\hat{C}\}$, *then* $x_2 = \varepsilon$ *and* $S \Rightarrow_G^* \pi(x)$;
(c) if $x_3 \in \{\hat{B}\hat{C}, \hat{C}\}$, *then* $x_0 = \hat{\$}$, $x_2 = \varepsilon$, *and* $S \Rightarrow_G^* \tau(x)$.

Proof. This claim is established by induction on $n \geq 1$.

Basis. For $n = 1$, this claim is clear [the only applicable rule to S' is $(S' \to \$S, \emptyset, \emptyset) \in P_L$, introduced in (1)].

Induction Hypothesis. Suppose that there exists $n \geq 1$ such that the claim holds for all derivations of length ℓ, where $1 \leq \ell \leq n$.

Induction Step. Consider any derivation of the form

$$S' \Rightarrow_H^{n+1} w \Rightarrow_H^* z$$

where $w \in V'^*$ and $z \in T^*$. Since $n + 1 \geq 1$, this derivation can be expressed as

$$S' \Rightarrow_H^n x \Rightarrow_H w \Rightarrow_H^* z$$

for some $x \in V'^+$. By the induction hypothesis, $x = x_0x_1x_2x_3x_4x_5$, where $x_0 \in \{\varepsilon, \$, \hat{\$}\}$, $x_1 \in \{\bar{A}, \bar{B}\}^*$, $x_2 \in \{A, B\}^*$, $x_3 \in \{S, \hat{A}BC, \hat{A}\hat{B}C, \hat{A}\hat{B}\hat{C}, \hat{B}\hat{C}, \hat{C}, \varepsilon\}$, $x_4 \in \{B, C\}^*$, and $x_5 \in (T \cup \{\#\})^*$. Furthermore, (a) through (c), stated in the claim, hold.

Next, we consider all possible forms of $x \Rightarrow_H w$, covered by the following six cases—(i) through (vi).

(i) *Application of a rule from (2).* Let $x_3 = S$, $x_1 = x_4 = \varepsilon$, and

$$(S \to uS\#a, \emptyset, \{\bar{A}, \bar{B}, \hat{A}, \hat{B}, \hat{C}, \#\}) \in P_L$$

introduced in (2), where $u \in \{A, AB\}^*$ and $a \in T$, so

$$x_0x_2Sx_5 \Rightarrow_H x_0x_2uS\#ax_5$$

Observe that if $x_4 \neq \varepsilon$, then $w \Rightarrow_H^* z$ does not hold. Indeed, if $x_4 \neq \varepsilon$, then to erase the nonterminals in x_4, there have to be As in x_2. However, the # symbol, introduced between x_2 and x_4, blocks the applicability of $(C \to \hat{C}, \{\hat{A}, \hat{B}\}, N \cup \{\hat{\$}, \hat{C}, \#\}) \in P_L$, introduced in (5), which is needed to erase the nonterminals in x_4. Since $(\# \to \varepsilon, \emptyset, N') \in P_L$, introduced in (11), requires that there are no nonterminals to the left of #, the derivation cannot be successfully finished. Hence, $x_4 = \varepsilon$. Since $u \in \{A, B\}^*$ and $\#a \in (T \cup \{\#\})^*$, $x_0x_2uS\#ax_5$ is of the required form. As $x_3 = S$, by (a), $S \Rightarrow_G^* \pi(x)$. Observe that $\pi(x) = \pi(x_0x_2)S\pi(x_5)$. By (2), $S \to uSa \in P$, so

$$\pi(x_0x_2)S\pi(x_5) \Rightarrow_G \pi(x_0x_2)uSa\pi(x_5)$$

Since $\pi(x_0x_2)uSa\pi(x_5) = \pi(x_0x_2uS\#ax_5)$ and both $\hat{B}\hat{C}$ and $\hat{C}$ are not substrings of $x_0x_2uS\#ax_5$, the induction step is completed for (i).

(ii) *Application of a rule from (3).* Make this part of the proof by analogy with (i).

(iii) *Application of a rule from (4).* Make this part of the proof by analogy with (i).

(iv) *Application of a rule from (5), (6), (9), (10), or (11).* Let $x \Rightarrow_H w$ by a rule from (5), (6), (9), (10), or (11). Then, $S \Rightarrow_G^* \pi(w)$ follows directly from the induction hypothesis. Observe that w is of the required form, and the induction step is completed for (iv).

(v) *Application of a rule from (7).* Let $x_3 \in \{\hat{A}\hat{B}\hat{C}, \hat{B}\hat{C}, \hat{C}\}$. By the induction hypothesis [more specifically, by (b) and (c)], $x_2 = \varepsilon$. Then, there are three subcases, depending on x_3, as demonstrated next.

(v.i) Let $x_3 = \hat{A}\hat{B}\hat{C}$. Then, $x_0x_1\hat{A}\hat{B}\hat{C}x_4x_5 \Rightarrow_H x_0x_1\hat{B}\hat{C}x_4x_5$ by $(\hat{A} \to \varepsilon, \{\hat{\$}\}, N \cup \{\hat{A}, \hat{B}, \hat{C}, \#\}) \in P_L$, introduced in (7). Observe that this is the only applicable rule from (7). By the induction hypothesis, $S \Rightarrow_G^* \pi(x)$. Since $\pi(x) = \pi(x_1)ABC\pi(x_4x_5)$,

$$\pi(x_1)ABC\pi(x_4x_5) \Rightarrow_G \pi(x_1)\pi(x_4x_5)$$

by $ABC \to \varepsilon$. As $w = x_0x_1\hat{A}\hat{C}x_4x_5$ is of the required form and $\pi(x_0x_1)\pi(x_4x_5) = \tau(w)$, the induction step is completed for (v.i).

(v.ii) Let $x_3 = \hat{B}\hat{C}$. Then, $x_0x_1\hat{B}\hat{C}x_4x_5 \Rightarrow_H x_0x_1\hat{C}x_4x_5$ by $(\hat{B} \to \varepsilon, \{\hat{\$}\}, N \cup \{\hat{A}, \hat{B}, \hat{C}, \#\}) \in P_L$, introduced in (7). Observe that this is the only applicable rule from (7). By the induction hypothesis, $S \Rightarrow_G^* \tau(x)$. As $w = x_0x_1\hat{C}x_4x_5$ is of the required form and $\tau(x) = \tau(w)$, the induction step is completed for (v.ii).

(v.iii) Let $x_3 = \hat{C}$. Then, $x_0x_1\hat{C}x_4x_5 \Rightarrow_H x_0x_1x_4x_5$ by $(\hat{C} \to \varepsilon, \{\hat{\$}\}, N \cup \{\hat{A}, \hat{B}, \hat{C}, \#\}) \in P_L$, introduced in (7). Observe that this is the only applicable rule from (7). By the induction hypothesis, $S \Rightarrow_G^* \tau(x)$. As $w = x_0x_1x_4x_5$ is of the required form and $\tau(x) = \tau(w)$, the induction step is completed for (v.iii).

(vi) *Application of a rule from (8).* Let $x \Rightarrow_H w$ by a rule from (8). Then, $x_3 \notin \{\hat{B}\hat{C}, \hat{C}\}$ has to hold; otherwise, no string of terminals can be obtained anymore. Indeed, the deletion of $\hat{B}$ and $\hat{C}$ requires that there are no symbols from N to the left of them, and to rewrite A or B to their barred versions, $\hat{\$}$ cannot be present to the left of them. However, by (c), it is there. Therefore, $S \Rightarrow_G^* \pi(w)$ follows directly from the induction hypothesis. Furthermore, w is of the required form; if not, then observe that no string of terminals can be obtained anymore. Hence, the induction step is completed for (vi).

Observe that cases (i) through (vi) cover all possible forms of $x \Rightarrow_H w$. Thus, the claim holds. □

We next prove that $L(H) = L(G)$. Consider Claim 1 when $x \in T^*$. Then, $S' \Rightarrow_H^* \$\varphi(x)$. By $(\$ \rightarrow \varepsilon, \emptyset, \emptyset) \in P_L$, introduced in (10),

$$\$\varphi(x) \Rightarrow_H \varphi(x)$$

Let $x = a_1a_2 \cdots a_k$, where $k = |x|$ (the case when $k = 0$ means that $x = \varepsilon$), so $\varphi(x) = \#a_1\#a_2 \cdots \#a_k$. By (11), $(\# \rightarrow \varepsilon, \emptyset, N') \in P_L$, so

$$\begin{aligned} \#a_1\#a_2 \cdots \#a_k &\Rightarrow_H a_1\#a_2 \cdots \#a_k \\ &\Rightarrow_H a_1a_2 \cdots \#a_k \\ &\vdots \\ &\Rightarrow_H a_2a_2 \cdots a_k \end{aligned}$$

Hence, $x \in L(G)$ implies that $x \in L(H)$, so $L(G) \subseteq L(H)$.

Consider Claim 2 when $x \in T^*$. Then, $S \Rightarrow_G^* \pi(x)$. Since $x \in T^*$, $\pi(x) = x$. Hence, $x \in L(H)$ implies that $x \in L(G)$, so $L(H) \subseteq L(G)$.

The two inclusions, $L(G) \subseteq L(H)$ and $L(H) \subseteq L(G)$, imply that $L(H) = L(G)$. Since $\text{card}(P_R) = 2$, the theorem holds. □

From Theorem 6.4.12 and its proof, we obtain the following corollary, which strengthens Theorem 6.4.7.

Corollary 6.4.13. *Let K be a recursively enumerable language. Then, there is a one-sided random context grammar, $H = (N, T, P_L, P_R, S)$, such that $L(H) = K$, $\text{card}(N) = 13$, $\text{nrrcn}(H) = 2$, and $\text{card}(P_R) = 2$.* □

We close this section by suggesting two important open problem areas.

Open Problem 6.4.14. Can the achieved results be improved? Especially, reconsider Theorem 6.4.7. By proving that every one-sided random context grammar G can be converted into an equivalent one-sided random context H with no right random context nonterminals, we would establish the generative power of left random context grammars (see Sect. 6.1). □

Open Problem 6.4.15. Recall that propagating one-sided random context grammars characterize the family of context-sensitive languages (see Theorem 6.2.3). Can we also limit the overall number of nonterminals in terms of this propagating version like in Theorem 6.4.1? □

6.5 Leftmost Derivations

The investigation of grammars that perform leftmost derivations is central to formal language theory as a whole. Indeed, from a practical viewpoint, leftmost derivations fulfill a crucial role in parsing, which represents a key application area of formal grammars (see [1, 2, 9, 36]). From a theoretical viewpoint, an effect of leftmost derivation restrictions to the power of grammars restricted in this way represents an intensively investigated area of this theory as clearly indicated by many studies on the subject. More specifically, [3, 4, 27, 33, 47] contain fundamental results concerning leftmost derivations in classical Chomsky grammars, [6, 21, 34, 44, 48] and Sect. 5.3 in [11] give an overview of the results concerning leftmost derivations in regulated grammars published until late 1980s, and [12–14, 35, 38, 40] together with Sect. 7.3 in [39] present several follow-up results. In addition, [24, 26, 46] cover language-defining devices introduced with some kind of leftmost derivations, and [5] discusses the recognition complexity of derivation languages of various regulated grammars with leftmost derivations. Finally, [26, 37, 41] study grammar systems working under the leftmost derivation restriction, and [18, 19, 42] investigates leftmost derivations in terms of P systems.

Considering the significance of leftmost derivations, it comes as no surprise that the present section pays a special attention to them. Indeed, it introduces three types of leftmost derivation restrictions placed upon one-sided random context grammars. In the *type*-1 *derivation restriction*, discussed in Sect. 6.5.1, during every derivation step, the leftmost occurrence of a nonterminal has to be rewritten. In the *type*-2 *derivation restriction*, covered in Sect. 6.5.2, during every derivation step, the leftmost occurrence of a nonterminal which can be rewritten has to be rewritten. In the *type*-3 *derivation restriction*, studied in Sect. 6.5.2, during every derivation step, a rule is chosen, and the leftmost occurrence of its left-hand side is rewritten.

In this section, we place the three above-mentioned leftmost derivation restrictions on one-sided random context grammars, and prove results (I) through (III), given next.

(I) One-sided random context grammars with type-1 leftmost derivations characterize **CF** (Theorem 6.5.4). An analogous result holds for propagating one-sided random context grammars (Theorem 6.5.5).

(II) One-sided random context grammars with type-2 leftmost derivations characterize **RE** (Theorem 6.5.9). Propagating one-sided random context grammars with type-2 leftmost derivations characterize **CS** (Theorem 6.5.11).

(III) One-sided random context grammars with type-3 leftmost derivations characterize **RE** (Theorem 6.5.15). Propagating one-sided random context grammars with type-3 leftmost derivations characterize **CS** (Theorem 6.5.17).

6.5.1 Type-1 Leftmost Derivations

In the first derivation restriction type, during every derivation step, the leftmost occurrence of a nonterminal has to be rewritten. This type of leftmost derivations corresponds to the well-known leftmost derivations in context-free grammars (see Sect. 3.3).

Definition 6.5.1. Let $G = (N, T, P_L, P_R, S)$ be a one-sided random context grammar. The *type*-1 *direct leftmost derivation relation* over V^*, symbolically denoted by ${}_{\rm lm}\!\overset{1}{\Rightarrow}_G$, is defined as follows. Let $u \in T^*$, $A \in N$ and $x, v \in V^*$. Then,

$$uAv \; {}_{\rm lm}\!\overset{1}{\Rightarrow}_G \; uxv$$

if and only if

$$uAv \Rightarrow_G uxv$$

Let ${}_{\text{lm}}\!\overset{1}{\Rightarrow}{}^{n}_{G}$ and ${}_{\text{lm}}\!\overset{1}{\Rightarrow}{}^{*}_{G}$ denote the nth power of ${}_{\text{lm}}\!\overset{1}{\Rightarrow}_{G}$, for some $n \geq 0$, and the reflexive-transitive closure of ${}_{\text{lm}}\!\overset{1}{\Rightarrow}_{G}$, respectively. The ${}_{\text{lm}}^{1}$-*language* of G is denoted by $L(G, {}_{\text{lm}}\!\overset{1}{\Rightarrow})$ and defined as

$$L\big(G, {}_{\text{lm}}\!\overset{1}{\Rightarrow}\big) = \big\{w \in T^* \mid S \;{}_{\text{lm}}\!\overset{1}{\Rightarrow}{}^{*}_{G}\; w\big\}$$

□

Notice that if the leftmost occurrence of a nonterminal cannot be rewritten by any rule, then the derivation is blocked.

The language families generated by one-sided random context grammars with type-1 leftmost derivations and propagating one-sided random context grammars with type-1 leftmost derivations are denoted by $\mathbf{ORC}({}_{\text{lm}}\!\overset{1}{\Rightarrow})$ and $\mathbf{ORC}^{-\varepsilon}({}_{\text{lm}}\!\overset{1}{\Rightarrow})$, respectively.

Next, we prove that $\mathbf{ORC}({}_{\text{lm}}\!\overset{1}{\Rightarrow}) = \mathbf{ORC}^{-\varepsilon}({}_{\text{lm}}\!\overset{1}{\Rightarrow}) = \mathbf{CF}$.

Lemma 6.5.2. *For every context-free grammar G, there is a one-sided random context grammar H such that $L(H, {}_{\text{lm}}\!\overset{1}{\Rightarrow}) = L(G)$. Furthermore, if G is propagating, so is H.*

Proof. Let $G = (N, T, P, S)$ be a context-free grammar. Construct the one-sided random context grammar

$$H = \big(N, T, P', P', S\big)$$

where

$$P' = \big\{(A \rightarrow x, \emptyset, \emptyset) \mid A \rightarrow x \in P\big\}$$

As the rules in P' have their permitting and forbidding contexts empty, any successful type-1 leftmost derivation in H is also a successful derivation in G, so the inclusion $L(H, {}_{\text{lm}}\!\overset{1}{\Rightarrow}) \subseteq L(G)$ holds. On the other hand, let $w \in L(G)$ be a string successfully generated by G. Then, it is well-known that there exists a successful leftmost derivation of w in G (see Theorem 3.3.13). Observe that such a leftmost derivation is also possible in H. Thus, the other inclusion $L(G) \subseteq L(H, {}_{\text{lm}}\!\overset{1}{\Rightarrow})$ holds as well. Finally, notice that whenever G is propagating, so is H. Hence, the theorem holds. □

Lemma 6.5.3. *For every one-sided random context grammar G, there is a context-free grammar H such that $L(H) = L(G, {}_{\text{lm}}\!\overset{1}{\Rightarrow})$. Furthermore, if G is propagating, so is H.*

Proof. Let $G = (N, T, P_L, P_R, S)$ be a one-sided random context grammar. In what follows, angle brackets $\langle$ and $\rangle$ are used to incorporate more symbols into a single compound symbol. Construct the context-free grammar

$$H = \left(N', T, P, \langle S, \emptyset\rangle\right)$$

in the following way. Initially, set

$$N' = \left\{\langle A, Q\rangle \mid A \in N, Q \subseteq N\right\}$$

and $P = \emptyset$. Without any loss of generality, assume that $N' \cap V = \emptyset$. Perform (1) and (2), given next.

(1) For each $(A \rightarrow y_0 Y_1 y_1 Y_2 y_2 \cdots Y_h y_h, U, W) \in P_R$, where $y_i \in T^*$, $Y_j \in N$, for all i and j, $0 \leq i \leq h$, $1 \leq j \leq h$, for some $h \geq 0$, and for each $\langle A, Q\rangle \in N'$ such that $U \subseteq Q$ and $W \cap Q = \emptyset$, extend P by adding

$$\begin{aligned}\langle A, Q\rangle \rightarrow y_0 &\langle Y_1, Q \cup \{Y_2, Y_3, \dots, Y_h\}\rangle y_1\\ &\langle Y_2, Q \cup \{Y_3, \dots, Y_h\}\rangle y_2\\ &\vdots\\ &\langle Y_h, Q\rangle y_h\end{aligned}$$

(2) For each $(A \rightarrow y_0 Y_1 y_1 Y_2 y_2 \cdots Y_h y_h, \emptyset, W) \in P_L$, where $y_i \in T^*$, $Y_j \in N$, for all i and j, $0 \leq i \leq h$, $1 \leq j \leq h$, for some $h \geq 0$, and for each $\langle A, Q\rangle \in N'$, extend P by adding

$$\begin{aligned}\langle A, Q\rangle \rightarrow y_0 &\langle Y_1, Q \cup \{Y_2, Y_3, \dots, Y_h\}\rangle y_1\\ &\langle Y_2, Q \cup \{Y_3, \dots, Y_h\}\rangle y_2\\ &\vdots\\ &\langle Y_h, Q\rangle y_h\end{aligned}$$

Before proving that $L(H) = L(G, {}_{\text{lm}}\!\overset{1}{\Rightarrow})$, let us give an insight into the construction. As G always rewrites the leftmost occurrence of a nonterminal, we use compound nonterminals of the form $\langle A, Q\rangle$ in H, where A is a nonterminal and Q is a set of nonterminals that appear to the right of this occurrence of A. When simulating rules from P_R, the check for the presence and absence of symbols is accomplished by using Q. Also, when rewriting A in $\langle A, Q\rangle$ to some y, the compound nonterminals from N' are generated instead of nonterminals from N.

Rules from P_L are simulated analogously; however, notice that if the permitting set of such a rule is nonempty, it is never applicable in G. Therefore, such rules are not introduced to P'. Furthermore, since there are no nonterminals to the left of the leftmost occurrence of a nonterminal, no check for their absence is done.

Clearly, $L(G, {}_{\text{lm}}\!\overset{1}{\Rightarrow}) \subseteq L(H)$. The opposite inclusion, $L(H) \subseteq L(G, {}_{\text{lm}}\!\overset{1}{\Rightarrow})$, can be proved by analogy with the proof of Lemma 6.5.2 by simulating the leftmost derivation of every $w \in L(H)$ by G. Observe that since the check for the presence and absence of symbols in H is done in the second components of the compound nonterminals, each rule introduced to P in (1) and (2) can be simulated by a rule from P_R and P_L from which it is constructed.

Since H is propagating whenever G is propagating, the theorem holds. □

Theorem 6.5.4. $\mathbf{ORC}({}_{\text{lm}}\!\overset{1}{\Rightarrow}) = \mathbf{CF}$

Proof. By Lemma 6.5.2, $\mathbf{CF} \subseteq \mathbf{ORC}({}_{\text{lm}}\!\overset{1}{\Rightarrow})$. By Lemma 6.5.3, $\mathbf{ORC}({}_{\text{lm}}\!\overset{1}{\Rightarrow}) \subseteq \mathbf{CF}$. Consequently, $\mathbf{ORC}({}_{\text{lm}}\!\overset{1}{\Rightarrow}) = \mathbf{CF}$, so the theorem holds. □

Theorem 6.5.5. $\mathbf{ORC}^{-\varepsilon}({}_{\text{lm}}\!\overset{1}{\Rightarrow}) = \mathbf{CF}$

Proof. Since it is well-known that any context-free grammar can be converted to an equivalent context-free grammar without any erasing rules (see Theorem 7.1.8), this theorem follows from Lemmas 6.5.2 and 6.5.3. □

6.5.2 Type-2 Leftmost Derivations

In the second derivation restriction type, during every derivation step, the leftmost occurrence of a nonterminal that can be rewritten has to be rewritten.

Definition 6.5.6. Let $G = (N, T, P_L, P_R, S)$ be a one-sided random context grammar. The *type-2 direct leftmost derivation relation* over V^*, symbolically denoted by ${}_{\text{lm}}\!\overset{2}{\Rightarrow}_G$, is defined as follows. Let $u, x, v \in V^*$ and $A \in N$. Then,

$$uAv \; {}_{\text{lm}}\!\overset{2}{\Rightarrow}_G \; uxv$$

if and only if $uAv \Rightarrow_G uxv$ and there is no $B \in N$ and $y \in V^*$ such that $u = u_1 B u_2$ and $u_1 B u_2 A v \Rightarrow_G u_1 y u_2 A v$.

Let ${}_{\text{lm}}\!\overset{2}{\Rightarrow}^n_G$ and ${}_{\text{lm}}\!\overset{2}{\Rightarrow}^*_G$ denote the nth power of ${}_{\text{lm}}\!\overset{2}{\Rightarrow}_G$, for some $n \geq 0$, and the reflexive-transitive closure of ${}_{\text{lm}}\!\overset{2}{\Rightarrow}_G$, respectively. The ${}_{\text{lm}}\!\overset{2}{}$*-language* of G is denoted by $L(G, {}_{\text{lm}}\!\overset{2}{\Rightarrow})$ and defined as

$$L\big(G, {}_{\text{lm}}\!\overset{2}{\Rightarrow}\big) = \big\{w \in T^* \mid S \; {}_{\text{lm}}\!\overset{2}{\Rightarrow}^*_G \; w\big\}$$ □

The language families generated by one-sided random context grammars with type-2 leftmost derivations and propagating one-sided random context grammars with type-2 leftmost derivations are denoted by $\mathbf{ORC}({}_{\text{lm}}\!\overset{2}{\Rightarrow})$ and $\mathbf{ORC}^{-\varepsilon}({}_{\text{lm}}\!\overset{2}{\Rightarrow})$, respectively.

Next, we prove that $\mathbf{ORC}({}_{\text{lm}}\!\overset{2}{\Rightarrow}) = \mathbf{RE}$ and $\mathbf{ORC}^{-\varepsilon}({}_{\text{lm}}\!\overset{2}{\Rightarrow}) = \mathbf{CS}$.

Lemma 6.5.7. *For every one-sided random context grammar G, there is a one-sided random context grammar H such that $L(H, {}_{\mathrm{lm}}{\overset{2}{\Rightarrow}}) = L(G)$. Furthermore, if G is propagating, so is H.*

Proof. Let $G = (N, T, P_L, P_R, S)$ be a one-sided random context grammar. We construct the one-sided random context grammar H in such a way that always allows it to rewrite an arbitrary occurrence of a nonterminal. Construct

$$H = \left(N', T, P'_L, P'_R, S\right)$$

as follows. Initially, set $\bar{N} = \{\bar{A} \mid A \in N\}$, $\hat{N} = \{\hat{A} \mid A \in N\}$, $N' = N \cup \bar{N} \cup \hat{N}$, and $P'_L = P'_R = \emptyset$. Without any loss of generality, assume that N, $\bar{N}$, and $\hat{N}$ are pairwise disjoint. Define the function ψ from 2^N to $2^{\bar{N}}$ as $\psi(\emptyset) = \emptyset$ and

$$\psi\left(\{A_1, A_2, \dots, A_n\}\right) = \left\{\bar{A}_1, \bar{A}_2, \dots, \bar{A}_n\right\}$$

Perform (1) through (3), given next.

(1) For each $A \in N$,

  (1.1) add $(A \rightarrow \bar{A}, \emptyset, N \cup \hat{N})$ to P'_L,
  (1.2) add $(\bar{A} \rightarrow \hat{A}, \emptyset, N \cup \bar{N})$ to P'_R,
  (1.3) add $(\hat{A} \rightarrow A, \emptyset, \bar{N} \cup \hat{N})$ to P'_R.

(2) For each $(A \rightarrow y, U, W) \in P_R$, add $(A \rightarrow y, U, W)$ to P'_R.
(3) For each $(A \rightarrow y, U, W) \in P_L$, add $(A \rightarrow y, \psi(U), \psi(W) \cup N \cup \hat{N})$ to P'_L.

Before proving that $L(H) = L(G)$, let us informally explain (1) through (3). Rules from (2) and (3) simulate the corresponding rules from P_R and P_L, respectively. Rules from (1) allow H to rewrite any occurrence of a nonterminal.

Consider a sentential form $x_1 A x_2$, where $x_1, x_2 \in (N \cup T)^*$ and $A \in N$. To rewrite A in H using type-2 leftmost derivations, all occurrences of nonterminals in x_1 are first rewritten to their barred versions by rules from (1.1). Then, A can be rewritten by a rule from (2) or (3). By rules from (1.1), every occurrence of a nonterminal in the current sentential form is then rewritten to its barred version. Rules from (1.2) then start rewriting barred nonterminals to hatted nonterminals, which is performed from the right to the left. Finally, hatted nonterminals are rewritten to their original versions by rules from (1.3). This is also performed from the right to the left.

To establish $L(H, {}_{\mathrm{lm}}{\overset{2}{\Rightarrow}}) = L(G)$, we prove two claims. First, Claim 1 shows how derivations of G are simulated by H. Then, Claim 2 demonstrates the converse—that is, it shows how derivations of H are simulated by G.

Claim 1. If $S \Rightarrow_G^n x$, where $x \in V^*$, for some $n \geq 0$, then $S \;{}_{\mathrm{lm}}{\overset{2}{\Rightarrow}}{}_H^* \; x$.

Proof. This claim is established by induction on $n \geq 0$.

Basis. For $n = 0$, this claim obviously holds.

Induction Hypothesis. Suppose that there exists $n \geq 0$ such that the claim holds for all derivations of length ℓ, where $0 \leq \ell \leq n$.

Induction Step. Consider any derivation of the form

$$S \Rightarrow_G^{n+1} w$$

where $w \in V^*$. Since $n + 1 \geq 1$, this derivation can be expressed as

$$S \Rightarrow_G^n x \Rightarrow_G w$$

for some $x \in V^+$. By the induction hypothesis, $S \,{}_{\mathrm{lm}}^{\;2}\!\!\Rightarrow_H^* x$. Next, we consider all possible forms of $x \Rightarrow_G w$, covered by the following two cases—(i) and (ii).

(i) *Application of* $(A \rightarrow y, U, W) \in P_R$. Let $x = x_1 A x_2$ and $r = (A \rightarrow y, U, W) \in P_R$, where $x_1, x_2 \in V^*$ such that $U \subseteq \mathrm{alph}(x_2)$ and $W \cap \mathrm{alph}(x_2) = \emptyset$, so

$$x_1 A x_2 \Rightarrow_G x_1 y x_2$$

If $x_1 \in T^*$, then $x_1 A x_2 \,{}_{\mathrm{lm}}^{\;2}\!\!\Rightarrow_H x_1 y x_2$ by the corresponding rule introduced in (2), and the induction step is completed for (i). Therefore, assume that $\mathrm{alph}(x_1) \cap N \neq \emptyset$. Let $x_1 = z_0 Z_1 z_1 Z_2 z_2 \cdots Z_h z_h$, where $z_i \in T^*$ and $Z_j \in N$, for all i and j, $0 \leq i \leq h$, $1 \leq j \leq h$, for some $h \geq 1$. By rules introduced in (1.1),

$$z_0 Z_1 z_1 Z_2 z_2 \cdots Z_h z_h A x_2 \,{}_{\mathrm{lm}}^{\;2}\!\!\Rightarrow_H^* z_0 \bar{Z}_1 z_1 \bar{Z}_2 z_2 \cdots \bar{Z}_h z_h A x_2$$

By the corresponding rule to r introduced in (2),

$$z_0 \bar{Z}_1 z_1 \bar{Z}_2 z_2 \cdots \bar{Z}_h z_h A x_2 \,{}_{\mathrm{lm}}^{\;2}\!\!\Rightarrow_H z_0 \bar{Z}_1 z_1 \bar{Z}_2 z_2 \cdots \bar{Z}_h z_h y x_2$$

By rules introduced in (1.1) through (1.3),

$$z_0 \bar{Z}_1 z_1 \bar{Z}_2 z_2 \cdots \bar{Z}_h z_h y x_2 \,{}_{\mathrm{lm}}^{\;2}\!\!\Rightarrow_H^* z_0 Z_1 z_1 Z_2 z_2 \cdots Z_h z_h y x_2$$

which completes the induction step for (i).

(ii) *Application of* $(A \rightarrow y, U, W) \in P_L$. Let $x = x_1 A x_2$ and $r = (A \rightarrow y, U, W) \in P_L$, where $x_1, x_2 \in V^*$ such that $U \subseteq \mathrm{alph}(x_1)$ and $W \cap \mathrm{alph}(x_1) = \emptyset$, so

$$x_1 A x_2 \Rightarrow_G x_1 y x_2$$

To complete the induction step for (ii), proceed by analogy with (i), but use a rule from (3) instead of a rule from (2).

Observe that cases (i) and (ii) cover all possible forms of $x \Rightarrow_G w$. Thus, the claim holds. □

Set $V = N \cup T$ and $V' = N' \cup T$. Define the homomorphism τ from V'^* to V^* as $\tau(A) = \tau(\bar{A}) = \tau(\hat{A}) = A$, for all $A \in N$, and $\tau(a) = a$, for all $a \in T$.

Claim 2. If $S \;_{\text{lm}}^{\;2}\!\!\Rightarrow^n_H x$, where $x \in V'^*$, for some $n \geq 0$, then $S \Rightarrow^*_G \tau(x)$, and either $x \in (\bar{N} \cup T)^* V^*$, $x \in (\bar{N} \cup T)^*(\hat{N} \cup T)^*$, or $x \in (\hat{N} \cup T)^* V^*$.

Proof. This claim is established by induction on $n \geq 0$.

Basis. For $n = 0$, this claim obviously holds.

Induction Hypothesis. Suppose that there exists $n \geq 0$ such that the claim holds for all derivations of length ℓ, where $0 \leq \ell \leq n$.

Induction Step. Consider any derivation of the form

$$S \;_{\text{lm}}^{\;2}\!\!\Rightarrow^{n+1}_H w$$

where $w \in V'^*$. Since $n + 1 \geq 1$, this derivation can be expressed as

$$S \;_{\text{lm}}^{\;2}\!\!\Rightarrow^n_H x \;_{\text{lm}}^{\;2}\!\!\Rightarrow_H w$$

for some $x \in V'^+$. By the induction hypothesis, $S \Rightarrow^*_G \tau(x)$, and either $x \in (\bar{N} \cup T)^* V^*$, $x \in (\bar{N} \cup T)^*(\hat{N} \cup T)^*$, or $x \in (\hat{N} \cup T)^* V^*$. Next, we consider all possible forms of $x \;_{\text{lm}}^{\;2}\!\!\Rightarrow_H w$, covered by the following five cases—(i) through (v).

(i) *Application of a rule introduced in (1.1).* Let $(A \to \bar{A}, \emptyset, N \cup \hat{N}) \in P'_L$ be a rule introduced in (1.1). Observe that this rule is applicable only if $x = x_1 A x_2$, where $x_1 \in (\bar{N} \cup T)^*$ and $x_2 \in V^*$. Then,

$$x_1 A x_2 \;_{\text{lm}}^{\;2}\!\!\Rightarrow_H x_1 \bar{A} x_2$$

Since $\tau(x_1 \bar{A} x_2) = \tau(x_1 A x_2)$ and $x_1 \bar{A} x_2 \in (\bar{N} \cup T)^* V^*$, the induction step is completed for (i).

(ii) *Application of a rule introduced in (1.2).* Let $(\bar{A} \to \hat{A}, \emptyset, N \cup \bar{N}) \in P'_R$ be a rule introduced in (1.2). Observe that this rule is applicable only if $x = x_1 \bar{A} x_2$, where $x_1 \in (\bar{N} \cup T)^*$ and $x_2 \in (\hat{N} \cup T)^*$. Then,

$$x_1 \bar{A} x_2 \;_{\text{lm}}^{\;2}\!\!\Rightarrow_H x_1 \hat{A} x_2$$

Since $\tau(x_1 \hat{A} x_2) = \tau(x_1 \bar{A} x_2)$ and $x_1 \hat{A} x_2 \in (\bar{N} \cup T)^*(\hat{N} \cup T)^*$, the induction step is completed for (ii).

(iii) *Application of a rule introduced in (1.3).* Let $(\hat{A} \to A, \emptyset, \bar{N} \cup \hat{N}) \in P'_R$ be a rule introduced in (1.3). Observe that this rule is applicable only if $x = x_1 \hat{A} x_2$, where $x_1 \in (\hat{N} \cup T)^*$ and $x_2 \in V^*$. Then,

$$x_1 \hat{A} x_2 \;_{\text{lm}}^{\;2}\!\!\Rightarrow_H x_1 A x_2$$

Since $\tau(x_1 A x_2) = \tau(x_1 \hat{A} x_2)$ and $x_1 A x_2 \in (\hat{N} \cup T)^* V^*$, the induction step is completed for (iii).

(iv) *Application of a rule introduced in (2).* Let $(A \to y, U, W) \in P'_R$ be a rule introduced in (2) from $(A \to y, U, W) \in P_R$, and let $x = x_1 A x_2$ such that $U \subseteq \text{alph}(x_2)$ and $W \cap \text{alph}(x_2) = \emptyset$. Then,

$$x_1 A x_2 \ {}_{\text{lm}}\!\overset{2}{\Rightarrow}_H \ x_1 y x_2$$

and

$$\tau(x_1) A \tau(x_2) \Rightarrow_G \tau(x_1) y \tau(x_2)$$

Clearly, $x_1 y x_2$ is of the required form, so the induction step is completed for (iv).

(v) *Application of a rule introduced in (3).* Let $(A \to y, \psi(U), \psi(W) \cup N \cup \hat{N}) \in P'_L$ be a rule introduced in (3) from $(A \to y, U, W) \in P_L$, and let $x = x_1 A x_2$ such that $\psi(U) \subseteq \text{alph}(x_1)$ and $(\psi(W) \cup N \cup \hat{N}) \cap \text{alph}(x_1) = \emptyset$. Then,

$$x_1 A x_2 \ {}_{\text{lm}}\!\overset{2}{\Rightarrow}_H \ x_1 y x_2$$

and

$$\tau(x_1) A \tau(x_2) \Rightarrow_G \tau(x_1) y \tau(x_2)$$

Clearly, $x_1 y x_2$ is of the required form, so the induction step is completed for (v).

Observe that cases (i) through (v) cover all possible forms of $x \ {}_{\text{lm}}\!\overset{2}{\Rightarrow}_H \ w$. Thus, the claim holds. □

We next prove that $L(H, {}_{\text{lm}}\!\overset{2}{\Rightarrow}) = L(G)$. Consider Claim 1 for $x \in T^*$. Then, $S \Rightarrow_G^* x$ implies that $S \ {}_{\text{lm}}\!\overset{2}{\Rightarrow}{}_H^* \ x$, so $L(G) \subseteq L(H, {}_{\text{lm}}\!\overset{2}{\Rightarrow})$. Consider Claim 2 for $x \in T^*$. Then, $S \ {}_{\text{lm}}\!\overset{2}{\Rightarrow}{}_H^* \ x$ implies that $S \Rightarrow_G^* x$, so $L(H, {}_{\text{lm}}\!\overset{2}{\Rightarrow}) \subseteq L(G)$. Consequently, $L(H, {}_{\text{lm}}\!\overset{2}{\Rightarrow}) = L(G)$.

Since H is propagating whenever G is propagating, the theorem holds. □

Lemma 6.5.8. $\mathbf{ORC}({}_{\text{lm}}\!\overset{2}{\Rightarrow}) \subseteq \mathbf{RE}$

Proof. This inclusion follows from Church's thesis. □

Theorem 6.5.9. $\mathbf{ORC}({}_{\text{lm}}\!\overset{2}{\Rightarrow}) = \mathbf{RE}$

Proof. Since $\mathbf{ORC} = \mathbf{RE}$ (see Theorem 6.2.4), Lemma 6.5.7 implies that $\mathbf{RE} \subseteq \mathbf{ORC}({}_{\text{lm}}\!\overset{2}{\Rightarrow})$. By Lemma 6.5.8, $\mathbf{ORC}({}_{\text{lm}}\!\overset{2}{\Rightarrow}) \subseteq \mathbf{RE}$. Consequently, $\mathbf{ORC}({}_{\text{lm}}\!\overset{2}{\Rightarrow}) = \mathbf{RE}$, so the theorem holds. □

Lemma 6.5.10. $\mathbf{ORC}^{-\varepsilon}({}_{\mathrm{lm}}^{\;2}\!\!\Rightarrow) \subseteq \mathbf{CS}$

Proof. Since the length of sentential forms in derivations of propagating one-sided random context grammars is nondecreasing, propagating one-sided random context grammars can be simulated by context-sensitive grammars. A rigorous proof of this lemma is left to the reader. □

Theorem 6.5.11. $\mathbf{ORC}^{-\varepsilon}({}_{\mathrm{lm}}^{\;2}\!\!\Rightarrow) = \mathbf{CS}$

Proof. Since $\mathbf{ORC}^{-\varepsilon} = \mathbf{CS}$ (see Theorem 6.2.3), Lemma 6.5.7 implies that $\mathbf{CS} \subseteq \mathbf{ORC}^{-\varepsilon}({}_{\mathrm{lm}}^{\;2}\!\!\Rightarrow)$. By Lemma 6.5.10, $\mathbf{ORC}^{-\varepsilon}({}_{\mathrm{lm}}^{\;2}\!\!\Rightarrow) \subseteq \mathbf{CS}$. Consequently, we have $\mathbf{ORC}^{-\varepsilon}({}_{\mathrm{lm}}^{\;2}\!\!\Rightarrow) = \mathbf{CS}$, so the theorem holds. □

6.5.3 Type-3 Leftmost Derivations

In the third derivation restriction type, during every derivation step, a rule is chosen, and the leftmost occurrence of its left-hand side is rewritten.

Definition 6.5.12. Let $G = (N, T, P_L, P_R, S)$ be a one-sided random context grammar. The *type-3 direct leftmost derivation relation* over V^*, symbolically denoted by ${}_{\mathrm{lm}}^{\;3}\!\!\Rightarrow_G$, is defined as follows. Let $u, x, v \in V^*$ and $A \in N$. Then,

$$uAv \;{}_{\mathrm{lm}}^{\;3}\!\!\Rightarrow_G\; uxv$$

if and only if $uAv \Rightarrow_G uxv$ and $\mathrm{alph}(u) \cap \{A\} = \emptyset$.

Let ${}_{\mathrm{lm}}^{\;3}\!\!\Rightarrow_G^n$ and ${}_{\mathrm{lm}}^{\;3}\!\!\Rightarrow_G^*$ denote the nth power of ${}_{\mathrm{lm}}^{\;3}\!\!\Rightarrow_G$, for some $n \geq 0$, and the reflexive-transitive closure of ${}_{\mathrm{lm}}^{\;3}\!\!\Rightarrow_G$, respectively. The ${}_{\mathrm{lm}}^{\;3}$*-language* of G is denoted by $L(G, {}_{\mathrm{lm}}^{\;3}\!\!\Rightarrow)$ and defined as

$$L\big(G, {}_{\mathrm{lm}}^{\;3}\!\!\Rightarrow\big) = \big\{w \in T^* \mid S \;{}_{\mathrm{lm}}^{\;3}\!\!\Rightarrow_G^*\; w\big\}$$ □

Notice the following difference between the second and the third type. In the former, the leftmost occurrence of a rewritable nonterminal is chosen first, and then, a choice of a rule with this nonterminal on its let-hand side is made. In the latter, a rule is chosen first, and then, the leftmost occurrence of its left-hand side is rewritten.

The language families generated by one-sided random context grammars with type-3 leftmost derivations and propagating one-sided random context grammars with type-3 leftmost derivations are denoted by $\mathbf{ORC}({}_{\mathrm{lm}}^{\;3}\!\!\Rightarrow)$ and $\mathbf{ORC}^{-\varepsilon}({}_{\mathrm{lm}}^{\;3}\!\!\Rightarrow)$, respectively.

Next, we prove that $\mathbf{ORC}({}_{\mathrm{lm}}^{\;3}\!\!\Rightarrow) = \mathbf{RE}$ and $\mathbf{ORC}^{-\varepsilon}({}_{\mathrm{lm}}^{\;3}\!\!\Rightarrow) = \mathbf{CS}$.

Lemma 6.5.13. *For every one-sided random context grammar G, there is a one-sided random context grammar H such that $L(H, {}_{\mathrm{lm}}^{\;3}\!\!\Rightarrow) = L(G)$. Furthermore, if G is propagating, so is H.*

Proof. Let $G = (N, T, P_L, P_R, S)$ be a one-sided random context grammar. We prove this lemma by analogy with the proof of Lemma 6.5.7. That is, we construct the one-sided random context grammar H in such a way that always allows it to rewrite an arbitrary occurrence of a nonterminal. Construct

$$H = \left(N', T, P'_L, P'_R, S\right)$$

as follows. Initially, set $\bar{N} = \{\bar{A} \mid A \in N\}$, $N' = N \cup \bar{N}$, and $P'_L = P'_R = \emptyset$. Without any loss of generality, assume that $N \cap \bar{N} = \emptyset$. Define the function ψ from 2^N to $2^{\bar{N}}$ as $\psi(\emptyset) = \emptyset$ and

$$\psi\left(\{A_1, A_2, \dots, A_n\}\right) = \left\{\bar{A}_1, \bar{A}_2, \dots, \bar{A}_n\right\}$$

Perform (1) through (3), given next.

(1) For each $A \in N$,

 (1.1) add $(A \to \bar{A}, \emptyset, N)$ to P'_L;
 (1.2) add $(\bar{A} \to A, \emptyset, \bar{N})$ to P'_R.

(2) For each $(A \to y, U, W) \in P_R$, add $(A \to y, U, W)$ to P'_R.
(3) For each $(A \to y, U, W) \in P_L$, let $U = \{X_1, X_2, \dots, X_k\}$, and for each

$$U' \in \left\{\{Y_1, Y_2, \dots, Y_k\} \mid Y_i \in \{X_i, \bar{X}_i\}, 1 \leq i \leq k\right\}$$

add $(A \to y, U', W \cup \Psi(W))$ to P'_L ($U' = \emptyset$ if and only if $U = \emptyset$).

Before proving that $L(G) = L(H, {}_{\text{lm}}^{\;3}\!\Rightarrow)$, let us give an insight into the construction. Rules introduced in (1) allow H to rewrite an arbitrary occurrence of a nonterminal. Rules from (2) and (3) simulate the corresponding rules from P_R and P_L, respectively.

Consider a sentential form $x_1 A x_2$, where $x_1, x_2 \in (N \cup T)^*$ and $A \in N$, and a rule, $r = (A \to y, U, W) \in P'_L \cup P'_R$, introduced in (2) or (3). If $A \in \text{alph}(x_1)$, all occurrences of nonterminals in x_1 are rewritten to their barred versions by rules from (1). Then, r is applied, and all barred nonterminals are rewritten back to their non-barred versions. Since not all occurrences of nonterminals in x_1 need to be rewritten to their barred versions before r is applied, all combinations of barred and non-barred nonterminals in the left permitting contexts of the resulting rules in (3) are considered.

The identity $L(H, {}_{\text{lm}}^{\;3}\!\Rightarrow) = L(G)$ can be established by analogy with the proof given in Lemma 6.5.7, and we leave its proof to the reader. Finally, notice that whenever G is propagating, so is H. Hence, the theorem holds. □

Lemma 6.5.14. $\mathbf{ORC}({}_{\text{lm}}^{\;3}\!\Rightarrow) \subseteq \mathbf{RE}$

Proof. This inclusion follows from Church's thesis. □

Theorem 6.5.15. $\mathbf{ORC}({}_{\text{lm}}^{\;3}\!\Rightarrow) = \mathbf{RE}$

Proof. Since $\mathbf{ORC} = \mathbf{RE}$ (see Theorem 6.2.4), Lemma 6.5.13 implies that $\mathbf{RE} \subseteq \mathbf{ORC}({}_{\mathrm{lm}}^{3}\!\Rightarrow)$. By Lemma 6.5.14, $\mathbf{ORC}({}_{\mathrm{lm}}^{3}\!\Rightarrow) \subseteq \mathbf{RE}$. Consequently, $\mathbf{ORC}({}_{\mathrm{lm}}^{3}\!\Rightarrow) = \mathbf{RE}$, so the theorem holds. □

Lemma 6.5.16. $\mathbf{ORC}^{-\varepsilon}({}_{\mathrm{lm}}^{3}\!\Rightarrow) \subseteq \mathbf{CS}$

Proof. This lemma can be established by analogy with the proof of Lemma 6.5.10. □

Theorem 6.5.17. $\mathbf{ORC}^{-\varepsilon}({}_{\mathrm{lm}}^{3}\!\Rightarrow) = \mathbf{CS}$

Proof. Since $\mathbf{ORC}^{-\varepsilon} = \mathbf{CS}$ (see Theorem 6.2.3), Lemma 6.5.13 implies that $\mathbf{CS} \subseteq \mathbf{ORC}^{-\varepsilon}({}_{\mathrm{lm}}^{3}\!\Rightarrow)$. By Lemma 6.5.16, $\mathbf{ORC}^{-\varepsilon}({}_{\mathrm{lm}}^{3}\!\Rightarrow) \subseteq \mathbf{CS}$. Consequently, we have $\mathbf{ORC}^{-\varepsilon}({}_{\mathrm{lm}}^{3}\!\Rightarrow) = \mathbf{CS}$, so the theorem holds. □

In the conclusion of this section, we compare the achieved results with some well-known results of formal language theory. More specifically, we relate the language families generated by one-sided random context grammars with leftmost derivations to the language families generated by random context grammars with leftmost derivations.

The families of languages generated by random context grammars with type-1 leftmost derivations, random context grammars with type-2 leftmost derivations, and random context grammars with type-3 leftmost derivations are denoted by $\mathbf{RC}({}_{\mathrm{lm}}^{1}\!\Rightarrow)$, $\mathbf{RC}({}_{\mathrm{lm}}^{2}\!\Rightarrow)$, and $\mathbf{RC}({}_{\mathrm{lm}}^{3}\!\Rightarrow)$, respectively (see [11] for the definitions of all these families). The notation without ε stands for the corresponding propagating family. For example, $\mathbf{RC}^{-\varepsilon}({}_{\mathrm{lm}}^{1}\!\Rightarrow)$ denotes the language family generated by propagating random context grammars with type-1 leftmost derivations.

The fundamental relations between these families are summarized next.

Corollary 6.5.18. $\mathbf{CF} \subset \mathbf{RC}^{-\varepsilon} \subset \mathbf{ORC}^{-\varepsilon} = \mathbf{CS} \subset \mathbf{ORC} = \mathbf{RC} = \mathbf{RE}$

Proof. This corollary follows from Theorems 6.2.3 and 6.2.4 in this section and from Corollary 4.3.10. □

Considering type-1 leftmost derivations, we significantly decrease the power of both one-sided random context grammars and random context grammars.

Corollary 6.5.19. $\mathbf{ORC}({}_{\mathrm{lm}}^{1}\!\Rightarrow) = \mathbf{RC}({}_{\mathrm{lm}}^{1}\!\Rightarrow) = \mathbf{CF}$

Proof. This corollary follows from Theorem 6.5.4 in this section and from Theorem 1.4.1 in [11]. □

Type-2 leftmost derivations increase the generative power of propagating random context grammars, but the generative power of random context grammars remains unchanged.

Corollary 6.5.20.

(i) $\mathbf{ORC}^{-\varepsilon}({}_{\mathrm{lm}}^{2}\!\Rightarrow) = \mathbf{RC}^{-\varepsilon}({}_{\mathrm{lm}}^{2}\!\Rightarrow) = \mathbf{CS}$
(ii) $\mathbf{ORC}({}_{\mathrm{lm}}^{2}\!\Rightarrow) = \mathbf{RC}({}_{\mathrm{lm}}^{2}\!\Rightarrow) = \mathbf{RE}$

Proof. This corollary follows from Theorems 6.5.9 and 6.5.11 in this section and from Theorem 1.4.4 in [11]. □

Finally, type-3 leftmost derivations are not enough for propagating random context grammars to generate the family of context-sensitive languages, so one-sided random context grammars with type-3 leftmost derivations are more powerful.

Corollary 6.5.21.

(i) $\mathbf{RC}^{-\varepsilon}({}_{\mathrm{lm}}^{\;3}\!\Rightarrow) \subset \mathbf{ORC}^{-\varepsilon}({}_{\mathrm{lm}}^{\;3}\!\Rightarrow) = \mathbf{CS}$
(ii) $\mathbf{ORC}({}_{\mathrm{lm}}^{\;3}\!\Rightarrow) = \mathbf{RC}({}_{\mathrm{lm}}^{\;3}\!\Rightarrow) = \mathbf{RE}$

Proof. This corollary follows from Theorems 6.5.15 and 6.5.17 in the this section, from Theorem 1.4.5 in [11], and from Remarks 5.11 in [13]. □

We close this section by making a remark about rightmost derivations. Of course, we can define and study rightmost derivations in one-sided random context grammars by analogy with their leftmost counterparts, discussed above. We can also reformulate and establish the same results as above in terms of the rightmost derivations. All this discussion of rightmost derivations is so analogous with the above discussion of leftmost derivations that we leave it to the reader.

6.6 Generalized One-Sided Forbidding Grammars

In Sect. 4.4, we introduced and studied generalized forbidding grammars based upon context-free rules, each of which may be associated with finitely many *forbidding strings*. A rule like this can rewrite a nonterminal provided that none of its forbidding strings occur in the current sentential form; apart from this, these grammars work just like context-free grammars. As opposed to context-free grammars, however, they are computationally complete—that is, they generate the family of recursively enumerable languages (see Theorem 4.4.5), and this property obviously represents their crucially important advantage over ordinary context-free and forbidding grammars (see Theorem 4.3.9).

Taking a closer look at the rewriting process in generalized forbidding grammars, we see that they always verify the absence of forbidding strings within their entire sentential forms. To simplify and accelerate their rewriting process, it is obviously more than desirable to modify these grammars so they make this verification only within some prescribed portions of the rewritten sentential forms while remaining computationally complete. *Generalized one-sided forbidding grammars*, which are defined and studied in the present section, represent a modification satisfying these properties.

More precisely, in a generalized one-sided forbidding grammar, the set of rules is divided into the set of *left forbidding rules* and the set of *right forbidding rules*. When applying a left forbidding rule, the grammar checks the absence of its forbidding strings only in the prefix to the left of the rewritten nonterminal in the current sentential form. Similarly, when applying a right forbidding rule, it performs an analogous check to the right. Apart from this, it works like any generalized forbidding grammar.

Most importantly, the present section demonstrates that generalized one-sided forbidding grammars characterize the family of recursively enumerable languages. In fact, these grammars remain computationally complete even under the restriction that any of their forbidding strings is of length two or less. On the other hand, if a generalized one-sided forbidding grammar has all left forbidding rules without any forbidding strings, then it necessarily generates a context-free language; an analogous result holds in terms of right forbidding rules, too. Even more surprisingly, any generalized one-sided forbidding grammar that has the set of left forbidding rules coinciding with the set of right forbidding rules generates a context-free language.

This section is divided into two subsections. First, Sect. 6.6.1 defines generalized one-sided forbidding grammars and illustrate them by an example. Then, Sect. 6.6.2 establishes their generative power.

6.6.1 *Definitions and Examples*

In this section, we define generalized one-sided forbidding grammars and illustrate them by an example. Recall that for an alphabet N and a string $x \in N^*$, $\mathrm{sub}(x)$ denotes the set of all substrings of x, and $\mathrm{fin}(N)$ denotes the set of all finite languages over N (see Sect. 3.1).

Definition 6.6.1. A *generalized one-sided forbidding grammar* is a quintuple

$$G = \left(N, T, P_L, P_R, S\right)$$

where N and T are two disjoint alphabets, $S \in N$, and

$$P_L, P_R \subseteq N \times \left(N \cup T\right)^* \times \mathrm{fin}(N)$$

are two finite relations. Set $V = N \cup T$. The components V, N, T, P_L, P_R, and S are called the *total alphabet*, the alphabet of *nonterminals*, the alphabet of *terminals*, the set of *left forbidding rules*, the set of *right forbidding rules*, and the *start symbol*, respectively. Each $(A, x, F) \in P_L \cup P_R$ is written as $(A \to x, F)$ throughout this section. For $(A \to x, F) \in P_L$, F is called the *left forbidding context*. Analogously, for $(A \to x, F) \in P_R$, F is called the *right forbidding context*. The *direct derivation relation* over V^*, symbolically denoted by $\Rightarrow_G$, is defined as follows. Let $u, v \in V^*$ and $(A \to x, F) \in P_L \cup P_R$. Then,

$$uAv \Rightarrow_G uxv$$

if and only if

$$(A \rightarrow x, F) \in P_L \text{ and } F \cap \text{sub}(u) = \emptyset$$

or

$$(A \rightarrow x, F) \in P_R \text{ and } F \cap \text{sub}(v) = \emptyset$$

Let $\Rightarrow_G^n$ and $\Rightarrow_G^*$ denote the nth power of $\Rightarrow_G$, for some $n \geq 0$, and the reflexive-transitive closure of $\Rightarrow_G$, respectively. The *language* of G is denoted by $L(G)$ and defined as

$$L(G) = \{w \in T^* \mid S \Rightarrow_G^* w\}$$ □

Next, we introduce the notion of a degree of G. Informally, it is the length of the longest string in the forbidding contexts of the rules of G. Let V be an alphabet. For $L \in \text{fin}(V)$, max-len(L) denotes the length of the longest string in L. We set max-len$(\emptyset) = 0$.

Definition 6.6.2. Let $G = (N, T, P_L, P_R, S)$ be a generalized one-sided forbidding grammar. G is of *degree* (m, n), where $m, n \geq 0$, if $(A \rightarrow x, F) \in P_L$ implies that max-len$(F) \leq m$ and $(A \rightarrow x, F) \in P_R$ implies that max-len$(F) \leq n$. □

Next, we illustrate the previous definitions by an example.

Example 6.6.3. Consider the generalized one-sided forbidding grammar

$$G = (\{S, A, B, A', B', \bar{A}, \bar{B}\}, \{a, b, c\}, P_L, P_R, S)$$

where P_L contains the following five rules

$$(S \rightarrow AB, \emptyset) \quad (B \rightarrow bB'c, \{A, \bar{A}\}) \quad (B' \rightarrow B, \{A'\})$$
$$(B \rightarrow \bar{B}, \{A, A'\}) \quad (\bar{B} \rightarrow \varepsilon, \{\bar{A}\})$$

and P_R contains the following four rules

$$(A \rightarrow aA', \{B'\}) \quad (A' \rightarrow A, \{B\})$$
$$(A \rightarrow \bar{A}, \{B'\}) \quad (\bar{A} \rightarrow \varepsilon, \{B\})$$

Since the length of the longest string in the forbidding contexts of rules from P_L and P_R is 1, G is of degree $(1, 1)$. It is rather easy to see that every derivation that generates a nonempty string of $L(G)$ is of the form

$$
\begin{aligned}
S &\Rightarrow_G AB \\
&\Rightarrow_G aA'B \\
&\Rightarrow_G aA'bB'c \\
&\Rightarrow_G aAbB'c \\
&\Rightarrow_G aAbBc \\
&\Rightarrow_G^* a^n Ab^n Bc^n \\
&\Rightarrow_G a^n \bar{A}b^n Bc^n \\
&\Rightarrow_G a^n \bar{A}b^n \bar{B}c^n \\
&\Rightarrow_G a^n b^n \bar{B}c^n \\
&\Rightarrow_G a^n b^n c^n
\end{aligned}
$$

where $n \geq 1$. The empty string is generated by

$$S \Rightarrow_G AB \Rightarrow_G \bar{A}B \Rightarrow_G \bar{A}\bar{B} \Rightarrow_G \bar{B} \Rightarrow_G \varepsilon$$

Based on the previous observations, we see that G generates the non-context-free language $\{a^n b^n c^n \mid n \geq 0\}$. □

The language family generated by generalized one-sided forbidding grammars of degree (m, n) is denoted by $\mathbf{GOF}(m, n)$. Furthermore, set

$$\mathbf{GOF} = \bigcup_{m,n \geq 0} \mathbf{GOF}(m, n)$$

6.6.2 Generative Power

In this section, we establish the generative power of generalized one-sided forbidding grammars. More specifically, we prove results (I) through (IV), given next.

(I) Generalized one-sided forbidding grammars of degrees $(n, 0)$ or $(0, n)$, for any non-negative integer n, characterize only the family of context-free languages (Theorem 6.6.6).

(II) Generalized one-sided forbidding grammars of degree $(1, 1)$ generate a proper superfamily of the family of context-free languages (Theorem 6.6.7).

(III) Generalized one-sided forbidding grammars of degrees $(1, 2)$ or $(2, 1)$ characterize the family of recursively enumerable languages (Theorem 6.6.10).

(IV) Generalized one-sided forbidding grammars with the set of left forbidding rules coinciding with the set of right forbidding rules characterize only the family of context-free languages (Theorem 6.6.16).

First, we consider generalized one-sided forbidding grammars of degrees $(n, 0)$ and $(0, n)$, where $n \geq 0$.

Lemma 6.6.4. $\mathbf{GOF}(n, 0) = \mathbf{CF}$ *for every* $n \geq 0$.

Proof. Let n be a non-negative integer. As any context-free grammar is also a generalized one-sided forbidding grammar in which the empty sets are attached to each of its rules, the inclusion $\mathbf{CF} \subseteq \mathbf{GOF}(n, 0)$ holds. To establish the other inclusion, $\mathbf{GOF}(n, 0) \subseteq \mathbf{CF}$, let $G = (N, T, P_L, P_R, S)$ be a generalized one-sided forbidding grammar of degree $(n, 0)$, and let

$$H = (N, T, P', S)$$

be a context-free grammar with

$$P' = \{A \to x \mid (A \to x, F) \in P_L \cup P_R\}$$

As any successful derivation in G is also a successful derivation in H, the inclusion $L(G) \subseteq L(H)$ holds. On the other hand, let $w \in L(H)$ be a string successfully generated by H. Then, it is well-known that there exists a successful leftmost derivation of w in H (see Theorem 3.3.13). Such a leftmost derivation is, however, also possible in G because the leftmost nonterminal can always be rewritten. Thus, the other inclusion $L(H) \subseteq L(G)$ holds as well, which completes the proof. □

Lemma 6.6.5. $\mathbf{GOF}(0, n) = \mathbf{CF}$ *for every* $n \geq 0$.

Proof. This lemma can be proved by analogy with the proof of Lemma 6.6.4. The only difference is that instead of leftmost derivations, we use rightmost derivations. □

Theorem 6.6.6. $\mathbf{GOF}(n, 0) = \mathbf{GOF}(0, n) = \mathbf{CF}$ *for every* $n \geq 0$.

Proof. This theorem follows from Lemmas 6.6.4 and 6.6.5. □

Next, we consider generalized one-sided forbidding grammars of degree $(1, 1)$.

Theorem 6.6.7. $\mathbf{CF} \subset \mathbf{GOF}(1, 1)$

Proof. This theorem follows from Example 6.6.3. □

In what follows, we prove that generalized one-sided forbidding grammars of degrees $(1, 2)$ and $(2, 1)$ are computationally complete—that is, they characterize the family of recursively enumerable languages.

Lemma 6.6.8. $\mathbf{GOF}(2, 1) = \mathbf{RE}$

Proof. The inclusion $\mathbf{GOF}(2, 1) \subseteq \mathbf{RE}$ follows from Church's thesis, so we only prove that $\mathbf{RE} \subseteq \mathbf{GOF}(2, 1)$.

Let $K \in \mathbf{RE}$. By Theorem 4.1.4, there is a phrase-structure grammar $G = (N, T, P, S)$ in the Penttonen normal form such that $L(G) = K$. We next construct a generalized one-sided forbidding grammar H of degree $(2, 1)$ such that $L(H) = L(G)$. Set

$$W = \{\langle r, i\rangle \mid r = (AB \to AC) \in P, A, B, C \in N, i = 1, 2\}$$

Let S' and # be two new symbols. Without any loss of generality, assume that N, W, and $\{S', \#\}$ are pairwise disjoint. Construct

$$H = \left(N', T, P_L, P_R, S'\right)$$

as follows. Initially, set $N' = N \cup W \cup \{S', \#\}$, $P_L = \emptyset$, and $P_R = \emptyset$. Perform (1) through (5), given next.

(1) Add $(S' \to \#S, \emptyset)$ to P_L.
(2) For each $A \to a \in P$, where $A \in N$ and $a \in T$, add $(A \to a, N')$ to P_R.
(3) For each $A \to y \in P$, where $A \in N$ and $y \in \{\varepsilon\} \cup NN$, add $(A \to y, \emptyset)$ to P_L.
(4) For each $r = (AB \to AC) \in P$, where $A, B, C \in N$,

  (4.1) add $(B \to \langle r, 1\rangle\langle r, 2\rangle, W)$ to P_L;
  (4.2) add $(\langle r, 2\rangle \to C, N'W - \{A\langle r, 1\rangle\})$ to P_L;
  (4.3) add $(\langle r, 1\rangle \to \varepsilon, W)$ to P_R.

(5) Add $(\# \to \varepsilon, N')$ to P_R.

Before proving that $L(H) = L(G)$, let us informally describe (1) through (5). G generates each string of $L(G)$ by simulating the corresponding derivations of H as follows. Every derivation is started by $(S' \to \#S, \emptyset) \in P_L$, introduced in (1). Context-free rules of the form $A \to y$, where $A \in N$ and $y \in T \cup \{\varepsilon\} \cup NN$, are simulated by rules from (2) and (3). Since rules introduced in (2) forbid the presence of nonterminals to the right of the rewritten symbol, every sentential form of H is of the form xy, where $x \in N'^*$ and $y \in T^*$—that is, it begins with a string of nonterminals and ends with a string of terminals. In this way, no terminal is followed by a nonterminal. This is needed to properly simulate context-sensitive rules, described next. Rules of the form $AB \to AC$, where $A, B, C \in N$, are simulated in a three-step way by rules from (4). Observe that the forbidding context of rules from (4.2) ensures that the rewritten symbol B is directly preceded by A. Indeed, if B is not directly preceded by A, then a string different from $A\langle r, 1\rangle$, where $r = (AB \to AC)$, occurs to the left of $\langle r, 2\rangle$ (recall that # is at the beginning of every sentential form having at least one nonterminal). The end-marker # is erased at the end of every successful derivation by $(\# \to \varepsilon, N') \in P_R$, introduced in (5).

To establish $L(H) = L(G)$, we prove three claims. Claim 1 demonstrates that every $y \in L(G)$ can be generated by G in two stages; first, only nonterminals are generated, and then, all nonterminals are rewritten to terminals. Claim 2 shows how such derivations of G are simulated by H. Finally, Claim 3 shows how derivations of H are simulated by G.

Claim 1. For every $y \in L(G)$, there exists a derivation of the form $S \Rightarrow_G^* x \Rightarrow_G^* y$, where $x \in N^+$, and during $x \Rightarrow_G^* y$, only rules of the form $A \to a$, where $A \in N$ and $a \in T$, are applied.

Proof. Let $y \in L(G)$. Since there are no rules in P with symbols from T on their left-hand sides, we can always rearrange all the applications of the rules occurring in $S \Rightarrow_G^* y$ so the claim holds. □

Claim 2. If $S \Rightarrow_G^n x$, where $x \in N^*$, for some $n \geq 0$, then $S' \Rightarrow_H^* \#x$.

Proof. This claim is established by induction on $n \geq 0$.

Basis. Let $n = 0$. Then, for $S \Rightarrow_G^0 S$, there is $S' \Rightarrow_H \#S$ by the rule from (1), so the basis holds.

Induction Hypothesis. Suppose that there exists $n \geq 0$ such that the claim holds for all derivations of length ℓ, where $0 \leq \ell \leq n$.

Induction Step. Consider any derivation of the form

$$S \Rightarrow_G^{n+1} w$$

where $w \in N^*$. Since $n + 1 \geq 1$, this derivation can be expressed as

$$S \Rightarrow_G^n x \Rightarrow_G w$$

for some $x \in N^+$. By the induction hypothesis, $S' \Rightarrow_H^* \#x$.

Next, we consider all possible forms of $x \Rightarrow_G w$, covered by the following two cases—(i) and (ii).

(i) Let $A \rightarrow y \in P$ and $x = x_1 A x_2$, where $A \in N$, $x_1, x_2 \in N^*$, and $y \in \{\varepsilon\} \cup NN$. Then,

$$x_1 A x_2 \Rightarrow_G x_1 y x_2$$

By (2), $(A \rightarrow y, \emptyset) \in P_L$, so

$$\#x_1 A x_2 \Rightarrow_H \#x_1 y x_2$$

which completes the induction step for (i).

(ii) Let $AB \rightarrow AC \in P$ and $x = x_1 AB x_2$, where $A, B, C \in N$ and $x_1, x_2 \in N^*$. Then,

$$x_1 AB x_2 \Rightarrow_G x_1 BC x_2$$

Let $r = (AB \rightarrow AC)$. By (4.1), $(B \rightarrow \langle r, 1\rangle\langle r, 2\rangle, W) \in P_L$. Since $\mathrm{sub}(\#x_1 A) \cap W = \emptyset$,

$$\#x_1 AB x_2 \Rightarrow_H \#x_1 A\langle r, 1\rangle\langle r, 2\rangle x_2$$

By (4.2), $(\langle r, 2\rangle \rightarrow C, N'W - \{A\langle r, 1\rangle\}) \in P_L$. Since $\mathrm{sub}(\#x_1 A\langle r, 1\rangle) \cap (N'W - \{A\langle r, 1\rangle\}) = \emptyset$,

$$\#x_1 A\langle r, 1\rangle\langle r, 2\rangle x_2 \Rightarrow_H \#x_1 A\langle r, 1\rangle C x_2$$

By (4.3), $(\langle r, 1\rangle \rightarrow \varepsilon, W) \in P_R$. Since $\text{sub}(Cx_2) \cap W = \emptyset$,

$$\#x_1 A\langle r, 1\rangle C x_2 \Rightarrow_H \#x_1 A C x_2$$

which completes the induction step for (ii).

Observe that cases (i) and (ii) cover all possible forms of $x \Rightarrow_G w$. Thus, the claim holds. □

Set $V = N \cup T$ and $V' = N' \cup T$. Define the homomorphism τ from V'^* to V^* as $\tau(X) = X$ for all $X \in V$, $\tau(S') = S$, $\tau(\#) = \varepsilon$, $\tau(\langle r, 1\rangle) = \varepsilon$ and $\tau(\langle r, 2\rangle) = B$ for all $r = (AB \rightarrow AC) \in P$.

Claim 3. If $S' \Rightarrow_H^n x$, where $x \in V'^*$, for some $n \geq 1$, then $S \Rightarrow_G^* \tau(x)$ and x is of the form uv, where $u \in \{\varepsilon\} \cup \{\#\}(N \cup W)^*$, $v \in T^*$, and if $\langle r, 2\rangle \in \text{sub}(u)$, then this occurrence of $\langle r, 2\rangle$ is directly preceded by $\langle r, 1\rangle$.

Proof. This claim is established by induction on $n \geq 0$.

Basis. Let $n = 1$. Then, for $S' \Rightarrow_H \#S$ by the rule from (1), there is $S \Rightarrow_G^0 S$. Since $\#S$ is of the required form, the basis holds.

Induction Hypothesis. Suppose that there exists $n \geq 1$ such that the claim holds for all derivations of length ℓ, where $1 \leq \ell \leq n$.

Induction Step. Consider any derivation of the form

$$S' \Rightarrow_H^{n+1} w$$

where $w \in V'^*$. Since $n + 1 \geq 1$, this derivation can be expressed as

$$S' \Rightarrow_H^n x \Rightarrow_H w$$

for some $x \in V'^+$. By the induction hypothesis, $S \Rightarrow_G^* \tau(x)$ and x is of the form uv, where $u \in \{\varepsilon\} \cup \{\#\}(N \cup W)^*$, $v \in T^*$, and if $\langle r, 2\rangle \in \text{sub}(u)$, then this occurrence of $\langle r, 2\rangle$ is directly preceded by $\langle r, 1\rangle$.

Next, we consider all possible forms of $x \Rightarrow_H w$, covered by the following six cases—(i) through (vi).

(i) *Application of a rule from (2).* Let $x = x_1 A x_2$ and $(A \rightarrow a, N') \in P_R$ so that $N' \cap \text{sub}(x_2) = \emptyset$, where $x_1, x_2 \in V'^*$, $A \in N$, and $a \in T$. Then,

$$x_1 A x_2 \Rightarrow_H x_1 a x_2$$

Clearly, $x_1 a x_2$ is of the required form. By the induction hypothesis, $\tau(x) = \tau(x_1)A\tau(x_2)$. By (2), $A \rightarrow a \in P$, so

$$\tau(x_1)A\tau(x_2) \Rightarrow_G \tau(x_1)a\tau(x_2)$$

which completes the induction step for (i).

(ii) *Application of a rule from (3).* Let $x = x_1 A x_2$ and $(A \rightarrow y, \emptyset) \in P_L$, where $x_1, x_2 \in V'^*$, $A \in N$, and $y \in \{\varepsilon\} \cup NN$. Then,

$$x_1 A x_2 \Rightarrow_H x_1 y x_2$$

Clearly, $x_1 y x_2$ is of the required form. By the induction hypothesis, $\tau(x) = \tau(x_1) A \tau(x_2)$. By (3), $A \rightarrow y \in P$, so

$$\tau(x_1) A \tau(x_2) \Rightarrow_G \tau(x_1) y \tau(x_2)$$

which completes the induction step for (ii).

(iii) *Application of a rule from (4.1).* Let $x = x_1 B x_2$ and $(B \rightarrow \langle r, 1\rangle\langle r, 2\rangle, W) \in P_L$ so that $W \cap \text{sub}(x_1) = \emptyset$, where $x_1, x_2 \in V'^*$, $B \in N$, and $r = (AB \rightarrow AC) \in P$. Then,

$$x_1 B x_2 \Rightarrow_H x_1 \langle r, 1\rangle\langle r, 2\rangle x_2$$

Clearly, $x_1 \langle r, 1\rangle\langle r, 2\rangle x_2$ is of the required form, and since $\tau(x_1 \langle r, 1\rangle\langle r, 2\rangle x_2) = \tau(x_1) B \tau(x_2)$, the induction step for (iii) follows directly from the induction hypothesis.

(iv) *Application of a rule from (4.2).* Let $x = x_1 \langle r, 2\rangle x_2$ and $(\langle r, 2\rangle \rightarrow C, N'W - \{A\langle r, 1\rangle\}) \in P_L$ so that $(N'W - \{A\langle r, 1\rangle\}) \cap \text{sub}(x_1) = \emptyset$, where $x_1, x_2 \in V'^*$, $C \in N$, and $r = (AB \rightarrow AC) \in P$. By the induction hypothesis, since $\langle r, 2\rangle \in \text{sub}(x)$, x is of the form $\#x_1' \langle r, 1\rangle\langle r, 2\rangle x_2$, where $x_1' \in V'^*$. Furthermore, since $(N'W - \{A\langle r, 1\rangle\}) \cap \text{sub}(x_1) = \emptyset$, x_1' is of the form $x_1'' A$. So,

$$\#x_1'' A \langle r, 1\rangle\langle r, 2\rangle x_2 \Rightarrow_H \#x_1'' A \langle r, 1\rangle C x_2$$

Clearly, $\#x_1'' A \langle r, 1\rangle C x_2$ is of the required form. By the induction hypothesis, $\tau(x) = \tau(x_1'') AB \tau(x_2)$. By (4.2), $AB \rightarrow AC \in P$, so

$$\tau(x_1'') AB \tau(x_2) \Rightarrow_G \tau(x_1'') AC \tau(x_2)$$

which completes the induction step for (iv).

(v) *Application of a rule from (4.3).* Let $x = x_1 \langle r, 1\rangle x_2$ and $(\langle r, 1\rangle \rightarrow \varepsilon, W) \in P_R$ so that $W \cap \text{sub}(x_2) = \emptyset$, where $x_1, x_2 \in V'^*$ and $r = (AB \rightarrow AC) \in P$. Then,

$$x_1 \langle r, 1\rangle x_2 \Rightarrow_H x_1 x_2$$

Clearly, $x_1 x_2$ is of the required form. Since $\tau(x_1 \langle r, 1\rangle x_2) = \tau(x_1 x_2)$, the induction step for (v) follows directly from the induction hypothesis.

(vi) *Application of a rule from (5).* Let $x = \#x'$ and $(\# \rightarrow \varepsilon, N') \in P_R$ so that $N' \cap \text{sub}(x') = \emptyset$ (this implies that $x' \in T^*$). Then, $\#x' \Rightarrow_H x'$. Clearly, x' is of the required form. Since $x' \in T^*$, $\tau(x) = x$, so the induction step for (vi) follows directly from the induction hypothesis.

Observe that cases (i) through (vi) cover all possible forms of $x \Rightarrow_H w$. Thus, the claim holds. □

We next establish $L(H) = L(G)$. Let $y \in L(G)$. Then, by Claim 1, there exists a derivation $S \Rightarrow_G^* x \Rightarrow_G^* y$ such that $x \in N^+$ and during $x \Rightarrow_G^* y$, G uses only rules of the form $A \rightarrow a$, where $A \in N$ and $a \in T$. By Claim 2, $S' \Rightarrow_H^* \#x$. Let $x = X_1X_2 \cdots X_k$ and $y = a_1a_2 \cdots a_k$, where $k = |x|$. Since $x \Rightarrow_G^* y$, $X_i \rightarrow a_i \in P$ for $i = 1, 2, \dots, k$. By (2), $(X_i \rightarrow a_i, N') \in P_R$ for $i = 1, 2, \dots, k$. Then,

$$\begin{aligned}\#X_1 \cdots X_{k-1}X_k &\Rightarrow_H \#X_1 \cdots X_{k-1}a_k \\ &\Rightarrow_H \#X_1 \cdots a_{k-1}a_k \\ &\vdots \\ &\Rightarrow_H \#a_1a_2 \cdots a_k\end{aligned}$$

By (5), $(\# \rightarrow \varepsilon, N') \in P_R$. Since $y \in T^*$, $\#y \Rightarrow_H y$. Consequently, $y \in L(G)$ implies that $y \in L(H)$, so $L(G) \subseteq L(H)$.

Consider Claim 3 for $x \in T^*$. Then, $x \in L(H)$ implies that $\tau(x) = x \in L(G)$, so $L(H) \subseteq L(G)$. As $L(G) \subseteq L(H)$ and $L(H) \subseteq L(G)$, $L(H) = L(G)$. Since H is of degree $(2, 1)$, the theorem holds. □

Lemma 6.6.9. $\textbf{GOF}(1, 2) = \textbf{RE}$

Proof. This lemma can be proved by analogy with the proof of Lemma 6.6.8. First, by modifying the proofs given in [45], we can convert any phrase-structure grammar into an equivalent phrase-structure grammar $G = (N, T, P, S)$, where every rule in P is in one of the following four forms:

(i) $BA \rightarrow CA$,
(ii) $A \rightarrow BC$,
(iii) $A \rightarrow a$,
(iv) $A \rightarrow \varepsilon$, where $A, B, C \in N$, and $a \in T$.

Notice that this normal form differs from the Penttonen normal only by the form of context-sensitive rules. Then, in the proof Lemma 6.6.8, we accordingly modify the rules introduced to P_L and P_R so that the resulting grammar is of degree $(1, 2)$ instead of $(2, 1)$. A rigorous proof of this theorem is left to the reader. □

Theorem 6.6.10. $\textbf{GOF}(1, 2) = \textbf{GOF}(2, 1) = \textbf{RE}$

Proof. This theorem follows from Lemmas 6.6.8 and 6.6.9. □

From Theorem 6.6.10, we obtain the following three corollaries.

Corollary 6.6.11. **GOF**(m, n) = **RE** *for every* $m \geq 2$ *and* $n \geq 1$. □

Corollary 6.6.12. **GOF**(m, n) = **RE** *for every* $m \geq 1$ *and* $n \geq 2$. □

Corollary 6.6.13. **GOF** = **RE** □

We next turn our attention to generalized one-sided forbidding grammars with the set of left forbidding rules coinciding with the set of right forbidding rules.

Lemma 6.6.14. *Let K be a context-free language. Then, there exists a generalized one-sided forbidding grammar, $G = (N, T, P_L, P_R, S)$, satisfying $P_L = P_R$ and $L(G) = K$.*

Proof. Let K be a context-free language. Then, there exists a context-free grammar, $H = (N, T, P, S)$, such that $L(H) = K$. Define the generalized one-sided forbidding grammar

$$G = \big(N, T, P', P', S\big)$$

with

$$P' = \big\{(A \to x, \emptyset, \emptyset) \mid A \to x \in P\big\}$$

Clearly, $L(G) = L(H) = K$, so the lemma holds. □

Lemma 6.6.15. *Let $G = (N, T, P_L, P_R, S)$ be a generalized one-sided forbidding grammar satisfying $P_L = P_R$. Then, $L(G)$ is context-free.*

Proof. Let $G = (N, T, P_L, P_R, S)$ be a generalized one-sided forbidding grammar satisfying $P_L = P_R$. Define the context free grammar $H = (N, T, P', S)$ with

$$P' = \big\{A \to x \mid (A \to x, \emptyset, F) \in P_L\big\}$$

Observe that since $P_L = P_R$, it is sufficient to consider just the rules from P_L. As any successful derivation in G is also a successful derivation in H, the inclusion $L(G) \subseteq L(H)$ holds. On the other hand, let $w \in L(H)$ be a string successfully generated by H. Then, it is well-known that there exists a successful leftmost derivation of w in H (see Theorem 3.3.13). Observe that such a leftmost derivation is also possible in G because the leftmost nonterminal can always be rewritten. Indeed, P' contains only rules originating from the rules in P_L and all rules in P_L are applicable to the leftmost nonterminal. Thus, the other inclusion $L(H) \subseteq L(G)$ holds as well, which completes the proof. □

Theorem 6.6.16. *A language K is context-free if and only if there is a generalized one-sided forbidding grammar, $G = (N, T, P_L, P_R, S)$, satisfying $K = L(G)$ and $P_L = P_R$.*

Proof. This theorem follows from Lemmas 6.6.14 and 6.6.15. □

In the conclusion of this section, we first describe relations of generalized one-sided forbidding grammars to other variants of forbidding grammars. Then, we state several open problems related to the achieved results.

We begin by considering generalized forbidding grammars (see Sect. 4.4).

Corollary 6.6.17. $\mathbf{GF} = \mathbf{GOF}$

Proof. This corollary follows from Corollary 6.6.13 in the previous section and from Theorem 4.4.5, which says that $\mathbf{GF} = \mathbf{RE}$. □

Next, we move to forbidding grammars (see Sect. 4.3).

Corollary 6.6.18. $\mathbf{For} \subset \mathbf{GOF}$

Proof. This corollary follows from Corollary 6.6.13 in the previous section and from Theorem 4.3.9, which says that $\mathbf{For} \subset \mathbf{RE}$. □

From the definition of a one-sided forbidding grammar, we immediately obtain the following corollary.

Corollary 6.6.19. $\mathbf{GOF}(1, 1) = \mathbf{OFor}$ □

The next three open problem areas are related to the achieved results.

Open Problem 6.6.20. By Theorem 6.6.19, $\mathbf{GOF}(1, 1) = \mathbf{OFor}$. However, recall that it is not known whether $\mathbf{OFor} = \mathbf{RE}$ or $\mathbf{OFor} \subset \mathbf{RE}$ (see Sect. 6.2). Are generalized one-sided forbidding grammars of degree $(1, 1)$ capable of generating all recursively enumerable languages? □

Open Problem 6.6.21. Let $G = (N, T, P_L, P_R, S)$ be a generalized one-sided forbidding grammar. If $(A \rightarrow x, F) \in P_L \cup P_R$ implies that $|x| \geq 1$, then G is said to be *propagating*. What is the generative power of propagating generalized one-sided forbidding grammars? Do they characterize the family of context-sensitive languages? □

Open Problem 6.6.22. By Theorem 6.6.10, the degrees $(2, 1)$ or $(1, 2)$ suffice to characterize the family of recursively enumerable languages. Can we also place a limitation on the number of nonterminals or on the number of rules with nonempty forbidding contexts? Recall that in terms of generalized forbidding grammars, a limitation like this has been achieved (see Sect. 4.4.2). □

6.7 LL One-Sided Random Context Grammars

In this chapter, have introduced and studied one-sided random context grammars from a purely theoretical viewpoint. From a more practical viewpoint, however, it is also desirable to make use of them in such grammar-based application-oriented fields as syntax analysis (see [1, 2]). An effort like this obviously gives rise to introducing and investigating their parsing-related variants, such as LL versions—the subject of the present section.

LL one-sided random context grammars, introduced in this section, represent ordinary one-sided random context grammars restricted by analogy with LL requirements placed upon LL context-free grammars. That is, for every positive integer k, (1) LL(k) one-sided random context grammars always rewrite the leftmost nonterminal in the current sentential form during every derivation step, and (2) if there are two or more applicable rules with the same nonterminal on their left-hand sides, then the sets of all terminal strings of length k that can begin a string obtained by a derivation started by using these rules are disjoint. The class of LL grammars is the union of all LL(k) grammars, for every $k \geq 1$.

Recall that one-sided random context grammars characterize the family of recursively enumerable languages (see Theorem 6.2.4). Of course, it is natural to ask whether LL one-sided random context grammars generate the family of LL context-free languages or whether they are more powerful. As its main result, this section proves that the families of LL one-sided random context languages and LL context-free languages coincide. Indeed, it describes transformations that convert any LL(k) one-sided random context grammar to an equivalent LL(k) context-free grammar and conversely.

In fact, we take a closer look at the generation of languages by both versions of LL grammars. That is, we demonstrate an advantage of LL one-sided random context grammars over LL context-free grammars. More precisely, for every $k \geq 1$, we present a specific LL(k) one-sided random context grammar G and prove that every equivalent LL(k) context-free grammar has necessarily more nonterminals or rules than G. Thus, to rephrase this result more broadly and pragmatically, we actually show that LL(k) one-sided random context grammars can possibly allow us to specify LL(k) languages more succinctly and economically than LL(k) context-free grammars do.

This section is divided into three subsections. First, Sect. 6.7.1 defines LL one-sided forbidding grammars. Then, Sect. 6.7.2 gives a motivational example. After that, Sect. 6.7.3 proves the main result sketched above, and formulates three open problems.

6.7.1 Definitions

In this section, we define LL context-free grammars and LL one-sided random context grammars. Since we pay a principal attention to context-free and one-sided random context grammars working in the leftmost way, in what follows, by a context-free and one-sided random context grammar, respectively, we always mean a context-free and one-sided random context grammar working in the leftmost way, respectively (see Sects. 3.3 and 6.5). In terms of one-sided random context grammars, by this leftmost way, we mean the type-1 leftmost derivations (see Sect. 6.5.1).

We begin by defining the LL(k) property of context-free grammars, for every $k \geq 1$. To simplify the definition, we end all sentential forms by k end-markers, denoted by \$, and we extend the derivation relation to $V^*\{\$\}^k$ in the standard way—that is, $u\$^k \Rightarrow v\$^k$ if and only if $u \Rightarrow v$.

Definition 6.7.1 (see [1]). Let $G = (N, T, P, S)$ be a context-free grammar and $\$ \notin N \cup T$ be a symbol. For every $r = (A \rightarrow x) \in P$ and $k \geq 1$, define

$$\mathrm{Predict}_k(r) \subseteq T^*\{\$\}^*$$

as follows: $\gamma \in \mathrm{Predict}_k(r)$ if and only if $|\gamma| = k$ and

$$S\$^k \;{}_{\mathrm{lm}}\!\Rightarrow_G^* \; uAv\$^k \;{}_{\mathrm{lm}}\!\Rightarrow_G \; uxv\$^k \;{}_{\mathrm{lm}}\!\Rightarrow_G^* \; u\gamma w$$

where $u \in T^*$, $v, x \in V^*$, and $w \in V^*\{\$\}^*$. □

Using the above definition, we next define LL context-free grammars.

Definition 6.7.2 (see [1]). Let $G = (N, T, P, S)$ be a context-free grammar. G is an *LL(k) context-free grammar*, where $k \geq 1$, if it satisfies the following condition: if $p = (A \rightarrow x) \in P$ and $r = (A \rightarrow y) \in P$ such that $x \neq y$, then

$$\mathrm{Predict}_k(p) \cap \mathrm{Predict}_k(r) = \emptyset$$

If there exists $k \geq 1$ such that G is an LL(k) context-free grammar, then G is an *LL context-free grammar*. □

Next, we move to the definition of LL one-sided random context grammars. To simplify this definition, we first introduce the notion of leftmost applicability. Informally, a random context rule r is *leftmost-applicable* to a sentential form y if the leftmost nonterminal in y can be rewritten by applying r.

Definition 6.7.3. Let $G = (N, T, P_L, P_R, S)$ be a one-sided random context grammar. A rule $(A \rightarrow x, U, W) \in P_L \cup P_R$ is *leftmost-applicable* to $y \in V^*$ if and only if $y = uAv$, where $u \in T^*$ and $v \in V^*$, and

$$(A \rightarrow x, U, W) \in P_L, U \subseteq \mathrm{alph}(u) \text{ and } W \cap \mathrm{alph}(u) = \emptyset$$

or

$$(A \rightarrow x, U, W) \in P_R, U \subseteq \mathrm{alph}(v) \text{ and } W \cap \mathrm{alph}(v) = \emptyset$$ □

Let us note that the leftmost property of the direct derivation relation has significant consequences to the applicability of rules from P_L. We point out these consequences later in Lemma 6.7.7.

By analogy with the Predict set in context-free grammars, we introduce such a set to one-sided random context grammars. It is then used to define LL one-sided random context grammars. Notice that as opposed to context-free grammars, in the current sentential form, the applicability of a random context rule $(A \rightarrow x, U, W)$

depends not only on the presence of A but also on the presence and absence of symbols from U and W, respectively. This has to be properly reflected in the definition.

Definition 6.7.4. Let $G = (N, T, P_L, P_R, S)$ be a one-sided random context grammar and $\$ \notin N \cup T$ be a symbol. For every $r = (A \rightarrow x, U, W) \in P_L \cup P_R$ and $k \geq 1$, define

$$\text{Predict}_k(r) \subseteq T^*\{\$\}^*$$

as follows: $\gamma \in \text{Predict}_k(r)$ if and only if $|\gamma| = k$ and

$$S\$^k \ {}_{\text{lm}}{}^{1}\!\!\Rightarrow^*_G uAv\$^k \ {}_{\text{lm}}{}^{1}\!\!\Rightarrow_G uxv\$^k \ {}_{\text{lm}}{}^{1}\!\!\Rightarrow^*_G u\gamma w$$

where $u \in T^*$, $v, x \in V^*$, $w \in V^*\{\$\}^*$, and r is leftmost-applicable to uAv. □

Making use of the above definition, we next define LL one-sided random context grammars.

Definition 6.7.5. Let $G = (N, T, P_L, P_R, S)$ be a one-sided random context grammar. G is an *LL(k) one-sided random context grammar*, where $k \geq 1$, if it satisfies the following condition: for any $p = (A \rightarrow x, U, W), r = (A \rightarrow x', U', W') \in P_L \cup P_R$ such that $p \neq r$, if $\text{Predict}_k(p) \cap \text{Predict}_k(r) \neq \emptyset$, then there is no $w \in V^*$ such that $S \ {}_{\text{lm}}{}^{1}\!\!\Rightarrow^*_G w$ with both p and r being leftmost-applicable to w.

If there exists $k \geq 1$ such that G is an LL(k) one-sided random context grammar, then G is an *LL one-sided random context grammar.* □

6.7.2 A Motivational Example

In this short section, we give an example of an LL(k) one-sided random context grammar, for every $k \geq 1$. In this example, we argue that LL(k) one-sided random context grammars can describe some languages more succinctly than LL(k) context-free grammars.

Example 6.7.6. Let k be a positive integer and $G = (N, T, \emptyset, P_R, S)$ be a one-sided random context grammar, where $N = \{S\}$, $T = \{a, b, c, d\}$, and

$$P_R = \{(S \rightarrow d^{k-1}c, \emptyset, \emptyset), (S \rightarrow d^{k-1}aSS, \emptyset, \{S\}), (S \rightarrow d^{k-1}bS, \{S\}, \emptyset)\}$$

Notice that G is an LL(k) one-sided random context grammar. Observe that the second rule can be applied only to a sentential form containing exactly one occurrence of S,while the third rule can be applied only to a sentential form containing at least two occurrences of S. The generated language $L(G)$ can be described by the following expression

$$\left(d^{k-1}a(d^{k-1}b)^*d^{k-1}c\right)^*d^{k-1}c$$

Next, we argue that $L(G)$ cannot be generated by any LL(k) context-free grammar having a single nonterminal and at most three rules. This shows us that for some languages, LL(k) one-sided random context grammars need fewer rules or nonterminals than LL(k) context-free grammars do to describe them.

We proceed by contradiction. Suppose that there exists an LL(k) context-free grammar

$$H = \left(\{S\}, T, P', S\right)$$

such that $L(H) = L(G)$ and $\text{card}(P') \leq 3$. Observe that since there is only a single nonterminal, to satisfy the LL(k) property, the right-hand side of each rule in P' has to start with a string of terminals. Furthermore, since there is only a single nonterminal and all the strings in $L(G)$ begin with either $d^{k-1}a$ or $d^{k-1}c$, each rule has to begin with $d^{k-1}a$ or $d^{k-1}c$. Therefore, to satisfy the LL(k) property, there can be at most two rules. However, then at least one of these rules has to have b somewhere on its right-hand side, so the number of occurrences of b depends on the number of occurrences of a or c. Thus, $L(H) \neq L(G)$, which contradicts $L(H) = L(G)$. Hence, there is no LL(k) context-free grammar that generates $L(G)$ with only a single nonterminal and at most three rules. □

6.7.3 *Generative Power*

In this section, we prove that LL one-sided random context grammars characterize the family of LL context-free languages.

First, we establish a normal form for LL one-sided random context grammars, which greatly simplifies the proof of the subsequent Lemma 6.7.8. In this normal form, an LL one-sided random context grammar does not have any left random context rules.

Lemma 6.7.7. *For every LL(k) one-sided random context grammar G, where $k \geq 1$, there is an LL(k) one-sided random context grammar $H = (N, T, \emptyset, P_R, S)$ such that $L(H) = L(G)$.*

Proof. Let $H = (N, T, P_L, P_R, S)$ be an LL(k) one-sided random context grammar, where $k \geq 1$. Construct the one-sided random context grammar

$$H = \left(N, T, \emptyset, P'_R, S\right)$$

where

$$P'_R = P_R \cup \{(A \to x, \emptyset, \emptyset) \mid (A \to x, \emptyset, W) \in P_L\}$$

Notice that rules from P_L are simulated by right random context rules from P'_R. In a greater detail, let $r = (A \to x, U, W) \in P_L$. Observe that if $U \neq \emptyset$, then r is never applicable in G, so if this is the case, we do not add a rule corresponding to r to P'_R. Furthermore, observe that we do not have to check the absence of nonterminals from W because there are no nonterminals to the left of the leftmost nonterminal in any sentential form.

Clearly, $L(H) = L(G)$ and H is an LL(k) one-sided random context grammar. Hence, the lemma holds. □

To establish the equivalence between LL(k) one-sided random context grammars and LL(k) context-free grammars, we first show how to transform any LL(k) one-sided random context grammar into an equivalent LL(k) context-free grammar. Our transformation is based on the construction used in the proof of Lemma 6.5.3.

Let G be a one-sided random context grammar. By analogy with context-free grammars, in the remainder of this section, we write $x \Rightarrow_G y\ [r]$ to denote that in this derivation step, rule r was used.

Lemma 6.7.8. *For every LL(k) one-sided random context grammar G, where $k \geq 1$, there is an LL(k) context-free grammar H such that $L(H) = L(G)$.*

Proof. Let $G = (N, T, P_L, P_R, S)$ be an LL(k) one-sided random context grammar, where $k \geq 1$. Without any loss of generality, making use of Lemma 6.7.7, we assume that $P_L = \emptyset$. In what follows, angle brackets $\langle$ and $\rangle$ are used to incorporate more symbols into a single compound symbol. Construct the context-free grammar

$$H = (N', T, P', \langle S, \emptyset \rangle)$$

in the following way. Initially, set

$$N' = \{\langle A, Q \rangle \mid A \in N, Q \subseteq N\}$$

and $P' = \emptyset$. Without any loss of generality, we assume that $N' \cap (N \cup T) = \emptyset$. Next, for each

$$(A \to y_0 Y_1 y_1 Y_2 y_2 \cdots Y_h y_h, U, W) \in P_R$$

where $y_i \in T^*$, $Y_j \in N$, for all i and j, $0 \leq i \leq h$, $1 \leq j \leq h$, for some $h \geq 0$, and for each $\langle A, Q \rangle \in N'$ such that $U \subseteq Q$ and $W \cap Q = \emptyset$, extend P' by adding

$$\begin{aligned}\langle A, Q\rangle \to y_0 &\langle Y_1, Q \cup \{Y_2, Y_3, \dots, Y_h\}\rangle y_1\\ &\langle Y_2, Q \cup \{Y_3, \dots, Y_h\}\rangle y_2\\ &\vdots\\ &\langle Y_h, Q\rangle y_h\end{aligned}$$

Before proving that $L(H) = L(G)$, let us give an insight into the construction. As G always rewrites the leftmost occurrence of a nonterminal, we use compound nonterminals of the form $\langle A, Q\rangle$ in H, where A is a nonterminal, and Q is a set of nonterminals that appear to the right of this occurrence of A. When simulating rules from P_R, the check for the presence and absence of symbols is accomplished by using Q. Also, when rewriting A in $\langle A, Q\rangle$ to some y, the compound nonterminals from N' are generated instead of nonterminals from N.

To establish $L(H) = L(G)$, we prove two claims. First, Claim 1 shows how derivations of G are simulated by H. Then, Claim 2 demonstrates the converse—that is, it shows how G simulates derivations of H.

Set $V = N \cup T$ and $V' = N' \cup T$. Define the homomorphism τ from V'^* to V^* as $\tau(\langle A, Q\rangle) = A$ for all $A \in N$ and $Q \subseteq N$, and $\tau(a) = a$ for all $a \in T$.

Claim 1. If $S \;{}_{\mathrm{lm}}^{\;1}\!\!\Rightarrow_G^m x$, *where* $x \in V^*$ *and* $m \geq 0$, *then* $\langle S, \emptyset\rangle \;{}_{\mathrm{lm}}\!\!\Rightarrow_H^* x'$, *where* $\tau(x') = x$ *and* x' *is of the form*

$$x' = x_0\langle X_1, \{X_2, X_3, \dots, X_n\}\rangle x_1\langle X_2, \{X_3, \dots, X_n\}\rangle x_2 \cdots \langle X_n, \emptyset\rangle x_n$$

where $X_i \in N$ *for* $i = 1, 2, \dots, n$ *and* $x_j \in T^*$ *for* $j = 0, 1, \dots, n$, *for some* $n \geq 0$.

Proof. This claim is established by induction on $m \geq 0$.

Basis. Let $m = 0$. Then, for $S \;{}_{\mathrm{lm}}^{\;1}\!\!\Rightarrow_G^0 S$, $\langle S, \emptyset\rangle \;{}_{\mathrm{lm}}\!\!\Rightarrow_H^0 \langle S, \emptyset\rangle$, so the basis holds.

Induction Hypothesis. Suppose that there exists $m \geq 0$ such that the claim holds for all derivations of length ℓ, where $0 \leq \ell \leq m$.

Induction Step. Consider any derivation of the form

$$S \;{}_{\mathrm{lm}}^{\;1}\!\!\Rightarrow_G^{m+1} w$$

where $w \in V^*$. Since $m + 1 \geq 1$, this derivation can be expressed as

$$S \;{}_{\mathrm{lm}}^{\;1}\!\!\Rightarrow_G^m x \;{}_{\mathrm{lm}}^{\;1}\!\!\Rightarrow_G w\ [r]$$

for some $x \in V^+$ and $r \in P_R$. By the induction hypothesis, $\langle S, \emptyset\rangle \;{}_{\mathrm{lm}}\!\!\Rightarrow_H^* x'$, where $\tau(x') = x$ and x' is of the form

$$x' = x_0\langle X_1, \{X_2, X_3, \dots, X_n\}\rangle x_1\langle X_2, \{X_3, \dots, X_n\}\rangle x_2 \cdots \langle X_n, \emptyset\rangle x_n$$

where $X_i \in N$ for $i = 1, 2, \ldots, n$ and $x_j \in T^*$ for $j = 0, 1, \ldots, n$, for some $n \geq 1$. As $x \ {}_{\text{lm}}\!\overset{1}{\Rightarrow}_G\ w$ $[r]$, $x = x_0 X_1 x_1 X_2 x_2 \cdots X_n x_n$, $r = (X_1 \rightarrow y, U, W)$, $U \subseteq \{X_2, X_3, \ldots, X_n\}$, $W \cap \{X_2, X_3, \ldots, X_n\} = \emptyset$, and $w = x_0 y x_1 X_2 x_2 \cdots X_n x_n$. By the construction of H, there is

$$r' = \big(\langle X_1, \{X_2, X_3, \ldots, X_n\}\rangle \rightarrow y'\big) \in P'$$

where $\tau(y') = y$. Then,

$$x' \ {}_{\text{lm}}\!\Rightarrow_H\ x_0 y' x_1 \langle X_2, \{X_3, \ldots, X_n\}\rangle x_2 \cdots \langle X_n, \emptyset\rangle x_n \ [r']$$

Since $w' = x_0 y' x_1 \langle X_2, \{X_3, \ldots, X_n\}\rangle x_2 \cdots \langle X_n, \emptyset\rangle x_n$ is of the required form and, moreover, $\tau(w') = w$, the induction step is completed. □

Claim 2. If $\langle S, \emptyset\rangle \ {}_{\text{lm}}\!\Rightarrow_H^m\ x$*, where* $x \in V'^*$ *and* $m \geq 0$*, then* $S \ {}_{\text{lm}}\!\overset{1}{\Rightarrow}{}_G^*\ \tau(x)$ *and* x *is of the form*

$$x = x_0 \langle X_1, \{X_2, X_3, \ldots, X_n\}\rangle x_1 \langle X_2, \{X_3, \ldots, X_n\}\rangle x_2 \cdots \langle X_n, \emptyset\rangle x_n$$

where $X_i \in N$ *for* $i = 1, 2, \ldots, n$ *and* $x_j \in T^*$ *for* $j = 0, 1, \ldots, n$*, for some* $n \geq 0$.

Proof. This claim is established by induction on $m \geq 0$.

Basis. Let $m = 0$. Then, for $\langle S, \emptyset\rangle \ {}_{\text{lm}}\!\Rightarrow_H^0\ \langle S, \emptyset\rangle$, $S \ {}_{\text{lm}}\!\overset{1}{\Rightarrow}{}_G^0\ S$, so the basis holds.

Induction Hypothesis. Suppose that there exists $m \geq 0$ such that the claim holds for all derivations of length ℓ, where $0 \leq \ell \leq m$.

Induction Step. Consider any derivation of the form

$$\langle S, \emptyset\rangle \ {}_{\text{lm}}\!\Rightarrow_H^{m+1}\ w$$

where $w \in V'^*$. Since $m + 1 \geq 1$, this derivation can be expressed as

$$\langle S, \emptyset\rangle \ {}_{\text{lm}}\!\Rightarrow_H^m\ x \ {}_{\text{lm}}\!\Rightarrow_H\ w \ [r']$$

for some $x \in V^+$ and $r' \in P'$. By the induction hypothesis, $S \ {}_{\text{lm}}\!\overset{1}{\Rightarrow}{}_G^*\ \tau(x)$ and x is of the form

$$x = x_0 \langle X_1, \{X_2, X_3, \ldots, X_n\}\rangle x_1 \langle X_2, \{X_3, \ldots, X_n\}\rangle x_2 \cdots \langle X_n, \emptyset\rangle x_n$$

where $X_i \in N$ for $i = 1, 2, \ldots, n$ and $x_j \in T^*$ for $j = 0, 1, \ldots, n$, for some $n \geq 0$. As $x \ {}_{\text{lm}}\!\Rightarrow_H\ w$ $[r']$,

$$r' = (\langle X_1, \{X_2, X_3, \ldots, X_n\}\rangle \rightarrow y') \in P'$$

where $y' \in V'^*$, and there is $r = (X_1 \rightarrow y, U, W) \in P_R$, where $U \subseteq \{X_2, X_3, \ldots, X_n\}$, $W \cap \{X_2, X_3, \ldots, X_n\} = \emptyset$, and $\tau(y') = y$. Then,

$$x_0 X_1 x_1 X_2 x_2 \cdots X_n x_n \ {}_{\mathrm{lm}}\!\overset{1}{\Rightarrow}_G\ x_0 y x_1 X_2 x_2 \cdots X_n x_n\ [r]$$

Since $x_0 y x_1 X_2 x_2 \cdots X_n x_n$ is of the required form and it equals $\tau(w)$, the induction step is completed. □

Consider Claim 1 for $x \in T^*$. Then, $S\ {}_{\mathrm{lm}}\!\overset{1}{\Rightarrow}{}_G^*\ x$ implies that $S\ {}_{\mathrm{lm}}\!\Rightarrow_H^*\ x$, so $L(G) \subseteq L(H)$. Consider Claim 2 for $x \in T^*$. Then, $\langle S, \emptyset\rangle\ {}_{\mathrm{lm}}\!\Rightarrow_H^*\ x$ implies that $S\ {}_{\mathrm{lm}}\!\overset{1}{\Rightarrow}{}_G^*\ x$, so $L(H) \subseteq L(G)$. Hence, $L(H) = L(G)$.

Finally, we argue that H is an LL(k) context-free grammar. To simplify the argumentation, we establish another claim. It represents a slight modification of Claim 2. Let $\$ \notin V' \cup V$ be an end-marker.

Claim 3. If $\langle S, \emptyset\rangle\$^k\ {}_{\mathrm{lm}}\!\Rightarrow_H^m\ x\$^k$, *where* $x \in V'^*$ *and* $m \geq 0$, *then* $S\$^k\ {}_{\mathrm{lm}}\!\overset{1}{\Rightarrow}{}_G^*\ \tau(x)\$^k$ *and x is of the form specified in Claim 2.*

Proof. This claim can be established by analogy with the proof of Claim 2, so we leave its proof to the reader. □

For the sake of contradiction, suppose that H is not an LL(k) context-free grammar—that is, assume that there are $p' = (X \rightarrow y_1) \in P'$ and $r' = (X \rightarrow y_2) \in P'$ such that $y_1 \neq y_2$ and $\mathrm{Predict}_k(p) \cap \mathrm{Predict}_k(r) \neq \emptyset$. Let γ be a string from $\mathrm{Predict}_k(p') \cap \mathrm{Predict}_k(r')$. By the construction of P', X is of the form $X = \langle A, Q\rangle$, for some $A \in N$ and $Q \subseteq N$, and there are $p = (A \rightarrow \tau(y_1), U_1, W_1) \in P_R$ and $r = (A \rightarrow \tau(y_2), U_2, W_2) \in P_R$ such that $U_1 \subseteq Q$, $U_2 \subseteq Q$, $W_1 \cap Q = \emptyset$, and $W_2 \cap Q = \emptyset$. Since $\gamma \in \mathrm{Predict}_k(p') \cap \mathrm{Predict}_k(r')$,

$$\langle S, \emptyset\rangle\$^k\ {}_{\mathrm{lm}}\!\Rightarrow_H^*\ u\langle A, Q\rangle v\$^k\ {}_{\mathrm{lm}}\!\Rightarrow_H\ u y_1 v\$^k\ [p']\ {}_{\mathrm{lm}}\!\Rightarrow_H^*\ u\gamma w_1$$

and

$$\langle S, \emptyset\rangle\$^k\ {}_{\mathrm{lm}}\!\Rightarrow_H^*\ u\langle A, Q\rangle v\$^k\ {}_{\mathrm{lm}}\!\Rightarrow_H\ u y_2 v\$^k\ [r']\ {}_{\mathrm{lm}}\!\Rightarrow_H^*\ u\gamma w_2$$

for some $u \in T^*$, $v \in V'^*$ such that $\mathrm{alph}(\tau(v)) = Q$ (see Claim 3), and $\gamma, w_1, w_2 \in V'^*\{\$\}^*$. Then, by Claim 3,

$$S\$^k\ {}_{\mathrm{lm}}\!\overset{1}{\Rightarrow}{}_G^*\ uA\tau(v)\$^k\ {}_{\mathrm{lm}}\!\overset{1}{\Rightarrow}_G\ u\tau(y_1 v)\$^k\ [p]\ {}_{\mathrm{lm}}\!\overset{1}{\Rightarrow}{}_G^*\ u\gamma\tau(w_1)$$

and

$$S\$^k\ {}_{\mathrm{lm}}\!\overset{1}{\Rightarrow}{}_G^*\ uA\tau(v)\$^k\ {}_{\mathrm{lm}}\!\overset{1}{\Rightarrow}_G\ u\tau(y_1 v)\$^k\ [r]\ {}_{\mathrm{lm}}\!\overset{1}{\Rightarrow}{}_G^*\ u\gamma\tau(w_2)$$

However, by Definition 6.7.4, $\gamma \in \mathrm{Predict}_k(p)$ and $\gamma \in \mathrm{Predict}_k(r)$, so

$$\mathrm{Predict}_k(p) \cap \mathrm{Predict}_k(r) \neq \emptyset$$

Since both p and r have the same left-hand side and since both are leftmost-applicable to $uA\tau(v)$, we have a contradiction with the fact that G is an LL(k) one-sided random context grammar. Hence, H is an LL(k) context-free grammar, and the lemma holds. □

Next, we show how to transform any LL(k) context-free grammar into an equivalent LL(k) one-sided random context grammar, for every $k \geq 1$.

Lemma 6.7.9. *For every LL(k) context-free grammar G, where $k \geq 1$, there is an LL(k) one-sided random context grammar H such that $L(H) = L(G)$.*

Proof. Let $G = (N, T, P, S)$ be an LL(k) context-free grammar, where $k \geq 1$. Then, the one-sided random context grammar $H = (N, T, P', \emptyset, S)$, where

$$P' = \{(A \rightarrow x, \emptyset, \emptyset) \mid A \rightarrow x \in P\}$$

is clearly an LL(k) one-sided random context grammar that satisfies $L(H) = L(G)$. Hence, the lemma holds. □

For every $k \geq 1$, let **LL - CF**(k) and **LL - ORC**(k) denote the families of languages generated by LL(k) context-free grammars and LL(k) one-sided random context grammars, respectively.

The following theorem represents the main result of this section.

Theorem 6.7.10. **LL - ORC**(k) = **LL - CF**(k) *for* $k \geq 1$.

Proof. This theorem follows from Lemmas 6.7.8 and 6.7.9. □

Define the language families **LL - CF** and **LL - ORC** as

$$\mathbf{LL\text{-}CF} = \bigcup_{k \geq 1} \mathbf{LL\text{-}CF}(k)$$

$$\mathbf{LL\text{-}ORC} = \bigcup_{k \geq 1} \mathbf{LL\text{-}ORC}(k)$$

From Theorem 6.7.10, we obtain the following corollary.

Corollary 6.7.11. **LL - ORC** = **LL - CF** □

We conclude this section by proposing three open problem areas as suggested topics of future investigations related to the topic of the present section.

Open Problem 6.7.12. Is the LL property of LL one-sided random context grammars decidable? Reconsider the construction in the proof of Lemma 6.7.8. Since it is decidable whether a given context-free grammar is an LL context-free grammar (see [1]), we might try to convert a one-sided random context grammar G into a context-free grammar H, and show that G is an LL one-sided random context grammar if and only if H is an LL context-free grammar. The only-if part of this equivalence follows from Lemma 6.7.8. However, the if part does not hold. Indeed, we give an example of a one-sided random context grammar that is not LL, but which the construction in the proof of Lemma 6.7.8 turns into a context-free grammar that is LL. Consider the one-sided random context grammar

$$G = \big(\{S, F_1, F_2\}, \{a\}, \emptyset, P_R, S\big)$$

where

$$P_R = \big\{(S \rightarrow a, \emptyset, \{F_1\}), (S \rightarrow a, \emptyset, \{F_2\})\big\}$$

Clearly, G is not an LL one-sided random context grammar. However, observe that the construction converts G into the LL context-free grammar

$$H = \big(N, \{a\}, P', \langle S, \emptyset\rangle\big)$$

where P' contains $\langle S, \emptyset\rangle \rightarrow a$ (other rules are never applicable so we do not list them here). Hence, the construction used in the proof of Lemma 6.7.8 cannot be straightforwardly used for deciding the LL property of one-sided random context grammars. □

Open Problem 6.7.13. Reconsider the proof of Lemma 6.7.8. Observe that for a single right random context rule from P_R, the construction introduces several rules to P', depending on the number of nonterminals of G. Hence, H contains many more rules than G. Obviously, we may eliminate all useless nonterminals and rules from H by using standard methods. However, does an elimination like this always result into the most economical context-free grammar? In other words, is there an algorithm which, given G, finds an equivalent LL(k) context-free grammar H such that there is no other equivalent LL(k) context-free grammar with fewer nonterminals or rules than H has? □

Open Problem 6.7.14. Given an LL(k) context-free grammar G, where $k \geq 1$, is there an algorithm which converts G into an equivalent LL(k) one-sided random context grammar that contains fewer rules than G? □

References

1. Aho, A.V., Ullman, J.D.: The Theory of Parsing, Translation and Compiling, Volume I: Parsing. Prentice-Hall, New Jersey (1972)
2. Aho, A.V., Lam, M.S., Sethi, R., Ullman, J.D.: Compilers: Principles, Techniques, and Tools, 2nd edn. Addison-Wesley, Boston (2006)
3. Baker, B.S.: Non-context-free grammars generating context-free languages. Inf. Control **24**(3), 231–246 (1974)
4. Cannon, R.L.: Phrase structure grammars generating context-free languages. Inf. Control **29**(3), 252–267 (1975)

5. Cojocaru, L., Mäkinen, E.: On the complexity of Szilard languages of regulated grammars. Technical report, Department of Computer Sciences, University of Tampere, Tampere, Finland (2010)
6. Cremers, A.B., Maurer, H.A., Mayer, O.: A note on leftmost restricted random context grammars. Inf. Process. Lett. **2**(2), 31–33 (1973)
7. Csuhaj-Varjú, E., Vaszil, G.: Scattered context grammars generate any recursively enumerable language with two nonterminals. Inf. Process. Lett. **110**(20), 902–907 (2010)
8. Csuhaj-Varjú, E., Masopust, T., Vaszil, G.: Cooperating distributed grammar systems with permitting grammars as components. Rom. J. Inf. Sci. Technol. **12**(2), 175–189 (2009)
9. Cytron, R., Fischer, C., LeBlanc, R.: Crafting a Compiler. Addison-Wesley, Boston (2009)
10. Czeizler, E., Czeizler, E., Kari, L., Salomaa, K.: On the descriptional complexity of Watson-Crick automata. Theor. Comput. Sci. **410**(35), 3250–3260 (2009)
11. Dassow, J., Păun, G.: Regulated Rewriting in Formal Language Theory. Springer, New York (1989)
12. Dassow, J., Fernau, H., Păun, G.: On the leftmost derivation in matrix grammars. Int. J. Found. Comput. Sci. **10**(1), 61–80 (1999)
13. Fernau, H.: Regulated grammars under leftmost derivation. Grammars **3**(1), 37–62 (2000)
14. Fernau, H.: Nonterminal complexity of programmed grammars. Theor. Comput. Sci. **296**(2), 225–251 (2003)
15. Fernau, H., Meduna, A.: A simultaneous reduction of several measures of descriptional complexity in scattered context grammars. Inf. Process. Lett. **86**(5), 235–240 (2003)
16. Fernau, H., Meduna, A.: On the degree of scattered context-sensitivity. Theor. Comput. Sci. **290**(3), 2121–2124 (2003)
17. Fernau, H., Freund, R., Oswald, M., Reinhardt, K.: Refining the nonterminal complexity of graph-controlled, programmed, and matrix grammars. J. Autom. Lang. Comb. **12**(1–2), 117–138 (2007)
18. Ferretti, C., Mauri, G., Păun, G., Zandron, C.: On three variants of rewriting P systems. Theor. Comput. Sci. **301**(1–3), 201–215 (2003)
19. Freund, R., Oswald, M.: P systems with activated/prohibited membrane channels. In: Membrane Computing. Lecture Notes in Computer Science, vol. 2597, pp. 261–269. Springer, Berlin/Heidelberg (2003)
20. Geffert, V.: Normal forms for phrase-structure grammars. Theor. Inf. Appl. **25**(5), 473–496 (1991)
21. Ginsburg, S., Spanier, E.H.: Control sets on grammars. Theory Comput. Syst. **2**(2), 159–177 (1968)
22. Goldefus, F., Masopust, T., Meduna, A.: Left-forbidding cooperating distributed grammar systems. Theor. Comput. Sci. **20**(3), 1–11 (2010)
23. Holzer, M., Kutrib, M.: Nondeterministic finite automata: recent results on the descriptional and computational complexity. In: Implementation and Applications of Automata. Lecture Notes in Computer Science, vol. 5148, pp. 1–16. Springer (2008)
24. Kasai, T.: An hierarchy between context-free and context-sensitive languages. J. Comput. Syst. Sci. **4**, 492–508 (1970)
25. Kleijn, H.C.M., Rozenberg, G.: Sequential, continuous and parallel grammars. Inf. Control **48**(3), 221–260 (1981)
26. Lukáš, R., Meduna, A.: Multigenerative grammar systems. Schedae Informaticae **2006**(15), 175–188 (2006)
27. Luker, M.: A generalization of leftmost derivations. Theory Comput. Syst. **11**(1), 317–325 (1977)
28. Madhu, M.: Descriptional complexity of rewriting P systems. J. Autom. Lang. Comb. **9**(2–3), 311–316 (2004)
29. Masopust, T.: Descriptional complexity of multi-parallel grammars. Inf. Process. Lett. **108**(2), 68–70 (2008)
30. Masopust, T.: On the descriptional complexity of scattered context grammars. Theor. Comput. Sci. **410**(1), 108–112 (2009)

31. Masopust, T., Meduna, A.: On descriptional complexity of partially parallel grammars. Fundamenta Informaticae **87**(3), 407–415 (2008)
32. Masopust, T., Meduna, A.: Descriptional complexity of three-nonterminal scattered context grammars: an improvement. In: Proceedings of 11th International Workshop on Descriptional Complexity of Formal Systems, pp. 235–245. Otto-von-Guericke-Universität Magdeburg (2009)
33. Matthews, G.H.: A note on asymmetry in phrase structure grammars. Inf. Control **7**, 360–365 (1964)
34. Maurer, H.A.: Simple matrix languages with a leftmost restriction. Inf. Control **23**(2), 128–139 (1973)
35. Meduna, A.: On the number of nonterminals in matrix grammars with leftmost derivations. In: New Trends in Formal Languages: Control, Cooperation, and Combinatorics (to Jürgen Dassow on the occasion of his 50th birthday), pp. 27–38. Springer, New York (1997)
36. Meduna, A.: Elements of Compiler Design. Auerbach Publications, Boston (2007)
37. Meduna, A., Goldefus, F.: Weak leftmost derivations in cooperative distributed grammar systems. In: 5th Doctoral Workshop on Mathematical and Engineering Methods in Computer Science, pp. 144–151. Brno University of Technology, Brno, CZ (2009)
38. Meduna, A., Techet, J.: Canonical scattered context generators of sentences with their parses. Theor. Comput. Sci. **2007**(389), 73–81 (2007)
39. Meduna, A., Techet, J.: Scattered Context Grammars and Their Applications. WIT Press, Southampton (2010)
40. Meduna, A., Škrkal, O.: Combined leftmost derivations in matrix grammars. In: Proceedings of 7th International Conference on Information Systems Implementation and Modelling (ISIM'04), pp. 127–132. Ostrava, CZ (2004)
41. Mihalache, V.: Matrix grammars versus parallel communicating grammar systems. In: Mathematical Aspects of Natural and Formal Languages, pp. 293–318. World Scientific Publishing, River Edge (1994)
42. Mutyam, M., Krithivasan, K.: Tissue P systems with leftmost derivation. Ramanujan Math. Soc. Lect. Notes Series **3**, 187–196 (2007)
43. Okubo, F.: A note on the descriptional complexity of semi-conditional grammars. Inf. Process. Lett. **110**(1), 36–40 (2009)
44. Păun, G.: A variant of random context grammars: semi-conditional grammars. Theor. Comput. Sci. **41**(1), 1–17 (1985)
45. Penttonen, M.: One-sided and two-sided context in formal grammars. Inf. Control **25**(4), 371–392 (1974)
46. Rosenkrantz, D.J.: Programmed grammars and classes of formal languages. J. ACM **16**(1), 107–131 (1969)
47. Rozenberg, G., Salomaa, A. (eds.): Handbook of Formal Languages, Volumes 1 through 3. Springer, New York (1997)
48. Salomaa, A.: Matrix grammars with a leftmost restriction. Inf. Control **20**(2), 143–149 (1972)
49. Vaszil, G.: On the descriptional complexity of some rewriting mechanisms regulated by context conditions. Theor. Comput. Sci. **330**(2), 361–373 (2005)

Chapter 7
On Erasing Rules and Their Elimination

Abstract The present three-section chapter studies how to eliminate erasing rules, having the empty string on their right-hand sides, from context-free grammars and their regulated versions. The chapter points out that this important topic still represents a largely open problem area in the theory of regulated grammars. Section 7.1 gives two methods of eliminating erasing rules from ordinary context-free grammars. One method is based upon a well-known technique, but the other represents a completely new algorithm that performs this elimination. Section 7.2 establishes workspace theorems for regular-controlled grammars. In essence, these theorems give derivation conditions under which erasing rules can be removed from these grammars. Section 7.3 discusses the topic of this chapter in terms of scattered context grammars. First, it points out that scattered context grammars with erasing rules characterize the family of recursively enumerable languages while their propagating versions do not. In fact, propagating scattered context grammars cannot generate any non-context-sensitive language, so some scattered context grammars with erasing rules are necessarily unconvertible to equivalent propagating scattered context grammars. This section establishes a sufficient condition under which this conversion is always possible.

Keywords Elimination of erasing rules • New methods of elimination for ordinary context-free grammars • Workspace theorems for regular-controlled grammars • Elimination of erasing rules in scattered context grammars

This chapter, consisting of three sections, discusses the elimination of erasing rules from context-free grammars and their regulated versions. This elimination often fulfills a key role in formal grammars and their applications. From a practical viewpoint, some grammar-based methods, including many parsing algorithms, strictly require the absence of any erasing rules in the applied grammars. From a theoretical viewpoint, the absence of any erasing rules frequently simplifies the achievement of various results concerning grammars and their languages.

A. Meduna and P. Zemek, *Regulated Grammars and Automata*,
DOI 10.1007/978-1-4939-0369-6_7,

Section 7.1 describes two methods that eliminate erasing rules from ordinary context-free grammars. One method is based upon a well-known technique while the other represents a brand new algorithm that does this job. We also discuss how these algorithms might be adapted for regulated grammars in the future.

Section 7.2 concerns workspace theorems for regular-controlled grammars. In essence, these theorems give derivation conditions under which erasing rules can be removed from these grammars.

Finally, Sect. 7.3 discusses erasing rules and their effect on the generative power of scattered context grammars. It points out that scattered context grammars with erasing rules characterize the family of recursively enumerable languages while propagating scattered context grammars cannot generate any non-context-sensitive language. Consequently, some scattered context grammars with erasing rules can be converted to equivalent propagating scattered context grammars while others cannot. It is thus highly desirable to have a sufficient condition under which this conversion is always possible. Section 7.3 establishes a condition like this.

7.1 Elimination of Erasing Rules from Context-Free Grammars

In this section, we first recall the well-known technique that removes all erasing rules from context-free grammars without affecting their generative power (see, for instance, Theorem 5.1.3.2.4 in [7]). Apart from this technique, we add a brand new algorithm that does the same job. In the conclusion, we discuss how these algorithms may be adapted for regulated grammars, which are also based upon context-free rules, but their rules are applied in a regulated way as opposed to a completely unregulated application of rules in context-free grammars.

The present section consists of three subsections. Section 7.1.1 describes the classical algorithm that removes erasing rules from context-free grammars. Then, Sect. 7.1.2 presents a completely new algorithm that performs this elimination. Finally, Sect. 7.1.3 discusses how these algorithms might be adapted for regulated grammars in the future.

7.1.1 The Standard Algorithm

This section describes the classical algorithm that removes erasing rules from context-free grammars. This algorithm is based upon a predetermination of ε-nonterminals—that is, the nonterminals from which the empty string can be derived.

Definition 7.1.1. Let $G = (N, T, P, S)$ be a context-free grammar. A nonterminal $A \in N$ is said to be an *ε-nonterminal* if and only if $A \Rightarrow_G^* \varepsilon$. □

The following algorithm determines the set of all ε-nonterminals in a given context-free grammar (see Algorithm 5.1.3.2.1 in [7]).

Algorithm 7.1.2. *Determination of ε-nonterminals in a context-free grammar.*

Input: A context-free grammar, $G = (N, T, P, S)$.
Output: The set of all ε-nonterminals in G, $N_\varepsilon = \{A \mid A \in N \text{ and } A \Rightarrow_G^* \varepsilon\}$.
Method: Initially, set $N_\varepsilon = \{A \mid A \to \varepsilon \in P\}$. Next, apply the following step until N_ε cannot be extended

if $A \to X_1X_2 \cdots X_n \in P$, where $X_i \in N_\varepsilon$, for all i, $1 \leq i \leq n$, for some $n \geq 1$
then $N_\varepsilon = N_\varepsilon \cup \{A\}$. □

First, N_ε is set to nonterminals from which the empty string can be derived in a single step. Then, if there is a rule which have only nonterminals from N_ε on its right-hand side, include the left-hand side of this rule to N_ε. This step is repeated until N_ε cannot be extended.

Example 7.1.3. Let $G = (N, T, P, S)$ be a context-free grammar, where $N = \{S, A, B, C\}, T = \{b, c\}$, and P consists of the following six rules

$$S \to AB \qquad A \to \varepsilon \qquad B \to b$$
$$S \to BC \qquad B \to A \qquad C \to c$$

With G on its input, Algorithm 7.1.2 produces $N_\varepsilon = \{S, A, B\}$. □

For a proof of the following lemma, see Lemma 5.1.3.2.2 in [7].

Lemma 7.1.4. *Algorithm 7.1.2 is correct—that is, with a context-free grammar G on its input, it halts and correctly produces the set of all ε-nonterminals in G.*

By using Algorithm 7.1.2, we can define the standard algorithm for elimination of erasing rules from context-free grammars (see Sect. 7.3.1 in [6]).

Algorithm 7.1.5. *Standard elimination of erasing rules from context-free grammars.*

Input: A context-free grammar, $G = (N, T, P, S)$.
Output: A propagating context-free grammar, $H = (N, T, P', S)$, such that $L(H) = L(G)$.
Method: Use Algorithm 7.1.2 to compute N_ε from G. Initially, set $P' = \emptyset$. To complete P', apply the following step until P' cannot be extended

if $A \to x_0X_1x_1X_2x_2 \cdots X_nx_n \in P$, where $x_i \in N_\varepsilon^*, X_j \in V$, for all i and $j, 0 \leq i \leq n, 1 \leq j \leq n$, for some $n \geq 1$
then add $A \to X_1X_2 \cdots X_n$ to P'. □

Consider the context-free grammar $G = (N, T, P, S)$, where $A \to aAbB, A \to \varepsilon, B \to \varepsilon \in P, A, B \in N$, and $a, b \in T^*$. Notice that $A, B \in N_\varepsilon$. The idea behind Algorithm 7.1.5 is that if some nonterminal A is erased in a derivation (in an arbitrary number of steps), then it does not need to be present in the sentential form, so there is no reason of deriving it. Hence, H additionally has rules $A \rightarrow abB$, $A \rightarrow aAb$, and $A \rightarrow ab$. It also has to contain the original rule $A \rightarrow aAbB$ if there are some non-erasing rules $A \rightarrow x$ and $B \rightarrow y$ in P, which can lead to a derivation of a string of terminal symbols. This procedure is performed for all rules in P.

Notice that if we transform a rule $A \rightarrow BB$ with $B \in N_\varepsilon$, then H has only $A \rightarrow BB$ and $A \rightarrow B$ because it does not make a difference if the first occurrence of B is erased or the second one. We do not want to include the erasing rule $A \rightarrow \varepsilon$ either (by the if condition, there has to be at least one symbol that is chosen as a not-to-be-erased symbol).

Example 7.1.6. Let $G = (N, T, P, S)$ be a context-free grammar, where $N = \{S, A, B, C\}$, $T = \{a, b, c\}$, and P consists of the following seven rules

$S \rightarrow ABC$	$A \rightarrow a$	$B \rightarrow bB$	$C \rightarrow cBcC$
	$A \rightarrow BB$	$B \rightarrow \varepsilon$	$C \rightarrow \varepsilon$

With G on its input, Algorithm 7.1.5 produces the propagating context-free grammar, $H = (N, T, P', S)$, where P' contains these rules

$S \rightarrow ABC$	$A \rightarrow a$	$B \rightarrow bB$	$C \rightarrow cBcC$
$S \rightarrow AB$	$A \rightarrow BB$	$B \rightarrow b$	$C \rightarrow cBc$
$S \rightarrow BC$	$A \rightarrow B$		$C \rightarrow ccC$
$S \rightarrow AC$			$C \rightarrow cc$

Observe that $L(H) = L(G)$. □

For a proof of the following lemma, see Theorem 7.9 in [6].

Lemma 7.1.7. *Algorithm 7.1.5 is correct—that is, with a context-free grammar G on its input, it halts and correctly produces a propagating context-free grammar H such that $L(H) = L(G)$.*

Theorem 7.1.8. *Let G be a context-free grammar. Then, there is a propagating context-free grammar H such that $L(H) = L(G)$.*

Proof. This theorem follows from Algorithm 7.1.5 and Lemma 7.1.7. □

7.1.2 A New Algorithm

As a matter of fact, the elimination of erasing rules described in the previous section works in two phases. First, it determines all ε-nonterminals in a given context-free grammar. Then, having determined these nonterminals, it performs the desired elimination. In the present section, we explain how to eliminate erasing rules in a one-phase way. That is, we present a brand new algorithm that performs this elimination without any predetermination of ε-nonterminals, which obviously represents a significant advantage over the two-phase method given in Sect. 7.1.1.

To give a more detailed insight into this algorithm, consider an arbitrary context-free grammar, $G = (N, T, P, S)$, and $Y \in N$. If Y derives ε in G, then a derivation like this can be expressed in the following step-by-step way

$$Y \Rightarrow_G y_1 \Rightarrow_G y_2 \Rightarrow_G \cdots \Rightarrow_G y_n \Rightarrow_G \varepsilon$$

where $y_i \in V^*$, for all $i, 1 \leq i \leq n$, for some $n \geq 1$. If a sentential form contains several occurrences of Y, each of them can be erased in this way although there may exist many alternative ways of erasing Y. Based upon these observations, the next algorithm introduces compound nonterminals of the form $\langle X, U \rangle$, in which X is a symbol that is not erased during the derivation, and U is a set of symbols that are erased. Within the compound nonterminal, the algorithm simulates the erasure of symbols in U in the way sketched above. Observe that since U is a set, U contains no more than one occurrence of any symbol because there is no need to record several occurrences of the same symbol; indeed, as already pointed out, all these occurrences can be erased in the same way.

Algorithm 7.1.9. *Elimination of erasing rules from context-free grammars without any predetermination of ε-nonterminals.*

Input: A context-free grammar, $G = (N, T, P, S)$.
Output: A propagating context-free grammar, $H = (N', T, P', S')$, such that $L(H) = L(G)$.
Method: Initially, set

$$\begin{aligned} N' &= \{\langle X, U \rangle \mid X \in V, U \subseteq N\} \\ S' &= \langle S, \emptyset \rangle \\ P' &= \{\langle a, \emptyset \rangle \to a \mid a \in T\} \end{aligned}$$

Apply the following two steps until P' cannot be extended.

(1) **If** $A \to x_0 X_1 x_1 X_2 x_2 \cdots X_n x_n \in P$, where $x_i \in N^*, X_j \in V$, for all i and $j, 0 \leq i \leq n, 1 \leq j \leq n$, for some $n \geq 1$
then for each $U \subseteq N$, add
$\langle A, U \rangle \to \langle X_1, U \cup \text{alph}(x_0 x_1 x_2 \cdots x_n) \rangle \langle X_2, \emptyset \rangle \cdots \langle X_n, \emptyset \rangle$ to P'.
(2) **If** $\langle X, U \rangle \in N'$ and $A \to x \in P$, where $A \in U$ and $x \in N^*$
then add $\langle X, U \rangle \to \langle X, (U - \{A\}) \cup \text{alph}(x) \rangle$ to P'. □

Example 7.1.10. Consider the context-free grammar $G = (N, T, P, S)$, where $N = \{S\}, T = \{a, b\}$, and $P = \{S \to aSb, S \to \varepsilon\}$.

The generated language is $L(G) = \{a^n b^n \mid n \geq 0\}$. Algorithm 7.1.9 produces a propagating context-free grammar

$$H = (N', T, P', S')$$

where

$$N' = \{\langle S, \emptyset\rangle, \langle a, \emptyset\rangle, \langle b, \emptyset\rangle, \langle S, \{S\}\rangle, \langle a, \{S\}\rangle, \langle b, \{S\}\rangle\}$$

$S' = \langle S, \emptyset\rangle$, and P' contains the following six rules

$$\begin{array}{ll} \langle S, \emptyset\rangle \rightarrow \langle a, \emptyset\rangle\langle S, \emptyset\rangle\langle b, \emptyset\rangle & \langle b, \{S\}\rangle \rightarrow \langle b, \emptyset\rangle \\ \langle S, \emptyset\rangle \rightarrow \langle a, \{S\}\rangle\langle b, \emptyset\rangle & \langle a, \emptyset\rangle \rightarrow a \\ \langle a, \{S\}\rangle \rightarrow \langle a, \emptyset\rangle & \langle b, \emptyset\rangle \rightarrow b \end{array}$$

For example, for

$$\begin{aligned} S &\Rightarrow_G aSb \\ &\Rightarrow_G aaSbb \\ &\Rightarrow_G aabb \end{aligned}$$

there is

$$\begin{aligned} \langle S, \emptyset\rangle &\Rightarrow_H \langle a, \emptyset\rangle\langle S, \emptyset\rangle\langle b, \emptyset\rangle \\ &\Rightarrow_H \langle a, \emptyset\rangle\langle a, \{S\}\rangle\langle b, \emptyset\rangle\langle b, \emptyset\rangle \\ &\Rightarrow_H \langle a, \emptyset\rangle\langle a, \emptyset\rangle\langle b, \emptyset\rangle\langle b, \emptyset\rangle \\ &\Rightarrow_H^* aabb \end{aligned}$$

Clearly, $L(H) = L(G)$. □

Lemma 7.1.11. *Algorithm 7.1.9 is correct—that is, with a context-free grammar G on its input, it halts and correctly produces a propagating context-free grammar H such that $L(H) = L(G)$.*

Proof. Clearly, the algorithm always halts. Since P' does not contain any erasing rules, H is propagating. To establish $L(H) = L(G)$, we prove three claims.

The first claim shows how derivations of G are simulated by H. It is then used to prove $L(G) \subseteq L(H)$ later in the proof. Recall the meaning of the $^{\varepsilon}$ and $^{\not\varepsilon}$notation from Definition 3.3.26 because this notation is frequently used in the rest of the proof.

Claim 1. If

$$S \Rightarrow_G^m {}^{\varepsilon}x_0{}^{\not\varepsilon}X_1{}^{\varepsilon}x_1{}^{\not\varepsilon}X_2{}^{\varepsilon}x_2 \cdots {}^{\not\varepsilon}X_h{}^{\varepsilon}x_h \Rightarrow_G^* z$$

where $x_i \in N^*$, $X_j \in V$, for all i and j, $0 \leq i \leq h$, $1 \leq j \leq h$, and $z \in L(G)$, for some $m \geq 0$ and $h \geq 1$, then

$$\langle S, \emptyset\rangle \Rightarrow_H^* \langle X_1, U_1\rangle\langle X_2, U_2\rangle \cdots \langle X_h, U_h\rangle$$

where $\bigcup_{1 \leq i \leq h} U_i \subseteq \bigcup_{0 \leq i \leq h} \mathrm{alph}(x_i)$.

Proof. This claim is established by induction on $m \geq 0$.

Basis. The basis for $m = 0$ is clear.

Induction Hypothesis. Suppose that there exists $m \geq 0$ such that the claim holds for all derivations of length ℓ, where $0 \leq \ell \leq m$.

Induction Step. Consider any derivation of the form

$$S \Rightarrow_G^{m+1} w \Rightarrow_G^* z$$

where $w \in V^*$ and $z \in L(G)$. Since $m + 1 \geq 1$, this derivation can be expressed as

$$S \Rightarrow_G^m x \Rightarrow_G w \Rightarrow_G^* z$$

where $x \in V^+$. Let $x = {}^{\varepsilon}x_0{}^{\not\varepsilon}X_1{}^{\varepsilon}x_1{}^{\not\varepsilon}X_2{}^{\varepsilon}x_2 \cdots {}^{\not\varepsilon}X_h{}^{\varepsilon}x_h$, where $x_i \in N^*$, $X_j \in V$, for all i and j, $0 \leq i \leq h$, $1 \leq j \leq h$, for some $h \geq 1$. Then, by the induction hypothesis,

$$\langle S, \emptyset \rangle \Rightarrow_H^* \langle X_1, U_1 \rangle \langle X_2, U_2 \rangle \cdots \langle X_h, U_h \rangle$$

where $\bigcup_{1 \leq i \leq h} U_i \subseteq \bigcup_{0 \leq i \leq h} \mathrm{alph}(x_i)$, for some $n \geq 0$.

Now, we consider all possible forms of $x \Rightarrow_G w$, covered by the next two cases—(i) and (ii).

(i) Let $X_j \rightarrow y_0 Y_1 y_1 \cdots Y_q y_q \in P$, where $y_i \in N^*$, for all $i, 0 \leq i \leq q$, $Y_i \in V$, for all $i, 1 \leq i \leq 1$, for some $j, 1 \leq j \leq h$, and $q \geq 1$, so

$$\begin{aligned} &{}^{\varepsilon}x_0{}^{\not\varepsilon}X_1{}^{\varepsilon}x_1 \cdots {}^{\not\varepsilon}X_j{}^{\varepsilon}x_j \cdots {}^{\not\varepsilon}X_h{}^{\varepsilon}x_h \Rightarrow_G \\ &{}^{\varepsilon}x_0 X_1{}^{\varepsilon}x_1 \cdots {}^{\not\varepsilon}X_{j-1}{}^{\varepsilon}x_{j-1}{}^{\varepsilon}y_0{}^{\not\varepsilon}Y_1{}^{\varepsilon}y_1 \cdots {}^{\not\varepsilon}Y_q{}^{\varepsilon}y_q{}^{\varepsilon}x_j{}^{\not\varepsilon}X_{j+1}{}^{\varepsilon}x_{j+1} \cdots {}^{\not\varepsilon}X_h{}^{\varepsilon}x_h \end{aligned}$$

By (1) in the algorithm,

$$\langle X_j, U_j \rangle \rightarrow \langle Y_1, U_j \cup \mathrm{alph}(y_0 y_1 \cdots y_q) \rangle \langle Y_2, \emptyset \rangle \cdots \langle Y_q, \emptyset \rangle \in P'$$

so

$$\begin{aligned} &\langle X_1, U_1 \rangle \langle X_2, U_2 \rangle \cdots \langle X_j, U_j \rangle \cdots \langle X_h, U_h \rangle \Rightarrow_H \\ &\langle X_1, U_1 \rangle \langle X_2, U_2 \rangle \cdots \langle X_{j-1}, U_{j-1} \rangle \langle Y_1, U_j \cup \mathrm{alph}(y_0 y_1 \cdots y_q) \rangle \langle Y_2, \emptyset \rangle \cdots \langle Y_q, \emptyset \rangle \\ &\quad \langle X_{j+1}, U_{j+1} \rangle \cdots \langle X_h, U_h \rangle \end{aligned}$$

Clearly,

$$\Big(\bigcup_{1 \leq i \leq h} U_i\Big) \cup \Big(\bigcup_{0 \leq i \leq q} \mathrm{alph}(y_i)\Big) \subseteq \Big(\bigcup_{0 \leq i \leq h} \mathrm{alph}(x_i)\Big) \cup \Big(\bigcup_{0 \leq i \leq q} \mathrm{alph}(y_i)\Big)$$

which completes the induction step for (i).

(ii) Let $x_j = x'_j A x''_j$ and $A \to y \in P$, where $y \in N^*$ and $x'_j, x''_j \in N^*$, so

$$
\begin{aligned}
&{}^{\varepsilon}x_0 {}^{\not\varepsilon}X_1 {}^{\varepsilon}x_1 \cdots {}^{\not\varepsilon}X_j {}^{\varepsilon}x_j \cdots {}^{\not\varepsilon}X_h {}^{\varepsilon}x_h \Rightarrow_G \\
&{}^{\varepsilon}x_0 {}^{\not\varepsilon}X_1 {}^{\varepsilon}x_1 \cdots {}^{\not\varepsilon}X_{j-1} {}^{\varepsilon}x_{j-1} {}^{\not\varepsilon}X_j {}^{\varepsilon}x'_j {}^{\varepsilon}y {}^{\varepsilon}x''_j {}^{\not\varepsilon}X_{j+1} {}^{\varepsilon}x_{j+1} \cdots {}^{\not\varepsilon}X_h {}^{\varepsilon}x_h
\end{aligned}
$$

If $A \notin \bigcup_{1 \leq i \leq h} U_i$, then

$$\langle X_1, U_1 \rangle \langle X_2, U_2 \rangle \cdots \langle X_h, U_h \rangle \Rightarrow_H^0 \langle X_1, U_1 \rangle \langle X_2, U_2 \rangle \cdots \langle X_h, U_h \rangle$$

and clearly

$$\bigcup_{1 \leq i \leq h} U_i \subseteq \Big(\bigcup_{0 \leq i \leq h, i \neq j} \text{alph}(x_i) \Big) \cup \text{alph}(x'_j y x''_j)$$

Therefore, assume that $A \in \bigcup_{1 \leq i \leq h} U_i$. By (2) in the algorithm,

$$\langle X_k, U_k \rangle \to \langle X_k, (U_k - \{A\}) \cup \text{alph}(y) \rangle \in P'$$

where $U_k = U'_k \cup \{A\}$, $U'_k \subseteq N$, for some k, $1 \leq k \leq h$, so

$$
\begin{aligned}
&\langle X_1, U_1 \rangle \langle X_2, U_2 \rangle \cdots \langle X_k, U_k \rangle \cdots \langle X_h, U_h \rangle \Rightarrow_H \\
&\langle X_1, U_1 \rangle \langle X_2, U_2 \rangle \cdots \langle X_k, (U_k - \{A\}) \cup \text{alph}(y) \rangle \cdots \langle X_h, U_h \rangle
\end{aligned}
$$

Clearly,

$$\Big(\bigcup_{1 \leq i \leq h, i \neq k} U_i \Big) \cup (U'_k \cup \text{alph}(y)) \subseteq \Big(\bigcup_{0 \leq i \leq h, i \neq j} \text{alph}(x_i) \Big) \cup \text{alph}(x'_j y x''_j)$$

which completes the induction step for (ii).

Observe that these two cases cover all possible derivations of the form $x \Rightarrow_G w$. Thus, the claim holds. □

The second claim shows that in H, every derivation of any $z \in L(H)$ can be expressed as a two-part derivation. In the first part, every occurring symbol is a two-component nonterminal. In the second part, only the rules of the form $\langle a, \emptyset \rangle \to a$, where $a \in T$, are used.

Claim 2. Let $z \in L(H)$*. Then, there exists a derivation* $\langle S, \emptyset \rangle \Rightarrow_H^* x \Rightarrow_H^* z$*, where* $x \in N'^+$*, and during* $x \Rightarrow_H^* z$*, only rules of the form* $\langle a, \emptyset \rangle \to a$*, where* $a \in T$*, are used.*

Proof. Since H is a context-free grammar, we can always rearrange all the applications of rules so the claim holds. □

The third claim shows how derivations of H are simulated by G. It is used to prove $L(H) \subseteq L(G)$ later in the proof.

Claim 3. If

$$\langle S, \emptyset \rangle \Rightarrow_H^n \langle X_1, U_1 \rangle \langle X_2, U_2 \rangle \cdots \langle X_h, U_h \rangle$$

where $X_i \in V$ and $U_i \subseteq N$, for all i, $1 \leq i \leq h$, for some $n \geq 0$ and $h \geq 1$, then

$$S \Rightarrow_G^* x_0 X_1 x_1 X_2 x_2 \cdots X_h x_h$$

where $x_i \in N^*$, for all $i, 0 \leq i \leq h$, and $\bigcup_{1 \leq i \leq h} U_i \subseteq \bigcup_{0 \leq i \leq h} \mathrm{alph}(x_i)$.

Proof. This claim is established by induction on $n \geq 0$.

Basis. The basis for $n = 0$ is clear.

Induction Hypothesis. Suppose that there exists $n \geq 0$ such that the claim holds for all derivations of length ℓ, where $0 \leq \ell \leq n$.

Induction Step. Consider any derivation of the form

$$\langle S, \emptyset \rangle \Rightarrow_H^{n+1} w$$

where $w \in N'^+$. Since $n + 1 \geq 1$, this derivation can be expressed as

$$\langle S, \emptyset \rangle \Rightarrow_H^n x \Rightarrow_H w$$

where $x \in N'^+$. Let $x = \langle X_1, U_1 \rangle \langle X_2, U_2 \rangle \cdots \langle X_h, U_h \rangle$, where $X_i \in V, U_i \in N^*$, for all $i, 1 \leq i \leq h$, for some $h \geq 1$. By the induction hypothesis,

$$S \Rightarrow_G^* x_0 X_1 x_1 X_2 x_2 \cdots X_h x_h$$

where $x_i \in N^*$, for all $i, 1 \leq i \leq h$, such that $\bigcup_{1 \leq i \leq h} U_i \subseteq \bigcup_{0 \leq i \leq h} \mathrm{alph}(x_i)$.

Now, we consider all possible forms of $x \Rightarrow_H w$, covered by the next two cases—(i) and (ii).

(i) Let $\langle X_j, U_j \rangle \to \langle Y_1, W \rangle \langle Y_2, \emptyset \rangle \cdots \langle Y_q, \emptyset \rangle \in P'$, where $W \subseteq N$ and $Y_i \in N$, for all $i, 1 \leq i \leq q$, for some $q \geq 1$, so

$$\begin{aligned} &\langle X_1, U_1 \rangle \langle X_2, U_2 \rangle \cdots \langle X_j, U_j \rangle \cdots \langle X_h, U_h \rangle \Rightarrow_H \\ &\langle X_1, U_1 \rangle \langle X_2, U_2 \rangle \cdots \langle X_{j-1}, U_{j-1} \rangle \langle Y_1, W \rangle \langle Y_2, \emptyset \rangle \cdots \\ &\quad \cdots \langle Y_q, \emptyset \rangle \langle X_{j+1}, U_{j+1} \rangle \cdots \langle X_h, U_h \rangle \end{aligned}$$

By (1) in the algorithm, W is of the form $W = U_j \cup \mathrm{alph}(y_0 y_1 \cdots y_q)$, where $y_i \in N^*$, for all $i, 1 \leq i \leq q$, so $X_j \to y_0 Y_1 y_1 \cdots Y_q y_q \in P$. Therefore,

$$x_0 X_1 x_1 \cdots X_j x_j \cdots X_h x_h \Rightarrow_G$$
$$x_0 X_1 x_1 \cdots X_{j-1} x_{j-1} y_0 Y_1 y_1 \cdots Y_q y_q x_j X_{j+1} x_{j+1} \cdots X_h x_h$$

Clearly,

$$\left(\bigcup_{1 \le i \le h} U_i\right) \cup \left(\bigcup_{0 \le i \le q} \mathrm{alph}(y_i)\right)$$
$$\subseteq$$
$$\left(\bigcup_{0 \le i \le h} \mathrm{alph}(x_i)\right) \cup \left(\bigcup_{0 \le i \le q} \mathrm{alph}(y_i)\right)$$

(ii) Let $\langle X_j, U_j \rangle \to \langle X_j, W \rangle \in P'$, for some $j, 1 \le j \le h$, where $W \subseteq N$, so

$$\langle X_1, U_1 \rangle \langle X_2, U_2 \rangle \cdots \langle X_j, U_j \rangle \cdots \langle X_h, U_h \rangle \Rightarrow_H$$
$$\langle X_1, U_1 \rangle \langle X_2, U_2 \rangle \cdots \langle X_j, W \rangle \cdots \langle X_h, U_h \rangle$$

By (2) in the algorithm, W is of the form $W = (U_j - \{A\}) \cup \mathrm{alph}(y)$, where $A \in N$ and $y \in N^*$, and $A \to y \in P$. Recall that $\bigcup_{1 \le i \le h} U_i \subseteq \bigcup_{0 \le i \le h} \mathrm{alph}(x_i)$ by the induction hypothesis. Since $A \in \bigcup_{1 \le i \le h} U_i$, some x_k has to be of the form $x_k = x'_k A x''_k$, where $x'_k, x''_k \in N^*$, so

$$x_0 X_1 x_1 \cdots X_k x_k \cdots X_h x_h \Rightarrow_G x_0 X_1 x_1 \cdots X_{k-1} x_{k-1} X_k x'_k y x''_k X_{k+1} x_{k+1} \cdots X_h x_h$$

Clearly,

$$\left(\bigcup_{1 \le i \le h, i \ne j} U_i\right) \cup \left(U_j - \{A\}\right) \cup \mathrm{alph}(y)$$
$$=$$
$$\left(\bigcup_{1 \le i \le h, i \ne k} \mathrm{alph}(x_i)\right) \cup \left(\mathrm{alph}(x_k) - \{A\}\right) \cup \mathrm{alph}(y)$$

Observe that these two cases cover all possible derivations of the form $x \Rightarrow_H w$. Thus, the claim holds. □

We next prove that $L(H) = L(G)$. Consider Claim 1 with $x_i = \varepsilon$ and $X_j \in T$, for all i and $j, 0 \le i \le h, 1 \le j \le h$, for some $h \ge 1$. Then,

$$S \Rightarrow_G^m X_1 X_2 \cdots X_h$$

implies

$$\langle S, \emptyset \rangle \Rightarrow_H^* \langle X_1, \emptyset \rangle \langle X_2, \emptyset \rangle \cdots \langle X_h, \emptyset \rangle$$

By the initialization part of the algorithm, $\langle X_i, \emptyset \rangle \to X_i \in P'$, for all i, so

$$\begin{aligned}\langle S, \emptyset\rangle &\Rightarrow_H X_1\langle X_2, \emptyset\rangle \cdots \langle X_h, \emptyset\rangle \\ &\Rightarrow_H X_1 X_2 \cdots \langle X_h, \emptyset\rangle \\ &\ \ \vdots \\ &\Rightarrow_H X_1 X_2 \cdots \langle X_h, \emptyset\rangle\end{aligned}$$

Therefore, $L(G) \subseteq L(H)$. Let $z \in L(H)$. By Claim 2,

$$\langle S, \emptyset\rangle \Rightarrow_H^* x \Rightarrow_H^* z$$

where $x \in N'^+$. Observe that by Claim 3,

$$S \Rightarrow_G^* z$$

Therefore, $L(H) \subseteq L(G)$. As $L(G) \subseteq L(H)$ and $L(H) \subseteq L(G)$, $L(H) = L(G)$, so the lemma holds. □

Based upon the achieved results from this section, we obtain another way of proving Theorem 7.1.8.

Theorem 7.1.12. *Let G be a context-free grammar. Then, there is a propagating context-free grammar H such that $L(H) = L(G)$.*

Proof. This theorem follows from Algorithm 7.1.9 and Lemma 7.1.11. □

Open Problem 7.1.13. Compared to its input grammar, the output grammar produced by Algorithm 7.1.9 has many more symbols and rules. Can we improve this algorithm so it works in a more economical way? □

7.1.3 *Can Erasing Rules Be Eliminated from Regulated Grammars?*

Considering the subject of this book, we are obviously tempted to adapt Algorithms 7.1.5 or 7.1.9 for regulated grammars. Unfortunately, an adaptation like this is far from being straightforward. Let us take a closer look at difficulties we run into when trying to make this adaptation for regular-controlled grammars—that is, one of the major types of regulated grammars (see Sect. 5.1).

Let $H = (G, \Xi)$ be a regular-controlled grammar, where $G = (N, T, \Psi, P, S)$. Let us try to modify either of the two algorithms given earlier in this chapter for H.

(I) Take the standard algorithm—that is, Algorithms 7.1.5. Recall that this algorithm works by introducing more rules for a single rule, depending on the number of occurrences of ε-nonterminals present on the right-hand side of that rule. For instance, assume that $r: A \rightarrow bBcC \in P$ and $B, C \in N_\varepsilon$;

under this assumption, the algorithm introduces rules $r_1: A \to bBcC, r_2: A \to bcC, r_3: A \to bBc$, and $r_4: A \to bc$. Then, we substitute the use of r in the original control language with $\{r_1, r_2, r_3, r_4\}$. However, if we apply r_2 and, later on, the control language prescribes the application of a rule with B on its left-hand side, there might be no B in the sentential form. If we use r_2, we should also substitute every rule with B on its left-hand side from the control language, and so on. However, a modification like this is obviously extremely difficult, if not impossible, under the strict requirement that the resulting propagating grammar is equivalent with the original grammar.

(II) Take the new algorithm—that is, Algorithm 7.1.9. This algorithm introduces compound nonterminals of the form $\langle X, U\rangle$, where $X \in V$ and $U \subseteq N$. Since U is a set, several occurrences of a nonterminal are stored as a single occurrence. For instance, let $r: A \to BBBB$ and $s: B \to \varepsilon$ be rules in P, and let $\langle C, \{A\}\rangle$ be a nonterminal in the resulting propagating grammar. If A is rewritten in $\langle C, \{A\}\rangle$ by r—that is, $\langle C, \{A\}\rangle \Rightarrow \langle C, \{B\}\rangle$ $[r]$, there might be a sequence of s rules in the control language, but there is only one B in the second component. Based upon a similar argument like in the conclusion of (I), we see that the equivalence of the original grammar and the resulting propagating grammar can hardly be guaranteed in this straightforward modification of Algorithm 7.1.9.

Observe that the above arguments are applicable to any other major type of grammatical regulation (see Chaps. 4 and 5). Consequently, as these significant difficulties suggest, it is likely that *no* is the answer to the question used as the title of this section.

From Theorems 5.1.6, 5.2.5, and 5.3.4, we obtain the following corollary.

Corollary 7.1.14.

(i) $\mathbf{rC}^{-\varepsilon} = \mathbf{rC}$ *if and only if* $\mathbf{M}^{-\varepsilon} = \mathbf{M}$
(ii) $\mathbf{M}^{-\varepsilon} = \mathbf{M}$ *if and only if* $\mathbf{P}^{-\varepsilon} = \mathbf{P}$
(iii) $\mathbf{P}^{-\varepsilon} = \mathbf{P}$ *if and only if* $\mathbf{rC}^{-\varepsilon} = \mathbf{rC}$ □

Is any of the inclusions $\mathbf{rC}^{-\varepsilon} \subseteq \mathbf{rC}$, $\mathbf{M}^{-\varepsilon} \subseteq \mathbf{M}$, and $\mathbf{P}^{-\varepsilon} \subseteq \mathbf{P}$ proper? If so, then so are the other inclusions. Similarly, if any of the inclusions $\mathbf{rC}^{-\varepsilon} \subseteq \mathbf{rC}$, $\mathbf{M}^{-\varepsilon} \subseteq \mathbf{M}$, and $\mathbf{P}^{-\varepsilon} \subseteq \mathbf{P}$ is, in fact, an identity, then so are the others. No identity like this has been proved or disproved so far, however.

To close Sect. 7.1, we see that the exact effect of erasing rules to the generative power of regulated grammars represents a challenging open problem area in formal language theory as a whole. Under these difficult circumstances, rather than determine this effect in general, it is highly advisable to undertake a more modest approach and solve this problem in terms of some special cases of regulated grammars, which brings us straight to the topic of the next two sections.

7.2 Workspace Theorems for Regular-Controlled Grammars

Indisputably, the workspace theorem for phrase-structure grammars fulfills a crucially important role in the grammatically oriented theory of formal languages as a whole (see Theorem 3.3.15). Indeed, it represents a powerful tool to demonstrate that if a phrase-structure grammar H generates each of its sentences by a derivation satisfying a linear length-limited condition (specifically, this condition requires that there is a positive integer k such that H generates every sentence $y \in L(H)$ by a derivation in which every sentential form x satisfies $|x| \leq k|y|$), then $L(H)$ is context sensitive. That is, $L(H)$ is generated by a context-sensitive grammar, which has no erasing rules and each of its derivations of nonempty strings has the property that the length of strings increases monotonically as follows from its definition (see Definitions 3.3.5). Considering this significant theorem, it is obvious that establishing similar workspace theorems for other types of grammar may be useful to formal language theory as well (for instance, regarding accepting programmed grammars, some workspace arguments have been considered in [1]). This section presents two new workspace theorems of this kind in terms of regular-controlled grammars (see Sect. 5.1).

More specifically, let H be a regular-controlled grammar. If there is a positive integer k such that H generates every sentence $y \in L(H)$ by a derivation in which every sentential form x satisfies a linear length-limited condition, then $L(H)$ is generated by a propagating regular-controlled grammar H'. To be exact, if for every sentence $y \in L(H)$, there exists a constant k and a derivation such that every sentential form x in the derivation contains at most $(k-1)|x|/k$ occurrences of nonterminals that are erased throughout the rest of the derivation, then $L(H)$ is generated by a propagating regular-controlled grammar H'. Consequently, the language family of all the languages generated by regular-controlled grammars satisfying the above condition is properly included in the family of context-sensitive languages because the latter properly contains the language family generated by propagating regular-controlled grammars (see Theorem 5.1.6). This section also gives an analogical workspace theorem in terms of regular-controlled grammars with appearance checking.

Let us sketch a significance of these workspace theorems in terms the theory of regulated grammars, which is central to this book. As demonstrated by the present book, this theory has already established many fundamental properties concerning regulated grammars. Nevertheless, there still remain some crucially significant open problems concerning them, including the exact effect of erasing rules to the generative power of some regulated grammars (see Chap. 5 and the conclusion of the previous section). Observe that the workspace theorems imply that erasing rules do not effect the power of regular-controlled grammars satisfying the workspace condition described above, so this theory can narrow the future investigation concerning this problem only to the grammars in which this condition is unsatisfied.

First, we define the key notion of this section and illustrate it by examples. For any rational number i, let floor(i) denote the greatest integer smaller than or equal to i.

Definition 7.2.1. Let $H = (G, \Xi, W)$ be a regular-controlled grammar with appearance checking, where $G = (N, T, \Psi, P, S)$. Let

$$D: S \Rightarrow_{(G,W)} x_1 \Rightarrow_{(G,W)} x_2 \Rightarrow_{(G,W)} \cdots \Rightarrow_{(G,W)} x_n = y\ [\rho]$$

be a derivation in G, where $x_i \in V^*$, for all $i, 1 \leq i \leq n$, for some $n \geq 1, y \in L(H)$, and $\rho \in \Xi$; then, the *workspace* of D is denoted by $WS_H(D)$ and defined as the smallest integer k satisfying that in $\Delta(D)$, there are at most

$$\text{floor}\left(\frac{(k-1)|x|}{k}\right)$$

ε-subtrees rooted at the symbols of x_i, for all i. For $y \in L(H)$, the *workspace* of y is denoted by $WS_H(y)$ and defined as the smallest integer in

$$\{WS_H(D) \mid D: S \Rightarrow^*_{(G,W)} y\ [\rho], \rho \in \Xi\}$$

If there exists a positive integer k such that for all $y \in L(H), WS_H(y) \leq k$, then H *generates* $L(H)$ *within a* k*-limited workspace*, symbolically written as

$$_{k\text{-lim}}WS_H(L(H))$$

□

Next, we illustrate Definition 7.2.1 by two examples of regular-controlled grammars without appearance checking.

Example 7.2.2. Let $H = (G, \Xi)$ be a regular-controlled grammar, where

$$\begin{aligned}
G &= (\{S, A, B, C\}, \{a, b, c\}, \{r_1, r_2, \ldots, r_7\}, P, S) \\
P &= \{r_1: S \to ABC, r_2: A \to aA, r_3: B \to bB, r_4: C \to cC, \\
&\qquad r_5: A \to \varepsilon, r_6: B \to \varepsilon, r_7: C \to \varepsilon\} \\
\Xi &= \{r_1\}\{r_2r_3r_4\}^*\{r_5r_6r_7\}
\end{aligned}$$

In every successful derivation, r_1 is applied first. Then, r_2, r_3, and r_4 are applied n times, for some $n \geq 0$. The derivation is completed by applying r_5, r_6, and r_7. Clearly, H generates the non-context-free language

$$\{a^n b^n c^n \mid n \geq 1\}$$

Observe that for every $a^n b^n c^n \in L(H)$, where $n \geq 1$, there exists a derivation D of the form

$$D: S \Rightarrow^*_G a^n A b^n B c^n C \Rightarrow^3_G a^n b^n c^n\ [\rho]$$

with $\rho \in \Xi$. Notice that $WS_H(D)$ cannot be equal to 1 because this would imply that no erasures are ever performed. However, observe that $WS_H(D) = 2$ and, furthermore, ${}_{2\text{-lim}}WS_H(L(H))$ because every sentential form x in D satisfies that in $\Delta(D)$, there are at most

$$\text{floor}\left(\frac{(2-1)|x|}{2}\right) = \text{floor}\left(\frac{|x|}{2}\right)$$

ε-subtrees rooted at the symbols of x. In other words, no more than half of all symbols in any sentential form in D are erased. □

Example 7.2.3. Let $H = (G, \Xi)$ be a regular-controlled grammar, where

$$\begin{aligned} G &= (\{S\}, \{a\}, \{r_1, r_2, r_3\}, P, S) \\ P &= \{r_1\colon S \to SS, r_2\colon S \to a, r_3\colon S \to \varepsilon\} \\ \Xi &= \{r_1\}^*\{r_2\}^*\{r_3\}^* \end{aligned}$$

Clearly, $L(H) = \{a^n \mid n \geq 1\}$. Observe that even though there is a derivation of the form

$$S \Rightarrow_G^* S^{m+p+1} \Rightarrow_G S^m a S^p \Rightarrow_G^* a \ [\rho]$$

where $\rho \in \Xi$, for any $m, p \geq 0$, it holds that ${}_{1\text{-lim}}WS_H(L(H))$. Indeed, for every $a^n \in L(H)$, where $n \geq 1$, there exists a derivation of the form

$$S \Rightarrow_G^* S^n \Rightarrow_G^* a^n \ [\rho]$$

with $\rho \in \Xi$, where no erasing rules are used. □

The following algorithm converts any regular-controlled grammar with appearance checking H satisfying ${}_{k\text{-lim}}WS_H(L(H))$, for some $k \geq 1$, to an equivalent propagating regular-controlled grammar with appearance checking H'. To give an insight into this conversion, we first explain the fundamental idea underlying this algorithm. H' uses two-component nonterminals of the form $\langle Z, z\rangle$, in which Z is a symbol and z is a string of no more than $k + g$ nonterminals, where g is the length of the longest right-hand side of a rule. To explain the meaning of these two components in $\langle Z, z\rangle$, Z is a symbol that is not erased during the derivation while z is a string of nonterminals that are erased. H' selects Z and z in a non-deterministic way so all possible selections of these two components are covered.

At any time, H' can move the nonterminals between the second components because these nonterminals are to be erased anyway, so it is completely irrelevant where they occur in the sentential forms. In addition, by Definition 7.2.1, there is always enough space to accommodate all these to-be-erased nonterminals in the second components of all nonterminals. Otherwise, as already pointed out, the algorithm uses the first components to simulate rewriting symbols that H does not erase during derivation while the simulation of rewriting nonterminals that H erases is performed in the second components.

Let us point out that the algorithm makes no predetermination of nonterminals from which ε can be derived as opposed to most standard methods of removing erasing rules, including the standard removal of erasing rules from context-free grammars (see Sect. 7.1.1). Indeed, if the output grammar improperly extends the second component of a two-component nonterminal by a nonterminal that is not erased throughout the rest of the derivation, then this occurrence of the nonterminal never disappears in this component, so a terminal string cannot be generated under this improper selection.

Algorithm 7.2.4. *Elimination of erasing rules from any regular-controlled grammar with appearance checking H satisfying ${}_{k\text{-lim}}WS_H(L(H))$, for some $k \geq 1$.*

Input: A context-free grammar, $G = (N, T, \Psi, P, S)$, a deterministic finite automaton, $M = (Q, \Psi, R, s, F)$, and an appearance checking set, $W \subseteq \Psi$, such that ${}_{k\text{-lim}}WS_H(L(H))$, for some $k \geq 1$, where $H = (G, L(M), W)$ is a regular-controlled grammar with appearance checking.

Output: A propagating context-free grammar, $G' = (N', T, \Psi', P', S')$, a deterministic finite automaton, $M' = (Q', \Psi', R', s, F)$, and an appearance checking set, $W' \subseteq \Psi'$, such that $L(H') = L(H)$, where $H' = (G', L(M'), W')$ is a propagating regular-controlled grammar with appearance checking.

Note: Let us note that in what follows, brackets $\langle, \rangle$, $\lfloor, \rfloor$, $\lceil$, and $\rceil$ are used to incorporate more symbols into a single compound symbol. Without any loss of generality, we assume that $\bot \notin Q$.

Method: Set $V = N \cup T$ and $k' = k + \max(\{|y| \mid A \rightarrow y \in P\})$. Initially, set

$$
\begin{aligned}
N' &= \{\langle Z, z\rangle \mid Z \in V, z \in N^*, 0 \leq |z| \leq k'\} \cup N \\
S' &= \langle S, \varepsilon\rangle \\
\Psi' &= \{\lfloor\langle a, \varepsilon\rangle \rightarrow a\rfloor \mid a \in T\} \\
P' &= \{\lfloor\langle a, \varepsilon\rangle \rightarrow a\rfloor\colon \langle a, \varepsilon\rangle \rightarrow a \mid a \in T\} \\
Q' &= Q \cup \{\bot\} \\
R' &= \{f\lfloor\langle a, \varepsilon\rangle \rightarrow a\rfloor \rightarrow f \mid f \in F, a \in T\} \\
W' &= \emptyset
\end{aligned}
$$

Repeat (1) through (4), given next, until none of the sets Ψ', P', Q', R', and W' can be extended in this way.

(1) **If** $r\colon A \rightarrow y_0Y_1y_1Y_2y_2 \cdots Y_my_m \in P$ and $\langle A, z\rangle, \langle Y_1, zy_0y_1 \cdots y_m\rangle \in N'$, where $y_i \in N^*$, $Y_j \in V$, for all i and j, $0 \leq i \leq m$, $1 \leq j \leq m$, for some $m \geq 1$

then

(1.1) add $t = \lfloor r, z, y_0, Y_1y_1, Y_2y_2, \ldots, Y_my_m\rfloor$ to Ψ';

(1.2) add $t: \langle A, z\rangle \to \langle Y_1, zy_0y_1 \cdots y_m\rangle\langle Y_2, \varepsilon\rangle \cdots \langle Y_m, \varepsilon\rangle$ to P';
(1.3) for each $pr \to q \in R$, add $pt \to q$ to R'.

(2) **If** $r: A \to y \in P$ and $\langle X, uAv\rangle, \langle X, uyv\rangle \in N'$, where $u, v, y \in N^*$
then

(2.1) add $t = \lfloor X, uAv, uyv, r\rfloor$ to Ψ';
(2.2) add $t: \langle X, uAv\rangle \to \langle X, uyv\rangle$ to P';
(2.3) for each $pr \to q \in R$, add $pt \to q$ to R'.

(3) **If** $\langle X, uAv\rangle, \langle Y, y\rangle, \langle Y, yA\rangle \in N'$, where $A \in N, u, v, y \in N^*$
then

(3.1) add $t_1 = \lfloor 1, X, uA, v, Yy\rfloor$ and $t_2 = \lfloor 2, X, uA, v, Yy\rfloor$ to Ψ';
(3.2) add $t_1: \langle X, uAv\rangle \to \langle X, uv\rangle$ and $t_2: \langle Y, y\rangle \to \langle Y, yA\rangle$ to P';
(3.3) for each $p \in Q$, add $q = \lceil p, t_1, t_2\rceil$ to Q' and add $pt_1 \to q$ and $qt_2 \to p$ to R'.

(4) **If** $r: A \to y \in P$ such that $r \in W$ and $pr \to q \in R, y \in V^*, p, q \in Q$
then

(4.1) add $t_1, t_2, \ldots, t_m$ to Ψ' and to W', where $t_i = \lfloor r, A_i\rfloor$, for all $i, 1 \leq i \leq m$, and A_1 through A_m are all nonterminals from N' of the form $\langle X, uAv\rangle$ or $\langle A, u'\rangle, X \in V, u, v, u' \in N^*$, for some $m \geq 1$;
(4.2) add $t_1: A_1 \to Ay, t_2: A_2 \to Ay, \ldots, t_m: A_m \to Ay$ to P';
(4.3) for every $j, 1 \leq j \leq m-1$, add $q_j = \lceil r, p, t_j, q\rceil$ to Q', and add $pt_1 \to q_1, q_1t_2 \to q_2, \ldots, q_{m-1}t_m \to q$ to R'.

For each $p \in Q'$ and $t \in \Psi'$ such that $pt \to q \notin R'$ for any $q \in Q'$, add $pt \to \bot$ to R'. For each $t \in \Psi'$, add $\bot t \to \bot$ to R'. □

Before verifying Algorithm 7.2.4 rigorously, we informally sketch the meaning of rules introduced in (1) through (4) and illustrate the algorithm by an example.

Rules introduced in (1) are used to simulate rewriting nonterminals that G does not erase during derivation. The simulation of rewriting nonterminals that G erases is performed by rules introduced in (2). The movement of nonterminals between the second components is done by rules introduced in (3). Rules from (4) handle the simulation of rules in the appearance checking mode as follows. To simulate $r: A \to y \in P, r \in W, G'$ has to make sure that A does not appear in any of the two component nonterminals in the current sentential form. If it does, then no string of terminals can be obtained. Indeed, if a rule from (4) is not applied in the appearance checking mode, the generated nonterminals cannot be rewritten to terminals because there are no rules with nonterminals from N on their left-hand sides in P'.

Finally, notice that the very last step of Algorithm 7.2.4 makes M' complete.

Example 7.2.5. Consider $H = (G, \Xi)$ from Example 7.2.2. Let $M = (Q, \Psi, R, s, F)$ be a deterministic finite automaton such that $L(M) = \Xi$. Set $W = \emptyset$.

With G, M, and W as input, Algorithm 7.2.4 produces a propagating context-free grammar G', a deterministic finite automaton M', and appearance checking set W', whose definition is left to the reader (notice that $W' = \emptyset$). We only describe the derivations of abc in G and in G'. That is, for

$$\begin{aligned} S &\Rightarrow_G ABC && [r_1] \\ &\Rightarrow_G aABC && [r_2] \\ &\Rightarrow_G aAbBC && [r_3] \\ &\Rightarrow_G aAbBcC && [r_4] \\ &\Rightarrow_G abBcC && [r_5] \\ &\Rightarrow_G abcC && [r_6] \\ &\Rightarrow_G abc && [r_7] \end{aligned}$$

one of the corresponding derivations in G' is

$$\begin{aligned} \langle S, \varepsilon \rangle &\Rightarrow_{G'} \langle A, \varepsilon \rangle \langle B, \varepsilon \rangle \langle C, \varepsilon \rangle && [\lfloor r_1, \varepsilon, \varepsilon, A, B, C \rfloor] \\ &\Rightarrow_{G'} \langle a, A \rangle \langle B, \varepsilon \rangle \langle C, \varepsilon \rangle && [\lfloor r_2, \varepsilon, \varepsilon, aA \rfloor] \\ &\Rightarrow_{G'} \langle a, A \rangle \langle b, B \rangle \langle C, \varepsilon \rangle && [\lfloor r_3, \varepsilon, \varepsilon, bB \rfloor] \\ &\Rightarrow_{G'} \langle a, A \rangle \langle b, B \rangle \langle c, C \rangle && [\lfloor r_4, \varepsilon, \varepsilon, cC \rfloor] \\ &\Rightarrow_{G'} \langle a, \varepsilon \rangle \langle b, B \rangle \langle c, C \rangle && [\lfloor a, A, \varepsilon, r_5 \rfloor] \\ &\Rightarrow_{G'} \langle a, \varepsilon \rangle \langle b, \varepsilon \rangle \langle c, C \rangle && [\lfloor b, B, \varepsilon, r_6 \rfloor] \\ &\Rightarrow_{G'} \langle a, \varepsilon \rangle \langle b, \varepsilon \rangle \langle c, \varepsilon \rangle && [\lfloor c, C, \varepsilon, r_7 \rfloor] \\ &\Rightarrow_{G'} a \langle b, \varepsilon \rangle \langle c, \varepsilon \rangle && [\lfloor \langle a, \varepsilon \rangle \rightarrow a \rfloor] \\ &\Rightarrow_{G'} ab \langle c, \varepsilon \rangle && [\lfloor \langle b, \varepsilon \rangle \rightarrow b \rfloor] \\ &\Rightarrow_{G'} abc && [\lfloor \langle c, \varepsilon \rangle \rightarrow c \rfloor] \end{aligned}$$

Clearly, both grammars generate the same language. □

Lemma 7.2.6. *Algorithm 7.2.4 is correct.*

Proof. Clearly, the algorithm always halts. Since P' does not contain any erasing rules, G' is propagating. To show that $L(H') = L(H)$, we first introduce some mathematical notions needed later in the proof.

Define the function $\mathfrak{D}$ from Ψ^* to Q as

$$\mathfrak{D}(w) = p \text{ if and only if } (s, w) \vdash_M^* (p, \varepsilon)$$

where s is the start state of M. Define the function $\mathfrak{D}'$ from Ψ'^* to Q' as

$$\mathfrak{D}'(w) = p \text{ if and only if } (s, w) \vdash_{M'}^* (p, \varepsilon)$$

where s is the start state of M'. Set $V' = N' \cup T$.

Next, we prove six claims. Claims 1, 2, and 4 are used to improve the readability of the rest of the proof. Claim 5 shows how derivations of G are simulated by G'.

Claim 6 shows the converse—that is, it demonstrates how derivations of G' are simulated by G. Claims 3, 5, and 6 are then used to establish $L(H') = L(H)$.

The first claim explains the role of $\bot$. That is, once M' occurs in $\bot$, M' cannot accept its input.

Claim 1. *Let*

$$\langle S, \varepsilon\rangle \Rightarrow^*_{(G,W')} x\ [\gamma]$$

where $x \in V'^+$, *such that* $\mathfrak{D}'(\gamma) = \bot$. *Then, there is no* $\chi \in \Psi'^*$ *such that*

$$\langle S, \varepsilon\rangle \Rightarrow^*_{(G,W')} x\ [\gamma] \Rightarrow^*_{(G,W')} y\ [\chi]$$

with $y \in L(H')$.

Proof. Let

$$\langle S, \varepsilon\rangle \Rightarrow^*_{(G,W')} x\ [\gamma]$$

where $x \in V'^+$, such that $\mathfrak{D}'(\gamma) = \bot$. Since $\bot \notin F$ and $\bot t \to t \in R'$, for all $t \in \Psi'$, by the very last step of the algorithm, this claim holds. □

The second claim shows that G' can always move nonterminals from N between the second components of its compound nonterminals.

Claim 2. *Let*

$$\langle S, \varepsilon\rangle \Rightarrow^*_{(G,W')} \langle X_1, u_1\rangle\langle X_2, u_2\rangle \cdots \langle X_h, u_h\rangle\ [\gamma]$$

where $u_i \in N^*, X_i \in V$, *for all* $i, 1 \leq i \leq h$, *for some* $h \geq 1$, *such that* $|u_1u_2\cdots u_h| < hk'$ *and* $\mathfrak{D}'(\gamma) \in Q$. *Then,*

$$\langle X_1, u_1\rangle\langle X_2, u_2\rangle \cdots \langle X_h, u_h\rangle \Rightarrow^*_{(G,W')} \langle X_1, v_1\rangle\langle X_2, v_2\rangle \cdots \langle X_h, v_h\rangle\ [\chi]$$

for any $v_1v_2\cdots v_h \in \mathrm{perm}(u_1u_2\cdots u_h)$ *such that* $\mathfrak{D}'(\gamma\chi) = \mathfrak{D}'(\gamma)$.

Proof. Let

$$\langle S, \varepsilon\rangle \Rightarrow^*_{(G,W')} \langle X_1, u_1\rangle\langle X_2, u_2\rangle \cdots \langle X_h, u_h\rangle\ [\gamma]$$

where $u_i \in N^*, X_i \in V$, for all $i, 1 \leq i \leq h$, for some $h \geq 1$, such that $|u_1u_2\cdots u_h| < hk'$ and $\mathfrak{D}'(\gamma) \in Q$. Since $|u_1u_2\cdots u_h| < hk'$, there has to be at least one nonterminal $\langle X_j, u_j\rangle$ with $|u_j| < k'$, for some $j, 1 \leq j \leq h$. By (3.2) in the algorithm, P' contains rules of the form

$$t_1\colon \langle Y, y_1Ay_2\rangle \to \langle Y, y_1y_2\rangle$$

and

$$t_2\colon \langle Z, z\rangle \rightarrow \langle Z, zA\rangle$$

where $Y, Z \in V, A \in N$, and $y_1, y_2, z \in N^*$. Observe that by using rules of this form, it is possible to consecutively derive

$$\langle X_1, u_1\rangle\langle X_2, u_2\rangle \cdots \langle X_h, u_h\rangle \Rightarrow^*_{(G,W')} \langle X_1, v_1\rangle\langle X_2, v_2\rangle \cdots \langle X_h, v_h\rangle\ [\chi]$$

for any $v_1 v_2 \cdots v_h \in \mathrm{perm}(u_1 u_2 \cdots u_h)$. Let $p = \mathfrak{D}'(\gamma)$ and $q = \lceil p, t_1, t_2\rceil$. By (3.3) in the algorithm, R' contains $pt_1 \rightarrow q$ and $qt_2 \rightarrow p$, so $\mathfrak{D}'(\gamma\chi) = \mathfrak{D}'(\gamma)$. Therefore, the claim holds. □

The third claim shows that in G', every derivation of any $y \in L(H')$ can be expressed as a two-part derivation. In the first part, every occurring symbol is a two-component nonterminal. In the second part, only the rules of the form $\langle a, \varepsilon\rangle \rightarrow a$, where $a \in T$, are used.

Claim 3. Let

$$\langle S, \varepsilon\rangle \Rightarrow^*_{(G,W')} y\ [\gamma]$$

where $y \in T^+$, *such that* $\gamma \in L(M')$. *Then,*

$$\langle S, \varepsilon\rangle \Rightarrow^*_{(G,W')} x\ [\gamma_1] \Rightarrow^*_{(G,W')} y\ [\gamma_2]$$

where $x \in (N' - N)^+$, *such that* $\gamma_1\gamma_2 \in L(M')$ *and during*

$$x \Rightarrow^*_{(G,W')} y\ [\gamma_2]$$

only rules of the form $\langle a, \varepsilon\rangle \rightarrow a$ *are used, where* $a \in T$.

Proof. Let

$$\langle S, \varepsilon\rangle \Rightarrow^*_{(G,W')} y\ [\gamma]$$

where $y \in T^+$, such that $\gamma \in L(M')$. By the initialization part of the algorithm,

$$\lfloor\langle a, \varepsilon\rangle \rightarrow a\rfloor\colon \langle a, \varepsilon\rangle \rightarrow a \in P'$$

and

$$f\,\lfloor\langle a, \varepsilon\rangle \rightarrow a\rfloor \rightarrow f \in R'$$

for all $f \in F$ and for all $a \in T$. Let $p = \mathfrak{D}'(\gamma)$. Since $\gamma \in L(M'), p \in F$. Based on these observations, we can always rearrange all the applications of the rules occurring in

$$\langle S, \varepsilon \rangle \Rightarrow^*_{(G,W')} y$$

so the claim holds. A fully rigorous proof is left to the reader. □

The fourth claim shows that after the first part of the derivation described in Claim 3 is completed, G' can generate $y \in L(H')$ if and only if the sequence of rules used during the derivation satisfies the condition described in Claim 4.

Claim 4. Let

$$\langle S, \varepsilon \rangle \Rightarrow^*_{(G,W')} \langle X_1, \varepsilon \rangle \langle X_2, \varepsilon \rangle \cdots \langle X_h, \varepsilon \rangle \; [\gamma]$$

such that $X_i \in T$, for all $i, 1 \leq i \leq h$, for some $h \geq 1$. Then, $X_1 X_2 \cdots X_h \in L(H')$ if and only if $\mathfrak{D}'(\gamma) \in F$.

Proof. Let

$$\langle S, \varepsilon \rangle \Rightarrow^*_{(G,W')} \langle X_1, \varepsilon \rangle \langle X_2, \varepsilon \rangle \cdots \langle X_h, \varepsilon \rangle \; [\gamma]$$

such that $X_i \in T$, for all $i, 1 \leq i \leq h$, for some $h \geq 1$. By the initialization part of the algorithm,

$$\lfloor \langle X_i, \varepsilon \rangle \rightarrow X_i \rfloor \colon \langle X_i, \varepsilon \rangle \rightarrow X_i \in P'$$

for all $i, 1 \leq i \leq h$, so

$$\langle X_1, \varepsilon \rangle \langle X_2, \varepsilon \rangle \cdots \langle X_h, \varepsilon \rangle \Rightarrow^h_{(G,W')} X_1 X_2 \cdots X_h \; [\chi]$$

Let $p = \mathfrak{D}'(\gamma)$. Since $f \lfloor \langle X_i, \varepsilon \rangle \rightarrow X_i \rfloor \rightarrow f \in R$ by the initialization part of the algorithm, for all $i, 1 \leq i \leq h$, and for all $f \in F$, and there is no $t \colon \langle a, \varepsilon \rangle \rightarrow a \in P'$, where $a \in T$, such that $pt \rightarrow o \in R'$ with $o \neq p, \mathfrak{D}'(\gamma\chi) \in F$ if and only if $\mathfrak{D}'(\gamma) \in F$. Therefore, $X_1 X_2 \cdots X_h \in L(H')$ if and only if $\mathfrak{D}'(\gamma) \in F$, so the claim holds. □

The fifth claim shows how G' simulates G. Since ${}_{k\text{-lim}}WS_H(L(H))$, we only describe the simulation of derivations D of $z \in L(H)$ satisfying $WS_H(D) \leq k$. This claim is used to prove that $L(H) \subseteq L(H')$ later in this proof. Recall the meaning of the ${}^{\varepsilon}$ and ${}^{\not\varepsilon}$ notation from Definition 3.3.26 because this notation is frequently used in the rest of the proof.

Claim 5. If

$$D \colon S \Rightarrow^n_{(G,W)} {}^{\varepsilon}x_0 {}^{\not\varepsilon}X_1 {}^{\varepsilon}x_1 {}^{\not\varepsilon}X_2 {}^{\varepsilon}x_2 \cdots {}^{\not\varepsilon}X_h {}^{\varepsilon}x_h \; [\alpha] \Rightarrow^*_{(G,W)} z$$

is a derivation in G satisfying $WS_H(D) \leq k$, *where* $z \in L(H), x_i \in N^*, X_j \in V$, *for all i and* $j, 0 \leq i \leq h, 1 \leq j \leq h$, *for some* $n \geq 0$ *and* $h \geq 1$, *then*

$$\langle S, \varepsilon \rangle \Rightarrow^*_{(G,W')} \langle X_1, u_1 \rangle \langle X_2, u_2 \rangle \cdots \langle X_h, u_h \rangle \; [\gamma]$$

where $u_1 u_2 \cdots u_h \in \mathrm{perm}(x_0 x_1 \cdots x_h)$ *and* $\mathfrak{D}'(\gamma) = \mathfrak{D}(\alpha)$.

Proof. This claim is established by induction on $n \geq 0$.

Basis. Let $n = 0$. Then, for

$$S \Rightarrow^0_{(G,W)} S \Rightarrow^*_{(G,W)} z$$

where $z \in L(H)$, there is

$$\langle S, \varepsilon \rangle \Rightarrow^0_{(G,W')} \langle S, \varepsilon \rangle$$

satisfying both conditions of the claim, so the basis holds.

Induction Hypothesis. Suppose that there exists $n \geq 0$ such that the claim holds for all derivations of length ℓ, where $0 \leq \ell \leq n$.

Induction Step. Consider any derivation of the form

$$D\colon S \Rightarrow^{n+1}_{(G,W)} w \Rightarrow^*_{(G,W)} z$$

where $w \in V^+$ and $z \in L(H)$, satisfying $WS_H(D) \leq k$. Since $n + 1 \geq 1$, this derivation can be expressed as

$$S \Rightarrow^n_{(G,W)} x \; [\alpha] \Rightarrow_{(G,W)} w \; [r] \Rightarrow^*_{(G,W)} z$$

for some $x \in V^+$. Let $x = {}^{\varepsilon}x_0 {}^{\not\varepsilon}X_1 {}^{\varepsilon}x_1 {}^{\not\varepsilon}X_2 {}^{\varepsilon}x_2 \cdots {}^{\not\varepsilon}X_h {}^{\varepsilon}x_h$, where $x_i \in N^*, X_j \in V$, for all i and $j, 0 \leq i \leq h, 1 \leq j \leq h$, for some $h \geq 1$. Then, by the induction hypothesis,

$$\langle S, \varepsilon \rangle \Rightarrow^*_{(G,W')} \langle X_1, u_1 \rangle \langle X_2, u_2 \rangle \cdots \langle X_h, u_h \rangle \; [\gamma]$$

where $u_1 u_2 \cdots u_h \in \mathrm{perm}(x_0 x_1 \cdots x_h)$ and $\mathfrak{D}'(\gamma) = \mathfrak{D}(\alpha)$. Let $p = \mathfrak{D}(\alpha)$ and $pr \to q \in R$.

Now, we consider all possible forms of $x \Rightarrow_{(G,W)} w \; [r]$, covered by the next three cases—(i) through (iii).

(i) Let $r\colon X_j \to y_0 Y_1 y_1 Y_2 y_2 \cdots Y_m y_m \in P$, where $y_i \in N^*$, for all $i, 0 \leq i \leq h, Y_i \in V$, for all $i, 1 \leq i \leq h$, for some $j, 1 \leq j \leq h$, and $m \geq 0$, so

$$
\begin{aligned}
& {}^{\varepsilon}x_0 {}^{\not\varepsilon}X_1 {}^{\varepsilon}x_1 \cdots {}^{\not\varepsilon}X_j {}^{\varepsilon}x_j \cdots {}^{\not\varepsilon}X_h {}^{\varepsilon}x_h \\
\Rightarrow_{(G,W)} & {}^{\varepsilon}x_0 {}^{\not\varepsilon}X_1 {}^{\varepsilon}x_1 \cdots {}^{\varepsilon}y_0 {}^{\not\varepsilon}Y_1 {}^{\varepsilon}y_1 {}^{\not\varepsilon}Y_2 {}^{\varepsilon}y_2 \cdots {}^{\not\varepsilon}Y_m {}^{\varepsilon}y_m {}^{\varepsilon}x_j \cdots {}^{\not\varepsilon}X_h {}^{\varepsilon}x_h \; [r] \\
\Rightarrow^*_{(G,W)} & z
\end{aligned}
$$

Let $y = y_0 y_1 \cdots y_m$. By (1.2) in the algorithm, there is

$$t\colon \langle X_j, u_j\rangle \to \langle Y_1, u_j y\rangle\langle Y_2, \varepsilon\rangle \cdots \langle Y_m, \varepsilon\rangle \in P'$$

(without any loss of generality, based on Claim 2, on Definition 7.2.1, and on the definition of k', we assume that $|u_j y| \leq k'$). Therefore,

$$
\begin{aligned}
& \langle X_1, u_1\rangle\langle X_2, u_2\rangle \cdots \langle X_j, u_j\rangle \cdots \langle X_h, u_h\rangle \\
\Rightarrow_{(G,W')} & \langle X_1, u_1\rangle\langle X_2, u_2\rangle \cdots \langle Y_1, u_j y\rangle\langle Y_2, \varepsilon\rangle \cdots \langle Y_m, \varepsilon\rangle \cdots \langle X_h, u_h\rangle \; [t]
\end{aligned}
$$

Recall that

$$u_1 u_2 \cdots u_h \in \operatorname{perm}(x_0 x_1 \cdots x_h)$$

by the induction hypothesis. Obviously,

$$u_1 u_2 \cdots u_j y u_{j+1} \cdots u_h \in \operatorname{perm}(x_0 x_1 \cdots x_{j-1} y x_j \cdots x_h)$$

Since $\mathfrak{D}'(\gamma) = \mathfrak{D}(\alpha) = p$ and $pr \to q \in R$, $pt \to q \in R'$ by (1.3) in the algorithm, so $\mathfrak{D}'(\gamma t) = \mathfrak{D}(\alpha r) = q$. Thus, the induction step is completed for (i).

(ii) Let $r\colon A \to y \in P$ such that $x_j = x'_j A x''_j$, for some $j, 0 \leq j \leq h$, where $y, x'_j, x''_j \in N^*$, so

$$
\begin{aligned}
& {}^{\varepsilon}x_0 {}^{\not\varepsilon}X_1 {}^{\varepsilon}x_1 \cdots {}^{\not\varepsilon}X_j {}^{\varepsilon}x'_j {}^{\varepsilon}A {}^{\varepsilon}x''_j \cdots {}^{\not\varepsilon}X_h {}^{\varepsilon}x_h \\
\Rightarrow_{(G,W)} & {}^{\varepsilon}x_0 {}^{\not\varepsilon}X_1 {}^{\varepsilon}x_1 \cdots {}^{\not\varepsilon}X_j {}^{\varepsilon}x'_j {}^{\varepsilon}y {}^{\varepsilon}x''_j \cdots {}^{\not\varepsilon}X_h {}^{\varepsilon}x_h \; [r] \\
\Rightarrow^*_{(G,W)} & z
\end{aligned}
$$

By the induction hypothesis, some u_i has to be of the form $u'_i A u''_i$, where $u'_i, u''_i \in N^*$, for some $i, 1 \leq i \leq h$. By (2.2) in the algorithm, there is

$$t\colon \langle X_i, u'_i A u''_i\rangle \to \langle X_i, u'_i y u''_i\rangle \in P'$$

(without any loss of generality, based on Claim 2, on Definition 7.2.1, and on the definition of k', we assume that $|u'_i y u''_i| \leq k'$). Therefore,

$$
\begin{aligned}
& \langle X_1, u_1\rangle\langle X_2, u_2\rangle \cdots \langle X_i, u'_i A u''_i\rangle \cdots \langle X_h, u_h\rangle \\
\Rightarrow_{(G,W')} & \langle X_1, u_1\rangle\langle X_2, u_2\rangle \cdots \langle X_i, u'_i y u''_i\rangle \cdots \langle X_h, u_h\rangle \; [t]
\end{aligned}
$$

Recall that

$$u_1u_2\cdots u_i'Au_i''\cdots u_h \in \mathrm{perm}(x_0x_1\cdots x_j'Ax_j''\cdots x_h)$$

by the induction hypothesis. Obviously,

$$u_1u_2\cdots u_i'yu_i''\cdots u_h \in \mathrm{perm}(x_0x_1\cdots x_j'yx_j''\cdots x_h)$$

Since $\mathfrak{D}'(\gamma) = \mathfrak{D}(\alpha) = p$ and $pr \to q \in R, pt \to q \in R'$ by (2.3) in the algorithm, so $\mathfrak{D}'(\gamma t) = \mathfrak{D}(\alpha r) = q$. Thus, the induction step is completed for (ii).

(iii) Let $r\colon A \to y \in P$ such that $A \notin \mathrm{alph}(x)$ and $r \in W$, so

$$x \Rightarrow_{(G,W)} x\ [r] \Rightarrow^*_{(G,W)} z$$

Since $r \in W$, by (4.2) in the algorithm, there are

$$t_1\colon A_1 \to Ay, t_2\colon A_2 \to Ay, \dots, t_m\colon A_m \to Ay \in P'$$

where A_1 through A_m are all nonterminals from N' of the form $\langle X, uAv\rangle$ or $\langle A, u'\rangle, X \in V, u, v, u' \in N^*$, for some $m \geq 1$, and $t_1, t_2, \dots, t_m \in W'$ by (4.1) in the algorithm. Let $x' = \langle X_1, u_1\rangle\langle X_2, u_2\rangle \cdots \langle X_h, u_h\rangle$. By the induction hypothesis, there is no $i, 1 \leq i \leq h$, such that $X_i = A$ or $u_i = u_i'Au_i''$. Therefore,

$$x' \Rightarrow^m_{(G,W')} x'\ [t_1t_2\cdots t_m]$$

Obviously,

$$u_1u_2\cdots u_h \in \mathrm{perm}(x_0x_1\cdots x_h)$$

By (4.3) in the algorithm, $(p, t_1, q_1), (q_1, t_2, q_2), \dots, q_{m-1}t_m \to q \in R'$, where $q_j = \lceil r, p, t_j, q\rceil$, for all $j, 1 \leq j \leq m-1$. As $\mathfrak{D}'(\gamma) = \mathfrak{D}(\alpha) = p, \mathfrak{D}'(\gamma t_1t_2\cdots t_m) = \mathfrak{D}(\alpha r) = q$. Thus, the induction step is completed for (iii).

Observe that cases (i) through (iii) cover all possible forms of $x \Rightarrow_{(G,W)} w\ [r]$. Thus, the claim holds. □

The following claim explains how G simulates derivations that satisfy a prescribed form in G'. This claim is used to prove that $L(H') \subseteq L(H)$ later in this proof.

Claim 6. If

$$\langle S, \varepsilon\rangle \Rightarrow^n_{(G,W')} \langle X_1, u_1\rangle\langle X_2, u_2\rangle \cdots \langle X_h, u_h\rangle\ [\gamma]$$

such that $\mathfrak{D}'(\gamma) \in Q$, *where* $u_i \in N^*, X_j \in V$, *for all* i *and* $j, 0 \leq i \leq h, 1 \leq j \leq h$, *for some* $n \geq 0$ *and* $h \geq 1$, *then*

$$S \Rightarrow^*_{(G,W)} x_0 X_1 x_1 X_2 x_2 \cdots X_h x_h \ [\alpha]$$

where $x_0 x_1 \cdots x_h \in \mathrm{perm}(u_1 u_2 \cdots u_h)$ *and* $\mathfrak{D}(\alpha) = \mathfrak{D}'(\gamma)$.

Proof. This claim is established by induction on $n \geq 0$.

Basis. Let $n = 0$. Then, for

$$\langle S, \varepsilon \rangle \Rightarrow^0_{(G,W')} \langle S, \varepsilon \rangle$$

there is

$$S \Rightarrow^0_{(G,W)} S$$

satisfying both conditions of the claim, so the basis holds.

Induction Hypothesis. Suppose that there exists $n \geq 0$ such that the claim holds for all derivations of length ℓ, where $0 \leq \ell \leq n$.

Induction Step. Consider any derivation of the form

$$\langle S, \varepsilon \rangle \Rightarrow^{n+1}_{(G,W')} w$$

where $w \in (N' - N)^+$. Since $n + 1 \geq 1$, this derivation can be expressed as

$$\langle S, \varepsilon \rangle \Rightarrow^n_{(G,W')} x \ [\gamma] \Rightarrow_{(G,W')} w \ [t]$$

for some $x \in (N' - N)^+$ such that $\mathfrak{D}'(\gamma) \in Q$. Let $x = \langle X_1, u_1 \rangle \langle X_2, u_2 \rangle \cdots \langle X_h, u_h \rangle$, where $u_i \in N^*, X_j \in V$, for all i and $j, 0 \leq i \leq h, 1 \leq j \leq h$, for some $h \geq 1$. By the induction hypothesis,

$$S \Rightarrow^*_{(G,W)} x_0 X_1 x_1 X_2 x_2 \cdots X_h x_h \ [\alpha]$$

where $x_0 x_1 \cdots x_h \in \mathrm{perm}(u_1 u_2 \cdots u_h)$ and $\mathfrak{D}(\alpha) = \mathfrak{D}'(\gamma)$.

Now, we consider all possible forms of $x \Rightarrow_{(G,W')} w \ [t]$, covered by the next four cases—(i) through (iv).

(i) Let $t\colon \langle X_j, u_j \rangle \to \langle Y_1, u_j y_0 y_1 \cdots y_m \rangle \langle Y_2, \varepsilon \rangle \cdots \langle Y_m, \varepsilon \rangle \in P'$ be introduced in (1.2) in the algorithm from $r\colon X_j \to y_0 Y_1 y_1 Y_2 y_2 \cdots Y_m y_m \in P$, where $y_i \in N^*$, for all $i, 0 \leq i \leq m, Y_i \in V$, for all $i, 1 \leq i \leq m$, for some $j, 1 \leq j \leq m$ and $m \geq 1$. Let $y = y_0 y_1 \cdots y_m$. Then,

$$\begin{aligned}&\langle X_1, u_1\rangle\langle X_2, u_2\rangle \cdots \langle X_j, u_j\rangle \cdots \langle X_h, u_h\rangle\\ \Rightarrow_{(G,W')}\ &\langle X_1, u_1\rangle\langle X_2, u_2\rangle \cdots \langle Y_1, u_j y\rangle\langle Y_2, \varepsilon\rangle \cdots \langle Y_m, \varepsilon\rangle \cdots \langle X_h, u_h\rangle\ [t]\end{aligned}$$

Since $r: X_j \rightarrow y_0 Y_1 y_1 Y_2 y_2 \cdots Y_m y_m \in P$,

$$\begin{aligned}&x_0 X_1 x_1 X_2 x_2 \cdots X_j x_j \cdots X_h x_h\\ \Rightarrow_{(G,W)}\ &x_0 X_1 x_1 X_2 x_2 \cdots y_0 Y_1 y_1 Y_2 y_2 \cdots Y_m y_m x_j \cdots X_h x_h\ [r]\end{aligned}$$

Recall that

$$u_1 u_2 \cdots u_h \in \operatorname{perm}(x_0 x_1 \cdots x_h)$$

by the induction hypothesis. Obviously,

$$u_1 u_2 \cdots u_j y u_{j+1} \cdots u_h \in \operatorname{perm}(x_0 x_1 \cdots x_{j-1} y x_j \cdots x_h)$$

Let $p = \mathfrak{D}'(\gamma)$ and $pt \rightarrow q \in R'$, for some $q \in Q'$. Since $\mathfrak{D}(\alpha) = \mathfrak{D}'(\gamma) = p$ by the induction hypothesis and $pr \rightarrow q \in R$ by (1.3) in the algorithm, $\mathfrak{D}(\alpha r) = \mathfrak{D}'(\gamma t) = q$. Thus, the induction step is completed for (i).

(ii) Let $t: \langle X_j, u'_j A u''_j\rangle \rightarrow \langle X_j, u'_j y u''_j\rangle \in P'$ be introduced in (2.2) in the algorithm from $r: A \rightarrow y \in P$ such that $u_j = u'_j A u''_j$, where $u'_j, u''_j, y \in N^*$, for some $j, 1 \leq j \leq h$. Then,

$$\begin{aligned}&\langle X_1, u_1\rangle\langle X_2, u_2\rangle \cdots \langle X_j, u'_j A u''_j\rangle \cdots \langle X_h, u_h\rangle\\ \Rightarrow_{(G,W')}\ &\langle X_1, u_1\rangle\langle X_2, u_2\rangle \cdots \langle X_j, u'_j y u''_j\rangle \cdots \langle X_h, u_h\rangle\ [t]\end{aligned}$$

By the induction hypothesis, $x_i = x'_i A x''_i$, for some $i, 0 \leq i \leq h$, where $x'_i, x''_i \in N^*$. Since $r: A \rightarrow y \in P$,

$$\begin{aligned}&x_0 X_1 x_1 X_2 x_2 \cdots X_i x'_i A x''_i \cdots X_h x_h\\ \Rightarrow_{(G,W)}\ &x_0 X_1 x_1 X_2 x_2 \cdots X_i x'_i y x''_i \cdots X_h x_h\ [r]\end{aligned}$$

Recall that

$$u_1 u_2 \cdots u'_j A u''_j \cdots u_h \in \operatorname{perm}(x_0 x_1 \cdots x'_i A x''_i \cdots x_h)$$

by the induction hypothesis. Obviously,

$$u_1 u_2 \cdots u'_i y u''_i \cdots u_h \in \operatorname{perm}(x_0 x_1 \cdots x'_j y x''_j \cdots x_h)$$

Let $p = \mathfrak{D}'(\gamma)$ and $pt \rightarrow q \in R'$, for some $q \in Q'$. Since $\mathfrak{D}(\alpha) = \mathfrak{D}'(\gamma) = p$ by the induction hypothesis and $pr \rightarrow q \in R$ by (2.3) in the algorithm, $\mathfrak{D}(\alpha r) = \mathfrak{D}'(\gamma t) = q$. Thus, the induction step is completed for (ii).

(iii) Let $t = t_1 \in \Psi'$ be introduced in (4.1) in the algorithm from $r: A \to y \in P$ such that $t \in W'$ and there is no i, $1 \leq i \leq h$, such that $X_i = A$ or $u_i = u'_i A u''_i$. Then,

$$x \Rightarrow_{(G,W')} x\ [t_1]$$

By (4.2) in the algorithm, there are

$$t_1: A_1 \to Ay, t_2: A_2 \to Ay, \dots, t_m: A_m \to Ay \in P'$$

where A_1 through A_m are all nonterminals from N' of the form $\langle X, uAv \rangle$ or $\langle A, u' \rangle$, $X \in V$, $u, v, u' \in N^*$, for some $m \geq 1$, and $t_1, t_2, \dots, t_m \in W'$ by (4.1) in the algorithm. Observe that now t_2 through t_m have to be applied in G' in the appearance checking mode; otherwise, if some other rules are applied, then by Claim 1, there is no sequence of rules that would lead to a derivation of a sentence from $L(H')$. Therefore,

$$x \Rightarrow^{m-1}_{(G,W')} x\ [t_2 \cdots t_m]$$

Let $x' = x_0 X_1 x_1 X_2 x_2 \cdots X_h x_h$. By the induction hypothesis, $A \notin \text{alph}(x')$. By (4) in the algorithm, $r \in W$, so

$$x' \Rightarrow_{(G,W)} x'\ [r]$$

Obviously,

$$u_1 u_2 \cdots u_h \in \text{perm}(x_0 x_1 \cdots x_h)$$

Let $p = \mathfrak{D}'(\gamma)$ and $pr \to q \in R$, for some $q \in Q$. By (4.3) in the algorithm, $(p, t_1, q_1), (q_1, t_2, q_2), \dots, q_{m-1} t_m \to q \in R'$, where $q_j = \lceil r, p, t_j, q \rceil$, for all j, $1 \leq j \leq m-1$. Since $\mathfrak{D}(\alpha) = \mathfrak{D}'(\gamma) = p$ by the induction hypothesis , $\mathfrak{D}(\alpha r) = \mathfrak{D}'(\gamma t_1 t_2 \cdots t_m) = q$. Thus, the induction step is completed for (iii).

(iv) Let $t_1: \langle X_j, u'_j A u''_j \rangle \to \langle X_j, u'_j u''_j \rangle \in P'$ be introduced in (3.2) in the algorithm such that $u_j = u'_j A u''_j$, where $u'_j, u''_j \in N^*$, for some j, $1 \leq j \leq h$. Then,

$$\begin{aligned} &\langle X_1, u_1 \rangle \langle X_2, u_2 \rangle \cdots \langle X_j, u'_j A u''_j \rangle \cdots \langle X_h, u_h \rangle \\ \Rightarrow_{(G,W')} &\langle X_1, u_1 \rangle \langle X_2, u_2 \rangle \cdots \langle X_j, u'_j u''_j \rangle \cdots \langle X_h, u_h \rangle\ [t_1] \end{aligned}$$

Let $p = \mathfrak{D}'(\gamma)$ and $pt_1 \to q \in R'$, for some $q \in Q'$. By (3.3) in the algorithm, $q = \lceil p, t_1, t_2 \rceil$, where $t_2: \langle X_i, z \rangle \to \langle X_i, zA \rangle \in P'$ by (3.2) in the algorithm, for some i, $1 \leq i \leq h$, and $z \in N^*$ (such i surely exists because $|u'_j u''_j| < k'$). Observe that now t_2 has to be applied in G'; otherwise, if a rule of a different form than the form of t_2 is applied, then $qt' \to \bot \in R$, where t' is that rule, and, consequently, by Claim 1, there is no sequence of rules that will lead to a derivation of a sentence from $L(H')$. Therefore,

$$\langle X_1, u_1\rangle\langle X_2, u_2\rangle \cdots \langle X_h, u_h\rangle \Rightarrow^2_{(G,W')} \langle X_1, v_1\rangle\langle X_2, v_2\rangle \cdots \langle X_h, v_h\rangle\ [t_1 t_2]$$

where $v_1 v_2 \cdots v_h \in \text{perm}(u_1 u_2 \cdots u_h)$. By (3.3) in the algorithm, $qt_2 \rightarrow p \in R'$, so $\mathfrak{D}'(\gamma t_1 t_2) = \mathfrak{D}'(\gamma)$. To complete the induction step, reconsider cases (i) through (iv).

Observe that cases (i) through (iv) cover all possible forms of $x \Rightarrow_{(G,W')} w\ [t]$. Thus, the claim holds. □

Next, we prove that $L(H') = L(H)$. Recall that ${}_{k\text{-lim}}WS_H(L(H))$ and consider Claim 5 when $x_i = \varepsilon, X_j \in T$, for all i and $j, 0 \leq i \leq h, 1 \leq j \leq h$, for some $h \geq 1$, and let $x = X_1 X_2 \cdots X_h$. Then,

$$S \Rightarrow^*_{(G,W)} x\ [\alpha]$$

and

$$\langle S, \varepsilon\rangle \Rightarrow^*_{(G,W')} \langle X_1, \varepsilon\rangle\langle X_2, \varepsilon\rangle \cdots \langle X_h, \varepsilon\rangle\ [\gamma]$$

such that $\mathfrak{D}'(\gamma) = \mathfrak{D}(\alpha)$. Let $p = \mathfrak{D}(\alpha)$. Since $x \in T^+, x \in L(H)$ if and only if $p \in F$. By Claim 4, $x \in L(H')$ if and only if $p \in F$. Hence, $L(H) \subseteq L(H')$.

Let $y \in L(H')$. By Claim 3,

$$\langle S, \varepsilon\rangle \Rightarrow^*_{(G,W')} x\ [\gamma_1] \Rightarrow^*_{(G,W')} y\ [\gamma_2]$$

where $x \in (N' - N)^+$, such that $\gamma_1\gamma_2 \in L(M')$ and during

$$x \Rightarrow^*_{(G,W')} y\ [\gamma_2]$$

only rules of the form $\langle a, \varepsilon\rangle \rightarrow a, a \in T$, are used. Therefore, let

$$x = \langle X_1, \varepsilon\rangle\langle X_2\ \varepsilon\rangle \cdots \langle X_h, \varepsilon\rangle$$

where $X_i \in T$, for all $i, 1 \leq i \leq h$, for some $h \geq 1$. Let $p = \mathfrak{D}'(\gamma_1)$. Since $\gamma_1\gamma_2 \in L(M'), p \neq \bot$ by Claim 1, so considering Claim 3, $p \in Q$. Consequently, by Claim 6,

$$S \Rightarrow^*_{(G,W)} X_1 X_2 \cdots X_h\ [\alpha]$$

such that $\mathfrak{D}(\alpha) = p$. Let $z = X_1 X_2 \cdots X_h$. Clearly, $z \in L(H)$ if and only if $p \in F$. By Claim 4, $z \in L(H')$ if and only if $p \in F$. Hence, $L(H') \subseteq L(H)$.

As $L(H) \subseteq L(H')$ and $L(H') \subseteq L(H), L(H') = L(H)$, so Lemma 7.2.6 holds. □

Theorem 7.2.7. *Let H be a regular-controlled grammar with appearance checking satisfying ${}_{k\text{-lim}}WS_H(L(H))$, for some $k \geq 1$. Then, $L(H) \in \mathbf{rC}_{ac}^{-\varepsilon}$.*

Proof. This theorem follows from Algorithm 7.2.4 and Lemma 7.2.6. □

Theorem 7.2.8. *Let H be a regular-controlled grammar satisfying ${}_{k\text{-lim}}WS_H(L(H))$, for some $k \geq 1$. Then, $L(H) \in \mathbf{rC}^{-\varepsilon}$.*

Proof. Reconsider Algorithm 7.2.4. Observe that if the input appearance checking set W is empty, then so is the output appearance checking set W' (in symbols, $W = \emptyset$ implies that $W' = \emptyset$). Therefore, Theorem 7.2.8 holds. □

The next theorem points out that the previous two theorems are achieved effectively.

Theorem 7.2.9. *Let $H = (G, L(M), W)$ be a regular-controlled grammar with appearance checking satisfying ${}_{k\text{-lim}}WS_H(L(H))$, for some $k \geq 1$, where M is a deterministic finite automaton. Then, there is an algorithm which converts H to a propagating regular-controlled grammar with appearance checking, $H' = (G', L(M'), W')$, such that $L(H') = L(H)$, where M' is a deterministic finite automaton. Furthermore, if $W = \emptyset$, then $W' = \emptyset$.*

Proof. This theorem follows from Algorithm 7.2.4, Lemma 7.2.6, and the observation made in the proof of Theorem 7.2.8. □

We next establish a denotation for the language families generated by regular-controlled grammars satisfying the workspace condition (see Definition 7.2.1). We define the following two families

$$\begin{aligned} \mathbf{rC}^{ws} &= \{L(H) \mid H \text{ is a regular-controlled grammar satisfying} \\ &\qquad {}_{k\text{-lim}}WS_H(L(H)), \text{ for some } k \geq 1\} \\ \mathbf{rC}_{ac}^{ws} &= \{L(H) \mid H \text{ is a regular-controlled grammar with appearance} \\ &\qquad \text{checking satisfying} {}_{k\text{-lim}}WS_H(L(H)), \text{ for some } k \geq 1\} \end{aligned}$$

Corollary 7.2.10. $\mathbf{CF} \subset \mathbf{rC}^{ws} = \mathbf{rC}^{-\varepsilon} \subset \mathbf{rC}_{ac}^{ws} = \mathbf{rC}_{ac}^{-\varepsilon} \subset \mathbf{CS} \subset \mathbf{RE} = \mathbf{rC}_{ac}$

Proof. Clearly, $\mathbf{rC}^{-\varepsilon} \subseteq \mathbf{rC}^{ws}$ and $\mathbf{rC}_{ac}^{-\varepsilon} \subseteq \mathbf{rC}_{ac}^{ws}$. From Theorems 7.2.7 and 7.2.8, $\mathbf{rC}^{ws} = \mathbf{rC}^{-\varepsilon}$ and $\mathbf{rC}_{ac}^{ws} = \mathbf{rC}_{ac}^{-\varepsilon}$. The rest of this corollary follows from Theorem 5.1.6. □

Regarding the relation of $\mathbf{rC}$ to other language families, it holds that $\mathbf{rC}^{-\varepsilon} \subseteq \mathbf{rC} \subset \mathbf{RE}$ (see Theorem 5.1.6), but it is an open question whether $\mathbf{rC}^{-\varepsilon} \subset \mathbf{rC}$.

We close this section by pointing out several remarks regarding the proved results, and by proposing two open problems.

(I) By analogy with establishing workspace theorems for regular-controlled grammars (see Theorems 7.2.7 and 7.2.8), we can obtain analogical theorems in terms of other regulated grammars. Specifically, these theorems can be straightforwardly reformulated in terms of matrix or programmed grammars.

(II) Consider accepting regulated grammars discussed in [2]. These grammars work, in essence, from terminal strings back towards the start symbol so they reduce the right-hand sides of rules to the corresponding left-hand sides. Surprisingly, Algorithm 7.2.4 cannot be straightforwardly adapted for them. Indeed, consider any derivation that reduces ε to a nonterminal by making several reductions. Of course, any proper simulation of these reductions requires recording the context in which the reduced symbols occur in the sentential forms. However, Algorithm 7.2.4 does not record these symbols in context at all. In fact, it does just the opposite: it arbitrarily moves these nonterminals between the second components of nonterminals throughout the sentential forms. Therefore, obtaining an algorithm that does the job of Algorithm 7.2.4 in terms of accepting regulated grammars represents a topic of the future research concerning the subject of the present section.

Open Problem 7.2.11. Consider Definition 7.2.1. Modify this definition strictly according to the derivation condition used in the workspace theorem for phrase-structure grammars (see Theorem 3.3.15). Reconsider Algorithm 7.2.4 and all the results obtained in the present section. Do they hold in terms of this modified definition, too? □

Open Problem 7.2.12. Establish workspace theorems for some more general framework, such as *grammars controlled by a bicolored digraph* (see p. 115 in [3]). Recall that regular-controlled grammars, matrix grammars, programmed grammars, and some other regulated grammars are special cases of this framework [4]. □

7.3 Generalized Restricted Erasing in Scattered Context Grammars

Considering the subject of this chapter, the present section deals with erasing rules in scattered context grammars (see Sect. 4.7). Specifically, it investigates their effect on the generative power of these grammars. On the one hand, scattered context grammars, which may contain erasing rules, characterize the family of recursively enumerable languages (see Theorem 4.7.6). On the other hand, all languages generated by propagating scattered context grammars, which do not contain erasing rules at all, are non-context-sensitive (see Theorem 4.7.6). As a result, some scattered context grammars with erasing rules cannot be converted to equivalent propagating scattered context grammars. Therefore, establishing conditions under which this conversion is performable represents a central topic of this investigation, and the present section pays a special attention to it. More specifically, this section demonstrates that this conversion is always possible if a scattered context grammar *erases its nonterminals in a generalized* k*-restricted way*, where $k \geq 0$: in every sentential form of a derivation, each of its substrings consisting of nonterminals from which the grammar derives empty strings later in the derivation is of length k or less. Consequently, the scattered context grammars that have erasing rules, but apply them in a generalized k-restricted way, are equivalent to the scattered context grammars that do not have erasing rules at all.

In [5], it was demonstrated that the family of all propagating scattered context languages is closed under restricted homomorphism. Note that our definition of generalized k-restricted erasing differs significantly from the way symbols are erased by k-restricted homomorphism. Under k-restricted homomorphism, a language can be generated by a propagating scattered context grammar if at most k symbols are deleted between any two consecutive terminals in a sentence (see [5]). However, under generalized k-restricted erasing, an unlimited number of symbols can be deleted between any two consecutive terminals in a sentence if the grammar erases its nonterminals in a generalized k-restricted way. In this sense, the result achieved in the present section generalizes the original result concerning k-restricted homomorphism.

To define the generalized k-restricted erasing formally, some auxiliary notions are needed.

Definition 7.3.1. The *core grammar underlying a scattered context grammar* $G = (V, T, P, S)$ is denoted by $\text{core}(G)$ and defined as the context-free grammar $\text{core}(G) = (V, T, P', S)$ with

$$P' = \{A_i \to x_i \mid (A_1, \dots, A_i, \dots, A_n) \to (x_1, \dots, x_i, \dots, x_n) \in P, 1 \leq i \leq n\}$$

□

Definition 7.3.2. Let $G = (V, T, P, S)$ be a scattered context grammar, and let $\text{core}(G)$ be the core grammar underlying G. Let

$$\begin{aligned} u_1 A_1 \cdots u_n A_n u_{n+1} &= v \\ \Rightarrow_G u_1 x_1 \cdots u_n x_n u_{n+1} &= w\ [(A_1, \dots, A_n) \to (x_1, \dots, x_n)] \end{aligned}$$

where $u_i \in V^*$, for all i, $1 \leq i \leq n+1$. The *partial m-step context-free simulation* of this derivation step by $\text{core}(G)$ is denoted by

$$\text{cf-sim}(v \Rightarrow_G w)_m$$

and defined as an m-step derivation of $\text{core}(G)$ of the form

$$\begin{aligned} & u_1 A_1 u_2 A_2 \cdots u_n A_n u_{n+1} \\ \Rightarrow_{\text{core}(G)} & u_1 x_1 u_2 A_2 \cdots u_n A_n u_{n+1} \\ \Rightarrow_{\text{core}(G)}^{m-1} & u_1 x_1 u_2 x_2 \cdots u_m x_m u_{m+1} A_{m+1} \cdots u_n A_n u_{n+1} \end{aligned}$$

where $m \leq n$. The *context-free simulation* of a derivation step of G, denoted by

$$\text{cf-sim}(v \Rightarrow_G w)$$

is the partial n-step context-free simulation of this step. Let $v = v_1 \Rightarrow_G^* v_n = w$ be of the form

$$v_1 \Rightarrow_G v_2 \Rightarrow_G \cdots \Rightarrow_G v_n$$

The context-free simulation of $v \Rightarrow_G^* w$ by core(G) is denoted by cf-sim$(v \Rightarrow_G^* w)$ and defined as

$$v_1 \Rightarrow_{\text{core}(G)}^* v_2 \Rightarrow_{\text{core}(G)}^* \cdots \Rightarrow_{\text{core}(G)}^* v_n$$

such that for all $i, 1 \leq i \leq n-1$, $v_i \Rightarrow_{\text{core}(G)}^* v_{i+1}$ is the context-free simulation of $v_i \Rightarrow_G v_{i+1}$. □

We next straightforwardly extend the notation of Definition 3.3.26 to derivations in scattered context grammars.

Definition 7.3.3. Let $G = (V, T, P, S)$ be a scattered context grammar, and let core(G) be the core grammar underlying G. Let $S \Rightarrow_G^* w$ be of the form $S \Rightarrow_G^* uXv \Rightarrow_G^* w$, where $u, v \in V^*$ and $X \in V$. Let cf-sim$(S \Rightarrow_G^* w)$ be the context-free simulation of $S \Rightarrow_G^* w$.

If $X \in V - T$ and the frontier of the subtree rooted at this occurrence of X in $\Delta(\text{cf-sim}(S \Rightarrow_G^* w))$ is ε, then G *erases* this occurrence of X in $S \Rightarrow_G^* w$, symbolically written as

$$^{\varepsilon}X$$

If either $X \in T$, or $X \in V - T$ and the frontier of the subtree rooted at this occurrence of X in $\Delta(\text{cf-sim}(S \Rightarrow_G^* w))$ differs from ε, then G *does not erase* this occurrence of X in $S \Rightarrow_G^* w$, symbolically written as

$$^{\not\varepsilon}X$$

By analogy with Definition 3.3.26, we define the notation $^{\varepsilon}x$ and $^{\not\varepsilon}x$, where $x \in V^*$. □

Generalized k-restricted erasing is defined next.

Definition 7.3.4. Let $G = (V, T, P, S)$ be a scattered context grammar, and let $k \geq 0$. G *erases its nonterminals in a generalized k-restricted way* if for every $y \in L(G)$, there exists a derivation $S \Rightarrow_G^* y$ such that every sentential form x of the derivation satisfies the following two properties

1. every $x = uAvBw$ such that $^{\not\varepsilon}A$, $^{\not\varepsilon}B$, and $^{\varepsilon}v$ satisfies $|v| \leq k$;
2. every $x = uAw$ such that $^{\not\varepsilon}A$ satisfies: if $^{\varepsilon}u$, then $|u| \leq k$, and if $^{\varepsilon}w$, then $|w| \leq k$. □

We illustrate these definitions by the following example.

Example 7.3.5. Let $G_1 = (N_1, T, P_1, S_1)$ and $G_2 = (N_2, T, P_2, S_2)$ be two right-linear grammars that satisfy $N_1 \cap N_2 = \emptyset$. The following example demonstrates the use of generalized k-restricted erasing to construct the scattered context grammar G satisfying

$$L(G) = \{ww \mid w \in L(G_1) \backslash L(G_2)\}$$

Suppose that $S, X, Y \notin N_1 \cup N_2 \cup T$. Define the homomorphism α from T^* to $\{\bar{a} \mid a \in T\}^*$ as $\alpha(a) = \bar{a}$, for all $a \in T$. Construct the scattered context grammar

$$G = \big(N_1 \cup N_2 \cup \{\bar{a} \mid a \in T\} \cup \{S, X, Y, Z\} \cup T, T, P, S\big)$$

with P defined as follows:

(1) add $(S) \to (XS_1XS_2XS_1XS_2)$ to P;
(2) for each $A \to xB \in P_1 \cup P_2$, add $(A, A) \to \big(\alpha(x)B, \alpha(x)B\big)$ to P;
(3) for each $A \to x \in P_1 \cup P_2$, add $(A, A) \to \big(\alpha(x)Z, \alpha(x)Z\big)$ to P.
(4) for each $a \in T$, add $(X, \bar{a}, X, \bar{a}, X, \bar{a}, X, \bar{a}) \to (\varepsilon, X, \varepsilon, X, \varepsilon, X, \varepsilon, X)$ to P;
(5) add $(X, Z, X, Z, X, Z, X, Z) \to (\varepsilon, \varepsilon, Y, \varepsilon, \varepsilon, \varepsilon, Y, \varepsilon)$ to P;
(6) for each $a \in T$, add $(Y, \bar{a}, Y, \bar{a}) \to (a, Y, a, Y)$ to P;
(7) add $(Y, Y) \to (\varepsilon, \varepsilon)$ to P.

Let $y \in T^*$ be the longest string such that $A \to yB \in P_1$ or $A \to y \in P_1$. Similarly, let $z \in T^*$ be the longest string such that $A \to zB \in P_2$ or $A \to z \in P_2$. Then, observe that G erases its nonterminals in a generalized $\big(|y| + |z| + 3\big)$-restricted way, for all $w \in L, w \neq \varepsilon$.

For instance, consider

$$G_1 = \big(\{S_1, a\}, \{a, b\}, \{S_1 \to aaS_1, S_1 \to \varepsilon\}, S_1\big)$$

with $L(G_1) = \{a^{2n} \mid n \geq 0\}$, and

$$G_2 = \big(\{S_2, a, b\}, \{a, b\}, \{S_2 \to aaaaS_2, S_2 \to b\}, S_2\big)$$

with $L(G_2) = \{a^{4n}b \mid n \geq 0\}$. We construct the scattered context grammar

$$G = \big(\{S_1, S_2, S, X, Y, \bar{a}, \bar{b}, a, b\}, \{a, b\}, P, S\big)$$

with P containing the following rules

$$
\begin{aligned}
1 &: (S) \rightarrow (XS_1XS_2XS_1XS_2) \\
2_a &: (S_1, S_1) \rightarrow (\bar{a}\bar{a}S_1, \bar{a}\bar{a}S_1) \\
2_b &: (S_2, S_2) \rightarrow (\bar{a}\bar{a}\bar{a}\bar{a}S_2, \bar{a}\bar{a}\bar{a}\bar{a}S_2) \\
3_a &: (S_1, S_1) \rightarrow (Z, Z) \\
3_b &: (S_2, S_2) \rightarrow (\bar{b}Z, \bar{b}Z) \\
4_a &: (X, \bar{a}, X, \bar{a}, X, \bar{a}, X, \bar{a}) \rightarrow (\varepsilon, X, \varepsilon, X, \varepsilon, X, \varepsilon, X) \\
4_b &: (X, \bar{b}, X, \bar{b}, X, \bar{b}, X, \bar{b}) \rightarrow (\varepsilon, X, \varepsilon, X, \varepsilon, X, \varepsilon, X) \\
5 &: (X, Z, X, Z, X, Z, X, Z) \rightarrow (\varepsilon, \varepsilon, Y, \varepsilon, \varepsilon, \varepsilon, Y, \varepsilon) \\
6_a &: (Y, \bar{a}, Y, \bar{a}) \rightarrow (a, Y, a, Y) \\
6_b &: (Y, \bar{b}, Y, \bar{b}) \rightarrow (b, Y, b, Y) \\
7 &: (Y, Y) \rightarrow (\varepsilon, \varepsilon)
\end{aligned}
$$

This grammar generates the language

$$L(G) = L = \{ww \mid w \in L(G_1) \backslash L(G_2)\} = \{a^{2n}ba^{2n}b \mid n \geq 0\}$$

In addition, G erases its nonterminals in a generalized 9-restricted way in every derivation; to illustrate, consider the following derivation

$$
\begin{aligned}
S &\Rightarrow_G XS_1XS_2XS_1XS_2 \ [1] \Rightarrow_G X\bar{a}\bar{a}S_1XS_2X\bar{a}\bar{a}S_1XS_2 \ [2_a] \\
&\Rightarrow_G X\bar{a}\bar{a}S_1X\bar{a}\bar{a}\bar{a}\bar{a}S_2X\bar{a}\bar{a}S_1X\bar{a}\bar{a}\bar{a}\bar{a}S_2 \ [2_b] \\
&\Rightarrow_G X\bar{a}S_1X\bar{a}\bar{a}\bar{a}S_2X\bar{a}S_1X\bar{a}\bar{a}\bar{a}S_2 \ [4_a] \Rightarrow_G XS_1X\bar{a}\bar{a}S_2XS_1X\bar{a}\bar{a}S_2 \ [4_a] \\
&\Rightarrow_G XZX\bar{a}\bar{a}S_2XZX\bar{a}\bar{a}S_2 \ [3_a] \Rightarrow_G XZX\bar{a}\bar{a}\bar{b}ZXZX\bar{a}\bar{a}\bar{b}Z \ [3_b] \\
&\Rightarrow_G Y\bar{a}\bar{a}\bar{b}Y\bar{a}\bar{a}\bar{b} \ [5] \Rightarrow_G aY\bar{a}\bar{b}aY\bar{a}\bar{b} \ [6_a] \Rightarrow_G aaY\bar{b}aaY\bar{b} \ [6_a] \\
&\Rightarrow_G aabYaabY \ [6_b] \Rightarrow_G aabaab \ [7]
\end{aligned}
$$

Notice that S_2 is not deleted in a successful derivation of $w \in L$, and the derivation can always proceed so that there are at most four $\bar{a}$s in front of S_2 and two $\bar{a}$s in front of S_1. Therefore, in the case when all $\bar{a}$s, S_1 and both Xs are deleted in front of $S_2, k = 9$, so the erasing is performed in a generalized 9-restricted way. In fact, the two $\bar{a}$s preceding S_2 cannot be deleted in a successful derivation of G, so the erasing is 7-restricted. □

As the previous example illustrates, erasing rules fulfill a useful role during the verification of a relation between substrings of a sentential form, so their involvement in a scattered context grammar is often desirable. Of course, this involvement does not rule out the existence of a propagating scattered context grammar equivalent to the scattered context grammar with erasing rules.

For instance, the existence of erasing rules in the grammar in Example 7.3.5 might suggest that no propagating scattered context grammar generates the resulting language. However, this is not the case at all. For instance, $\{a^{2n}ba^{2n}b \mid n \geq 0\}$ is clearly context-sensitive, and can be easily generated by a propagating scattered context grammar. This observation brings us to the following theorem, which says that if a scattered context grammar erases its nonterminals in a generalized k-restricted way, then it can be converted to an equivalent propagating scattered context grammar.

Theorem 7.3.6. *For every scattered context grammar G that erases its nonterminals in a generalized k-restricted way, there exists a propagating scattered context grammar $\bar{G}$ such that $L(G) = L(\bar{G})$.*

Proof. Let $G = (V, T, P, S)$ be a scattered context grammar. For each

$$p = (A_1, \ldots, A_i, \ldots, A_n) \to (x_1, \ldots, x_i, \ldots, x_n) \in P$$

let $\llparenthesis p, i \rrparenthesis$ denote $A_i \to x_i$, for all i, $1 \leq i \leq n$. Let

$$\begin{aligned} \Psi &= \{\llparenthesis p, i \rrparenthesis \mid p \in P, 1 \leq i \leq \mathrm{len}(p)\} \\ \Psi' &= \{\llparenthesis p, i \rrparenthesis' \mid \llparenthesis p, i \rrparenthesis \in \Psi\} \end{aligned}$$

Set

$$\bar{N}_1 = \{\langle x \rangle \mid x \in (V - T)^* \cup (V - T)^* T (V - T)^*, |x| \leq 2k + 1\}$$

For each $\langle x \rangle \in \bar{N}_1$ and $\llparenthesis p, i \rrparenthesis \in \Psi$, define

$$\text{lhs-replace}\big(\langle x \rangle, \llparenthesis p, i \rrparenthesis\big) = \Big\{\langle x_1 \llparenthesis p, i \rrparenthesis x_2 \rangle \mid x_1 \,\mathrm{lhs}\big(\llparenthesis p, i \rrparenthesis\big) x_2 = x\Big\}$$

Set

$$\bar{N}_2 = \Big\{\langle x \rangle \mid \langle x \rangle \in \text{lhs-replace}\big(\langle y \rangle, \llparenthesis p, i \rrparenthesis\big), \langle y \rangle \in \bar{N}_1, \llparenthesis p, i \rrparenthesis \in \Psi\Big\}$$

For each $\langle x \rangle \in \bar{N}_1$ and $\llparenthesis p, i \rrparenthesis' \in \Psi'$, define

$$\text{insert}\big(\langle x \rangle, \llparenthesis p, i \rrparenthesis'\big) = \Big\{\langle x_1 \llparenthesis p, i \rrparenthesis' x_2 \rangle \mid x_1 x_2 = x\Big\}$$

Set

$$\bar{N}_2' = \Big\{\langle x \rangle \mid \langle x \rangle \in \text{insert}\big(\langle y \rangle, \llparenthesis p, i \rrparenthesis'\big), \langle y \rangle \in \bar{N}_1, \llparenthesis p, i \rrparenthesis' \in \Psi'\Big\}$$

For each $x = \langle x_1 \rangle \cdots \langle x_n \rangle \in (\bar{N}_1 \cup \bar{N}_2 \cup \bar{N}_2')^*$, for some $n \geq 1$, define

$$\mathrm{join}(x) = x_1 \cdots x_n$$

For each $x \in \bar{N}_1 \cup \bar{N}_2 \cup \bar{N}_2'$, define

$$\text{split}(x) = \{y \mid x = \text{join}(y)\}$$

Set

$$\bar{V} = T \cup \bar{N}_1 \cup \bar{N}_2 \cup \bar{N}_2' \cup \{\bar{S}\}$$

Define the propagating scattered context grammar $\bar{G} = (\bar{V}, T, \bar{P}, \bar{S})$ with $\bar{P}$ constructed in the following way.

(1) For each $p = (S) \rightarrow (x) \in P$, add $(\bar{S}) \rightarrow (\langle (\!| p, 1 |\!) \rangle)$ to $\bar{P}$.
(2) For each $\langle x \rangle \in \bar{N}_1$, each $X \in \text{insert}(\langle x \rangle, (\!| p, n |\!)')$, where $p \in P, \text{len}(p) = n$, each $\langle y \rangle \in \bar{N}_1$, and each $Y \in \text{lhs-replace}(\langle y \rangle, (\!| q, 1 |\!))$, where $q \in P$, add

  (2.1) $(X, \langle y \rangle) \rightarrow (\langle x \rangle, Y)$, and
  (2.2) $(\langle y \rangle, X) \rightarrow (Y, \langle x \rangle)$ to $\bar{P}$;
  (2.3) if $\langle x \rangle = \langle y \rangle$, add
  $(X) \rightarrow (Y)$ to $\bar{P}$;
  (2.4) add $(X) \rightarrow (\langle x \rangle)$ to $\bar{P}$.

(3) For each $\langle x \rangle \in \bar{N}_1$, each $X \in \text{insert}(\langle x \rangle, (\!| p, i |\!)')$, where $p \in P, i < \text{len}(p)$, each $\langle y \rangle \in \bar{N}_1$, and each $Y \in \text{lhs-replace}(\langle y \rangle, (\!| p, i+1 |\!))$, add

  (3.1) $(X, \langle y \rangle) \rightarrow (\langle x \rangle, Y)$ to $\bar{P}$;
  (3.2) if $\langle x \rangle = \langle y \rangle$, and $X = x_1 (\!| p, i |\!)' x_2, Y = y_1 (\!| p, i+1 |\!) y_2$ satisfy $|x_1 (\!| p, i |\!)'| < |y_1 (\!| p, i+1 |\!)|$, add $(X) \rightarrow (Y)$ to $\bar{P}$.

(4) For each $\langle x_1 (\!| p, i |\!) x_2 \rangle \in \text{lhs-replace}(\langle x \rangle, (\!| p, i |\!))$, $\langle x \rangle \in \bar{N}_1, (\!| p, i |\!) \in \Psi$, and each $Y \in \text{split}(x_1 \, \text{rhs}((\!| p, i |\!)) (\!| p, i |\!)' x_2)$, add $(\langle x_1 (\!| p, i |\!) x_2 \rangle) \rightarrow (Y)$ to $\bar{P}$.
(5) For each $a \in T$, add $(\langle a \rangle) \rightarrow (a)$ to $\bar{P}$.

The propagating scattered context grammar $\bar{G}$ simulates G by using nonterminals of the form $\langle \dots \rangle$. In every nonterminal of this form, during every simulated derivation step, $\bar{G}$ records a substring corresponding to the current sentential form of G. The grammar $\bar{G}$ performs its derivation so that each of the nonterminals $\langle \dots \rangle$ contains at least one symbol not erased later in the derivation.

The rule constructed in (1) only initializes the simulation process. By rules introduced in (2) through (4), $\bar{G}$ simulates the application of a scattered context rule $p \in P$ in a left-to-right way. In a greater detail, by using a rule from (2), $\bar{G}$ nondeterministically selects a scattered context rule $p \in P$. Suppose that p consists of context-free rules $r_1, \dots, r_{i-1}, r_i, \dots, r_n$. By using rules from (3) and (4), $\bar{G}$ simulates the application of r_1 through r_n one by one. To explain this in a greater detail, suppose that $\bar{G}$ has just completed the simulation of r_{i-1}. Then, to the right of this simulation, $\bar{G}$ selects $\text{lhs}(r_i)$ by using a rule from (3). That is, this selection takes place either in a nonterminal of G in which the simulation of r_{i-1} has been performed or in one of the nonterminals appearing to the right of this nonterminal. After this selection, by using a rule from (4), $\bar{G}$ performs the replacement of the selected symbol $\text{lhs}(r_i)$ with $\text{rhs}(r_i)$. If a terminal occurs inside of a nonterminal of $\bar{G}$, then a rule from (5) allows $\bar{G}$ to change this nonterminal to the terminal contained in it.

Next, we establish $L(G) = L(\bar{G})$. In what follows, we denote the set of rules introduced in step i of the construction by $\bar{P}_i$, for $i = 1, 2, \dots, 5$.

Claim 1. Every successful derivation in $\bar{G}$ can be expressed in the following way

$$\begin{aligned} \bar{S} &\Rightarrow_{\bar{G}} \langle (\!| p, 1 |\!) \rangle \; [p_1] \\ &\Rightarrow_{\bar{G}}^{+} u \qquad [\Phi] \\ &\Rightarrow_{\bar{G}}^{+} v \qquad [\Theta] \end{aligned}$$

where $p_1 = (\bar{S}) \to (\langle (\!| p, 1 |\!) \rangle) \in \bar{P}_1, u = \langle a_1 \rangle \cdots \langle a_n \rangle, n \geq 1, a_1, \dots, a_n \in T, v = a_1 \cdots a_n$, Φ and Θ are sequences of rules from $\bar{P}_2 \cup \bar{P}_3 \cup \bar{P}_4$, and $\bar{P}_5$, respectively.

Proof. Rules from $\bar{P}_1$ are the only rules with $\bar{S}$ on their left-hand sides. Therefore, the derivation starts with a step made by one of these rules, so

$$\begin{aligned} \bar{S} &\Rightarrow_{\bar{G}} \langle (\!| p, 1 |\!) \rangle \; [p_1] \\ &\Rightarrow_{\bar{G}}^{+} v \end{aligned}$$

Observe that $\bar{S}$ does not appear on the right-hand side of any of the introduced rules. Therefore, rules from $\bar{P}_1$ are not used during the rest of the derivation.

The rules from $\bar{P}_5$ are the only rules with their right-hand sides over T; therefore, they have to be used to complete the derivation. Observe that none of the rules from $\bar{P}_1 \cup \bar{P}_2 \cup \bar{P}_3 \cup \bar{P}_4$ rewrites nonterminals of the form $\langle a \rangle$, where $a \in T$. As all rules from $\bar{P}_5$ are context-free, they can be applied whenever $\langle a \rangle$ appears in the sentential form. Without loss of generality, we assume that they are applied at the end of the derivation process; therefore,

$$\begin{aligned} \bar{S} &\Rightarrow_{\bar{G}} \langle (\!| p, 1 |\!) \rangle \; [p_1] \\ &\Rightarrow_{\bar{G}}^{+} u \qquad [\Phi] \\ &\Rightarrow_{\bar{G}}^{+} v \qquad [\Theta] \end{aligned}$$

so the claim holds. □

The following notation describes the strings that may appear in a successful derivation of $\bar{G}$. Let $S \Rightarrow_{\bar{G}}^{*} y \Rightarrow_{\bar{G}}^{*} w, w \in L(\bar{G}), y \in (\bar{N}_1 \cup \bar{N}_2 \cup \bar{N}_2')^*$, and let for each $\langle z \rangle \in \text{alph}(y)$ there exists

1. $A \in \text{alph}(z)$ such that ${}^{\not g}\!A$, or
2. $(\!| p, i |\!) \in \text{alph}(z), (\!| p, i |\!) \in \Psi, A = \text{lhs}((\!| p, i |\!))$ such that ${}^{\not g}\!A$.

Then, we write $\check{y}$.

Claim 2. Let

$$w_1 \in \text{split}\big(u_1 (\!|p, 1|\!) u_2 A_2 \cdots u_n A_n u_{n+1}\big)$$
$$\tau = u_1 A_1 u_2 A_2 \cdots u_n A_n u_{n+1}$$

where $u_1, \ldots, u_{n+1} \in V^*, A_1, \ldots, A_n \in V - T, A_1 = \text{lhs}\big((\!|p, 1|\!)\big), p = (A_1, \ldots, A_n) \rightarrow (x_1, \ldots, x_n) \in P$, *and* $\check{w}_1$*; then, every partial h-step context-free simulation*

$$\text{cf-sim}\big(\big(\tau = u_1 A_1 u_2 A_2 \cdots u_n A_n u_{n+1} \Rightarrow_G u_1 x_1 u_2 x_2 \cdots u_n x_n u_{n+1}\ [p]\big)\big)_h$$

of the form

$$\begin{aligned}
&u_1 A_1 u_2 A_2 u_3 A_3 \cdots u_n A_n u_{n+1} = \tau \\
\Rightarrow_{\text{core}(G)} &u_1 x_1 u_2 A_2 u_3 A_3 \cdots u_n A_n u_{n+1} \\
\Rightarrow_{\text{core}(G)} &u_1 x_1 u_2 x_2 u_3 A_3 \cdots u_n A_n u_{n+1} \\
\Rightarrow^{h-2}_{\text{core}(G)} &u_1 x_1 u_2 x_2 u_3 x_3 \cdots u_h x_h u_{h+1} A_{h+1} \cdots u_n A_n u_{n+1}
\end{aligned}$$

is performed in core(G) *if and only if*

$$\begin{aligned}
&w_1 \\
\Rightarrow_{\bar{G}}\ &w'_1\ [p_1^4] \\
\Rightarrow_{\bar{G}}\ &w_2\ [p_2^3] \\
\Rightarrow_{\bar{G}}\ &w'_2\ [p_2^4] \\
&\vdots \\
\Rightarrow_{\bar{G}}^{2h-5}\ &w_h\ [p_h^3] \\
\Rightarrow_{\bar{G}}\ &w'_h\ [p_h^4]
\end{aligned}$$

is performed in $\bar{G}$, *where* $p_2^3, \ldots, p_h^3 \in \bar{P}_3, p_1^4, \ldots, p_h^4 \in \bar{P}_4$, *and*

$$\begin{aligned}
w'_1 &\in \text{split}\big(u_1 x_1 (\!|p, 1|\!)' u_2 A_2 \cdots u_n A_n u_{n+1}\big) \\
w_2 &\in \text{split}\big(u_1 x_1 u_2 (\!|p, 2|\!) u_3 \cdots u_n A_n u_{n+1}\big) \\
w'_2 &\in \text{split}\big(u_1 x_1 u_2 x_2 (\!|p, 2|\!)' u_3 \cdots u_n A_n u_{n+1}\big) \\
&\vdots \\
w_h &\in \text{split}\big(u_1 x_1 u_2 x_2 \cdots u_h (\!|p, h|\!) u_{h+1} A_{h+1} \cdots u_n A_n u_{n+1}\big) \\
w'_h &\in \text{split}\big(u_1 x_1 u_2 x_2 \cdots u_h x_h (\!|p, h|\!)' u_{h+1} A_{h+1} \cdots u_n A_n u_{n+1}\big)
\end{aligned}$$

in addition, every $w \in \{w_2, \ldots, w_h, w'_1, \ldots, w'_h\}$ *satisfies* $\check{w}$.

Proof. We start by proving the only-if part of the claim.

Only If. The only-if part is proved by induction on g for all partial g-step context-free simulations, for $g \geq 1$.

Basis. Let $g = 1$. Then,

$$u_1 A_1 u_2 A_2 \cdots u_n A_n u_{n+1} \Rightarrow_{\text{core}(G)} u_1 x_1 u_2 A_2 \cdots u_n A_n u_{n+1}$$

and $w_1 \in \text{split}\big(u_1 \llparenthesis p, 1 \rrparenthesis u_2 A_2 \cdots u_n A_n u_{n+1}\big)$. Notice that every rule from $\bar{P}_2 \cup \bar{P}_3 \cup \bar{P}_4$ contains one symbol from

$$\text{insert}\big(\langle x \rangle, \llparenthesis p, i \rrparenthesis'\big) \cup \text{lhs-replace}\big(\langle x \rangle, \llparenthesis p, i \rrparenthesis\big)$$

where $\langle x \rangle \in \bar{N}_1$, $\llparenthesis p, i \rrparenthesis' \in \Psi'$, $\llparenthesis p, i \rrparenthesis \in \Psi$, both on its left-hand side and right-hand side. Therefore, only one of these symbols occurs in every sentential form of $\bar{G}$. As

$$w_1 \in \text{split}\big(u_1 \llparenthesis p, 1 \rrparenthesis u_2 A_2 \cdots u_n A_n u_{n+1}\big)$$

only a rule from $\bar{P}_4$ can be used in $\bar{G}$. The sentential form w_1 can be expressed as

$$w_1 \in \text{split}(u_{11})\big\langle u_{12} \llparenthesis p, 1 \rrparenthesis u_{21} \big\rangle\, \text{split}(u_{22})$$

where $u_{11}u_{12} = u_1$, $u_{21}u_{22} = u_2 A_2 \cdots u_n A_n u_{n+1}$. Consider the following two forms of $\big\langle u_{12} \llparenthesis p, 1 \rrparenthesis u_{21} \big\rangle$.

1. Let $\big\langle E_1^1 \cdots E_{l^1}^1 B^1 \cdots E_1^\alpha \cdots \hat{E}_m^\alpha \cdots E_{l^\alpha}^\alpha B^\alpha \cdots E_1^\beta \cdots E_{l^\beta}^\beta B^\beta E_1^{\beta+1} \cdots E_{l^{\beta+1}}^{\beta+1} \big\rangle$ with $\hat{E}_m^\alpha = \llparenthesis p, 1 \rrparenthesis$, $E_m^\alpha = \text{lhs}\big(\llparenthesis p, 1 \rrparenthesis\big)$, ${}^{\not\varepsilon}B^i$, for all i, $1 \leq i \leq \beta$, ${}^{\varepsilon}E_1^i, \ldots, {}^{\varepsilon}E_{l^i}^i$, for all i, $1 \leq i \leq \beta + 1$. If G erases nonterminals in a generalized k-restricted way, then $\bar{G}$ satisfies

$$\Big|\text{rhs}\big(\llparenthesis p, 1 \rrparenthesis\big)\Big| + l^\alpha - 1 \leq k$$

   By using the corresponding simulating rule $p_1^4 \in \bar{P}_4$, $\big\langle u_{12} \llparenthesis p, 1 \rrparenthesis u_{21} \big\rangle$ can be rewritten and splitted so that we obtain $\langle y_1 \rangle \cdots \langle y_o \rangle$, for some $o \geq 1$, satisfying

$$\begin{aligned} y_1 \cdots y_o = {} & E_1^1 \cdots E_{l^1}^1 B^1 \cdots E_1^\alpha \cdots \text{rhs}\big(\llparenthesis p, 1 \rrparenthesis\big) \llparenthesis p, 1 \rrparenthesis' \cdots \\ & \cdots E_{l^\alpha}^\alpha B^\alpha \cdots E_1^\beta \cdots E_{l^\beta}^\beta B^\beta E_1^{\beta+1} \cdots E_{l^{\beta+1}}^{\beta+1} \end{aligned}$$

   where for all j, $1 \leq j \leq o$, there exists $B \in \{B^1, \ldots, B^\beta\}$ such that $B \in \text{alph}(y_j)$. If G does not erase nonterminals in a generalized k-restricted way, then some y_j, $1 \leq j \leq o$, may satisfy

$$\text{alph}(y_j) \cap \big\{B^1, \ldots, B^\beta\big\} = \emptyset$$

However, in this case, $\langle y_j \rangle$ will be eventually rewritten to $\langle \varepsilon \rangle$, which cannot be rewritten by any rule, so the derivation would be unsuccessful.

2. Let $\langle E_1^1 \cdots E_{l^1}^1 B^1 \cdots E_1^\alpha \cdots E_{l^\alpha}^\alpha \hat{B}^\alpha \cdots E_1^\beta \cdots E_{l^\beta}^\beta B^\beta E_1^{\beta+1} \cdots E_{l^{\beta+1}}^{\beta+1} \rangle$ with $\hat{B}^\alpha = (\!| p, 1 |\!)$, $B^\alpha = \mathrm{lhs}((\!| p, 1 |\!))$, ${}^{\not\varepsilon}B^i$, for all $i, 1 \leq i \leq \beta$, and ${}^{\varepsilon}E_1^i, \dots, {}^{\varepsilon}E_{l^i}^i$, for all $i, 1 \leq i \leq \beta + 1$. Consider the simulated rule, denoted by $(\!| p, 1 |\!)$, of the form

$$B^\alpha \rightarrow C_1^1 \cdots C_{p^1}^1 D^1 \cdots C_1^\gamma \cdots C_{p^\gamma}^\gamma D^\gamma C_1^{\gamma+1} \cdots C_{p^{\gamma+1}}^{\gamma+1}$$

with ${}^{\not\varepsilon}D^i$, for all $i, 1 \leq i \leq \gamma$, and ${}^{\varepsilon}C_1^i, \dots, {}^{\varepsilon}C_{p^i}^i$, for all $i, 1 \leq i \leq \gamma + 1$. If G erases nonterminals in a generalized k-restricted way, the rule satisfies p^2, $\dots$, $p^\gamma \leq k$, and $p^1 + l^\alpha \leq k$, $p^{\gamma+1} + l^{\alpha+1} \leq k$. By using the corresponding simulating rule $p_1^4 \in \bar{P}_4$, $\langle u_{12} (\!| p, 1 |\!) u_{21} \rangle$ can be rewritten and splitted so that we obtain $\langle y_1 \rangle \cdots \langle y_o \rangle$, for some $o \geq 1$, satisfying

$$\begin{aligned} y_1 \cdots y_o = {} & E_1^1 \cdots E_{l^1}^1 B^1 \cdots E_1^\alpha \cdots E_{l^\alpha}^\alpha \\ & C_1^1 \cdots C_{p^1}^1 D^1 \cdots C_1^\gamma \cdots C_{p^\gamma}^\gamma D^\gamma C_1^{\gamma+1} \cdots C_{p^{\gamma+1}}^{\gamma+1} (\!| p, 1 |\!)' \cdots \\ & \cdots E_1^\beta \cdots E_{l^\beta}^\beta B^\beta E_1^{\beta+1} \cdots E_{l^{\beta+1}}^{\beta+1} \end{aligned}$$

where for all $j, 1 \leq j \leq o$, there exists $B \in \{B^1, \dots, B^\beta, D^1, \dots, D^\gamma\}$ such that $B \in \mathrm{alph}(y_j)$. If G does not erase nonterminals in a generalized k-restricted way, then some y_j, $1 \leq j \leq o$, may satisfy

$$\mathrm{alph}(y_j) \cap \{B^1, \dots, B^\beta, D^1, \dots, D^\gamma\} = \emptyset$$

Similarly to the form (1), $\langle y_j \rangle$ will be eventually rewritten to $\langle \varepsilon \rangle$, so the derivation would be unsuccessful.

Therefore, some rule $p_1^4 \in \bar{P}_4$ is applicable, so we obtain

$$w_1 \Rightarrow_{\bar{G}} w_1' \; [p_1^4]$$

with

$$w_1' \in \mathrm{split}(u_{11})\, \mathrm{split}\big(u_{12} x_1 (\!| p, 1 |\!)' u_{21}\big)\, \mathrm{split}(u_{22})$$

so $w_1' \in \mathrm{split}\big(u_{11} u_{12} x_1 (\!| p, 1 |\!)' u_{21} u_{22}\big)$, and

$$w_1' \in \mathrm{split}\big(u_1 x_1 (\!| p, 1 |\!)' u_2 A_2 \cdots u_n A_n u_{n+1}\big)$$

Induction Hypothesis. Suppose that the claim holds for all partial k-step context-free simulations, where $k \leq g$, for some $g \geq 1$.

Induction Step. Consider a partial $(g + 1)$-step context-free simulation

$$x \Rightarrow^{g+1}_{\text{core}(G)} z$$

where $z \in V^*$. As $g + 1 \geq 2$, there exists $y \in V^*$ such that

$$x \Rightarrow^{g}_{\text{core}(G)} y \Rightarrow_{\text{core}(G)} z \; [A_{g+1} \to x_{g+1}]$$

where $y = u_1 x_1 u_2 x_2 \cdots u_g x_g u_{g+1} A_{g+1} \cdots u_n A_n u_{n+1}$, and by the induction hypothesis,

$$w_1 \Rightarrow^{2g-1}_{\bar{G}} w'_g$$

We perform the partial $(g + 1)$-step context-free simulation of a rule p, so $g < \text{len}(p)$ and

$$w'_g \in \text{split}\big(u_1 x_1 u_2 x_2 \cdots u_g x_g (\!| p, g |\!)' u_{g+1} A_{g+1} \cdots u_n A_n u_{n+1}\big)$$

therefore, only rules from $\bar{P}_3$ can be used. The sentential form w'_g has either of the following two forms

- $\text{split}(r_1)\big\langle r_2 (\!| p, g |\!)' r_3 \big\rangle \, \text{split}(r_4 A_{g+1} r_5)$ such that $y = r_1 r_2 r_3 r_4 A_{g+1} r_5$. In this case, a rule from (3.1) can be used and after its application, we obtain

$$w_{g+1} = \text{split}(r_1)\langle r_2 r_3 \rangle \, \text{split}\big(r_4 (\!| p, g + 1 |\!) r_5\big)$$

- $\text{split}(r_1)\big\langle r_2 (\!| p, g |\!)' u_{g+1} A_{g+1} r_3 \big\rangle \, \text{split}(r_4)$ such that $y = r_1 r_2 u_{g+1} A_{g+1} r_3 r_4$. In this case, a rule from (3.1) can be used and after its application, we obtain

$$w_{g+1} = \text{split}(r_1)\big\langle r_2 u_{g+1} (\!| p, g + 1 |\!) r_3 \big\rangle \, \text{split}(r_4)$$

Thus, this derivation step can be expressed as

$$w'_g \Rightarrow_{\bar{G}} w_{g+1} \; [p^3_{g+1}]$$

where $p^3_{g+1} \in \bar{P}_3$, and

$$w_{g+1} \in \text{split}\big(u_1 x_1 u_2 x_2 \cdots u_g x_g u_{g+1} (\!| p, g + 1 |\!) \cdots u_n A_n u_{n+1}\big)$$

If $\check{w}'_g$, the form of the rules from $\bar{P}_3$ implies that w_{g+1} satisfies $\check{w}_{g+1}$ as well. At this point, only rules from $p^4_{g+1} \in \bar{P}_4$ can be used. The proof of

$$w_{g+1} \Rightarrow_{\bar{G}} w'_{g+1} \; [p^4_{g+1}]$$

is analogous to the proof of

$$w_1 \Rightarrow_{\bar{G}} w_1' \; [p_1^4]$$

described in the basis and is left to the reader. As a result,

$$w'_{g+1} \in \text{split}\big(u_1 x_1 u_2 x_2 \cdots u_{g+1} x_{g+1} \llparenthesis p, g+1 \rrparenthesis' u_{g+2} A_{g+2} \cdots u_n A_n u_{n+1}\big)$$

satisfying $\breve{w}'_{g+1}$.

If. The if part is proved by induction on g for all g-step derivations in $\bar{G}$, for $g \geq 1$.

Basis. Let $g = 1$. Then, $w_1 \Rightarrow_{\bar{G}} w_1' \; [p_1^4]$ and $\tau = u_1 A_1 \cdots u_n A_n u_{n+1}$. Clearly,

$$u_1 A_1 \cdots u_n A_n u_{n+1} \Rightarrow_{\text{core}(G)} u_1 x_1 \cdots u_n A_n u_{n+1} \; [A_1 \to x_1]$$

Induction Hypothesis. Suppose that the claim holds for all k-step derivations, where $k \leq g$, for some $g \geq 1$.

Induction Step. Consider any derivation of the form

$$w_1 \Rightarrow_{\bar{G}}^{2(g+1)-1} w'_{g+1}$$

where

$$w'_{g+1} \in \text{split}\big(u_1 x_1 \cdots u_{g+1} x_{g+1} \llparenthesis p, g+1 \rrparenthesis' u_{g+2} A_{g+2} \cdots u_n A_n u_{n+1}\big)$$

Since $2(g+1) - 1 \geq 3$, there exists a derivation

$$w_1 \Rightarrow_{\bar{G}}^{2g-1} w'_g \Rightarrow_{\bar{G}} w_{g+1} \Rightarrow_{\bar{G}} w'_{g+1}$$

where

$$w_{g+1} \in \text{split}\big(u_1 x_1 \cdots u_g x_g u_{g+1} \llparenthesis p, g+1 \rrparenthesis u_{g+2} A_{g+2} \cdots u_n A_n u_{n+1}\big)$$

By the induction hypothesis, there is a derivation

$$u_1 A_1 \cdots u_n A_n u_{n+1} \Rightarrow_{\text{core}(G)}^{g} u_1 x_1 \cdots u_g x_g u_{g+1} A_{g+1} \cdots u_n A_n u_{n+1}$$

Clearly,

$$\begin{aligned} & u_1 x_1 \cdots u_g x_g u_{g+1} A_{g+1} u_{g+2} A_{g+2} \cdots u_n A_n u_{n+1} \\ \Rightarrow_{\text{core}(G)} & u_1 x_1 \cdots u_g x_g u_{g+1} x_{g+1} u_{g+2} A_{g+2} \cdots u_n A_n u_{n+1} \; [A_{g+1} \to x_{g+1}] \end{aligned}$$

Hence, the claim holds. □

Claim 3. The result from Claim 2 holds for a context-free simulation.

Proof. As the context-free simulation of the whole scattered context rule

$$(A_1, \ldots, A_n) \to (x_1, \ldots, x_n)$$

is a partial context-free simulation of the length n, Claim 2 holds for a context-free simulation as well. □

We use the following notation to describe the simulation of a derivation step in G by rules of $\bar{G}$ as demonstrated by Claims 2 and 3. Let

$$\begin{aligned} & u_1 A_1 \cdots u_n A_n u_{n+1} \\ \Rightarrow_G & u_1 x_1 \cdots u_n x_n u_{n+1} \; [p] \end{aligned}$$

for some $p = (A_1, \ldots, A_n) \to (x_1, \ldots, x_n) \in P$. Then, $\Rrightarrow$ denotes the simulation of this derivation step in $\bar{G}$. We write

$$w_1 \Rrightarrow w'_n \; [p]$$

or, shortly, $w_1 \Rrightarrow w'_n$. Therefore, $w_1 \Rightarrow_{\bar{G}}^{2n-1} w'_n$ from Claim 2 is equal to $w_1 \Rrightarrow w'_n$.

Claim 4. Let $x_1 \in V^*$ *and* $\bar{x}'_1 \in \mathrm{split}(x'_{11} (\!| p_1, 1 |\!) x'_{12})$, *where* $x_1 = x'_{11} \, \mathrm{lhs}((\!| p_1, 1 |\!)) x'_{12}$, $(\!| p_1, 1 |\!) \in \Psi$, *and* $\check{\bar{x}}'_1$; *then, every derivation*

$$\begin{aligned} & x_1 \\ \Rightarrow_G & x_2 \quad [p_1] \\ & \vdots \\ \Rightarrow_G & x_{m+1} \; [p_m] \end{aligned}$$

is performed in G if and only if

$$\begin{aligned} & \bar{x}'_1 \\ \Rrightarrow \; & \bar{x}_1 \quad [p_1] \\ \Rightarrow_{\bar{G}} & \bar{x}'_2 \quad [p'_2] \\ \Rrightarrow \; & \bar{x}_2 \quad [p_2] \\ \Rightarrow_{\bar{G}} & \bar{x}'_3 \quad [p'_3] \\ & \vdots \\ \Rrightarrow \; & \bar{x}_m \quad [p_m] \\ \Rightarrow_{\bar{G}} & \bar{x}'_{m+1} \; [p'_{m+1}] \end{aligned}$$

is performed in $\bar{G}$, *where* $x_2, \ldots, x_{m+1} \in V^*$, $p_1, \ldots, p_m \in P$, $p'_2, \ldots, p'_{m+1} \in \bar{P}_2$,

$$\begin{aligned}\bar{x}_i &\in \operatorname{split}\big(x_{i1}\llparenthesis p_i, \operatorname{len}(p_i)\rrparenthesis' x_{i2}\big)\\ \bar{x}'_j &\in \operatorname{split}\big(x'_{j1}\llparenthesis p_j, 1\rrparenthesis x'_{j2}\big)\end{aligned}$$

for all i *and* j, $1 \leq i \leq m, 2 \leq j \leq m$, *and*

- $\bar{x}'_{m+1} \in \operatorname{split}\big(x'_{(m+1)1}\llparenthesis p_{m+1}, 1\rrparenthesis x'_{(m+1)2}\big)$, *for* $x_{m+1} \notin T^*$, *or*
- $\bar{x}'_{m+1} \in \operatorname{split}(x_{m+1})$, *for* $x_{m+1} \in T^*$,

where $x_{i1}x_{i2} = x_i$, *for all* $i, 1 \leq i \leq m$, $x'_{j1} \operatorname{lhs}(\llparenthesis p_j, 1\rrparenthesis) x'_{j2} = x_j$, *for all* $j, 2 \leq j \leq m+1$, *and every* $\bar{x} \in \{\bar{x}_1, \ldots, \bar{x}_m, \bar{x}'_2, \ldots, \bar{x}'_{m+1}\}$ *satisfies* $\check{\bar{x}}$.

Proof. We start by proving the only-if part of the claim.

Only If. The only-if part is proved by induction on g for all g-step derivations in G, for $g \geq 0$.

Basis. Let $g = 0$. Then, $x_1 \Rightarrow_G^0 x_1$ and $\bar{x}'_1 \Rightarrow_{\bar{G}}^0 \bar{x}'_1$.

Induction Hypothesis. Suppose that the claim holds for all k-step derivations, where $k \leq g$, for some $g \geq 0$.

Induction Step. Consider any derivation of the form

$$x_1 \Rightarrow_G^{g+1} x_{g+2}$$

Since $g + 1 \geq 1$, there exists a derivation

$$x_1 \Rightarrow_G^{g} x_{g+1} \Rightarrow_G x_{g+2}$$

By the induction hypothesis, there is a derivation

$$\bar{x}'_1 \Rrightarrow \bar{x}_1 \Rightarrow_{\bar{G}} \bar{x}'_2 \Rrightarrow \bar{x}_2 \Rightarrow_{\bar{G}} \cdots \Rrightarrow \bar{x}_g \Rightarrow_{\bar{G}} \bar{x}'_{g+1}$$

Because we are performing $(g + 1)$-step derivation, $x_{g+1} \notin T^*$. The sentential forms $\bar{x}_g$ and $\bar{x}'_{g+1}$ in the g-step simulation have to satisfy

$$\begin{aligned}\bar{x}_g &\in \operatorname{split}\big(x_{g1}\llparenthesis p_g, \operatorname{len}(p_g)\rrparenthesis' x_{g2}\big)\\ \bar{x}'_{g+1} &\in \operatorname{split}\big(x'_{(g+1)1}\llparenthesis p_{g+1}, 1\rrparenthesis x'_{(g+1)2}\big)\end{aligned}$$

where $x_g = x_{g1}x_{g2}$,

$$x_{g+1} = x'_{(g+1)1} \operatorname{lhs}(\llparenthesis p_{g+1}, 1\rrparenthesis) x'_{(g+1)2}$$

and $\check{\bar{x}}'_{g+1}$. Then, by Claim 3,

$$\bar{x}'_{g+1} \Rightarrow \bar{x}_{g+1}$$

where

$$\bar{x}_{g+1} \in \text{split}\big(x_{(g+1)1} (\!| p_{g+1}, \text{len}(p_{g+1}) |\!)' x_{(g+1)2}\big)$$

$x_{(g+1)1} x_{(g+1)2} = x_{g+1}$, and $\check{\bar{x}}_{g+1}$. As

$$\bar{x}_{g+1} \in \text{split}\big(x_{(g+1)1} (\!| p_{g+1}, \text{len}(p_{g+1}) |\!)' x_{(g+1)2}\big)$$

$(\!| p, i |\!)' = (\!| p_{g+1}, \text{len}(p_{g+1}) |\!)' \in \Psi'$, and $i = \text{len}(p)$, only rules from $\bar{P}_2$ can be used. Let

$$x_{g+2} = u_{(g+2)1} A_{(g+2)1} \cdots u_{(g+2)j} A_{(g+2)j} u_{(g+2)(j+1)}$$

where $u_{(g+2)1}, \ldots, u_{(g+2)(j+1)} \in V^*$, $A_{(g+2)1}, \ldots, A_{(g+2)j} \in V - T$, for some $j \geq 1$. We distinguish the following three forms of $\bar{x}_{g+1}$.

1. Let $\text{split}(r_1)\big\langle r_2 (\!| p_{g+1}, \text{len}(p_{g+1}) |\!)' r_3 \big\rangle \, \text{split}(r_4 A_{(g+2)1} r_5)$ such that

$$x_{g+2} = r_1 r_2 r_3 r_4 A_{(g+2)1} r_5$$

   Then, a rule constructed in (2.1) can be used, and by its application, we obtain a sentential form

$$\bar{x}'_{g+2} = \text{split}(r_1) \langle r_2 r_3 \rangle \, \text{split}\big(r_4 (\!| p_{g+2}, 1 |\!) r_5\big)$$

2. Let $\text{split}(r_1 A_{(g+2)1} r_2)\big\langle r_3 (\!| p_{g+1}, \text{len}(p_{g+1}) |\!)' r_4 \big\rangle \, \text{split}(r_5)$ such that

$$x_{g+2} = r_1 A_{(g+2)1} r_2 r_3 r_4 r_5$$

   Then, a rule constructed in (2.2) can be used, and by its application, we obtain a sentential form

$$\bar{x}'_{g+2} = \text{split}\big(r_1 (\!| p_{g+2}, 1 |\!) r_2\big) \langle r_3 r_4 \rangle \, \text{split}(r_5)$$

3. Consider 3.a and 3.b, given next.

   3.a Let $\text{split}(r_1)\big\langle r_2 (\!| p_{g+1}, \text{len}(p_{g+1}) |\!)' r_3 A_{(g+2)1} r_4 \big\rangle \, \text{split}(r_5)$ such that

$$x_{g+2} = r_1 r_2 r_3 A_{(g+2)1} r_4 r_5$$

Then, a rule constructed in (2.3) can be used, and by its application, we obtain a sentential form

$$\bar{x}'_{g+2} = \text{split}(r_1)\langle r_2 r_3 (\!| p_{g+2}, 1 |\!) r_4 \rangle \, \text{split}(r_5)$$

3.b Let $\text{split}(r_1)\langle r_2 A_{(g+2)1} r_3 (\!| p_{g+1}, \text{len}(p_{g+1}) |\!)' r_4 \rangle \, \text{split}(r_5)$ such that

$$x_{g+2} = r_1 r_2 A_{(g+2)1} r_3 r_4 r_5$$

Then, a rule constructed in (2.3) can be used, and by its application, we obtain a sentential form

$$\bar{x}'_{g+2} = \text{split}(r_1)\langle r_2 (\!| p_{g+2}, 1 |\!) r_3 r_4 \rangle \, \text{split}(r_5)$$

In the case when $x_{g+2} \in T^*$, $\bar{x}_{g+1}$ has the form

$$\text{split}(r_1)\langle r_2 (\!| p_{g+1}, \text{len}(p_{g+1}) |\!)' r_3 \rangle \, \text{split}(r_4)$$

where $x_{g+2} = r_1 r_2 r_3 r_4 \in T^*$. Then, a rule constructed in (2.4) can be used, and by its application, we obtain a sentential form

$$\bar{x}'_{g+2} = \text{split}(r_1)\langle r_2 r_3 \rangle \, \text{split}(r_4)$$

Notice that a rule from (2.4) is applicable also in the case when $x_{g+2} \notin T^*$ (cases 1 through 3 above). This rule removes the symbol from Ψ' from the sentential form. However, as all rules from (2) through (4) require a symbol from Ψ or Ψ', its application for $x_{g+2} \notin T^*$ does not lead to a successful derivation.

As a result, there is a derivation

$$\bar{x}_{g+1} \Rightarrow_{\bar{G}} \bar{x}'_{g+2} \; [p'_{g+1}]$$

where

- $\bar{x}'_{g+2} \in \text{split}\big(x'_{(g+2)1} (\!| p_{g+2}, 1 |\!) x'_{(g+2)2}\big)$, for $x_{g+2} \notin T^*$, or
- $\bar{x}'_{g+2} \in \text{split}(x_{g+2})$, for $x_{g+2} \in T^*$,

with $x'_{(g+2)1} \, \text{lhs}\big((\!| p_{g+2}, 1 |\!)\big) x'_{(g+2)2} = x_{g+2}$, satisfying $\check{\bar{x}}'_{g+2}$.

If. The if part is proved by induction on the number g of simulation steps in a derivation of $\bar{G}$, for $g \geq 0$.

Basis. Let $g = 0$. Then, $\bar{x}'_1 \Rightarrow^0_{\bar{G}} \bar{x}'_1$ and $x_1 \Rightarrow^0_G x_1$.

Induction Hypothesis. Suppose that the claim holds for all derivations of $\bar{G}$ containing k simulation steps, where $k \leq g$, for some $g \geq 0$.

Induction Step. Consider any derivation of the form

$$\bar{x}'_1 \Rrightarrow \bar{x}_1 \Rightarrow_{\bar{G}} \cdots \Rrightarrow \bar{x}_{g+1} \Rightarrow_{\bar{G}} \bar{x}'_{g+2}$$

Since $g + 1 \geq 1$, there exists a derivation

$$\bar{x}'_1 \Rrightarrow \bar{x}_1 \Rightarrow_{\bar{G}} \cdots \Rrightarrow \bar{x}_g \Rightarrow_{\bar{G}} \bar{x}'_{g+1} \Rrightarrow \bar{x}_{g+1} \Rightarrow_{\bar{G}} \bar{x}'_{g+2}$$

By the induction hypothesis, there is a derivation

$$x_1 \Rightarrow_G^g x_{g+1}$$

As

$$\begin{aligned} \bar{x}'_{g+1} &\in \mathrm{split}(x'_{(g+1)1} (\!| p_{g+1}, 1 |\!) x'_{(g+1)2}) \\ x_{g+1} &= x'_{(g+1)1} \,\mathrm{lhs}((\!| p_{g+1}, 1 |\!)) x'_{(g+1)2} \end{aligned}$$

and

$$\bar{x}'_{g+1} \Rrightarrow \bar{x}_{g+1} \Rightarrow_{\bar{G}} \bar{x}'_{g+2}$$

then, by Claim 3, there is also a derivation

$$x_{g+1} \Rightarrow_G x_{g+2}$$

such that

$$\begin{aligned} \bar{x}'_{g+2} &\in \mathrm{split}(x'_{(g+2)1} (\!| p_{g+2}, 1 |\!) x'_{(g+2)2}) \\ x_{g+2} &= x'_{(g+2)1} \,\mathrm{lhs}((\!| p_{g+2}, 1 |\!)) x'_{(g+2)2} \end{aligned}$$

or $\bar{x}'_{g+2} \in \mathrm{split}(x_{g+2})$. □

From Claim 1,

$$\bar{S} \Rightarrow_{\bar{G}} \langle (\!| p, 1 |\!) \rangle$$

As $\langle (\!| p, 1 |\!) \rangle \in \mathrm{split}((\!| p, 1 |\!))$ and $S = \mathrm{lhs}((\!| p, 1 |\!))$, the simulation of G as described in Claim 4 can be performed, so

$$\langle (\!| p, 1 |\!) \rangle \Rightarrow_{\bar{G}}^* u\ [\Phi]$$

where Φ is a sequence of rules from $\bar{P}_2 \cup \bar{P}_3 \cup \bar{P}_4$. If a successful derivation is simulated, then we obtain $u = \langle a_1 \rangle \cdots \langle a_n \rangle$, where $n \geq 1$ and $a_1, \ldots, a_n \in T$. Finally, by using rules from $\bar{P}_5$, we obtain

$$u \Rightarrow_{\bar{G}}^+ v$$

where $v = a_1 \cdots a_n$. Therefore, every string during whose generation G erases nonterminals in a generalized k-restricted way can be generated by a propagating scattered context grammar $\bar{G}$. □

The family of languages generated by scattered context grammars that erase their nonterminals in a generalized k-restricted way is denoted by $\mathbf{SCAT}_k$.

Corollary 7.3.7. $\mathbf{SCAT}^{-\varepsilon} = \mathbf{SCAT}_k$, *for all* $k \geq 0$. □

References

1. Bordihn, H.: A grammatical approach to the LBA problem. In: New Trends in Formal Languages – Control, Cooperation, and Combinatorics, pp. 1–9. Springer, London (1997)
2. Bordihn, H., Fernau, H.: Accepting grammars with regulation. Int. J. Comput. Math. **53**(1), 1–18 (1994)
3. Dassow, J., Păun, G.: Regulated Rewriting in Formal Language Theory. Springer, New York (1989)
4. Fernau, H.: Unconditional transfer in regulated rewriting. Acta Inform. **34**(11), 837–857 (1997)
5. Greibach, S.A., Hopcroft, J.E.: Scattered context grammars. J. Comput. Syst. Sci. **3**(3), 233–247 (1969)
6. Hopcroft, J.E., Motwani, R., Ullman, J.D.: Introduction to Automata Theory, Languages, and Computation, 3rd edn. Addison-Wesley, Boston (2006)
7. Meduna, A.: Automata and Languages: Theory and Applications. Springer, London (2000)

Chapter 8
Extension of Languages Resulting from Regulated Grammars

Abstract The present chapter, consisting of two sections, studies modified regulated grammars so they generate their languages extended by some extra symbols that represent useful information related to the generated languages. Section 8.1 describes a transformation of any regular-controlled grammar with appearance checking G to a propagating regular-controlled with appearance checking H whose language $L(H)$ has every sentence of the form $w\rho$, where w is a string of terminals in G and ρ is a sequence of rules in H, so that (1) $w\rho \in L(H)$ if and only if $w \in L(G)$ and (2) ρ is a parse of w in H. Consequently, for every recursively enumerable language K, there exists a propagating regular-controlled grammar with appearance checking H with $L(H)$ of the above-mentioned form so K results from $L(H)$ by erasing all rules in $L(H)$. Analogical results are established (a) for regular-controlled grammars without appearance checking and (b) for these grammars that make only leftmost derivations. Section 8.2 studies a language operation referred to as *coincidental extension*, which extend strings by inserting some symbols into the languages generated by propagating scattered context grammars.

Keywords Extension of languages generated by regulated grammars • Regular-controlled grammars • Scattered context grammars • Appearance checking • Leftmost derivations • Erasing rules • Language operations

Regulated grammars are sometimes modified so they generate their languages extended by some extra symbols that represent useful information related to the generated languages. For instance, these symbols may represent information used during parsing directed by these grammars. To give another example, shuffling or inserting some symbols into the generated languages may correspond to language operations that fulfill an important role in various modern fields of informatics, ranging from cryptography through various text algorithms to DNA computation (see [2, 28, 29]). Therefore, it comes as no surprise that the theory of regulated grammars has payed a special attention to their investigation (see [15, 16, 19]).

A. Meduna and P. Zemek, *Regulated Grammars and Automata*,
DOI 10.1007/978-1-4939-0369-6_8,

The present chapter consists of two sections on this subject. Section 8.1 explains how to transform any regular-controlled grammar with appearance checking (see Sect. 5.1), $H = (G, \Xi, W)$, to a propagating regular-controlled with appearance checking, $H' = (G', \Xi', W')$, whose language $L(H')$ has every sentence of the form $w\rho$, where w is a string of terminals in G and ρ is a sequence of rules in G', so that (1) $w\rho \in L(H')$ if and only if $w \in L(H)$ and (2) ρ is a parse of w in G'. Consequently, for every recursively enumerable language K, there exists a propagating regular-controlled grammar with appearance checking H' with $L(H')$ of the above form so K results from $L(H')$ by erasing all rules in $L(H')$. In addition, analogical results are established (a) in terms of these grammars without appearance checking and (b) in terms of these grammars that make only leftmost derivations. In the conclusion, Sect. 8.1 points out some consequences implied by the achieved results.

Section 8.2 studies a language operation referred to as *coincidental extension*, which extend strings by inserting some symbols into the languages generated by propagating scattered context grammars (see Sect. 4.7).

8.1 Regular-Controlled Generators

Indisputably, parsing represents an important application area of grammatically oriented formal language theory (see [1, 3, 6, 7, 22, 25, 26]). Indeed, this area fulfills a crucial role in most computer science fields that grammatically analyze and process languages, ranging from compilers through computational linguistics up to bioinformatics. As demonstrated by several studies, such as [4, 8, 12–14, 17, 24, 27, 30], regulated grammars are important to this application area as well.

Sentences followed by their parses, which represent sequences of rules according to which the corresponding sentences are generated, frequently represent the goal information resulting from parsing because the parses allow us to construct the grammatical derivation of sentences whenever needed. Of course, we are particularly interested in achieving this information by using some simplified versions of regulated grammars, such as propagating versions. The present section discusses this topic in terms of regular-controlled grammars (see Sect. 5.1).

As the main result of this section, we demonstrate how to transform every regular-controlled grammar with appearance checking, $H = (G, \Xi, W)$, to a propagating regular-controlled grammar with appearance checking, $H' = (G', \Xi', W')$, whose language $L(H')$ has every sentence of the form $w\rho$, where w is a string of terminals in G and ρ is a sequence of rules in G', and in addition, each $w\rho \in L(H')$ satisfies (1) $w\rho \in L(H')$ if and only if $w \in L(H)$, and (2) ρ is a parse of w in G'. Since every recursively enumerable language K is generated by a regular-controlled grammar with appearance checking H, we actually demonstrate that for every recursively enumerable language K, there exists a propagating regular-controlled grammar with appearance checking H' such that $L(H')$ is of the form described above and by deleting all the parses in $L(H')$, we obtain precisely K.

This representation of the family of recursively enumerable languages is of some interest because propagating regular-controlled grammars with appearance checking define only a proper subfamily of the family of context-sensitive languages, which is properly contained in the family of recursively enumerable languages (see Theorem 5.1.6).

Apart from the main result described above, we add two closely related results. First, we explain how to modify the above transformation of G to G' so G' makes only leftmost derivations. Second, we prove that if H does not have any appearance checking set, then neither does H'.

It is worth noting that the generation of sentences followed by their parses is systematically studied in terms of matrix grammars in Sect. 7.2 of [8], which refers to these languages as *extended Szilard languages* (as pointed out in Sect. 3.3, the notion of a *parse* represents a synonym of several other notions, including a *Szilard word*). However, there exist three important differences between both studies. First, the study given in our section is based upon regular-controlled grammars rather than matrix grammars. Second, in the present section, the output grammar G' is always propagating even if G is not while in the other study, the output matrix grammar contains erasing rules whenever the input grammar does. Finally, in our study, the output grammar G' generates sentences followed by their parses in G' while in the other study, this generation is made in terms of the input matrix grammar. A similar result to the presented one is achieved in [23] in terms of scattered context grammars (see also Chap. 7 of [24]). Recall that Szilard languages and their properties have been vividly studied in recent four decades (see [18], references given therein, and [5]). Finally, see [11] for an extension of the notion of a Szilard language in terms of P systems.

First, we give an insight into Algorithm 8.1.1, which represents the main achievement of this section. The algorithm converts a regular-controlled grammar with appearance checking, $H = (G, \Xi, W)$, to a propagating regular-controlled grammar with appearance checking, $H' = (G', \Xi', W')$, so a terminal string w is in $L(H)$ if and only if $w\rho$ is in $L(H')$ and $S' \Rightarrow^*_{G'} w\ [\rho]$, where S' is the start symbol of G'.

G' simulates the derivation of w in G so that during every step, the sentential form of G has the three-part form $x\bar{Z}u\gamma Z$, in which the first part x consists of symbols that are not erased during the rest of the derivation, the second part u consists of nonterminals that are erased in the rest of the derivation, and the third and last part γ is a string of labels of rules used to generate x in G'. The two special nonterminals $\bar{Z}$ and Z mark the beginning of the second part and the end of the third part, respectively.

A single derivation step in G is simulated by G' as follows. First, G' non-deterministically selects a rule $r: A \rightarrow y$ from P with $y = y_0Y_1y_1Y_2y_2 \cdots Y_f y_f$, for some $f \geq 1$, where $Y_1Y_2 \cdots Y_f$ is a string of symbols that G does not erase during the rest of the derivation after applying r while all the symbols of $y_0y_1 \cdots y_f$ are erased. If G' makes an improper non-deterministic selection of this kind—that is, the selection does not correspond to a derivation in G, then G' is not able to generate a terminal string as explained in the notes following Algorithm 8.1.1.

In (I) through (III), given next, we describe three possible ways of simulating the application of r by G'.

(I) If $Y_1 Y_2 \cdots Y_f$ is nonempty and, therefore, y is not completely erased during the rest of the derivation, G' simulates the application of r in the following three-phase way

(i) it rewrites A to $Y_1 Y_2 \cdots Y_f$ in x by using a rule labeled with t,
(ii) adds the symbols of $y_0 y_1 \cdots y_f$ to u, and
(iii) writes t behind the end of γ.

(II) If $y = y_0 y_1 \cdots y_f$—in other words, y is completely erased during the rest of the derivation, then A is rewritten in u and no label is appended to γ.
(III) If $y = \varepsilon$, then the rule is, in fact, an erasing rule of the form $r\colon A \to \varepsilon$, whose application G simulates so it rewrites an occurrence of A in u to ℓ_1, where $\ell_1\colon S \to S$ is a special rule in G' (S is the start symbol of G).

Eventually, all symbols in the second part of sentential forms are rewritten to ℓ_1. At the end of every successful derivation, $\bar{Z}$ is rewritten to ℓ_0, where $\ell_0\colon S' \to S$ is a special rule in G'. Therefore, every parse begins with a prefix from $\{\ell_0\}\{\ell_1\}^*$. Observe that $S' \Rightarrow^*_{G'} S\ [\ell_0 \ell_1^k]$, for any $k \geq 0$.

Finally, let us note that in order to distinguish between symbols from the first part of the sentential form and its second part, the latter part has its symbols barred (for example, $\bar{A}$ is introduced instead of A).

Algorithm 8.1.1.

Input: A context-free grammar, $G = (N, T, \Psi, P, S)$, a complete finite automaton, $M = (Q, \Psi, R, s, F)$, and an appearance checking set, $W \subseteq \Psi$.
Output: A propagating context-free grammar, $G' = (N', T', \Psi', P', S')$, a complete finite automaton, $M' = (Q', \Psi', R', s', F')$, and an appearance checking set, $W' \subseteq \Psi'$, such that

$$L(H') = \{w \mid w\rho \in L(H'), S' \Rightarrow^*_{G'} w\ [\rho]\}$$

where $H = (G, L(M), W)$ and $H' = (G', L(M'), W')$ are two regular-controlled grammars with appearance checking.
Note: Let us note that in what follows, symbols $\lfloor$, $\rfloor$, $\lceil$, and $\rceil$ are used to clearly incorporate more symbols into a single compound symbol. Without any loss of generality, we assume that $s', \$, \bot \notin Q$, and $S', \#, Z, \bar{Z}, \ell_0, \ell_1, \ell_2, \ell_\$ \notin (\Psi \cup N \cup T)$. If $r\colon S \to S \in P$, then we set $\ell_1 = r$.
Method: Set $V = N \cup T$ and $\bar{N} = \{\bar{A} \mid A \in N\}$. Define the homomorphism τ from N^* to $\bar{N}^*$ as $\tau(A) = \bar{A}$, for all $A \in N$. Initially, set

$$\begin{aligned}
\Psi' &= \{\ell_0, \ell_1, \ell_2, \ell_\$\} \\
T' &= T \cup \Psi' \\
N' &= N \cup \bar{N} \cup \{S', \#, Z, \bar{Z}\} \\
P' &= \{\ell_0 \colon S' \to S, \ell_1 \colon S \to S, \ell_2 \colon S' \to S\bar{Z}Z, \ell_\$ \colon \bar{Z} \to \ell_0\} \\
W' &= \emptyset \\
Q' &= Q \cup \{s', \$, \bot\} \\
R' &= \{s'\ell_2 \to s\} \cup \{p\ell_\$ \to \$ \mid p \in F\} \\
F' &= \{\$\}
\end{aligned}$$

Repeat (1) through (3), given next, until none of the sets Ψ', T', P', W', Q', and R' can be extended in this way. The order of the execution of these three steps is irrelevant.

(1) **If** $r \colon A \to y_0 Y_1 y_1 Y_2 y_2 \cdots Y_f y_f \in P$, where $y_i \in N^*$, $Y_j \in V$, for all i and j, $0 \le i \le f$, $1 \le j \le f$, for some $f \ge 1$
then

(1.1) for each k, $1 \le k \le 4$, add $t_k = \lfloor k, r, y_0, Y_1 y_1, Y_2 y_2, \ldots, Y_f y_f \rfloor$ to Ψ' and T';
(1.2) add the following four rules to P':
$t_1 \colon A \to Y_1 Y_2 \cdots Y_f$,
$t_2 \colon \bar{Z} \to \bar{Z}z$, where $z = \tau(y_0 y_1 \cdots y_f)$,
$t_3 \colon Z \to t_1 Z$,
$t_4 \colon Z \to t_1$;
(1.3) for each $pr \to q \in R$, add $o_1 = \lceil p, t_1, q \rceil$, $o_2 = \lceil p, t_2, q \rceil$ to Q' and add $pt_1 \to o_1$, $o_1 t_2 \to o_2$, $o_2 t_3 \to q$, $o_2 t_4 \to q$ to R'.

(2) **If** $r \colon A \to y \in P$, where $y \in N^*$
then

(2.1) add $\bar{r}$ to Ψ' and T';
(2.2) add $\bar{r} \colon \bar{A} \to z$ to P', where $z = \tau(y)$ if and only if $|y| > 0$ and $z = \ell_1$ if and only if $|y| = 0$;
(2.3) for each $pr \to q \in R$, add $p\bar{r} \to q$ to R'.

(3) **If** $r \colon A \to y \in P$, where $y \in V^*$ and $r \in W$
then

(3.1) add r' and r'' to Ψ', T', and W';
(3.2) add $r' \colon A \to y\#$ and $r'' \colon \bar{A} \to y\#$ to P';
(3.3) for each $pr \to q \in R$, add $q' = \lceil p, r', q \rceil$ to Q' and add $pr' \to q'$, $q'r'' \to q$ to R'.

For each $p \in Q'$ and $a \in \Psi'$ such that $pa \to q \notin R'$ for any $q \in Q'$, add $pa \to \bot$ to R'. For each $a \in \Psi'$, add $\bot a \to \bot$ to R'. □

Before proving that this algorithm is correct, we make some informal comments concerning the rules of G'. The rules introduced in (1) are used to simulate an application of a rule where some of the symbols on its right-hand side are erased throughout the rest of the derivation while some others are not erased. The rules introduced in (2) are used to simulate an application of a rule which erases a nonterminal by making one or more derivation steps. The rules introduced in (3) are used to simulate an application of a rule in the appearance checking mode. Observe that when simulating $r: A \to y \in P$, where $r \in W$, in the appearance checking mode, we have to check the absence of both A and $\bar{A}$. If either of these two nonterminals appear in the current sentential form, the introduced nonterminal # is never rewritten because there is no rule in P' with # on its left-hand side. Finally, notice that the last step of Algorithm 8.1.1 makes M' complete.

Reconsider (1). Notice that the algorithm works correctly although it makes no predetermination of nonterminals from which ε can be derived. Indeed, if the output grammar improperly selects a nonterminal that is not erased throughout the rest of the derivation, then this occurrence of the nonterminal never disappears, so a terminal string cannot be generated under this improper selection.

Next, we prove that Algorithm 8.1.1 is correct.

Lemma 8.1.2. *Let $G = (N, T, \Psi, P, S)$ be a context-free grammar, $M = (Q, \Psi, R, s, F)$ be a complete finite automaton, and $W \subseteq \Psi$ be an appearance checking set. Algorithm 8.1.1 is correct—that is, it converts G, M, and W to a propagating context-free grammar, $G' = (N', T', \Psi', P', S')$, a complete finite automaton, $M' = (Q', \Psi', R', s', F')$, and an appearance checking set, $W' \subseteq \Psi'$, such that*

$$L(H') = \{w \mid w\rho \in L(H'), S' \Rightarrow^*_{G'} w\ [\rho]\}$$

where $H = (G, L(M), W)$ and $H' = (G', L(M'), W')$ are two regular-controlled grammars with appearance checking.

Proof. Clearly, the algorithm always halts. Since P' does not contain any erasing rules, G' is propagating. To establish

$$L(H) = \{w \mid w\rho \in L(H'), S' \Rightarrow^*_{G'} w\ [\rho]\}$$

we first introduce some notions needed later in the proof.

First, we define a function which, given a string of rule labels ρ, gives the state of the deterministic finite automaton M in which it appears after reading ρ. Define the function $\mathfrak{D}$ from Ψ^* to Q as

$$\mathfrak{D}(\rho) = p \text{ if and only if } (s, \rho) \vdash^*_M (p, \varepsilon)$$

where s is the start state of M. Next, we define an analogical function for M'. Define the function $\mathfrak{D}'$ from Ψ'^* to Q' as

$$\mathfrak{D}'(\rho) = p \text{ if and only if } (s', \rho) \vdash^*_{M'} (p, \varepsilon)$$

where s' is the start state of M'.

Let $\Omega \subseteq \Psi'$ be the set of all rule labels introduced in (1.1) in the algorithm which are of the form $\lfloor 1, r, y_0, Y_1 y_1, Y_2 y_2, \dots, Y_f y_f \rfloor$ (that is, $k = 1$). Define the homomorphism ω over Ψ'^* as $\omega(t) = t$, for all $t \in \Omega$, and $\omega(t) = \varepsilon$, for all $t \in \Psi' - \Omega$. Informally, given a string of rule labels, this homomorphism erases all occurrences of rule labels that are not in Ω.

Next, define the regular substitution π from N^* to $2^{(\bar{N} \cup \{\ell_1\})^*}$ as

$$\pi(A) = \{\ell_1\}^* \{\bar{A}\} \{\ell_1\}^*$$

for all $A \in N$. Informally, using π, strings formed by occurrences of ℓ_1 may be added so they precede and follow any occurrence of a nonterminal from N.

First, we show how derivations of G are simulated by G'. The following claim is used to prove that

$$L(H) \subseteq \{w \mid w\rho \in L(H'), S' \Rightarrow^*_{G'} w\ [\rho]\}$$

later in the proof. Recall the meaning of the ${}^{\varepsilon}$ and ${}^{\not\varepsilon}$ notation from Definition 3.3.26.

Claim 1. Let

$$S \Rightarrow^m_{(G,W)} {}^{\varepsilon}x_0 {}^{\not\varepsilon}X_1 {}^{\varepsilon}x_1 {}^{\not\varepsilon}X_2 {}^{\varepsilon}x_2 \cdots {}^{\not\varepsilon}X_h {}^{\varepsilon}x_h\ [\alpha] \Rightarrow^*_{(G,W)} w$$

where $w \in L(H)$, $x_i \in N^*$, $X_j \in V$, for all i and j, $0 \leq i \leq h$, $1 \leq j \leq h$, for some $h \geq 1$ and $m \geq 0$. Then,

$$S' \Rightarrow^*_{(G',W')} X_1 X_2 \cdots X_h \bar{Z} u \gamma v\ [\beta]$$

where

(i) $u \in \pi(u')\{\ell_1\}^*$, *for some* $u' \in \mathrm{perm}(x_0 x_1 \cdots x_h)$;
(ii) $\gamma = \omega(\beta)$;
(iii) $v = \varepsilon$ *if and only if* $X_1 X_2 \cdots X_h \in T^*$ *and* $v = Z$ *if and only if* $X_1 X_2 \cdots X_h \notin T^*$;
(iv) $\mathfrak{D}(\alpha) = \mathfrak{D}'(\beta)$.

Proof. This claim is established by induction on $m \geq 0$.

Basis. Let $m = 0$. Then,

$$S \Rightarrow^0_{(G,W)} S\ [\varepsilon] \Rightarrow^*_{(G,W)} w$$

where $w \in L(H)$. By the initialization part of the algorithm, $\ell_2\colon S' \to S\bar{Z}Z \in P'$, so

$$S' \Rightarrow_{(G',W')} S\bar{Z}Z\ [\ell_2]$$

Clearly, (i)–(iii) hold. Since $\mathfrak{D}(\varepsilon) = s$ and $s'\ell_2 \rightarrow s \in R'$ by the initialization part of the algorithm, $\mathfrak{D}(\varepsilon) = \mathfrak{D}'(\ell_2)$, so (iv) also holds. Thus, the basis holds.

Induction Hypothesis. Suppose that the claim holds for all derivations of length ℓ or less, where $0 \leq \ell \leq m$, for some $m \geq 0$.

Induction Step. Consider any derivation of the form

$$S \Rightarrow^{m+1}_{(G,W)} \chi \Rightarrow^{*}_{(G,W)} w$$

where $\chi \in V^+$ and $w \in L(H)$. Since $m + 1 \geq 1$, this derivation can be expressed as

$$S \Rightarrow^{m}_{(G,W)} x\ [\alpha] \Rightarrow_{(G,W)} \chi\ [r] \Rightarrow^{*}_{(G,W)} w$$

for some $x \in V^+$, $\alpha \in \Psi^*$, and $r \in \Psi$. Let $x = {}^{\varepsilon}x_0{}^{\not\varepsilon}X_1{}^{\varepsilon}x_1{}^{\not\varepsilon}X_2{}^{\varepsilon}x_2 \cdots {}^{\not\varepsilon}X_h{}^{\varepsilon}x_h$, where $x_i \in N^*$, $X_j \in V$, for all i and j, $0 \leq i \leq h$, $1 \leq j \leq h$, for some $h \geq 1$. Then, by the induction hypothesis,

$$S' \Rightarrow^{*}_{(G',W')} X_1X_2 \cdots X_h\bar{Z}u\gamma v\ [\beta]$$

such that (i) through (iv) hold. Let $\mathfrak{D}(\alpha) = \mathfrak{D}'(\beta) = p$ and $pr \rightarrow q \in R$.

Now, we consider all possible forms of $x \Rightarrow_{(G,W)} \chi\ [r]$, covered by the next three cases—(a), (b), and (c).

(a) *Application of a rule rewriting a not-to-be-erased nonterminal.* Let $r\colon A \rightarrow y \in P$, where $y \in V^+$, such that $A = X_k$, for some k, $1 \leq k \leq h$. Let $y = {}^{\varepsilon}y_0{}^{\not\varepsilon}Y_1{}^{\varepsilon}y_1{}^{\not\varepsilon}Y_2{}^{\varepsilon}y_2 \cdots {}^{\not\varepsilon}Y_f{}^{\varepsilon}y_f$, where $y_i \in N^*$, $Y_j \in V$, for all i and j, $0 \leq i \leq h$, $1 \leq j \leq h$, for some $f \geq 1$. Then,

$$\begin{gathered} {}^{\varepsilon}x_0{}^{\not\varepsilon}X_1{}^{\varepsilon}x_1{}^{\not\varepsilon}X_2{}^{\varepsilon}x_2 \cdots {}^{\not\varepsilon}X_k{}^{\varepsilon}x_k \cdots {}^{\not\varepsilon}X_h{}^{\varepsilon}x_h \Rightarrow_{(G,W)} \\ {}^{\varepsilon}x_0{}^{\not\varepsilon}X_1{}^{\varepsilon}x_1{}^{\not\varepsilon}X_2{}^{\varepsilon}x_2 \cdots {}^{\not\varepsilon}X_{k-1}{}^{\varepsilon}x_{k-1}y^{\varepsilon}x_k{}^{\not\varepsilon}X_{k+1}{}^{\varepsilon}x_{k+1} \cdots {}^{\not\varepsilon}X_h{}^{\varepsilon}x_h\ [r] \end{gathered}$$

By (1.1) in the algorithm, there are $t_i = \lfloor i, r, y_0, Y_1y_1, Y_2y_2, \cdots, Y_fy_f \rfloor \in \Psi'$, for all i, $1 \leq i \leq 4$, and by (1.2), P' contains

$$\begin{aligned} &t_1\colon A \rightarrow Y_1Y_2 \cdots Y_f \\ &t_2\colon \bar{Z} \rightarrow \bar{Z}z \\ &t_3\colon Z \rightarrow t_1Z \\ &t_4\colon Z \rightarrow t_1 \end{aligned}$$

where $z = \tau(y_0 y_1 \cdots y_n)$. Also, by (1.3), there are $o_1 = \lceil p, t_1, q \rceil$, $o_2 = \lceil p, t_2, q \rceil \in Q'$, and R' contains

$$\begin{aligned} pt_1 &\to o_1 \\ o_1 t_2 &\to o_2 \\ o_2 t_3 &\to q \\ o_2 t_4 &\to q \end{aligned}$$

As a result,

$$\begin{array}{ll} X_1 X_2 \cdots X_k \cdots X_h \bar{Z} u \gamma v & \Rightarrow_{(G',W')} \\ X_1 X_2 \cdots X_{k-1} Y_1 Y_2 \cdots Y_f X_{k+1} \cdots X_h \bar{Z} u \gamma v \ [t_1] & \Rightarrow_{(G',W')} \\ X_1 X_2 \cdots X_{k-1} Y_1 Y_2 \cdots Y_f X_{k+1} \cdots X_h \bar{Z} z u \gamma v \ [t_2] & \end{array}$$

Since $X_k \in N$, $v = Z$ by the induction hypothesis [see (iii)]. Let

$$g = X_1 X_2 \cdots X_{k-1} Y_1 Y_2 \cdots Y_f X_{k+1} \cdots X_h$$

If $g \in T^*$, then

$$g \bar{Z} z u \gamma Z \Rightarrow_{(G',W')} g \bar{Z} z u \gamma t_1 \ [t_4]$$

Otherwise, if $g \notin T^*$, then

$$g \bar{Z} z u \gamma Z \Rightarrow_{(G',W')} g \bar{Z} z u \gamma t_1 Z \ [t_3]$$

so (iii) holds. Since $u \in \pi(u')\{\ell_1\}^*$ by the induction hypothesis, for some $u' \in \mathrm{perm}(x_0 x_1 \cdots x_h)$, and $z \in \pi(y_0 y_1 \cdots y_f)$, (i) also holds. Since R' contains

$$\begin{aligned} pt_1 &\to o_1 \\ o_1 t_2 &\to o_2 \\ o_2 t_3 &\to q \\ o_2 t_4 &\to q \end{aligned}$$

by (1.3), (iv) holds. Finally, since $\gamma = \omega(\beta)$ by the induction hypothesis, and $\omega(t_1 t_2 t_i) = t_1$, $\omega(\beta t_1 t_2 t_i) = \gamma t_1$, for each $i \in \{3, 4\}$, which proves (ii).

(b) *Application of a rule rewriting a to-be-erased nonterminal.* Let $r\colon A \to y \in P$, where $y \in N^*$, such that $x_k = x'_k A x''_k$, where $x'_k, x''_k \in N^*$, for some k, $0 \le k \le h$. Then,

$$\begin{array}{l} {}^{\varepsilon}x_0 {}^{\not g}X_1 {}^{\varepsilon}x_1 {}^{\not g}X_2 {}^{\varepsilon}x_2 \cdots {}^{\not g}X_k {}^{\varepsilon}x'_k {}^{\varepsilon}A {}^{\varepsilon}x''_k \cdots {}^{\not g}X_h {}^{\varepsilon}x_h \Rightarrow_{(G,W)} \\ {}^{\varepsilon}x_0 {}^{\not g}X_1 {}^{\varepsilon}x_1 {}^{\not g}X_2 {}^{\varepsilon}x_2 \cdots {}^{\not g}X_k {}^{\varepsilon}x'_k {}^{\varepsilon}y {}^{\varepsilon}x''_k \cdots {}^{\not g}X_h {}^{\varepsilon}x_h \ [r] \end{array}$$

By (2.2) in the algorithm, $\bar{r}\colon \bar{A} \to z \in P'$, where $z = \tau(y)$ if and only if $|y| > 0$ and $z = \ell_1$ if and only if $|y| = 0$. By (2.3), $p\bar{r} \to q \in R'$. By the induction hypothesis [see (i)], $u = u_1\bar{A}u_2$, for some $u_1, u_2 \in (\bar{N} \cup \{\ell_1\})^*$, so

$$X_1X_2\cdots X_h\bar{Z}u_1\bar{A}u_2\gamma v \Rightarrow_{(G',W')} X_1X_2\cdots X_h\bar{Z}u_1zu_2\gamma v\ [\bar{r}]$$

Clearly, (ii) through (iv) hold. Since $u \in \pi(u')\{\ell_1\}^*$ by the induction hypothesis, for some $u' \in \mathrm{perm}(x_0x_1\cdots x_h)$, and $z \in \pi(y)$, (i) also holds.

(c) *Application of a rule in the appearance checking mode.* Let $r\colon A \to y \in P$, where $y \in V^*$, $A \notin \mathrm{alph}(x)$, and $r \in W$. Then, $x \Rightarrow_{(G,W)} x\ [r]$. Let $z = X_1X_2\cdots X_h\bar{Z}u\gamma v$. Since $A \notin \mathrm{alph}(x)$, $A, \bar{A} \notin \mathrm{alph}(z)$. By (3.2), P' contains

$$r'\colon A \to y\#$$
$$r''\colon \bar{A} \to y\#$$

where $r', r'' \in W'$. Therefore, $z \Rightarrow^2_{(G',W')} z\ [r'r'']$. By (3.3), R' contains

$$pr' \to q'$$
$$q'r'' \to q$$

where $q' = \lceil p, r', q\rceil \in Q'$. Since $\mathfrak{D}(\alpha) = \mathfrak{D}'(\beta) = p$ by the induction hypothesis [see (iv)], $\mathfrak{D}(\alpha r) = \mathfrak{D}'(\beta r'r'') = q$, so (iv) holds. Clearly, (i) through (iii) also hold.

Observe that cases (a) through (c) cover all possible forms of $x \Rightarrow_{(G,W)} \chi\ [r]$. Thus, the claim holds. □

Set $V' = N' \cup T'$. The next claim explains the role of $\bot$. That is, once M' occurs in $\bot$, it cannot accept its input.

Claim 2. If $S' \Rightarrow^*_{(G',W')} x\ [\alpha]$, where $x \in V'^*$, such that $\mathfrak{D}'(\alpha) = \bot$, then there is no $\beta \in \Psi'^*$ satisfying $\mathfrak{D}'(\alpha\beta) \in F'$ such that $S' \Rightarrow^*_{(G',W')} x\ [\alpha] \Rightarrow^*_{(G',W')} w\ [\beta]$, where $w \in T^*$.

Proof. By the last step of the algorithm, $\bot a \to \bot \in R'$, for all $a \in \Psi'$. Since $\bot \notin F'$, the claim holds. □

The following claim says that every sentence in $L(H')$ is generated without using the unit rule $\ell_0\colon S' \to S \in P'$.

Claim 3. Let $w \in V'^+$. Then, $w \in L(H')$ if and only if $S' \Rightarrow^*_{(G',W')} w\ [\alpha]$ such that $\mathfrak{D}'(\alpha) \in F'$ and $\ell_0 \notin \mathrm{alph}(\alpha)$.

Proof. The proof is divided into the only-if part and the if part.

Only If. Observe that by the last step of the algorithm, $p\ell_0 \to \bot \in R'$, for all $p \in Q'$. Therefore, the only-if part of this claim follows from Claim 2.

If. Trivial. □

We next show how derivations of G' are simulated by G. The following claim is used to prove that

$$\{w \mid w\rho \in L(H'), S' \Rightarrow_{G'}^{*} w\ [\rho]\} \subseteq L(H')$$

later in the proof.

Claim 4. Let

$$S' \Rightarrow_{(G',W')}^{n} x\ [\beta]$$

where $x \in V'^{+}$ and $\mathfrak{D}'(\beta) \in Q$, for some $n \geq 1$. Then,

$$S \Rightarrow_{(G,W)}^{*} x_0 X_1 x_1 X_2 x_2 \cdots X_h x_h\ [\alpha]$$

and x is of the form $x = X_1 X_2 \cdots X_h \bar{Z} u \gamma v$, where $x_i \in N^*$, $X_j \in V$, for all i and j, $0 \leq i \leq h$, $1 \leq j \leq h$, for some $h \geq 1$, and

(i) $u \in \pi(u')\{\ell_1\}^*$, *for some* $u' \in \mathrm{perm}(x_0 x_1 \cdots x_h)$;
(ii) $\gamma = \omega(\beta)$;
(iii) $v = \varepsilon$ *if and only if* $X_1 X_2 \cdots X_h \in T^*$ *and* $v = Z$ *if and only if* $X_1 X_2 \cdots X_h \notin T^*$;
(iv) $\mathfrak{D}(\alpha) = \mathfrak{D}'(\beta)$.

Proof. This claim is established by induction on $n \geq 1$.

Basis. Let $n = 1$. Considering Claim 3, the only applicable rule to S' is ℓ_2: $S' \rightarrow S\bar{Z}Z \in P'$, which is introduced in the initialization part of the algorithm, so

$$S' \Rightarrow_{(G',W')} S\bar{Z}Z\ [\ell_2]$$

Then, $S \Rightarrow_G^0 S\ [\varepsilon]$. Clearly, (i)–(iii) hold. Since $s'\ell_2 \rightarrow s \in R'$ by the initialization part of the algorithm and $\mathfrak{D}(\varepsilon) = s$, $\mathfrak{D}'(\ell_2) = \mathfrak{D}(\varepsilon)$, so (iv) also holds. Thus, the basis holds.

Induction Hypothesis. Suppose that the claim holds for all derivations of length ℓ or less, where $1 \leq \ell \leq n$, for some $n \geq 1$.

Induction Step. Consider any derivation of the form

$$S' \Rightarrow_{(G',W')}^{n+1} \chi$$

where $\chi \in V'^{+}$. Since $n + 1 > 1$, this derivation can be expressed as

$$S' \Rightarrow_{(G',W')}^{n} x\ [\beta] \Rightarrow_{(G',W')} \chi\ [t]$$

for some $x \in V'^{+}$, $\beta \in \Psi'^{*}$, and $t \in \Psi'$. By the induction hypothesis, $\mathfrak{D}'(\beta) \in Q$, x is of the form $x = X_1X_2 \cdots X_h\bar{Z}u\gamma v$, where $X_i \in V$, for all i, $1 \leq i \leq h$, for some $h \geq 1$, $u \in (\bar{N} \cup \{\ell_1\})^{*}$, $\gamma \in \Psi'^{*}$, and $v \in \{\varepsilon, Z\}$, such that (i) through (iv) hold. Furthermore,

$$S \Rightarrow^{*}_{(G,W)} x_0X_1x_1X_2x_2 \cdots X_hx_h$$

Let $\mathfrak{D}'(\beta) = \mathfrak{D}(\alpha) = p$ and $pt \rightarrow q \in R'$.

Next, we consider all possible forms of $x \Rightarrow_{(G',W')} \chi$ $[t]$, covered by the following four cases—(a) through (d).

(a) *Application of a rule from (1).* Let $t = \lfloor 1, r, y_0, Y_1y_1, Y_2y_2, \ldots, Y_fy_f \rfloor$ be a label introduced in (1.1) in the algorithm and $t: X_k \rightarrow Y_1Y_2 \cdots Y_f \in P'$ be a rule introduced in (1.1) from $r: X_k \rightarrow y_0Y_1y_1Y_2y_2 \cdots Y_fy_f \in P$, where $y_i \in N^{*}$, $Y_j \in V$, for all i and j, $0 \leq i \leq f$, $1 \leq j \leq f$, for some $f \geq 1$ and k, $1 \leq k \leq h$. Since $X_k \in N$, $v = Z$ by the induction hypothesis. Then,

$$X_1X_2 \cdots X_{k-1}X_kX_{k+1} \cdots X_h\bar{Z}u\gamma Z \Rightarrow_{(G',W')}$$
$$X_1X_2 \cdots X_{k-1}Y_1Y_2 \cdots Y_fX_{k+1} \cdots X_h\bar{Z}u\gamma Z \; [t]$$

By (1.1), there are $t_i = \lfloor i, r, y_0, Y_1y_1, \ldots, Y_fy_f \rfloor \in \Psi'$, for all $i \in \{2, 3, 4\}$, and by (1.2), P' contains

$$t_2: \bar{Z} \rightarrow \bar{Z}z$$
$$t_3: Z \rightarrow tZ$$
$$t_4: Z \rightarrow t$$

where $z = \tau(y_0y_1 \cdots y_f)$. Also, by (1.3), there are $o_1 = \lceil p, t, e \rceil, o_2 = \lceil p, t_2, e \rceil \in Q'$ and $pt_1 \rightarrow o_1$, $o_1t_2 \rightarrow o_2$, $o_2t_3 \rightarrow e$, $o_2t_4 \rightarrow e \in R'$, for some $e \in Q$. Let $g = X_1X_2 \cdots X_{k-1}Y_1Y_2 \cdots Y_fX_{k+1} \cdots X_h$. Considering Claim 2, observe that the only applicable rules in G' are now t_2 followed by either t_3 or t_4. Therefore,

$$g\bar{Z}u\gamma Z \Rightarrow_{(G',W')} g\bar{Z}zu\gamma Z \; [t_2]$$

There are the following two cases depending on whether g is in T^{*}.

(a.1) Assume that $g \notin T^{*}$. Then, we have to apply t_3; otherwise, there is no way of rewriting the remaining nonterminals in g to terminals. Indeed, observe that nonterminals in g can be rewritten to terminals only by rules introduced in (1.2), but every such sequence of rules has to be finished by a rewrite of Z. Thus,

$$g\bar{Z}zu\gamma Z \Rightarrow_{(G',W')} g\bar{Z}zu\gamma tZ \; [t_3]$$

(a.2) Assume that $g \in T^*$. Then, we have to apply t_4; otherwise, there is no way of rewriting Z to a terminal (since $g \in T^*$, no rule introduced in (1.2) is applicable in the rest of the derivation). Thus,

$$g\bar{Z}zu\gamma Z \Rightarrow_{(G',W')} g\bar{Z}zu\gamma t \ [t_4]$$

In either case,

$$x_0X_1x_1X_2x_2\cdots X_kx_k\cdots X_hx_h \Rightarrow_{(G,W)}$$
$$x_0X_1x_1X_2x_2\cdots X_{k-1}x_{k-1}y_0Y_1y_1\cdots Y_fy_fx_kX_{k+1}x_{k+1}\cdots X_hx_h \ [r]$$

By the induction hypothesis, $u \in \pi(u')\{\ell_1\}^*$, for some $u' \in \text{perm}(x_0x_1\cdots \cdots x_h)$. Obviously, $z \in \pi(y_0y_1\cdots y_f)$, so (i) holds. By the induction hypothesis, $\gamma = \omega(\beta)$. Since $t \in \Omega$, $\omega(t) = t$. Therefore, $\omega(\beta tt_2t') = \gamma t$, where either $t' = t_3$ or $t' = t_4$, depending on whether $g \notin T^*$ or $g \in T^*$ (see (a.1) and (a.2) above). Thus, (ii) holds. Clearly, by (a.1) and (a.2) above, (iii) holds. By (1.3), since t is constructed from r, $pr \to e \in R$, so $\mathfrak{D}(\alpha r) = e$. Since R' contains

$$\begin{aligned} pt &\to o_1 \\ o_1t_2 &\to o_2 \\ o_2t' &\to e \end{aligned}$$

$\mathfrak{D}'(\beta tt_2t') = e$. Thus, $\mathfrak{D}(\alpha r) = \mathfrak{D}'(\beta tt_2t')$, which proves that (iv) also holds.

(b) *Application of a rule from (2).* Let $t = \bar{r}$ and $\bar{r}: \bar{A} \to z \in P'$ be a rule introduced in (2.2) in the algorithm from $r: A \to y \in P$, where $z = \tau(y)$ if and only if $|y| > 0$ and $z = \ell_1$ if and only if $|y| = 0$. Let $g = x_0X_1x_1X_2x_2\cdots X_hx_h$. Let $u = u_1\bar{A}u_2$, for some $u_1, u_2 \in (\bar{N} \cup \{\ell_1\})^*$, so

$$X_1X_2\cdots X_h\bar{Z}u_1\bar{A}u_2\gamma v \Rightarrow_{(G',W')} X_1X_2\cdots X_h\bar{Z}u_1zu_2\gamma v \ [\bar{r}]$$

By the induction hypothesis [see (i)], $x_k = x'_kAx''_k$, where $x'_k, x''_k \in N^*$, for some k, $0 \leq k \leq h$, so

$$x_0X_1x_1X_2x_2\cdots X_kx'_kAx''_k\cdots X_hx_h \Rightarrow_{(G,W)} x_0X_1x_1X_2x_2\cdots X_kx'_kyx''_k\cdots X_hx_h$$

Clearly, (ii) and (iii) hold. By (2.3), $pr \to q \in R$, so (iv) holds. Since $u \in \pi(u')\{\ell_1\}^*$ by the induction hypothesis, for some $u' \in \text{perm}(x_0x_1\cdots x_h)$, and $z \in \pi(y)$, (i) also holds.

(c) *Application of a rule from (3).* Let $t = r'$ and $r': A \to y\# \in P'$ be introduced in (3.2) in the algorithm from $r: A \to y \in P$, where $y \in V^*$ and $r \in W$. To apply r' and to be able to derive a string of terminals from $L(H')$, observe that $A, \bar{A} \notin \text{alph}(x)$ and $pr \to q \in R$ have to hold. Therefore, by (3.1)–(3.3), $r', r'' \in W'$, P' contains

$$r'\colon A \to y\#$$
$$r''\colon \bar{A} \to y\#$$

$q' = \lceil p, r', q \rceil \in Q'$, and R' contains

$$pr' \to q'$$
$$q'r'' \to q$$

so $x \Rightarrow^2_{(G',W')} x\ [r'r'']$. Let $g = x_0 X_1 x_1 X_2 x_2 \cdots X_h x_h$. Since $A \notin \text{alph}(x)$, $A \notin \text{alph}(g)$, so $g \Rightarrow_{(G,W)} g\ [r]$. Since $\mathfrak{D}(\alpha) = \mathfrak{D}'(\beta) = p$, $\mathfrak{D}(\alpha r) = \mathfrak{D}'(\beta r' r'') = q$, so (iv) holds. Clearly, (i) through (iii) also hold.

(d) *Application of* $\ell_\$\colon \bar{Z} \to \ell_0$. Let $t = \ell_\$$, $p \in F$, $g = X_1 X_2 \cdots X_h$, $\ell_\$\colon \bar{Z} \to \ell_0 \in P'$ be the rule introduced in the initialization part of the algorithm, and $p\ell_\$ \to \$ \in R'$, so

$$g\bar{Z}u\gamma v \Rightarrow_{(G',W')} g\ell_0 u\gamma v\ [\ell_\$]$$

There is no $\sigma \in \Psi'^{+}$ such that $\mathfrak{D}(\beta\ell_\$\sigma) \in F'$. Thus, if $g \notin T^*$ or $u \notin \{\ell_1\}^*$, then there is no way of deriving a sentence from $L(H')$. On the other hand, if $g \in T^*$ and $u \in \{\ell_1\}^*$, then $g\ell_0 u\gamma \in L(H')$ [recall that since $g \in T^*$, $v = \varepsilon$ by the induction hypothesis, see (iii)].

We next prove that Lemma 8.1.2 holds in the case when $g \in T^*$ and $u \in \{\ell_1\}^*$. Since $u \in \{\ell_1\}^*$, $x_i = \varepsilon$, for all i, $0 \le i \le h$, and since $p \in F$ and $g \in T^*$, $g \in L(H)$. Clearly, $S' \Rightarrow^*_{G'} S\ [\ell_0 u]$. By the induction hypothesis, $\gamma = \omega(\beta)$. By the definition of ω and by (1.2), γ consists only of rules used to generate g. Thus, $S \Rightarrow^*_G w\ [\gamma]$, which proves

$$\{w \mid w\rho \in L(H'), S' \Rightarrow^*_{G'} w\ [\rho]\} \subseteq L(H)$$

By Claim 5, stated below, $L(H) \subseteq \{w \mid w\rho \in L(H'),\ S' \Rightarrow^*_{G'} w\ [\rho]\}$. Therefore, Lemma 8.1.2 holds in this case.

Observe that cases (a) through (d) cover all possible forms of $x \Rightarrow_{(G',W')} \chi\ [t]$. Thus, the claim holds. □

Let $\Gamma = \{w \mid w\rho \in L(H'),\ S' \Rightarrow^*_{G'} w\ [\rho]\}$. We prove that $L(H) = \Gamma$. This proof consists of the following two claims.

Claim 5. $L(H) \subseteq \Gamma$.

Proof. Consider Claim 1 for $x_i = \varepsilon$, $X_j \in T$, for all i and j, $0 \le i \le h$, $1 \le j \le h$, for some $h \ge 1$, so

$$S \Rightarrow^*_{(G,W)} X_1 X_2 \cdots X_h\ [\alpha] \Rightarrow^*_{(G,W)} w$$

where $w \in L(H)$. Since $X_1X_2 \cdots X_h \in T^+$ and $w \in L(H)$, $w = X_1X_2 \cdots X_h$ and $\mathfrak{D}(\alpha) \in F$. Then,

$$S' \Rightarrow^*_{(G',W')} w\bar{Z}\ell_1^k\gamma \ [\beta]$$

where $k \geq 0$, $\gamma \in \omega(\beta)$, and $\mathfrak{D}(\alpha) = \mathfrak{D}'(\beta)$, so $\mathfrak{D}'(\beta) \in F$. Let $p = \mathfrak{D}'(\beta)$. By the initialization part of the algorithm, $\ell_\$: \bar{Z} \rightarrow \ell_0 \in P'$ and $p\ell_\$ \rightarrow \$ \in R'$, so

$$w\bar{Z}\ell_1^k\gamma \Rightarrow_{(G',W')} w\ell_0\ell_1^k\gamma$$

Let $\rho = \ell_0\ell_1^k\gamma$. Since $\mathfrak{D}'(\beta\ell_\$) = \$$ and $\$ \in F'$ by the initialization part of the algorithm, $w\rho \in L(H')$.

Finally, it remains to prove that $S' \Rightarrow^*_{G'} w \ [\rho]$. Clearly,

$$S' \Rightarrow^{k+1}_{G'} S \ [\ell_0\ell_1^k]$$

By (ii) in Claim 1, $\gamma = \omega(\beta)$. By the definition of ω and by (1.2) in the algorithm, γ consists only of rules used to generate w, so $S \Rightarrow^*_G w \ [\gamma]$. Therefore, $S' \Rightarrow^*_{G'} w \ [\rho]$. Consequently, we have $L(H) \subseteq \Gamma$, so the claim holds. □

Claim 6. $\Gamma \subseteq L(H)$.

Proof. Consider Claim 4 with $x = X_1X_2 \cdots X_h\bar{Z}\ell_1^k\gamma$ and $\mathfrak{D}'(\beta) \in F$, where $k \geq 0$, $\gamma \in \omega(\beta)$, and $X_j \in T$, for all j, $1 \leq j \leq h$, for some $h \geq 1$, so

$$S' \Rightarrow^*_{(G',W')} X_1X_2 \cdots X_h\bar{Z}\ell_1^k\gamma \ [\beta]$$

Let $w = X_1X_2 \cdots X_h$. Then,

$$S \Rightarrow^*_{(G,W)} w \ [\alpha]$$

such that $\mathfrak{D}(\alpha) = \mathfrak{D}'(\beta)$. Let $p = \mathfrak{D}'(\beta)$. By the initialization part of the algorithm, $\ell_\$: \bar{Z} \rightarrow \ell_0 \in P'$ and $p\ell_\$ \rightarrow \$ \in R'$, so

$$w\bar{Z}\ell_1^k\gamma \Rightarrow_{(G',W')} w\ell_0\ell_1^k\gamma$$

Let $\rho = \ell_0\ell_1^k\gamma$. Since $\mathfrak{D}'(\beta\ell_\$) = \$$ and $\$ \in F'$ by the initialization part of the algorithm, $w\rho \in L(H')$. The derivation

$$S' \Rightarrow^*_{G'} w \ [\rho]$$

could be demonstrated in the same way as in the proof of Claim 5. Since $p \in F$, $w \in L(H)$. Consequently, we have $\Gamma \subseteq L(H)$, so the claim holds. □

Since $L(H) \subseteq \Gamma$ and $\Gamma \subseteq L(H)$, $L(H) = \Gamma$. Thus, Lemma 8.1.2 holds. □

The next theorem represents the main result of this section.

Theorem 8.1.3. *Let K be a recursively enumerable language. Then, there is a propagating regular-controlled grammar with appearance checking, $H' = (G', \Xi, W')$, where $G' = (N, T, \Psi, P, S)$, such that*

$$K = \{w \mid w\rho \in L(H), S \Rightarrow^*_{G'} w\ [\rho]\}$$

Proof. By Theorem 5.1.6, there are a context-free grammar G, a complete finite automaton M, and an appearance checking set W, such that $L(H) = K$, where $H = (G, L(M), W)$ is a regular-controlled grammar with appearance checking. Let $G' = (N, T, \Psi, P, S)$, M', and W' be the propagating context-free grammar, the complete finite automaton, and the appearance checking set, respectively, constructed from G, M, and W by Algorithm 8.1.1. Set $\Xi = L(M')$. By Lemma 8.1.2,

$$L(H) = \{w \mid w\rho \in L(H'), S \Rightarrow^*_{G'} w\ [\rho]\}$$

where $H' = (G', \Xi, W')$ is the requested propagating regular-controlled grammar with appearance checking. Therefore, the theorem holds. □

Let $G = (N, T, \Psi, P, S)$ be an arbitrary context-free grammar, and let $S \Rightarrow^*_G w\ [\rho]$ be a derivation in G, where $w \in T^*$. Observe that since G is a context-free grammar, we can always rearrange rules from ρ so that $S\ {}_{\mathrm{lm}}\!\Rightarrow^*_G w\ [\sigma]$, for some $\sigma \in \mathrm{perm}(\rho)$. Therefore, we obtain the following corollary of Theorem 8.1.3.

Corollary 8.1.4. *Let K be a recursively enumerable language. Then, there is a propagating regular-controlled grammar with appearance checking, $H' = (G', \Xi, W')$, where $G' = (N, T, \Psi, P, S)$, such that*

$$K = \{w \mid w\rho \in L(H'), S\ {}_{\mathrm{lm}}\!\Rightarrow^*_{G'} w\ [\sigma], \textit{for some } \sigma \in \mathrm{perm}(\rho)\}$$ □

Observe that Corollary 8.1.4 can be straightforwardly rephrased in terms of rightmost derivations and right parses.

Corollary 8.1.5. *Let K be a recursively enumerable language. Then, there is a propagating regular-controlled grammar with appearance checking, $H' = (G', \Xi, W')$, where $G' = (N, T, \Psi, P, S)$, such that*

$$K = \{w \mid w\rho \in L(H'), S\ {}_{\mathrm{rm}}\!\Rightarrow^*_{G'} w\ [\sigma], \textit{for some } \sigma \in \mathrm{perm}(\rho)\}$$ □

Reconsider Algorithm 8.1.1. Observe that if the input appearance checking set W is empty, then so is the output appearance checking set W' (in symbols, $W = \emptyset$ implies that $W' = \emptyset$). Therefore, we obtain the following corollary.

Corollary 8.1.6. *Let $H = (G, \Xi)$ be a regular-controlled grammar. Then, there is a propagating regular-controlled grammar, $H' = (G', \Xi')$, where $G' = (N, T, \Psi, P, S)$, such that*

$$L(H) = \{w \mid w\rho \in L(H'), S \Rightarrow^*_{G'} w\ [\rho]\}$$

□

Reconsider Algorithm 8.1.1. The language $L(H')$ in the output of the algorithm contains strings followed by their parses. Set $T' = T' - T$, $N' = N' \cup T$, and introduce rules of the form $a \rightarrow \ell_{-1}$, for all $a \in T$, where $\ell_{-1}\colon S' \rightarrow S'$ is a new rule in P'. Then, observe that the output grammar directly generates only parses with no preceding strings of terminals from T. This generation is particularly useful when we are interested only in parses.

Reconsider Algorithm 8.1.1 and the proof of Lemma 8.1.2. Let $w\rho \in L(H')$, where $w \in T^*$, $\rho \in \Psi'^*$, and $S' \Rightarrow^*_{G'} w\ [\rho]$. Observe that the only nonterminals that appear in the sentential forms of this derivation are from $N \cup \{S'\}$. Furthermore, for each $t \in \text{alph}(\rho)$, the rule labeled with t, $t\colon A \rightarrow Y_1Y_2 \cdots Y_f \in P'$, introduced in (1.2) in the algorithm, has its right-hand side no longer than the right-hand side of the original rule, $r\colon A \rightarrow y_0Y_1y_1Y_2y_2 \cdots Y_fy_f \in P$, from which it was constructed, where $y_i \in N^*$, $Y_j \in V$, for all i and j, $0 \leq i \leq f$, $1 \leq j \leq f$, for some $f \geq 1$.

Next, we modify Theorem 8.1.3 using symbol-exhaustive right and left quotients (see Sect. 3.3).

Theorem 8.1.7. *Let K be a recursively enumerable language and let § be a symbol such that $\S \notin \text{alph}(K)$. Then, there is a propagating regular-controlled grammar with appearance checking H' such that $K = L(H') \,/\!\!/\, \{\S\}^+$.*

Proof. Reconsider Algorithm 8.1.1. Without any loss of generality, we assume that $\S \notin V'$, where $V' = N' \cup T'$. Let ν be a homomorphism from V'^* to $(V' \cup \{\S\})^*$ defined as $\nu(t) = \S$, for all $t \in \Psi'$, and $\nu(X) = X$, for all $X \in V' - \Psi'$. Observe that by replacing each $t\colon A \rightarrow x \in P'$ with $t\colon A \rightarrow \nu(x)$, we can establish this theorem by analogy with Theorem 8.1.3. □

By a straightforward modification of Algorithm 8.1.1, we can demonstrate the next theorem, too. It represents a reformulation of Theorem 8.1.7 in terms of symbol-exhaustive left quotient. We leave its proof to the reader.

Theorem 8.1.8. *Let K be a recursively enumerable language and let § be a symbol such that $\S \notin \text{alph}(K)$. Then, there is a propagating regular-controlled grammar with appearance checking H' such that $K = \{\S\}^+ \setminus\!\!\setminus L(H') /\!\!/$.* □

Let us note that in terms of scattered context grammars, [9] gives a similar characterization to the characterization result presented in Theorem 8.1.8.

We close this section by suggesting a new open problem area. Reconsider G and G' in Algorithm 8.1.1. Consider any derivation D in G and the corresponding derivation D' by which G' simulates D. Finally, consider the derivation D'' that results from D' by eliminating all applications of the unit rules $\ell_0\colon S' \rightarrow S$ and $\ell_1\colon S \rightarrow S$. Observe that D'' consists of no more derivation steps than D does.

Informally and more generally speaking, in this sense, G' works at least as fast as G does if the applications of $\ell_0\colon S' \to S$ and $\ell_1\colon S \to S$ are ignored in the derivations in G'.

Open Problem 8.1.9. Can we make an alternative algorithm such that it does the same job as Algorithm 8.1.1, and simultaneously it satisfies the same valuable property concerning the number of derivation steps in G' without the necessity of ignoring the applications of any rules? □

8.2 Coincidental Extension of Scattered Context Languages

As already pointed out in the beginning of this chapter, language operations that extend strings by shuffling or inserting some symbols into them fulfill an important role in several modern computer science fields. The present section defines and discusses another operation of this kind—*coincidental extension*—in terms of languages generated by propagating scattered context grammars (see Sect. 4.7).

To give an insight into this operation, consider an alphabet V, a symbol $\# \notin V$, and two languages $K \subseteq (V \cup \{\#\})^*$ and $L \subseteq V^*$. For a string $x = a_1a_2\cdots a_{n-1}a_n \in L$, any string of the form $\#^i a_1 \#^i a_2 \#^i \cdots \#^i a_{n-1} \#^i a_n \#^i$, where $i \geq 0$, is a coincidental #-extension of x. K is a coincidental #-extension of L if every string of K represents a coincidental extension of a string in L and the deletion of all #s in K results in L.

We next formalize the coincidental extension of languages and illustrate this extension by an example.

Definition 8.2.1. Let V be an alphabet and $\# \notin V$ be a symbol. A *coincidental #-extension* of ε is any string of the form

$$\#^i$$

where $i \geq 0$. A coincidental #-extension of $a_1a_2\ldots a_{n-1}a_n \in V^+$ is any string of the form

$$\#^i a_1 \#^i a_2 \#^i \cdots \#^i a_{n-1} \#^i a_n \#^i$$

where $n \geq 1$, $a_j \in V$, for all j, $1 \leq j \leq n$, for some $i \geq 0$. For any language, $L \subseteq V^*$, ${}_{\#}CE(L)$ denotes the set of all coincidental #-extensions of strings in L. Define the almost identity ω from $(V \cup \{\#\})^*$ to $(V \cup \{\#\})^*$ as $\omega(a) = a$, for all $a \in (V - \{\#\})$, and $\omega(\#) = \varepsilon$. Let $K \subseteq (V \cup \{\#\})^*$. K is a *coincidental #-extension* of L, symbolically written as $L \;{}_{\#}\!\blacktriangleleft\; K$, if $K \subseteq {}_{\#}CE(L)$ and $L = \{\omega(x) \mid x \in K\}$. □

Example 8.2.2. Consider the following three languages K, L, and M.

- $K = \{\#^i a \#^i b \#^i \mid i \geq 5\} \cup \{\#^i c^n \#^i d^n \#^i \mid n, i \geq 0\}$
- $L = \{ab\} \cup \{c^n d^n \mid n \geq 0\}$
- $M = \{\#^i a \#^i b \#^i \mid i \geq 5\} \cup \{\#^i c^n \#^i d^n \#^{i+1} \mid n, i \geq 0\}$

Observe that $L \,_{\#}\!\blacktriangleleft\, K$, but there exists no language N such that $N \,_{\#}\!\blacktriangleleft\, M$. □

Next, we prove that for every recursively enumerable language K, there exists a propagating scattered context language that represents a coincidental extension of K. The use of this result is sketched in the conclusion of this section.

Theorem 8.2.3. *Let $K \in$ **RE**. Then, there exists a propagating scattered context grammar H such that $K \,_{\#}\!\blacktriangleleft\, L(H)$.*

Proof. Let $K \in$ **RE**. Express K as $K = J \cup L$, where $J = \{x \mid x \in K$ and $|x| \leq 2\}$ and $L = K - J$.

First, we construct a propagating scattered context grammar G such that $L \,_{\#}\!\blacktriangleleft\, L(G)$. By Theorem 4.7.6, there exists a scattered context grammar, $\Gamma = (V, T, \Pi, S)$, such that $L = L(\Gamma)$. Without any loss of generality, we assume that $\{X, Y, Z, \#, \$, \S\} \cap V = \emptyset$. Set $U = \{\langle a \rangle \mid a \in T\}$ and $N = V - T$. Define the homomorphism γ from V to $(V \cup U \cup \{Y\})^*$ as $\gamma(a) = \langle a \rangle Y$, for all $a \in T$, and $\gamma(A) = AY$, for all $A \in N$. Extend the domain of γ to V^+ in the standard manner; non-standardly, however, define $\gamma(\varepsilon) = Y$ rather than $\gamma(\varepsilon) = \varepsilon$. Set

$$W = \{\langle ja \rangle \mid j \in \{0, 1, 2, 3\} \text{ and } a \in T\}$$

and

$$V' = \{X, Y, Z, \#, \$, \S\} \cup V \cup U \cup W$$

Define the propagating scattered context grammar

$$G = (V', T \cup \{\#\}, P, Z)$$

with P constructed by performing the next six steps

(1) add $(Z) \rightarrow (YS\$)$ to P;
(2) for every $(A_1, A_2, \ldots, A_n) \rightarrow (x_1, x_2, \ldots, x_n) \in \Pi$,
add $(A_1, A_2, \ldots, A_n, \$) \rightarrow (\gamma(x_1), \gamma(x_2), \ldots, \gamma(x_n), \$)$ to P;
(3) add $(Y, \$) \rightarrow (YY, \$)$ to P;
(4) for every $a, b, c \in T$,
add $(\langle a \rangle, \langle b \rangle, \langle c \rangle, \$) \rightarrow (\langle 0a \rangle, \langle 0b \rangle, \langle 0c \rangle, \S)$ to P;
(5) for every $a, b, c, d \in T$, add
$(Y, \langle 0a \rangle, Y, \langle 0b \rangle, Y, \langle 0c \rangle, \S) \rightarrow (\#, \langle 0a \rangle, X, \langle 0b \rangle, Y, \langle 0c \rangle, \S)$,
$(\langle 0a \rangle, \langle 0b \rangle, \langle 0c \rangle, \S) \rightarrow (\langle 4a \rangle, \langle 1b \rangle, \langle 2c \rangle, \S)$,
$(\langle 4a \rangle, X, \langle 1b \rangle, Y, \langle 2c \rangle, \S) \rightarrow (\langle 4a \rangle, \#, \langle 1b \rangle, X, \langle 2c \rangle, \S)$,
$(\langle 4a \rangle, \langle 1b \rangle, \langle 2c \rangle, \langle d \rangle, \S) \rightarrow (a, \langle 4b \rangle, \langle 1c \rangle, \langle 2d \rangle, \S)$,
$(\langle 4a \rangle, \langle 1b \rangle, \langle 2c \rangle, \S) \rightarrow (a, \langle 1b \rangle, \langle 3c \rangle, \S)$,
$(\langle 1a \rangle, X, \langle 3b \rangle, Y, \S) \rightarrow (\langle 1a \rangle, \#, \langle 3b \rangle, \#, \S)$
to P;
(6) for every $a, b \in T$, add
$(\langle 1a \rangle, X, \langle 3b \rangle, \S) \rightarrow (a, \#, b, \#)$ to P.

For a proof of $L \;_{\#}\!\blacktriangleleft\; L(G)$, see [20, 21]. Construct a propagating scattered context grammar, $\bar{G} = (\bar{V}, T, \bar{P}, \bar{Z})$, such that $(\bar{V} - T) \cap (V' - T) = \emptyset$ and $J \;_{\#}\!\blacktriangleleft\; L(G')$ (recall that $J = \{x \mid x \in K \text{ and } |x| \leq 2\}$ and $L = K - J$). Let $\hat{Z} \notin V' \cup \bar{V}$. Consider the propagating scattered context grammar

$$H = \left(V' \cup \bar{V} \cup \{\hat{Z}\}, T \cup \{\#\}, P \cup \bar{P} \cup \{(\hat{Z}) \to (Z), (\hat{Z}) \to (\bar{Z})\}, \hat{Z}\right)$$

Observe that $K \;_{\#}\!\blacktriangleleft\; L(H)$, so the theorem holds. □

To illustrate the use of the previous result, observe that Theorem 8.2.3 straightforwardly implies some results about $\mathbf{SCAT}^{-\varepsilon}$, which were originally proved in a rather complicated way in formal language theory. For instance, Theorem 8.2.3 straightforwardly implies the next corollary, which was established in [19] in a rather complex way.

Corollary 8.2.4. *For every $K \in$ **RE**, there exists an almost identity ω and a language $J \in \mathbf{SCAT}^{-\varepsilon}$ such that $K = \omega(J)$.* □

As any almost identity is a special case of a homomorphism, Corollary 8.2.4 also implies the following fundamental result concerning $\mathbf{SCAT}^{-\varepsilon}$ (see the second corollary on p. 245 in [10]).

Corollary 8.2.5. *For every $K \in$ **RE**, there exists a homomorphism η and a language $J \in \mathbf{SCAT}^{-\varepsilon}$ such that $K = \eta(J)$.* □

We close this section by formulating and open problem area.

Open Problem 8.2.6. Investigate coincidental extension in terms of other regulated grammars, such as regular-controlled grammars, matrix grammars, and programmed grammars (see Chap. 5). □

References

1. Aho, A.V., Lam, M.S., Sethi, R., Ullman, J.D.: Compilers: Principles, Techniques, and Tools, 2nd edn. Addison-Wesley, Boston (2006)
2. Amos, M.: DNA computation. Ph.D. thesis, University of Warwick, England (1997)
3. Bal, H., Grune, D., Jacobs, C., Langendoen, K.: Modern Compiler Design. Wiley, Hoboken (2000)
4. Bravo, C., Neto, J.J.: Building context-sensitive parsers from CF grammars with regular control language. In: Implementation and Application of Automata 8th International Conference, pp. 306–308. Springer (2003)
5. Cojocaru, L., Mäkinen, E., Tiplea, F.L.: Classes of Szilard languages in NC^1. In: Symbolic and Numeric Algorithms for Scientific Computing (SYNASC), 11th International Symposium, pp. 299–306 (2009)
6. Cooper, K.D., Torczon, L.: Engineering a Compiler. Morgan Kaufmann Publishers, San Francisco (2004)
7. Cytron, R., Fischer, C., LeBlanc, R.: Crafting a Compiler. Addison-Wesley, Boston (2009)

8. Dassow, J., Păun, G.: Regulated Rewriting in Formal Language Theory. Springer, New York (1989)
9. Ehrenfeucht, A., Rozenberg, G.: An observation on scattered grammars. Inform. Process. Lett. **9**(2), 84–85 (1979)
10. Greibach, S.A., Hopcroft, J.E.: Scattered context grammars. J. Comput. Syst. Sci. **3**(3), 233–247 (1969)
11. Gutiérrez-Naranjo, M.A., Pérez-Jiménez, M.J., Riscos-Núñez, A.: Multidimensional Sevilla carpets associated with P systems. In: Proceedings of the ESF Exploratory Workshop on Cellular Computing (Complexity Aspects), pp. 225–236 (2005)
12. Jirák, O.: Delayed execution of scattered context grammar rules. In: Proceedings of the 15th Conference and Competition EEICT 2009, pp. 405–409. Brno University of Technology, Brno, CZ (2009)
13. Jirák, O.: Table-driven parsing of scattered context grammar. In: Proceedings of the 16th Conference and Competition EEICT 2010, pp. 171–175. Brno University of Technology, Brno, CZ (2010)
14. Jirák, O., Kolář, D.: Derivation in scattered context grammar via lazy function evaluation. In: 5th Doctoral Workshop on Mathematical and Engineering Methods in Computer Science, pp. 118–125. Masaryk University (2009)
15. Kari, L.: Power of controlled insertion and deletion. In: Results and Trends in Theoretical Computer Science. Lecture Notes in Computer Science, vol. 812, pp. 197–212. Springer, Berlin (1994)
16. Kari, L.: On insertion and deletion in formal languages. Ph.D. thesis, University of Turku, Finland (1997)
17. Kolář, D.: Scattered context grammars parsers. In: Proceedings of the 14th International Congress of Cybernetics and Systems of WOCS, pp. 491–500. Wroclaw University of Technology (2008)
18. Mäkinen, E.: A bibliography on Szilard languages. Department of Computer Sciences, University of Tampere. Available on http://www.cs.uta.fi/reports/pdf/Szilard.pdf
19. Meduna, A.: Syntactic complexity of scattered context grammars. Acta Inform. **1995**(32), 285–298 (1995)
20. Meduna, A.: Coincidental extension of scattered context languages. Acta Inform. **39**(5), 307–314 (2003)
21. Meduna, A.: Erratum: Coincidental extension of scattered context languages. Acta Inform. **39**(9), 699 (2003)
22. Meduna, A.: Elements of Compiler Design. Auerbach Publications, Boston (2007)
23. Meduna, A., Techet, J.: Canonical scattered context generators of sentences with their parses. Theor. Comput. Sci. **2007**(389), 73–81 (2007)
24. Meduna, A., Techet, J.: Scattered Context Grammars and Their Applications. WIT Press, Southampton (2010)
25. Muchnick, S.S.: Advanced Compiler Design and Implementation. Morgan Kaufmann Publishers, San Francisco (1997)
26. Parsons, T.W.: Introduction to Compiler Construction. Computer Science Press, New York (1992)
27. Rußmann, A.: Dynamic LL(k) parsing. Acta Inform. **34**(4), 267–289 (1997)
28. Rytter, W., Crochemore, M.: Text Algorithms. Oxford University Press, New York (1994)
29. Seberry, J., Pieprzyk, J.: Cryptography: An Introduction to Computer Security. Prentice-Hall, New Jersey (1989)
30. Sebesta, R.W.: On context-free programmed grammars. Comput. Lang. **14**(2), 99–108 (1989)

Chapter 9
Sequential Rewriting Over Word Monoids

Abstract Concerning grammars, formal language theory standardly defines the relation of a direct derivation over free monoids generated by an alphabet—that is, finitely many symbols. The present chapter modifies this standard definition so it introduces this relation over free monoids generated by finitely many strings. It explains that this modification can be seen as a very natural context-based grammatical regulation; indeed, a derivation step is performed on the condition that the rewritten sentential form occurs in the free monoids generated in this modified way. This chapter consists of two sections. Section 9.1 defines the above modification rigorously and explains its relation to the subject of this book. Section 9.2 demonstrates that this modification results into a large increase of the generative power of context-free grammars. In fact, even if the free monoids are generated by strings consisting of no more than two symbols, the resulting context-free grammars are as powerful as phrase-structure grammars.

Keywords Algebraic approach to grammatical regulation • Derivations over string-generated free monoids • Context conditions • Generative power • Computational completeness

The present chapter approaches the subject of this book from a strictly mathematical viewpoint. In formal language theory, the relation of a direct derivation $\Rightarrow$ is introduced over V^*, where V is the total alphabet of a grammar; to rephrase this in terms of algebra, $\Rightarrow$ is thus defined over the free monoid whose generators are symbols. We modify this definition by using strings rather than symbols as the generators. More precisely, we introduce this relation over the free monoid generated by a finite set of strings; in symbols, $\Rightarrow$ is defined over W^*, where W is a finite language. As a result, this modification represents a very natural context condition: a derivation step is performed on the condition that the rewritten sentential form occurs in W^*. This context condition results into a large increase

A. Meduna and P. Zemek, *Regulated Grammars and Automata*,
DOI 10.1007/978-1-4939-0369-6_9,

of generative power of context-free grammars. In fact, even if W contains strings consisting of no more than two symbols, the resulting power of these grammars is equal to the power of phrase-structure grammars.

This chapter consists of two sections. Section 9.1 defines context-free grammars over word monoids. Then, Sect. 9.2 establishes their generative power.

9.1 Definitions

Without further ado, we next define context-free grammars over word monoids.

Definition 9.1.1. A *context-free grammar over word monoid* (a *wm-grammar* for short) is a pair

$$(G, W)$$

where

$$G = (V, T, P, S)$$

is a context-free grammar, and W, called the *set of generators*, is a finite language over V. (G, W) is of *degree i*, where i is a natural number, if $y \in W$ implies that $|y| \leq i$. (G, W) is said to be *propagating* if $A \to x \in P$ implies that $x \neq \varepsilon$.

The *direct derivation* $\Rightarrow_{(G,W)}$ on W^* is defined as follows: if $p = A \to y \in P$, $xAz, xyz \in W^*$ for some $x, z \in V^*$, then xAz directly derives xyz, written as

$$xAz \Rightarrow_{(G,W)} xyz \; [p]$$

In the standard way, we define $\Rightarrow^k_{(G,W)}$ for $k \geq 0$, $\Rightarrow^+_{(G,W)}$, and $\Rightarrow^*_{(G,W)}$. The *language* of (G, W), symbolically denoted by $L(G, W)$, is defined as

$$L(G, W) = \left\{ w \in T^* \mid S \Rightarrow^*_{(G,W)} w \right\}$$ □

Let **WM** denote the family of languages generated by wm-grammars. The family of languages generated by wm-grammars of degree i is denoted by $\mathbf{WM}(i)$. The families languages generated by propagating wm-grammars of degree i and propagating wm-grammars of any degree are denoted by $\mathbf{WM}^{-\varepsilon}(i)$ and $\mathbf{WM}^{-\varepsilon}$, respectively.

9.2 Generative Power

Making use of the language families introduced in the previous section, we establish the generative power of wm-grammar by demonstrating that

$$\begin{array}{c}\mathbf{WM}^{-\varepsilon}(1) = \mathbf{WM}(1) = \mathbf{CF} \\ \subset \\ \mathbf{WM}^{-\varepsilon}(2) = \mathbf{WM}^{-\varepsilon} = \mathbf{CS} \\ \subset \\ \mathbf{WM}(2) = \mathbf{WM} = \mathbf{RE}\end{array}$$

Theorem 9.2.1. $\mathbf{WM}^{-\varepsilon}(0) = \mathbf{WM}(0) = \emptyset$ *and* $\mathbf{WM}^{-\varepsilon}(1) = \mathbf{WM}(1) = \mathbf{CF}$

Proof. This theorem follows immediately from the definitions. □

Next, we prove that (1) a language is context-sensitive if and only if it is generated by a propagating wm-grammar of degree 2 and (2) a language is recursively enumerable if and only if it is generated by a wm-grammar of degree 2.

Theorem 9.2.2. $\mathbf{WM}^{-\varepsilon}(2) = \mathbf{CS}$

Proof. It is straightforward to prove that $\mathbf{WM}^{-\varepsilon}(2) \subseteq \mathbf{CS}$; hence, it is sufficient to prove the converse inclusion.

Let L be a context-sensitive language. Without any loss of generality, we assume that L is generated by a context-sensitive grammar

$$G = \big(N_{CF} \cup N_{CS} \cup T, T, P, S\big)$$

of the form described in Theorem 4.1.6. Let

$$V = N_{CS} \cup N_{CF} \cup T$$

The propagating wm-grammar (G', W) of degree 2 is defined as follows:

$$G' = \big(V', T, P', S\big)$$

where

$$\begin{aligned} V' &= V \cup Q \\ Q &= \big\{\langle A, B, C\rangle \mid AB \to AC \in P,\ A, C \in N_{CF},\ B \in N_{CS}\big\} \end{aligned}$$

Clearly, without loss of generality, we assume that $Q \cap V = \emptyset$. The set of rules P' is defined in the following way

(1) If $A \to x \in P$, $A \in N_{CF}$, $x \in N_{CS} \cup T \cup N_{CF}^2$, then add $A \to x$ to P';
(2) If $AB \to AC \in P$, $A, C \in N_{CF}$, $B \in N_{CS}$, then add $B \to \langle A, B, C\rangle$ and $\langle A, B, C\rangle \to C$ to P'.

The set of generators W is defined as follows:

$$W = \big\{A\langle A, B, C\rangle \mid \langle A, B, C\rangle \in Q,\ A \in N_{CF}\big\} \cup V$$

Obviously (G', W) is a propagating wm-grammar of degree 2. Next, let h be a finite substitution from V'^* into V^* defined as

(1) For all $D \in V$, $h(D) = D$;
(2) For all $\langle X, D, Z\rangle \in Q$, $h(\langle X, D, Z\rangle) = D$.

Let h^{-1} be the inverse of h. To show that $L(G) = L(G', W)$, we first prove that

$$S \Rightarrow_G^m w \quad \text{if and only if} \quad S \Rightarrow_{(G'W)}^n v$$

where $v \in W^* \cap h^{-1}(w)$, $w \in V^+$, for some $m, n \geq 0$.

Only If. This is established by induction on $m \geq 0$.

Basis. Let $m = 0$. The only w is S because $S \Rightarrow_G^0 S$. Clearly, $S \Rightarrow_{(G',W)}^0 S$ and $S \in h^{-1}(S)$.

Induction Hypothesis. Let us suppose that the claim holds for all derivations of length m or less, for some $m \geq 0$.

Induction Step. Consider any derivation of the form

$$S \Rightarrow_G^{m+1} x$$

where $x \in V^+$. Since $m + 1 \geq 1$, there is some $y \in V^+$ and $p \in P$ such that

$$S \Rightarrow_G^m y \Rightarrow_G x\ [p]$$

and by the induction hypothesis, there is also a derivation

$$S \Rightarrow_{(G',W)}^n y''$$

for some $y'' \in W^* \cap h^{-1}(y)$, $n \geq 0$.

(i) Let us assume that $p = D \rightarrow y_2$, $D \in N_{CF}$, $y_2 \in N_{CS} \cup T \cup N_{CF}^2$, $y = y_1 D y_3$, $y_1, y_3 \in V^*$, and $x = y_1 y_2 y_3$. Since from the definition of h^{-1} it is clear that $h^{-1}(Z) = \{Z\}$ for all $Z \in N_{CF}$, we can write $y'' = z_1 D z_3$, where $z_1 \in h^{-1}(y_1)$ and $z_3 \in h^{-1}(y_3)$. It is clear that $D \rightarrow y_2 \in P'$ (see the definition of P').

Let $z_3 \notin QV'^*$. Then,

$$S \Rightarrow_{(G',W)}^n z_1 D z_3 \Rightarrow_{(G',W)} z_1 y_2 z_3$$

and clearly, $z_1 y_2 z_3 \in h^{-1}(y_1 y_2 y_3) \cap W^*$.

Let $z_3 \in QV'^*$; that is, $z_3 = Yr$ for some $Y \in Q$, $r \in V'^*$. Thus, $Dh(Y) \rightarrow DC \in P$ (for some $C \in N_{CF}$), $y_3 = h(Y)s$, where $r \in h^{-1}(s)$ and $s \in V^*$. Hence, we have $h(Y) \rightarrow Y \in P'$ (see (2) in the definition of P'). Observe that $h(Y) \rightarrow Y$ is the only rule in P' that has Y appearing on its right-hand side. Also it is clear that r is not in QV'^{*} (see the definition of W). Thus, $\{z_1 Dh(Y)r, z_1 y_2 h(Y)r\} \subseteq W^*$, and since

$$S \Rightarrow^{n}_{(G',W)} z_1 DYr$$

there is also the following derivation in (G', W)

$$S \Rightarrow^{n-1}_{(G',W)} z_1 Dh(Y)r \Rightarrow_{(G',W)} z_1 DYr \; [h(Y) \rightarrow Y]$$

Thus,

$$S \Rightarrow^{n-1}_{(G',W)} z_1 Dh(Y)r \Rightarrow_{(G',W)} z_1 y_2 h(Y)r \; [D \rightarrow y_2]$$

such that $z_1 y_2 h(Y)r$ is in $h^{-1}(x) \cap W^*$.

(ii) Let $p = AB \rightarrow AC$, $A, C \in N_{CF}$, $B \in N_{CS}$, $y = y_1 ABy_2$, $y_1, y_2 \in V^*$, $x = y_1 ACy_2$, $y'' = z_1 AYz_2$, $z_i \in h^{-1}(y_i)$, $i \in \{1, 2\}$, and $Y \in h^{-1}(B)$. Clearly,

$$\{B \rightarrow \langle A, B, C\rangle, \langle A, B, C\rangle \rightarrow C\} \subseteq P'$$

and $A\langle A, B, C\rangle \in W$.

Let $Y = B$. Since $B \in N_{CS}$, $z_2 \notin QV'^{*}$; therefore, $z_1 A\langle A, B, C\rangle z_2 \in W^*$ (see the definition of W). Thus,

$$\begin{aligned} S &\Rightarrow^{n}_{(G',W)} z_1 ABz_2 \\ &\Rightarrow_{(G',W)} z_1 A\langle A, B, C\rangle z_2 \; [B \rightarrow \langle A, B, C\rangle] \\ &\Rightarrow_{(G',W)} z_1 ACz_2 \qquad [\langle A, B, C\rangle \rightarrow C] \end{aligned}$$

and $z_1 ACz_2 \in h^{-1}(x) \cap W^*$.

Let $Y \in Q$. Clearly, $h(Y) = B$ and by the definitions of Q and P', we have $B \rightarrow Y \in P'$. Thus, we can express the derivation

$$S \Rightarrow^{n}_{(G',W)} z_1 AYz_2$$

as

$$S \Rightarrow^{n-1}_{(G',W)} z_1 ABz_2 \Rightarrow_{(G',W)} z_1 AYz_2 \; [B \rightarrow Y]$$

Since $z_1 A\langle A, B, C\rangle z_2 \in W^*$, we have

$$\begin{aligned} S &\Rightarrow^{n-1}_{(G',W)} z_1 ABz_2 \\ &\Rightarrow_{(G',W)} z_1 A\langle A, B, C\rangle z_2 \\ &\Rightarrow_{(G',W)} z_1 ACz_2 \end{aligned}$$

where $z_1 ACz_2 \in h^{-1}(x) \cap W^*$.

If. This is established by induction on $n \geq 0$.

Basis. For $n = 0$ the only v is S because $S \Rightarrow^0_{(G',W)} S$. Since $S \in h^{-1}(S)$, we have $w = S$. Clearly, $S \Rightarrow^0_G S$.

Induction Hypothesis. Let us assume the claim holds for all derivations of length at most n, for some $n \geq 0$.

Induction Step. Consider any derivation of the form

$$S \Rightarrow^{n+1}_{(G',W)} u$$

where $u \in h^{-1}(x) \cap W^*$ and $x \in V^+$. Since $n + 1 \geq 1$, there are $p \in P'$, $y \in V^+$, and $v \in h^{-1}(y) \cap W^*$ such that

$$S \Rightarrow^n_{(G',W)} v \Rightarrow_{(G',W)} u\ [p]$$

and by the induction hypothesis,

$$S \Rightarrow^*_G y$$

Let $v = r'Ds'$, $y = rBs$, $r' \in h^{-1}(r)$, $s' \in h^{-1}(s)$, $r, s \in V^*$, $D \in h^{-1}(B)$, $u = r'z's'$, and $p = D \rightarrow z' \in P'$. Moreover, let us consider the following three cases.

(i) Let $h(z') = B$ [see (2)]. Then, $u = r'z's' \in h^{-1}(rBs)$; that is, $x = rBs$. By the induction hypothesis, we have

$$S \Rightarrow^*_G rBs$$

(ii) Let $z' \in T \cup N_{CS} \cup N^2_{CF}$. Then, there is a rule $B \rightarrow z' \in P$. Since $z' \in h^{-1}(z')$, we have $x = rz's$. Clearly,

$$S \Rightarrow^*_G rBs \Rightarrow_G rz's\ [B \rightarrow z']$$

(iii) Let $z' = C \in N_{CF}$, $D = \langle A, B, C \rangle \in Q$. By the definition of W, we have $r' = t'A$, $r = tA$, where $t' \in h^{-1}(t)$, $t \in V^*$; therefore, $x = tACs$. By the definition of Q, there is a rule $AB \rightarrow AC \in P$. Thus,

$$S \Rightarrow^*_G tABs \Rightarrow_G tACs\ [AB \rightarrow AC]$$

By the inspection of P', we have considered all possible derivations of the form

$$S \Rightarrow^n_{(G',W)} v \Rightarrow_{(G',W)} u$$

in (G', W). Thus, by the principle of induction, we have established that

$$S \Rightarrow^n_{(G',W)} u$$

for some $n \geq 0$ and $u \in W^*$ implies

$$S \Rightarrow^*_G x$$

where $x \in V^*$ and $u \in h^{-1}(x)$. Hence,

$$S \Rightarrow^m_G w \quad \text{if and only if} \quad S \Rightarrow^n_{(G',W)} v$$

where $v \in W^* \cap h^{-1}(w)$ and $w \in V^*$, for some $m, n \geq 0$.

A proof of the equivalence of G and (G', W) can be derived from the above. Indeed, by the definition of h^{-1}, we have $h^{-1}(a) = \{a\}$ for all $a \in T$. Thus, by the statement above and by the definition of W, we have for any $x \in T^*$

$$S \Rightarrow^*_G x \quad \text{if and only if} \quad S \Rightarrow^*_{(G',W)} x$$

That is, $L(G) = L(G', W)$. Thus, $\mathbf{WM}^{-\varepsilon}(2) = \mathbf{CS}$, which proves the theorem. □

Observe that the form of the wm-grammar in the proof of Theorem 9.2.2 implies the following corollary.

Corollary 9.2.3. *Let L be a context-sensitive language over an alphabet T. Then, L is generated by a propagating wm-grammar (G, W) of degree 2, where $G = (V, T, P, S)$ satisfies*

(i) $T \subseteq W$ and $(W - V) \subseteq (V - T)^2$;
(ii) If $A \to x$ and $|x| > 1$, then $x \in (V - T)^2$. □

Next, we study the wm-grammars of degree 2 with erasing rules. We prove that these grammars generate precisely **RE**.

Theorem 9.2.4. $\mathbf{WM}(2) = \mathbf{RE}$

Proof. Clearly, we have $\mathbf{WM}(2) \subseteq \mathbf{RE}$; hence it suffices to show $\mathbf{RE} \subseteq \mathbf{WM}(2)$. The inclusion $\mathbf{RE} \subseteq \mathbf{WM}(2)$ can be proved by the techniques given in the proof of Theorem 9.2.2 because every language $L \in \mathbf{RE}$ can be generated by a grammar $G = (V, T, P, S)$ of the form of Theorem 4.1.7. The details are left to the reader. □

Since the form of the resulting wm-grammar in the proof of Theorem 9.2.4 is analogous to the wm-grammar in the proof of Theorem 9.2.2 (except that the former may contain some erasing rules), we obtain the next corollary.

Corollary 9.2.5. *Let L be a recursively enumerable language over an alphabet T, Then, L can be generated by a wm-grammar (G, W) of degree 2, where $G = (V, T, P, S)$ such that*

(i) $T \subseteq W$ *and* $(W - V) \subseteq (V - T)^2$;
(ii) If $A \to x$ *and* $|x| > 1$, *then* $x \in (V - T)^2$. □

Summing up the results from the present section, we also obtain the following corollary.

Corollary 9.2.6.

$$\begin{array}{c} \mathbf{WM}^{-\varepsilon}(1) = \mathbf{WM}(1) = \mathbf{CF} \\ \subset \\ \mathbf{WM}^{-\varepsilon}(2) = \mathbf{WM}^{-\varepsilon} = \mathbf{CS} \\ \subset \\ \mathbf{WM}(2) = \mathbf{WM} = \mathbf{RE} \end{array}$$

Proof. This corollary follows from Theorems 9.2.1, 9.2.2, and 9.2.4. □

So far, we have demonstrated that propagating wm-grammars of degree 2 and wm-grammars of degree 2 characterize **CS** and **RE**, respectively. Next, we show that the characterization of **RE** can be further improved in such a way that even some reduced versions of wm-grammars suffice to generate all the family of recursively enumerable languages. More specifically, we can simultaneously reduce the number of nonterminals and the number of strings of length two occurring in the set of generators without any decrease of the generative power.

Theorem 9.2.7. *Every $L \in$ **RE** can be defined by a ten-nonterminal context-free grammar over a word monoid generated by an alphabet and six strings of length 2.*

Proof. Let $L \in$ **RE**. By Theorem 4.1.11, $L = L(G)$, where G is a phrase-structure grammar of the form

$$G = \big(V, T, P \cup \{AB \to \varepsilon, CD \to \varepsilon\}, S\big)$$

such that P contains only context-free rules and

$$V - T = \{S, A, B, C, D\}$$

Let us define a wm-grammar (G', W) of degree 2, where

$$G' = \big(V', T, P', S\big)$$

and

$$\begin{aligned} V' &= \{S, A, B, C, D, \langle AB\rangle, \langle CD\rangle, \langle left\rangle, \langle right\rangle, \langle empty\rangle\} \cup T, \\ P' &= P \cup \{B \to \langle AB\rangle, \langle AB\rangle \to \langle right\rangle, \\ &\quad D \to \langle CD\rangle, \langle CD\rangle \to \langle right\rangle, \\ &\quad A \to \langle left\rangle, C \to \langle left\rangle, \\ &\quad \langle left\rangle \to \langle empty\rangle, \langle right\rangle \to \langle empty\rangle, \langle empty\rangle \to \varepsilon\} \end{aligned}$$

The set of generators is defined as

$$\begin{aligned} W = \big\{&A\langle AB\rangle, C\langle CD\rangle, \langle left\rangle\langle AB\rangle, \langle left\rangle\langle CD\rangle, \\ &\langle left\rangle\langle right\rangle, \langle empty\rangle\langle right\rangle, \langle empty\rangle\big\} \cup T \cup \big\{S, A, B, C, D\big\} \end{aligned}$$

Clearly, (G', W) is a wm-grammar with the required properties. To establish $L(G) \subseteq L(G', W)$, we first prove the following claim.

Claim 1. $S \Rightarrow_G^m w$ *implies that* $S \Rightarrow_{(G',W)}^* w$*, where* $w \in V^*$ *for some* $m \geq 0$.

Proof. This claim is established by induction on $m \geq 0$.

Basis. Let $m = 0$. The only w is S because $S \Rightarrow_G^0 S$. Clearly, $S \Rightarrow_{(G',W)}^0 S$.

Induction Hypothesis. Suppose that the claim holds for all derivations of length m or less, for some $m \geq 0$.

Induction Step. Consider any derivation of the form

$$S \Rightarrow_G^{m+1} w$$

with $w \in V^*$. As $m + 1 \geq 1$, there exists $y \in W^+$ and $p \in P$ such that

$$S \Rightarrow_G^m y \Rightarrow_G w\ [p]$$

By the induction hypothesis, there also exists a derivation

$$S \Rightarrow_{(G',W)}^n y$$

Observe that $y \in W^*$ because $V \subseteq W$. The rule p has one of these three forms

(i) p is a context-free rule in P;
(ii) p has the form $AB \to \varepsilon$;
(iii) p has the form $CD \to \varepsilon$.

Next, we consider these three possibilities.

(i) Let $p = E \to y_2$, $y = y_1 E y_3$, $E \in \{S, A, B, C, D\}$, $y_1, y_3 \in V^*$, and $w = y_1 y_2 y_3$. By the construction of P', $E \to y_2 \in P'$. Thus,

$$S \Rightarrow_{(G',W)}^n y_1 E y_3 \Rightarrow_{(G',W)} y_1 y_2 y_3\ [E \to y_2]$$

(ii) Let $p = AB \rightarrow \varepsilon$, $y = y_1ABy_2$, $y_1, y_2 \in V^*$, $w = y_1y_2$. At this point, we construct the following derivation in (G', W)

$$\begin{aligned}
S &\Rightarrow^n_{(G',W)} y_1ABy_2 \\
&\Rightarrow_{(G',W)} y_1A\langle AB\rangle y_2 && [B \rightarrow \langle AB\rangle] \\
&\Rightarrow_{(G',W)} y_1\langle left\rangle\langle AB\rangle y_2 && [A \rightarrow \langle left\rangle] \\
&\Rightarrow_{(G',W)} y_1\langle left\rangle\langle right\rangle y_2 && [\langle AB\rangle \rightarrow \langle right\rangle] \\
&\Rightarrow_{(G',W)} y_1\langle empty\rangle\langle right\rangle y_2 && [\langle left\rangle \rightarrow \langle empty\rangle] \\
&\Rightarrow_{(G',W)} y_1\langle empty\rangle\langle empty\rangle y_2 && [\langle right\rangle \rightarrow \langle empty\rangle] \\
&\Rightarrow_{(G',W)} y_1\langle empty\rangle y_2 && [\langle empty\rangle \rightarrow \varepsilon] \\
&\Rightarrow_{(G',W)} y_1y_2 && [\langle empty\rangle \rightarrow \varepsilon]
\end{aligned}$$

(iii) Let $p = CD \rightarrow \varepsilon$, $y = y_1CDy_2$, $y_1, y_2 \in V^*$, $w = y_1y_2$. By analogy with (ii), we can prove that

$$S \Rightarrow^*_{(G',W)} y_1y_2$$

Thus, Claim 1 now follows by the principle of induction. □

Next, we sketch how to verify $L(G', W) \subseteq L(G)$. First, we make two observations, which follow from the definition of W.

Let

$$\begin{aligned}
S &\Rightarrow^*_{(G',W)} y_1ABy_2 \\
&\Rightarrow_{(G',W)} y_1A\langle AB\rangle y_2 \ [B \rightarrow \langle AB\rangle] \\
&\Rightarrow^*_{(G',W)} w
\end{aligned}$$

where $w \in T^*$. Then, during the derivation

$$y_1A\langle AB\rangle y_2 \Rightarrow^*_{(G',W)} w$$

the following six derivation steps necessarily occur

(1) A is rewritten according to $A \rightarrow \langle left\rangle$, so $\langle left\rangle\langle AB\rangle$ is produced.
(2) $\langle AB\rangle$ is rewritten according to $\langle AB\rangle \rightarrow \langle right\rangle$, so $\langle left\rangle\langle right\rangle$ is produced.
(3) $\langle left\rangle$ is rewritten according to $\langle left\rangle \rightarrow \langle empty\rangle$, so $\langle empty\rangle\langle right\rangle$ is produced.
(4) $\langle right\rangle$ is rewritten according to $\langle right\rangle \rightarrow \langle empty\rangle$, so $\langle empty\rangle\langle empty\rangle$ is produced.
(5) In $\langle empty\rangle\langle empty\rangle$, one $\langle empty\rangle$ is erased according to $\langle empty\rangle \rightarrow \varepsilon$.
(6) The other $\langle empty\rangle$ is erased according to $\langle empty\rangle \rightarrow \varepsilon$.

On the other hand, let

$$S \Rightarrow^*_{(G',W)} y_1 C D y_2$$
$$\Rightarrow_{(G',W)} y_1 C \langle CD \rangle y_2 \ [D \to \langle CD \rangle]$$
$$\Rightarrow^*_{(G',W)} w$$

where $w \in T^*$. Then, during the derivation

$$y_1 C \langle CD \rangle y_2 \Rightarrow^*_{(G',W)} w$$

the following six derivation steps necessarily occur

(1) C is rewritten according to $C \to \langle left \rangle$, so $\langle left \rangle \langle CD \rangle$ is produced.
(2) $\langle CD \rangle$ is rewritten according to $\langle CD \rangle \to \langle right \rangle$, so $\langle left \rangle \langle right \rangle$ is produced.
(3) $\langle left \rangle$ is rewritten according to $\langle left \rangle \to \langle empty \rangle$, so $\langle empty \rangle \langle right \rangle$ is produced.
(4) $\langle right \rangle$ is rewritten according to $\langle right \rangle \to \langle empty \rangle$, so $\langle empty \rangle \langle empty \rangle$ is produced.
(5) One $\langle empty \rangle$ in $\langle empty \rangle \langle empty \rangle$ is erased according to $\langle empty \rangle \to \varepsilon$.
(6) The other $\langle empty \rangle$ is erased according to $\langle empty \rangle \to \varepsilon$.

Considering the previous two observations, we can easily prove the following claim.

Claim 2. $S \Rightarrow^m_{(G',W)} w$ *implies that* $S \Rightarrow^*_G w$, *where* $w \in T^*$, *for some* $m \geq 0$.

Proof. This proof is left to the reader. □

By Claim 1, $L(G) \subseteq L(G', W)$. From Claim 2, it follows that $L(G', W) \subseteq L(G)$. Therefore, $L(G) = L(G', W)$, and Theorem 9.2.7 holds. □

Recall that for ordinary context-free grammars (which coincide with the wm-grammars of degree 1 in terms of the present chapter), Gruska in [1] proved that for every natural number $n \geq 1$, the context-free grammars with $n + 1$ nonterminals are more powerful that the context-free grammars with n nonterminals. Consequently, if we reduce the number of nonterminals in context-free grammars over letter monoids, then we also reduce the power of these grammars. On the other hand, by Theorem 9.2.7, context-free grammars defined over word monoids keep their power even if we reduce their number of nonterminals to 10.

References

1. Gruska, J.: On a classification of context-free languages. Kybernetika **13**, 22–29 (1967)

Part IV
Regulated Grammars: Parallelism

This part studies parallel grammars regulated similarly to the regulation of sequential grammars, discussed earlier in this book. It consists of Chaps. 10 through 12.

Chapter 10 bases its discussion on ET0L grammars, which can be viewed, in essence, as generalized parallel versions of context-free grammars. It concentrates its attention on their context-conditional versions, including some of their special versions, such as *forbidding ET0L grammars*, *simple semi-conditional ET0L grammars*, and *left random context ET0L grammars*.

Chapter 11 studies how to perform the parallel generation of languages in a uniform way with respect to the rewritten sentential forms during the generation process. Specifically, it transforms parallel regulated grammars so they produce only strings that have a uniform permutation-based form.

Chapter 12 reconsiders the topic of Chap. 9 in terms of grammatical parallelism, represented by E0L grammars. That is, it studies algebraic regulation based upon E0L grammars with the relation of a direct derivation defined over free monoids generated by finite sets of strings.

Chapter 10
Regulated ET0L Grammars

Abstract ET0L grammars can be seen as generalized parallel versions of context-free grammars. More precisely, there exist three main conceptual differences between them and context-free grammars. First, instead of a single set of rules, they have finitely many sets of rules. Second, the left-hand side of a rule may be formed by any grammatical symbol, including a terminal. Third, all symbols of a string are simultaneously rewritten during a single derivation step. The present chapter studies ET0L grammars regulated in a context-conditional way. Specifically, by analogy with sequential context-conditional grammars, this chapter discusses *context-conditional ET0L grammars* that capture this dependency so each of their rules may be associated with finitely many strings representing *permitting conditions* and, in addition, finitely many strings representing *forbidding conditions*. A rule like this can rewrite a symbol if all its permitting conditions occur in the rewritten current sentential form and, simultaneously, all its forbidding conditions do not. Otherwise, these grammars work just like ordinary ET0L grammars. The chapter consists of four sections. Section 10.1 defines the basic version of context-conditional ET0L grammars. The other sections investigate three variants of the basic version—*forbidding ET0L grammars* (Sect. 10.2), *simple semi-conditional ET0L grammars* (Sect. 10.3), and *left random context ET0L grammars* (Sect. 10.4). All these sections concentrate their attention on establishing the generative power of the ET0L grammars under investigation.

Keywords Grammatical parallelism • Regulated ET0L grammars • Context-conditional ET0L grammars and their variants • Forbidding ET0L grammars • Simple semi-conditional ET0L grammars • Left random context ET0L grammars • Generative power

Extended tabled zero-sided Lindenmayer grammars or, more briefly and customarily, ET0L grammars (see [10–13]) represent the first type of parallel grammars considered in this part of the book. Originally, these grammars were introduced in connection with a theory proposed for the development of filamentous

A. Meduna and P. Zemek, *Regulated Grammars and Automata*,
DOI 10.1007/978-1-4939-0369-6_10,

organisms. Developmental stages of cellular arrays are described by strings with each symbol being a cell. Rules correspond to developmental instructions with which organisms can be produced. They are applied simultaneously to all cell-representing symbols because in a growing organism development proceeds simultaneously everywhere. That is, all symbols, including terminals, are always rewritten to adequately reflect the development of real organisms that contain no dead cells which would remain permanently fixed in their place in the organism; disappearing cells are represented by ε. Instead of a single set of rules, ET0L grammars have a finite set of sets containing rules. Each of them contains rules that describe developmental instructions corresponding to a specific biological circumstances, such as environmental conditions concerning temperature or coexistence of other organisms. Naturally, during a single derivation step, rules from only one of these sets can be applied to the rewritten string.

Considering these biologically motivated features of ET0L grammars, we see the following three main conceptual differences between them and the previously discussed sequential grammars, based upon context-free grammars.

(I) Instead of a single set of rules, they have finitely many sets of rules.
(II) The left-hand side of a rule may be formed by any grammatical symbol, including a terminal.
(III) All symbols of a string are simultaneously rewritten during a single derivation step.

In many cell organisms, developmental instructions according to which cells are changed depend on the absence or, on the contrary, existence of some other cells in the organisms. A formalization and exploration of this dependency by means of regulated ET0L grammars represents the principal subject of this chapter. Specifically, the chapter discusses *context-conditional ET0L grammars* that capture this dependency so each of their rules may be associated with finitely many strings representing *permitting conditions* and, in addition, finitely many strings representing *forbidding conditions*. A rule like this can rewrite a symbol if all its permitting conditions occur in the rewritten current sentential form and, simultaneously, all its forbidding conditions do not; otherwise, these grammars work just like ET0L grammars as sketched above.

In this chapter, by analogy with sequential context-conditional grammars (see Sect. 4.2), we first define context-conditional ET0L grammars quite generally (Definition 10.1.1). Then, we investigate their three variants—*forbidding ET0L grammars* (Sect. 10.2), *simple semi-conditional ET0L grammars* (Sect. 10.3), and *left random context ET0L grammars* (Sect. 10.4).

10.1 Context-Conditional ET0L Grammars

In the present section, we demonstrate that context-conditional ET0L grammars characterize the family of recursively enumerable languages (see Theorem 10.1.8), and, without erasing rules, they characterize the family of context-sensitive languages (see Theorem 10.1.6).

This section consists of Sects. 10.1.1 and 10.1.2. The former defines and illustrates context-conditional ET0L grammars. The latter establishes their generative power.

10.1.1 Definitions

In this section, we define context-conditional ET0L grammars.

Definition 10.1.1. A *context-conditional ET0L grammar* (a *C-ET0L grammar* for short) is a $(t+3)$-tuple

$$G = \big(V, T, P_1, \dots, P_t, S\big)$$

where $t \geq 1$, and V, T, and S are the *total alphabet*, the *terminal alphabet* ($T \subset V$), and the *start symbol* ($S \in V - T$), respectively. Every P_i, where $1 \leq i \leq t$, is a finite set of rules of the form

$$\big(a \to x, Per, For\big)$$

with $a \in V$, $x \in V^*$, and $Per, For \subseteq V^+$ are finite languages. If every $(a \to x, Per, For) \in P_i$ for $i = 1, 2, \dots, t$ satisfies that $|x| \geq 1$, then G is said to be *propagating* (a *C-EPT0L grammar* for short). G has *degree* (r, s), where r and s are natural numbers, if for every $i = 1, \dots, t$ and $(a \to x, Per, For) \in P_i$, max-len$(Per) \leq r$ and max-len$(For) \leq s$.

Let $u, v \in V^*$, $u = a_1 a_2 \cdots a_q$, $v = v_1 v_2 \cdots v_q$, $q = |u|$, $a_j \in V$, $v_j \in V^*$, and $p_1, p_2, \dots, p_q$ be a sequence of rules $p_j = (a_j \to v_j, Per_j, For_j) \in P_i$ for all $j = 1, \dots, q$ and some $i \in \{1, \dots, t\}$. If for every p_j, $Per_j \subseteq \text{sub}(u)$ and $For_j \cap \text{sub}(u) = \emptyset$, then u *directly derives* v according to $p_1, p_2, \dots, p_q$ in G, denoted by

$$u \Rightarrow_G v \; [p_1, p_2, \dots, p_q]$$

In the standard way, define $\Rightarrow_G^k$ for $k \geq 0$, $\Rightarrow_G^*$, and $\Rightarrow_G^+$. The *language* of G is denoted by $L(G)$ and defined as

$$L(G) = \big\{x \in T^* \mid S \Rightarrow_G^* x\big\}$$

□

Definition 10.1.2. Let $G = (V, T, P_1, \ldots, P_t, S)$ be a C-ET0L grammar, for some $t \geq 1$. If $t = 1$, then G is called a *context-conditional E0L* grammar (a *C-E0L grammar* for short). If G is a propagating C-E0L grammar, then G is said to be a *C-EP0L grammar*. □

The language families defined by C-EPT0L, C-ET0L, C-EP0L, and C-E0L grammars of degree (r, s) are denoted by $\mathbf{C\text{-}EPT0L}(r, s)$, $\mathbf{C\text{-}ET0L}(r, s)$, $\mathbf{C\text{-}EP0L}(r, s)$, and $\mathbf{C\text{-}E0L}(r, s)$, respectively. Set

$$\mathbf{C\text{-}EPT0L} = \bigcup_{r=0}^{\infty} \bigcup_{s=0}^{\infty} \mathbf{C\text{-}EPT0L}(r, s) \qquad \mathbf{C\text{-}ET0L} = \bigcup_{r=0}^{\infty} \bigcup_{s=0}^{\infty} \mathbf{C\text{-}ET0L}(r, s)$$

$$\mathbf{C\text{-}EP0L} = \bigcup_{r=0}^{\infty} \bigcup_{s=0}^{\infty} \mathbf{C\text{-}EP0L}(r, s) \qquad \mathbf{C\text{-}E0L} = \bigcup_{r=0}^{\infty} \bigcup_{s=0}^{\infty} \mathbf{C\text{-}E0L}(r, s)$$

10.1.2 Generative Power

In this section, we discuss the generative power of context-conditional grammars.

Lemma 10.1.3. $\mathbf{C\text{-}EP0L} \subseteq \mathbf{C\text{-}EPT0L} \subseteq \mathbf{C\text{-}ET0L}$ *and* $\mathbf{C\text{-}EP0L} \subseteq \mathbf{C\text{-}E0L} \subseteq \mathbf{C\text{-}ET0L}$. *For any* $r, s \geq 0$, $\mathbf{C\text{-}EP0L}(r, s) \subseteq \mathbf{C\text{-}EPT0L}(r, s) \subseteq \mathbf{C\text{-}ET0L}(r, s)$, *and* $\mathbf{C\text{-}EP0L}(r, s) \subseteq \mathbf{C\text{-}E0L}(r, s) \subseteq \mathbf{C\text{-}ET0L}(r, s)$.

Proof. This lemma follows from Definitions 10.1.1 and 10.1.2. □

Theorem 10.1.4.

$$\begin{array}{c} \mathbf{CF} \\ \subset \\ \mathbf{C\text{-}E0L}(0, 0) = \mathbf{C\text{-}EP0L}(0, 0) = \mathbf{E0L} = \mathbf{EP0L} \\ \subset \\ \mathbf{C\text{-}ET0L}(0, 0) = \mathbf{C\text{-}EPT0L}(0, 0) = \mathbf{ET0L} = \mathbf{EPT0L} \\ \subset \\ \mathbf{CS} \end{array}$$

Proof. Clearly, C-EP0L and C-E0L grammars of degree $(0, 0)$ are ordinary EP0L and E0L grammars, respectively. Analogously, C-EPT0L and C-ET0L grammars of degree $(0, 0)$ are EPT0L and ET0L grammars, respectively. Since $\mathbf{CF} \subset \mathbf{E0L} = \mathbf{EP0L} \subset \mathbf{ET0L} = \mathbf{EPT0L} \subset \mathbf{CS}$ (see Theorem 3.3.23), we get $\mathbf{CF} \subset \mathbf{C\text{-}E0L}(0, 0) = \mathbf{C\text{-}EP0L}(0, 0) = \mathbf{E0L} \subset \mathbf{C\text{-}ET0L}(0, 0) = \mathbf{C\text{-}EPT0L}(0, 0) = \mathbf{ET0L} \subset \mathbf{CS}$; therefore, the theorem holds. □

Lemma 10.1.5. **C - EPT0L**$(r, s) \subseteq$ **CS**, *for any* $r \geq 0$, $s \geq 0$.

Proof. For $r = 0$ and $s = 0$, we have

$$\mathbf{C\text{-}EPT0L}(0, 0) = \mathbf{EPT0L} \subset \mathbf{CS}$$

The following proof demonstrates that the inclusion holds for any r and s such that $r + s \geq 1$.

Let L be a language generated by a C-EPT0L grammar

$$G = (V, T, P_1, \ldots, P_t, S)$$

of degree (r, s), for some $r, s \geq 0$, $r + s \geq 1$, $t \geq 1$. Let k be the greater number of r and s. Set

$$M = \{x \in V^+ \mid |x| \leq k\}$$

For every P_i, where $1 \leq i \leq t$, define

$$\text{cf-rules}(P_i) = \{a \to z \mid (a \to z, Per, For) \in P_i,\ a \in V,\ z \in V^+\}$$

Then, set

$$\begin{aligned} N_F &= \{\lfloor X, x \rfloor \mid X \subseteq M,\ x \in M \cup \{\varepsilon\}\} \\ N_T &= \{\langle X \rangle \mid X \subseteq M\} \\ N_B &= \{\lceil Q \rceil \mid Q \subseteq \text{cf-rules}(P_i)\ 1 \leq i \leq t\} \\ V' &= V \cup N_F \cup N_T \cup N_B \cup \{\rhd, \lhd, \$, S', \#\} \\ T' &= T \cup \{\#\} \end{aligned}$$

Construct the context-sensitive grammar

$$G' = (V', T', P', S')$$

with the finite set of rules P' constructed by performing (1) through (7), given next.

(1) Add $S' \to \rhd \lfloor \emptyset, \varepsilon \rfloor S \lhd$ to P'
(2) For all $X \subseteq M$, $x \in (V^k \cup \{\varepsilon\})$ and $y \in V^k$, extend P' by adding

$$\lfloor X, x \rfloor y \to y \lfloor X \cup \text{sub}(xy, k), y \rfloor$$

(3) For all $X \subseteq M$, $x \in (V^k \cup \{\varepsilon\})$ and $y \in V^+$, $|y| \leq k$, extend P' by adding

$$\lfloor X, x \rfloor y \lhd \to y \langle X \cup \text{sub}(xy, k) \rangle \lhd$$

(4) For all $X \subseteq M$ and $Q \subseteq$ cf-rules(P_i), where $i \in \{1, \ldots, t\}$, such that for every $a \to z \in Q$, there exists $(a \to z, Per, For) \in P_i$ satisfying $Per \subseteq X$ and $For \cap X = \emptyset$, extend P' by adding

$$\langle X \rangle \lhd \to \lceil Q \rceil \lhd$$

(5) For every $Q \subseteq$ cf-rules(P_i) for some $i \in \{1, \ldots, t\}$, $a \in V$ and $z \in V^+$ such that $a \to z \in Q$, extend P' by adding

$$a \lceil Q \rceil \to \lceil Q \rceil z$$

(6) For all $Q \subseteq$ cf-rules(P_i) for some $i = \{1, \ldots, t\}$, extend P' by adding

$$\rhd \lceil Q \rceil \to \rhd \lfloor \emptyset, \varepsilon \rfloor$$

(7) Add $\rhd \lfloor \emptyset, \varepsilon \rfloor \to \#\$$, $\$\lhd \to \#\#$, and $\$a \to a\$$, for all $a \in T$, to P'

To prove that $L(G) = L(G')$, we first establish Claims 1 through 3.

Claim 1. Every successful derivation in G' has the form

$$\begin{aligned} S' &\Rightarrow_{G'} \rhd \lfloor \emptyset, \varepsilon \rfloor S \lhd \\ &\Rightarrow_{G'}^{+} \rhd \lfloor \emptyset, \varepsilon \rfloor x \lhd \\ &\Rightarrow_{G'} \#\$x \lhd \\ &\Rightarrow_{G'}^{|x|} \#x\$ \lhd \\ &\Rightarrow_{G'} \#x\#\# \end{aligned}$$

such that $x \in T^+$ and during $\rhd \lfloor \emptyset, \varepsilon \rfloor S \lhd \Rightarrow_{G'}^{+} \rhd \lfloor \emptyset, \varepsilon \rfloor x \lhd$, every sentential form w satisfies $w \in \{\rhd\} H^+ \{\lhd\}$, where $H \subseteq V' - \{\rhd, \lhd, \#, \$, S'\}$.

Proof. The only rule that can rewrite the start symbol is $S' \to \rhd \lfloor \emptyset, \varepsilon \rfloor S \lhd$; thus,

$$S' \Rightarrow_{G'} \rhd \lfloor \emptyset, \varepsilon \rfloor S \lhd$$

After that, every sentential form that occurs in

$$\rhd \lfloor \emptyset, \varepsilon \rfloor S \lhd \Rightarrow_{G'}^{+} \rhd \lfloor \emptyset, \varepsilon \rfloor x \lhd$$

can be rewritten by using any of the rules introduced in (2) through (6) from the construction of P'. By the inspection of these rules, it is obvious that the edge symbols $\rhd$ and $\lhd$ remain unchanged and no other occurrences of them appear inside the sentential form. Moreover, there is no rule generating a symbol from $\{\#, \$, S'\}$. Therefore, all these sentential forms belong to $\{\rhd\} H^+ \{\lhd\}$.

Next, let us explain how G' generates a string from $L(G')$. Only $\rhd \lfloor \emptyset, \varepsilon \rfloor \to \#\$$ can rewrite $\rhd$ to a symbol from T (see (7) in the definition of P'). According to the left-hand side of this rule, we obtain

$$S' \Rightarrow_{G'} \rhd \lfloor \emptyset, \varepsilon \rfloor S \lhd \Rightarrow_{G'}^{*} \rhd \lfloor \emptyset, \varepsilon \rfloor x \lhd \Rightarrow_{G'} \#\$x\lhd$$

where $x \in H^+$. To rewrite $\lhd$, G' uses $\$\lhd \rightarrow \#\#$. Thus, G' needs $\$$ as the left neighbor of $\lhd$. Suppose that $x = a_1a_2 \cdots a_q$, where $q = |x|$ and $a_i \in T$, for all $i \in \{1, \ldots, q\}$. Since for every $a \in T$ there is $\$a \rightarrow a\$ \in P'$ [see (7)], we can construct

$$\begin{aligned} \#\$a_1a_2 \cdots a_n\lhd \Rightarrow_{G'} & \ \#a_1\$a_2 \cdots a_n\lhd \\ \Rightarrow_{G'} & \ \#a_1a_2\$ \cdots a_n\lhd \\ \Rightarrow_{G'}^{|x|-2} & \ \#a_1a_2 \cdots a_n\$\lhd \end{aligned}$$

Notice that this derivation can be constructed only for x that belong to T^+. Then, $\$\lhd$ is rewritten to ##. As a result,

$$S' \Rightarrow_{G'} \rhd \lfloor \emptyset, \varepsilon \rfloor S \lhd \Rightarrow_{G'}^{+} \rhd \lfloor \emptyset, \varepsilon \rfloor x \lhd \Rightarrow_{G'} \#\$x\lhd \Rightarrow_{G'}^{|x|} \#x\$\lhd \Rightarrow_{G'} \#x\#\#$$

with the required properties. Thus, the claim holds. □

The following claim demonstrates how G' simulates a direct derivation from G—the heart of the construction.

Let $x \Rightarrow_{G'}^{\oplus} y$ denote the derivation $x \Rightarrow_{G'}^{+} y$ such that $x = \rhd \lfloor \emptyset, \varepsilon \rfloor u \lhd$, $y = \rhd \lfloor \emptyset, \varepsilon \rfloor v \lhd$, $u, v \in V^+$, and during $x \Rightarrow_{G'}^{+} y$, there is no other occurrence of a string of the form $\rhd \lfloor \emptyset, \varepsilon \rfloor z \lhd$, $z \in V^*$.

Claim 2. For every $u, v \in V^*$,

$$\rhd \lfloor \emptyset, \varepsilon \rfloor u \lhd \Rightarrow_{G'}^{\oplus} \rhd \lfloor \emptyset, \varepsilon \rfloor v \lhd \quad \text{if and only if} \quad u \Rightarrow_G v$$

Proof. The proof is divided into the only-if part and the if part.

Only If. Let us show how G' rewrites $\rhd \lfloor \emptyset, \varepsilon \rfloor u \lhd$ to $\rhd \lfloor \emptyset, \varepsilon \rfloor v \lhd$ by performing a derivation consisting of a forward phase and a backward phase.

During the first, forward phase, G' scans u to obtain all nonempty substrings of length k or less. By repeatedly using rules

$$\lfloor X, x \rfloor y \rightarrow y \lfloor X \cup \mathrm{sub}(xy, k), y \rfloor$$

where $X \subseteq M$, $x \in (V^k \cup \{\varepsilon\})$, $y \in V^k$ (see (2) in the definition of P'), the occurrence of a symbol with form $\lfloor X, x \rfloor$ is moved toward the end of the sentential form. Simultaneously, the substrings of u are collected in X. The forward phase is finished by

$$\lfloor X, x \rfloor y \lhd \rightarrow y \langle X \cup \mathrm{sub}(xy, k) \rangle \lhd$$

where $x \in (V^k \cup \{\varepsilon\})$, $y \in V^+$, $|y| \leq k$ [see (3)]; the rule reaches the end of u and completes $X = \mathrm{sub}(u, k)$. Formally,

$$\triangleright\lfloor\emptyset,\varepsilon\rfloor u\triangleleft \Rightarrow_{G'}^{+} \triangleright u\langle X\rangle\triangleleft$$

such that $X = \text{sub}(u, k)$. Then, $\langle X\rangle$ is changed to $\lceil Q\rceil$, where

$$Q = \{a \to z \mid (a \to z, Per, For) \in P_i,\ a \in V,\ z \in V^+,\\ Per, For \subseteq M,\ Per \subseteq X,\ For \cap X = \emptyset\}$$

for some $i \in \{1, \ldots, t\}$, by

$$\langle X\rangle\triangleleft \to \lceil Q\rceil\triangleleft$$

[see (4)]. In other words, G' selects a subset of rules from P_i that could be used to rewrite u in G.

The second, backward phase simulates rewriting of all symbols in u in parallel. Since

$$a\lceil Q\rceil \to \lceil Q\rceil z \in P'$$

for all $a \to z \in Q, a \in V, z \in V^+$ [see (5)],

$$\triangleright u\lceil Q\rceil\triangleleft \Rightarrow_{G'}^{|u|} \triangleright\lceil Q\rceil v\triangleleft$$

such that $\lceil Q\rceil$ moves left and every symbol $a \in V$ in u is rewritten to some z provided that $a \to z \in Q$. Finally, $\lceil Q\rceil$ is rewritten to $\lfloor\emptyset,\varepsilon\rfloor$ by

$$\triangleright\lceil Q\rceil \to \triangleright\lfloor\emptyset,\varepsilon\rfloor$$

As a result, we obtain

$$\begin{aligned}\triangleright\lfloor\emptyset,\varepsilon\rfloor u\triangleleft &\Rightarrow_{G'}^{+} \triangleright u\langle X\rangle\triangleleft \Rightarrow_{G'} \triangleright u\lceil Q\rceil\triangleleft\\ &\Rightarrow_{G'}^{|u|} \triangleright\lceil Q\rceil v\triangleleft \Rightarrow_{G'} \triangleright\lfloor\emptyset,\varepsilon\rfloor v\triangleleft\end{aligned}$$

Observe that this is the only way of deriving

$$\triangleright\lfloor\emptyset,\varepsilon\rfloor u\triangleleft \Rightarrow_{G'}^{\oplus} \triangleright\lfloor\emptyset,\varepsilon\rfloor v\triangleleft$$

Let us show that $u \Rightarrow_G v$. Indeed, because we have $(a \to z, Per, For) \in P_i$ for every $a\lceil Q\rceil \to \lceil Q\rceil z \in P$ used in the backward phase, where $Per \subseteq \text{sub}(u, k)$ and $For \cap \text{sub}(u, k) = \emptyset$ (see the construction of Q), there exists a derivation

$$u \Rightarrow_G v\ [p_1 \cdots p_q]$$

where $|u| = q$, and $p_j = (a \to z, Per, For) \in P_i$ such that $a\lceil Q\rceil \to \lceil Q\rceil z$ has been applied in the $(q - j + 1)$th derivation step in

$$\triangleright u\lceil Q\rceil\triangleleft \Rightarrow_{G'}^{|u|} \triangleright\lceil Q\rceil v\triangleleft$$

where $a \in V$, $z \in V^+$, $1 \leq j \leq q$.

If. The converse implication can be proved similarly to the only-if part, so we leave it to the reader. □

Claim 3. $S' \Rightarrow_{G'}^+ \triangleright\lfloor\emptyset, \varepsilon\rfloor x\triangleleft$ *if and only if* $S \Rightarrow_G^* x$, *for all* $x \in V^+$.

Proof. The proof is divided into the only-if part and the if part.

Only If. The only-if part is proved by induction on the ith occurrence of the sentential form w satisfying $w = \triangleright\lfloor\emptyset, \varepsilon\rfloor u\triangleleft$, $u \in V^+$, during the derivation in G'.

Basis. Let $i = 1$. Then, $S' \Rightarrow_{G'} \triangleright\lfloor\emptyset, \varepsilon\rfloor S\triangleleft$ and $S \Rightarrow_G^0 S$.

Induction Hypothesis. Suppose that the claim holds for all $i \leq h$, for some $h \geq 1$.

Induction Step. Let $i = h + 1$. Since $h + 1 \geq 2$, we can express

$$S' \Rightarrow_{G'}^+ \triangleright\lfloor\emptyset, \varepsilon\rfloor x_i\triangleleft$$

as

$$S' \Rightarrow_{G'}^+ \triangleright\lfloor\emptyset, \varepsilon\rfloor x_{i-1}\triangleleft \Rightarrow_{G'}^{\oplus} \triangleright\lfloor\emptyset, \varepsilon\rfloor x_i\triangleleft$$

where $x_{i-1}, x_i \in V^+$. By the induction hypothesis,

$$S \Rightarrow_G^* x_{i-1}$$

Claim 2 says that

$$\triangleright\lfloor\emptyset, \varepsilon\rfloor x_{i-1}\triangleleft \Rightarrow_{G'}^{\oplus} \triangleright\lfloor\emptyset, \varepsilon\rfloor x_i\triangleleft \quad \text{if and only if} \quad x_{i-1} \Rightarrow_G x_i$$

Hence,

$$S \Rightarrow_G^* x_{i-1} \Rightarrow_G x_i$$

and the only-if part holds.

If. By induction on n, we prove that

$$S \Rightarrow_G^n x \quad \text{implies that} \quad S' \Rightarrow_{G'}^+ \triangleright\lfloor\emptyset, \varepsilon\rfloor x\triangleleft$$

for all $n \geq 0$, $x \in V^+$.

Basis. For $n = 0$, $S \Rightarrow_G^0 S$ and $S' \Rightarrow_{G'} \triangleright\lfloor\emptyset, \varepsilon\rfloor S\triangleleft$.

Induction Hypothesis. Assume that the claim holds for all n or less, for some $n \geq 0$.

Induction Step. Consider any derivation of the form

$$S \Rightarrow_G^{n+1} x$$

where $x \in V^+$. Since $n + 1 \geq 1$, there exists $y \in V^+$ such that

$$S \Rightarrow_G^n y \Rightarrow_G x$$

and by the induction hypothesis, there is also a derivation

$$S' \Rightarrow_{G'}^+ \rhd\lfloor\emptyset, \varepsilon\rfloor y\lhd$$

From Claim 2, we have

$$\rhd\lfloor\emptyset, \varepsilon\rfloor y\lhd \Rightarrow_{G'}^{\oplus} \rhd\lfloor\emptyset, \varepsilon\rfloor x\lhd$$

Therefore,

$$S' \Rightarrow_{G'}^+ \rhd\lfloor\emptyset, \varepsilon\rfloor y\lhd \Rightarrow_{G'}^{\oplus} \rhd\lfloor\emptyset, \varepsilon\rfloor x\lhd$$

and the converse implication holds as well. □

From Claims 1 and 3, we see that any successful derivation in G' is of the form

$$S' \Rightarrow_{G'}^+ \rhd\lfloor\emptyset, \varepsilon\rfloor x\lhd \Rightarrow_{G'}^+ \#x\#\#$$

such that

$$S \Rightarrow_G^* x, \ x \in T^+$$

Therefore, we have for each $x \in T^+$,

$$S' \Rightarrow_{G'}^+ \#x\#\# \quad \text{if and only if} \quad S \Rightarrow_G^* x$$

Define the homomorphism h over $(T \cup \{\#\})^*$ as $h(\#) = \varepsilon$ and $h(a) = a$ for all $a \in T$. Observe that h is 4-linear erasing with respect to $L(G')$. Furthermore, notice that $h(L(G')) = L(G)$. Since **CS** is closed under linear erasing (see Theorem 10.4 on p. 98 in [16]), $L \in$ **CS**. Thus, Lemma 10.1.5 holds. □

Theorem 10.1.6. C - EPT0L = CS

Proof. By Lemma 10.1.5, **C - EPT0L** $\subseteq$ **CS**. Later in this chapter, we define two special cases of C-EPT0L grammars and prove that they generate all the family of context-sensitive languages (see Theorems 10.2.11 and 10.3.7). Therefore, **CS** $\subseteq$ **C - EPT0L**, and hence **C - EPT0L** = **CS**. □

Lemma 10.1.7. $\mathbf{C\text{-}ET0L} \subseteq \mathbf{RE}$

Proof. This lemma follows from Church's thesis. To obtain an algorithm converting any C-ET0L grammar to an equivalent phrase-structure grammar, use the technique presented in Lemma 10.1.5. □

Theorem 10.1.8. $\mathbf{C\text{-}ET0L} = \mathbf{RE}$

Proof. By Lemma 10.1.7, $\mathbf{C\text{-}ET0L} \subseteq \mathbf{RE}$. In Sects. 10.2 and 10.3, we introduce two special cases of C-ET0L grammars and demonstrate that even these grammars generate $\mathbf{RE}$ (see Theorems 10.2.14 and 10.3.4); therefore, $\mathbf{RE} \subseteq \mathbf{C\text{-}ET0L}$. As a result, $\mathbf{C\text{-}ET0L} = \mathbf{RE}$. □

10.2 Forbidding ET0L Grammars

Forbidding ET0L grammars, discussed in the present section, represent context-conditional ET0L grammars in which no rule has any permitting condition. First, this section defines and illustrates them (see Sect. 10.2.1). Then, it establishes their generative power and reduces their degree without affecting the power (see Sect. 10.2.2).

10.2.1 Definitions and Examples

In this section, we define forbidding ET0L grammars.

Definition 10.2.1. Let $G = (V, T, P_1, \ldots, P_t, S)$ be a C-ET0L grammar. If every $p = (a \rightarrow x, Per, For) \in P_i$, where $i = 1, \ldots, t$, satisfies $Per = \emptyset$, then G is said to be *forbidding ET0L grammar* (an *F-ET0L grammar* for short). If G is a propagating F-ET0L grammar, then G is said to be an *F-EPT0L grammar*. If $t = 1$, G is called an *F-E0L grammar*. If G is a propagating F-E0L grammar, G is called an *F-EP0L grammar*. □

Let $G = (V, T, P_1, \ldots, P_t, S)$ be an F-ET0L grammar of degree (r, s). From the above definition, $(a \rightarrow x, Per, For) \in P_i$ implies that $Per = \emptyset$ for all $i = 1, \ldots, t$. By analogy with sequential forbidding grammars, we thus omit the empty set in the rules. For simplicity, we also say that the degree of G is s instead of (r, s).

The families of languages generated by F-E0L grammars, F-EP0L grammars, F-ET0L grammars, and F-EPT0L grammars of degree s are denoted by $\mathbf{F\text{-}E0L}(s)$, $\mathbf{F\text{-}EP0L}(s)$, $\mathbf{F\text{-}ET0L}(s)$, and $\mathbf{F\text{-}EPT0L}(s)$, respectively. Moreover, set

$$\mathbf{F\text{-}EPT0L} = \bigcup_{s=0}^{\infty} \mathbf{F\text{-}EPT0L}(s) \qquad \mathbf{F\text{-}ET0L} = \bigcup_{s=0}^{\infty} \mathbf{F\text{-}ET0L}(s)$$

$$\mathbf{F\text{-}EP0L} = \bigcup_{s=0}^{\infty} \mathbf{F\text{-}EP0L}(s) \qquad \mathbf{F\text{-}E0L} = \bigcup_{s=0}^{\infty} \mathbf{F\text{-}E0L}(s)$$

Example 10.2.2. Let

$$G = \big(\{S, A, B, C, a, \bar{a}, b\}, \{a, b\}, P, S\big)$$

be an F-EP0L grammar, where

$$\begin{aligned} P = \{&(S \to ABA, \emptyset), \\ &(A \to aA, \{\bar{a}\}), \\ &(B \to bB, \emptyset), \\ &(A \to \bar{a}, \{\bar{a}\}), \\ &(\bar{a} \to a, \emptyset), \\ &(B \to C, \emptyset), \\ &(C \to bC, \{A\}), \\ &(C \to b, \{A\}), \\ &(a \to a, \emptyset), \\ &(b \to b, \emptyset)\} \end{aligned}$$

Obviously, G is an F-EP0L grammar of degree 1. Observe that for every string from $L(G)$, there exists a derivation of the form

$$\begin{aligned} S &\Rightarrow_G ABA \\ &\Rightarrow_G aAbBaA \\ &\Rightarrow_G^+ a^{m-1}Ab^{m-1}Ba^{m-1}A \\ &\Rightarrow_G a^{m-1}\bar{a}b^{m-1}Ca^{m-1}\bar{a} \\ &\Rightarrow_G a^m b^m C a^m \\ &\Rightarrow_G^+ a^m b^{n-1} C a^m \\ &\Rightarrow_G a^m b^n a^m \end{aligned}$$

with $1 \leq m \leq n$. Hence,

$$L(G) = \big\{a^m b^n a^m \mid 1 \leq m \leq n\big\}$$

Note that $L(G) \notin \mathbf{E0L}$ (see p. 268 in [14]); however, $L(G) \in \mathbf{F\text{-}EP0L}(1)$. As a result, F-EP0L grammars of degree 1 are more powerful than ordinary E0L grammars. □

10.2.2 Generative Power and Reduction

Next, we investigate the generative power of F-ET0L grammars of all degrees.

Theorem 10.2.3. **F - EPT0L(0)=EPT0L**, **F - ET0L(0)=ET0L**, **F - EP0L(0) = EP0L**, *and* **F - E0L(0) = E0L**

Proof. This theorem follows from Definition 10.2.1. □

Lemmas 10.2.4–10.2.7, given next, inspect the generative power of forbidding ET0L grammars of degree 1. As a conclusion, in Theorem 10.2.8, we demonstrate that both F-EPT0L(1) and F-ET0L(1) grammars generate precisely the family of ET0L languages.

Lemma 10.2.4. **EPT0L ⊆ F - EP0L(1)**

Proof. Let

$$G = (V, T, P_1, \dots, P_t, S)$$

be an EPT0L grammar, where $t \geq 1$. Set

$$W = \{\langle a, i\rangle \mid a \in V,\ i = 1, \dots, t\}$$

and

$$F(i) = \{\langle a, j\rangle \in W \mid j \neq i\}$$

Then, construct an F-EP0L grammar of degree 1

$$G' = (V', T, P', S)$$

where $V' = V \cup W$, $(V \cap W = \emptyset)$ and the set of rules P' is defined as follows:

(1) For each $a \in V$ and $i = 1, \dots, t$, add $(a \rightarrow \langle a, i\rangle, \emptyset)$ to P';
(2) If $a \rightarrow z \in P_i$ for some $i \in \{1, \dots, t\}$, $a \in V$, $z \in V^+$, add $(\langle a, i\rangle \rightarrow z, F(i))$ to P'.

Next, to demonstrate that $L(G) = L(G')$, we prove Claims 1 and 2.

Claim 1. For each derivation $S \Rightarrow^n_{G'} x$, $n \geq 0$,

(I) If $n = 2k + 1$ *for some* $k \geq 0$, $x \in W^+$;
(II) If $n = 2k$ *for some* $k \geq 0$, $x \in V^+$.

Proof. The claim follows from the definition of P'. Indeed, every rule in P' is either of the form $(a \rightarrow \langle a, i\rangle, \emptyset)$ or $(\langle a, i\rangle \rightarrow z, F(i))$, where $a \in V$, $\langle a, i\rangle \in W$, $z \in V^+$, $i \in \{1, \dots, t\}$. Since $S \in V$,

$$S \Rightarrow_{G'}^{2k+1} x \quad \text{implies} \quad x \in W^+$$

and

$$S \Rightarrow_{G'}^{2k} x \quad \text{implies} \quad x \in V^+$$

Thus, the claim holds. □

Define the finite substitution g from V^* to V'^* such that for every $a \in V$,

$$g(a) = \{a\} \cup \{\langle a, i\rangle \in W \mid i = 1, \dots, t\}$$

Claim 2. $S \Rightarrow_G^* x$ *if and only if* $S \Rightarrow_{G'}^* x'$ *for some* $x' \in g(x)$, $x \in V^+$, $x' \in V'^+$.

Proof. The proof is divided into the only-if part and the if part.

Only If. By induction on $n \geq 0$, we show that for all $x \in V^+$,

$$S \Rightarrow_G^n x \quad \text{implies} \quad S \Rightarrow_{G'}^{2n} x$$

Basis. Let $n = 0$. Then, the only x is S; therefore, $S \Rightarrow_G^0 S$ and also $S \Rightarrow_{G'}^0 S$.

Induction Hypothesis. Suppose that

$$S \Rightarrow_G^n x \quad \text{implies} \quad S \Rightarrow_{G'}^{2n} x$$

for all derivations of length n or less, for some $n \geq 0$.

Induction Step. Consider any derivation of the form

$$S \Rightarrow_G^{n+1} x$$

Since $n + 1 \geq 1$, this derivation can be expressed as

$$S \Rightarrow_G^n y \Rightarrow_G x\ [p_1, p_2, \dots, p_q]$$

such that $y \in V^+$, $q = |y|$, and $p_j \in P_i$ for all $j = 1, \dots, q$ and some $i \in \{1, \dots, t\}$. By the induction hypothesis,

$$S \Rightarrow_{G'}^{2n} y$$

Suppose that $y = a_1 a_2 \cdots a_q$, $a_j \in V$. Let

$$\begin{aligned} S &\Rightarrow_{G'}^{2n} a_1 a_2 \cdots a_q \\ &\Rightarrow_{G'} \langle a_1, i\rangle\langle a_2, i\rangle \cdots \langle a_q, i\rangle\ [p'_1, p'_2, \cdots, p'_q] \\ &\Rightarrow_{G'} z_1 z_2 \cdots z_q \qquad\qquad\quad [p''_1, p''_2, \cdots, p''_q] \end{aligned}$$

where $p'_j = (a_j \to \langle a_j, i\rangle, \emptyset)$ and $p''_j = (\langle a_j, i\rangle \to z_j, F(i))$ such that $p_j = a_j \to z_j, z_j \in V^+$, for all $j = 1, \dots, q$. Then, $z_1 z_2 \cdots z_q = x$; therefore,

$$S \Rightarrow_{G'}^{2(n+1)} x$$

If. The converse implication is established by induction on $n \geq 0$. That is, we prove that

$$S \Rightarrow_{G'}^{n} x' \quad \text{implies} \quad S \Rightarrow_{G}^{*} x$$

for some $x' \in g(x)$, $n \geq 0$.

Basis. For $n = 0$, $S \Rightarrow_{G'}^{0} S$ and $S \Rightarrow_{G}^{0} S$; clearly, $S \in g(S)$.

Induction Hypothesis. Assume that there exists a natural number m such that the claim holds for every n, where $0 \leq n \leq m$.

Induction Step. Consider any derivation of the form

$$S \Rightarrow_{G'}^{m+1} x'$$

Express this derivation as

$$S \Rightarrow_{G'}^{m} y' \Rightarrow_{G'} x' \; [p'_1, p'_2, \dots, p'_q]$$

where $y' \in V'^+$, $q = |y'|$, and $p'_1, p'_2, \dots, p'_q$ is a sequence of rules from P'. By the induction hypothesis,

$$S \Rightarrow_{G}^{*} y$$

where $y \in V^+$, $y' \in g(y)$. Claim 1 says that there exist the following two cases—(i) and (ii).

(i) Let $m = 2k$ for some $k \geq 0$. Then, $y' \in V^+$, $x' \in W^+$, and every rule

$$p'_j = (a_j \to \langle a_j, i\rangle, \emptyset)$$

where $a_j \in V$, $\langle a_j, i\rangle \in W$, $i \in \{1, \dots, t\}$. In this case, $\langle a_j, i\rangle \in g(a_j)$ for every a_j and any i (see the definition of g); hence, $x' \in g(y)$ as well.

(ii) Let $m = 2k + 1$. Then, $y' \in W^+$, $x' \in V^+$, and each p'_j is of the form

$$p'_j = (\langle a_j, i\rangle \to z_j, F(i))$$

where $\langle a_j, i\rangle \in W$, $z_j \in V^+$. Moreover, according to the forbidding conditions of p'_j, all $\langle a_j, i\rangle$ in y' have the same i. Thus, $y' = \langle a_1, i\rangle\langle a_2, i\rangle \cdots \langle a_q, i\rangle$ for some $i \in \{1, \ldots, t\}$, $y = g^{-1}(y') = a_1 a_2 \cdots a_q$, and $x' = z_1 z_2 \cdots z_q$. By the definition of P',

$$(\langle a_j, i\rangle \to z_j, F(i)) \in P' \quad \text{implies} \quad a_j \to z_j \in P_i$$

Therefore,

$$S \Rightarrow_G^* a_1 a_2 \cdots a_q \Rightarrow_G z_1 z_2 \cdots z_q \; [p_1, p_2, \ldots, p_q]$$

where $p_j = a_j \to z_j \in P_i$ such that $p'_j = (\langle a_j, i\rangle \to z_j, F(i))$. Obviously, $x' = x = z_1 z_2 \cdots z_q$.

This completes the induction and establishes Claim 2. □

By Claim 2, for any $x \in T^+$,

$$S \Rightarrow_G^* x \quad \text{if and only if} \quad S \Rightarrow_{G'}^* x$$

Therefore, $L(G) = L(G')$, so the lemma holds. □

In order to simplify the notation in the proof of the following lemma, for every subset of rules

$$P \subseteq \{(a \to z, F) \mid a \in V,\ z \in V^*,\ F \subseteq V\}$$

define

$$\text{left}(P) = \{a \mid (a \to z, F) \in P\}$$

Informally, left(P) denotes the set of the left-hand sides of all rules in P.

Lemma 10.2.5. F - EPT0L(1) $\subseteq$ EPT0L

Proof. Let

$$G = (V, T, P_1, \ldots, P_t, S)$$

be an F-EPT0L grammar of degree 1, $t \geq 1$. Let Q be the set of all subsets $O \subseteq P_i$, $1 \leq i \leq t$, such that every $(a \to z, F) \in O$, $a \in V$, $z \in V^+$, $F \subseteq V$, satisfies $F \cap \text{left}(O) = \emptyset$. Introduce a new set Q' so that for each $O \in Q$, add

$$\{a \to z \mid (a \to z, F) \in O\}$$

to Q'. Express

$$Q' = \{Q'_1, \ldots, Q'_m\}$$

where m is the cardinality of Q'. Then, construct the EPT0L grammar

$$G' = \left(V, T, Q'_1, \ldots, Q'_m, S\right)$$

To see the basic idea behind the construction of G', consider a pair of rules $p_1 = (a_1 \rightarrow z_1, F_1)$ and $p_2 = (a_2 \rightarrow z_2, F_2)$ from P_i, for some $i \in \{1, \ldots, t\}$. During a single derivation step, p_1 and p_2 can concurrently rewrite a_1 and a_2 provided that $a_2 \notin F_1$ and $a_1 \notin F_2$, respectively. Consider any $O \subseteq P_i$ containing no pair of rules $(a_1 \rightarrow z_1, F_1)$ and $(a_2 \rightarrow z_2, F_2)$ such that $a_1 \in F_2$ or $a_2 \in F_1$. Observe that for any derivation step based on O, no rule from O is blocked by its forbidding conditions; thus, the conditions can be omitted. A formal proof is given next.

Claim 1. $S \Rightarrow_G^n x$ *if and only if* $S \Rightarrow_{G'}^n x$, $x \in V^*$, $n \geq 0$.

Proof. The claim is proved by induction on $n \geq 0$.

Only If. By induction $n \geq 0$, we prove that

$$S \Rightarrow_G^n x \quad \text{implies} \quad S \Rightarrow_{G'}^n x$$

for all $x \in V^*$.

Basis. Let $n = 0$. As obvious, $S \Rightarrow_G^0 S$ and $S \Rightarrow_{G'}^0 S$.

Induction Hypothesis. Suppose that the claim holds for all derivations of length n or less, for some $n \geq 0$.

Induction Step. Consider any derivation of the form

$$S \Rightarrow_G^{n+1} x$$

Since $n + 1 \geq 1$, there exists $y \in V^+$, $q = |y|$, and a sequence $p_1, \ldots, p_q$, where $p_j \in P_i$ for all $j = 1, \ldots, q$ and some $i \in \{1, \ldots, t\}$, such that

$$S \Rightarrow_G^n y \Rightarrow_G x\ [p_1, \ldots, p_q]$$

By the induction hypothesis,

$$S \Rightarrow_{G'}^n y$$

Set

$$O = \left\{p_j \mid 1 \leq j \leq q\right\}$$

Observe that

$$y \Rightarrow_G x\ [p_1, \ldots, p_q]$$

implies that $\text{alph}(y) = \text{left}(O)$. Moreover, every $p_j = (a \to z, F) \in O$, $a \in V$, $z \in V^+$, $F \subseteq V$, satisfies $F \cap \text{alph}(y) = \emptyset$. Hence, $(a \to z, F) \in O$ implies $F \cap \text{left}(O) = \emptyset$. Inspect the definition of G' to see that there exists

$$Q'_r = \{a \to z \mid (a \to z, F) \in O\}$$

for some r, $1 \leq r \leq m$. Therefore,

$$S \Rightarrow^n_{G'} y \Rightarrow_{G'} x\ [p'_1, \ldots, p'_q]$$

where $p'_j = a \to z \in Q'_r$ such that $p_j = (a \to z, F) \in O$, for all $j = 1, \ldots, q$.

If. The if part demonstrates for every $n \geq 0$,

$$S \Rightarrow^n_{G'} x \text{ implies that } S \Rightarrow^n_G x$$

where $x \in V^*$.

Basis. Suppose that $n = 0$. As obvious, $S \Rightarrow^0_{G'} S$ and $S \Rightarrow^0_G S$.

Induction Hypothesis. Assume that the claim holds for all derivations of length n or less, for some $n \geq 0$.

Induction Step. Consider any derivation of the form

$$S \Rightarrow^{n+1}_{G'} x$$

As $n + 1 \geq 1$, there exists a derivation

$$S \Rightarrow^n_{G'} y \Rightarrow_{G'} x\ [p'_1, \ldots, p'_q]$$

such that $y \in V^+$, $q = |y|$, each $p'_i \in Q'_r$ for some $r \in \{1, \ldots, m\}$, and by the induction hypothesis,

$$S \Rightarrow^n_G y$$

Then, by the definition of Q'_r, there exists P_i and $O \subseteq P_i$ such that every $(a \to z, F) \in O$, $a \in V$, $z \in V^+$, $F \subseteq V$, satisfies $a \to z \in Q'_r$ and $F \cap \text{left}(O) = \emptyset$. Since $\text{alph}(y) \subseteq \text{left}(O)$, $(a \to z, F) \in O$ implies that $F \cap \text{alph}(y) = \emptyset$. Hence,

$$S \Rightarrow^n_G y \Rightarrow_G x\ [p_1, \ldots, p_q]$$

where $p_j = (a \to z, F) \in O$ for all $j = 1, \ldots, q$. □

From the claim above,

$$S \Rightarrow^*_G x \quad \text{if and only if} \quad S \Rightarrow^*_{G'} x$$

for all $x \in T^*$. Consequently, $L(G) = L(G')$, and the lemma holds. □

The following two lemmas can be proved by analogy with Lemmas 10.2.4 and 10.2.5. The details are left to the reader.

Lemma 10.2.6. $\mathbf{ET0L} \subseteq \mathbf{F\text{-}E0L}(1)$ □

Lemma 10.2.7. $\mathbf{F\text{-}ET0L}(1) \subseteq \mathbf{ET0L}$ □

Theorem 10.2.8.

$$\mathbf{F\text{-}EP0L}(1) = \mathbf{F\text{-}EPT0L}(1) = \mathbf{F\text{-}E0L}(1) = \mathbf{F\text{-}ET0L}(1) = \mathbf{ET0L} = \mathbf{EPT0L}$$

Proof. By Lemmas 10.2.4 and 10.2.5, $\mathbf{EPT0L} \subseteq \mathbf{F\text{-}EP0L}(1)$ and $\mathbf{F\text{-}EPT0L}(1) \subseteq \mathbf{EPT0L}$, respectively. Since $\mathbf{F\text{-}EP0L}(1) \subseteq \mathbf{F\text{-}EPT0L}(1)$, we get $\mathbf{F\text{-}EP0L}(1) = \mathbf{F\text{-}EPT0L}(1) = \mathbf{EPT0L}$. Analogously, from Lemmas 10.2.6 and 10.2.7, we have $\mathbf{F\text{-}E0L}(1) = \mathbf{F\text{-}ET0L}(1) = \mathbf{ET0L}$. Theorem 3.3.23 implies that $\mathbf{EPT0L} = \mathbf{ET0L}$. Therefore,

$$\mathbf{F\text{-}EP0L}(1) = \mathbf{F\text{-}EPT0L}(1) = \mathbf{F\text{-}E0L}(1) = \mathbf{F\text{-}ET0L}(1) = \mathbf{EPT0L} = \mathbf{ET0L}$$

Thus, the theorem holds. □

Next, we investigate the generative power of F-EPT0L grammars of degree 2. The following lemma establishes a normal form for context-sensitive grammars so that the grammars satisfying this form generate only sentential forms containing no nonterminal from N_{CS} as the leftmost symbol of the string. We make use of this normal form in Lemma 10.2.10.

Lemma 10.2.9. *Every context-sensitive language $L \in \mathbf{CS}$ can be generated by a context-sensitive grammar, $G = (N_1 \cup N_{CF} \cup N_{CS} \cup T, T, P, S_1)$, where N_1, N_{CF}, N_{CS}, and T are pairwise disjoint alphabets, $S_1 \in N_1$, and in P, every rule has one of the following forms*

(i) $AB \to AC$, *where* $A \in (N_1 \cup N_{CF})$, $B \in N_{CS}$, $C \in N_{CF}$;
(ii) $A \to B$, *where* $A \in N_{CF}$, $B \in N_{CS}$;
(iii) $A \to a$, *where* $A \in (N_1 \cup N_{CF})$, $a \in T$;
(iv) $A \to C$, *where* $A, C \in N_{CF}$;
(v) $A_1 \to C_1$, *where* $A_1, C_1 \in N_1$;
(vi) $A \to DE$, *where* $A, D, E \in N_{CF}$;
(vii) $A_1 \to D_1E$, *where* $A_1, D_1 \in N_1$, $E \in N_{CF}$.

Proof. Let

$$G' = \big(N_{CF} \cup N_{CS} \cup T, T, P', S\big)$$

be a context-sensitive grammar of the form defined in Theorem 4.1.6. From this grammar, we construct a grammar

$$G = \big(N_1 \cup N_{CF} \cup N_{CS} \cup T, T, P, S_1\big)$$

where

$$
\begin{aligned}
N_1 &= \{X_1 \mid X \in N_{CF}\} \\
P &= P' \cup \{A_1B \to A_1C \mid AB \to AC \in P', A, C \in N_{CF}, B \in N_{CS}, A_1 \in N_1\} \\
&\cup \{A_1 \to a \mid A \to a \in P', A \in N_{CF}, A_1 \in N_1, a \in T\} \\
&\cup \{A_1 \to C_1 \mid A \to C \in P', A, C \in N_{CF}, A_1, C_1 \in N_1\} \\
&\cup \{A_1 \to D_1E \mid A \to DE \in P', A, D, E \in N_{CF}, A_1, D_1 \in N_1\}
\end{aligned}
$$

G works by analogy with G' except that in G every sentential form starts with a symbol from $N_1 \cup T$ followed by symbols that are not in N_1. Notice, however, that by $AB \to AC$, G' can never rewrite the leftmost symbol of any sentential form. Based on these observations, it is rather easy to see that $L(G) = L(G')$; a formal proof of this identity is left to the reader. As G is of the required form, Lemma 10.2.9 holds. □

Lemma 10.2.10. **CS** $\subseteq$ **F - EP0L(2)**

Proof. Let L be a context-sensitive language generated by a grammar

$$G = \big(N_1 \cup N_{CF} \cup N_{CS} \cup T, T, P, S_1\big)$$

of the form of Lemma 10.2.9. Set

$$
\begin{aligned}
V &= N_1 \cup N_{CF} \cup N_{CS} \cup T \\
P_{CS} &= \{AB \to AC \mid AB \to AC \in P, A \in (N_1 \cup N_{CF}), B \in N_{CS}, C \in N_{CF}\} \\
P_{CF} &= P - P_{CS}
\end{aligned}
$$

Informally, P_{CS} and P_{CF} are the sets of context-sensitive and context-free rules in P, respectively, and V denotes the total alphabet of G.

Let f be an arbitrary bijection from V to $\{1, \ldots, m\}$, where m is the cardinality of V, and let f^{-1} be the inverse of f.

Construct an F-EP0L grammar of degree 2,

$$G' = \big(V', T, P', S_1\big)$$

with V' defined as

$$
\begin{aligned}
W_0 &= \{\langle A, B, C\rangle \mid AB \to AC \in P_{CS}\} \\
W_S &= \{\langle A, B, C, j\rangle \mid AB \to AC \in P_{CS}, 1 \le j \le m + 1\} \\
W &= W_0 \cup W_S \\
V' &= V \cup W
\end{aligned}
$$

where V, W_0, and W_S are pairwise disjoint alphabets. The set of rules P' is constructed by performing (1) through (3), given next.

(1) For every $X \in V$, add $(X \to X, \emptyset)$ to P'.
(2) For every $A \to u \in P_{CF}$, add $(A \to u, W)$ to P'.
(3) For every $AB \to AC \in P_{CS}$, extend P' by adding

(3.1) $(B \to \langle A, B, C \rangle, W)$;
(3.2) $(\langle A, B, C \rangle \to \langle A, B, C, 1 \rangle, W - \{\langle A, B, C \rangle\})$;
(3.3) $(\langle A, B, C, j \rangle \to \langle A, B, C, j + 1 \rangle, \{f^{-1}(j)\langle A, B, C, j \rangle\})$ for all $1 \leq j \leq m$ such that $f(A) \neq j$;
(3.4) $(\langle A, B, C, f(A) \rangle \to \langle A, B, C, f(A) + 1 \rangle, \emptyset)$;
(3.5) $(\langle A, B, C, m + 1 \rangle \to C, \{\langle A, B, C, m + 1 \rangle^2\})$.

Let us informally explain how G' simulates the non-context-free rules of the form $AB \to AC$ (see rules of (3) in the construction of P'). First, chosen occurrences of B are rewritten with $\langle A, B, C \rangle$ by $(B \to \langle A, B, C \rangle, W)$. The forbidding condition of this rule guarantees that there is no simulation already in process. After that, left neighbors of all occurrences of $\langle A, B, C \rangle$ are checked not to be any symbols from $V - \{A\}$. In a greater detail, G' rewrites $\langle A, B, C \rangle$ with $\langle A, B, C, i \rangle$ for $i = 1$. Then, in every $\langle A, B, C, i \rangle$, G' increments i by one as long as i is less or equal to the cardinality of V; simultaneously, it verifies that the left neighbor of every $\langle A, B, C, i \rangle$ differs from the symbol that f maps to i except for the case when $f(A) = i$. Finally, G' checks that there are no two adjoining symbols $\langle A, B, C, m + 1 \rangle$. At this point, the left neighbors of $\langle A, B, C, m+1 \rangle$ are necessarily equal to A, so every occurrence of $\langle A, B, C, m + 1 \rangle$ is rewritten to C.

Observe that the other symbols remain unchanged during the simulation. Indeed, by the forbidding conditions, the only rules that can rewrite symbols $X \notin W$ are of the form $(X \to X, \emptyset)$. Moreover, the forbidding condition of $(\langle A, B, C \rangle \to \langle A, B, C, 1 \rangle, W - \{\langle A, B, C \rangle\})$ implies that it is not possible to simulate two different non-context-free rules at the same time.

To establish that $L(G) = L(G')$, we first prove Claims 1 through 5.

Claim 1. $S_1 \Rightarrow_{G'}^{n} x'$ *implies that* $\mathrm{lms}(x') \in (N_1 \cup T)$ *for every* $n \geq 0$, $x' \in V'^*$.

Proof. The claim is proved by induction on $n \geq 0$.

Basis. Let $n = 0$. Then, $S_1 \Rightarrow_{G'}^{0} S_1$ and $S_1 \in N_1$.

Induction Hypothesis. Assume that the claim holds for all derivations of length n or less, for some $n \geq 0$.

Induction Step. Consider any derivation of the form

$$S_1 \Rightarrow_{G'}^{n+1} x'$$

where $x' \in V'^*$. Since $n + 1 \geq 1$, there is a derivation

$$S_1 \Rightarrow_{G'}^{n} y' \Rightarrow_{G'} x' \; [p_1, \ldots, p_q]$$

$y' \in V'^*$, $q = |y'|$, and by the induction hypothesis, $\text{lms}(y') \in (N_1 \cup T)$. Inspect P' to see that the rule p_1 that rewrites the leftmost symbol of y' is one of the following forms $(A_1 \to A_1, \emptyset)$, $(a \to a, \emptyset)$, $(A_1 \to a, W)$, $(A_1 \to C_1, W)$, or $(A_1 \to D_1 E, W)$, where $A_1, C_1, D_1 \in N_1$, $a \in T$, $E \in N_{CF}$ (see (1) and (2) in the definition of P' and Lemma 10.2.9). It is obvious that the leftmost symbols of the right-hand sides of these rules belong to $(N_1 \cup T)$. Hence, $\text{lms}(x') \in (N_1 \cup T)$, so the claim holds. □

Claim 2. $S_1 \Rightarrow_{G'}^n y_1' X y_3'$*, where* $X \in W_S$*, implies that* $y_1' \in V'^+$ *for any* $y_3' \in V'^*$.

Proof. Informally, the claim says that every occurrence of a symbol from W_S has always a left neighbor. Clearly, this claim follows from the statement of Claim 1. Since $W_S \cap (N_1 \cup T) = \emptyset$, X cannot be the leftmost symbol in a sentential form and the claim holds. □

Claim 3. $S_1 \Rightarrow_{G'}^n x'$, $n \geq 0$*, implies that* x' *has one of the following three forms*

(I) $x' \in V^*$;
(II) $x' \in (V \cup W_0)^*$ *and* $\text{occur}(x', W_0) > 0$;
(III) $x' \in (V \cup \{\langle A, B, C, j\rangle\})^*$, $\text{occur}(x', \{\langle A, B, C, j\rangle\}) > 0$, *and* $\{f^{-1}(k)\langle A, B, C, j\rangle \mid 1 \leq k < j, k \neq f(A)\} \cap \text{sub}(x') = \emptyset$*, where* $\langle A, B, C, j\rangle \in W_S$, $A \in (N_1 \cup N_{CF})$, $B \in N_{CS}$, $C \in N_{CF}$, $1 \leq j \leq m+1$.

Proof. We prove the claim by induction on $n \geq 0$.

Basis. Let $n = 0$. Clearly, $S_1 \Rightarrow_{G'}^0 S_1$ and S_1 is of type (I).

Induction Hypothesis. Suppose that the claim holds for all derivations of length n or less, for some $n \geq 0$.

Induction Step. Consider any derivation of the form

$$S_1 \Rightarrow_{G'}^{n+1} x'$$

Since $n + 1 \geq 1$, there exists $y' \in V'^*$ and a sequence of rules $p_1, \dots, p_q$, where $p_i \in P'$, $1 \leq i \leq q$, $q = |y'|$, such that

$$S_1 \Rightarrow_{G'}^n y' \Rightarrow_{G'} x' \; [p_1, \dots, p_q]$$

Let $y' = a_1 a_2 \dots a_q$, $a_i \in V'$.

By the induction hypothesis, y' can only be of forms (I) through (III). Thus, the following three cases cover all possible forms of y'.

(i) Let $y' \in V^*$ [form (I)]. In this case, every rule p_i can be either of the form $(a_i \to a_i, \emptyset)$, $a_i \in V$, or $(a_i \to u, W)$ such that $a_i \to u \in P_{CF}$, or $(a_i \to \langle A, a_i, C\rangle, W)$, $a_i \in N_{CS}$, $\langle A, a_i, C\rangle \in W_0$ (see the definition of P').

Suppose that for every $i \in \{1, \dots, q\}$, p_i has one of the first two listed forms. According to the right-hand sides of these rules, we obtain $x' \in V^*$; that is, x' is of form (I).

If there exists i such that $p_i = (a_i \to \langle A, a_i, C\rangle, W)$ for some $A \in (N_1 \cup N_{CF})$, $a_i \in N_{CS}$, $C \in N_{CF}$, $\langle A, a_i, C\rangle \in W_0$, we get $x' \in (V \cup W_0)^*$ with $\text{occur}(x', W_0) > 0$. Thus, x' belongs to (II).

(ii) Let $y' \in (V \cup W_0)^*$ and $\text{occur}(y', W_0) > 0$ [form (II)]. At this point, p_i is either $(a_i \to a_i, \emptyset)$ (rewriting $a_i \in V$ to itself) or $(\langle A, B, C\rangle \to \langle A, B, C, 1\rangle, W - \{\langle A, B, C\rangle\})$ rewriting $a_i = \langle A, B, C\rangle \in W_0$ to $\langle A, B, C, 1\rangle \in W_S$, where $A \in (N_1 \cup N_{CF})$, $B \in N_{CS}$, $C \in N_{CF}$. Since $\text{occur}(y', W_0) > 0$, there exists at least one i such that $a_i = \langle A, B, C\rangle \in W_0$. The corresponding rule p_i can be used provided that $\text{occur}(y', W - \{\langle A, B, C\rangle\}) = 0$. Therefore, $y' \in (V \cup \{\langle A, B, C\rangle\})^*$, so $x' \in (V \cup \{\langle A, B, C, 1\rangle\})^*$, $\text{occur}(x', \{\langle A, B, C, 1\rangle\}) > 0$. That is, x' is of type (III).

(iii) Assume that $y' \in (V \cup \{\langle A, B, C, j\rangle\})^*$, $\text{occur}(y', \{\langle A, B, C, j\rangle\}) > 0$, and

$$\text{sub}(y') \cap \{f^{-1}(k)\langle A, B, C, j\rangle \mid 1 \le k < j, k \ne f(A)\} = \emptyset$$

where $\langle A, B, C, j\rangle \in W_S$, $A \in (N_1 \cup N_{CF})$, $B \in N_{CS}$, $C \in N_{CF}$, $1 \le j \le m + 1$ [form (III)]. By the inspection of P', we see that the following four forms of rules can be used to rewrite y' to x'

(a) $(a_i \to a_i, \emptyset)$, $a_i \in V$;
(b) $(\langle A, B, C, j\rangle \to \langle A, B, C, j + 1\rangle, \{f^{-1}(j)\langle A, B, C, j\rangle\})$, $1 \le j \le m$, $j \ne f(A)$;
(c) $(\langle A, B, C, f(A)\rangle \to \langle A, B, C, f(A) + 1\rangle, \emptyset)$;
(d) $(\langle A, B, C, m + 1\rangle \to C, \{\langle A, B, C, m + 1\rangle^2\})$.

Let $1 \le j \le m$, $j \ne f(A)$. Then, symbols from V are rewritten to themselves [case (a)] and every occurrence of $\langle A, B, C, j\rangle$ is rewritten to $\langle A, B, C, j + 1\rangle$ by (b). Clearly, we obtain $x' \in (V \cup \{\langle A, B, C, j + 1\rangle\})^*$ such that $\text{occur}(x', \{\langle A, B, C, j + 1\rangle\}) > 0$. Furthermore, (b) can be used only when $f^{-1}(j)\langle A, B, C, j\rangle \notin \text{sub}(y')$. As

$$\text{sub}(y') \cap \{f^{-1}(k)\langle A, B, C, j\rangle \mid 1 \le k < j, k \ne f(A)\} = \emptyset$$

it holds that

$$\text{sub}(y') \cap \{f^{-1}(k)\langle A, B, C, j\rangle \mid 1 \le k \le j, k \ne f(A)\} = \emptyset$$

Since every occurrence of $\langle A, B, C, j\rangle$ is rewritten to $\langle A, B, C, j + 1\rangle$ and other symbols are unchanged,

$$\text{sub}(x') \cap \{f^{-1}(k)\langle A, B, C, j + 1\rangle \mid 1 \le k < j + 1, k \ne f(A)\} = \emptyset$$

Therefore, x' is of form (III).

Next, assume that $j = f(A)$. Then, all occurrences of $\langle A, B, C, j\rangle$ are rewritten to $\langle A, B, C, j + 1\rangle$ by (c), and symbols from V are rewritten to themselves. As before, we obtain $x' \in (V \cup \{\langle A, B, C, j + 1\rangle\})^*$ and $\text{occur}(x', \{\langle A, B, C, j + 1\rangle\}) > 0$. Moreover, because

$$\mathrm{sub}(y') \cap \{f^{-1}(k)\langle A, B, C, j\rangle \mid 1 \leq k < j,\ k \neq f(A)\} = \emptyset$$

and j is $f(A)$,

$$\mathrm{sub}(x') \cap \{f^{-1}(k)\langle A, B, C, j+1\rangle \mid 1 \leq k < j+1,\ k \neq f(A)\} = \emptyset$$

and x' belongs to (III) as well.

Finally, let $j = m + 1$. Then, every occurrence of $\langle A, B, C, j\rangle$ is rewritten to C [case (d)]. Therefore, $x' \in V^*$, so x' has form (I).

In (i), (ii), and (iii), we have considered all derivations that rewrite y' to x', and in each of these cases, we have shown that x' has one of the requested forms. Therefore, Claim 3 holds. □

To prove the following claims, we need a finite letter-to-letters substitution g from V^* into V'^* defined as

$$\begin{aligned} g(X) = \{X\} &\cup \{\langle A, X, C\rangle \mid \langle A, X, C\rangle \in W_0\} \\ &\cup \{\langle A, X, C, j\rangle \mid \langle A, X, C, j\rangle \in W_S, 1 \leq j \leq m+1\} \end{aligned}$$

for all $X \in V$, $A \in (N_1 \cup N_{CF})$, $C \in N_{CF}$. Let g^{-1} be the inverse of g.

Claim 4. Let $y' = a_1 a_2 \cdots a_q$, $a_i \in V'$, $q = |y'|$, and $g^{-1}(a_i) \Rightarrow_G^{h_i} g^{-1}(u_i)$ for all $i \in \{1, \ldots, q\}$ and some $h_i \in \{0, 1\}$, $u_i \in V'^+$. Then, $g^{-1}(y') \Rightarrow_G^{r} g^{-1}(x')$ such that $x' = u_1 u_2 \cdots u_q$, $r = \sum_{i=1}^{q} h_i$, $r \leq q$.

Proof. First, consider any derivation of the form

$$g^{-1}(X) \Rightarrow_G^h g^{-1}(u)$$

where $X \in V'$, $u \in V'^+$, $h \in \{0, 1\}$. If $h = 0$, then $g^{-1}(X) = g^{-1}(u)$. Let $h = 1$. Then, there surely exists a rule $p = g^{-1}(X) \to g^{-1}(u) \in P$ such that

$$g^{-1}(X) \Rightarrow_G g^{-1}(u)\ [p]$$

Return to the statement of this claim. We can construct

$$\begin{aligned} g^{-1}(a_1)g^{-1}(a_2)\cdots g^{-1}(a_q) &\Rightarrow_G^{h_1} g^{-1}(u_1)g^{-1}(a_2)\cdots g^{-1}(a_q) \\ &\Rightarrow_G^{h_2} g^{-1}(u_1)g^{-1}(u_2)\cdots g^{-1}(a_q) \\ &\ \ \vdots \\ &\Rightarrow_G^{h_q} g^{-1}(u_1)g^{-1}(u_2)\cdots g^{-1}(u_q) \end{aligned}$$

where

$$g^{-1}(y') = g^{-1}(a_1)\cdots g^{-1}(a_q)$$

and

$$g^{-1}(u_1)\cdots g^{-1}(u_q) = g^{-1}(u_1 \cdots u_q) = g^{-1}(x')$$

In such a derivation, each $g^{-1}(a_i)$ is either left unchanged (if $h_i = 0$) or rewritten to $g^{-1}(u_i)$ by the corresponding rule $g^{-1}(a_i) \rightarrow g^{-1}(u_i)$. Obviously, the length of this derivation is $\sum_{i=1}^{q} h_i$. □

Claim 5. $S_1 \Rightarrow_G^* x$ *if and only if* $S_1 \Rightarrow_{G'}^* x'$, *where* $x \in V^*$, $x' \in V'^*$, $x' \in g(x)$.

Proof. The proof is divided into the only-if part and the if part.

Only If. The only-if part is established by induction on $n \geq 0$. That is, we show that

$$S_1 \Rightarrow_G^n x \quad \text{implies} \quad S_1 \Rightarrow_{G'}^* x$$

where $x \in V^*$, for $n \geq 0$.

Basis. Let $n = 0$. Then, $S_1 \Rightarrow_G^0 S_1$ and $S_1 \Rightarrow_{G'}^0 S_1$ as well.

Induction Hypothesis. Assume that the claim holds for all derivations of length n or less, for some $n \geq 0$.

Induction Step. Consider any derivation of the form

$$S_1 \Rightarrow_G^{n+1} x$$

Since $n + 1 > 0$, there exists $y \in V^*$ and $p \in P$ such that

$$S_1 \Rightarrow_G^n y \Rightarrow_G x \ [p]$$

and by the induction hypothesis, there is also a derivation

$$S_1 \Rightarrow_{G'}^* y$$

Let $y = a_1 a_2 \cdots a_q$, $a_i \in V$, $1 \leq i \leq q$, $q = |y|$. The following cases (i) and (ii) cover all possible forms of p.

(i) Let $p = A \rightarrow u \in P_{CF}$, $A \in (N_1 \cup N_{CF})$, $u \in V^*$. Then, $y = y_1 A y_3$ and $x = y_1 u y_3$, $y_1, y_3 \in V^*$. Let $s = |y_1| + 1$. Since we have $(A \rightarrow u, W) \in P'$, we can construct a derivation

$$S_1 \Rightarrow_{G'}^* y \Rightarrow_{G'} x \ [p_1, \cdots, p_q]$$

such that $p_s = (A \rightarrow u, W)$ and $p_i = (a_i \rightarrow a_i, \emptyset)$ for all $i \in \{1, \cdots, q\}$, $i \neq s$.

(ii) Let $p = AB \rightarrow AC \in P_{CS}$, $A \in (N_1 \cup N_{CF})$, $B \in N_{CS}$, $C \in N_{CF}$. Then, $y = y_1 AB y_3$ and $x = y_1 AC y_3$, $y_1, y_3 \in V^*$. Let $s = |y_1| + 2$. In this case, there is the following derivation

$$
\begin{aligned}
S_1 &\Rightarrow^*_{G'} y_1 A B y_3 \\
&\Rightarrow_{G'} y_1 A \langle A, B, C\rangle y_3 && [p_s = (B \rightarrow \langle A, B, C\rangle, W)] \\
&\Rightarrow_{G'} y_1 A \langle A, B, C, 1\rangle y_3 && [p_s = (\langle A, B, C\rangle \rightarrow \langle A, B, C, 1\rangle, \\
& && \quad W - \{\langle A, B, C\rangle\})] \\
&\Rightarrow_{G'} y_1 A \langle A, B, C, 2\rangle y_3 && [p_s = (\langle A, B, C, 1\rangle \rightarrow \langle A, B, C, 2\rangle, \\
& && \quad \{f^{-1}(1)\langle A, B, C, j\rangle\})] \\
&\ \vdots \\
&\Rightarrow_{G'} y_1 A \langle A, B, C, f(A)\rangle y_3 && [p_s = (\langle A, B, C, f(A) - 1\rangle \rightarrow \\
& && \quad \langle A, B, C, f(A)\rangle, \{f^{-1}(f(A) - 1) \\
& && \quad \langle A, B, C, f(A) - 1\rangle\})] \\
&\Rightarrow_{G'} y_1 A \langle A, B, C, f(A) + 1\rangle y_3 && [p_s = (\langle A, B, C, f(A)\rangle \rightarrow \\
& && \quad \langle A, B, C, f(A) + 1\rangle, \emptyset)] \\
&\Rightarrow_{G'} y_1 A \langle A, B, C, f(A) + 2\rangle y_3 && [p_s = (\langle A, B, C, f(A) + 1\rangle \rightarrow \\
& && \quad \langle A, B, C, f(A) + 2\rangle, \{f^{-1}(f(A) + 1) \\
& && \quad \langle A, B, C, f(A) + 1\rangle\})] \\
&\ \vdots \\
&\Rightarrow_{G'} y_1 A \langle A, B, C, m + 1\rangle y_3 && [p_s = (\langle A, B, C, m\rangle \rightarrow \langle A, B, C, m + 1\rangle, \\
& && \quad \{f^{-1}(m)\langle A, B, C, m\rangle\})] \\
&\Rightarrow_{G'} y_1 A C y_3 && [p_s = (\langle A, B, C, m + 1\rangle \rightarrow C, \\
& && \quad \{\langle A, B, C, m + 1\rangle^2\})]
\end{aligned}
$$

such that $p_i = (a_i \rightarrow a_i, \emptyset)$ for all $i \in \{1, \dots, q\}, i \neq s$.

If. By induction on $n \geq 0$, we prove that

$$S_1 \Rightarrow^n_{G'} x' \quad \text{implies} \quad S_1 \Rightarrow^*_G x$$

where $x' \in V'^*$, $x \in V^*$ and $x' \in g(x)$.

Basis. Let $n = 0$. The only x' is S_1 because $S_1 \Rightarrow^0_{G'} S_1$. Obviously, $S_1 \Rightarrow^0_G S_1$ and $S_1 \in g(S_1)$.

Induction Hypothesis. Suppose that the claim holds for any derivation of length n or less, for some $n \geq 0$.

Induction Hypothesis. Consider any derivation of the form

$$S_1 \Rightarrow^{n+1}_{G'} x'$$

Since $n + 1 \geq 1$, there exists $y' \in V'^*$ and a sequence of rules $p_1, \dots, p_q$ from P', $q = |x'|$, such that

$$S_1 \Rightarrow^n_{G'} y' \Rightarrow_{G'} x' \ [p_1, \dots, p_q]$$

Let $y' = a_1a_2 \cdots a_q$, $a_i \in V'$, $1 \leq i \leq q$. By the induction hypothesis, we have

$$S_1 \Rightarrow_G^* y$$

where $y \in V^*$ such that $y' \in g(y)$.

From Claim 3, y' has one of the following forms (i), (ii), or (iii), described next.

(i) Let $y' \in V'^*$ (see (I) in Claim 3). Inspect P' to see that there are three forms of rules rewriting symbols a_i in y':

(i.a) $p_i = (a_i \rightarrow a_i, \emptyset) \in P'$, $a_i \in V$. In this case,

$$g^{-1}(a_i) \Rightarrow_G^0 g^{-1}(a_i)$$

(i.b) $p_i = (a_i \rightarrow u_i, W) \in P'$ such that $a_i \rightarrow u_i \in P_{CF}$. Since $a_i = g^{-1}(a_i)$, $u_i = g^{-1}(u_i)$ and $a_i \rightarrow u_i \in P$,

$$g^{-1}(a_i) \Rightarrow_G g^{-1}(u_i) \ [a_i \rightarrow u_i]$$

(i.c) $p_i = (a_i \rightarrow \langle A, a_i, C\rangle, W) \in P'$, $a_i \in N_{CS}$, $A \in (N_1 \cup N_{CF})$, $C \in N_{CF}$. Since $g^{-1}(a_i) = g^{-1}(\langle A, a_i, C\rangle)$, we have

$$g^{-1}(a_i) \Rightarrow_G^0 g^{-1}(\langle A, a_i, C\rangle)$$

We see that for all a_i, there exists a derivation

$$g^{-1}(a_i) \Rightarrow_G^{h_i} g^{-1}(z_i)$$

for some $h_i \in \{0, 1\}$, where $z_i \in V'^+$, $x' = z_1z_2 \cdots z_q$. Therefore, by Claim 4, we can construct

$$S_1 \Rightarrow_G^* y \Rightarrow_G^r x$$

where $0 \leq r \leq q$, $x = g^{-1}(x')$.

(ii) Let $y' \in (V \cup W_0)^*$ and $\text{occur}(y', W_0) > 0$ [see (II)]. At this point, the following two forms of rules can be used to rewrite a_i in y'—(ii.a) or (ii.b).

(ii.a) $p_i = (a_i \rightarrow a_i, \emptyset) \in P'$, $a_i \in V$. As in case (i.a),

$$g^{-1}(a_i) \Rightarrow_G^0 g^{-1}(a_i)$$

(ii.b) $p_i = (\langle A, B, C\rangle \rightarrow \langle A, B, C, 1\rangle, W - \{\langle A, B, C\rangle\})$, $a_i = \langle A, B, C\rangle \in W_0$, $A \in (N_1 \cup N_{CF})$, $B \in N_{CS}$, $C \in N_{CF}$. Since $g^{-1}(\langle A, B, C\rangle) = g^{-1}(\langle A, B, C, 1\rangle)$,

$$g^{-1}(\langle A, B, C\rangle) \Rightarrow_G^0 g^{-1}(\langle A, B, C, 1\rangle)$$

Thus, there exists a derivation

$$S_1 \Rightarrow_G^* y \Rightarrow_G^0 x$$

where $x = g^{-1}(x')$.

(iii) Let $y' \in (V \cup \{\langle A, B, C, j\rangle\})^*$, $\text{occur}(y', \{\langle A, B, C, j\rangle\}) > 0$, and

$$\text{sub}(y') \cap \{f^{-1}(k)\langle A, B, C, j\rangle \mid 1 \leq k < j,\ k \neq f(A)\} = \emptyset$$

where $\langle A, B, C, j\rangle \in W_S$, $A \in (N_1 \cup N_{CF})$, $B \in N_{CS}$, $C \in N_{CF}$, $1 \leq j \leq m + 1$ [see (III)]. By the inspection of P', the following four forms of rules can be used to rewrite y' to x':

(iii.a) $p_i = (a_i \rightarrow a_i, \emptyset)$, $a_i \in V$;
(iii.b) $p_i = (\langle A, B, C, j\rangle \rightarrow \langle A, B, C, j + 1\rangle, \{f^{-1}(j)\langle A, B, C, j\rangle\})$, $1 \leq j \leq m$, $j \neq f(A)$;
(iii.c) $p_i = (\langle A, B, C, f(A)\rangle \rightarrow \langle A, B, C, f(A) + 1\rangle, \emptyset)$;
(iii.d) $p_i = (\langle A, B, C, m + 1\rangle \rightarrow C, \{\langle A, B, C, m + 1\rangle^2\})$.

Let $1 \leq j \leq m$. G' can rewrite such y' using only the rules (iii.a) through (iii.c). Since $g^{-1}(\langle A, B, C, j\rangle) = g^{-1}(\langle A, B, C, j + 1\rangle)$ and $g^{-1}(a_i) = g^{-1}(a_i)$, by analogy with (ii), we obtain

$$S_1 \Rightarrow_G^* y \Rightarrow_G^0 x$$

such that $x = g^{-1}(x')$.

Let $j = m + 1$. In this case, only the rules (iii.a) and (iii.d) can be used. Since $\text{occur}(y', \{\langle A, B, C, j\rangle\}) > 0$, there is at least one occurrence of $\langle A, B, C, m + 1\rangle$ in y', and by the forbidding condition of the rule (iii.c), $\langle A, B, C, m + 1\rangle^2 \notin \text{sub}(y')$. Observe that for $j = m + 1$,

$$\begin{aligned}&\{f^{-1}(k)\langle A, B, C, m + 1\rangle \mid 1 \leq k < j,\ k \neq f(A)\}\\ &= \{X\langle A, B, C, m + 1\rangle \mid X \in V,\ X \neq A\}\end{aligned}$$

and thus

$$\text{sub}(y') \cap \{X\langle A, B, C, m + 1\rangle \mid X \in V,\ X \neq A\} = \emptyset$$

According to Claim 2, $\langle A, B, C, m + 1\rangle$ has always a left neighbor in y'. As a result, the left neighbor of every occurrence of $\langle A, B, C, m+1\rangle$ is A. Therefore, we can express y', y, and x' as follows:

$$\begin{aligned} y' &= y_1 A\langle A, B, C, m{+}1\rangle y_2 A\langle A, B, C, m{+}1\rangle y_3 \cdots y_r A\langle A, B, C, m{+}1\rangle y_{r+1} \\ y &= g^{-1}(y_1)ABg^{-1}(y_2)ABg^{-1}(y_3)\cdots g^{-1}(y_r)ABg^{-1}(y_{r+1}) \\ x' &= y_1 ACy_2ACy_3 \cdots y_r ACy_{r+1} \end{aligned}$$

where $r \geq 1$, $y_s \in V^*$, $1 \leq s \leq r + 1$. Since we have $p = AB \rightarrow AC \in P$, there is a derivation

$$\begin{aligned} S_1 &\Rightarrow_G^* g^{-1}(y_1)ABg^{-1}(y_2)ABg^{-1}(y_3)\cdots g^{-1}(y_r)ABg^{-1}(y_{r+1}) \\ &\Rightarrow_G g^{-1}(y_1)ACg^{-1}(y_2)ABg^{-1}(y_3)\cdots g^{-1}(y_r)ABg^{-1}(y_{r+1}) \; [p] \\ &\Rightarrow_G g^{-1}(y_1)ACg^{-1}(y_2)ACg^{-1}(y_3)\cdots g^{-1}(y_r)ABg^{-1}(y_{r+1}) \; [p] \\ &\;\;\vdots \\ &\Rightarrow_G g^{-1}(y_1)ACg^{-1}(y_2)ACg^{-1}(y_3)\cdots g^{-1}(y_r)ACg^{-1}(y_{r+1}) \; [p] \end{aligned}$$

where $g^{-1}(y_1)ACg^{-1}(y_2)ACg^{-1}(y_3)\cdots g^{-1}(y_r)ACg^{-1}(y_{r+1}) = g^{-1}(x') = x$.

Since cases (i), (ii), and (iii) cover all possible forms of y', we have completed the induction and established Claim 5. □

The equivalence of G and G' follows from Claim 5. Indeed, observe that by the definition of g, we have $g(a) = \{a\}$ for all $a \in T$. Therefore, by Claim 5, we have for any $x \in T^*$,

$$S_1 \Rightarrow_G^* x \quad \text{if and only if} \quad S_1 \Rightarrow_{G'}^* x$$

Thus, $L(G) = L(G')$, and the lemma holds. □

Theorem 10.2.11. **CS** $=$ **F - EP0L(2)** $=$ **F - EPT0L(2)** $=$ **F - EP0L** $=$ **F - EPT0L**

Proof. By Lemma 10.2.10, **CS** $\subseteq$ **F - EP0L(2)** $\subseteq$ **F - EPT0L(2)** $\subseteq$ **F - EPT0L**. From Lemma 10.1.5 and the definition of F-ET0L grammars, we have **F - EPT0L**(s) $\subseteq$ **F - EPT0L** $\subseteq$ **C - EPT0L** $\subseteq$ **CS** for any $s \geq 0$. Moreover, **F - EP0L**(s) $\subseteq$ **F - EP0L** $\subseteq$ **F - EPT0L**. Thus, **CS** $=$ **F - EP0L(2)** $=$ **F - EPT0L(2)** $=$ **F - EP0L** $=$ **F - EPT0L**, and the theorem holds. □

Return to the proof of Lemma 10.2.10. Observe the form of the rules in the F-EP0L grammar G'. This observation gives rise to the next corollary.

Corollary 10.2.12. *Every context-sensitive language can be generated by an F-EP0L grammar $G = (V, T, P, S)$ of degree 2 such that every rule from P has one of the following forms*

(i) $(a \rightarrow a, \emptyset)$, $a \in V$;
(ii) $(X \rightarrow x, F)$, $X \in V - T$, $|x| \in \{1, 2\}$, $\text{max-len}(F) = 1$;
(iii) $(X \rightarrow Y, \{z\})$, $X, Y \in V - T$, $z \in V^2$. □

Next, we demonstrate that the family of recursively enumerable languages is generated by the forbidding E0L grammars of degree 2.

Lemma 10.2.13. **RE $\subseteq$ F - E0L(2)**

Proof. Let L be a recursively enumerable language generated by a phrase structure grammar

$$G = (V, T, P, S)$$

having the form defined in Theorem 4.1.7, where

$$\begin{aligned} V &= N_{CF} \cup N_{CS} \cup T \\ P_{CS} &= \{AB \to AC \in P \mid A, C \in N_{CF}, B \in N_{CS}\} \\ P_{CF} &= P - P_{CS} \end{aligned}$$

Let \$ be a new symbol and m be the cardinality of $V \cup \{\$\}$. Furthermore, let f be an arbitrary bijection from $V \cup \{\$\}$ onto $\{1, \dots, m\}$, and let f^{-1} be the inverse of f.

Define the F-E0L grammar

$$G' = (V', T, P', S')$$

of degree 2 as follows:

$$\begin{aligned} W_0 &= \{\langle A, B, C\rangle \mid AB \to AC \in P\} \\ W_S &= \{\langle A, B, C, j\rangle \mid AB \to AC \in P, 1 \le j \le m\} \\ W &= W_0 \cup W_S \\ V' &= V \cup W \cup \{S', \$\} \end{aligned}$$

where $A, C \in N_{CF}$, $B \in N_{CS}$, and V, W_0, W_S, and $\{S', \$\}$ are pairwise disjoint alphabets. The set of rules P' is constructed by performing (1) through (4), given next.

(1) Add $(S' \to \$S, \emptyset)$, $(\$ \to \$, \emptyset)$ and $(\$ \to \varepsilon, V' - T - \{\$\})$ to P'.
(2) For all $X \in V$, add $(X \to X, \emptyset)$ to P'.
(3) For all $A \to u \in P_{CF}$, $A \in N_{CF}$, $u \in \{\varepsilon\} \cup N_{CS} \cup T \cup (\bigcup_{i=1}^{2} N_{CF}^{i})$, add $(A \to u, W)$ to P'.
(4) If $AB \to AC \in P_{CS}$, $A, C \in N_{CF}$, $B \in N_{CS}$, then add the rules defined in (4.1) through (4.5) into P'.

(4.1) $(B \to \langle A, B, C\rangle, W)$;
(4.2) $(\langle A, B, C\rangle \to \langle A, B, C, 1\rangle, W - \{\langle A, B, C\rangle\})$;
(4.3) $(\langle A, B, C, j\rangle \to \langle A, B, C, j+1\rangle, \{f^{-1}(j)\langle A, B, C, j\rangle\})$ for all $1 \le j \le m$ such that $f(A) \neq j$;
(4.4) $(\langle A, B, C, f(A)\rangle \to \langle A, B, C, f(A)+1\rangle, \emptyset)$;
(4.5) $(\langle A, B, C, m+1\rangle \to C, \{\langle A, B, C, m+1\rangle^2\})$.

Let us only give a gist of the reason why $L(G) = L(G')$. The construction above resembles the construction in Lemma 10.2.10 very much. Indeed, to simulate the non-context-free rules $AB \to AC$ in F-E0L grammars, we use the same technique as in F-EP0L grammars from Lemma 10.2.10. We only need to guarantee that no sentential form begins with a symbol from N_{CS}. This is solved by an auxiliary nonterminal \$ in the definition of G'. The symbol is always generated in the first derivation step by $(S' \rightarrow \$S, \emptyset)$ (see (1) in the definition of P'). After that, it appears as the leftmost symbol of all sentential forms containing some nonterminals. The only rule that can erase it is $(\$ \rightarrow \varepsilon, V' - T - \{\$\})$.

Therefore, by analogy with the technique used in Lemma 10.2.10, we can establish

$$S \Rightarrow_G^* x \quad \text{if and only if} \quad S' \Rightarrow_{G'}^+ \$x'$$

such that $x \in V^*$, $x' \in (V' - \{S', \$\})^*$, $x' \in g(x)$, where g is a finite substitution from V^* into $(V' - \{S', \$\})^*$ defined as

$$\begin{aligned} g(X) = \{X\} &\cup \{\langle A, X, C\rangle \mid \langle A, X, C\rangle \in W_0\} \\ &\cup \{\langle A, X, C, j\rangle \mid \langle A, X, C, j\rangle \in W_S, 1 \leq j \leq m + 1\} \end{aligned}$$

for all $X \in V$, $A, C \in N_{CF}$. The details are left to the reader.

As in Lemma 10.2.10, we have $g(a) = \{a\}$ for all $a \in T$; hence, for all $x \in T^*$,

$$S \Rightarrow_G^* x \quad \text{if and only if} \quad S' \Rightarrow_{G'}^+ \$x$$

Since

$$\$x \Rightarrow_{G'} x \; [(\$ \rightarrow \varepsilon, V' - T - \{\$\})]$$

we obtain

$$S \Rightarrow_G^* x \quad \text{if and only if} \quad S' \Rightarrow_{G'}^+ x$$

Consequently, $L(G) = L(G')$; thus, $\mathbf{RE} \subseteq \mathbf{F\text{-}E0L}(2)$. □

Theorem 10.2.14. $\mathbf{RE} = \mathbf{F\text{-}E0L}(2) = \mathbf{F\text{-}ET0L}(2) = \mathbf{F\text{-}E0L} = \mathbf{F\text{-}ET0L}$

Proof. By Lemma 10.2.13, we have $\mathbf{RE} \subseteq \mathbf{F\text{-}E0L}(2) \subseteq \mathbf{F\text{-}ET0L}(2) \subseteq \mathbf{F\text{-}ET0L}$. From Lemma 10.1.7, it follows that $\mathbf{F\text{-}ET0L}(s) \subseteq \mathbf{F\text{-}ET0L} \subseteq \mathbf{C\text{-}ET0L} \subseteq \mathbf{RE}$, for any $s \geq 0$. Therefore, $\mathbf{RE} = \mathbf{F\text{-}E0L}(2) = \mathbf{F\text{-}ET0L}(2) = \mathbf{F\text{-}E0L} = \mathbf{F\text{-}ET0L}$, so the theorem holds. □

By analogy with Corollary 10.2.12, we obtain the following normal form.

Corollary 10.2.15. *Every recursively enumerable language can be generated by an F-E0L grammar $G = (V, T, P, S)$ of degree 2 such that every rule from P has one of the following forms*

(i) $(a \rightarrow a, \emptyset)$, $a \in V$;
(ii) $(X \rightarrow x, F)$, $X \in V - T$, $|x| \leq 2$, *and* $F \neq \emptyset$ *implies that* $\text{max-len}(F) = 1$;
(iii) $(X \rightarrow Y, \{z\})$, $X, Y \in V - T$, $z \in V^2$. □

Moreover, we obtain the following relations between F-ET0L language families.

Corollary 10.2.16.

$$\begin{gathered}\mathbf{CF}\\ \subset\\ \mathbf{F\text{-}EP0L}(0)=\mathbf{F\text{-}E0L}(0)=\mathbf{EP0L}=\mathbf{E0L}\\ \subset\\ \mathbf{F\text{-}EP0L}(1)=\mathbf{F\text{-}EPT0L}(1)=\mathbf{F\text{-}E0L}(1)=\mathbf{F\text{-}ET0L}(1)\\ =\mathbf{F\text{-}EPT0L}(0)=\mathbf{F\text{-}ET0L}(0)=\mathbf{EPT0L}=\mathbf{ET0L}\\ \subset\\ \mathbf{F\text{-}EP0L}(2)=\mathbf{F\text{-}EPT0L}(2)=\mathbf{F\text{-}EP0L}=\mathbf{F\text{-}EPT0L}=\mathbf{CS}\\ \subset\\ \mathbf{F\text{-}E0L}(2)=\mathbf{F\text{-}ET0L}(2)=\mathbf{F\text{-}E0L}=\mathbf{F\text{-}ET0L}=\mathbf{RE}\end{gathered}$$

Proof. This corollary follows from Theorems 10.2.3, 10.2.8, 10.2.11, and 10.2.14. □

10.3 Simple Semi-Conditional ET0L Grammars

Simple semi-conditional ET0L grammars represent another variant of context-conditional ET0L grammars with restricted sets of context conditions. By analogy with sequential simple semi-conditional grammars (see Sect. 4.6), these grammars are context-conditional ET0L grammars in which every rule contains no more than one context condition. This section defines them (see Sect. 10.3.1), and establishes their power and reduces their degree (see Sect. 10.3.2).

10.3.1 Definitions

In this section, we define simple semi-conditional ET0L grammars.

Definition 10.3.1. Let $G = (V, T, P_1, \dots, P_t, S)$ be a context-conditional ET0L grammar, for some $t \geq 1$. If for all $p = (a \rightarrow x, Per, For) \in P_i$ for every $i = 1, \dots, t$ holds that $\text{card}(Per) + \text{card}(For) \leq 1$, G is said to be a *simple semi-conditional ET0L grammar* (*SSC-ET0L grammar* for short). If G is a propagating SSC-ET0L grammar, then G is called an *SSC-EPT0L grammar*. If $t = 1$, then G is called an *SSC-E0L grammar*; if, in addition, G is a propagating SSC-E0L grammar, G is said to be an *SSC-EP0L grammar*. □

Let $G = (V, T, P_1, \dots, P_t, S)$ be an SSC-ET0L grammar of degree (r, s). By analogy with ssc-grammars, in each rule $(a \rightarrow x, Per, For) \in P_i$, $i = 1, \dots, t$, we omit braces and instead of $\emptyset$, we write 0. For example, we write $(a \rightarrow x, EF, 0)$ instead of $(a \rightarrow x, \{EF\}, \emptyset)$.

Let $\mathbf{SSC\text{-}EPT0L}(r, s)$, $\mathbf{SSC\text{-}ET0L}(r, s)$, $\mathbf{SSC\text{-}EP0L}(r, s)$, and $\mathbf{SSC\text{-}E0L}(r, s)$ denote the families of languages generated by SSC-EPT0L, SSC-ET0L, SSC-EP0L, and SSC-E0L grammars of degree (r, s), respectively. Furthermore, the families of languages generated by SSC-EPT0L, SSC-ET0L, SSC-EP0L, and SSC-E0L grammars of any degree are denoted by $\mathbf{SSC\text{-}EPT0L}$, $\mathbf{SSC\text{-}ET0L}$, $\mathbf{SSC\text{-}EP0L}$, and $\mathbf{SSC\text{-}E0L}$, respectively. Moreover, set

$$\mathbf{SSC\text{-}EPT0L} = \bigcup_{r=0}^{\infty} \bigcup_{s=0}^{\infty} \mathbf{SSC\text{-}EPT0L}(r, s)$$

$$\mathbf{SSC\text{-}ET0L} = \bigcup_{r=0}^{\infty} \bigcup_{s=0}^{\infty} \mathbf{SSC\text{-}ET0L}(r, s)$$

$$\mathbf{SSC\text{-}EP0L} = \bigcup_{r=0}^{\infty} \bigcup_{s=0}^{\infty} \mathbf{SSC\text{-}EP0L}(r, s)$$

$$\mathbf{SSC\text{-}E0L} = \bigcup_{r=0}^{\infty} \bigcup_{s=0}^{\infty} \mathbf{SSC\text{-}E0L}(r, s)$$

10.3.2 Generative Power and Reduction

Next, let us investigate the generative power of SSC-ET0L grammars. The following lemma proves that every recursively enumerable language can be defined by an SSC-E0L grammar of degree $(1, 2)$.

Lemma 10.3.2. $\mathbf{RE} \subseteq \mathbf{SSC\text{-}E0L}(1, 2)$

Proof. Let

$$G = \big(N_{CF} \cup N_{CS} \cup T, T, P, S\big)$$

be a phrase-structure grammar of the form of Theorem 4.1.7. Then, let $V = N_{CF} \cup N_{CS} \cup T$ and m be the cardinality of V. Let f be an arbitrary bijection from V to $\{1, \dots, m\}$, and f^{-1} be the inverse of f. Set

$$\begin{aligned} M = {} & \{\#\} \cup \\ & \{\langle A, B, C\rangle \mid AB \to AC \in P, A, C \in N_{CF}, B \in N_{CS}\} \cup \\ & \{\langle A, B, C, i\rangle \mid AB \to AC \in P, A, C \in N_{CF}, B \in N_{CS}, 1 \le i \le m+2\} \end{aligned}$$

and

$$W = \big\{[A, B, C] \mid AB \to AC \in P, A, C \in N_{CF}, B \in N_{CS}\big\}$$

Next, construct an SSC-E0L grammar of degree $(1, 2)$

$$G' = (V', T, P', S')$$

where

$$V' = V \cup M \cup W \cup \{S'\}$$

Without any loss of generality, we assume that V, M, W, and $\{S'\}$ are pairwise disjoint. The set of rules P' is constructed by performing (1) through (5), given next.

(1) Add $(S' \to \#S, 0, 0)$ to P'.
(2) For all $A \to x \in P$, $A \in N_{CF}$, $x \in \{\varepsilon\} \cup N_{CS} \cup T \cup N_{CF}^2$, add $(A \to x, \#, 0)$ to P'.
(3) For every $AB \to AC \in P$, $A, C \in N_{CF}$, $B \in N_{CS}$, add the following rules to P'

(3a) $(\# \to \langle A, B, C\rangle, 0, 0)$;
(3b) $(B \to [A, B, C], \langle A, B, C\rangle, 0)$;
(3c) $(\langle A, B, C\rangle \to \langle A, B, C, 1\rangle, 0, 0)$;
(3d) $([A, B, C] \to [A, B, C], 0, \langle A, B, C, m + 2\rangle)$;
(3e) $(\langle A, B, C, i\rangle \to \langle A, B, C, i + 1\rangle, 0, f^{-1}(i)[A, B, C])$ for all $1 \le i \le m$, $i \neq f(A)$;
(3f) $(\langle A, B, C, f(A)\rangle \to \langle A, B, C, f(A) + 1\rangle, 0, 0)$;
(3g) $(\langle A, B, C, m + 1\rangle \to \langle A, B, C, m + 2\rangle, 0, [A, B, C]^2)$;
(3h) $(\langle A, B, C, m + 2\rangle \to \#, 0, \langle A, B, C, m + 2\rangle[A, B, C])$;
(3i) $([A, B, C] \to C, \langle A, B, C, m + 2\rangle, 0)$.

(4) For all $X \in V$, add $(X \to X, 0, 0)$ to P'.
(5) Add $(\# \to \#, 0, 0)$ and $(\# \to \varepsilon, 0, 0)$ to P'.

Let us explain how G' works. During the simulation of a derivation in G, every sentential form starts with an auxiliary symbol from M, called the master. This symbol determines the current simulation mode and controls the next derivation step. Initially, the master is set to # (see (1) in the definition of P'). In this mode, G' simulates context-free rules [see (2)]; notice that symbols from V can always be rewritten to themselves by (4). To start the simulation of a non-context-free rule of the form $AB \to AC$, G' rewrites the master to $\langle A, B, C\rangle$. In the following step, chosen occurrences of B are rewritten to $[A, B, C]$; no other rules can be used except rules introduced in (4). At the same time, the master is rewritten to $\langle A, B, C, i\rangle$ with $i = 1$ [see (3c)]. Then, i is repeatedly incremented by one until i is greater than the cardinality of V [see rules (3e) and (3f)]. Simultaneously, the master's conditions make sure that for every i such that $f^{-1}(i) \neq A$, no $f^{-1}(i)$ appears as the left neighbor of any occurrence of $[A, B, C]$. Finally, G' checks that there are no two adjoining $[A, B, C]$ [see (3g)] and that $[A, B, C]$ does not appear as the right neighbor of the master [see (3h)]. At this point, the left neighbors of $[A, B, C]$ are necessarily equal to A and every occurrence of $[A, B, C]$ is rewritten to C. In the same derivation step, the master is rewritten to #.

Observe that in every derivation step, the master allows G' to use only a subset of rules according to the current mode. Indeed, it is not possible to combine context-free and non-context-free simulation modes. Furthermore, no two different non-context-free rules can be simulated at the same time. The simulation ends when # is erased by $(\# \to \varepsilon, 0, 0)$. After this erasure, no other rule can be used.

The following three claims demonstrate some important properties of derivations in G' to establish $L(G) = L(G')$.

Claim 1. *$S' \Rightarrow^+_{G'} w'$ implies that $w' \in M(V \cup W)^*$ or $w' \in (V \cup W)^*$. Furthermore, if $w' \in M(V \cup W)^*$, every v' such that $S' \Rightarrow^+_{G'} v' \Rightarrow^*_{G'} w'$ belongs to $M(V \cup W)^*$ as well.*

Proof. When deriving w', G' first rewrites S' to $\#S$ by using $(S' \to \#S, 0, 0)$, where $\# \in M$ and $S \in V$. Next, inspect P' to see that every symbol from M is always rewritten to a symbol belonging to M or, in the case of #, erased by $(\# \to \varepsilon, 0, 0)$. Moreover, there are no rules generating new occurrences of symbols from $(M \cup \{S'\})$. Thus, all sentential forms derived from S' belong either to $M(V \cup W)^*$ or to $(V \cup W)^*$. In addition, if a sentential form belongs to $M(V \cup W)^*$, all previous sentential forms (except for S') are also from $M(V \cup W)^*$. □

Claim 2. *Every successful derivation in G' is of the form*

$$S' \Rightarrow_{G'} \#S \Rightarrow^+_{G'} \#u' \Rightarrow_{G'} w' \Rightarrow^*_{G'} w'$$

where $u' \in V^*$, $w' \in T^*$.

Proof. From Claim 1 and its proof, every successful derivation has the form

$$S' \Rightarrow_{G'} \#S \Rightarrow^+_{G'} \#u' \Rightarrow_{G'} v' \Rightarrow^*_{G'} w'$$

where $u', v' \in (V \cup W)^*$, $w' \in T^*$. This claim shows that

$$\#u' \Rightarrow_{G'} v' \Rightarrow^*_{G'} w'$$

implies that $u' \in V$ and $v' = w'$. Consider

$$\#u' \Rightarrow_{G'} v' \Rightarrow^*_{G'} w'$$

where $u', v' \in (V \cup W)^*$, $w' \in T^*$. Assume that u' contains a nonterminal $[A, B, C] \in W$. There are two rules rewriting $[A, B, C]$:

$$p_1 = ([A, B, C] \to [A, B, C], 0, \langle A, B, C, m + 2\rangle)$$

and

$$p_2 = ([A, B, C] \to C, \langle A, B, C, m + 2\rangle, 0)$$

Because of its permitting condition, p_2 cannot be applied during $\#u' \Rightarrow_{G'} v'$. If $[A, B, C]$ is rewritten by p_1—that is, $[A, B, C] \in \text{alph}(v')$—$[A, B, C]$ necessarily occurs in all sentential forms derived from v'. Thus, no u' containing a nonterminal from W results in a terminal string; hence, $u' \in V^*$. By analogical considerations, establish that also $v' \in V^*$. Next, assume that v' contains some $A \in N_{CF}$ or $B \in N_{CS}$. The first one can be rewritten by $(A \to z, \#, 0)$, $z \in V^*$, and the second one by $(B \to [A, B, C], \langle A, B, C\rangle, 0)$, $[A, B, C] \in W$, $\langle A, B, C\rangle \in M$. In both cases, the permitting condition forbids an application of the rule. Consequently, $v' \in T^*$. It is sufficient to show that $v' = w'$. Indeed, every rule rewriting a terminal is of the form $(a \to a, 0, 0)$, $a \in T$. □

Claim 3. Let $S' \Rightarrow^n_{G'} Zx'$, $Z \in M$, $x' \in (V \cup W)^*$, $n \geq 1$. *Then,* Zx' *has one of the following forms*

(I) $Z = \#$, $x' \in V^*$;
(II) $Z = \langle A, B, C\rangle$, $x' \in V^*$, *for some* $A, C \in N_{CF}$, $B \in N_{CS}$;
(III) $Z = \langle A, B, C, i\rangle$, $x' \in (V \cup \{[A, B, C]\})^*$, $1 \leq i \leq m + 1$, *and* $\{f^{-1}(j)[A, B, C] \mid 1 \leq j < i,\ j \neq f(A)\} \cap \text{sub}(x') = \emptyset$ *for some* $A, C \in N_{CF}$, $B \in N_{CS}$;
(IV) $Z = \langle A, B, C, m + 2\rangle$, $x' \in (V \cup \{[A, B, C]\})^*$, $\{X[A, B, C] \mid X \in V,\ X \neq A\} \cap \text{sub}(x') = \emptyset$, *and* $[A, B, C]^2 \notin \text{sub}(x')$ *for some* $A, C \in N_{CF}$, $B \in N_{CS}$.

Proof. This claim is proved by induction on $n \geq 1$.

Basis. Let $n = 1$. Then, $S' \Rightarrow_{G'} \#S$, where $\#S$ is of type (I).

Induction Hypothesis. Suppose that the claim holds for all derivations of length n or less, for some $n \geq 1$.

Induction Step. Consider any derivation of the form

$$S' \Rightarrow^{n+1}_{G'} Qx'$$

where $Q \in M$, $x' \in (V \cup W)^*$. Since $n + 1 \geq 2$, by Claim 1, there exists $Zy' \in M(V \cup W)^*$ and a sequence of rules $p_0, p_1, \ldots, p_q$, where $p_i \in P'$, $0 \leq i \leq q$, $q = |y'|$, such that

$$S' \Rightarrow^n_{G'} Zy' \Rightarrow_{G'} Qx'\ [p_0, p_1, \ldots, p_q]$$

Let $y' = a_1 a_2 \cdots a_q$, where $a_i \in (V \cup W)$ for all $i = 1, \ldots, q$. By the induction hypothesis, the following cases (i) through (iv) cover all possible forms of Zy'.

(i) Let $Z = \#$ and $y' \in V^*$ [form (I)]. According to the definition of P', p_0 is either $(\# \to \langle A, B, C\rangle, 0, 0)$, $A, C \in N_{CF}$, $B \in N_{CS}$, or $(\# \to \#, 0, 0)$, or $(\# \to \varepsilon, 0, 0)$, and every p_i is either of the form $(a_i \to z, \#, 0)$, $z \in \{\varepsilon\} \cup N_{CS} \cup T \cup N^2_{CF}$, or $(a_i \to a_i, 0, 0)$. Obviously, y' is always rewritten to a string $x' \in V^*$. If # is rewritten to $\langle A, B, C\rangle$, we get $\langle A, B, C\rangle x'$ that is of form (II). If # remains unchanged, $\#x'$ is of type (I). In case that # is erased, the resulting sentential form does not belong to $M(V \cup W)^*$ required by this claim [which also holds for all strings derived from x' (see Claim 1)].

(ii) Let $Z = \langle A, B, C\rangle$, $y' \in V^*$, for some $A, C \in N_{CF}$, $B \in N_{CS}$ [form (II)]. In this case, $p_0 = (\langle A, B, C\rangle \rightarrow \langle A, B, C, 1\rangle, 0, 0)$ and every p_i is either $(a_i \rightarrow [A, B, C], \langle A, B, C\rangle, 0)$ or $(a_i \rightarrow a_i, 0, 0)$ (see the definition of P'). It is easy to see that $\langle A, B, C, 1\rangle x'$ belongs to (III).

(iii) Let $Z = \langle A, B, C, j\rangle$, $y' \in (V \cup \{[A, B, C]\})^*$, and y' satisfies

$$\{f^{-1}(k)[A, B, C] \mid 1 \leq k < j,\ k \neq f(A)\} \cap \text{sub}(y') = \emptyset$$

$1 \leq j \leq m + 1$, for some $A, C \in N_{CF}$, $B \in N_{CS}$ [form (III)]. The only rules rewriting symbols from y' are $(a_i \rightarrow a_i, 0, 0)$, $a_i \in V$, and $([A, B, C] \rightarrow [A, B, C], 0, \langle A, B, C, m+2\rangle)$; thus, y' is rewritten to itself. By the inspection of P', p_0 can be of the following three forms.

(a) If $j \neq f(A)$ and $j < m + 1$,

$$p_0 = \left(\langle A, B, C, j\rangle \rightarrow \langle A, B, C, j + 1\rangle, 0, f^{-1}(j)[A, B, C]\right)$$

Clearly, p_0 can be used only when $f^{-1}(j)[A, B, C] \notin \text{sub}(Zy')$. As

$$\{f^{-1}(k)[A, B, C] \mid 1 \leq k < j,\ k \neq f(A)\} \cap \text{sub}(y') = \emptyset$$

it also

$$\{f^{-1}(k)[A, B, C] \mid 1 \leq k \leq j,\ k \neq f(A)\} \cap \text{sub}(y') = \emptyset$$

Since $\langle A, B, C, j\rangle$ is rewritten to $\langle A, B, C, j + 1\rangle$ and y' is unchanged, we get $\langle A, B, C, j + 1\rangle y'$ with

$$\{f^{-1}(k)[A, B, C] \mid 1 \leq k < j + 1,\ k \neq f(A)\} \cap \text{sub}(y') = \emptyset$$

which is of form (III).

(b) If $j = f(A)$,

$$p_0 = \left(\langle A, B, C, f(A)\rangle \rightarrow \langle A, B, C, f(A) + 1\rangle, 0, 0\right)$$

As before, $Qx' = \langle A, B, C, j + 1\rangle y'$. Moreover, because

$$\{f^{-1}(k)[A, B, C] \mid 1 \leq k < j,\ k \neq f(A)\} \cap \text{sub}(y') = \emptyset$$

and $j = f(A)$,

$$\{f^{-1}(k)[A, B, C] \mid 1 \leq k < j + 1,\ k \neq f(A)\} \cap \text{sub}(x') = \emptyset$$

Consequently, Qx' belongs to (III) as well.

(c) If $j = m + 1$,

$$p_0 = \big(\langle A, B, C, m + 1\rangle \to \langle A, B, C, m + 2\rangle, 0, [A, B, C]^2\big)$$

Then, $Qx' = \langle A, B, C, m + 2\rangle y'$. The application of p_0 implies that $[A, B, C]^2 \notin \mathrm{sub}(x')$. In addition, observe that for $j = m + 1$,

$$\begin{aligned}&\{f^{-1}(k)[A, B, C] \mid 1 \leq k < j,\ k \neq f(A)\}\\&= \{X[A, B, C] \mid X \in V,\ X \neq A\}\end{aligned}$$

Hence,

$$\{X[A, B, C] \mid X \in V,\ X \neq A\} \cap \mathrm{sub}(x') = \emptyset$$

As a result, Qx' is of form (IV).

(iv) Let $Z = \langle A, B, C, m + 2\rangle$, $y' \in (V \cup \{[A, B, C]\})^*$, $[A, B, C]^2 \notin \mathrm{sub}(y')$, and

$$\{X[A, B, C] \mid X \in V,\ X \neq A\} \cap \mathrm{sub}(y') = \emptyset$$

for some $A, C \in N_{CF}$, $B \in N_{CS}$ [form (IV)]. Inspect P' to see that

$$p_0 = \big(\langle A, B, C, m + 2\rangle \to \#, 0, \langle A, B, C, m + 2\rangle[A, B, C]\big)$$

and p_i is either

$$(a_i \to a_i, 0, 0),\ a_i \in V$$

or

$$\big([A, B, C] \to C, \langle A, B, C, m + 2\rangle, 0\big)$$

where $1 \leq i \leq q$. According to the right-hand sides of these rules, $Qx' \in \{\#\}V^*$; that is, Qx' belongs to (I).

In cases (i) through (iv), we have demonstrated that every sentential form obtained in $n + 1$ derivation steps satisfies the statement of this claim. Therefore, we have finished the induction step and established Claim 3. □

To prove the following claims, define the finite substitution g from V^* into $(V \cup W)^*$ as

$$g(X) = \{X\} \cup \{[A, B, C] \in W \mid A, C \in N_{CF},\ B \in N_{CS}\}$$

for all $X \in V$. Let g^{-1} be the inverse of g.

Claim 4. *Let* $y' = a_1 a_2 \cdots a_q$, $a_i \in (V \cup W)^*$, $q = |y'|$, *and* $g^{-1}(a_i) \Rightarrow_G^{h_i} g^{-1}(x'_i)$ *for all* $i \in \{1, \ldots, q\}$ *and some* $h_i \in \{0, 1\}$, $x'_i \in (V \cup W)^*$. *Then,* $g^{-1}(y') \Rightarrow_G^{h} g^{-1}(x')$ *such that* $x' = x'_1 x'_2 \cdots x'_q$, $h = \sum_{i=1}^{q} h_i$, $h \leq q$.

Proof. Consider any derivation of the form

$$g^{-1}(X) \Rightarrow_G^{l} g^{-1}(u)$$

$X \in (V \cup W)$, $u \in (V \cup W)^*$, $l \in \{0, 1\}$. If $l = 0$, $g^{-1}(X) = g^{-1}(u)$. Let $l = 1$. Then, there surely exists a rule $p = g^{-1}(X) \rightarrow g^{-1}(u) \in P$ such that

$$g^{-1}(X) \Rightarrow_G g^{-1}(u) \; [p]$$

Return to the statement of this claim. We can construct this derivation

$$\begin{aligned} g^{-1}(a_1)g^{-1}(a_2)\cdots g^{-1}(a_q) &\Rightarrow_G^{h_1} g^{-1}(x'_1)g^{-1}(a_2)\cdots g^{-1}(a_q) \\ &\Rightarrow_G^{h_2} g^{-1}(x'_1)g^{-1}(x'_2)\cdots g^{-1}(a_q) \\ &\;\vdots \\ &\Rightarrow_G^{h_q} g^{-1}(x'_1)g^{-1}(x'_2)\cdots g^{-1}(x'_q) \end{aligned}$$

where

$$g^{-1}(y') = g^{-1}(a_1)\cdots g^{-1}(a_q)$$

and

$$g^{-1}(x'_1)\cdots g^{-1}(x'_q) = g^{-1}(x'_1 \cdots x'_q) = g^{-1}(x')$$

In such a derivation, each $g^{-1}(a_i)$ is either left unchanged (if $h_i = 0$) or rewritten to $g^{-1}(x'_i)$ by the corresponding rule $g^{-1}(a_i) \rightarrow g^{-1}(x'_i)$. Obviously, the length of this derivation is $\sum_{i=1}^{q} h_i$. □

Claim 5. $S \Rightarrow_G^* x$ *if and only if* $S' \Rightarrow_{G'}^+ Qx'$, *where* $g^{-1}(x') = x$, $Q \in M$, $x \in V^*$, $x' \in (V \cup W)^*$.

Proof. The proof is divided into the only-if part and the if part.

Only If. By induction on $n \geq 0$, we show that

$$S \Rightarrow_G^n x \quad \text{implies} \quad S' \Rightarrow_{G'}^+ \#x$$

where $x \in V^*$, $n \geq 0$. Clearly, $g^{-1}(x) = x$.

Basis. Let $n = 0$. Then, $S \Rightarrow_G^0 S$. In G', $S' \Rightarrow_{G'} \#S$ by using $(S' \rightarrow \#S, 0, 0)$.

Induction Hypothesis. Assume that the claim holds for all derivations of length n or less, for some $n \geq 0$.

Induction Step. Consider any derivation of the form

$$S \Rightarrow_G^{n+1} x$$

As $n + 1 \geq 1$, there exists $y \in V^*$ and $p \in P$ such that

$$S \Rightarrow_G^n y \Rightarrow_G x \; [p]$$

Let $y = a_1 a_2 \cdots a_q$, $a_i \in V$ for all $1 \leq i \leq q$, where $q = |y|$. By the induction hypothesis,

$$S' \Rightarrow_{G'}^+ \#y$$

The following cases investigate all possible forms of p.

(i) Let $p = A \to z$, $A \in N_{CF}$, $z \in \{\varepsilon\} \cup N_{CS} \cup T \cup N_{CF}^2$. Then, $y = y_1 A y_3$ and $x = y_1 z y_3$, $y_1, y_3 \in V^*$. Let $l = |y_1| + 1$. In this case, we can construct

$$S' \Rightarrow_{G'}^+ \#y \Rightarrow_{G'} \#x \; [p_0, p_1, \ldots, p_q]$$

such that $p_0 = (\# \to \#, 0, 0)$, $p_l = (A \to z, \#, 0)$, and $p_i = (a_i \to a_i, 0, 0)$ for all $1 \leq i \leq q, i \neq l$.

(ii) Let $p = AB \to AC$, $A, C \in N_{CF}$, $B \in N_{CS}$. Then, $y = y_1 A B y_3$ and $x = y_1 A C y_3$, $y_1, y_3 \in V^*$. Let $l = |y_1| + 2$. At this point, there exists the following derivation

$$\begin{aligned}
S' &\Rightarrow_{G'}^+ \#y_1 A B y_3 \\
&\Rightarrow_{G'} \langle A, B, C\rangle y_1 A B y_3 \\
&\Rightarrow_{G'} \langle A, B, C, 1\rangle y_1 A[A, B, C] y_3 \\
&\Rightarrow_{G'} \langle A, B, C, 2\rangle y_1 A[A, B, C] y_3 \\
&\;\vdots \\
&\Rightarrow_{G'} \langle A, B, C, f(A)\rangle y_1 A[A, B, C] y_3 \\
&\Rightarrow_{G'} \langle A, B, C, f(A)+1\rangle y_1 A[A, B, C] y_3 \\
&\;\vdots \\
&\Rightarrow_{G'} \langle A, B, C, m+1\rangle y_1 A[A, B, C] y_3 \\
&\Rightarrow_{G'} \langle A, B, C, m+2\rangle y_1 A[A, B, C] y_3 \\
&\Rightarrow_{G'} \#y_1 A C y_3
\end{aligned}$$

If. The if part establishes that

$$S' \Rightarrow_{G'}^n Qx' \quad \text{implies} \quad S \Rightarrow_{G'}^* x$$

where $g^{-1}(x') = x$, $Q \in M$, $x' \in (V \cup W)^*$, $x \in V^*$, $n \geq 1$. This claim is proved by induction on $n \geq 0$.

Basis. Assume that $n = 1$. Since the only rule that can rewrite S' is $(S' \rightarrow \#S, 0, 0)$, $S' \Rightarrow_{G'} \#S$. Clearly, $S \Rightarrow_G^0 S$ and $g^{-1}(S) = S$.

Induction Hypothesis. Suppose that the claim holds for any derivation of length n or less, for some $n \geq 1$.

Induction Step. Consider any derivation of the form

$$S' \Rightarrow_{G'}^{n+1} Qx'$$

where $Qx' \in M(V \cup W)^*$. Since $n + 1 \geq 2$, by Claim 1, there exists a derivation

$$S' \Rightarrow_{G'}^{+} Zy' \Rightarrow_{G'} Qx' \; [p_0, p_1, \ldots, p_q]$$

where $Zy' \in M(V \cup W)^*$, and $p_i \in P'$ for all $i \in \{0, 1, \ldots, q\}$, $q = |y'|$. By the induction hypothesis, there is also a derivation

$$S \Rightarrow_{G'}^{*} y$$

where $y \in V^*$, $g^{-1}(y') = y$. Let $y' = a_1 a_2 \cdots a_q$. Claim 3 says that Zy' has one of the following forms.

(i) Let $Z = \#$ and $y' \in V^*$. Then, there are the following two forms of rules rewriting a_i in y'.

(i.a) Let $(a_i \rightarrow a_i, 0, 0)$, $a_i \in V$. In this case,

$$g^{-1}(a_i) \Rightarrow_G^0 g^{-1}(a_i)$$

(i.b) Let $(a_i \rightarrow x_i, \#, 0)$, $x_i \in \{\varepsilon\} \cup N_{CS} \cup T \cup N_{CF}^2$. Since $a_i = g^{-1}(a_i)$, $x_i = g^{-1}(x_i)$ and $a_i \rightarrow x_i \in P$,

$$g^{-1}(a_i) \Rightarrow_G g^{-1}(x_i) \; [a_i \rightarrow x_i]$$

We see that for all a_i, there exists a derivation

$$g^{-1}(a_i) \Rightarrow_G^{h_i} g^{-1}(x_i)$$

for some $h_i \in \{0, 1\}$, where $x_i \in V^*$, $x' = x_1 x_2 \cdots x_q$. Therefore, by Claim 4, we can construct

$$S' \Rightarrow_G^{*} y \Rightarrow_G^{h} x$$

where $0 \leq h \leq q$, $x = g^{-1}(x')$.

(ii) Let $Z = \langle A, B, C \rangle$, $y' \in V^*$, for some $A, C \in N_{CF}$, $B \in N_{CS}$. At this point, the following two forms of rules can be used to rewrite a_i in y'.

(ii.a) Let $(a_i \to a_i, 0, 0)$, $a_i \in V$. As in case (i.a),

$$g^{-1}(a_i) \Rightarrow_G^0 g^{-1}(a_i)$$

(ii.b) Let $(a_i \to [A, B, C], \langle A, B, C \rangle, 0)$, $a_i = B$. Since $g^{-1}([A, B, C]) = g^{-1}(B)$, we have

$$g^{-1}(a_i) \Rightarrow_G^0 g^{-1}([A, B, C])$$

Thus, there exists the derivation

$$S \Rightarrow_G^* y \Rightarrow_G^0 x, \; x = g^{-1}(x')$$

(iii) Let $Z = \langle A, B, C, j \rangle$, $y' \in (V \cup \{[A, B, C]\})^*$, and

$$\{f^{-1}(k)[A, B, C] \mid 1 \le k < j, \; k \ne f(A)\} \cap \mathrm{sub}(y') = \emptyset$$

$1 \le j \le m + 1$, for some $A, C \in N_{CF}$, $B \in N_{CS}$. Then, the only rules rewriting symbols from y' are

$$(a_i \to a_i, 0, 0), \; a_i \in V$$

and

$$([A, B, C] \to [A, B, C], 0, \langle A, B, C, m + 2 \rangle)$$

Hence, $x' = y'$. Since we have

$$S \Rightarrow_G^* y, \; g^{-1}(y') = y$$

it also holds that $g^{-1}(x') = y$.

(iv) Let $Z = \langle A, B, C, m + 2 \rangle$, $y' \in (V \cup \{[A, B, C]\})^*$, $[A, B, C]^2 \notin \mathrm{sub}(y')$,

$$\{X[A, B, C] \mid X \in V, \; X \ne A\} \cap \mathrm{sub}(y') = \emptyset$$

for some $A, C \in N_{CF}$, $B \in N_{CS}$. G' rewrites $\langle A, B, C, m + 2 \rangle$ by using

$$(\langle A, B, C, m + 2 \rangle \to \#, 0, \langle A, B, C, m + 2 \rangle[A, B, C])$$

which forbids $\langle A, B, C, m + 2 \rangle[A, B, C]$ as a substring of Zy'. As a result, the left neighbor of every occurrence of $[A, B, C]$ in $\langle A, B, C, m + 2 \rangle y'$ is A.

Inspect P' to see that a_i can be rewritten either by $(a_i \to a_i, 0, 0)$, $a_i \in V$, or by $([A, B, C] \to C, \langle A, B, C, m+2 \rangle, 0)$. Therefore, we can express

$$\begin{aligned} y' &= y_1 A[A,B,C] y_2 A[A,B,C] y_3 \cdots y_l A[A,B,C] y_{l+1} \\ y &= y_1 AB y_2 AB y_3 \cdots y_l AB y_{l+1} \\ x' &= y_1 AC y_2 AC y_3 \cdots y_l AC y_{l+1} \end{aligned}$$

where $l \geq 0$, $y_k \in V^*$, $1 \leq k \leq l+1$. Since we have $p = AB \to AC \in P$, there is a derivation

$$\begin{aligned} S &\Rightarrow_G^* y_1 AB y_2 AB y_3 \cdots y_l AB y_{l+1} \\ &\Rightarrow_G y_1 AC y_2 AB y_3 \cdots y_l AB y_{l+1} \; [p] \\ &\Rightarrow_G y_1 AC y_2 AC y_3 \cdots y_l AB y_{l+1} \; [p] \\ &\;\;\vdots \\ &\Rightarrow_G y_1 AC y_2 AC y_3 \cdots y_l AC y_{l+1} \; [p] \end{aligned}$$

Since cases (i) through (iv) cover all possible forms of y', we have completed the induction and established Claim 5. □

Let us finish the proof of Lemma 10.3.2. Consider any derivation of the form

$$S \Rightarrow_G^* w, \; w \in T^*$$

From Claim 5, it follows that

$$S' \Rightarrow_{G'}^+ \#w$$

because $g(a) = \{a\}$ for every $a \in T$. Then, as shown in Claim 2,

$$S' \Rightarrow_{G'}^+ \#w \Rightarrow_{G'} w$$

and hence,

$$S \Rightarrow_G^* w \quad \text{implies} \quad S' \Rightarrow_{G'}^+ w$$

for all $w \in T^*$. To prove the converse implication, consider a successful derivation of the form

$$S' \Rightarrow_{G'}^+ \#u \Rightarrow_{G'} w \Rightarrow_{G'}^* w$$

$u \in V^*$, $w \in T^*$ (see Claim 2). Observe that by the definition of P', for every

$$S' \Rightarrow_{G'}^+ \#u \Rightarrow_{G'} w$$

there also exists a derivation

$$S' \Rightarrow_{G'}^{+} \#u \Rightarrow_{G'}^{*} \#w \Rightarrow_{G'} w$$

Then, according to Claim 5, $S \Rightarrow_G^* w$. Consequently, we get for every $w \in T^*$,

$$S \Rightarrow_G^* w \quad \text{if and only if} \quad S' \Rightarrow_{G'}^* w$$

Therefore, $L(G) = L(G')$. □

Lemma 10.3.3. $\mathbf{SSC\text{-}ET0L}(r, s) \subseteq \mathbf{RE}$ *for any* $r, s \geq 0$.

Proof. By Lemma 10.1.7, $\mathbf{C\text{-}ET0L} \subseteq \mathbf{RE}$. Since $\mathbf{SSC\text{-}ET0L}(r, s) \subseteq \mathbf{C\text{-}ET0L}$ for all $r, s \geq 0$ (see Definition 10.3.1), $\mathbf{SSC\text{-}ET0L}(r, s) \subseteq \mathbf{RE}$ for all $r, s \geq 0$ as well. □

Inclusions established in Lemmas 10.3.2 and 10.3.3 imply the following theorem.

Theorem 10.3.4.

$$\mathbf{SSC\text{-}E0L}(1, 2) = \mathbf{SSC\text{-}ET0L}(1, 2) = \mathbf{SSC\text{-}E0L} = \mathbf{SSC\text{-}ET0L} = \mathbf{RE}$$

Proof. From Lemmas 10.3.2 and 10.3.3, we have that $\mathbf{RE} \subseteq \mathbf{SSC\text{-}E0L}(1, 2)$ and $\mathbf{SSC\text{-}ET0L}(r, s) \subseteq \mathbf{RE}$ for any $r, s \geq 0$. By the definitions, $\mathbf{SSC\text{-}E0L}(1, 2) \subseteq \mathbf{SSC\text{-}ET0L}(1, 2) \subseteq \mathbf{SSC\text{-}ET0L}$ and $\mathbf{SSC\text{-}E0L}(1, 2) \subseteq \mathbf{SSC\text{-}E0L} \subseteq \mathbf{SSC\text{-}ET0L}$. Hence, $\mathbf{SSC\text{-}E0L}(1, 2) = \mathbf{SSC\text{-}ET0L}(1, 2) = \mathbf{SSC\text{-}E0L} = \mathbf{SSC\text{-}ET0L} = \mathbf{RE}$. □

Next, let us investigate the generative power of propagating SSC-ET0L grammars.

Lemma 10.3.5. $\mathbf{CS} \subseteq \mathbf{SSC\text{-}EP0L}(1, 2)$

Proof. We can base this proof on the same technique as in Lemma 10.3.2. However, we have to make sure that the construction produces no erasing rules. This requires some modifications of the original algorithm; in particular, we have to eliminate the rule $(\# \rightarrow \varepsilon, 0, 0)$.

Let L be a context-sensitive language generated by a context-sensitive grammar

$$G = \big(V, T, P, S\big)$$

of the normal form of Theorem 4.1.6, where

$$V = N_{CF} \cup N_{CS} \cup T$$

Let m be the cardinality of V. Define a bijection f from V to $\{1, \dots, m\}$. Let f^{-1} be the inverse of f. Set

$$
\begin{aligned}
M &= \{\langle \# \mid X\rangle \mid X \in V\} \cup \\
&\quad \{\langle A, B, C \mid X\rangle \mid AB \to AC \in P,\ X \in V\} \cup \\
&\quad \{\langle A, B, C, i \mid X\rangle \mid AB \to AC \in P,\ 1 \le i \le m+2,\ X \in V\} \\
W &= \{[A, B, C, X] \mid AB \to AC \in P,\ X \in V\}, \text{ and} \\
V' &= V \cup M \cup W
\end{aligned}
$$

where V, M, and W are pairwise disjoint. Then, construct the SSC-EP0L grammar of degree $(1, 2)$,

$$G' = \big(V', T, P', \langle \# \mid S\rangle\big)$$

with the set of rules P' constructed by performing (1) through (4), given next.

(1) For all $A \to x \in P$, $A \in N_{CF}$, $x \in T \cup N_{CS} \cup N_{CF}^2$,

(1a) for all $X \in V$, add $(A \to x, \langle \# \mid X\rangle, 0)$ to P';
(1b) if $x \in T \cup N_{CS}$, add $(\langle \# \mid A\rangle \to \langle \# \mid x\rangle, 0, 0)$ to P';
(1c) if $x = YZ$, $YZ \in N_{CF}^2$, add $(\langle \# \mid A\rangle \to \langle \# \mid Y\rangle Z, 0, 0)$ to P'.

(2) For all $X \in V$ and for every $AB \to AC \in P$, $A, C \in N_{CF}$, $B \in N_{CS}$, extend P' by adding

(2a) $(\langle \# \mid X\rangle \to \langle A, B, C \mid X\rangle, 0, 0)$;
(2b) $(B \to [A, B, C, X], \langle A, B, C \mid X\rangle, 0)$;
(2c) $(\langle A, B, C \mid X\rangle \to \langle A, B, C, 1 \mid X\rangle, 0, 0)$;
(2d) $([A, B, C, X] \to [A, B, C, X], 0, \langle A, B, C, m+2\rangle X)$;
(2e) $(\langle A, B, C, i \mid X\rangle \to \langle A, B, C, i+1 \mid X\rangle, 0, f^{-1}(i)[A, B, C, X])$ for all $1 \le i \le m$, $i \ne f(A)$;
(2f) $(\langle A, B, C, f(A) \mid X\rangle \to \langle A, B, C, f(A)+1 \mid X\rangle, 0, 0)$;
(2g) $(\langle A, B, C, m+1 \mid X\rangle \to \langle A, B, C, m+2 \mid X\rangle, 0, [A, B, C, X]^2)$;
(2h) $(\langle A, B, C, m+2 \mid X\rangle \to \langle \# \mid X\rangle, 0, 0)$ for $X = A$, $(\langle A, B, C, m+2 \mid X\rangle \to \langle \# \mid X\rangle, 0, \langle A, B, C, m+2 \mid X\rangle[A, B, C, X])$ otherwise;
(2i) $([A, B, C, X] \to C, \langle A, B, C, m+2 \mid X\rangle, 0)$.

(3) For all $X \in V$, add $(X \to X, 0, 0)$ to P'.
(4) For all $X \in V$, add $(\langle \# \mid X\rangle \to \langle \# \mid X\rangle, 0, 0)$ and $(\langle \# \mid X\rangle \to X, 0, 0)$ to P'.

Consider the construction above and the construction used in the proof of Lemma 10.3.2. Observe that the present construction does not attach the master as an extra symbol before sentential forms. Instead, the master is incorporated with its right neighbor into one composite symbol. For example, if G generates $AabCadd$, the corresponding sentential form in G' is $\langle \# \mid A\rangle abCadd$, where $\langle \# \mid A\rangle$ is one symbol. At this point, we need no rule erasing #; the master is simply rewritten to the symbol with which it is incorporated [see rules of (4)]. In addition, this modification involves some changes to the algorithm: First, G' can rewrite symbols incorporated with the master [see rules of (1b) and (1c)]. Second, conditions of the rules depending on the master refer to the composite symbols. Finally, G' can make context-sensitive rewriting of the composite master's right neighbor [see rules of (2h)]. For instance, if

$$ABadC \Rightarrow_G ACadC\ [AB \rightarrow AC]$$

in G, G' derives

$$\langle \# \mid A \rangle BadC \Rightarrow_{G'}^{+} \langle \# \mid A \rangle CadC$$

Based on the observations above, the reader can surely establish $L(G) = L(G')$ by analogy with the proof of Lemma 10.3.2. Thus, the fully rigorous version of this proof is omitted. □

Lemma 10.3.6. $\mathbf{SSC\text{-}EPT0L}(r, s) \subseteq \mathbf{CS}$, *for all* $r, s \geq 0$.

Proof. By Lemma 10.1.5, $\mathbf{C\text{-}EPT0L}(r, s) \subseteq \mathbf{CS}$, for any $r \geq 0$, $s \geq 0$. Since every SSC-EPT0L grammar is a special case of a C-EPT0L grammar (see Definition 10.3.1), we obtain $\mathbf{SSC\text{-}EPT0L}(r, s) \subseteq \mathbf{CS}$, for all $r, s \geq 0$. □

Theorem 10.3.7.

$$\mathbf{CS} = \mathbf{SSC\text{-}EP0L}(1,2) = \mathbf{SSC\text{-}EPT0L}(1,2) = \mathbf{SSC\text{-}EP0L} = \mathbf{SSC\text{-}EPT0L}$$

Proof. By Lemma 10.3.5, we have $\mathbf{CS} \subseteq \mathbf{SSC\text{-}EP0L}(1,2)$. Lemma 10.3.6 says that $\mathbf{SSC\text{-}EPT0L}(r, s) \subseteq \mathbf{CS}$ for all $r, s \geq 0$. From the definitions, it follows that $\mathbf{SSC\text{-}EP0L}(1,2) \subseteq \mathbf{SSC\text{-}EPT0L}(1,2) \subseteq \mathbf{SSC\text{-}EPT0L}$ and $\mathbf{SSC\text{-}EP0L}(1,2) \subseteq \mathbf{SSC\text{-}EP0L} \subseteq \mathbf{SSC\text{-}EPT0L}$. Hence, we have the identity $\mathbf{SSC\text{-}EP0L}(1,2) = \mathbf{SSC\text{-}EPT0L}(1,2) = \mathbf{SSC\text{-}EP0L} = \mathbf{SSC\text{-}EPT0L} = \mathbf{CS}$, so the theorem holds. □

The following corollary summarizes the established relations between the language families generated by SSC-ET0L grammars.

Corollary 10.3.8.

$$\mathbf{CF}$$
$$\subset$$
$$\mathbf{SSC\text{-}EP0L}(0,0) = \mathbf{SSC\text{-}E0L}(0,0) = \mathbf{EP0L} = \mathbf{E0L}$$
$$\subset$$
$$\mathbf{SSC\text{-}EPT0L}(0,0) = \mathbf{SSC\text{-}ET0L}(0,0) = \mathbf{EPT0L} = \mathbf{ET0L}$$
$$\subset$$
$$\mathbf{SSC\text{-}EP0L}(1,2) = \mathbf{SSC\text{-}EPT0L}(1,2) = \mathbf{SSC\text{-}EP0L} = \mathbf{SSC\text{-}EPT0L} = \mathbf{CS}$$
$$\subset$$
$$\mathbf{SSC\text{-}E0L}(1,2) = \mathbf{SSC\text{-}ET0L}(1,2) = \mathbf{SSC\text{-}E0L} = \mathbf{SSC\text{-}ET0L} = \mathbf{RE}$$

□

Open Problem 10.3.9. Notice that Corollary 10.3.8 does not include some related language families. For instance, it contains no language families generated by SSC-ET0L grammars with degrees $(1, 1)$, $(1, 0)$, and $(0, 1)$. What is their generative power? What is the generative power of SSC-ET0L grammars of degree $(2, 1)$? Are they as powerful as SSC-ET0L grammars of degree $(1, 2)$? □

10.4 Left Random Context ET0L Grammars

As their name indicates, *left random context ET0L grammars* (*LRC-ET0L grammars* for short) represent another variant of context-conditional ET0L grammars. In this variant, a set of *permitting symbols* and a set of *forbidding symbols* are attached to each of their rules, just like in random context grammars (see Sect. 4.3). A rule like this can rewrite a symbol if each of its permitting symbols occurs to the left of the rewritten symbol in the current sentential form while each of its forbidding symbols does not occur there. LRC-ET0L grammars represent the principal subject of this section.

In the present section, we demonstrate that LRC-ET0L grammars are computationally complete—that is, they characterize the family of recursively enumerable languages (see Theorem 10.4.8). In fact, we prove that the family of recursively enumerable languages is characterized even by LRC-ET0L grammars with a limited number of nonterminals (see Theorem 10.4.10). We also demonstrate how to characterize the family of context-sensitive languages by these grammars without erasing rules (see Theorem 10.4.7).

In addition, we study a variety of special cases of LRC-ET0L grammars. First, we introduce *left random context E0L grammars* (*LRC-E0L grammars* for short), which represent LRC-ET0L grammars with a single set of rules. We prove that the above characterizations hold in terms of LRC-E0L grammars as well. Second, we introduce *left permitting E0L grammars* (*LP-E0L grammars* for short), which represent LRC-E0L grammars where each rule has only a set of permitting symbols. Analogously, we define *left forbidding E0L grammars* (*LF-E0L grammars* for short) as LRC-E0L grammars where each rule has only a set of forbidding symbols. We demonstrate that LP-E0L grammars are more powerful than ordinary E0L grammars and that LF-E0L grammars are at least as powerful as ordinary ET0L grammars.

This section consists of two subsections. Section 10.4.1 defines LRC-ET0L grammars and their variants. Section 10.4.2 studies their generative power and reduction.

10.4.1 Definitions and Examples

In this section, we define LRC-ET0L grammars and their variants. In addition, we illustrate them by examples.

Definition 10.4.1. A *left random context ET0L grammar* (a *LRC-ET0L grammar* for short) is an $(n + 3)$-tuple

$$G = \left(V, T, P_1, P_2, \ldots, P_n, w\right)$$

where V, T, and w are defined as in an ET0L grammar, $N = V - T$ is the alphabet of *nonterminals*, and $P_i \subseteq V \times V^* \times 2^N \times 2^N$ is a finite relation, for all i, $1 \leq i \leq n$, for some $n \geq 1$. By analogy with phrase-structure grammars, elements of P_i are called *rules* and instead of $(X, y, U, W) \in P_i$, we write $(X \to y, U, W)$ throughout this section. The *direct derivation relation* over V^*, symbolically denoted by $\Rightarrow_G$, is defined as follows:

$$u \Rightarrow_G v$$

if and only if

- $u = X_1 X_2 \cdots X_k$,
- $v = y_1 y_2 \cdots y_k$,
- $(X_i \to y_i, U_i, W_i) \in P_h$,
- $U_i \subseteq \text{alph}(X_1 X_2 \cdots X_{i-1})$, and
- $\text{alph}(X_1 X_2 \cdots X_{i-1}) \cap W_i = \emptyset$,

for all i, $1 \leq i \leq k$, for some $k \geq 1$ and $h \leq n$. For $(X \to y, U, W) \in P_i$, U and W are called the *left permitting context* and the *left forbidding context*, respectively. Let $\Rightarrow_G^m$, $\Rightarrow_G^*$, and $\Rightarrow_G^+$ denote the mth power of $\Rightarrow_G$, for $m \geq 0$, the reflexive-transitive closure of $\Rightarrow_G$, and the transitive closure of $\Rightarrow_G$, respectively. The *language* of G is denoted by $L(G)$ and defined as

$$L(G) = \left\{x \in T^* \mid w \Rightarrow_G^* x\right\}$$

□

Definition 10.4.2. Let $G = (V, T, P_1, P_2, \ldots, P_n, w)$ be a LRC-ET0L grammar, for some $n \geq 1$. If every $(X \to y, U, W) \in P_i$ satisfies that $W = \emptyset$, for all i, $1 \leq i \leq n$, then G is a *left permitting ET0L grammar* (a *LP-ET0L grammar* for short). If every $(X \to y, U, W) \in P_i$ satisfies that $U = \emptyset$, for all i, $1 \leq i \leq n$, then G is a *left forbidding ET0L grammar* (a *LF-ET0L grammar* for short). □

By analogy with ET0L grammars (see their definition in Sect. 3.3), we define *LRC-EPT0L, LP-EPT0L, LF-EPT0L, LRC-E0L, LP-E0L, LF-E0L, LRC-EP0L, LP-EP0L*, and *LF-EP0L grammars*.

The language families that are generated by LRC-ET0L, LP-ET0L, LF-ET0L, LRC-EPT0L, LP-EPT0L and LF-EPT0L grammars are denoted by **LRC - ET0L**, **LP - ET0L**, **LF - ET0L**, **LRC - EPT0L**, **LP - EPT0L**, and **LF - EPT0L**, respectively. The language families generated by LRC-E0L, LP-E0L, LF-E0L, LRC-EP0L, LP-EP0L, and LF-EP0L grammars are denoted by **LRC - E0L**, **LP - E0L**, **LF - E0L**, **LRC - EP0L**, **LP - EP0L**, and **LF - EP0L**, respectively.

Next, we illustrate the above-introduced notions by two examples.

Example 10.4.3. Consider $K = \{a^m b^n a^m \mid 1 \leq m \leq n\}$. This language is generated by the LF-EP0L grammar

$$G = \big(\{A, B, B', \bar{a}, a, b\}, \{a, b\}, P, ABA\big)$$

with P containing the following nine rules

$$\begin{array}{lll} (A \to aA, \emptyset, \{\bar{a}\}) & (a \to a, \emptyset, \emptyset) & (B' \to bB', \emptyset, \{A\}) \\ (A \to \bar{a}, \emptyset, \{\bar{a}\}) & (B \to bB, \emptyset, \emptyset) & (B' \to b, \emptyset, \{A\}) \\ (\bar{a} \to a, \emptyset, \{A\}) & (B \to B', \emptyset, \emptyset) & (b \to b, \emptyset, \emptyset) \end{array}$$

To rewrite A to a string not containing A, $(A \to \bar{a}, \emptyset, \{\bar{a}\})$ has to be used. Since the only rule which can rewrite $\bar{a}$ is $(\bar{a} \to a, \emptyset, \{A\})$, and the rules that can rewrite A have $\bar{a}$ in their forbidding contexts, it is guaranteed that both As are rewritten to $\bar{a}$ simultaneously; otherwise, the derivation is blocked. The rules $(B' \to bB', \emptyset, \{A\})$ and $(B' \to b, \emptyset, \{A\})$ are applicable only if there is no A to the left of B'. Therefore, after these rules are applied, no more as can be generated. Consequently, we see that for every string from $L(G)$, there exists a derivation of the form

$$\begin{aligned} ABA &\Rightarrow_G^* a^{m-1}Ab^{m-1}Ba^{m-1}A \\ &\Rightarrow_G a^{m-1}\bar{a}b^{m-1}B'a^{m-1}\bar{a} \\ &\Rightarrow_G^+ a^m b^n a^m \end{aligned}$$

with $1 \leq m \leq n$. Hence, $L(G) = K$. □

Recall that $K \notin$ **E0L** (see p. 268 in [14]); however, $K \in$ **LF-EP0L**. As a result, LF-EP0L grammars are more powerful than ordinary E0L grammars. This is in contrast to left forbidding context-free grammars (see Sect. 6.2.2), which characterize only the family of context-free languages (see Theorem 6.2.13).

The next example shows how to generate K by an LP-EP0L grammar, which implies that LP-E0L grammars have greater expressive power than E0L grammars.

Example 10.4.4. Consider the LRC-EP0L grammar

$$H = \big(\{S, A, A', B, B', \bar{a}, a, b\}, \{a, b\}, P, S\big)$$

with P containing the following fourteen rules

$$\begin{array}{lll} (S \to ABA', \emptyset, \emptyset) & (A' \to aA', \{A\}, \emptyset) & (B \to bB, \emptyset, \emptyset) \\ (S \to \bar{a}B'\bar{a}, \emptyset, \emptyset) & (A' \to \bar{a}, \{\bar{a}\}, \emptyset) & (B \to bB', \emptyset, \emptyset) \\ (A \to aA, \emptyset, \emptyset) & (\bar{a} \to \bar{a}, \emptyset, \emptyset) & (B' \to bB', \{\bar{a}\}, \emptyset) \\ (A \to a\bar{a}, \emptyset, \emptyset) & (\bar{a} \to a, \emptyset, \emptyset) & (B' \to b, \emptyset, \{\bar{a}\}, \emptyset) \\ & (a \to a, \emptyset, \emptyset) & (b \to b, \emptyset, \emptyset) \end{array}$$

If the first applied rule is $(S \to \bar{a}B'\bar{a}, \emptyset, \emptyset)$, then the generated string of terminals clearly belongs to K from Example 10.4.3. By using this rule, we can obtain a string with only two as, which is impossible if $(S \to ABA', \emptyset, \emptyset)$ is used instead. Therefore, we assume that $(S \to ABA', \emptyset, \emptyset)$ is applied as the first rule. Observe that $(A' \to aA', \{A\}, \emptyset)$ can be used only when there is A present to the left of A' in the current sentential form. Also, $(A' \to a, \{\bar{a}\}, \emptyset)$ can be applied only after $(A \to a\bar{a}, \emptyset, \emptyset)$ is used. Finally, note that $(B' \to bB', \{\bar{a}\}, \emptyset)$ and $(B' \to b, \emptyset, \{\bar{a}\}, \emptyset)$ can be applied only if there is $\bar{a}$ to the left of B'. Therefore, after these rules are used, no more as can be generated. Consequently, we see that for every string from $L(G)$ with more than two as, there exists a derivation of the form

$$\begin{aligned} S &\Rightarrow_H ABA' \\ &\Rightarrow_H^* a^{m-2}Ab^{m-2}Ba^{m-2}A' \\ &\Rightarrow_H a^{m-1}\bar{a}b^{m-1}B'a^{m-1}A' \\ &\Rightarrow_H^+ a^{m-1}\bar{a}b^{n-1}B'a^{m-1}\bar{a} \\ &\Rightarrow_H a^m b^n a^m \end{aligned}$$

with $2 \leq m \leq n$. Hence, $L(H) = K$. □

10.4.2 Generative Power and Reduction

In this section, we establish the generative power of LRC-ET0L grammars and their special variants. More specifically, we prove that **LRC - EPT0L** $=$ **LRC - EP0L** $=$ **CS** (Theorem 10.4.7), **LRC - ET0L** $=$ **LRC - E0L** $=$ **RE** (Theorem 10.4.8), **ET0L** $\subseteq$ **LF - EP0L** (Theorem 10.4.12), and **E0L** $\subset$ **LP - EP0L** (Theorem 10.4.13).

First, we consider LRC-EPT0L and LRC-EP0L grammars.

Lemma 10.4.5. CS $\subseteq$ LRC - EP0L

Proof. Let $G = (N, T, P, S)$ be a context-sensitive grammar. Without any loss of generality, making use of Theorem 4.1.5, we assume that G is in the Penttonen normal form. Next, we construct a LRC-EP0L grammar H such that $L(H) = L(G)$. Set

$$\begin{aligned} \bar{N} &= \{\bar{A} \mid A \in N\} \\ \hat{N} &= \{\hat{A} \mid A \in N\} \\ N' &= N \cup \bar{N} \cup \hat{N} \end{aligned}$$

Without any loss of generality, we assume that $\bar{N}$, $\hat{N}$, N, and T are pairwise disjoint. Construct

$$H = \left(V', T, P', S\right)$$

as follows. Initially, set $V' = N' \cup T$ and $P' = \emptyset$. Perform (1) through (5), given next.

(1) For each $A \rightarrow a \in P$, where $A \in N$ and $a \in T$, add $(A \rightarrow a, \emptyset, N')$ to P'.
(2) For each $A \rightarrow BC \in P$, where $A, B, C \in N$, add $(A \rightarrow BC, \emptyset, \bar{N} \cup \hat{N})$ to P'.
(3) For each $AB \rightarrow AC \in P$, where $A, B, C \in N$,

 (3.1) add $(B \rightarrow C, \{\hat{A}\}, N \cup (\hat{N} - \{\hat{A}\}))$ to P';
 (3.2) for each $D \in N$, add $(D \rightarrow D, \{\hat{A}, B\}, \hat{N} - \{\hat{A}\})$ to P'.

(4) For each $D \in N$, add $(D \rightarrow \bar{D}, \emptyset, \bar{N} \cup \hat{N})$, $(D \rightarrow \hat{D}, \emptyset, \bar{N} \cup \hat{N})$, $(\bar{D} \rightarrow D, \emptyset, N \cup \hat{N})$, and $(\hat{D} \rightarrow D, \emptyset, N \cup \hat{N})$ to P'.
(5) For each $a \in T$ and each $D \in N$, add $(a \rightarrow a, \emptyset, N')$ and $(D \rightarrow D, \emptyset, \bar{N} \cup \hat{N})$ to P'.

Before proving that $L(H) = L(G)$, let us give an insight into the construction. The simulation of context-free rules of the form $A \rightarrow BC$, where $A, B, C \in N$, is done by rules introduced in (2). Rules from (5) are used to rewrite all the remaining symbols.

H simulates context-sensitive rules—that is, rules of the form $AB \rightarrow AC$, where $A, B, C \in N$—as follows. First, it rewrites all nonterminals to the left of A to their barred versions by rules from (4), A to $\hat{A}$ by $(A \rightarrow \hat{A}, \emptyset, \bar{N} \cup \hat{N})$ from (4), and all the remaining symbols by passive rules from (5). Then, it rewrites B to C by $(B \rightarrow C, \{\hat{A}\}, N \cup (\hat{N} - \{\hat{A}\}))$ from (3.1), barred nonterminals to non-barred nonterminals by rules from (4), $\hat{A}$ back to A by $(\hat{A} \rightarrow A, \emptyset, N \cup \hat{N})$ from (4), all other nonterminals by passive rules from (3.2), and all terminals by passive rules from (5). For example, for

$$abXYABZ \Rightarrow_G abXYACZ$$

there is

$$abXYABZ \Rightarrow_H ab\bar{X}\bar{Y}\hat{A}BZ \Rightarrow_H abXYACZ$$

Observe that if H makes an improper selection of the symbols rewritten to their barred and hatted versions, like in $AXB \Rightarrow_H \hat{A}\bar{X}B$, then the derivation is blocked because every rule of the form $(\bar{D} \rightarrow D, \emptyset, N \cup \hat{N})$, where $D \in D$, requires that there are no hatted nonterminals to the left of $\bar{D}$.

To prevent $AAB \Rightarrow_H AaB \Rightarrow_H \hat{A}aB \Rightarrow_H AaC$, rules simulating $A \rightarrow a$, where $A \in N$ and $a \in T$, introduced in (1), can be used only if there are no nonterminals to the left of A. Therefore, a terminal can never appear between two nonterminals, and so every sentential form generated by H is of the form x_1x_2, where $x_1 \in T^*$ and $x_2 \in N'^*$.

To establish $L(H) = L(G)$, we prove three claims. Claim 1 demonstrates that every $y \in L(G)$ can be generated in two stages; first, only nonterminals are generated, and then, all nonterminals are rewritten to terminals. Claim 2 shows how such derivations of every $y \in L(G)$ in G are simulated by H. Finally, Claim 3 shows how derivations of H are simulated by G.

Claim 1. Let $y \in L(G)$. Then, in G, there exists a derivation $S \Rightarrow_G^* x \Rightarrow_G^* y$, where $x \in N^+$, and during $x \Rightarrow_G^* y$, only rules of the form $A \rightarrow a$, where $A \in N$ and $a \in T$, are applied.

Proof. Let $y \in L(G)$. Since there are no rules in P with symbols from T on their left-hand sides, we can always rearrange all the applications of the rules occurring in $S \Rightarrow_G^* y$ so the claim holds. □

Claim 2. If $S \Rightarrow_G^n x$, where $x \in N^+$, for some $n \geq 0$, then $S \Rightarrow_H^* x$.

Proof. This claim is established by induction on $n \geq 0$.

Basis. For $n = 0$, this claim obviously holds.

Induction Hypothesis. Suppose that there exists $n \geq 0$ such that the claim holds for all derivations of length ℓ, where $0 \leq \ell \leq n$.

Induction Step. Consider any derivation of the form

$$S \Rightarrow_G^{n+1} w$$

where $w \in N^+$. Since $n + 1 \geq 1$, this derivation can be expressed as

$$S \Rightarrow_G^n x \Rightarrow_G w$$

for some $x \in N^+$. By the induction hypothesis, $S \Rightarrow_H^* x$.

Next, we consider all possible forms of $x \Rightarrow_G w$, covered by the following two cases—(i) and (ii).

(i) Let $A \rightarrow BC \in P$ and $x = x_1Ax_2$, where $A, B, C \in N$ and $x_1, x_2 \in N^*$. Then, $x_1Ax_2 \Rightarrow_G x_1BCx_2$. By (2), $(A \rightarrow BC, \emptyset, \bar{N} \cup \hat{N}) \in P'$, and by (5), $(D \rightarrow D, \emptyset, \bar{N} \cup \hat{N}) \in P'$, for each $D \in N$. Since $\text{alph}(x_1Ax_2) \cap (\bar{N} \cup \hat{N}) = \emptyset$,

$$x_1Ax_2 \Rightarrow_H x_1BCx_2$$

which completes the induction step for (i).

(ii) Let $AB \rightarrow AC \in P$ and $x = x_1ABx_2$, where $A, B, C \in N$ and $x_1, x_2 \in N^*$. Then, $x_1ABx_2 \Rightarrow_G x_1BCx_2$. Let $x_1 = X_1X_2 \cdots X_k$, where $X_i \in N$, for all i, $1 \leq i \leq k$, for some $k \geq 1$. By (4), $(X_i \rightarrow \bar{X}_i, \emptyset, \bar{N} \cup \hat{N}) \in P'$, for all i, $1 \leq i \leq k$, and $(A \rightarrow \hat{A}, \emptyset, \bar{N} \cup \hat{N}) \in P'$. By (5), $(D \rightarrow D, \emptyset, \bar{N} \cup \hat{N}) \in P'$, for all $D \in \text{alph}(Bx_2)$. Since $\text{alph}(x_1Ax_2) \cap (\bar{N} \cup \hat{N}) = \emptyset$,

$$X_1X_2 \cdots X_kABx_2 \Rightarrow_H \bar{X}_1\bar{X}_2 \cdots \bar{X}_k\hat{A}Bx_2$$

By (3.1), $(B \rightarrow C, \{\hat{A}\}, N \cup (\hat{N} - \{\hat{A}\})) \in P'$. By (4), $(\bar{X}_i \rightarrow X_i, \emptyset, N \cup \hat{N}) \in P'$, for all i, $1 \leq i \leq k$, and $(\hat{A} \rightarrow A, \emptyset, N \cup \hat{N}) \in P'$. By (3.2), $(D \rightarrow D, \{\hat{A}, B\}, \hat{N} - \{\hat{A}\}) \in P'$, for all $D \in \text{alph}(x_2)$. Since $\text{alph}(\bar{X}_1\bar{X}_2 \cdots \bar{X}_k) \cap (N \cup \hat{N}) = \emptyset$,

$$\bar{X}_1\bar{X}_2\cdots\bar{X}_k\hat{A}Bx_2 \Rightarrow_H X_1X_2\cdots X_kACx_2$$

which completes the induction step for (ii).

Observe that cases (i) and (ii) cover all possible forms of $x \Rightarrow_G w$. Thus, the claim holds. □

Set $V = N \cup T$. Define the homomorphism τ from V'^* to V^* as $\tau(\bar{A}) = \tau(\hat{A}) = \tau(A) = A$, for all $A \in N$, and $\tau(a) = a$, for all $a \in T$.

Claim 3. If $S \Rightarrow_H^n x$, where $x \in V'^+$, for some $n \geq 0$, then $S \Rightarrow_G^* \tau(x)$, and x is of the form x_1x_2, where $x_1 \in T^*$ and $x_2 \in N'^*$.

Proof. This claim is established by induction on $n \geq 0$.

Basis. For $n = 0$, this claim obviously holds.

Induction Hypothesis. Suppose that there exists $n \geq 0$ such that the claim holds for all derivations of length ℓ, where $0 \leq \ell \leq n$.

Induction Step. Consider any derivation of the form

$$S \Rightarrow_H^{n+1} w$$

Since $n + 1 \geq 1$, this derivation can be expressed as

$$S \Rightarrow_H^n x \Rightarrow_H w$$

for some $x \in V'^+$. By the induction hypothesis, $S \Rightarrow_G^* \tau(x)$, and x is of the form x_1x_2, where $x_1 \in T^*$ and $x_2 \in N'^*$.

Next, we make the following four observations regarding the possible forms of $x \Rightarrow_H w$.

(i) A rule from (1) can be applied only to the leftmost occurrence of a nonterminal in x_2. Therefore, w is always of the required form.
(ii) Rules from (1) and (2) can be applied only if $\text{alph}(x) \cap (\bar{N} \cup \hat{N}) = \emptyset$. Furthermore, every rule from (1) and (2) is constructed from some $A \to a \in P$ and $A \to BC \in P$, respectively, where $A, B, C \in N$ and $a \in T$. If two or more rules are applied at once, G can apply them sequentially.
(iii) When a rule from (3.1)—that is, $(B \to C, \{\hat{A}\}, N \cup (\hat{N} - \{\hat{A}\}))$—is applied, $\hat{A}$ has to be right before the occurrence of B that is rewritten to C. Otherwise, the symbols between $\hat{A}$ and that occurrence of B cannot be rewritten by any rule and, therefore, the derivation is blocked. Furthermore, H can apply only a single such rule. Since every rule in (3.1) is constructed from some $AB \to AC \in P$, where $A, B, C \in N$, G applies $AB \to AC$ to simulate this rewrite.
(iv) If rules introduced in (3.2), (4), or (5) are applied, the induction step follows directly from the induction hypothesis.

Based on these observations, we see that the claim holds. □

Next, we establish $L(H) = L(G)$. Let $y \in L(G)$. Then, by Claim 1, in G, there exists a derivation $S \Rightarrow_G^* x \Rightarrow_G^* y$ such that $x \in N^+$ and during $x \Rightarrow_G^* y$, G uses only rules of the form $A \rightarrow a$, where $A \in N$ and $a \in T$. By Claim 2, $S \Rightarrow_H^* x$. Let $x = X_1X_2 \cdots X_k$ and $y = a_1a_2 \cdots a_k$, where $X_i \in N$, $a_i \in T$, $X_i \rightarrow a_i \in P$, for all i, $1 \leq i \leq k$, for some $k \geq 1$. By (1), $(X_i \rightarrow a_i, \emptyset, N') \in P'$, for all i. By (5), $(a_i \rightarrow a_i, \emptyset, N') \in P'$ and $(X_i \rightarrow X_i, \emptyset, \bar{N} \cup \hat{N}) \in P'$, for all i. Therefore,

$$\begin{aligned} X_1X_2 \cdots X_k &\Rightarrow_H a_1X_2 \cdots X_k \\ &\Rightarrow_H a_1a_2 \cdots X_k \\ &\vdots \\ &\Rightarrow_H a_1a_2 \cdots a_k \end{aligned}$$

Consequently, $y \in L(G)$ implies that $y \in L(H)$, so $L(G) \subseteq L(H)$.

Consider Claim 3 with $x \in T^+$. Then, $x \in L(H)$ implies that $\tau(x) = x \in L(G)$, so $L(H) \subseteq L(G)$. As $L(G) \subseteq L(H)$ and $L(H) \subseteq L(G)$, $L(H) = L(G)$, so the lemma holds. □

Lemma 10.4.6. LRC - EPT0L $\subseteq$ CS

Proof. Let $G = (V, T, P_1, P_2, \ldots, P_n, w)$ be a LRC-EPT0L grammar, for some $n \geq 1$. From G, we can construct a phrase-structure grammar, $H = (N', T, P', S)$, such that $L(G) = L(H)$ and if $S \Rightarrow_H^* x \Rightarrow_H^* z$, where $x \in (N' \cup T)^+$ and $z \in T^+$, then $|x| \leq 4|z|$. Consequently, by the workspace theorem (see Theorem 3.3.15), $L(H) \in$ **CS**. Since $L(G) = L(H)$, $L(G) \in$ **CS**, so the lemma holds. □

Theorem 10.4.7. LRC - EPT0L = LRC - EP0L = CS

Proof. **LRC - EP0L $\subseteq$ LRC - EPT0L** follows from the definition of a LRC-EP0L grammar. By Lemma 10.4.5, we have **CS $\subseteq$ LRC - EP0L**, which implies that **CS $\subseteq$ LRC - EP0L $\subseteq$ LRC - EPT0L**. Since **LRC - EPT0L $\subseteq$ CS** by Lemma 10.4.6, **LRC - EP0L $\subseteq$ LRC - EPT0L $\subseteq$ CS**. Hence, **LRC - EPT0L = LRC - EP0L = CS**, so the theorem holds. □

Hence, LRC-EP0L grammars characterize **CS**. Next, we focus on LRC-ET0L and LRC-E0L grammars.

Theorem 10.4.8. LRC - ET0L = LRC - E0L = RE

Proof. The inclusion **LRC - E0L $\subseteq$ LRC - ET0L** follows from the definition of a LRC-E0L grammar. The inclusion **LRC - ET0L $\subseteq$ RE** follows from Church's thesis. The inclusion **RE $\subseteq$ LRC - E0L** can be proved by analogy with the proof of Lemma 10.4.5. Observe that by Theorem 4.1.4, G can additionally contain rules of the form $A \rightarrow \varepsilon$, where $A \in N$. We can simulate these context-free rules in the same way we simulate $A \rightarrow BC$, where $A, B, C \in N$—that is, for each $A \rightarrow \varepsilon \in P$, we introduce $(A \rightarrow \varepsilon, \emptyset, \bar{N} \cup \hat{N})$ to P'. As **LRC - E0L $\subseteq$ LRC - ET0L $\subseteq$ RE** and **RE $\subseteq$ LRC - E0L $\subseteq$ LRC - ET0L**, **LRC - ET0L = LRC - E0L = RE**, so the theorem holds. □

The following corollary compares the generative power of LRC-E0L and LRC-ET0L grammars to the power of E0L and ET0L grammars.

Corollary 10.4.9.

$$\begin{array}{c}\mathbf{CF} \subset \mathbf{E0L} = \mathbf{EP0L} \subset \mathbf{ET0L} = \mathbf{EPT0L} \\ \subset \\ \mathbf{LRC\text{-}EPT0L} = \mathbf{LRC\text{-}EP0L} = \mathbf{CS} \\ \subset \\ \mathbf{LRC\text{-}ET0L} = \mathbf{LRC\text{-}E0L} = \mathbf{RE}\end{array}$$

Proof. This corollary follows from Theorem 3.3.23 in Sect. 3.3 and from Theorems 10.4.7 and 10.4.8 above. □

Next, we show that the family of recursively enumerable languages is characterized even by LRC-E0L grammars with a limited number of nonterminals. Indeed, we prove that every recursively enumerable language can be generated by a LRC-E0L grammar with seven nonterminals.

Theorem 10.4.10. *Let K be a recursively enumerable language. Then, there is a LRC-E0L grammar, $H = (V, T, P, w)$, such that $L(H) = K$ and* $\mathrm{card}(V - T) = 7$.

Proof. Let K be a recursively enumerable language. By Theorem 4.1.9, there is a phrase-structure grammar in the Geffert normal form

$$G = \big(\{S, A, B, C\}, T, P \cup \{ABC \to \varepsilon\}, S\big)$$

satisfying $L(G) = K$. Next, we construct a LRC-E0L grammar H such that $L(H) = L(G)$. Set $N = \{S, A, B, C\}$, $V = N \cup T$, and $N' = N \cup \{\bar{A}, \bar{B}, \#\}$ (without any loss of generality, we assume that $V \cap \{\bar{A}, \bar{B}, \#\} = \emptyset$). Construct

$$H = \big(V', T, P', S\#\big)$$

as follows. Initially, set $V' = N' \cup T$ and $P' = \emptyset$. Perform (1) through (8), given next.

(1) For each $a \in T$,
add $(a \to a, \emptyset, \emptyset)$ to P'.
(2) For each $X \in N$,
add $(X \to X, \emptyset, \{\bar{A}, \bar{B}, \#\})$ and $(X \to X, \{\bar{A}, \bar{B}, C\}, \{S, \#\})$ to P'.
(3) Add $(\# \to \#, \emptyset, \{\bar{A}, \bar{B}\})$, $(\# \to \#, \{\bar{A}, \bar{B}, C\}, \{S\})$, and $(\# \to \varepsilon, \emptyset, N' - \{\#\})$ to P'.
(4) For each $S \to uSa \in P$, where $u \in \{A, AB\}^*$ and $a \in T$,
add $(S \to uS\#a, \emptyset, \{\bar{A}, \bar{B}, \#\})$ to P'.
(5) For each $S \to uSv \in P$, where $u \in \{A, AB\}^*$ and $v \in \{BC, C\}^*$,
add $(S \to uSv, \emptyset, \{\bar{A}, \bar{B}, \#\})$ to P'.

(6) For each $S \rightarrow uv \in P$, where $u \in \{A, AB\}^*$ and $v \in \{BC, C\}^*$, add $(S \rightarrow uv, \emptyset, \{\bar{A}, \bar{B}, \#\})$ to P'.
(7) Add $(A \rightarrow \bar{A}, \emptyset, \{S, \bar{A}, \bar{B}, \#\})$ and $(B \rightarrow \bar{B}, \emptyset, \{S, \bar{A}, \bar{B}, \#\})$ to P'.
(8) Add $(\bar{A} \rightarrow \varepsilon, \emptyset, \{S, \bar{A}, \bar{B}, C, \#\})$, $(\bar{B} \rightarrow \varepsilon, \{\bar{A}\}, \{S, \bar{B}, C, \#\})$, and $(C \rightarrow \varepsilon, \{\bar{A}, \bar{B}\}, \{S, C, \#\})$ to P'.

Before proving that $L(H) = L(G)$, let us informally explain (1) through (8). H simulates the derivations of G that satisfy the form described in Theorem 4.1.9. Since H works in a parallel way, rules from (1) through (3) are used to rewrite symbols that are not actively rewritten. The context-free rules in P are simulated by rules from (4) through (6). The context-sensitive rule $ABC \rightarrow \varepsilon$ is simulated in a two-step way. First, rules introduced in (7) rewrite A and B to $\bar{A}$ and $\bar{B}$, respectively. Then, rules from (8) erase $\bar{A}$, $\bar{B}$, and C; for example,

$$AABCBC\#a\# \Rightarrow_H A\bar{A}\bar{B}CBC\#a\# \Rightarrow_H ABC\#a\#$$

The role of # is twofold. First, it ensures that every sentential form of H is of the form w_1w_2, where $w_1 \in (N' - \{\#\})^*$ and $w_2 \in (T \cup \{\#\})^*$. Since left permitting and left forbidding contexts cannot contain terminals, a mixture of symbols from T and N in H could produce a terminal string out of $L(G)$. For example, observe that $AaBC \Rightarrow_H^* a$, but such a derivation does not exist in G. Second, if any of $\bar{A}$ and $\bar{B}$ are present, $ABC \rightarrow \varepsilon$ has to be simulated. Therefore, it prevents derivations of the form $Aa \Rightarrow_H \bar{A}a \Rightarrow_H a$ (notice that the start string of H is $S\#$). Since H works in a parallel way, if rules from (7) are used improperly, the derivation is blocked, so no partial erasures are possible.

Observe that every sentential form of G and H contains at most one occurrence of S. In every derivation step of H, only a single rule from $P \cup \{ABC \rightarrow \varepsilon\}$ can be simulated at once. $ABC \rightarrow \varepsilon$ can be simulated only if there is no S. #s can be eliminated by an application of rules from (7); however, only if no nonterminals occur to the left of # in the current sentential form. Consequently, all #s are erased at the end of every successful derivation. Based on these observations and on Theorem 4.1.9, we see that every successful derivation in H is of the form

$$\begin{aligned} S\# &\Rightarrow_H^* w_1w_2\#a_1\#a_2\cdots\#a_n\# \\ &\Rightarrow_H^* \#a_1\#a_2\cdots\#a_n\# \\ &\Rightarrow_H^* a_1a_2\cdots a_n \end{aligned}$$

where $w_1 \in \{A, AB\}^*$, $w_2 \in \{BC, C\}^*$, and $a_i \in T$ for all $i = 1, \dots, n$, for some $n \geq 0$.

To establish $L(H) = L(G)$, we prove two claims. First, Claim 1 shows how derivations of G are simulated by H. Then, Claim 2 demonstrates the converse—that is, it shows how derivations of H are simulated by G.

Define the homomorphism φ from V^* to V'^* as $\varphi(X) = X$ for all $X \in N$, and $\varphi(a) = \#a$ for all $a \in T$.

Claim 1. If $S \Rightarrow_G^n x \Rightarrow_G^* z$, for some $n \geq 0$, where $x \in V^*$ and $z \in T^*$, then $S\# \Rightarrow_H^* \varphi(x)\#$.

Proof. This claim is established by induction on $n \geq 0$.

Basis. For $n = 0$, this claim obviously holds.

Induction Hypothesis. Suppose that there exists $n \geq 0$ such that the claim holds for all derivations of length ℓ, where $0 \leq \ell \leq n$.

Induction Step. Consider any derivation of the form

$$S \Rightarrow_G^{n+1} w \Rightarrow_G^* z$$

where $w \in V^*$ and $z \in T^*$. Since $n + 1 \geq 1$, this derivation can be expressed as

$$S \Rightarrow_G^n x \Rightarrow_G w \Rightarrow_G^* z$$

for some $x \in V^+$. Without any loss of generality, we assume that $x = x_1x_2x_3x_4$, where $x_1 \in \{A, AB\}^*$, $x_2 \in \{S, \varepsilon\}$, $x_3 \in \{BC, C\}^*$, and $x_4 \in T^*$ (see Theorem 4.1.9 and the form of rules in P). Next, we consider all possible forms of $x \Rightarrow_G w$, covered by the following four cases—(i) through (iv).

(i) *Application of $S \rightarrow uSa \in P$.* Let $x = x_1Sx_4$, $w = x_1uSax_4$, and $S \rightarrow uSa \in P$, where $u \in \{A, AB\}^*$ and $a \in T$. Then, by the induction hypothesis,

$$S\# \Rightarrow_H^* \varphi(x_1Sx_4)\#$$

By (4), $r = (S \rightarrow uS\#a, \emptyset, \{\bar{A}, \bar{B}, \#\}) \in P'$. Since $\varphi(x_1Sx_4)\# = x_1S\varphi(x_4)\#$ and $\mathrm{alph}(x_1S) \cap \{\bar{A}, \bar{B}, \#\} = \emptyset$, by (1)–(3), and by r,

$$x_1S\varphi(x_4)\# \Rightarrow_H x_1uS\#a\varphi(x_4)\#$$

As $\varphi(x_1uSax_4)\# = x_1uS\#a\varphi(x_4)\#$, the induction step is completed for (i).

(ii) *Application of $S \rightarrow uSv \in P$.* Let $x = x_1Sx_3x_4$, $w = x_1uSvx_3x_4$, and $S \rightarrow uSv \in P$, where $u \in \{A, AB\}^*$ and $v \in \{BC, C\}^*$. To complete the induction step for (ii), proceed by analogy with (i), but use a rule from (5) instead of a rule from (4).

(iii) *Application of $S \rightarrow uv \in P$.* Let $x = x_1Sx_3x_4$, $w = x_1uvx_3x_4$, and $S \rightarrow uv \in P$, where $u \in \{A, AB\}^*$ and $v \in \{BC, C\}^*$. To complete the induction step for (iii), proceed by analogy with (i), but use a rule from (6) instead of a rule from (4).

(iv) *Application of $ABC \rightarrow \varepsilon$.* Let $x = x_1'ABCx_3'x_4$, $w = x_1'x_3'x_4$, where $x_1x_2x_3 = x_1'ABCx_3'$, so $x \Rightarrow_G w$ by $ABC \rightarrow \varepsilon$. Then, by the induction hypothesis,

$$S\# \Rightarrow_H^* \varphi(x_1'ABCx_3'x_4)\#$$

Since $\varphi(x_1' ABCx_3' x_4)\# = x_1' ABCx_3' \varphi(x_4)\#$ and $\text{alph}(x_1' ABCx_3') \cap \{\bar{A}, \bar{B}, \#\} = \emptyset$,

$$x_1' ABCx_3' \varphi(x_4)\# \Rightarrow_H x_1' \bar{A}\bar{B}Cx_3' \varphi(x_4)\#$$

by rules from (1), (2), (3), and (7). Since $\text{alph}(x_1') \cap \{S, \bar{A}, \bar{B}, C, \#\} = \emptyset$, $\{\bar{A}\} \subseteq \text{alph}(x_1' \bar{A})$, $\text{alph}(x_1' \bar{A}) \cap \{S, \bar{B}, C, \#\} = \emptyset$, $\{\bar{A}, \bar{B}\} \subseteq \text{alph}(x_1' \bar{A}\bar{B})$, and $\{S, C, \#\} \cap \text{alph}(x_1' \bar{A}\bar{B}) = \emptyset$,

$$x_1' \bar{A}\bar{B}Cx_3' \varphi(x_4)\# \Rightarrow_H x_1' x_3' \varphi(x_4)\#$$

by rules from (1), (2), (3), and (8). As $\varphi(x_1' x_3' x_4)\# = x_1' x_3' \varphi(x_4)\#$, the induction step is completed for (iv).

Observe that cases (i) through (iv) cover all possible forms of $x \Rightarrow_G w$, so the claim holds. □

Define the homomorphism τ from V'^* to V^* as $\tau(X) = X$ for all $X \in N$, $\tau(a) = a$ for all $a \in T$, and $\tau(\bar{A}) = A$, $\tau(\bar{B}) = B$, $\tau(\#) = \varepsilon$.

Claim 2. If $S\# \Rightarrow_H^n x \Rightarrow_H^* z$, for some $n \geq 0$, where $x \in V'^*$ and $z \in T^*$, then $S \Rightarrow_G^* \tau(x)$.

Proof. This claim is established by induction on $n \geq 0$.

Basis. For $n = 0$, this claim obviously holds.

Induction Hypothesis. Suppose that there exists $n \geq 0$ such that the claim holds for all derivations of length ℓ, where $0 \leq \ell \leq n$.

Induction Step. Consider any derivation of the form

$$S\# \Rightarrow_H^{n+1} w \Rightarrow_H^* z$$

where $w \in V'^*$ and $z \in T^*$. Since $n + 1 \geq 1$, this derivation can be expressed as

$$S\# \Rightarrow_H^n x \Rightarrow_H w \Rightarrow_H^* z$$

for some $x \in V'^+$. By the induction hypothesis, $S \Rightarrow_G^* \tau(x)$. Next, we consider all possible forms of $x \Rightarrow_H w$, covered by the following five cases—(i) through (v).

(i) Let $x = x_1 S x_2$ and $w = x_1 uS\#a x_2$, where $x_1, x_2, \in V'^*$, such that $x_1 S x_2 \Rightarrow_H x_1 uS\#a x_2$ by $(S \to uS\#a, \emptyset, \{\bar{A}, \bar{B}, \#\})$—introduced in (4) from $S \to uSa \in P$, where $u \in \{A, AB\}^*$, $a \in T$—and by the rules introduced in (1), (2), and (3). Since $\tau(x_1 S x_2) = \tau(x_1) S \tau(x_2)$,

$$\tau(x_1) S \tau(x_2) \Rightarrow_G \tau(x_1) uSa\tau(x_2)$$

As $\tau(x_1) uSa\tau(x_2) = \tau(x_1 uS\#a x_2)$, the induction step is completed for (i).

(ii) Let $x = x_1 S x_2$ and $w = x_1 u S v x_2$, where $x_1, x_2, \in V'^*$, such that $x_1 S x_2 \Rightarrow_H x_1 u S v x_2$ by $(S \to uSv, \emptyset, \{\bar{A}, \bar{B}, \#\})$—introduced in (5) from $S \to uSv \in P$, where $u \in \{A, AB\}^*$, $v \in \{BC, C\}^*$—and by the rules introduced in (1), (2), and (3). Proceed by analogy with (i).

(iii) Let $x = x_1 S x_2$ and $w = x_1 u v x_2$, where $x_1, x_2, \in V'^*$, such that $x_1 S x_2 \Rightarrow_H x_1 u v x_2$ by $(S \to uv, \emptyset, \{\bar{A}, \bar{B}, \#\})$—introduced in (6) from $S \to uv \in P$, where $u \in \{A, AB\}^*$, $v \in \{BC, C\}^*$—and by the rules introduced in (1), (2), and (3). Proceed by analogy with (i).

(iv) Let $x = x_1 \bar{A}\bar{B}C x_2$ and $w = x_1' x_2$, where $x_1, x_2 \in V'^*$ and $\tau(x_1') = x_1$, such that $x_1 \bar{A}\bar{B}C x_2 \Rightarrow_H x_1' x_2$ by rules introduced in (1)–(3), (7), and (8). Since $\tau(x_1 \bar{A}\bar{B}C x_2) = \tau(x_1) ABC \tau(x_2)$,

$$\tau(x_1) ABC \tau(x_2) \Rightarrow_G \tau(x_1)\tau(x_2)$$

by $ABC \to \varepsilon$. As $\tau(x_1)\tau(x_2) = \tau(x_1' x_2)$, the induction step is completed for (iv).

(v) Let $x \Rightarrow_H w$ only by rules from (1)–(3), and from (7). As $\tau(x) = \tau(w)$, the induction step is completed for (v).

Observe that cases (i) through (v) cover all possible forms of $x \Rightarrow_H w$, so the claim holds. □

Next, we prove that $L(H) = L(G)$. Consider Claim 1 with $x \in T^*$. Then, $S \Rightarrow_G^* x$ implies that $S\# \Rightarrow_H^* \varphi(x)\#$. By (3), $(\# \to \varepsilon, \emptyset, N' - \{\#\}) \in P'$, and by (1), $(a \to a, \emptyset, \emptyset) \in P'$ for all $a \in T$. Since $\text{alph}(\varphi(x)\#) \cap (N' - \{\#\}) = \emptyset$, $\varphi(x)\# \Rightarrow_H x$. Hence, $L(G) \subseteq L(H)$. Consider Claim 2 with $x \in T^*$. Then, $S\# \Rightarrow_H^* x$ implies that $S \Rightarrow_G^* x$. Hence, $L(H) \subseteq L(G)$. Since $\text{card}(N') = 7$, the theorem holds. □

We turn our attention to LRC-E0L grammars containing only forbidding conditions.

Lemma 10.4.11. EPT0L $\subseteq$ LF - EP0L

Proof. Let $G = (V, T, P_1, P_2, \ldots, P_t, w)$ be an EPT0L grammar, for some $t \geq 1$. Set

$$R = \big\{\langle X, i\rangle \mid X \in V, 1 \leq i \leq t\big\}$$

and

$$F(i) = \big\{\langle X, j\rangle \in R \mid j \neq i\big\} \text{ for } i = 1, 2, \ldots, t$$

Without any loss of generality, we assume that $V \cap R = \emptyset$. Define the LF-EP0L grammar

$$H = \left(V', T, P', w\right)$$

where $V' = V \cup R$, and P' is constructed by performing the following two steps:

(1) For each $X \in V$ and each $i \in \{1, 2, \ldots, t\}$, add $(X \to \langle X, i\rangle, \emptyset, \emptyset)$ to P';
(2) For each $X \to y \in P_i$, where $1 \le i \le t$, add $(\langle X, i\rangle \to y, \emptyset, F(i))$ to P'.

To establish $L(H) = L(G)$, we prove three claims. Claim 1 points out that the every sentential form in H is formed either by symbols from R or from V, depending on whether the length of the derivation is even or odd. Claim 2 shows how derivations of G are simulated by H. Finally, Claim 3 demonstrates the converse—that is, it shows how derivations of H are simulated by G.

Claim 1. For every derivation $w \Rightarrow_H^n x$, where $n \ge 0$,

(i) If $n = 2k + 1$, for some $k \ge 0$, then $x \in R^+$;
(ii) If $n = 2k$, for some $k \ge 0$, then $x \in V^+$.

Proof. The claim follows from the construction of P'. Indeed, every rule in P' is either of the form $(X \to \langle X, i\rangle, \emptyset, \emptyset)$ or $(\langle X, i\rangle \to y, \emptyset, F(i))$, where $X \in V$, $1 \le i \le t$, and $y \in V^+$. Since $w \in V^+$, $w \Rightarrow_H^{2k+1} x$ implies that $x \in R^+$, and $w \Rightarrow_H^{2k} x$ implies that $x \in V^+$. Thus, the claim holds. □

Claim 2. If $w \Rightarrow_G^n x$, where $x \in V^+$, for some $n \ge 0$, then $w \Rightarrow_H^* x$.

Proof. This claim is established by induction on $n \ge 0$.

Basis. For $n = 0$, this claim obviously holds.

Induction Hypothesis. Suppose that there exists $n \ge 0$ such that the claim holds for all derivations of length ℓ, where $0 \le \ell \le n$.

Induction Step. Consider any derivation of the form

$$w \Rightarrow_G^{n+1} y$$

where $y \in V^+$. Since $n + 1 \ge 1$, this derivation can be expressed as

$$w \Rightarrow_G^n x \Rightarrow_G y$$

for some $x \in V^+$. Let $x = X_1X_2\cdots X_h$ and $y = y_1y_2\cdots y_h$, where $h = |x|$. As $x \Rightarrow_G y$, $X_i \to y_i \in P_m$, for all i, $1 \le i \le h$, for some $m \le t$.

By the induction hypothesis, $w \Rightarrow_H^* x$. By (1), $(X_i \to \langle X, m\rangle, \emptyset, \emptyset) \in P'$, for all i, $1 \le i \le h$. Therefore,

$$X_1X_2\cdots X_h \Rightarrow_H \langle X_1, m\rangle\langle X_2, m\rangle\cdots\langle X_h, m\rangle$$

By (2), $(\langle X_i, m\rangle \to y_i, \emptyset, F(m)) \in P'$, for all i, $1 \le i \le h$. Since $\mathrm{alph}(\langle X_1, m\rangle\langle X_2, m\rangle\cdots\langle X_h, m\rangle) \cap F(m) = \emptyset$,

$$\langle X_1, m\rangle\langle X_2, m\rangle \cdots \langle X_h, m\rangle \Rightarrow_H y_1 y_2 \cdots y_h$$

which proves the induction step. □

Define the homomorphism ψ from V'^* to V^* as $\psi(X) = \psi(\langle X, i\rangle) = X$, for all $X \in V$ and all i, $1 \leq i \leq t$.

Claim 3. If $w \Rightarrow_H^n x$, where $x \in V'^+$, for some $n \geq 0$, then $w \Rightarrow_G^* \psi(x)$.

Proof. This claim is established by induction on $n \geq 0$.

Basis. For $n = 0$, this claim obviously holds.

Induction Hypothesis. Suppose that there exists $n \geq 0$ such that the claim holds for all derivations of length ℓ, where $0 \leq \ell \leq n$.

Induction Step. Consider any derivation of the form

$$w \Rightarrow_H^{n+1} y$$

where $y \in V'^+$. Since $n + 1 \geq 1$, this derivation can be expressed as

$$w \Rightarrow_H^n x \Rightarrow_H y$$

for some $x \in V'^+$. By the induction hypothesis, $w \Rightarrow_G^* \psi(x)$. By Claim 1, there exist the following two cases—(i) and (ii).

(i) Let $n = 2k + 1$, for some $k \geq 0$. Then, $x \in R^+$, so let $x = \langle X_1, m_1\rangle\langle X_2, m_2\rangle \cdots \langle X_h, m_h\rangle$, where $h = |x|$, $X_i \in V$, for all i, $1 \leq i \leq h$, and $m_j \in \{1, 2, \dots, t\}$, for all j, $1 \leq j \leq h$. The only possible derivation in H is

$$\langle X_1, m_1\rangle\langle X_2, m_2\rangle \cdots \langle X_h, m_h\rangle \Rightarrow_H y_1 y_2 \cdots y_h$$

by rules introduced in (2), where $y_i \in V^*$, for all i, $1 \leq i \leq h$. Observe that $m_1 = m_2 = \cdots = m_h$; otherwise, $\langle X_h, m_h\rangle$ cannot be rewritten [see the form of left forbidding contexts of the rules introduced to P' in (2)]. By (2), $X_j \rightarrow y_j \in P_{m_h}$, for all j, $1 \leq j \leq h$. Since $\psi(x) = X_1 X_2 \cdots X_h$,

$$X_1 X_2 \cdots X_h \Rightarrow_G y_1 y_2 \cdots y_h$$

which proves the induction step for (i).

(ii) Let $n = 2k$, for some $k \geq 0$. Then, $x \in V^+$, so let $x = X_1 X_2 \cdots X_h$, where $h = |x|$. The only possible derivation in H is

$$X_1 X_2 \cdots X_h \Rightarrow_H \langle X_1, m_1\rangle\langle X_2, m_2\rangle \cdots \langle X_h, m_h\rangle$$

by rules introduced in (1), where $m_j \in \{1, 2, \ldots, t\}$, for all j, $1 \leq j \leq h$. Since $\psi(y) = \psi(x)$, where $y = \langle X_1, m_1\rangle\langle X_2, m_2\rangle \cdots \langle X_h, m_h\rangle$, the induction step for (ii) follows directly from the induction hypothesis.

Hence, the claim holds. □

Next, we establish $L(H) = L(G)$. Consider Claim 2 with $x \in T^+$. Then, $w \Rightarrow_G^* x$ implies that $w \Rightarrow_H^* x$, so $L(G) \subseteq L(H)$. Consider Claim 3 with $x \in T^+$. Then, $w \Rightarrow_H^* x$ implies that $w \Rightarrow_G^* \psi(x) = x$, so $L(H) \subseteq L(G)$. Hence, $L(H) = L(G)$, so the lemma holds. □

Theorem 10.4.12. E0L = EP0L ⊂ ET0L = EPT0L ⊆ LF - EP0L ⊆ LF - E0L

Proof. The inclusions **E0L** = **EP0L**, **ET0L** = **EPT0L**, and **E0L** ⊂ **ET0L** follow from Theorem 3.3.23. From Lemma 10.4.11, we have **EPT0L** ⊆ **LF - EP0L**. The inclusion **LF - EP0L** ⊆ **LF - E0L** follows directly from the definition of an LF-E0L grammar. □

Next, we briefly discuss LRC-E0L grammars containing only permitting conditions.

Theorem 10.4.13. E0L = EP0L ⊂ LP - EP0L ⊆ LP - E0L

Proof. The identity **E0L** = **EP0L** follows from Theorem 3.3.23. The inclusions **EP0L** ⊆ **LP - EP0L** ⊆ **LP - E0L** follow directly from the definition of an LP-E0L grammar. The properness of the inclusion **EP0L** ⊂ **LP - EP0L** follows from Example 10.4.4. □

To conclude this section, we compare LRC-ET0L grammars and their special variants to a variety of conditional ET0L grammars with respect to their generative power. Then, we formulate some open problem areas.

Consider *random context ET0L grammars* (abbreviated *RC-ET0L grammars*), see [15, 17] and Chap. 8 in [8]. These grammars have been recently discussed in connection to various grammar systems (see [1–7, 9]) and membrane systems (P systems, see [18]). Recall that as a generalization of LRC-ET0L grammars, they check the occurrence of symbols in the entire sequential form. Notice, however, that contrary to our definition of LRC-ET0L grammars, in [15, 17] and in other works, RC-ET0L grammars are defined so that they have permitting and forbidding conditions attached to whole sets of rules rather than to each single rule. Since we also study LRC-E0L grammars, which contain just a single set of rules, attachment to rules is more appropriate in our case, just like in terms of other types of regulated ET0L grammars discussed in this chapter.

The language families generated by RC-ET0L grammars and propagating RC-ET0L grammars are denoted by **RC - ET0L** and **RC - EPT0L**, respectively (for the definitions of these families, see [7]).

Theorem 10.4.14 (See [3]). RC - EPT0L ⊂ CS *and* **RC - ET0L ⊆ RE**

Let us point out that it is not known whether the inclusion **RC - ET0L** ⊆ **RE** is, in fact, proper (see [1, 7, 18]).

Corollary 10.4.15. $\mathbf{RC\text{-}EPT0L} \subset \mathbf{LRC\text{-}EP0L}$ *and* $\mathbf{RC\text{-}ET0L} \subseteq \mathbf{LRC\text{-}E0L}$

Proof. This corollary follows from Theorems 10.4.7, 10.4.8, and 10.4.14. □

Corollary 10.4.15 is of some interest because LRC-E0L grammars (i) have only a single set of rules and (ii) they check only prefixes of sentential forms.

A generalization of LF-ET0L grammars, called forbidding ET0L grammars (abbreviated F-ET0L grammars), is introduced and discussed in Sect. 10.2. Recall that as opposed to LF-ET0L grammars, these grammars check the absence of forbidding symbols in the entire sentential form. Furthermore, recall that $\mathbf{F\text{-}ET0L}(1)$ denotes the family of languages generated by F-ET0L grammars whose forbidding strings are of length one.

Corollary 10.4.16. $\mathbf{F\text{-}ET0L}(1) \subseteq \mathbf{LF\text{-}EP0L}$

Proof. This corollary follows from Lemma 10.4.11 and from Theorem 10.2.8, which says that $\mathbf{F\text{-}ET0L}(1) = \mathbf{ET0L}$. □

This result is also of some interest because LF-EP0L grammars (i) have only a single set of rules, (ii) have no rules of the form $(A \rightarrow \varepsilon, \emptyset, W)$, and (iii) they check only prefixes of sentential forms.

Furthermore, consider conditional ET0L grammars (C-ET0L grammars for short) and simple semi-conditional ET0L grammars (SSC-ET0L grammars for short) from Sects. 10.1 and 10.3, respectively. Recall that these grammars differ from RC-ET0L grammars by the form of their permitting and forbidding sets. In C-ET0L grammars, these sets contain strings rather than single symbols. SSC-ET0L grammars are C-ET0L grammars in which every rule can either forbid or permit the occurrence of a single string.

Recall that $\mathbf{C\text{-}ET0L}$ and $\mathbf{C\text{-}EPT0L}$ denote the language families generated by C-ET0L grammars and propagating C-ET0L grammars, respectively. The language families generated by SSC-ET0L grammars and propagating SSC-ET0L grammars are denoted by $\mathbf{SSC\text{-}ET0L}$ and $\mathbf{SSC\text{-}EPT0L}$, respectively.

Corollary 10.4.17.

$$\begin{array}{c}\mathbf{C\text{-}EPT0L} = \mathbf{SSC\text{-}EPT0L} = \mathbf{LRC\text{-}EP0L} \\ \subset \\ \mathbf{C\text{-}ET0L} = \mathbf{SSC\text{-}ET0L} = \mathbf{LRC\text{-}E0L}\end{array}$$

Proof. This corollary follows from Theorems 10.4.7 and 10.4.8 and from Theorems 10.1.6, 10.1.8, 10.3.7, and 10.3.4, which say that $\mathbf{C\text{-}EPT0L} = \mathbf{SSC\text{-}EPT0L} = \mathbf{CS}$ and $\mathbf{C\text{-}ET0L} = \mathbf{SSC\text{-}ET0L} = \mathbf{RE}$. □

We close this section by formulating several open problem areas suggested as topics of future investigation related to the present study.

Open Problem 10.4.18. By Theorem 10.4.12, $\mathbf{ET0L} \subseteq \mathbf{LF\text{-}E0L}$. Is this inclusion, in fact, an identity? □

Open Problem 10.4.19. ET0L and EPT0L grammars have the same generative power (see Theorem 3.3.23). Are LF-E0L and LF-EP0L grammars equally powerful? Are LP-E0L and LP-EP0L grammars equally powerful? □

Open Problem 10.4.20. What is the relation between the language families generated by ET0L grammars and by LP-E0L grammars? □

Open Problem 10.4.21. Establish the generative power of LP-ET0L and LF-ET0L grammars. □

Open Problem 10.4.22. Theorem 10.4.10 has proved that every recursively enumerable language can be generated by a LRC-E0L grammar with seven nonterminals. Can this result be improved? □

Open Problem 10.4.23. Recall that LRC-E0L grammars without erasing rules characterize the family of context-sensitive languages (see Theorem 10.4.7). Can we establish this characterization based upon these grammars with a limited number of nonterminals? □

References

1. Beek, M., Csuhaj-Varjú, E., Holzer, M., Vaszil, G.: On competence in CD grammar systems. In: Developments in Language Theory. Lecture Notes in Computer Science, vol. 3340, pp. 3–14. Springer, Berlin (2005)
2. Beek, M., Csuhaj-Varjú, E., Holzer, M., Vaszil, G.: On competence in CD grammar systems with parallel rewriting. Int. J. Found. Comput. Sci. **18**(6), 1425–1439 (2007)
3. Bordihn, H., Holzer, M.: Grammar systems with negated conditions in their cooperation protocols. J. Univers. Comput. Sci. **6**(12), 1165–1184 (2000)
4. Bordihn, H., Holzer, M.: Random context in regulated rewriting versus cooperating distributed grammar systems. In: LATA'08: Proceedings of the 2nd International Conference on Language and Automata Theory and Applications, pp. 125–136. Springer, New York (2008)
5. Csuhaj-Varjú, E., Păun, G., Salomaa, A.: Conditional tabled eco-grammar systems. J. Univers. Comput. Sci. **1**(5), 252–268 (1995)
6. Csuhaj-Varjú, E., Dassow, J., Vaszil, G.: Some new modes of competence-based derivations in CD grammar systems. In: Developments in Language Theory. Lecture Notes in Computer Science, vol. 5257, pp. 228–239. Springer, Berlin (2008)
7. Dassow, J.: On cooperating distributed grammar systems with competence based start and stop conditions. Fundamenta Informaticae **76**, 293–304 (2007)
8. Dassow, J., Păun, G.: Regulated Rewriting in Formal Language Theory. Springer, New York (1989)
9. Fernau, H., Holzer, M., Freund, R.: Hybrid modes in cooperating distributed grammar systems: internal versus external hybridization. Theor. Comput. Sci. **259**(1–2), 405–426 (2001)
10. Prusinkiewicz, P., Lindenmayer, A.: The Algorithmic Beauty of Plants. Springer, New York (1990)
11. Prusinkiewicz, P., Hammel, M., Hanan, J., Měch, R.: L-systems: From the theory to visual models of plants. In: Proceedings of the 2nd CSIRO Symposium on Computational Challenges in Life Sciences. CSIRO Publishing, Collingwood (1996)
12. Rozenberg, G., Salomaa, A.: Mathematical Theory of L Systems. Academic Press, Orlando (1980)

13. Rozenberg, G., Salomaa, A.: The Book of L. Springer, New York (1986)
14. Rozenberg, G., Salomaa, A. (eds.): Handbook of Formal Languages: Word, Language, Grammar, vol. 1. Springer, New York (1997)
15. Rozenberg, G., Solms, S.H.: Priorities on context conditions in rewriting systems. Inform. Sci. **14**(1), 15–50 (1978)
16. Salomaa, A.: Formal Languages. Academic, London (1973)
17. Solms, S.H.: Some notes on ET0L languages. Int. J. Comput. Math. **5**, 285–296 (1976)
18. Sosík, P.: The power of catalysts and priorities in membrane systems. Grammars **6**(1), 13–24 (2003)

Chapter 11
Uniform Regulated Rewriting in Parallel

Abstract The present chapter discusses how to perform the parallel generation of languages in a uniform way with respect to the rewritten strings. More precisely, it transform grammars that work in parallel so they produce only strings that have a uniform permutation-based form. In fact, this two-section chapter makes the regulated rewriting uniform in terms of both partially and totally parallel grammars. Indeed, Sect. 11.1 represents the semi-parallel language generation by scattered context grammars, which belong to the most important types of regulated grammars (see Sect. 4.7). It demonstrates how to transform scattered context grammars so they produce only strings that have a uniform permutation-based form. Then, Sect. 11.2 represents the totally parallel generation of languages by EIL grammars (see Sect. 3.3) and presents an analogical transformation for them.

Keywords Uniform grammatical rewriting in parallel • Rewritten strings of a uniform permutation-based form • Scattered context grammars • EIL grammars • Generative power • Computational completeness

The present chapter loosely continues with the topic opened in Sect. 4.1.2. Indeed, just like sequential grammars, parallel grammars can, in a general case, produce a very broad variety of quite different strings during the generation of their languages, and this diversity concerning the language generation often represents an undesirable grammatical phenomenon both theoretically and practically. In theory, the demonstration of properties concerning languages generated in this highly inconsistent way usually leads to unbearably lengthy and tedious proofs. In practice, this kind of language generation is obviously difficult to analyze. Therefore, the present chapter explains how to make the parallel generation of languages more uniform.

As a matter of fact, this two-section chapter makes the regulated rewriting uniform in terms of both partially and totally parallel grammars. Indeed, Sect. 11.1 represents the semi-parallel language generation by scattered context grammars, which belong to the most important types of regulated grammars (see Sect. 4.7).

A. Meduna and P. Zemek, *Regulated Grammars and Automata*, 431
DOI 10.1007/978-1-4939-0369-6_11, © Springer Science+Business Media New York 2014

It demonstrates how to transform scattered context grammars so they produce only strings that have a uniform permutation-based form. Then, Sect. 11.2 represents the totally parallel generation of languages by EIL grammars (see Sect. 3.3) and presents an analogical transformation for them.

11.1 Semi-Parallel Uniform Rewriting

In this section, we discuss the uniform generation of languages by scattered context grammars (see Sect. 4.7). More precisely, we demonstrate that for every recursively enumerable language L, there exists a scattered context grammar G and two equally long strings $z_1 \in \{A, B, C\}^*$ and $z_2 \in \{A, B, D\}^*$, where A, B, C, and D are nonterminals of G, so that G generates L and every string appearing in a generation of a sentence from L has the form $y_1 \cdots y_m u$, where u is a string of terminals and each y_i is a permutation of z_j, where $j \in \{1, 2\}$. Furthermore, we achieve an analogical result so that u precedes $y_1 \cdots y_m$.

Note that by analogy with Sect. 4.1.2, every transformation presented in this section assumes that its input grammar contains neither pseudoterminals nor useless nonterminals. Recall that **SCAT** denotes the family of languages generated by scattered context grammars. For a scattered context grammar G, define $F(G)$ and $\Delta(G)$ by analogy with Definition 3.3.2, and set

$$\mathbf{SCAT}[.i/j] = \{L \mid L = L(G), \text{ where } G = (V, T, P, S) \text{ is a scattered context grammar such that } \Delta(G) \subseteq T^* \mathrm{perm}(K)^*, \text{ where } K \text{ is a finite language consisting of equally long strings with } \mathrm{card}(K) = i \text{ and } \mathrm{card}(\mathrm{alph}(K)) = j\}$$

and

$$\mathbf{SCAT}[i/j.] = \{L \mid L = L(G), \text{ where } G = (V, T, P, S) \text{ is a scattered context grammar such that } \Delta(G) \subseteq \mathrm{perm}(K)^* T^*, \text{ where } K \text{ is a finite language consisting of equally long strings with } \mathrm{card}(K) = i \text{ and } \mathrm{card}(\mathrm{alph}(K)) = j\}$$

Lemma 11.1.1. *Let $L \in$ **RE**. Then, there exists a queue grammar $Q = (V, T, W, F, R, g)$ satisfying these two properties*

(i) $L = L(G)$;
(ii) Q derives every $w \in L(Q)$ in this way

$$\begin{aligned} g &\Rightarrow_Q^i a_1 u_1 b_1 \\ &\Rightarrow_Q u_1 x_1 y_1 c_1 \; [(a_1, b_1, x_1 y_1, c_1)] \\ &\Rightarrow_Q^j y_1 z_1 d \end{aligned}$$

where $i, j \geq 1$, $w = y_1 z_1$, $x_1, u_1 \in V^*$, $y_1, z_1 \in T^*$, $b_1, c_1 \in W$ and $d \in F$.

Proof. Let L be a recursively enumerable language. By Theorem 2.1 in [1], there exists a queue grammar

$$Q' = (V, T, W, F, R, g)$$

such that Q' derives every $w \in L(Q')$ as

$$\begin{aligned} g &\Rightarrow^{i}_{Q'} a_1 u_1 b_1 \\ &\Rightarrow_{Q'} u_1 x_1 y_1 c_1 \; [(a_1, b_1, x_1 y_1, c_1)] \\ &\Rightarrow^{j}_{Q'} y_1 z_1 d \end{aligned}$$

where $i, j \geq 0$, $w = y_1 z_1$, $x_1, u_1 \in V^*$, $y_1, z_1 \in T^*$, $b_1, c_1 \in W$, and $d \in F$ ($i = 0$ implies that $a_1 u_1 b_1 = u_1 x_1 y_1 c_1$ and $j = 0$ implies that $u_1 x_1 y_1 c_1 = y_1 z_1 d$). Transform Q' to an equivalent queue grammar Q so that Q generates every $w \in L(Q')$ by a derivation of the form above, where $i \geq 1$ and $j \geq 1$. A detailed version of this simple modification is left to the reader. □

Lemma 11.1.2. *Let $L \in$ **RE**. Then, there exists a scattered context grammar $G = (\{A, B, C, D, S\} \cup T, T, P, S)$ so that $L(G) = \mathrm{rev}(L)$ and*

$$\Delta(G) \subseteq \mathrm{perm}(\{A^t B^{n-t} C, A^t B^{n-t} D\})^* T^*$$

for some $t, n \geq 1$.

Proof. Let $L \in$ **RE**. By Lemma 11.1.1, without any loss of generality, assume that there exists a queue grammar

$$Q = (V, T, W, F, R, g)$$

such that $L = L(Q)$ and Q derives every $w \in L(Q)$ in this way

$$\begin{aligned} g &\Rightarrow^{i}_{Q} a_1 u_1 b_1 \\ &\Rightarrow_{Q} u_1 x_1 y_1 c_1 \; [(a_1, b_1, x_1 y_1, c_1)] \\ &\Rightarrow^{j}_{Q} y_1 z_1 d \end{aligned}$$

where $i, j \geq 1$, $w = y_1 z_1$, $x_1, u_1 \in V^*$, $y_1, z_1 \in T^*$, $b_1, c_1 \in W$ and $d \in F$. The following construction produces a scattered context grammar

$$G = (\{A, B, C, D, S\} \cup T, T, P, S)$$

satisfying

$$L(G) = \mathrm{rev}(L(Q))$$

and

$$\Delta(Q) \subseteq \mathrm{perm}(\{A^t B^{n-t} C, A^t B^{n-t} D\})^* T^*$$

for some $t, n \geq 1$.

For some $n \geq 2^{\mathrm{card}(V \cup W)}$ and $t \in \{1, \dots, n-1\}$, introduce an injective homomorphism β from $(V \cup W)$ to Z, where

$$Z = \{w \mid w \in (\{A, B\}^n - (\{A\}^t \{B\}^{n-t} \cup \{B\}^t \{A\}^{n-t})),\ \mathrm{occur}(w, A) = t\}$$

Intuitively, β represents $(V \cup W)$ in binary. Furthermore, let χ be the homomorphism from $(V \cup W)$ to $Z\{D\}$ defined as $\chi(a) = \beta(a)\{D\}$ for all $a \in (V \cup W)$. Extend the domain of β and χ to $(V \cup W)^*$ in the standard manner. Define the scattered context grammar

$$G = (\{A, B, C, D, S\} \cup T, T, P, S)$$

with P constructed by performing the next six steps.

(1) For $a \in V - T$ and $b \in W - F$ such that $ab = g$, add

$$(S \to A^t B^{n-t} C b_1 \cdots b_n C a_1 \cdots a_n C C A^t B^{n-t})$$

to P, where $b_i, a_i \in \{A, B\}$ for $i = 1, \dots, n$, $b_1 \cdots b_n = \beta(b)$, $a_1 \cdots a_n = \beta(a)$.

(2) For every $(a, b, x, c) \in R$, add

$$\begin{aligned}&(d_1, \dots, d_n, C, b_1, \dots, b_n, C, a_1, \dots, a_n, C, C, d_1, \dots, d_n) \to \\ &(d_1, \dots, d_n, C, e_1, \dots, e_n, \varepsilon, e_1, \dots, e_n, \beta(c) C A^t B^{n-t} C, \chi(x) C, d_1, \dots, d_n)\end{aligned}$$

to P, where $e_i = \varepsilon$, $d_i, b_i, a_i \in \{A, B\}$ for $i = 1, \dots, n$, $d_1 \cdots d_n = A^t B^{n-t}$, $b_1 \cdots b_n = \beta(b)$, $a_1 \cdots a_n = \beta(a)$.

(3) For every $(a, b, xy, c) \in R$ with $x \in V^+$ and $y \in T^*$, add

$$\begin{aligned}&(d_1, \dots, d_n, C, b_1, \dots, b_n, C, a_1, \dots, a_n, C, C, d_1, \dots, d_n) \to \\ &(f_1, \dots, f_n, C, e_1, \dots, e_n, \varepsilon, e_1, \dots, e_n, \beta(c) C A^t B^{n-t} C, \\ &\chi(x) A^t B^{n-t} C\, \mathrm{rev}(y), e_1, \dots, e_n)\end{aligned}$$

to P, where $e_i = \varepsilon$, $d_i, f_i, b_i, a_i \in \{A, B\}$ for $i = 1, \dots, n$, $d_1 \cdots d_n = A^t B^{n-t}$, $f_1 \cdots f_n = B^t A^{n-t}$, $b_1 \cdots b_n = \beta(b)$, $a_1 \cdots a_n = \beta(a)$.

(4) For every $(a, b, y, c) \in R$ with $y \in T^*$ and $c \in W - F$, add

$$\begin{aligned}&(f_1, \dots, f_n, C, b_1, \dots, b_n, C, a_1, \dots, a_n, C, C) \to \\ &(f_1, \dots, f_n, C, e_1, \dots, e_n, \varepsilon, e_1, \dots, e_n, \beta(c) C A^t B^{n-t} C, C\, \mathrm{rev}(y))\end{aligned}$$

to P, where $e_i = \varepsilon$, $f_i, b_i, a_i \in \{A, B\}$ for $i = 1, \dots, n$, $f_1 \cdots f_n = B^t A^{n-t}$, $b_1 \cdots b_n = \beta(b)$, $a_1 \cdots a_n = \beta(a)$.

(5) For every $(a, b, y, c) \in R$ with $y \in T^*$ and $c \in F$, add

$$\begin{aligned}&(f_1, \dots, f_n, C, b_1, \dots, b_n, C, a_1, \dots, a_n, C, d_1, \dots, d_n, C) \to \\ &(e_1, \dots, e_n, \varepsilon, e_1, \dots, e_n, \varepsilon, e_1, \dots, e_n, \varepsilon, e_1, \dots, e_n, \mathrm{rev}(y))\end{aligned}$$

to P, where $e_i = \varepsilon$, $f_i, b_i, a_i, d_i \in \{A, B\}$ for $i = 1, \dots, n$, $d_1 \cdots d_n = A^t B^{n-t}$, $f_1 \cdots f_n = B^t A^{n-t}$, $b_1 \cdots b_n = \beta(b)$, $a_1 \cdots a_n = \beta(a)$.

(6) Add

$$(C, C, d_1, \ldots, d_n, C, f, C) \rightarrow (C, C, e_1, \ldots, e_n, \varepsilon, fC, C)$$

to P, where $e_i = \varepsilon$, $f, d_i \in \{A, B\}$ for $i = 1, \ldots, n$, $d_1 \cdots d_n = A^t B^{n-t}$.

Next, we prove that $\Delta(G) \subseteq \text{perm}(\{A^t B^{n-t} C, A^t B^{n-t} D\})^* T^*$ and $L(G) = \text{rev}(L)$. For brevity, we omit some details in this proof; a complete version of this proof is left to the reader.

Consider any $z \in L(G)$. G generates z in this way

$$\begin{aligned} S &\Rightarrow_G A^t B^{n-t} C b_{1_1} \cdots b_{1_n} C a_{1_1} \cdots a_{1_n} CCA^t B^{n-t} \; [p_1] \\ &\Rightarrow_G^j u \\ &\Rightarrow_G v \\ &\Rightarrow_G^k w \\ &\Rightarrow_G \text{rev}(w_5) \; [p_5] \end{aligned}$$

where $j, k \geq 0$, $z = \text{rev}(w_5)$, and the five subderivations satisfy the following properties, (i) through (iv).

(i) In

$$S \Rightarrow_G A^t B^{n-t} C b_{1_1} \cdots b_{1_n} C a_{1_1} \cdots a_{1_n} CCA^t B^{n-t} \; [p_1]$$

p_1 is of the form

$$(S \rightarrow A^t B^{n-t} C b_{1_1} \cdots b_{1_n} C a_{1_1} \cdots a_{1_n} CCA^t B^{n-t})$$

where $a_{1_i}, b_{1_i} \in \{A, B\}$ for $i = 1, \ldots, n$, $b_{1_1} \cdots b_{1_n} = \beta(b_1)$ with $b_1 \in W$, $a_{1_1} \cdots a_{1_n} = \beta(a_1)$ with $a_1 \in V$, and $a_1 b_1 = g$ (see (1) in the construction of P).

(ii) In

$$A^t B^{n-t} C b_{1_1} \cdots b_{1_n} C a_{1_1} \cdots a_{1_n} CCA^t B^{n-t} \Rightarrow_G^j u$$

every derivation step that is not made by a rule introduced in (6) has the form

$$\begin{aligned} &A^t B^{n-t} C b_{2_1} \cdots b_{2_n} C a_{2_1} \cdots a_{2_n} C \chi(u_2) CA^t B^{n-t} \Rightarrow_G \\ &A^t B^{n-t} C c_{2_1} \cdots c_{2_n} CA^t B^{n-t} C \chi(u_2 x_2) CA^t B^{n-t} \; [p_2] \end{aligned}$$

where p_2 is of the form

$$\begin{aligned} &(d_{2_1}, \ldots, d_{2_n}, C, b_{2_1}, \ldots, b_{2_n}, C, a_{2_1}, \ldots, a_{2_n}, C, C, d_{2_1}, \ldots, d_{2_n}) \rightarrow \\ &(d_{2_1}, \ldots, d_{2_n}, C, e_{2_1}, \ldots, e_{2_n}, \varepsilon, e_{2_1}, \ldots, e_{2_n}, \beta(c) CA^t B^{n-t} C, \\ &\chi(x_2) C, d_{2_1}, \ldots, d_{2_n}) \end{aligned}$$

where $e_{2_i} = \varepsilon, a_{2_i}, b_{2_i}, d_{2_i} \in \{A, B\}$ for $i = 1, \dots, n$, $a_{2_1} \cdots a_{2_n} = \beta(a_2)$ with $a_2 \in V$, $b_{2_1} \cdots b_{2_n} = \beta(b_2)$ with $b_2 \in W$, $d_{2_1} \cdots b_{2_n} = A^t B^{n-t}$ (see (2) in the construction of P).

Thus,

$$A^t B^{n-t} C b_{1_1} \cdots b_{1_n} C a_{1_1} \cdots a_{1_n} C C A^t B^{n-t} \Rightarrow_G^j u$$

can be expressed as

$$\begin{aligned} & A^t B^{n-t} C b_{1_1} \cdots b_{1_n} C a_{1_1} \cdots a_{1_n} C C A^t B^{n-t} \Rightarrow_G \\ & \quad \vdots \\ & \Rightarrow_G A^t B^{n-t} C b_{2_1} \cdots b_{2_n} C a_{2_1} \cdots a_{2_n} C \chi(u_2) C A^t B^{n-t} \\ & \Rightarrow_G A^t B^{n-t} C c_{2_1} \cdots c_{2_n} C A^t B^{n-t} C \chi(u_2 x_2) C A^t B^{n-t} \\ & \quad \vdots \\ & \Rightarrow_G A^t B^{n-t} C b_{3_1} \cdots b_{3_n} C a_{3_1} \cdots a_{3_n} C A^t B^{n-t} C \chi(u_3) C A^t B^{n-t} \end{aligned}$$

where

$$u = A^t B^{n-t} C b_{3_1} \cdots b_{3_n} C a_{3_1} \cdots a_{3_n} C A^t B^{n-t} C \chi(u_3) C A^t B^{n-t}.$$

(iii) Step $u \Rightarrow_G v$ has the form

$$\begin{aligned} & A^t B^{n-t} C b_{3_1} \cdots b_{3_n} C a_{3_1} \cdots a_{3_n} C A^t B^{n-t} C \chi(u_3) C A^t B^{n-t} \Rightarrow_G \\ & B^t A^{n-t} C c_{3_1} \cdots c_{3_n} C A^t B^{n-t} C \chi(u_3 x_3) A^t B^{n-t} C \operatorname{rev}(y_3) \; [p_3] \end{aligned}$$

where

$$v = B^t A^{n-t} C c_{3_1} \cdots c_{3_n} C A^t B^{n-t} C \chi(u_3 x_3) A^t B^{n-t} C \operatorname{rev}(y_3)$$

and p_3 is of the form

$$\begin{aligned} & (d_{3_1}, \dots, d_{3_n}, C, b_{3_1}, \dots, b_{3_n}, C, a_{3_1}, \dots, a_{3_n}, C, C, d_{3_1}, \dots, d_{3_n}) \rightarrow \\ & (f_{3_1}, \dots, f_{3_n}, C, e_{3_1}, \dots, e_{3_n}, \varepsilon, e_{3_1}, \dots, e_{3_n}, \beta(c) C A^t B^{n-t} C, \\ & \chi(x_3) A^t B^{n-t} C \operatorname{rev}(y_3), e_{3_1}, \dots, e_{3_n}) \end{aligned}$$

where $e_{3_i} = \varepsilon, a_{3_i}, b_{3_i}, d_{3_i}, f_{3_i} \in \{A, B\}$ for $i = 1, \dots, n$, $a_{3_1} \cdots a_{3_n} = \beta(a_3)$ with $a_3 \in V$, $b_{3_1} \cdots b_{3_n} = \beta(b_3)$ with $b_3 \in W$, $d_{3_1} \cdots d_{3_n} = A^t B^{n-t}$, $f_{3_1} \cdots f_{3_n} = B^t A^{n-t}$ (see (3) in the construction of P).

(iv) In $v \Rightarrow_G^k w$, any derivation step that is not made by a rule introduced in (6) has the form

$$\begin{aligned} & B^t A^{n-t} C b_{4_1} \cdots b_{4_n} C a_{4_1} \cdots a_{4_n} C \chi(u_4) A^t B^{n-t} C \operatorname{rev}(v_4) \Rightarrow_G \\ & B^t A^{n-t} C c_{4_1} \cdots c_{4_n} C A^t B^{n-t} C \chi(u_4) A^t B^{n-t} C \operatorname{rev}(y_4) \operatorname{rev}(v_4) \; [p_4] \end{aligned}$$

where p_4 is of the form

$$(f_{4_1}, \ldots, f_{4_n}, C, b_{4_1}, \ldots, b_{4_n}, C, a_{4_1}, \ldots, a_{4_n}, C, C) \rightarrow$$
$$(f_{4_1}, \ldots, f_{4_n}, C, e_{4_1}, \ldots, e_{4_n}, \varepsilon, e_{4_1}, \ldots, e_{4_n}, \beta(c_4)CA^t B^{n-t} C, C \,\mathrm{rev}(y))$$

where $e_{4_i} = \varepsilon$, $a_{4_i}, b_{4_i}, f_{4_i} \in \{A, B\}$ for $i = 1, \ldots, n$, $f_{4_1} \cdots f_{4_n} = B^t A^{n-t}$, $b_{4_1} \cdots b_{4_n}$
$= \beta(b_4)$ with $b_4 \in W$, $a_{4_1} \cdots a_{4_n} = \beta(a_4)$ with $a_4 \in V$, $c_{4_1} \cdots c_{4_n} = \beta(c_4)$ with $c_4 \in W$.

As a result, $v \Rightarrow_G^k w$ can be expressed as

$$B^t A^{n-t} C c_{3_1} \cdots c_{3_n} C A^t B^{n-t} C \chi(u_3 x_3) A^t B^{n-t} C \,\mathrm{rev}(y_3)$$
$$\vdots$$
$$\Rightarrow_G B^t A^{n-t} C b_{4_1} \cdots b_{4_n} C a_{4_1} \cdots a_{4_n} C \chi(u_4) A^t B^{n-t} C \,\mathrm{rev}(v_4)$$
$$\Rightarrow_G B^t A^{n-t} C c_{4_1} \cdots c_{4_n} C A^t B^{n-t} C \chi(u_4) A^t B^{n-t} C \,\mathrm{rev}(y_4)\,\mathrm{rev}(v_4) \; [p_4]$$
$$\vdots$$
$$\Rightarrow_G B^t A^{n-t} C b_{5_1} \cdots b_{5_n} C a_{5_1} \cdots a_{5_n} C A^t B^{n-t} C \,\mathrm{rev}(w_5)$$

where

$$w = B^t A^{n-t} C b_{5_1} \cdots b_{5_n} C a_{5_1} \cdots a_{5_n} C A^t B^{n-t} C \,\mathrm{rev}(w_5)$$

and p_5 is of the form

$$(f_{5_1}, \ldots, f_{5_n}, C, b_{5_1}, \ldots, b_{5_n}, C, a_{5_1}, \ldots, a_{5_n}, C, d_{5_1}, \ldots, d_{5_n}, C) \rightarrow$$
$$(e_{5_1}, \ldots, e_{5_n}, \varepsilon, e_{5_1}, \ldots, e_{5_n}, \varepsilon, e_{5_1}, \ldots, e_{5_n}, \varepsilon, e_{5_1}, \ldots, e_{5_n}, \mathrm{rev}(y_5))$$

where $e_{5_i} = \varepsilon$, $a_{5_i}, b_{5_i}, d_{5_i}, f_{5_i} \in \{A, B\}$ for $i = 1, \ldots, n$, $a_{5_1} \cdots a_{5_n} = \beta(a_5)$ with $a_5 \in V$, $b_{5_1} \cdots b_{5_n} = \beta(b_5)$ with $b_5 \in W$, $d_{5_1} \cdots d_{5_n} = A^t B^{n-t}$, $f_{5_1} \cdots f_{5_n} = B^t A^{n-t}$ (see (5) in the construction of P').

In addition, during

$$A^t B^{n-t} C b_{1_1} \cdots b_{1_n} C a_{1_1} \cdots a_{1_n} C C A^t B^{n-t} \Rightarrow_G^j u$$

and

$$v \Rightarrow_G^k w$$

G uses a rule introduced in (6) to generate a sentential form that contains exactly n hs, where $h \in \{A, B\}$, between the second appearance of C and the third appearance of C, so G can use p_2 and p_4 as described above. Observe that in the previous generation of z by G, every sentential form belongs to $\mathrm{perm}(\{A^t B^{n-t} C, A^t B^{n-t} D\})^* T^*$, so

$$\Delta(G) \subseteq \mathrm{perm}(\{A^t B^{n-t} C, A^t B^{n-t} D\})^* T^*$$

Furthermore, the form of this generation and the construction of P imply that

$$g \Rightarrow_Q^* \text{rev}(z)d$$

with $d \in F$. Consequently, $L(Q)$ contains rev($L(G)$), so $L(G)$ is in rev($L(Q)$). Because $L = L(Q)$, $L(G) = \text{rev}(L)$. □

Lemma 11.1.3. **RE** $\subseteq$ **SCAT**[2/4.]

Proof. Let L be a recursively enumerable language. Set $L' = \text{rev}(L)$. As **RE** is closed under reversal, L' is a recursively enumerable language. By Lemma 11.1.2, there exists a scattered context grammar

$$G = \big(\{A, B, C, D, S\} \cup T, T, P, S\big)$$

so that

$$\Delta(G) \subseteq \text{perm}\big(\{A^t B^{n-t} C, A^t B^{n-t} D\}\big)^* T^*$$

and $L(G) = \text{rev}(L')$. Observe that $L(G)$, rev($L(Q)$), rev(L'), rev(rev(L)), and L coincide. As $L(G) \in$ **SCAT**[2/4.], this lemma holds. □

Theorem 11.1.4. **SCAT**[2/4.] $=$ **RE**

Proof. Clearly, **SCAT**[2/4.] $\subseteq$ **RE**. By Lemma 11.1.3, **RE** $\subseteq$ **SCAT**[2/4.]. Thus, **SCAT**[2/4.] $=$ **RE**. □

Lemma 11.1.5. **RE** $\subseteq$ **SCAT**[.2/4]

Proof. Let L be a recursively enumerable language. By Lemma 11.1.2, there exists a scattered context grammar

$$G' = \big(V, T, P', S\big)$$

satisfying $L(G') \in$ **SCAT**[2/4.] and $L(G') = \text{rev}(L)$. Introduce a scattered context grammar

$$G = \big(V, T, P, S\big)$$

where P is defined by the equivalence

$$(A_1, \dots, A_n) \to (x_1, \dots, x_n) \in P$$

if and only if

$$(A_n, \dots, A_1) \to (\text{rev}(x_n), \dots, \text{rev}(x_1)) \in P'$$

Observe that $L(G) \in$ **SCAT**[.2/4] and $L(G) = \text{rev}(\text{rev}(L))$. As $\text{rev}(\text{rev}(L)) = L$, this lemma holds. □

Theorem 11.1.6. $\mathbf{SCAT}[.2/4] = \mathbf{RE}$

Proof. Clearly, $\mathbf{SCAT}[.2/4] \subseteq \mathbf{RE}$. By Lemma 11.1.5, $\mathbf{RE} \subseteq \mathbf{SCAT}[.2/4]$. Thus, $\mathbf{SCAT}[.2/4] = \mathbf{RE}$. □

11.2 Parallel Uniform Rewriting

The present section converts any EIL grammar G to an equivalent EIL grammar $G' = (\{S, 0, 1\} \cup T, T, P, S)$ so that for every $x \in F(G')$,

$$x \in T^* \text{perm}(w)^*$$

where $w \in \{0, 1\}^*$. Then, it makes this conversion so that for every $x \in F(G')$,

$$x \in \text{perm}(w)^* T^*$$

Note that by analogy with Sect. 4.1.2, every transformation presented in this section assumes that its input grammar contains neither pseudoterminals nor useless nonterminals. Let $j \geq 0$. For an EIL grammar G, define $F(G)$ and $\Delta(G)$ by analogy with Definition 3.3.2, and set

$$\begin{aligned}\mathbf{EIL}[.j] = \big\{L \mid L = L(G),\ &\text{where } G = (V, T, P, S) \text{ is an EIL grammar} \\ &\text{such that } \text{card}(\text{alph}(F(G)) - T) = j \text{ and } F(G) \subseteq T^* \text{perm}(w)^*, \\ &\text{where } w \in (V - T)^*\big\}\end{aligned}$$

Analogously, define

$$\begin{aligned}\mathbf{EIL}[j.] = \big\{L \mid L = L(G),\ &\text{where } G = (V, T, P, S) \text{ is an EIL grammar} \\ &\text{such that } \text{card}(\text{alph}(F(G)) - T) = j \text{ and } F(G) \subseteq \text{perm}(w)^* T^*, \\ &\text{where } w \in (V - T)^*\big\}\end{aligned}$$

Lemma 11.2.1. *Let G be an E(1, 0)L grammar. Then, there exists an EIL grammar $G' = (\{S, 0, 1\} \cup T, T, P, S)$ such that $L(G) = L(G')$ and $F(G') \subseteq T^* \text{perm}(1^{n-2}00)^*$, for some $n \geq 1$.*

Proof. Let

$$G = \big(V, T, Q, \$\big)$$

be an E(1, 0)L grammar. For some natural numbers m and n such that $m \geq 3$ and $2m = n$, introduce an injective homomorphism β from V to

$$\big(\{1\}^m\{1\}^*\{0\}\{1\}^*\{0\} \cap \{0, 1\}^n\big) - \big\{1^{n-2}00\big\}$$

In addition, introduce an injective homomorphism χ from T to

$$(\{1\}^m\{1\}^*\{0\}\{1\}^*\{0\} \cap \{0,1\}^n) - \{1^{n-2}00\}$$

so that

$$\{\chi(a) \mid a \in T\} \cap \{\beta(A) \mid A \in V\} = \emptyset$$

Extend the domain of β and the domain of χ to V^* and T^*, respectively. Define the E$(2n-1,0)$L grammar

$$G' = (T \cup \{S,0,1\}, T, P, S)$$

with

$$P = P_\beta \cup P_\chi \cup P_\delta$$

where

$$\begin{aligned}
P_\beta = \{&S \to \beta(\$)\} \\
&\cup \{(\beta(X)x, 0) \to \beta(y) \mid X \in V \cup \{\varepsilon\},\ x \in \{0,1\}^{n-1},\ y \in V^*, \\
&\quad x0 = \beta(Y) \text{ for some } Y \in V \text{ such that } (X,Y) \to y \in Q\} \\
&\cup \{(\beta(a)x, 0) \to \chi(b) \mid a \in T \cup \{\varepsilon\},\ x \in \{0,1\}^{n-1}, \\
&\quad x0 = \beta(b) \text{ for some } b \in T\} \\
P_\chi = \{&(yx, 0) \to a \mid a \in T,\ y \in T^*,\ x \in \{0,1\}^*, \\
&\quad |yx| \le 2n-1,\ x0 = \chi(a)\} \\
&\cup \{(yx, y) \to \varepsilon \mid Y \in \{0,1\},\ y \in T^*,\ x \in \{0,1\}^*, \\
&\quad |x| \le n-2,\ |yx| \le 2n-1\} \\
&\cup \{(yx, Y) \to Y \mid Y \in \{0,1\},\ y \in T^*,\ x \in \{0,1\}^*, \\
&\quad |x| \ge n,\ |yx| \le 2n-1\} \\
&\cup \{(x, a) \to a \mid a \in T,\ |x| \le 2n-1\} \\
P_\delta = \{&(x, X) \to 1^{n-2}00 \mid x \in (T \cup \{0,1\})^{2n-1},\ X \in (T \cup \{0,1\}), \\
&\quad (P_\beta \cup P_\chi) \cap \{(x,X) \to z \mid z \in (T \cup \{0,1\})^*\} = \emptyset\}
\end{aligned}$$

To prove that $L(G) = L(G')$, we next establish Claims 1 through 4.

Claim 1. Let $S \Rightarrow_{G'}^m w$, where $w \in V^*$ and $m \ge 1$. Then, $w \in T^* \operatorname{perm}(1^{n-2}00)^*$.

Proof. The claim is proved by induction on $m \ge 1$.

Basis. Let $m = 1$. That is, $S \Rightarrow_{G'} \beta(\$)$ $[S \to \beta(\$)]$. As $T^* \operatorname{perm}(1^{n-2}00)^*$ contains $\beta(\$)$, the basis holds.

Induction Hypothesis. Suppose that for all $i = 1, \ldots, k$, where $k \ge 1$, if $S \Rightarrow_{G'}^i w$, then $w \in T^* \operatorname{perm}(1^{n-2}00)^*$.

Induction Step. Consider any derivation of the form

$$S \Rightarrow_{G'}^{k+1} w$$

where $w \in V^*$. Express $S \Rightarrow_{G'}^{k+1} w$ as

$$S \Rightarrow_{G'}^{k} u \Rightarrow_{G'} w \ [p]$$

where $p \in P$. By the induction hypothesis, $u \in T^* \operatorname{perm}(1^{n-2}00)^*$. Examine P to see that $w \in T^* \operatorname{perm}(1^{n-2}00)^*$ if $u \in T^* \operatorname{perm}(1^{n-2}00)^*$; the details are left to the reader. □

Claim 2. Let $\$ \Rightarrow_G^m w$, *for some* $m \geq 0$. *Then,* $S \Rightarrow_{G'}^{+} \beta(w)$.

Proof. This claim is proved by induction on $m \geq 0$.

Basis. Let $m = 0$. That is, $\$ \Rightarrow_G^0 \$$. Observe that $S \Rightarrow_{G'} \beta(\$)\ [S \rightarrow \beta(\$)]$, so the basis holds.

Induction Hypothesis. Suppose that for some $j \geq 1$, if $\$ \Rightarrow_G^i w$, where $i = 1, \ldots, j$, and $w \in V^*$, then $S \Rightarrow_{G'}^{*} \beta(w)$.

Induction Step. Consider any derivation of the form

$$\$ \Rightarrow_G^{j+1} y$$

Express $\$ \Rightarrow_G^{j+1} y$ as

$$\$ \Rightarrow_G^{j} x \Rightarrow_G y$$

Furthermore, express x as $x = X_1X_2 \cdots X_k$, where $k = |x|$ and $X_j \in V$, for $j = 1, \ldots, k$. Assume that G makes

$$X_1X_2 \cdots X_k \Rightarrow_G y$$

according to $(\varepsilon, X_1) \rightarrow y_1$, $(X_1, X_2) \rightarrow y_2, \ldots, (X_{k-1}, X_k) \rightarrow y_k$ so $y = y_1y_2 \cdots y_k$. By the induction hypothesis,

$$S \Rightarrow_{G'}^{+} \beta(x)$$

Express $\beta(x)$ as

$$\beta(x) = \beta(X_1)\beta(X_2) \cdots \beta(X_k)$$

where $X_j \in V$, for $j = 1, \ldots, k$. Return to P_β. Observe that P_β contains $(x_1, 0) \rightarrow \beta(y_1)$, where $x_10 = \beta(X_1)$, and $(\beta(X_{i-1})x_i, 0) \rightarrow \beta(y_i)$, where $x_i0 = \beta(X_i)$ for $i = 2, \ldots, k$. Thus,

$$\beta(X_1)\beta(X_2) \cdots \beta(X_k) \Rightarrow_{G'} \beta(y_1)\beta(y_2) \cdots \beta(y_k)$$

As $y = y_1 y_2 \cdots y_k$, $\beta(x) \Rightarrow_{G'} \beta(y)$. Consequently,

$$S \Rightarrow_{G'}^{+} \beta(y)$$

□

Claim 3. $L(G) \subseteq L(G')$

Proof. Let $w \in L(G')$. Thus, $S \Rightarrow_{G'}^{*} w$ and $w \in T^*$. By Claim 2, $S \Rightarrow_{G'}^{+} \beta(w)$. Recall that P_β contains

$$\{(\beta(a)x, 0) \to \chi(b) \mid a \in T,\ x \in \{0,1\}^{n-1},\ x0 = \beta(b) \text{ for some } b \in T\} \subseteq P_\beta$$

Therefore,

$$\beta(w) \Rightarrow_{G'} \chi(w)$$

Examine P_χ to see that

$$\chi(w) \Rightarrow_{G'}^{*} w$$

Hence, Claim 3 holds. □

Claim 4. $L(G') \subseteq L(G)$

Proof. Let $w \in L(G')$, and let $w = a_1 a_2 \cdots a_{n-1} a_n$ with $a_i \in T$ for $i = 1, \dots, n$, where n is a nonnegative integer ($w = \varepsilon$ if $n = 0$). Observe that

$$\begin{aligned}
S &\Rightarrow_{G'}^{*} \beta(a_1)\beta(a_2)\cdots\beta(a_{n-1})\beta(a_n) \\
&\Rightarrow_{G'} \chi(a_1)\chi(a_2)\cdots\chi(a_{n-1})\chi(a_n) \\
&\Rightarrow_{G'} a_1\chi(a_2)\cdots\chi(a_{n-1})\chi(a_n) \\
&\Rightarrow_{G'} a_1a_2\chi(a_3)\cdots\chi(a_{n-1})\chi(a_n) \\
&\ \vdots \\
&\Rightarrow_{G'} a_1a_2\cdots\chi(a_{n-1})\chi(a_n) \\
&\Rightarrow_{G'} a_1a_2\cdots a_{n-1}\chi(a_n) \\
&\Rightarrow_{G'} a_1a_2\cdots a_{n-1}a_n
\end{aligned}$$

In a greater detail, by using rules from P_β, G' makes

$$\begin{aligned}
S &\Rightarrow_{G'}^{*} \beta(a_1)\beta(a_2)\cdots\beta(a_{n-1})\beta(a_n) \\
&\Rightarrow_{G'} \chi(a_1)\chi(a_2)\cdots\chi(a_{n-1})\chi(a_n)
\end{aligned}$$

and by using rules from P_χ, G' makes the rest of this derivation. Examine P_β to see that if G' makes

$$\begin{aligned}
S &\Rightarrow_{G'}^{*} \beta(a_1)\beta(a_2)\cdots\beta(a_{n-1})\beta(a_n) \\
&\Rightarrow_{G'} \chi(a_1)\chi(a_2)\cdots\chi(a_{n-1})\chi(a_n)
\end{aligned}$$

by using rules from P_β, then $\$ \Rightarrow_G^* a_1a_2\cdots a_{n-1}a_n$. Because $w = a_1a_2\cdots a_{n-1}a_n$, $w \in L(G)$, so Claim 4 holds. □

By Claims 3 and 4, $L(G') = L(G)$, so Lemma 11.2.1 holds. □

Theorem 11.2.2. $\mathbf{EIL}[.2] = \mathbf{RE}$

Proof. Clearly, $\mathbf{EIL}[.2] \subseteq \mathbf{RE}$. By Theorem 6.1.3 in [2], for every $L \in \mathbf{RE}$, there exists an E(1, 0)L grammar G such that $L = L(G)$. Thus, by Lemma 11.2.1, $\mathbf{RE} \subseteq \mathbf{EIL}[.2]$. As $\mathbf{EIL}[.2] \subseteq \mathbf{RE}$ and $\mathbf{RE} \subseteq \mathbf{EIL}[.2]$, $\mathbf{RE} = \mathbf{EIL}[.2]$. □

Lemma 11.2.3. *Let G be an E(0, 1)L grammar. Then, there exists an EIL grammar $G' = (\{S, 0, 1\}, T, P, S)$ such that $L(G) = L(G')$ and $F(G') \subseteq \mathrm{perm}(1^{n-2}00)^*T^*$, for some $n \geq 6$.*

Proof. Let

$$G = \left(V, T, Q, \$\right)$$

be an E(0, 1)L grammar. For some natural numbers m and n such that $m \geq 3$ and $2m = n$, introduce an injective homomorphism β from V to

$$\left(\{0\}\{1\}^*\{0\}\{1\}^*\{1\}^m \cap \{0,1\}^n\right) - \left\{1^{n-2}00\right\}$$

In addition, introduce an injective homomorphism χ from T to

$$\left(\{0\}\{1\}^*\{0\}\{1\}^*\{1\}^m \cap \{0,1\}^n\right) - \left\{1^{n-2}00\right\}$$

so

$$\left\{\chi(a) \mid a \in T\right\} \cap \left\{\beta(A) \mid A \in V\right\} = \emptyset$$

Extend the domain of β and the domain of χ to V^* and T^*, respectively. Define the E(0, 2n − 1)L grammar

$$G' = \left(T \cup \{S, 0, 1\}, T, P, S\right)$$

with

$$P = P_\beta \cup P_\chi \cup P_\delta$$

where

$$\begin{aligned}
P_\beta = \{&S \to \beta(\$)\} \\
&\cup \{(0, x\beta(X)) \to \beta(y) \mid X \in V \cup \{\varepsilon\},\ x \in \{0,1\}^{n-1},\ y \in V^*, \\
&\quad 0x = \beta(Y) \text{ for some } Y \in V \text{ such that } (Y, X) \to y \in Q\} \\
&\cup \{(0, x\beta(a)) \to \chi(b) \mid a \in T \cup \{\varepsilon\},\ x \in \{0,1\}^{n-1}, \\
&\quad 0x = \beta(b) \text{ for some } b \in T\} \\
P_\chi = \{&(0, xy) \to a \mid a \in T,\ y \in T^*,\ x \in \{0,1\}^*, \\
&\quad |xy| \leq 2n-1,\ 0x = \chi(a)\} \\
&\cup \{(Y, xy) \to \varepsilon \mid Y \in \{0,1\},\ y \in T^*,\ x \in \{0,1\}^*, \\
&\quad |x| \leq n-2,\ |xy| \leq 2n-1\} \\
&\cup \{(Y, xy) \to Y \mid Y \in \{0,1\},\ y \in T^*,\ x \in \{0,1\}^*, \\
&\quad |x| \geq n,\ |xy| \leq 2n-1\} \\
&\cup \{(a, x) \to a \mid a \in T,\ |x| \leq 2n-1\} \\
P_\delta = \{&(X, x) \to 1^{n-2}00 \mid x \in (T \cup \{0,1\})^{2n-1},\ X \in (T \cup \{0,1\}), \\
&\quad (P_\beta \cup P_\chi) \cap \{(X, x) \to z \mid z \in (T \cup \{0,1\})^*\} = \emptyset\}
\end{aligned}$$

Complete this proof by analogy with the proof of Lemma 11.2.1. □

Theorem 11.2.4. EIL[2.] = RE

Proof. Clearly, **EIL**[2.] ⊆ **RE**. By Theorem 6.1.3 in [2], for every $L \in$ **RE** there exists an E(0, 1)L grammar G such that $L = L(G)$. Thus, by Lemma 11.2.3, **RE** ⊆ **EIL**[2.]. As **EIL**[2.] ⊆ **RE** and **RE** ⊆ **EIL**[2.], **EIL**[2.] = **RE**. □

From Theorems 11.2.2 and 11.2.4, we obtain the following corollary.

Corollary 11.2.5. EIL[.2] = EIL[.2] = RE □

We close the present chapter by formulating an open problem area.

Open Problem 11.2.6. All the uniform rewriting discussed in this chapter is obtained for grammars with erasing rules. In the techniques by which we achieved this uniform rewriting, these rules fulfill a crucial role. Therefore, we believe that these techniques cannot be straightforwardly adapted for grammars without erasing rules. Can we achieve some uniform rewriting for grammars without erasing rules by using completely different techniques? □

References

1. Kleijn, H.C.M., Rozenberg, G.: On the generative power of regular pattern grammars. Acta Inform. **20**, 391–411 (1983)
2. Rozenberg, G., Salomaa, A.: Mathematical Theory of L Systems. Academic, Orlando (1980)

Chapter 12
Parallel Rewriting Over Word Monoids

Abstract The present chapter studies the regulation of grammatical parallelism so it defines parallel derivations over free monoids generated by finitely many strings. The grammatical parallelism is represented by E0L grammars (an EOL grammar is an ET0L grammar with a single set of rules). The chapter demonstrates that this regulation results into a large increase of the generative power of ordinary E0L grammars, even if the strings that generate free monoids consist of no more than two symbols. In fact, the E0L grammars regulated in this way are computationally complete. This chapter consists of two sections. Section 12.1 defines E0L grammars over word monoids. Section 12.2 demonstrates their computational completeness.

Keywords Algebraic approach to grammatical regulation in parallel • E0L grammars • Parallel derivations over free monoids generated by strings • Context conditions • Computational completeness

In essence, the present chapter reconsiders the topic of Chap. 9 in terms of grammatical parallelism, represented by E0L grammars. That is, it defines E0L grammars with the relation of a direct derivation $\Rightarrow$ over the free monoid generated by a finite set of strings; in symbols, $\Rightarrow$ is defined over W^*, where W is a finite language. It demonstrates that this definition results into a large increase of generative power of ordinary E0L grammars, even if W contains strings consisting of no more than two symbols; in fact, E0L grammars defined in this way are computationally complete.

This chapter consists of two sections. First, Sect. 12.1 defines E0L grammars over word monoids. Then, Sect. 12.2 establishes their generative power.

12.1 Definitions

Without further ado, we next define E0L grammars over word monoids.

A. Meduna and P. Zemek, *Regulated Grammars and Automata*,

DOI 10.1007/978-1-4939-0369-6_12,

Definition 12.1.1. An *E0L grammar on word monoid* (a *WME0L grammar* for short) is a pair

$$(G, W)$$

where

$$G = (V, T, P, S)$$

is an E0L grammar. The set of generators W is a finite language over V. By analogy with wm-grammars, (G, W) has *degree* i, where i is a natural number, if every $y \in W$ satisfies $|y| \leq i$. If $A \to x \in P$ implies that $x \neq \varepsilon$, (G, W) is said to be propagating.

Let $x, y \in W^*$ be two strings such that $x = a_1 a_2 \cdots a_n$, $y = y_1 y_2 \cdots y_n$, $a_i \in V$, $y_i \in V^*$, $1 \leq i \leq n$, and $n \geq 0$. If $a_i \to y_i \in P$ for all $i = 1 \ldots n$, then x *directly derives* y according to rules $a_1 \to y_1, a_2 \to y_2, \ldots, a_n \to y_n$, symbolically written as

$$x \Rightarrow_{(G,W)} y \; [a_1 \to y_1, \ldots, a_n \to y_n]$$

As usual, the list of applied rules is omitted when no confusion arises. In the standard way, define $\Rightarrow^k_{(G,W)}$ for $k \geq 0$, $\Rightarrow^+_{(G,W)}$, and $\Rightarrow^*_{(G,W)}$.

The *language* of (G, W), denoted by $L(G, W)$, is defined as

$$L(G, W) = \left\{ w \in T^* \mid S \Rightarrow^*_{(G,W)} w \right\}$$

□

By **WME0L**(i), **WMEP0L**(i), **WME0L**, and **WMEP0L**, we denote the families of languages generated by WME0L grammars of degree i, propagating WME0L grammars of degree i, WME0L grammars, and propagating WME0L grammars, respectively.

Note that WME0L grammars of degree 2 are called *symbiotic E0L grammars* in [1]. The families of languages generated by symbiotic E0L grammars and propagating symbiotic E0L grammars are denoted by **SE0L** and **SEP0L**—that is, **SE0L** = **WME0L**(2).

12.2 Generative Power

In this section, we establish the power of WME0L grammars and their variants by demonstrating the following relations between the language families introduced in the previous section:

$$\begin{array}{c}\mathbf{CF}\\ \subset\\ \mathbf{WMEP0L}(1) = \mathbf{WME0L}(1) = \mathbf{EP0L} = \mathbf{E0L}\\ \subset\\ \mathbf{WMEP0L}(2) = \mathbf{CS}\\ \subset\\ \mathbf{WME0L}(2) = \mathbf{RE}\end{array}$$

Clearly,

$$\mathbf{WMEP0L}(0) = \mathbf{WME0L}(0) = \emptyset$$

Recall that for ordinary E0L grammars, $\mathbf{EP0L} = \mathbf{E0L}$ (see Theorem 3.3.23). Therefore, the following theorem follows immediately from the definitions.

Theorem 12.2.1. $\mathbf{WMEP0L}(1) = \mathbf{WME0L}(1) = \mathbf{EP0L} = \mathbf{E0L}$ □

Next, let us investigate WME0L grammars of degree 2 (symbiotic E0L grammars). In Theorems 12.2.2 and 12.2.4, we demonstrate that these grammars have remarkably higher generative power than WME0L grammars of degree 1. More specifically, propagating WME0L grammars of degree 2 generate precisely the family of context-sensitive languages and WME0L grammars of degree 2 generate all the family of recursively enumerable languages.

Theorem 12.2.2. $\mathbf{WMEP0L}(2) = \mathbf{CS}$

Proof. It is straightforward to prove that $\mathbf{WMEP0L}(2) \subseteq \mathbf{CS}$; hence, it is sufficient to prove the converse inclusion. Let L be a context-sensitive language generated by a context-sensitive grammar

$$G = \big(N_{CF} \cup N_{CS} \cup T, T, P, S\big)$$

of the form described in Theorem 4.1.6. Let

$$V = N_{CF} \cup N_{CS} \cup T$$

and

$$V' = V \cup Q$$

where

$$Q = \big\{\langle A, B, C\rangle \mid AB \to AC \in P,\ A, C \in N_{CF},\ B \in N_{CS}\big\}$$

Without any loss of generality, we assume that $Q \cap V = \emptyset$. The WMEP0L grammar of degree 2, (G', W), is defined as follows:

$$G' = \big(V', T, P', S\big)$$

where P' is constructed in the following way

(1) for all $A \in V'$, add $A \to A$ to P';
(2) if $A \to x \in P$, $A \in N_{CF}$, $x \in N_{CS} \cup T \cup N_{CF}^2$, then add $A \to x$ to P';
(3) if $AB \to AC \in P$, $A, C \in N_{CF}$, $B \in N_{CS}$, then add $B \to \langle A, B, C \rangle$ and $\langle A, B, C \rangle \to C$ to P'.

The set of generators, $W \subseteq (V \cup V^2)$, is defined as

$$W = \{A\langle A, B, C\rangle \mid \langle A, B, C\rangle \in Q,\ A \in N_{CF}\} \cup V$$

Obviously, (G', W) is a WMEP0L grammar of degree 2. Let us introduce a substitution from V'^* into V^* defined as

(a) for all $D \in V$, $h(D) = D$,
(b) for all $\langle X, D, Z\rangle \in Q$, $h(\langle X, D, Z\rangle) = D$.

Let h^{-1} be the inverse of h. To demonstrate that $L(G) = L(G', W)$, we first prove three claims.

Claim 1. If $S \Rightarrow_G^m w$, $w \in V^+$, for some $m \geq 0$, then $S \Rightarrow_{(G',W)}^* v$, where $v \in h^{-1}(w)$.

Proof. This claim is established by induction on $m \geq 0$.

Basis. Let $m = 0$. The only w is S because $S \Rightarrow_G^0 S$. Since $S \in W^*$, $S \Rightarrow_{(G',W)}^0 S$ and by the definition of h^{-1}, $S \in h^{-1}(S)$.

Induction Hypothesis. Let us suppose that the claim holds for all derivations of length m or less, for some $m \geq 0$.

Induction Step. Consider any derivation of the form

$$S \Rightarrow_G^{m+1} x$$

where $x \in V^*$. Since $m + 1 \geq 1$, there is some $y \in V^+$ and $p \in P$ such that

$$S \Rightarrow_G^m y \Rightarrow_G x\ [p]$$

and by the induction hypothesis, there is also a derivation

$$S \Rightarrow_{(G',W)}^n y'$$

for some $y' \in h^{-1}(y)$, $n \geq 0$. Thus, $y' \in W^*$.

(i) Let us first assume that $p = D \to y_2 \in P$, $D \in N_{CF}$, $y_2 \in N_{CS} \cup T \cup N_{CF}^2$, $y = y_1 D y_3$, and $x = y_1 y_2 y_3$, $y_1 = a_1 \cdots a_i$, $y_3 = b_1 \cdots b_j$, where $a_k, b_l \in V$, $1 \leq k \leq i$, $1 \leq l \leq j$, for some $i, j \geq 0$ ($i = 0$ implies that $y_1 = \varepsilon$ and $j = 0$ implies that $y_3 = \varepsilon$). From the definition of h^{-1}, it is clear that $h^{-1}(Z) = \{Z\}$

for all $Z \in N_{CF}$, $y' = z_1 D z_3$, where $z_1 \in h^{-1}(y_1)$ and $z_3 \in h^{-1}(y_3)$; in other words, $z_1 = c_1 \cdots c_i$, $z_3 = d_1 \cdots d_j$, where $c_k \in h^{-1}(a_k)$, $d_l \in h^{-1}(b_l)$, for $1 \le k \le i$, $1 \le l \le j$. It is clear that $D \to y_2 \in P'$.

Let $d_1 \notin Q$. Then, it is easy to see that $z_1 y_2 z_3 \in W^*$, so

$$z_1 D z_3 \Rightarrow_{(G',W)} z_1 y_2 z_3 \; [c_1 \to c_1, \dots, c_i \to c_i, D \to y_2, d_1 \to d_1, \dots, d_j \to d_j]$$

Therefore,

$$S \Rightarrow^n_{(G',W)} z_1 D z_3 \Rightarrow_{(G',W)} z_1 y_2 z_3$$

with $z_1 y_2 z_3 \in h^{-1}(y_1 y_2 y_3)$.

Let $d_1 \in Q$. That is, $Dh(d_1) \to DC \in P$ for some $C \in N_{CF}$ (see the definition of h). Consider (3) to see that $h(d_1) \to d_1 \in P'$ (observe that this rule is the only rule in P' that has d_1 appearing on its right-hand side). By the definition of W, $d_2 \notin Q$. Thus,

$$\{z_1 Dh(d_1)d_2 \cdots d_j, z_1 y_2 h(d_1) d_2 \cdots d_j\} \subseteq W^*$$

Since

$$S \Rightarrow^n_{(G',W)} z_1 D d_1 \cdots d_j$$

there exists the following derivation in (G', W)

$$\begin{aligned} S &\Rightarrow^{n-1}_{(G',W)} z_1 Dh(d_1)d_2 \cdots d_j \\ &\Rightarrow_{(G',W)} z_1 D d_1 d_2 \cdots d_j \; [c_1 \to c_1, \cdots, c_i \to c_i, D \to D, \\ &\qquad h(d_1) \to d_1, d_2 \to d_2, \dots, d_j \to d_j] \end{aligned}$$

Hence,

$$\begin{aligned} S &\Rightarrow^{n-1}_{(G',W)} z_1 Dh(d_1)d_2 \cdots d_j \\ &\Rightarrow_{(G',W)} z_1 y_2 h(d_1) d_2 \cdots d_j \; [c_1 \to c_1, \dots, c_i \to c_i, D \to y_2, \\ &\qquad h(d_1) \to h(d_1), d_2 \to d_2, \dots, d_j \to d_j] \end{aligned}$$

such that $z_1 y_2 h(d_1) d_2 \cdots d_j$ is in $h^{-1}(x)$.

(ii) Let $p = AB \to AC \in P$, $A, C \in N_{CF}$, $B \in N_{CS}$, $y = y_1 AB y_2$, $y_1, y_2 \in V^*$, $x = y_1 AC y_2$, $y' = z_1 A Y z_2$, $z_i \in h^{-1}(y_i)$, $i \in \{1, 2\}$, $Y \in h^{-1}(B)$, and $y_1 = a_1 \cdots a_i$, $y_3 = b_1 \cdots b_j$, $a_k, b_l \in V$, $1 \le k \le i$, $1 \le l \le j$, for some $i, j \ge 0$. Let $z_1 = c_1 \cdots c_i$, $z_3 = d_1 \cdots d_j$, $c_k \in h^{-1}(a_k)$, $d_l \in h^{-1}(b_l)$, $1 \le k \le i$, $1 \le l \le j$. Clearly,

$$\{B \to \langle A, B, C\rangle, \langle A, B, C\rangle \to C\} \subseteq P'$$

and $A\langle A, B, C\rangle \in W$ (see the definition of W).

Let $Y = B$. Since $y' \in W^*$ and $B \in N_{CS}$, we have $d_1 \notin Q$. Consequently, we have that $z_1 A\langle A, B, C\rangle z_2$ and $z_1 AC z_2$ are in W^* by the definition of W. Thus,

$$\begin{aligned} S &\Rightarrow^n_{(G',W)} z_1 ABz_2 \\ &\Rightarrow_{(G',W)} z_1 A\langle A, B, C\rangle z_2 \; [\pi_1] \\ &\Rightarrow_{(G',W)} z_1 AC z_2 \qquad\qquad [\pi_2] \end{aligned}$$

where

$$\pi_1 = c_1 \to c_1, \ldots, c_i \to c_i, A \to A, B \to \langle A, B, C\rangle, d_1 \to d_1, \ldots, d_j \to d_j$$

$$\pi_2 = c_1 \to c_1, \ldots, c_i \to c_i, A \to A, \langle A, B, C\rangle \to C, d_1 \to d_1, \ldots, d_j \to d_j$$

and $z_1 AC z_2 \in h^{-1}(x)$.

Let $Y \in Q$. Clearly, $h(Y) = B$. By (3) and by the definition of Q, we have $B \to Y \in P'$. As obvious, $z_1 AC z_2$ is in W^* for $d_1 \notin Q$. Thus, since

$$S \Rightarrow^n_{(G',W)} z_1 AY z_2$$

the string $z_1 AY z_2$ can be derived in (G', W) as follows:

$$\begin{aligned} S &\Rightarrow^{n-1}_{(G',W)} z_1 ABz_2 \\ &\Rightarrow_{(G',W)} z_1 AY z_2 \; [\pi] \end{aligned}$$

where

$$\pi = c_1 \to c_1, \ldots, c_i \to c_i, A \to A, B \to Y, d_1 \to d_1, \ldots, d_j \to d_j$$

Since $z_1 A\langle A, B, C\rangle z_2$ and $z_1 AC z_2$ belong to W^*, we have

$$\begin{aligned} S &\Rightarrow^{n-1}_{(G',W)} z_1 ABz_2 \\ &\Rightarrow_{(G',W)} z_1 A\langle A, B, C\rangle z_2 \; [\pi_1] \\ &\Rightarrow_{(G',W)} z_1 AC z_2 \qquad\qquad [\pi_2] \end{aligned}$$

where

$$\pi_1 = c_1 \to c_1, \ldots, c_i \to c_i, A \to A, B \to \langle A, B, C\rangle, d_1 \to d_1, \ldots, d_j \to d_j$$

$$\pi_2 = c_1 \to c_1, \ldots, c_i \to c_i, A \to A, \langle A, B, C\rangle \to C, d_1 \to d_1, \ldots, d_j \to d_j$$

and $z_1 AC z_2 \in h^{-1}(x)$.

Cases (i) and (ii) cover all possible rewriting of y in G. Thus, the claim holds. □

Claim 2. *Let* $S \Rightarrow^*_{(G',W)} v$, $v \in W^*$, $v = rDs$, *and* $p = D \rightarrow z \in P$. *Then,* $h(v) \Rightarrow^i_G h(r)h(z)h(s)$ *for some* $i = 0, 1$.

Proof. To verify this claim, consider the following three cases.

(i) Let $h(z) = h(D)$. Then,

$$h(v) \Rightarrow^0_G h(r)h(z)(s)$$

(ii) Let $z \in T \cup N_{CS} \cup N^2_{CF}$, $D \in N_{CF}$. Then, there is a rule $B \rightarrow z \in P$, and by the definition of h, we have $B \rightarrow z = h(B) \rightarrow h(z)$. Thus,

$$h(r)h(D)h(s) \Rightarrow_G h(r)h(z)h(s)\ [h(B) \rightarrow h(z)]$$

(iii) Let $z = C \in N_{CF}$ and $D = \langle A, B, C\rangle$ for some $\langle A, B, C\rangle \in Q$; see (3). By the definition of W, we have $r = tA$, where $t \in W^*$, and so $v = tACs$. By the definition of Q, there is a rule $AB \rightarrow AC \in P$. Thus,

$$tABs \Rightarrow_G tACs\ [AB \rightarrow AC]$$

where $tABs = h(tA)h(\langle A, B, C\rangle)h(s)$ and $tACs = h(tA)h(C)h(s)$.

By the inspection of P', cases (i) through (iii) cover all possible types of rules in P', which completes this proof. □

Claim 3. *If* $S \Rightarrow^n_{(G',W)} u$, $u \in W^*$, *for some* $n \geq 0$, *then* $S \Rightarrow^*_G h(u)$.

Proof. This claim is proved by induction on $n \geq 0$.

Basis. For $n = 0$, the only u is S because $S \Rightarrow^0_{(G',W)} S$. Since $S = h(S)$ we have $S \Rightarrow^0_G S$.

Induction Hypothesis. Let us assume that the claim holds for all derivations of length n or less, for some $n \geq 0$.

Induction Step. Consider any derivation of the form

$$S \Rightarrow^{n+1}_{(G',W)} u$$

where $u \in W^*$. Since $n + 1 \geq 1$, there is some $v \in W^*$ such that

$$S \Rightarrow^n_{(G',W)} v \Rightarrow_{(G',W)} u$$

and by the induction hypothesis

$$S \Rightarrow^*_G h(v)$$

Return to the proof of Claim 2. It should be clear that by using (i) through (iii) from Claim 3, we can construct a derivation of the form

$$h(v) \Rightarrow_G^i h(u)$$

for some $i \in \{0, \ldots, |u|\}$, in the following way. First, rewrite all occurrences of symbols corresponding to the case (iii). Then, rewrite all occurrences of symbols corresponding to (ii). The details of this construction are left to the reader.

Thus,

$$S \Rightarrow_G^* h(v) \Rightarrow_G^i h(u)$$

Hence, by the principle of induction, we have established Claim 3. □

A proof of the equivalence of G and (G', W) can be derived from Claims 1 and 3 in the following way. By the definition of h^{-1}, we have $h^{-1}(a) = \{a\}$ for all $a \in T$. Thus, by Claim 1, we have for any $x \in T^*$,

$$S \Rightarrow_G^* x \quad \text{implies} \quad S \Rightarrow_{(G',W)}^* x$$

That is, $L(G) \subseteq L(G', W)$.

Conversely, since $T^* \subseteq W^*$, we have, by the definition of h and Claim 3, for any $x \in T^*$,

$$S \Rightarrow_{(G',W)}^* x \quad \text{implies} \quad S \Rightarrow_G^* x$$

That is, $L(G', W) \subseteq L(G)$. As a result, $L(G) = L(G', W)$, so

$$\mathbf{WMEP0L}(2) = \mathbf{CS}$$

which proves the theorem. □

Observe that Theorem 12.2.2 and the definitions imply the following normal form.

Corollary 12.2.3. *Let L be a context-sensitive language over an alphabet T. Then, L can be generated by an WMEP0L grammar (G, W) of degree 2, $G = (V, T, P, S)$, where W is over an alphabet V such that $T \subseteq W$, $(W - V) \subseteq (V - T)^2$, and if $A \to x$ and $|x| > 1$, then $x \in (V - T)^2$.* □

Let us turn our attention to the investigation to WME0L grammars of degree 2 with erasing rules.

Theorem 12.2.4. WME0L(2) = RE

Proof. The inclusion **WME0L**(2) ⊆ **RE** follows from Church's thesis. Hence, it is sufficient to show that **RE** ⊆ **WME0L**(2). Each language $L \in$ **RE** can be generated by a phrase-structure grammar G having the form of Theorem 4.1.7. Thus, **RE** ⊆ **WME0L**(2) can be proved by analogy with the techniques used in the proof of Theorem 12.2.2. The details are left to the reader. □

Since the forms of the resulting WME0L(2) grammar in the proofs of Theorem 12.2.2 and Theorem 12.2.4 are analogous, we obtain the following corollary as an analogy to Corollary 12.2.3.

Corollary 12.2.5. *Let L be a recursively enumerable language over an alphabet T. Then, L can be generated by an WME0L grammar (G, W) of degree 2, $G = (V, T, P, S)$, where W is over an alphabet V such that $T \subseteq W$, $(W - V) \subseteq (V - T)^2$, and if $A \to x$ and $|x| > 1$, then $x \in (V - T)^2$.* □

Summing up the theorems established in this section, we obtain the following corollary, which closes the chapter.

Corollary 12.2.6.

$$\begin{array}{c}
\mathbf{CF} \\
\subset \\
\mathbf{WMEP0L}(1) = \mathbf{WME0L}(1) = \mathbf{EP0L} = \mathbf{E0L} \\
\subset \\
\mathbf{WMEP0L}(2) = \mathbf{CS} \\
\subset \\
\mathbf{WME0L}(2) = \mathbf{RE}
\end{array}$$

Proof. This corollary follows from Theorems 12.2.1, 12.2.2 and 12.2.4. □

References

1. Meduna, A.: Symbiotic E0L systems. Acta Cybern. **10**, 165–172 (1992)

Part V
Regulated Grammar Systems

This two-chapter part deals with regulated grammar systems, which represent, in essence, sets of mutually communicating grammars working under regulating restrictions. It covers regulated multigenerative grammar systems (Chap. 13) and controlled pure grammar systems (Chap. 14).

Chapter 13 studies *multigenerative grammar systems*, whose regulation is based upon a simultaneous generation of several strings composed together by some basic operation after the generation is completed.

Chapter 14 introduces *pure grammar systems*, which have only one type of symbols—terminals. The chapter regulates these systems by control languages and studies their generative power.

Chapter 13
Regulated Multigenerative Grammar Systems

Abstract This two-section chapter discusses regulated versions of grammar systems, referred to as *multigenerative grammar systems*, which consist of several components represented by context-free grammars. Their regulation is based upon a simultaneous generation of several strings, which are composed together by some basic operation, such as concatenation, after their generation is completed. Section 13.1 defines the basic versions of multigenerative grammar systems. During one generation step, each of their grammatical components rewrites a nonterminal in its sentential form. After this simultaneous generation is completed, all the generated strings are composed into a single string by some common string operation, such as union and concatenation. It is shown that these systems characterize the family of matrix languages. In addition, the section demonstrates that multigenerative grammar systems with any number of grammatical components can be transformed to equivalent two-component versions of these systems. Section 13.2 discusses leftmost versions of multigenerative grammar systems in which each generation step is performed in a leftmost manner. That is, all the grammatical components of these versions rewrite the leftmost nonterminal occurrence in their sentential forms; otherwise, they work as the basic versions. This section proves that leftmost multigenerative grammar systems are more powerful than their basic versions because they are computationally complete. It demonstrates that leftmost multigenerative grammar systems with any number of grammatical components can be transformed to equivalent two-component versions of these systems.

Keywords Grammar systems • Multigeneration • String operations • General and leftmost versions • Matrix grammars • Computational completeness

In essence, grammar systems consist of several grammatical components, usually represented by context-free grammars. These systems generate their languages by using all their components that work in a prescribed cooperative, distributed, or parallel way. The investigation of these systems represents a crucially important

A. Meduna and P. Zemek, *Regulated Grammars and Automata*,

DOI 10.1007/978-1-4939-0369-6_13, © Springer Science+Business Media New York 2014

trend in today's formal language theory (see Chap. 4 of [2] for an overview of the key concepts and results). Following the main subject of this book—regulated rewriting—we discuss regulated versions of these systems in the present two-section chapter. In fact, we narrow our attention only to multigenerative grammar systems, whose grammatical components generate their strings in a rule-controlled or nonterminal-controlled rewriting way.

In Sect. 13.1, we introduce the basic versions of multigenerative grammar systems. During one generation step, each of their grammatical components rewrites a nonterminal in its sentential form. After this simultaneous generation is completed, all the generated strings are composed into a single string by some common string operation, such as union and concatenation. More precisely, for a positive integer n, an n-generative grammar system works with n context-free grammatical components, each of which makes a derivation, and these n derivations are simultaneously controlled by a finite set of n-tuples consisting of rules. In this way, the grammar system generates n terminal strings, which are combined together by operation union, concatenation or the selection of the first generated string. We show that these systems characterize the family of matrix languages. In addition, we demonstrate that multigenerative grammar systems with any number of grammatical components can be transformed to equivalent two-component versions of these systems.

Section 13.2 discusses leftmost versions of multigenerative grammar systems in which each generation step is performed in the leftmost manner. That is, all the grammatical components of these versions rewrite the leftmost nonterminal occurrence in their sentential forms; otherwise, they work as the basic versions. We prove that leftmost multigenerative grammar systems are more powerful than their basic versions. Indeed, they generate the family of recursively enumerable languages, which properly contains the family of matrix languages (see Theorems 5.1.6 and 5.2.5). We also consider regulation by n-tuples of nonterminals, rather than rules, and prove that leftmost multigenerative grammar systems regulated by rules or nonterminals have the same generative power. In addition, like for the basic versions, we demonstrate that leftmost multigenerative grammar systems with any number of grammatical components can be transformed to equivalent two-component versions of these systems.

Observe that the results stated above are of some interest when compared to the corresponding results in terms of other grammatical models. Indeed, for instance, in terms of context-free grammars, their leftmost versions and their basic versions are equally powerful (see Theorem 3.3.13) while in terms of programmed grammars, the leftmost versions are less powerful than the general versions (see Theorem 1.4.1 in [1]).

13.1 Multigenerative Grammar Systems

In the present section, we define multigenerative grammar systems and demonstrate that they are as powerful as matrix grammars. We also show that any multigenerative grammar systems can be transformed to an equivalent two-component multigenerative grammar system.

Definition 13.1.1. An *n-generative rule-synchronized grammar system* (an *n-MGR* for short) is an $n+1$ tuple

$$\Gamma = \big(G_1, G_2, \dots, G_n, Q\big)$$

where

- $G_i = (N_i, T_i, P_i, S_i)$ is a context-free grammar, for each $i = 1, \dots, n$;
- Q is a finite set of n-tuples of the form $(p_1, p_2, \dots, p_n)$, where $p_i \in P_i$, for all $i = 1, \dots, n$.

A *sentential n-form* is an n-tuple of the form $\chi = (x_1, x_2, \dots, x_n)$, where $x_i \in (N_i \cup T_i)^*$, for all $i = 1, \dots, n$. Let $\chi = (u_1 A_1 v_1, u_2 A_2 v_2, \dots, u_n A_n v_n)$ and $\bar{\chi} = (u_1 x_1 v_1, u_2 x_2 v_2, \dots, u_n x_n v_n)$ be two sentential n-forms, where $A_i \in N_i$ and $u_i, v_i, x_i \in (N_i \cup T_i)^*$, for all $i = 1, \dots, n$. Let $(p_i \colon A_i \to x_i) \in P_i$, for all $i = 1, \dots, n$ and $(p_1, p_2, \dots, p_n) \in Q$. Then, χ directly derives $\bar{\chi}$ in Γ, denoted by

$$\chi \Rightarrow_\Gamma \bar{\chi}$$

In the standard way, we generalize $\Rightarrow_\Gamma$ to $\Rightarrow_\Gamma^k$, for all $k \geq 0$, $\Rightarrow_\Gamma^*$, and $\Rightarrow_\Gamma^+$.

The *n-language of Γ*, denoted by n-$L(\Gamma)$, is defined as

$$n\text{-}L(\Gamma) = \big\{(w_1, w_2, \dots, w_n) \mid (S_1, S_2, \dots, S_n) \Rightarrow_\Gamma^* (w_1, w_2, \dots, w_n),\ w_i \in T_i^*, \text{ for all } i = 1, \dots, n\big\}$$

The *language generated by Γ in the union mode*, $L_{union}(\Gamma)$, is defined as

$$L_{union}(\Gamma) = \bigcup_{i=1}^{n} \{w_i \mid (w_1, w_2, \dots, w_n) \in n\text{-}L(\Gamma)\}$$

The *language generated by Γ in the concatenation mode*, $L_{conc}(\Gamma)$, is defined as

$$L_{conc}(\Gamma) = \big\{w_1 w_2 \dots w_n \mid (w_1, w_2, \dots, w_n) \in n\text{-}L(\Gamma)\big\}$$

The *language generated by Γ in the first mode*, $L_{first}(\Gamma)$, is defined as

$$L_{first}(\Gamma) = \big\{w_1 \mid (w_1, w_2, \dots, w_n) \in n\text{-}L(\Gamma)\big\}$$

□

We illustrate the above definition by an example.

Example 13.1.2. Consider the 2-MGR $\Gamma = (G_1, G_2, Q)$, where

- $G_1 = (\{S_1, A_1\}, \{a, b, c\}, \{1\colon S_1 \to aS_1, 2\colon S_1 \to aA_1, 3\colon A_1 \to bA_1c, 4\colon A_1 \to bc\}, S_1)$,
- $G_2 = (\{S_2, A_2\}, \{d\}, \{1\colon S_2 \to S_2A_2, 2\colon S_2 \to A_2, 3\colon A_2 \to d\}, S_2)$,
- $Q = \{(1, 1), (2, 2), (3, 3), (4, 3)\}$.

Observe that

- $2\text{-}L(\Gamma) = \{(a^nb^nc^n, d^n) \mid n \geq 1\}$,
- $L_{union}(\Gamma) = \{a^nb^nc^n \mid n \geq 1\} \cup \{d^n \mid n \geq 1\}$,
- $L_{conc}(\Gamma) = \{a^nb^nc^nd^n \mid n \geq 1\}$, and
- $L_{first}(\Gamma) = \{a^nb^nc^n \mid n \geq 1\}$. □

Next, we prove that multigenerative grammar systems under all of the defined modes are equivalent to matrix grammars. Throughout the rest of this section, as indicated in Sect. 5.2, for brevity, we use rules and rule labels interchangeably.

Algorithm 13.1.3. *Conversion of an n-MGR in the union mode to an equivalent matrix grammar.*

Input: An n-MGR, $\Gamma = (G_1, G_2, \ldots, G_n, Q)$.
Output: A matrix grammar, $H = (G, M)$, satisfying $L_{union}(\Gamma) = L(H)$.
Method: Let $G_i = (N_i, T_i, P_i, S_i)$, for all $i = 1, \ldots, n$, and without any loss of generality, we assume that N_1 through N_n are pairwise disjoint. Let us choose arbitrary S satisfying $S \notin \bigcup_{j=1}^{n} N_j$. Then, construct

$$G = \bigl(N, T, P, S\bigr)$$

where

- $N = \{S\} \cup (\bigcup_{i=1}^{n} N_i) \cup (\bigcup_{i=1}^{n}\{\bar{A} \mid A \in N_i\})$;
- $T = \bigcup_{i=1}^{n} T_i$;
- $$\begin{aligned} P = \{&(s_1\colon S \to S_1h(S_2)\ldots h(S_n)),\\ &(s_2\colon S \to h(S_1)S_2\ldots h(S_n)),\\ &\quad\vdots\\ &(s_n\colon S \to h(S_1)h(S_2)\ldots S_n)\}\\ &\cup (\bigcup_{i=1}^{n} P_i)\\ &\cup (\bigcup_{i=1}^{n}\{h(A) \to h(x) \mid A \to x \in P_i\}), \end{aligned}$$
  where h is a homomorphism from $((\bigcup_{i=1}^{n} N_i) \cup (\bigcup_{i=1}^{n} T_i))^*$ to $(\bigcup_{j=1}^{n}\{\bar{A} \mid A \in N_i\})^*$, defined as $h(a) = \varepsilon$, for all $a \in \bigcup_{i=1}^{n} T_i$, and $h(A) = \bar{A}$, for all $A \in \bigcup_{i=1}^{n} N_i$.
- $$\begin{aligned} M = &\{s_1, s_2, \ldots, s_n\}\\ &\cup \{p_1\bar{p}_2\ldots\bar{p}_n \mid (p_1, p_2, \ldots, p_n) \in Q\}\\ &\cup \{\bar{p}_1p_2\ldots\bar{p}_n \mid (p_1, p_2, \ldots, p_n) \in Q\}\\ &\quad\vdots\\ &\cup \{\bar{p}_1\bar{p}_2\ldots p_n \mid (p_1, p_2, \ldots, p_n) \in Q\}. \end{aligned}$$ □

Theorem 13.1.4. *Let $\Gamma = (G_1, G_2, \ldots, G_n, Q)$ be an n-MGR. With Γ as its input, Algorithm 13.1.3 halts and correctly constructs a matrix grammar, $H = (G, M)$, such that $L_{union}(\Gamma) = L(H)$.*

Proof. Let $(p\colon A \to x)$ be a rule. Then, for simplicity and brevity, $\bar{p}$ denotes the rule $h(A) \to h(x)$. To prove this theorem, we first establish Claims 1 and 2.

Claim 1. Let $(S_1, S_2, \ldots, S_n) \Rightarrow_\Gamma^m (y_1, y_2, \ldots, y_n)$, where $m \geq 0$, $y_i \in (N_i \cup T_i)^*$, for all $i = 1, \ldots, n$. Then, $S \Rightarrow_H^{m+1} h(y_1)h(y_2)\ldots h(y_{j-1})y_j h(y_{j+1}) \ldots h(y_n)$, for any $j = 1, \ldots, n$.

Proof. This claim is proved by induction on $m \geq 0$.

Basis. Let $m = 0$. Then, $(S_1, S_2, \ldots, S_n) \Rightarrow_\Gamma^0 (S_1, S_2, \ldots, S_n)$. Notice that

$$S \Rightarrow_H h(S_1)h(S_2)\ldots h(S_{j-1})S_j h(S_{j+1})\ldots h(S_n)$$

for any $j = 1, \ldots, n$, because

$$\big(s_j\colon S \to h(S_1)h(S_2)\ldots h(S_{j-1})S_j h(S_{j+1})\ldots h(S_n)\big) \in M$$

Induction Hypothesis. Assume that the claim holds for all m-step derivations, where $m = 0, \ldots, k$, for some $k \geq 0$.

Induction Step. Consider any derivation of the form

$$(S_1, S_2, \ldots, S_n) \Rightarrow_\Gamma^{k+1} (y_1, y_2, \ldots, y_n)$$

Then, there exists a sentential n-form $(u_1 A_1 v_1, u_2 A_2 v_2, \ldots, u_n A_n v_n)$, where $u_i, v_i \in (T_i \cup N_i)^*$, $A_i \in N_i$ such that

$$\begin{aligned}(S_1, S_2, \ldots, S_n) &\Rightarrow_\Gamma^k (u_1 A_1 v_1, u_2 A_2 v_2, \ldots, u_n A_n v_n)\\ &\Rightarrow_\Gamma (u_1 x_1 v_1, u_2 x_2 v_2, \ldots, u_n x_n v_n)\end{aligned}$$

where $u_i x_i v_i = y_i$, for all $i = 1, \ldots, n$. First, observe that

$$(S_1, S_2, \ldots, S_n) \Rightarrow_\Gamma^k (u_1 A_1 v_1, u_2 A_2 v_2, \ldots, u_n A_n v_n)$$

implies that

$$\begin{aligned}S \Rightarrow_H^{k+1} &\; h(u_1 A_1 v_1)h(u_2 A_2 v_2)\ldots h(u_{j-1} A_{j-1} v_{j-1})\\ &\; u_j A_j v_j h(u_{j+1} A_{j+1} v_{j+1})\ldots h(u_n A_n v_n)\end{aligned}$$

for any $j = 1, \ldots, n$ by the induction hypothesis. Furthermore, let

$$\begin{aligned}(u_1 A_1 v_1, u_2 A_2 v_2, \ldots, u_n A_n v_n) \Rightarrow_\Gamma\\ (u_1 x_1 v_1, u_2 x_2 v_2, \ldots, u_n x_n v_n)\end{aligned}$$

Then, $((p_1: A_1 \to x_1), (p_2: A_2 \to x_2), \ldots, (p_n: A_n \to x_n)) \in Q$. Algorithm 13.1.3 implies that $\bar{p}_1\bar{p}_2 \ldots p_{\bar{j}-1} p_j p_{\bar{j}+1} \ldots \bar{p}_n \in M$, for any $j = 1, \ldots, n$. Hence,

$$h(u_1A_1v_1)h(u_2A_2v_2)\ldots h(u_{j-1}A_{j-1}v_{j-1})u_jA_jv_jh(u_{j+1}A_{j+1}v_{j+1})\ldots h(u_nA_nv_n) \Rightarrow_H h(u_1x_1v_1)h(u_2x_2v_2)\ldots h(u_{j-1}x_{j-1}v_{j-1})u_jx_jv_jh(u_{j+1}x_{j+1}v_{j+1})\ldots h(u_nx_nv_n)$$

by matrix $\bar{p}_1\bar{p}_2 \ldots p_{\bar{j}-1} p_j p_{\bar{j}+1} \ldots \bar{p}_n$, for any $j = 1, \ldots, n$. As a result, we obtain

$$S \Rightarrow_H^{k+2} h(u_1x_1v_1)h(u_2x_2v_2)\ldots h(u_{j-1}x_{j-1}v_{j-1}) u_jx_jv_jh(u_{j+1}x_{j+1}v_{j+1})\ldots h(u_nx_nv_n)$$

for any $j = 1, \ldots, n$. □

Claim 2. Let $S \Rightarrow_H^m y$, where $m \geq 1, y \in (N \cup T)^*$. Then, there exist $j \in \{1, \ldots, n\}$ and $y_i \in (N_i \cup T_i)^*$, for $i = 1, \ldots, n$, such that $(S_1, \ldots, S_n) \Rightarrow_\Gamma^{m-1} (y_1, \ldots, y_n)$ and $y = h(y_1)\ldots h(y_{j-1})y_jh(y_{j+1})\ldots h(y_n)$.

Proof. This claim is proved by induction on $m \geq 1$.

Basis. Let $m = 1$. Then, there exists exactly one of the following one-step derivations in H:

$S \Rightarrow_H S_1h(S_2)\ldots h(S_n)$ by matrix s_1, or

$S \Rightarrow_H h(S_1)S_2\ldots h(S_n)$ by matrix s_2, or

$\ldots$, or

$S \Rightarrow_H h(S_1)h(S_2)\ldots S_n$ by matrix s_n

Notice that trivially $(S_1, S_2, \ldots, S_n) \Rightarrow_\Gamma^0 (S_1, S_2, \ldots, S_n)$.

Induction Hypothesis. Assume that the claim holds for all m-step derivations, where $m = 1, \ldots, k$, for some $k \geq 1$.

Induction Step. Consider any derivation of the form

$$S \Rightarrow_H^{k+1} y$$

Then, there exists a sentential form w such that

$$S \Rightarrow_H^k w \Rightarrow_H y$$

where $w, y \in (N \cup T)^*$. As $w \Rightarrow_H y$, this derivation step can use only a matrix of the form $p_1p_2\ldots p_{j-1}p_jp_{j+1}\ldots p_n \in Q$, where p_j is a rule from P_j and $\bar{p}_i \in h(P_i)$, for $i = 1, \ldots, j-1, j+1, \ldots, n$. Hence, $w \Rightarrow_H y$ can be written as

$$h(w_i)\ldots h(w_{j-1})w_jh(w_{j+1})\ldots h(w_n) \Rightarrow_H z_1 \ldots z_n$$

where $w_j \Rightarrow_H z_j$ by the rule p_j and $h(w_i) \Rightarrow_H z_i$ by $\bar{p}_i$, for $i = 1, \dots, j-1, j+1, \dots, n$. Each rule $\bar{p}_i$ rewrites a barred nonterminal $\bar{A}_i \in h(N_i)$. Of course, then each rule p_i can be used to rewrite the respective occurrence of a non-barred nonterminal A_i in w_i in such a way that $w_i \Rightarrow_H y_i$ and $h(y_i) = z_i$, for all $i = 1, \dots, j-1, j+1, \dots, n$. By setting $y_j = z_j$, we obtain

$$(w_1, \dots, w_n) \Rightarrow_\Gamma (y_1, \dots, y_n)$$

and $y = h(y_1) \dots h(y_{j-1}) y_j h(y_{j+1}) \dots h(y_n)$. As a result, we obtain

$$(S_1, S_2, \dots, S_{j-1}, S_j, S_{j+1}, \dots, S_n) \Rightarrow_\Gamma^k$$
$$(u_1 x_1 v_1, u_2 x_2 v_2, \dots, u_{j-1} x_{j-1} v_{j-1}, u_j x_j v_j, u_{j+1} x_{j+1} v_{j+1}, \dots, u_n x_n j_n)$$

so $y = u_1 x_1 v_1 u_2 x_2 v_2 \dots u_{j-1} x_{j-1} v_{j-1} u_j x_j v_j u_{j+1} x_{j+1} v_{j+1} \dots u_n x_n v_n$. □

Consider Claim 1 for $y_i \in T_i^*$, for all $i = 1, \dots, n$. Notice that $h(a) = \varepsilon$, for all $a \in T_i$. We obtain an implication of the form

$$\text{if } (S_1, S_2, \dots, S_n) \Rightarrow_\Gamma^* (y_1, y_2, \dots, y_n)$$
$$\text{then } S \Rightarrow_H^* y_j, \text{ for any } j = 1, \dots, n$$

Hence, $L_{union}(\Gamma) \subseteq L(H)$. Consider Claim 1 for $y \in T^*$. Notice that $h(a) = \varepsilon$, for all $a \in T_i$. We obtain an implication of the form

$$\text{if } S \Rightarrow_H^* y$$
$$\text{then } (S_1, S_2, \dots, S_n) \Rightarrow_\Gamma^* (y_1, y_2, \dots, y_n)$$

and there exist an index $j = 1, \dots, n$ such that $y = y_j$. Hence, $L(H) \subseteq L_{union}(\Gamma)$. □

Algorithm 13.1.5. *Conversion of an n-MGR in the concatenation mode to an equivalent matrix grammar.*

Input: An n-MGR, $\Gamma = (G_1, G_2, \dots, G_n, Q)$.
Output: A matrix grammar, $H = (G, M)$, satisfying $L_{conc}(\Gamma) = L(H)$.
Method: Let $G_i = (N_i, T_i, P_i, S_i)$, for all $i = 1, \dots, n$, and without any loss of generality, we assume that N_1 through N_n are pairwise disjoint. Let us choose arbitrary S satisfying $S \notin \bigcup_{j=1}^{n} N_j$. Construct

$$G = \big(N, T, P, S\big)$$

where

- $N = \{S\} \cup (\bigcup_{i=1}^{n} N_i)$;
- $T = \bigcup_{i=1}^{n} T_i$;
- $P = \{(s\colon S \to S_1 S_2 \dots S_n)\} \cup (\bigcup_{i=1}^{n} P_i)$.

Finally, set $M = \{s\} \cup \{p_1 p_2 \dots p_n \mid (p_1, p_2, \dots, p_n) \in Q\}$. □

Theorem 13.1.6. *Let $\Gamma = (G_1, G_2, \ldots, G_n, Q)$ be an n-MGR. On input Γ, Algorithm 13.1.5 halts and correctly constructs a matrix grammar, $H = (G, M)$, such that $L_{conc}(\Gamma) = L(H)$.*

Proof. To prove this theorem, we first establish Claims 1 and 2.

Claim 1. Let $(S_1, S_2, \ldots, S_n) \Rightarrow_{\Gamma}^{m} (y_1, y_2, \ldots, y_n)$, where $m \geq 0$, $y_i \in (N_i \cup T_i)^*$, for all $i = 1, \ldots, n$. Then, $S \Rightarrow_{H}^{m+1} y_1 y_2 \ldots y_n$.

Proof. This claim is proved by induction on $m \geq 0$.

Basis. Let $m = 0$. Then, $(S_1, S_2, \ldots, S_n) \Rightarrow_{\Gamma}^{0} (S_1, S_2, \ldots, S_n)$. Notice that $S \Rightarrow_H S_1 S_2 \ldots S_n$, because $(s: S \to S_1 S_2 \ldots S_n) \in M$.

Induction Hypothesis. Assume that the claim holds for all m-step derivations, where $m = 0, \ldots, k$, for some $k \geq 0$.

Induction Step. Consider any derivation of the form

$$(S_1, S_2, \ldots, S_n) \Rightarrow_{\Gamma}^{k+1} (y_1, y_2, \ldots, y_n)$$

Then, there exists a sentential n-form $(u_1 A_1 v_1, u_2 A_2 v_2, \ldots, u_n A_n v_n)$, where $u_i, v_i \in (T_i \cup N_i)^*$, $A_i \in N_i$ such that

$$\begin{aligned}(S_1, S_2, \ldots, S_n) &\Rightarrow_{\Gamma}^{k} (u_1 A_1 v_1, u_2 A_2 v_2, \ldots, u_n A_n v_n)\\ &\Rightarrow_{\Gamma} (u_1 x_1 v_1, u_2 x_2 v_2, \ldots, u_n x_n v_n)\end{aligned}$$

where $u_i x_i v_i = y_i$, for all $i = 1, \ldots, n$. First, observe that

$$(S_1, S_2, \ldots, S_n) \Rightarrow_{\Gamma}^{k} (u_1 A_1 v_1, u_2 A_2 v_2, \ldots, u_n A_n v_n)$$

implies that

$$S \Rightarrow_{H}^{k+1} u_1 A_1 v_1 u_2 A_2 v_2 \ldots u_n A_n v_n$$

by the induction hypothesis. Furthermore, let

$$\begin{aligned}&(u_1 A_1 v_1, u_2 A_2 v_2, \ldots, u_n A_n v_n) \Rightarrow_{\Gamma}\\ &\quad (u_1 x_1 v_1, u_2 x_2 v_2, \ldots, u_n x_n v_n)\end{aligned}$$

Then, it holds that $((p_1: A_1 \to x_1), (p_2: A_2 \to x_2), \ldots, (p_n: A_n \to x_n)) \in Q$. Algorithm 13.1.5 implies that $p_1 p_2 \ldots p_n \in M$. Hence,

$$\begin{aligned}&u_1 A_1 v_1 u_2 A_2 v_2 \ldots u_n A_n v_n \Rightarrow_H\\ &\quad u_1 x_1 v_1 u_2 x_2 v_2 \ldots u_n x_n v_n\end{aligned}$$

by matrix $p_1 p_2 \ldots p_n$. As a result, we obtain

$$S \Rightarrow_{H}^{k+2} u_1 x_1 v_1 u_2 x_2 v_2 \ldots u_n x_n v_n$$

□

Claim 2. *Let $S \Rightarrow_H^m y$, where $m \geq 1, y \in (N \cup T)^*$. Then, $(S_1, S_2, \ldots, S_n) \Rightarrow_\Gamma^{m-1} (y_1, y_2, \ldots, y_n)$ such that $y = y_1 y_2 \ldots y_n$, where $y_i \in (N_i \cup T_i)^*$, for all $i = 1, \ldots, n$.*

Proof. This claim is proved by induction on $m \geq 1$.

Basis. Let $m = 1$. Then, there exists exactly one one-step derivation in H: $S \Rightarrow_H S_1 S_2 \ldots, S_n$ by matrix s. Notice that $(S_1, S_2, \ldots, S_n) \Rightarrow_\Gamma^0 (S_1, S_2, \ldots, S_n)$ trivially.

Induction Hypothesis. Assume that the claim holds for all m-step derivations, where $m = 1, \ldots, k$, for some $k \geq 1$.

Induction Step. Consider any derivation of the form

$$S \Rightarrow_H^{k+1} y$$

Then, there exists a sentential form w such that

$$S \Rightarrow_H^k w \Rightarrow_H y$$

where $w, y \in (N \cup T)^*$. First, observe that $S \Rightarrow_H^k w$ implies that

$$(S_1, S_2, \ldots, S_n) \Rightarrow_\Gamma^{k-1} (w_1, w_2, \ldots, w_n)$$

so that $w = w_1 w_2 \ldots w_n$, where $w_i \in (N_i \cup T_i)^*$, for all $i = 1, \ldots, n$, by the induction hypothesis. Furthermore, let $w \Rightarrow_H y$ by matrix $p_1 p_2 \ldots p_n \in M$, where $w = w_1 w_2 \ldots w_n$. Let p_i be a rule of the form $A_i \rightarrow x_i$. The rule p_i can be applied only inside substring w_i, for all $i = 1, \ldots, n$. Assume that $w_i = u_i A_i v_i$, where $u_i, v_i \in (N \cup T)^*$, $A_i \in N_i$, for all $i = 1, \ldots, n$. There exist a derivation step

$$\begin{aligned} &u_1 A_1 v_1 u_2 A_2 v_2 \ldots u_n A_n v_n \Rightarrow_H \\ &\quad u_1 x_1 v_1 u_2 x_2 v_2 \ldots u_n x_n v_n \end{aligned}$$

by matrix $p_1 p_2 \ldots p_n \in M$. Algorithm 13.1.5 implies that

$$\big((p_1: A_1 \rightarrow x_1), (p_2: A_2 \rightarrow x_2), \ldots, (p_n: A_n \rightarrow x_n)\big) \in Q$$

because $p_1 p_2 \ldots p_n \in M$. Hence,

$$\begin{aligned} &(u_1 A_1 v_1, u_2 A_2 v_2, \ldots, u_n A_n j_n \Rightarrow_\Gamma \\ &\quad (u_1 x_1 v_1, u_2 x_2 v_2, \ldots, u_n x_n j_n) \end{aligned}$$

As a result, we obtain

$$(S_1, S_2, \ldots, S_n) \Rightarrow_\Gamma^k (u_1 x_1 v_1, u_2 x_2 v_2, \ldots, u_n x_n j_n)$$

so that $y = u_1 x_1 v_1 u_2 x_2 v_2 \ldots u_n x_n v_n$. □

Consider Claim 1 for $y_i \in T_i^*$, for all $i = 1, \ldots, n$. We obtain an implication of this form

$$\text{if } (S_1, S_2, \ldots, S_n) \Rightarrow_\Gamma^* (y_1, y_2, \ldots, y_n) \\ \text{then } S \Rightarrow_H^* y_1 y_2 \ldots y_n$$

Hence, $L_{conc}(\Gamma) \subseteq L(H)$. Consider Claim 2 for $y \in T^*$. We obtain an implication of the form

$$\text{if } S \Rightarrow_H^* y \text{then } (S_1, S_2, \ldots, S_n) \Rightarrow_\Gamma^* (y_1, y_2, \ldots, y_n), \text{ such that } y = y_1 y_2 \ldots y_n$$

Hence, $L(H) \subseteq L_{conc}(\Gamma)$. □

Algorithm 13.1.7. *Conversion of an n-MGR in the first mode to an equivalent matrix grammar.*

Input: An n-MGR, $\Gamma = (G_1, G_2, \ldots, G_n, Q)$.
Output: A matrix grammar, $H = (G, M)$, satisfying $L_{first}(\Gamma) = L(H)$.
Method: Let $G_i = (N_i, T_i, P_i, S_i)$, for all $i = 1, \ldots, n$, and without any loss of generality, we assume that N_1 through N_n are pairwise disjoint. Let us choose arbitrary S satisfying $S \notin \bigcup_{j=1}^n N_j$. Construct

$$G = \big(N, T, P, S\big)$$

where

- $N = \{S\} \cup N_1 \cup (\bigcup_{i=2}^n \{\bar{A}: A \in N_i\})$;
- $T = T_1$;
- $P = \{(s: S \rightarrow S_1 h(S_2) \ldots h(S_n))\} \cup P_1$
  $\cup (\bigcup_{i=2}^n \{h(A) \rightarrow h(x) \mid A \rightarrow x \in P_i\})$,
  where h is a homomorphism from $((\bigcup_{i=2}^n N_i) \cup (\bigcup_{i=2}^n T_i))^*$ to $(\bigcup_{i=2}^n \{\bar{A} \mid A \in N_i\})^*$ defined as $h(a) = \varepsilon$, for all $a \in \bigcup_{i=2}^n T_i$ and $h(A) = \bar{A}$, for all $A \in \bigcup_{i=2}^n N_i$.

Finally, set $M = \{s\} \cup \{p_1 \bar{p}_2 \ldots \bar{p}_n \mid (p_1, p_2, \ldots, p_n) \in Q\}$. □

Theorem 13.1.8. *Let $\Gamma = (G_1, G_2, \ldots, G_n, Q)$ be an n-MGR. With Γ as its input, Algorithm 13.1.7 halts and correctly constructs a matrix grammar, $H = (G, M)$, such that $L_{first}(\Gamma) = L(H)$.*

Proof. Let $(p: A \rightarrow x)$ be a rule. Then, for simplicity and brevity, $\bar{p}$ denotes the rule $h(A) \rightarrow h(x)$. To prove this theorem, we first establish Claims 1 and 2.

Claim 1. Let $(S_1, S_2, \ldots, S_n) \Rightarrow_\Gamma^m (y_1, y_2, \ldots, y_n)$, where $m \geq 0$, $y_i \in (N_i \cup T_i)^*$, for all $i = 1, \ldots, n$. Then, $S \Rightarrow_H^{m+1} y_1 h(y_2) \ldots h(y_n)$.

Proof. This claim is proved by induction on $m \geq 0$.

Basis. Let $m = 0$. Then, $(S_1, S_2, \ldots, S_n) \Rightarrow_\Gamma^0 (S_1, S_2, \ldots, S_n)$. Notice that $S \Rightarrow_H S_1 h(S_2) \ldots h(S_n)$, because $(s: S \rightarrow S_1 h(S_2) \ldots h(S_n)) \in M$.

Induction Hypothesis. Assume that the claim holds for all m-step derivations, where $m = 0, \ldots, k$, for some $k \geq 0$.

Induction Step. Consider any derivation of the form

$$(S_1, S_2, \ldots, S_n) \Rightarrow_{\Gamma}^{k+1} (y_1, y_2, \ldots, y_n)$$

Then, there exists a sentential n-form $(u_1 A_1 v_1, u_2 A_2 v_2, \ldots, u_n A_n v_n)$, where $u_i, v_i \in (T_i \cup N_i)^*$, $A_i \in N_i$ such that

$$\begin{aligned}(S_1, S_2, \ldots, S_n) &\Rightarrow_{\Gamma}^{k} (u_1 A_1 v_1, u_2 A_2 v_2, \ldots, u_n A_n v_n)\\ &\Rightarrow_{\Gamma} (u_1 x_1 v_1, u_2 x_2 v_2, \ldots, u_n x_n v_n)\end{aligned}$$

where $u_i x_i v_i = y_i$, for all $i = 1, \ldots, n$. First, observe that

$$(S_1, S_2, \ldots, S_n) \Rightarrow_{\Gamma}^{k} (u_1 A_1 v_1, u_2 A_2 v_2, \ldots, u_n A_n v_n)$$

implies that

$$S \Rightarrow_{H}^{k+1} u_1 A_1 v_1 h(u_2 A_2 v_2) \ldots h(u_n A_n v_n)$$

by the induction hypothesis. Furthermore, let

$$\begin{aligned}&(u_1 A_1 v_1, u_2 A_2 v_2, \ldots, u_n A_n v_n) \Rightarrow_{\Gamma}\\ &\quad (u_1 x_1 v_1, u_2 x_2 v_2, \ldots, u_n x_n v_n)\end{aligned}$$

Then, it holds $((p_1{:}\ A_1 \rightarrow x_1), (p_2{:}\ A_2 \rightarrow x_2), \ldots, (p_n{:}\ A_n \rightarrow x_n)) \in Q$. Algorithm 13.1.7 implies that $p_1 \bar{p}_2 \ldots \ldots \bar{p}_n \in M$. Hence,

$$\begin{aligned}&u_1 A_1 v_1 h(u_2 A_2 v_2) \ldots h(u_n A_n v_n) \Rightarrow_H\\ &\quad u_1 x_1 v_1 h(u_2 x_2 v_2) \ldots h(u_n x_n v_n)\end{aligned}$$

by matrix $p_1 \bar{p}_2 \ldots \bar{p}_n$. As a result, we obtain

$$S \Rightarrow_{H}^{k+2} u_1 x_1 v_1 h(u_2 x_2 v_2) \ldots h(u_n x_n v_n)$$

□

Claim 2. Let $S \Rightarrow_H^m y$, where $m \geq 1, y \in (N \cup T)^*$. Then, $(S_1, S_2, \ldots, S_n) \Rightarrow_{\Gamma}^{m-1} (y_1, y_2, \ldots, y_n)$, where $y_i \in (N_i \cup T_i)^*$, for all $i = 1, \ldots, n$ so that $y = y_1 h(y_2) \ldots h(y_n)$.

Proof. This claim is proved by induction on $m \geq 1$.

Basis. Let $m = 1$. Then, there exists exactly one one-step derivation in H: $S \Rightarrow_H S_1 h(S_2) \ldots h(S_n)$ by matrix s. Notice that $(S_1, S_2, \ldots, S_n) \Rightarrow_{\Gamma}^{0} (S_1, S_2, \ldots, S_n)$ trivially.

Induction Hypothesis. Assume that the claim holds for all m-step derivations, where $m = 1, \ldots, k$, for some $k \geq 1$.

Induction Step. Consider any derivation of the form

$$S \Rightarrow_H^{k+1} y$$

Then, there is w such that

$$S \Rightarrow_H^{k} w \Rightarrow_H y$$

where $w, y \in (N \cup T)^*$. First, observe that $S \Rightarrow_H^k w$ implies that

$$(S_1, S_2, \ldots, S_n) \Rightarrow_\Gamma^{k-1} (w_1, w_2, \ldots, w_n)$$

so that $w = w_1 h(w_2) \ldots h(w_n)$, where $w_i \in (N_i \cup T_i)^*$, for all $i = 1, \ldots, n$, by the induction hypothesis. Furthermore, let $w \Rightarrow_H y$, where $w = w_1 h(w_2) \ldots h(w_n)$. Let p_1 be a rule of the form $A_1 \to x_1$. Let $\bar{p}_i$ be a rule of the form $h(A_i) \to h(x)$, for all $i = 2, \ldots, n$. The rule p_1 can be applied only inside substring w_1, the rule $\bar{p}_i$ can be applied only inside substring w_i, for all $i = 2, \ldots, n$. Assume that $w_i = u_i A_i v_i$, where $u_i, v_i \in (N_i \cup T_i)^*$, $A_i \in N_i$, for all $i = 1, \ldots, n$. There exists a derivation step

$$\begin{aligned} &u_1 A_1 v_1 h(u_2 A_2 v_2) \ldots h(u_n A_n v_n) \Rightarrow_H \\ &\quad u_1 x_1 v_1 h(u_2 x_2 v_2) \ldots h(u_n x_n v_n) \end{aligned}$$

by matrix $p_1 \bar{p}_2 \ldots \bar{p}_n \in M$. Algorithm 13.1.7 implies that

$$\big((p_1 \colon A_1 \to x_1), (p_2 \colon A_2 \to x_2), \ldots, (p_n \colon A_n \to x_n)\big) \in Q$$

because $p_1 \bar{p}_2 \ldots \bar{p}_n \in M$. Hence,

$$\begin{aligned} &(u_1 A_1 v_1, u_2 A_2 v_2, \ldots, u_n A_n j_n) \Rightarrow_\Gamma \\ &\quad (u_1 x_1 v_1, u_2 x_2 v_2, \ldots, u_n x_n j_n) \end{aligned}$$

As a result, we obtain

$$(S_1, S_2, \ldots, S_n) \Rightarrow_\Gamma^k (u_1 x_1 v_1, u_2 x_2 v_2, \ldots, u_n x_n j_n)$$

so that $y = u_1 x_1 v_1 h(u_2 x_2 v_2) \ldots h(u_n x_n v_n)$. □

Consider Claim 1 for $y_i \in T_i^*$, for all $i = 1, \ldots, n$. Notice that $h(a) = \varepsilon$, for all $a \in T_i$. We obtain an implication of the form

$$\begin{aligned} &\text{if } (S_1, S_2, \ldots, S_n) \Rightarrow_\Gamma^* (y_1, y_2, \ldots, y_n) \\ &\text{then } S \Rightarrow_H^* y_1 \end{aligned}$$

Hence, $L_{first}(\Gamma) \subseteq L(H)$. Consider Claim 2 for $y \in T^*$. Notice that $h(a) = \varepsilon$, for all $a \in T_i$. We obtain an implication of the form

if $S \Rightarrow_H^* y$
then $(S_1, S_2, \dots, S_n) \Rightarrow_\Gamma^* (y_1, y_2, \dots, y_n)$, such that $y = y_1$

Hence, $L(H) \subseteq L_{first}(\Gamma)$. Therefore, $L(H) = L_{first}(\Gamma)$. □

Algorithm 13.1.9. *Conversion of a matrix grammar to an equivalent 2-MGR.*

Input: A matrix grammar, $H = (G, M)$, and a string, $\bar{w} \in \bar{T}^*$, where $\bar{T}$ is any alphabet.

Output: A 2-MGR, $\Gamma = (G_1, G_2, Q)$, satisfying $\{w_1 \mid (w_1, \bar{w}) \in 2\text{-}L(\Gamma)\} = L(H)$.

Method: Let $G = (N, T, P, S)$. Then, set $G_1 = G$ and construct

$$G_2 = \big(N_2, T_2, P_2, S_2\big)$$

where

- $N_2 = \{\langle p_1 p_2 \dots p_k, j\rangle \mid p_1, \dots, p_k \in P, p_1 p_2 \dots p_k \in M, 1 \le j \le k-1\} \cup \{S_2\}$;
- $T_2 = \bar{T}$;
- $P_2 = \{S_2 \to \langle p_1 p_2 \dots p_k, 1\rangle \mid p_1, \dots, p_k \in P, p_1 p_2 \dots p_k \in M, k \ge 2\}$
  $\cup \{\langle p_1 p_2 \dots p_k, j\rangle \to \langle p_1 p_2 \dots p_k, j+1\rangle \mid p_1 p_2 \dots p_k \in M, k \ge 2, 1 \le j \le k-2\}$
  $\cup \{\langle p_1 p_2 \dots p_k, k-1\rangle \to S_2 \mid p_1, \dots, p_k \in P, p_1 p_2 \dots p_k \in M, k \ge 2\}$
  $\cup \{S_2 \to S_2 \mid p_1 \in M, |p_1| = 1\}$
  $\cup \{\langle p_1 p_2 \dots p_k, k-1\rangle \to \bar{w} \mid p_1, \dots, p_k \in P, p_1 p_2 \dots p_k \in M, k \ge 2\}$
  $\cup \{S_2 \to \bar{w} \mid p_1 \in M, |p_1| = 1\}$;
- $Q = \{(p_1, S_2 \to \langle p_1 p_2 \dots p_k, 1\rangle) \mid p_1, \dots, p_k \in P, p_1 p_2 \dots p_k \in M, k \ge 2\}$
  $\cup \{(p_{j+1}, \langle p_1 p_2 \dots p_k, j\rangle \to \langle p_1 p_2 \dots p_k, j+1\rangle) \mid p_1 p_2 \dots p_k \in M, k \ge 2, 1 \le j \le k-2\}$
  $\cup \{(p_k, \langle p_1 p_2 \dots p_k, k-1\rangle \to S_2) \mid p_1, \dots, p_k \in P, p_1 p_2 \dots p_k \in M, k \ge 2\}$
  $\cup \{(p_1, S_2 \to S_2) \mid p_1 \in M, |p_1| = 1\}$
  $\cup \{(p_k, \langle p_1 p_2 \dots p_k, k-1\rangle \to \bar{w}) \mid p_1, \dots, p_k \in P, p_1 p_2 \dots p_k \in M, k \ge 2\}$
  $\cup \{(p_1, S_2 \to \bar{w}) \mid p_1 \in M, |p_1| = 1\}$. □

Theorem 13.1.10. *Let H be a matrix grammar and $\bar{w}$ be a string. With H and $\bar{w}$ as its input, Algorithm 13.1.9 halts and correctly constructs a 2-MGR, $\Gamma = (G_1, G_2, Q)$, such that $\{w_1 \mid (w_1, \bar{w}) \in 2\text{-}L(\Gamma)\} = L(H)$.*

Proof. To prove this theorem, we first establish Claims 1 through 4.

Claim 1. Let $x \Rightarrow_H y$, *where* $x, y \in (N \cup T)^*$. *Then,* $(x, S_2) \Rightarrow_\Gamma^* (y, S_2)$ *and* $(x, S_2) \Rightarrow_\Gamma^* (y, \bar{w})$.

Proof. In this proof, we distinguish two cases—I and II. In I, we consider any derivation step of the form $x \Rightarrow_H y$ by a matrix consisting of a single rule. In II, we consider $x \Rightarrow_H y$ by a matrix consisting of several rules.

I. Consider any derivation step of the form $x \Rightarrow_H y$ by a matrix which contains only one rule $(p_1\colon A_1 \to x_1)$. It implies that $uA_1v \Rightarrow_G ux_1v\ [p_1]$, where $uA_1v = x, ux_1v = y$. Algorithm 13.1.9 implies that $(A_1 \to x_1, S_2 \to S_2) \in Q$ and $(A_1 \to x_1, S_2 \to \bar{w}) \in Q$. Hence, $(uA_1v, S_2) \Rightarrow_\Gamma (ux_1v, S_2)$ and $(uA_1v, S_2) \Rightarrow_\Gamma (ux_1v, \bar{w})$.

II. Let $x \Rightarrow_H y$ by a matrix of the form $p_1p_2 \dots p_k$, where $p_i, \dots, p_k \in P, k \geq 2$. It implies that

$$
\begin{aligned}
x &\Rightarrow_H y_1 & [p_1] \\
&\Rightarrow_H y_2 & [p_2] \\
&\ \vdots \\
&\Rightarrow_H y_{k-1} & [p_{k_1}] \\
&\Rightarrow_H y_k & [p_k]
\end{aligned}
$$

where $y_k = y$. Algorithm 13.1.9 implies that $(p_1, S_2 \to \langle p_1p_2 \dots p_k, 1\rangle) \in Q$, $(p_{j+1}, \langle p_1p_2 \dots p_k, j\rangle \to \langle p_1p_2 \dots p_k, j+1\rangle) \in Q$, where $j = 1, \dots, k-2$, $(p_k, \langle p_1p_2 \dots p_k, k-1\rangle \to S_2) \in Q$, $(p_k, \langle p_1p_2 \dots p_k, k-1\rangle \to \bar{w}) \in Q$. Hence,

$$
\begin{aligned}
(x, S_2) &\Rightarrow_\Gamma (y_1, \langle p_1p_2 \dots p_k, 1\rangle) \\
&\Rightarrow_\Gamma (y_2, \langle p_1p_2 \dots p_k, 2\rangle) \\
&\ \vdots \\
&\Rightarrow_\Gamma (y_{k-1}, \langle p_1p_2 \dots p_k, k-1\rangle) \\
&\Rightarrow_\Gamma (y_k, S_2)
\end{aligned}
$$

and

$$
\begin{aligned}
(x, S_2) &\Rightarrow_\Gamma (y_1, \langle p_1p_2 \dots p_k, 1\rangle) \\
&\Rightarrow_\Gamma (y_2, \langle p_1p_2 \dots p_k, 2\rangle) \\
&\ \vdots \\
&\Rightarrow_\Gamma (y_{k-1}, \langle p_1p_2 \dots p_k, k-1\rangle) \\
&\Rightarrow_\Gamma (y_k, \bar{w})
\end{aligned}
$$

where $y_k = y$. □

Claim 2. Let $x \Rightarrow_H^m y$, *where* $m \geq 1, y \in (N \cup T)^*$. *Then,* $(x, S_2) \Rightarrow_\Gamma^* (y, \bar{w})$.

Proof. This claim is proved by induction on $m \geq 1$.

Basis. Let $m = 1$ and let $x \Rightarrow_H y$. Claim 1 implies that $(x, S_2) \Rightarrow_\Gamma^* (y, \bar{w})$.

Induction Hypothesis. Assume that the claim holds for all m-step derivations, where $m = 1, \ldots, k$, for some $k \geq 1$.

Induction Step. Consider any derivation of the form

$$S \Rightarrow_H^{k+1} y$$

Then, there exists w such that

$$S \Rightarrow_H w \Rightarrow_H^k y$$

where $w, y \in (N \cup T)^*$. First, observe that $w \Rightarrow_H^k y$ implies that

$$(w, S_2) \Rightarrow_\Gamma^* (y, \bar{w})$$

by the induction hypothesis. Furthermore, let $x \Rightarrow_H w$. Claim 1 implies that

$$(x, S_2) \Rightarrow_\Gamma^* (w, S_2)$$

As a result, we obtain

$$(x, S_2) \Rightarrow_\Gamma^* (y, \bar{w})$$

□

Claim 3. Let $(y_0, S_2) \Rightarrow_\Gamma (y_1, z_1) \Rightarrow_\Gamma (y_2, z_2) \Rightarrow_\Gamma \cdots \Rightarrow_\Gamma (y_{k-1}, z_{k-1}) \Rightarrow_\Gamma (y_k, S_2)$ or $(y_0, S_2) \Rightarrow_\Gamma (y_1, z_1) \Rightarrow_\Gamma (y_2, z_2) \Rightarrow_\Gamma \cdots \Rightarrow_\Gamma (y_{k-1}, z_{k-1}) \Rightarrow_\Gamma (y_k, \bar{w})$, where $z_i \neq S_2$, for all $i = 1, \ldots, k-1$. Then, there exists a direct derivation step $y_0 \Rightarrow_H y_k$.

Proof. In this proof, we distinguish two cases—I and II. In I, we consider any derivation step of the form $x \Rightarrow_H y$ by a matrix consisting of a single rule. In II, we consider $x \Rightarrow_H y$ by a matrix consisting of several rules.

I. Consider any derivation step of the form

$$(uA_1v, S_2) \Rightarrow_\Gamma (ux_1v, S_2)$$

or

$$(uA_1v, S_2) \Rightarrow_\Gamma (ux_1v, \bar{w})$$

where $uA_1v = y_0, ux_1v = y_1$. Then, $(A_1 \rightarrow x_1, S_2 \rightarrow S_2) \in Q$ or $(A_1 \rightarrow x_1, S_2 \rightarrow \bar{w}) \in Q$. Algorithm 13.1.9 implies that there exists a matrix of the form $(p_1\colon A_1 \rightarrow x_1) \in M$. Hence,

$$uA_1v \Rightarrow_H ux_1v$$

II. Let

$$\begin{aligned}(y_0, S_2) &\Rightarrow_\Gamma (y_1, z_1)\\ &\Rightarrow_\Gamma (y_2, z_2)\\ &\vdots\\ &\Rightarrow_\Gamma (y_{k-1}, z_{k-1})\\ &\Rightarrow_\Gamma (y_k, S_2)\end{aligned}$$

or

$$\begin{aligned}(y_0, S_2) &\Rightarrow_\Gamma (y_1, z_1)\\ &\Rightarrow_\Gamma (y_2, z_2)\\ &\vdots\\ &\Rightarrow_\Gamma (y_{k-1}, z_{k-1})\\ &\Rightarrow_\Gamma (y_k, \bar{w})\end{aligned}$$

where $z_i \neq S_2$, for all $i = 1, \ldots, k - 1$ and $k \geq 2$. Algorithm 13.1.9 implies that there exists a matrix $p_1 p_2 \ldots p_k \in M$ and that $z_i = \langle p_1 p_2 \ldots p_k, i\rangle$, for all $i = 1, \ldots k - 1$. Hence,

$$y_0 \Rightarrow_H y_k$$

□

Claim 4. Let $(y_0, S_2) \Rightarrow_\Gamma (y_1, z_1) \Rightarrow_\Gamma (y_2, z_2) \Rightarrow_\Gamma \cdots \Rightarrow_\Gamma (y_{r-1}, z_{r-1}) \Rightarrow_\Gamma (y_r, \bar{w})$. *Set* $m = \mathrm{card}(\{i \mid 1 \leq i \leq r - 1, z_i = S_2\})$. *Then,* $y_0 \Rightarrow_H^{m+1} y_r$.

Proof. This claim is proved by induction on $m \geq 0$.

Basis. Let $m = 0$. Then, $z_i \neq S_2$, for all $i = 1, \ldots, k - 1$. Claim 3 implies that there exists a derivation step $y_0 \Rightarrow_H y_r$.

Induction Hypothesis. Assume that the claim holds for all m-step derivations, where $m = 0, \ldots, k$, for some $k \geq 0$.

Induction Step. Consider any derivation of the form

$$\begin{aligned}(y_0, S_2) &\Rightarrow_\Gamma (y_1, z_1)\\ &\Rightarrow_\Gamma (y_2, z_2)\\ &\vdots\\ &\Rightarrow_\Gamma (y_{r-1}, z_{r-1})\\ &\Rightarrow_\Gamma (y_r, \bar{w})\end{aligned}$$

where $\mathrm{card}(\{i \mid 1 \leq i \leq r - 1, z_i = S_2\}) = k + 1$. Then, there exists $p \in \{1, \ldots, r - 1\}$ such that $z_p = S_2$, $\mathrm{card}(\{i \mid 1 \leq i \leq p - 1, z_i = S_2\}) = 0$, $\mathrm{card}(\{i \mid p + 1 \leq i \leq r - 1, z_i = S_2\}) = k$, and

$$\begin{aligned}(y_0, z_0) &\Rightarrow_\Gamma (y_1, z_1)\\ &\vdots\\ &\Rightarrow_\Gamma (y_p, z_p)\\ &\vdots\\ &\Rightarrow_\Gamma (y_{r-1}, z_{r-1})\\ &\Rightarrow_\Gamma (y_r, \bar{w})\end{aligned}$$

First, observe that from

$$\begin{aligned}(y_p, z_p) &\Rightarrow_\Gamma (y_1, z_1)\\ &\vdots\\ &\Rightarrow_\Gamma (y_{r-1}, z_{r-1})\\ &\Rightarrow_\Gamma (y_r, \bar{w})\end{aligned}$$

where $z_p = S_2$ and $\mathrm{card}(\{i \mid p+1 \le i \le r-1, z_i = S_2\}) = k$, it follows

$$y_p \Rightarrow_H^{k+1} y_r$$

by the induction hypothesis. Furthermore, let

$$\begin{aligned}(y_p, z_p) &\Rightarrow_\Gamma (y_1, z_1)\\ &\vdots\\ &\Rightarrow_\Gamma (y_p, z_p)\end{aligned}$$

$\mathrm{card}(\{i \mid 1 \le i \le p-1, z_i = S_2\}) = 0$ implies that $z_i \ne S_2$, for all $i = 1, \dots, p$. Claim 3 implies that there exists a derivation step $y_0 \Rightarrow_H y_p$. As a result, we obtain

$$y_0 \Rightarrow_H^{k+2} y_r$$

□

We next prove the following two identities, (1) and (2).

(1) $\{w_1 \mid (w_1, \bar{w}) \in 2\text{-}L(\Gamma)\} = L(H)$.

Consider Claim 2 for $x = S$ and $y \in T^*$. We obtain an implication of the form

$$\begin{aligned}&\text{if } S \Rightarrow_H^* y\\ &\text{then } (S, S_2) \Rightarrow_\Gamma^* (y, \bar{w})\end{aligned}$$

Hence, $L(H) \subseteq \{w_1 \mid (w_1, \bar{w}) \in 2\text{-}L(\Gamma)\}$.

Consider Claim 4 for $y_0 = S$ and $y_r \in T^*$. We see that

$$\begin{aligned}&\text{if } (S, S_2) \Rightarrow_\Gamma^* (y_r, \bar{w})\\ &\text{then } S \Rightarrow_H^* y_r\end{aligned}$$

Hence, $\{w_1 \mid (w_1, \bar{w}) \in 2\text{-}L(\Gamma)\} \subseteq L(H)$.

(2) $\{(w_1, w_2) \mid (w_1, w_2) \in 2\text{-}L(\Gamma), w_2 \neq \bar{w}\} = \emptyset$. Notice that Algorithm 13.1.9 implies that $G_2 = (N_2, T_2, P_2, S_2)$ contains only rules of the form $A \to B$ and $A \to \bar{w}$, where $A, B \in N_2$. Hence, G_2 generates $\emptyset$ or $\{\bar{w}\}$. Γ contains G_2 as a second component; hence, $\{(w_1, w_2) \mid (w_1, w_2) \in 2\text{-}L(\Gamma), w_2 \neq \bar{w}\} = \emptyset$. □

Theorem 13.1.11. *For every matrix grammar H, there is a 2-MGR Γ such that $L(H) = L_{union}(\Gamma)$.*

Proof. To prove this theorem, we make use of Algorithm 13.1.9 with matrix grammar H and $\bar{w}$ as input, where $\bar{w}$ is any string in $L(H)$, provided that $L(H)$ is nonempty. Otherwise, if $L(H)$ is nonempty, let $\bar{w}$ be any string. We prove that $L(H) = L_{union}(\Gamma)$.

(1) If $L(H) = \emptyset$, take any string $\bar{w}$ and use Algorithm 13.1.9 to construct Γ. Observe that $L_{union}(\Gamma) = \emptyset = L(H)$.
(2) If $L(H) \neq \emptyset$, take any $\bar{w} \in L(H)$ and use Algorithm 13.1.9 to construct Γ. As obvious, $L_{union}(\Gamma) = L(H) \cup \bar{w} = L(H)$. □

Theorem 13.1.12. *For every matrix grammar H, there is a 2-MGR Γ such that $L(H) = L_{conc}(\Gamma)$.*

Proof. To prove this theorem, we make use of Algorithm 13.1.9 with matrix grammar H and $\bar{w} = \varepsilon$ as input. We prove that $L(H) = L_{conc}(\Gamma)$. Theorem 13.1.10 says that

$$\{w_1 \mid (w_1, \bar{w}) \in 2\text{-}L(\Gamma)\} = L(H)$$

and

$$\{(w_1, w_2) \mid (w_1, w_2) \in 2\text{-}L(\Gamma), w_2 \neq \bar{w}\} = \emptyset$$

Then,

$$\begin{aligned} L_{conc}(\Gamma) &= \{w_1 w_2 \mid (w_1, w_2) \in 2 - L(\Gamma)\} \\ &= \{w_1 w_2 \mid (w_1, w_2) \in 2 - L(\Gamma), w_2 = \bar{w}\} \\ &\quad \cup \{w_1 w_2 \mid (w_1, w_2) \in 2 - L(\Gamma), w_2 \neq \bar{w}\} \\ &= \{w_1 \bar{w} \mid (w_1, \bar{w}) \in 2 - L(\Gamma)\} \cup \emptyset \\ &= \{w_1 \bar{w} \mid (w_1, \bar{w}) \in 2 - L(\Gamma)\} \\ &= L(H) \end{aligned}$$

because $\bar{w} = \varepsilon$. □

Theorem 13.1.13. *For every matrix grammar H, there is a 2-MGR Γ such that $L(H) = L_{first}(\Gamma)$.*

Proof. To prove this theorem, we make use of Algorithm 13.1.9 with matrix grammar H and any $\bar{w}$ as input. We prove that $L(H) = L_{first}(\Gamma)$. Theorem 13.1.10 says that

$$\{w_1 \mid (w_1, \bar{w}) \in 2\text{-}L(\Gamma)\} = L(H)$$

and

$$\{(w_1, w_2) \mid (w_1, w_2) \in 2\text{-}L(\Gamma), w_2 \neq \bar{w}\} = \emptyset$$

Then,

$$\begin{aligned} L_{first}(\Gamma) &= \{w_1 \mid (w_1, w_2) \in n\text{-}L(\Gamma)\} \\ &= \{w_1 \mid (w_1, w_2) \in 2 - L(\Gamma), w_2 = \bar{w}\} \\ &\quad \cup \{w_1 \mid (w_1, w_2) \in 2 - L(\Gamma), w_2 \neq \bar{w}\} \\ &= \{w_1 \mid (w_1, \bar{w}) \in 2 - L(\Gamma)\} \cup \emptyset \\ &= \{w_1 \mid (w_1, \bar{w}) \in 2 - L(\Gamma)\} \\ &= L(H) \end{aligned}$$

Hence, the theorem holds. □

Let $\mathbf{MGR}_{n,X}$ denote the language families defined by n-MGRs in the X mode, where $X \in \{union, conc, first\}$. From the previous results, we obtain the following corollary.

Corollary 13.1.14. $\mathbf{M} = \mathbf{MGR}_{n,X}$, *where* $n \geq 2$, $X \in \{union, conc, first\}$. □

To summarize all the results, multigenerative grammar systems with any number of grammatical components are equivalent with two-component versions of these systems. Perhaps even more importantly, these systems are equivalent with matrix grammars, which generate a proper subfamily of the family of recursively enumerable languages (see Theorem 5.2.5).

We close this section by suggesting two open problem areas.

Open Problem 13.1.15. Consider other operations, like intersection, and study languages generated in this way by multigenerative grammars systems. □

Open Problem 13.1.16. Study multigenerative grammars systems that are based on other grammars than context-free grammars. Specifically, determine the generative power of multigenerative grammar systems with regular or right-linear grammars as components. □

13.2 Leftmost Multigenerative Grammar Systems

In this section, we study leftmost versions of multigenerative grammar systems, whose basic versions were defined and investigated in the previous section of this chapter. We prove that they characterize the family of recursively enumerable languages, which properly contains the family of matrix languages (see Theorems 5.1.6 and 5.2.5). We also consider regulation by n-tuples of nonterminals rather than rules, and we prove that leftmost multigenerative grammar systems regulated by rules or nonterminals have the same generative power. Just like for multigenerative grammar systems in the previous section, we explain how to reduce the number of grammatical components in leftmost multigenerative grammar systems to two.

Definition 13.2.1. A *leftmost n-generative rule-synchronized grammar system* (an *n-LMGR* for short) is an $n + 1$ tuple

$$\Gamma = \big(G_1, G_2, \ldots, G_n, Q\big)$$

where

- $G_i = (N_i, T_i, P_i, S_i)$ is a context-free grammar, for each $i = 1, \ldots, n$;
- Q is a finite set of n-tuples of the form $(p_1, p_2, \ldots, p_n)$, where $p_i \in P_i$, for all $i = 1, \ldots, n$.

A *sentential n-form* is an n-tuple of the form $\chi = (x_1, x_2, \ldots, x_n)$, where $x_i \in (N_i \cup T_i)^*$, for all $i = 1, \ldots, n$. Let $\chi = (u_1 A_1 v_1, u_2 A_2 v_2, \ldots, u_n A_n v_n)$ and $\bar{\chi} = (u_1 x_1 v_1, u_2 x_2 v_2, \ldots, u_n x_n v_n)$ be two sentential n-forms, where $A_i \in N_i$, $u_i \in T^*$, and $v_i, x_i \in (N_i \cup T_i)^*$, for all $i = 1, \ldots, n$. Let $(p_i \colon A_i \to x_i) \in P_i$, for all $i = 1, \ldots, n$ and $(p_1, p_2, \ldots, p_n) \in Q$. Then, χ directly derives $\bar{\chi}$ in Γ, denoted by

$$\chi \Rightarrow_\Gamma \bar{\chi}$$

In the standard way, we generalize $\Rightarrow_\Gamma$ to $\Rightarrow_\Gamma^k$, for all $k \geq 0$, $\Rightarrow_\Gamma^*$, and $\Rightarrow_\Gamma^+$.

The *n-language of* Γ, denoted by $n\text{-}L(\Gamma)$, is defined as

$$n\text{-}L(\Gamma) = \big\{(w_1, w_2, \ldots, w_n) \mid (S_1, S_2, \ldots, S_n) \Rightarrow_\Gamma^* (w_1, w_2, \ldots, w_n),\\ w_i \in T_i^*, \text{ for all } i = 1, \ldots, n\big\}$$

The *language generated by* Γ *in the union mode*, $L_{union}(\Gamma)$, is defined as

$$L_{union}(\Gamma) = \bigcup_{i=1}^{n} \big\{w_i \mid (w_1, w_2, \ldots, w_n) \in n\text{-}L(\Gamma)\big\}$$

The *language generated by* Γ *in the concatenation mode*, $L_{conc}(\Gamma)$, is defined as

$$L_{conc}(\Gamma) = \big\{w_1 w_2 \ldots w_n \mid (w_1, w_2, \ldots, w_n) \in n\text{-}L(\Gamma)\big\}$$

The *language generated by* Γ *in the first mode*, $L_{first}(\Gamma)$, is defined as

$$L_{first}(\Gamma) = \big\{w_1 \mid (w_1, w_2, \ldots, w_n) \in n\text{-}L(\Gamma)\big\}$$ □

Next, we illustrate the above definition by an example.

Example 13.2.2. Consider the 2-LMGR $\Gamma = (G_1, G_2, Q)$, where

- $G_1 = (\{S_1, A_1\}, \{a, b, c\}, \{1\colon S_1 \to aS_1, 2\colon S_1 \to aA_1, 3\colon A_1 \to bA_1c, 4\colon A_1 \to bc\}, S_1)$,
- $G_2 = (\{S_2, A_2\}, \{d\}, \{1\colon S_2 \to S_2A_2, 2\colon S_2 \to A_2, 3\colon A_2 \to d\}, S_2)$,
- $Q = \{(1, 1), (2, 2), (3, 3), (4, 3)\}$.

Observe that

- $2\text{-}L(\Gamma) = \{(a^n b^n c^n, d^n) \mid n \geq 1\}$,
- $L_{union}(\Gamma) = \{a^n b^n c^n \mid n \geq 1\} \cup \{d^n \mid n \geq 1\}$,
- $L_{conc}(\Gamma) = \{a^n b^n c^n d^n \mid n \geq 1\}$, and
- $L_{first}(\Gamma) = \{a^n b^n c^n \mid n \geq 1\}$. □

Next, we introduce regulation by n-tuples of nonterminals rather than rules.

Definition 13.2.3. A *leftmost* n*-generative nonterminal-synchronized grammar system* (an n*-LMGN* for short) is an $n + 1$ tuple

$$\Gamma = (G_1, G_2, \ldots, G_n, Q)$$

where

- $G_i = (N_i, T_i, P_i, S_i)$ is a context-free grammar, for each $i = 1, \ldots, n$;
- Q is a finite set of n-tuples of the form $(A_1, A_2, \ldots, A_n)$, where $A_i \in N_i$, for all $i = 1, \ldots, n$.

A *sentential* n*-form* is defined as a sentential n-form of an n-LMGR. Let $\chi = (u_1 A_1 v_1,\ u_2 A_2 v_2,\ \ldots,\ u_n A_n v_n)$ and $\bar{\chi} = (u_1 x_1 v_1,\ u_2 x_2 v_2,\ \ldots,\ u_n x_n v_n)$ be two sentential n-forms, where $A_i \in N_i$, $u_i \in T^*$, and $v_i, x_i \in (N_i \cup T_i)^*$, for all $i = 1, \ldots, n$. Let $(p_i\colon A_i \rightarrow x_i) \in P_i$, for all $i = 1, \ldots, n$ and $(A_1, A_2, \ldots, A_n) \in Q$. Then, χ directly derives $\bar{\chi}$ in Γ, denoted by

$$\chi \Rightarrow_\Gamma \bar{\chi}$$

In the standard way, we generalize $\Rightarrow_\Gamma$ to $\Rightarrow_\Gamma^k$, for all $k \geq 0$, $\Rightarrow_\Gamma^*$, and $\Rightarrow_\Gamma^+$.

An n*-language* for n-LMGN is defined as the n-language for n-LMGR, and a language generated by n-LMGN in the X mode, for each $X \in \{union,\ conc,\ first\}$, is defined as the language generated by n-LMGR in the X mode. □

Example 13.2.4. Consider the 2-LMGN $\Gamma = (G_1, G_2, Q)$, where

- $G_1 = (\{S_1, A_1\}, \{a, b, c\}, \{S_1 \rightarrow aS_1,\ S_1 \rightarrow aA_1,\ A_1 \rightarrow bA_1c,\ A_1 \rightarrow bc\}, S_1)$,
- $G_2 = (\{S_2, A_2\}, \{d\}, \{S_2 \rightarrow S_2A_2,\ S_2 \rightarrow A_2,\ A_2 \rightarrow d\}, S_2)$,
- $Q = \{(S_1, A_1), (S_2, A_2)\}$.

Observe that

- $2\text{-}L(\Gamma) = \{(a^n b^n c^n, d^n) \mid n \geq 1\}$,
- $L_{union}(\Gamma) = \{a^n b^n c^n \mid n \geq 1\} \cup \{d^n \mid n \geq 1\}$,
- $L_{conc}(\Gamma) = \{a^n b^n c^n d^n \mid n \geq 1\}$, and
- $L_{first}(\Gamma) = \{a^n b^n c^n \mid n \geq 1\}$. □

Lemma 13.2.5. *Let* Γ *be an* n*-LMGN and let* $\bar{\Gamma}$ *be an* n*-LMGR such that* $n\text{-}L(\Gamma) = n\text{-}L(\bar{\Gamma})$*. Then,* $L_X(\Gamma) = L_X(\bar{\Gamma})$*, for each* $X \in \{union,\ conc,\ first\}$*.*

Proof. I. First, we prove that $L_{union}(\Gamma) = L_{union}(\bar{\Gamma})$ as follows:

$$\begin{aligned} L_{union}(\Gamma) &= \{w \mid (w_1, \dots, w_n) \in n\text{-}L(\Gamma), w \in \{w_i \mid 1 \leq i \leq n\}\} \\ &= \{w \mid (w_1, \dots, w_n) \in n\text{-}L(\bar{\Gamma}), w \in \{w_i \mid 1 \leq i \leq n\}\} \\ &= L_{union}(\bar{\Gamma}) \end{aligned}$$

II. Second, we prove that $L_{conc}(\Gamma) = L_{conc}(\bar{\Gamma})$ as follows:

$$\begin{aligned} L_{conc}(\Gamma) &= \{w_1 \cdots w_n \mid (w_1, \dots, w_n) \in n\text{-}L(\Gamma)\} \\ &= \{w_1 \cdots w_n \mid (w_1, \dots, w_n) \in n\text{-}L(\bar{\Gamma})\} \\ &= L_{conc}(\bar{\Gamma}) \end{aligned}$$

III. Finally, we prove that $L_{first}(\Gamma) = L_{first}(\bar{\Gamma})$ as follows:

$$\begin{aligned} L_{first}(\Gamma) &= \{w_1 \mid (w_1, \dots, w_n) \in n\text{-}L(\Gamma)\} \\ &= \{w_1 \mid (w_1, \dots, w_n) \in n\text{-}L(\bar{\Gamma})\} \\ &= L_{first}(\bar{\Gamma}) \end{aligned}$$

□

Algorithm 13.2.6. *Conversion of an n-LMGN to an equivalent n-LMGR.*

Input: An n-LMGN, $\Gamma = (G_1, G_2, \dots, G_n, Q)$.
Output: An n-LMGR, $\bar{\Gamma} = (G_1, G_2, \dots, G_n, \bar{Q})$, such that $n\text{-}L(\Gamma) = n\text{-}L(\bar{\Gamma})$.
Method: Let $G_i = (N_i, T_i, P_i, S_i)$, for all $i = 1, \dots, n$, and set

$$\bar{Q} = \{(A_1 \rightarrow x_1, A_2 \rightarrow x_2, \dots, A_n \rightarrow x_n) \mid A_i \rightarrow x_i \in P_i, \text{ for all } i = 1, \dots, n, \text{ and } (A_1, A_2, \dots, A_n) \in Q\}.$$ □

Theorem 13.2.7. *Let $\Gamma = (G_1, G_2, \dots, G_n, Q)$ be an n-LMGN. With Γ as its input, Algorithm 13.2.6 halts and correctly constructs an n-LMGR, $\bar{\Gamma} = (G_1, G_2, \dots, G_n, \bar{Q})$, such that $n\text{-}L(\Gamma) = n\text{-}L(\bar{\Gamma})$, and $L_X(\Gamma) = L_X(\bar{\Gamma})$, for each $X \in \{union, conc, first\}$.*

Proof. The proof this theorem, we first establish Claims 1 and 2.

Claim 1. Let $(S_1, S_2, \dots, S_n) \Rightarrow^m_\Gamma (y_1, y_2, \dots, y_n)$, where $m \geq 0$, $y_i \in (N_i \cup T_i)^*$, for all $i = 1, \dots, n$. Then, $(S_1, S_2, \dots, S_n) \Rightarrow^m_{\bar{\Gamma}} (y_1, y_2, \dots, y_n)$.

Proof. This claim is proved by induction on $m \geq 0$.

Basis. The basis is clear.

Induction Hypothesis. Assume that Claim 1 holds for all m-step derivations, where $m = 0, \dots, k$, for some $k \geq 0$.

Induction Step. Consider any derivation of the form

$$(S_1, S_2, \dots, S_n) \Rightarrow^{k+1}_\Gamma (y_1, y_2, \dots, y_n)$$

Then, there exists a sentential n-form $(u_1A_1v_1, u_2A_2v_2, \ldots, u_nA_nv_n)$, where $u_i \in T_i^*$, $A_i \in N_i$, and $v_i \in (N_i \cup T_i)^*$, such that

$$\begin{aligned}(S_1, S_2, \ldots, S_n) &\Rightarrow_{\Gamma}^{k} (u_1A_1v_1, u_2A_2v_2, \ldots, u_nA_nv_n)\\ &\Rightarrow_{\Gamma} (u_1x_1v_1, u_2x_2v_2, \ldots, u_nx_nv_n)\end{aligned}$$

where $u_ix_iv_i = y_i$, for all $i = 1, \ldots, n$. Then, by the induction hypothesis, we have

$$(S_1, S_2, \ldots, S_n) \Rightarrow_{\bar{\Gamma}}^{k} (u_1A_1v_1, u_2A_2v_2, \ldots, u_nA_nv_n)$$

Since

$$\begin{aligned}&(u_1A_1v_1, u_2A_2v_2, \ldots, u_nA_nv_n) \Rightarrow_{\Gamma}\\ &\quad (u_1x_1v_1, u_2x_2v_2, \ldots, u_nx_nv_n)\end{aligned}$$

$(A_1, A_2, \ldots, A_n) \in Q$ and $A_i \to x_i \in P_i$, for all $i = 1, \ldots, n$. Algorithm 13.2.6 implies that $(A_1 \to x_1, A_2 \to x_2, \ldots, A_n \to x_n) \in \bar{Q}$, so

$$\begin{aligned}&(u_1A_1v_1, u_2A_2v_2, \ldots, u_nA_nv_n) \Rightarrow_{\bar{\Gamma}}\\ &\quad (u_1x_1v_1, u_2x_2v_2, \ldots, u_nx_nv_n)\end{aligned}$$

which proves the induction step. Therefore, Claim 1 holds. □

Claim 2. Let $(S_1, S_2, \ldots, S_n) \Rightarrow_{\bar{\Gamma}}^{m} (y_1, y_2, \ldots, y_n)$, *where* $m \geq 0$, $y_i \in (N_i \cup T_i)^*$, *for all* $i = 1, \ldots, n$. *Then,* $(S_1, S_2, \ldots, S_n) \Rightarrow_{\Gamma}^{m} (y_1, y_2, \ldots, y_n)$.

Proof. This claim is proved by induction on $m \geq 0$.

Basis. The basis is clear.

Induction Hypothesis. Assume that Claim 2 holds for all m-step derivations, where $m = 0, \ldots, k$, for some $k \geq 0$.

Induction Step. Consider any derivation of the form

$$(S_1, S_2, \ldots, S_n) \Rightarrow_{\bar{\Gamma}}^{k+1} (y_1, y_2, \ldots, y_n)$$

Then, there exists a sentential n-form $(u_1A_1v_1, u_2A_2v_2, \ldots, u_nA_nv_n)$, where $u_i \in T_i^*$, $A_i \in N_i$, $v_i \in (N_i \cup T_i)^*$, such that

$$\begin{aligned}(S_1, S_2, \ldots, S_n) &\Rightarrow_{\bar{\Gamma}}^{k} (u_1A_1v_1, u_2A_2v_2, \ldots, u_nA_nv_n)\\ &\Rightarrow_{\bar{\Gamma}} (u_1x_1v_1, u_2x_2v_2, \ldots, u_nx_nv_n)\end{aligned}$$

where $u_ix_iv_i = y_i$, for all $i = 1, \ldots, n$. Then, by the induction hypothesis, we have

$$(S_1, S_2, \ldots, S_n) \Rightarrow_{\Gamma}^{k} (u_1A_1v_1, u_2A_2v_2, \ldots, u_nA_nv_n)$$

Since

$$(u_1A_1v_1, u_2A_2v_2, \ldots, u_nA_nv_n) \Rightarrow_{\bar{\Gamma}} (u_1x_1v_1, u_2x_2v_2, \ldots, u_nx_nv_n)$$

$(A_1 \rightarrow x_1, A_2 \rightarrow x_2, \ldots, A_n \rightarrow x_n) \in \bar{Q}$, for all $i = 1, \ldots, n$. Algorithm 13.2.6 implies that $(A_1, A_2, \ldots, A_n) \in Q$ and $A_i \rightarrow x_i \in P_i$, so

$$(u_1A_1v_1, u_2A_2v_2, \ldots, u_nA_nv_n) \Rightarrow_{\Gamma} (u_1x_1v_1, u_2x_2v_2, \ldots, u_nx_nv_n)$$

which proves the induction step. Therefore, Claim 2 holds. □

Consider Claim 1 for $y_i \in T_i^*$, for all $i = 1, \ldots, n$. At this point, if

$$(S_1, S_2, \ldots, S_n) \Rightarrow_{\Gamma}^* (y_1, y_2, \ldots, y_n)$$

then

$$(S_1, S_2, \ldots, S_n) \Rightarrow_{\bar{\Gamma}}^* (y_1, y_2, \ldots, y_n)$$

Hence, $n\text{-}L(\Gamma) \subseteq n\text{-}L(\bar{\Gamma})$. Consider Claim 2 for $y_i \in T_i^*$, for all $i = 1, \ldots, n$. At this point, if

$$(S_1, S_2, \ldots, S_n) \Rightarrow_{\bar{\Gamma}}^* (y_1, y_2, \ldots, y_n)$$

then

$$(S_1, S_2, \ldots, S_n) \Rightarrow_{\Gamma}^* (y_1, y_2, \ldots, y_n)$$

Hence, $n\text{-}L(\bar{\Gamma}) \subseteq n\text{-}L(\Gamma)$. As $n\text{-}L(\Gamma) \subseteq n\text{-}L(\bar{\Gamma})$ and $n\text{-}L(\bar{\Gamma}) \subseteq n\text{-}L(\Gamma)$, $n\text{-}L(\Gamma) = n\text{-}L(\bar{\Gamma})$. By Lemma 13.2.5, this identity implies that $L_X(\Gamma) = L_X(\bar{\Gamma})$, for each $X \in \{union, conc, first\}$. Therefore, Theorem 13.2.7 holds. □

Algorithm 13.2.8. *Conversion of an n-LMGR to an equivalent n-LMGN.*

Input: An n-LMGR, $\Gamma = (G_1, G_2, \ldots, G_n, Q)$.
Output: An n-LMGN, $\bar{\Gamma} = (\bar{G}_1, \bar{G}_2, \ldots, \bar{G}_n, \bar{Q})$, such that $n\text{-}L(\Gamma) = n\text{-}L(\bar{\Gamma})$.
Method: Let $G_i = (N_i, T_i, P_i, S_i)$, for all $i = 1, \ldots, n$, and set

- $\bar{G}_i = (\bar{N}_i, T_i, \bar{P}_i, S_i)$, for all $i = 1, \ldots, n$, where

  $\bar{N}_i = \{\langle A, x\rangle \mid A \rightarrow x \in P_i\} \cup \{S_i\}$;
  $\bar{P}_i = \{\langle A, x\rangle \rightarrow y \mid A \rightarrow x \in P_i,\ y \in \tau_i(x)\} \cup \{S_i \rightarrow y \mid y \in \tau_i(S_i)\}$, where τ_i is a finite substitution from $(N_i \cup T_i)^*$ to $(\bar{N}_i \cup T_i)^*$ defined as $\tau_i(a) = \{a\}$, for all $a \in T$, and $\tau_i(A) = \{\langle A, x\rangle \mid A \rightarrow x \in P_i\}$, for all $A \in N_i$;

- $\bar{Q} = \{(\langle A_1, x_1\rangle, \langle A_2, x_2\rangle, \dots, \langle A_n, x_n\rangle) \mid (A_1 \rightarrow x_1, A_2 \rightarrow x_2, \dots, A_n \rightarrow x_n) \in Q\}$
  $\cup \{(S_1, S_2, \dots, S_n)\}$. □

Theorem 13.2.9. *Let $\Gamma = (G_1, G_2, \dots, G_n, Q)$ be an n-LMGR. With Γ as its input, Algorithm 13.2.8 halts and correctly constructs an n-LMGN, $\bar{\Gamma} = (\bar{G}_1, \bar{G}_2, \dots, \bar{G}_n, \bar{Q})$, such that $n\text{-}L(\Gamma) = n\text{-}L(\bar{\Gamma})$, and $L_X(\Gamma) = L_X(\bar{\Gamma})$, for each $X \in \{union, conc, first\}$.*

Proof. To prove this theorem, we first establish Claims 1 and 2.

Claim 1. Let $(S_1, S_2, \dots, S_n) \Rightarrow_{\Gamma}^{m} (z_1, z_2, \dots, z_n)$, where $m \geq 0$, $z_i \in (N_i \cup T_i)^*$, for all $i = 1, \dots, n$. Then, $(S_1, S_2, \dots, S_n) \Rightarrow_{\bar{\Gamma}}^{m+1} (\bar{z}_1, \bar{z}_2, \dots, \bar{z}_n)$, for any $\bar{z}_i \in \tau_i(z_i)$.

Proof. This claim is proved by induction on $m \geq 0$.

Basis. Let $m = 0$. Then,

$$(S_1, S_2, \dots, S_n) \Rightarrow_{\Gamma}^{0} (S_1, S_2, \dots, S_n)$$

Observe that

$$(S_1, S_2, \dots, S_n) \Rightarrow_{\bar{\Gamma}}^{1} (\bar{z}_1, \bar{z}_2, \dots, \bar{z}_n)$$

for any $\bar{z}_i \in \tau_i(z_i)$, because Algorithm 13.2.8 implies that $(S_1, S_2, \dots, S_n) \in \bar{Q}$ and $S_i \rightarrow \bar{z}_i \in \bar{P}_i$, for any $\bar{z}_i \in \tau_i(z_i)$, for all $i = 1, \dots, n$. Thus, the basis holds.

Induction Hypothesis. Assume that the claim holds for all m-step derivations, where $m = 0, \dots, k$, for some $k \geq 0$.

Induction Step. Consider any derivation of the form

$$(S_1, S_2, \dots, S_n) \Rightarrow_{\Gamma}^{k+1} (y_1, y_2, \dots, y_n)$$

Then, there exists a sentential n-form $(u_1 A_1 v_1, u_2 A_2 v_2, \dots, u_n A_n v_n)$, where $u_i \in T_i^*$, $A_i \in N_i$, $v_i \in (N_i \cup T_i)^*$, such that

$$\begin{aligned}(S_1, S_2, \dots, S_n) &\Rightarrow_{\Gamma}^{k} (u_1 A_1 v_1, u_2 A_2 v_2, \dots, u_n A_n v_n)\\ &\Rightarrow_{\Gamma} (u_1 x_1 v_1, u_2 x_2 v_2, \dots, u_n x_n v_n)\end{aligned}$$

where $u_i x_i v_i = y_i$, for all $i = 1, \dots, n$. Then, by the induction hypothesis, we have

$$(S_1, S_2, \dots, S_n) \Rightarrow_{\bar{\Gamma}}^{k+1} (\bar{w}_1, \bar{w}_2, \dots, \bar{w}_n)$$

for any $\bar{w}_i \in \tau_i(u_i A_i v_i)$, for all $i = 1, \dots, n$. Since

$$(u_1 A_1 v_1, u_2 A_2 v_2, \dots, u_n A_n v_n) \Rightarrow_{\Gamma} (u_1 x_1 v_1, u_2 x_2 v_2, \dots, u_n x_n v_n)$$

$(A_1 \rightarrow x_1, A_2 \rightarrow x_2, \ldots, A_n \rightarrow x_n) \in Q$. Algorithm 13.2.8 implies that $(\langle A_1, x_1\rangle, \langle A_2, x_2\rangle, \ldots, \langle A_n, x_n\rangle) \in \bar{Q}$ and $\langle A_i, x_i\rangle \rightarrow \bar{y}_i \in \bar{P}_i$, for any $\bar{y}_i \in \tau_i(x_i)$, for all $i = 1, \ldots, n$. Let $\bar{w}_i$ be any sentential form of the form $\bar{u}_i\langle A_i, x_i\rangle\bar{v}_i$, for all $i = 1, \ldots, n$, where $\bar{u}_i \in \tau_i(u_i)$ and $\bar{v}_i \in \tau_i(v_i)$. Then,

$$(\bar{u}_1\langle A_1, x_1\rangle\bar{v}_1, \bar{u}_2\langle A_2, x_2\rangle\bar{v}_2, \ldots, \bar{u}_n\langle A_n, x_n\rangle\bar{v}_n) \Rightarrow_{\bar{\Gamma}} (\bar{u}_1\bar{y}_1\bar{v}_1, \bar{u}_2\bar{y}_2\bar{v}_2, \ldots, \bar{u}_n\bar{y}_n\bar{v}_n)$$

where $\bar{u}_i\bar{y}_i\bar{v}_i$ is any sentential form, $\bar{u}_i\bar{y}_i\bar{v}_i \in \tau_i(u_iy_iv_i)$, for all $i = 1, \ldots, n$, which proves the induction step. Therefore, Claim 1 holds. □

Claim 2. Let $(S_1, S_2, \ldots, S_n) \Rightarrow^m_{\bar{\Gamma}} (\bar{z}_1, \bar{z}_2, \ldots, \bar{z}_n)$, *where* $m \geq 1$, $\bar{z}_i \in (T_i \cup \bar{N}_i)^*$, *for all* $i = 1, \ldots, n$. *Then,* $(S_1, S_2, \ldots, S_n) \Rightarrow^{m-1}_{\Gamma} (z_1, z_2, \ldots, z_n)$, *where* $\bar{z}_i \in \tau_i(z_i)$, *for all* $i = 1, \ldots, n$.

Proof. This claim is proved by induction on $m \geq 1$.

Basis. Let $m = 1$. Then,

$$(S_1, S_2, \ldots, S_n) \Rightarrow_{\bar{\Gamma}} (\bar{z}_1, \bar{z}_2, \ldots, \bar{z}_n)$$

implies that $S_i \rightarrow \bar{z}_i \in \bar{P}_i$, for all $i = 1, \ldots, n$. Algorithm 13.2.8 implies that $\bar{z}_i \in \tau_i(S_i)$, for all $i = 1, \ldots, n$, so

$$(S_1, S_2, \ldots, S_n) \Rightarrow^0_{\Gamma} (S_1, S_2, \ldots, S_n)$$

Since $\bar{z}_i \in \tau_i(S_i)$, for all $i = 1, \ldots, n$, the basis holds.

Induction Hypothesis. Assume that the claim holds for all m-step derivations, where $m = 1, \ldots, k$, for some $k \geq 1$.

Induction Step. Consider any derivation of the form

$$(S_1, S_2, \ldots, S_n) \Rightarrow^{k+1}_{\bar{\Gamma}} (\bar{y}_1, \bar{y}_2, \ldots, \bar{y}_n)$$

Then, there exists a sentential n-form $(\bar{u}_1\langle A_1, x_1\rangle\bar{v}_1, \bar{u}_2\langle A_2, x_2\rangle\bar{v}_2, \ldots, \bar{u}_n\langle A_n, x_n\rangle\bar{v}_n)$, where $\bar{u}_i \in T_i^*$, $\langle A_i, x_i\rangle \in \bar{N}_i$, $\bar{N}_i \in (\bar{N}_i \cup T_i)^*$, such that

$$\begin{aligned}(S_1, S_2, \ldots, S_n) &\Rightarrow^k_{\bar{\Gamma}} (\bar{u}_1\langle A_1, x_1\rangle\bar{v}_1, \bar{u}_2\langle A_2, x_2\rangle\bar{v}_2, \ldots, \bar{u}_n\langle A_n, x_n\rangle\bar{v}_n)\\ &\Rightarrow_{\bar{\Gamma}} (\bar{u}_1\bar{x}_1\bar{v}_1, \bar{u}_2\bar{x}_2\bar{v}_2, \ldots, \bar{u}_n\bar{x}_n\bar{v}_n)\end{aligned}$$

where $\bar{u}_i\bar{x}_i\bar{v}_i = \bar{y}_i$, for all $i = 1, \ldots, n$. Then, by the induction hypothesis, we have

$$(S_1, S_2, \ldots, S_n) \Rightarrow^{k-1}_{\Gamma}\Rightarrow_{\Gamma} (w_1, w_2, \ldots, w_n)$$

where $\bar{u}_i\langle A_i, x_i\rangle\bar{v}_i \in \tau_i(w_i)$, for all $i = 1, \ldots, n$. Since

$$(\bar{u}_1\langle A_1, x_1\rangle\bar{v}_1, \bar{u}_2\langle A_2, x_2\rangle\bar{v}_2, \ldots, \bar{u}_n\langle A_n, x_n\rangle\bar{v}_n) \Rightarrow_{\bar{\Gamma}} (\bar{u}_1\bar{x}_1\bar{v}_1, \bar{u}_2\bar{x}_2\bar{v}_2, \ldots, \bar{u}_n\bar{x}_n\bar{v}_n)$$

there are $(\langle A_n, x_n\rangle, \langle A_n, x_n\rangle, \ldots, \langle A_n, x_n\rangle) \in \bar{Q}$ and $\langle A_1, x_1\rangle \to \bar{x}_i \in \bar{P}_i$, for all $i = 1, \ldots, n$. Algorithm 13.2.8 implies that $(A_1 \to x_1, A_2 \to x_2, \ldots, A_n \to x_n) \in Q$ and $A_i \to x_i \in P_i$, where $\bar{x}_i \in \tau_i(x_i)$, for all $i = 1, \ldots, n$. We can express w_i as $w_i = u_i A_i v_i$, where $\bar{u}_i \in \tau_i(u_i)$, $\bar{v}_i \in \tau_i(v_i)$, and observe that $\langle A_i, x_i\rangle \in \tau_i(A_i)$ holds by the definition of τ_i, for all $i = 1, \ldots, n$. Then,

$$(u_1 A_1 v_1, u_2 A_2 v_2, \ldots, u_n A_n v_n) \Rightarrow_{\Gamma} (u_1 x_1 v_1, u_2 x_2 v_2, \ldots, u_n x_n v_n)$$

where $\bar{u}_i \in \tau_i(u_i)$, $\bar{v}_i \in \tau_i(v_i)$, and $\bar{x}_i \in \tau_i(x_i)$, for all $i = 1, \ldots, n$, which means that $\bar{u}_i \bar{x}_i \bar{v}_i \in \tau_i(u_i x_i v_i)$, for all $i = 1, \ldots, n$. Therefore,

$$\begin{aligned}(S_1, S_2, \ldots, S_n) &\Rightarrow_{\Gamma}^{k+1} (u_1 A_1 v_1, u_2 A_2 v_2, \ldots, u_n A_n v_n)\\ &\Rightarrow_{\Gamma} \quad (u_1 x_1 v_1, u_2 x_2 v_2, \ldots, u_n x_n v_n)\end{aligned}$$

where $\bar{u}_i \bar{x}_i \bar{v}_i \in \tau_i(u_i x_i v_i)$, for all $i = 1, \ldots, n$. Let $\bar{z}_i = \bar{u}_i \bar{x}_i \bar{v}_i$ and $z_i = u_i x_i v_i$, for all $i = 1, \ldots, n$. Then,

$$(S_1, S_2, \ldots, S_n) \Rightarrow_{\Gamma}^{k+2} (z_1, z_2, \ldots, z_n)$$

for all $\bar{z}_i \in \tau_i(z_i)$, which proves the induction step. Therefore, Claim 2 holds. □

Consider Claim 1 when $z_i \in T_i^*$, for all $i = 1, \ldots, n$. At this point, if

$$(S_1, S_2, \ldots, S_n) \Rightarrow_{\Gamma}^{*} (z_1, z_2, \ldots, z_n)$$

then

$$(S_1, S_2, \ldots, S_n) \Rightarrow_{\bar{\Gamma}}^{*} (\bar{z}_1, \bar{z}_2, \ldots, \bar{z}_n)$$

where $\bar{z}_i \in \tau_i(z_i)$, for all $i = 1, \ldots, n$. Since $\tau_i(a_i) = a_i$, for all $a_i \in T_i$, $\bar{z}_i = z_i$. Hence, $n\text{-}L(\Gamma) \subseteq n\text{-}L(\bar{\Gamma})$. Consider Claim 2 when $\bar{z}_i \in T_i^*$, for all $i = 1, \ldots, n$. At this point, if

$$(S_1, S_2, \ldots, S_n) \Rightarrow_{\bar{\Gamma}}^{m} (\bar{z}_1, \bar{z}_2, \ldots, \bar{z}_n)$$

then

$$(S_1, S_2, \ldots, S_n) \Rightarrow_{\Gamma}^{m-1} (z_1, z_2, \ldots, z_n)$$

where $\bar{z}_i \in \tau_i(z_i)$, for all $i = 1, \ldots, n$. Since $\tau_i(a_i) = a_i$, for all $a_i \in T_i$, $z_i = \bar{z}_i$. Hence, $n\text{-}L(\bar{\Gamma}) \subseteq n\text{-}L(\Gamma)$. As $n\text{-}L(\Gamma) \subseteq n\text{-}L(\bar{\Gamma})$ and $n\text{-}L(\bar{\Gamma}) \subseteq n\text{-}L(\Gamma)$, $n\text{-}L(\Gamma) = n\text{-}L(\bar{\Gamma})$. By Lemma 13.2.5, this identity implies that $L_X(\Gamma) = L_X(\bar{\Gamma})$, for each $X \in \{union, conc, first\}$. Therefore, Theorem 13.2.9 holds. □

From the achieved results, we immediately obtain the following corollary.

Corollary 13.2.10. *The family of languages generated by n-LMGN in the X mode coincides with the family of languages generated by n-LMGR in the X mode, where* $X \in \{union, conc, first\}$. □

Theorem 13.2.11. *For every recursively enumerable language L over some alphabet T, there exits a 2-LMGR,* $\Gamma = ((\bar{N}_1, T, \bar{P}_1, S_1), (\bar{N}_1, T, \bar{P}_1, S_1), Q)$, *such that*

(i) $\{w \in T^* \mid (S_1, S_2) \Rightarrow_{\Gamma}^{*} (w, w)\} = L$,
(ii) $\{w_1 w_2 \in T^* \mid (S_1, S_2) \Rightarrow_{\Gamma}^{*} (w_1, w_2), w_1 \neq w_2\} = \emptyset$.

Proof. Recall that for every recursive enumerable language L over some alphabet T, there exist two context-free grammars, $G_1 = (N_1, \bar{T}, P_1, S_1)$, $G_2 = (N_2, \bar{T}, P_2, S_2)$, and a homomorphism h from $\bar{T}^*$ to T^* such that $L = \{h(x) \mid x \in L(G_1) \cap L(G_2)\}$ (see Theorem 3.3.14). Furthermore, by Theorem 4.1.18, for every context-free grammar, there exists an equivalent context-free grammar in the Greibach normal form (see Definition 4.1.17). Hence, without any lost of generality, we assume that G_1 and G_2 are in the Greibach normal form. Consider the 2-LMGR

$$\Gamma = (G_1, G_2, Q)$$

where

- $G_i = (\bar{N}_i, T, \bar{P}_i, S_i)$, where

  $\bar{N}_i = N_i \cup \{\bar{a} \mid a \in \bar{T}\}$;
  $\bar{P}_i = \{A \rightarrow \bar{a}x \mid A \rightarrow ax \in P_i, a \in \bar{T}, x \in N_i^*\} \cup \{\bar{a} \rightarrow h(a) \mid a \in \bar{T}\}$,

  for $i = 1, 2$;
- $Q = \{(A_1 \rightarrow \bar{a}x_1, A_2 \rightarrow \bar{a}x_2) \mid A_1 \rightarrow \bar{a}x_1 \in P_1, A_2 \rightarrow \bar{a}x_2 \in P_2, a \in \bar{T}\}$
  $\cup \{(\bar{a} \rightarrow h(a), \bar{a} \rightarrow h(a)) \mid a \in \bar{T}\}$.

Consider properties (i) and (ii) in Theorem 13.2.11. Next, Claims 1 and 2 establish (i) and (ii), respectively.

Claim 1. $\{w \in T^* \mid (S_1, S_2) \Rightarrow_{\Gamma}^{*} (w, w)\} = L$

Proof. I. We prove that $L \subseteq \{w \in T^* \mid (S_1, S_2) \Rightarrow_{\Gamma}^{*} (w, w)\}$. Let w be any string. Then, there exists a string, $a_1 a_2 \cdots a_n \in \bar{T}^*$, such that

- $a_1 a_2 \cdots a_n \in L(G_1)$,
- $a_1 a_2 \cdots a_n \in L(G_2)$, and
- $h(a_1 a_2 \cdots a_n) = w$.

This means that there exist the following derivations in G_1 and G_2

$$\begin{aligned} S_1 &\Rightarrow_{G_1} a_1x_1 & [p_1] \\ &\Rightarrow_{G_1} a_1a_2x_2 & [p_2] \\ &\vdots \\ &\Rightarrow_{G_1} a_1a_2\cdots a_n & [p_n] \\ S_2 &\Rightarrow_{G_2} a_1y_1 & [r_1] \\ &\Rightarrow_{G_2} a_1a_2y_2 & [r_2] \\ &\vdots \\ &\Rightarrow_{G_2} a_1a_2\cdots a_n & [r_n] \end{aligned}$$

where $a_i \in \bar{T}$, $x_i \in N_i^*$, $y_i \in N_2^*$, $p_i \in P_1$, $r_i \in P_2$, for all $i = 1, \dots, n$. Observe that $\mathrm{sym}(\mathrm{rhs}(p_i), 1) = \mathrm{sym}(\mathrm{rhs}(r_i), 1) = a_i$, for all $i = 1, \dots, n$. The construction of Q implies the following two statements.

- Let $p_i\colon A_i \to a_iu_i \in \bar{P}_1$, $r_i\colon B_i \to a_iv_i \in \bar{P}_2$. Then, $(A_i \to \bar{a}_iu_i, B_i \to \bar{a}_iv_i) \in Q$, for all $i = 1, \dots, n$.
- Q contains $(\bar{a}_i \to h(a_i), \bar{a}_i \to h(a_i))$, for all $i = 1, \dots, n$.

Therefore, there exists

$$\begin{aligned} (S_1, S_2) &\Rightarrow_\Gamma (\bar{a}_1x_1, \bar{a}_1y_1) \\ &\Rightarrow_\Gamma (h(a_1)x_1, h(a_1)y_1) \\ &\Rightarrow_\Gamma (h(a_1)\bar{a}_2x_2, h(a_1)\bar{a}_2y_2) \\ &\Rightarrow_\Gamma (h(a_1)h(a_2)x_2, h(a_1)h(a_2)y_2) \\ &\vdots \\ &\Rightarrow_\Gamma (h(a_1)h(a_2)\cdots h(a_n), h(a_1)h(a_2)\cdots h(a_n)) \\ &= (h(a_1a_2\cdots a_n), h(a_1a_2\cdots a_n)) \\ &= (w, w) \end{aligned}$$

In brief, $(S_1, S_2) \Rightarrow_\Gamma^* (w, w)$. Hence, $L \subseteq \{w \in T^* \mid (S_1, S_2) \Rightarrow_\Gamma^* (w, w)\}$.

II. We prove that $\{w \in T^* \mid (S_1, S_2) \Rightarrow_\Gamma^* (w, w)\} \subseteq L$. Let $(S_1, S_2) \Rightarrow_\Gamma^* (w, w)$. Then, there exists

$$\begin{aligned} (S_1, S_2) &\Rightarrow_\Gamma (\bar{a}_1x_1, \bar{a}_1y_1) \\ &\Rightarrow_\Gamma (h(a_1)x_1, h(a_1)y_1) \\ &\Rightarrow_\Gamma (h(a_1)\bar{a}_2x_2, h(a_1)\bar{a}_2y_2) \\ &\Rightarrow_\Gamma (h(a_1)h(a_2)x_2, h(a_1)h(a_2)y_2) \\ &\vdots \\ &\Rightarrow_\Gamma (h(a_1)h(a_2)\cdots h(a_n), h(a_1)h(a_2)\cdots h(a_n)) \\ &= (h(a_1a_2\cdots a_n), h(a_1a_2\cdots a_n)) \\ &= (w, w) \end{aligned}$$

By analogy with part I, we can prove that there exist derivations in G_1 and G_2 of the forms

$$\begin{aligned} S_1 &\Rightarrow_{G_1} a_1x_1 && [p_1] \\ &\Rightarrow_{G_1} a_1a_2x_2 && [p_2] \\ &\vdots \\ &\Rightarrow_{G_1} a_1a_2\cdots a_n && [p_n] \\ S_2 &\Rightarrow_{G_2} a_1y_1 && [r_1] \\ &\Rightarrow_{G_2} a_1a_2y_2 && [r_2] \\ &\vdots \\ &\Rightarrow_{G_2} a_1a_2\cdots a_n && [r_n] \end{aligned}$$

This implies that $a_1a_2\cdots a_n \in L(G_1)$, $a_1a_2\cdots a_n \in L(G_2)$, and $h(a_1a_2\cdots a_n) = w$, so $w \in L$. Hence, $\{w \in T^* \mid (S_1, S_2) \Rightarrow_\Gamma^* (w, w)\} \subseteq L$. Therefore, Claim 1 holds. □

Claim 2. $\{w_1w_2 \in T^* \mid (S_1, S_2) \Rightarrow_\Gamma^* (w_1, w_2), w_1 \neq w_2\} = \emptyset$

Proof. By contradiction. Let $\{w_1w_2 \in T^* \mid (S_1, S_2) \Rightarrow_\Gamma^* (w_1, w_2), w_1 \neq w_2\} \neq \emptyset$. Then, there have to exist two different strings, $w_1 = h(a_1)h(a_2)\cdots h(a_n)$ and $w_2 = h(b_1)h(b_2)\cdots h(b_n)$, such that $(S_1, S_2) \Rightarrow_\Gamma^* (w_1, w_2)$.

I. Assume that $a_i = b_i$, for all $i = 1, \ldots, n$. Then, $w_1 = h(a_1)h(a_2)\cdots h(a_n) = h(b_1)h(b_2)\cdots h(b_n) = w_2$, which contradicts $w_1 \neq w_2$.

II. Assume that there exists some $k \leq n$ such that $a_k \neq b_k$. Then, there exists a derivation of the form

$$\begin{aligned} (S_1, S_2) &\Rightarrow_\Gamma (\bar{a}_1x_1, \bar{a}_1y_1) \\ &\Rightarrow_\Gamma (h(a_1)x_1, h(a_1)y_1) \\ &\Rightarrow_\Gamma (h(a_1)\bar{a}_2x_2, h(a_1)\bar{a}_2y_2) \\ &\Rightarrow_\Gamma (h(a_1)h(a_2)x_2, h(a_1)h(a_2)y_2) \\ &\vdots \\ &\Rightarrow_\Gamma (h(a_1)h(a_2)\cdots h(a_{k-1})x_{k-1}, h(a_1)h(a_2)\cdots h(a_{k-1})y_{k-1}) \end{aligned}$$

Then, there has to exist a derivation

$$(x_{k-1}, y_{k-1}) \Rightarrow_\Gamma (\bar{a}_kx_k, \bar{b}_ky_k)$$

where $\bar{a}_k \neq \bar{b}_k$. By the definition of Q, there has to be $(p, r) \in Q$ such that

$$\text{sym}\big(\text{rhs}(p), 1\big) = \text{sym}\big(\text{rhs}(r), 1\big)$$

Therefore, the next derivation has to be of the form

$$(x_{k-1}, y_{k-1}) \Rightarrow_\Gamma (\bar{a}_kx_k, \bar{b}_ky_k)$$

where $\bar{a}_k = \bar{b}_k$, which is a contradiction. Therefore, Claim 2 holds. □

Claims 1 and 2 imply that Theorem 13.2.11 holds. □

Theorem 13.2.12. *For any recursively enumerable language L over an alphabet T, there exists a 2-LMGR, $\Gamma = (G_1, G_2, Q)$, such that $L_{union}(\Gamma) = L$.*

Proof. By Theorem 13.2.11, for every recursively enumerable language L over an alphabet T, there exits a 2-LMGR

$$\bar{\Gamma} = \big((N_1, T, P_1, S_1), (N_1, T, P_1, S_1), Q\big)$$

such that

$$\big\{w \in T^* \mid (S_1, S_2) \Rightarrow_{\bar{\Gamma}}^* (w, w)\big\} = L$$

and

$$\big\{w_1 w_2 \in T^* \mid (S_1, S_2) \Rightarrow_{\bar{\Gamma}}^* (w_1, w_2), w_1 \neq w_2\big\} = \emptyset$$

Let $\Gamma = \bar{\Gamma}$. Then,

$$\begin{aligned} L_{union}(\Gamma) &= \{w \mid (S_1, S_2) \Rightarrow_{\Gamma}^* (w_1, w_2), w_i \in T^*, \text{for } i = 1, 2, \\ &\quad w \in \{w_1, w_2\}\} \\ &= \{w \mid (S_1, S_2) \Rightarrow_{\Gamma}^* (w, w)\} \cup \{w \mid (S_1, S_2) \Rightarrow_{\Gamma}^* (w_1, w_2), \\ &\quad w_i \in T^*, \text{for } i = 1, 2, w \in \{w_1, w_2\}, w_1 \neq w_2\} \\ &= \{w \mid (S_1, S_2) \Rightarrow_{\Gamma}^* (w, w)\} \cup \emptyset \\ &= \{w \mid (S_1, S_2) \Rightarrow_{\Gamma}^* (w, w)\} \\ &= L \end{aligned}$$

Therefore, Theorem 13.2.12 holds. □

Theorem 13.2.13. *For any recursively enumerable language L over an alphabet T, there exists a 2-LMGR, $\Gamma = (G_1, G_2, Q)$, such that $L_{first}(\Gamma) = L$.*

Proof. By Theorem 13.2.11, for every recursively enumerable language L over an alphabet T, there exits a 2-LMGR

$$\bar{\Gamma} = \big((N_1, T, P_1, S_1), (N_1, T, P_1, S_1), Q\big)$$

such that

$$\big\{w \in T^* \mid (S_1, S_2) \Rightarrow_{\bar{\Gamma}}^* (w, w)\big\} = L$$

and

$$\big\{w_1 w_2 \in T^* \mid (S_1, S_2) \Rightarrow_{\bar{\Gamma}}^* (w_1, w_2), w_1 \neq w_2\big\} = \emptyset$$

Let $\Gamma = \bar{\Gamma}$. Then,

$$\begin{aligned}
L_{first}(\Gamma) &= \{w_1 \mid (S_1, S_2) \Rightarrow_{\Gamma}^{*} (w_1, w_2), w_i \in T^*, \text{for } i = 1,2\} \\
&= \{w \mid (S_1, S_2) \Rightarrow_{\Gamma}^{*} (w, w)\} \cup \{w_1 \mid (S_1, S_2) \Rightarrow_{\Gamma}^{*} (w_1, w_2), \\
&\quad\; w_i \in T^*, \text{for } i = 1,2, w_1 \neq w_2\} \\
&= \{w \mid (S_1, S_2) \Rightarrow_{\Gamma}^{*} (w, w)\} \cup \emptyset \\
&= \{w \mid (S_1, S_2) \Rightarrow_{\Gamma}^{*} (w, w)\} \\
&= L
\end{aligned}$$

Therefore, Theorem 13.2.13 holds. □

Theorem 13.2.14. *For any recursively enumerable language L over an alphabet T, there exists a 2-LMGR, $\Gamma = (G_1, G_2, Q)$, such that $L_{conc}(\Gamma) = L$.*

Proof. By Theorem 13.2.11, we have that for every recursively enumerable language L over an alphabet T, there exits a 2-LMGR

$$\bar{\Gamma} = \big((N_1, T, P_1, S_1), (N_1, T, P_1, S_1), Q\big)$$

such that

$$\big\{w \in T^* \mid (S_1, S_2) \Rightarrow_{\bar{\Gamma}}^{*} (w, w)\big\} = L$$

and

$$\big\{w_1 w_2 \in T^* \mid (S_1, S_2) \Rightarrow_{\bar{\Gamma}}^{*} (w_1, w_2), w_1 \neq w_2\big\} = \emptyset$$

Let $G_1 = (N_1, T, P_1, S_1)$ and $G_2 = (N_2, \emptyset, \bar{P}_2, S_2)$, where $\bar{P}_2 = \{A \rightarrow g(x) \mid A \rightarrow x \in P_2\}$, where g is a homomorphism from $(N_2 \cup T)^*$ to N_2^* defined as $g(X) = X$, for all $X \in N_2$, and $g(a) = \varepsilon$, for all $a \in T$. We prove that $L_{conc}(\Gamma) = L$.

I. We prove that $L \subseteq L_{conc}(\Gamma)$. Let $w \in L$. Then, there exists a derivation of the form

$$(S_1, S_2) \Rightarrow_{\bar{\Gamma}}^{*} (w, w)$$

Thus, there exist a derivation of the form

$$(S_1, S_2) \Rightarrow_{\Gamma}^{*} (w, g(w))$$

Since $g(a) = \varepsilon$, for all $a \in T$, $g(w) = \varepsilon$, for all $w \in T^*$. Thus,

$$(S_1, S_2) \Rightarrow_{\Gamma}^{*} (w, \varepsilon)$$

Hence, $w\varepsilon = w$ and $w \in L_{conc}(\Gamma)$.

II. We prove that $L_{conc}(\Gamma) \subseteq L$. Let $w \in L$. Then, there exists a derivation of the form

$$(S_1, S_2) \Rightarrow_{\Gamma}^{*} (w, \varepsilon)$$

because $L(G_2) = \{\varepsilon\}$. Since $g(x) = \varepsilon$ in Γ, for all $x \in T^*$, there is a derivation of the form

$$(S_1, S_2) \Rightarrow_{\Gamma}^{*} (w, x)$$

where x is any string. Theorem 13.2.11 implies that $x = w$. Thus,

$$(S_1, S_2) \Rightarrow_{\Gamma}^{*} (w, w)$$

Hence, $w \in L$.

By I and II, Theorem 13.2.14 holds. □

We close this section by suggesting the next open problem area.

Open Problem 13.2.15. By analogy with leftmost n-generative nonterminal-synchronized grammar systems, discussed in this section, introduce n-generative nonterminal-synchronized grammar systems and study their generative power. □

References

1. Dassow, J., Păun, G.: Regulated Rewriting in Formal Language Theory. Springer, New York (1989)
2. Rozenberg, G., Salomaa, A. (eds.): Linear Modeling: Background and Application. Handbook of Formal Languages, vol. 2. Springer, New York (1997)

Chapter 14
Controlled Pure Grammar Systems

Abstract This chapter introduces pure grammar systems, which have only terminals. They generate their languages in the leftmost way, and in addition, this generative process is regulated by control languages over rule labels. The chapter concentrates its attention on investigating the generative power of these systems. It establishes three major results. First, without any control languages, these systems do not even generate some context-free languages. Second, with regular control languages, these systems characterize the family of recursively enumerable languages, and this result holds even if these systems have no more than two components. Finally, this chapter considers control languages as languages that are themselves generated by regular-controlled context-free grammars; surprisingly enough, with control languages of this kind, these systems over unary alphabets define nothing but regular languages. The chapter consists of two sections. First, Sect. 14.1 define controlled pure grammar systems and illustrate them by an example. Then, Sect. 14.2 rigorously establishes the results mentioned above and points out several open problems.

Keywords Grammar systems • Control languages • Pure versions • Leftmost derivations • Generative power • Language families

To grasp the discussion of the present chapter fully, we should realize that context-free grammars are quite central to formal language theory as a whole (see [10, 17, 23, 24]). It thus comes as no surprise that this theory has introduced a broad variety of their modified versions, ranging from simplified and restricted versions up to fundamentally generalized systems based upon these grammars. Grammar systems (see [9]), regulated context-free grammars (see Chaps. 4 and 5), pure context-free grammars (see [14–16] and p. 242 in [23]), and context-free grammars with leftmost derivations (see [19] and Sect. 5.1 in [17]) definitely belong to the key modifications of this kind. Next, we give an insight into these four modifications.

A. Meduna and P. Zemek, *Regulated Grammars and Automata*,
DOI 10.1007/978-1-4939-0369-6_14,

(I) Grammar systems consist of several context-free grammars, referred to as their components, which mutually cooperate and, in this way, generate the languages of the systems.
(II) Regulated context-free grammars prescribe the use of rules during derivations by some additional regulating mechanisms, such as control languages over the label of grammatical rules.
(III) Pure context-free grammars simplify ordinary context-free grammars by using only one type of symbols—terminals. There exist pure sequential versions of context-free grammars as well as pure parallel versions of context-free grammars, better known as 0L grammars (see Sect. 3.3).
(IV) Context-free grammars that perform only leftmost derivations fulfill a key role in a principal application area of these grammars—parsing (see [1, 19]).

Of course, formal language theory has also investigated various combinations of (I) through (IV). For instance, combining (I) and (III), pure grammar systems have been studied (see [2, 4, 5]). Similarly, based upon various combinations of (I) and (II), a number of regulated grammar systems were defined and discussed (see Chap. 13 and [3, 7, 8, 12, 13, 22]). Following this vivid investigation trend, the present chapter combines all the four modifications mentioned above.

More specifically, this chapter introduces pure grammar systems that generate their languages in the leftmost way, and in addition, this generative process is regulated by control languages over rule labels. The chapter concentrates its attention on investigating the generative power of these systems. It establishes three major results. First, without any control languages, these systems are not even able to generate all context-free languages (Theorem 14.2.4). Second, with regular control languages, these systems characterize the family of recursively enumerable languages, and this result holds even if these systems have no more than two components (Theorems 14.2.7 and 14.2.8). Finally, this chapter considers control languages as languages that are themselves generated by regular-controlled context-free grammars; surprisingly enough, with control languages of this kind, these systems over unary alphabets define nothing but regular languages (Theorem 14.2.9).

The present chapter is organized as follows. First, Sect. 14.1 define controlled pure grammar systems and illustrate them by an example. Then, Sect. 14.2 rigorously establishes the results mentioned above. A formulation of several open problems closes the chapter.

14.1 Definitions and Examples

In this section, we define controlled pure grammar systems and illustrate them by an example.

Informally, these systems are composed of n components, where $n \geq 1$, and a single alphabet. Every component contains a set of rewriting rules over the alphabet, each having a single symbol on its left-hand side, and a start string, from which these systems start their computation. Every rule is labeled with a unique label. Control languages for these systems are then defined over the set of all rule labels.

Definition 14.1.1. An *n-component pure grammar system* (an *n-pGS* for short), for some $n \geq 1$, is a $(2n + 2)$-tuple

$$\Gamma = \big(T, \Psi, P_1, w_1, P_2, w_2, \ldots, P_n, w_n\big)$$

where T and Ψ are two disjoint alphabets, $w_i \in T^*$, and $P_i \in \Psi \times T \times T^*$ for $i = 1, 2, \ldots, n$ are finite relations such that

(1) if $(r, a, x), (s, a, x) \in P_i$, then $r = s$;
(2) if $(r, a, x), (s, b, y) \in \bigcup_{1 \leq j \leq n} P_j$, where $a \neq b$ or $x \neq y$, then $r \neq s$.

The components Ψ, P_i, and w_i are called the alphabet of *rule labels*, the set of *rules* of the ith component, and the *start string* of the ith component, respectively. □

By analogy with context-free grammars, each rule (r, a, x) is written as $r\colon a \rightarrow x$ throughout this chapter.

A configuration of Γ is an n-tuple of strings. It represents an instantaneous description of Γ. The start configuration is formed by start strings.

Definition 14.1.2. Let $\Gamma = (T, \Psi, P_1, w_1, P_2, w_2, \ldots, P_n, w_n)$ be an n-pGS, for some $n \geq 1$. An n-tuple $(x_1, x_2, \ldots, x_n)$, where $x_i \in T^*$ for $i = 1, 2, \ldots, n$, is called a *configuration* of Γ. The configuration $(w_1, w_2, \ldots, w_n)$ is said to be the *start configuration*. □

At every computational step, a rule from some component i is selected, and it is applied to the leftmost symbol of the ith string in the current configuration. Other strings remain unchanged. Hence, these systems work in a sequential way.

Definition 14.1.3. Let $\Gamma = (T, \Psi, P_1, w_1, P_2, w_2, \ldots, P_n, w_n)$ be an n-pGS, for some $n \geq 1$, and let $(x_1, x_2, \ldots, x_n)$, $(z_1, z_2, \ldots, z_n)$ be two configurations of Γ. The *direct derivation relation* over $(T^*)^n$, symbolically denoted by $\Rightarrow_\Gamma$, is defined as

$$(x_1, x_2, \ldots, x_n) \Rightarrow_\Gamma (z_1, z_2, \ldots, z_n)\ [r]$$

if and only if $r\colon a \rightarrow y \in P_i$, $x_i = av$, $z_i = yv$, where $v \in T^*$, for some $i \in \{1, 2, \ldots, n\}$, and $z_j = x_j$ for every $j \neq i$; $(x_1, x_2, \ldots, x_n) \Rightarrow_\Gamma (z_1, z_2, \ldots, z_n)\ [r]$ is simplified to $(x_1, x_2, \ldots, x_n) \Rightarrow_\Gamma (z_1, z_2, \ldots, z_n)$ if r is immaterial.

Let $\chi_0, \chi_1, \chi_2, \ldots, \chi_m$ be $m + 1$ configurations such that

$$\chi_0 \Rightarrow_\Gamma \chi_1\ [r_1] \Rightarrow_\Gamma \chi_2\ [r_2] \Rightarrow_\Gamma \cdots \Rightarrow_\Gamma \chi_m\ [r_m]$$

by applying rules labeled with r_1 through r_m, for some $m \geq 1$. Then, we write

$$\chi_0 \Rightarrow_\Gamma^m \chi_m\ [r_1 \cdots r_m]$$

Moreover, for every configuration χ, we write $\chi \Rightarrow_{\Gamma}^{0} \chi$ $[\varepsilon]$. For any two configurations χ and χ', if $\chi \Rightarrow_{\Gamma}^{m} \chi'$ $[\rho]$ for $m \geq 0$ and $\rho \in \Psi^*$, then we write

$$\chi \Rightarrow_{\Gamma}^{*} \chi' \; [\rho]$$

□

In the language generated by Γ, we include every string z satisfying the following two conditions—(1) it appears in the first component in a configuration that can be computed from the start configuration, and (2) when it appears, all the other strings are empty.

Definition 14.1.4. Let $\Gamma = (T, \Psi, P_1, w_1, P_2, w_2, \dots, P_n, w_n)$ be an n-pGS, for some $n \geq 1$. The *language* of Γ is denoted by $L(\Gamma)$ and defined as

$$L(\Gamma) = \big\{z \in T^* \mid (w_1, w_2, \dots, w_n) \Rightarrow_{\Gamma}^{*} (z, \varepsilon, \dots, \varepsilon)\big\}$$

□

To control Γ, we define a language, Ξ, over its set of rule labels, and we require that every successful computation—that is, a computation leading to a string in the generated language—is made by a sequence of rules from Ξ.

Definition 14.1.5. Let $\Gamma = (T, \Psi, P_1, w_1, P_2, w_2, \dots, P_n, w_n)$ be an n-pGS, for some $n \geq 1$, and let $\Xi \subseteq \Psi^*$ be a *control language*. The *language generated by Γ with Ξ* is denoted by $L(\Gamma, \Xi)$ and defined as

$$L(\Gamma, \Xi) = \big\{z \in T^* \mid (w_1, w_2, \dots, w_n) \Rightarrow_{\Gamma}^{*} (z, \varepsilon, \dots, \varepsilon) \; [\rho] \text{ with } \rho \in \Xi\big\}$$

If Ξ is regular, then the pair (Γ, Ξ) is called a *regular-controlled n-pGS*. □

Next, we illustrate the previous definitions by an example.

Example 14.1.6. Consider the 4-pGS

$$\Gamma = \big(\{a, b, c\}, \{r_i \mid 1 \leq i \leq 11\}, P_1, c, P_2, a, P_3, a, P_4, a\big)$$

where

$$\begin{aligned}
P_1 &= \{r_1{:}\, c \to cc, r_2{:}\, c \to bc, r_3{:}\, b \to bb, r_4{:}\, b \to ab, r_5{:}\, a \to aa\} \\
P_2 &= \{r_6{:}\, a \to aa, r_7{:}\, a \to \varepsilon\} \\
P_3 &= \{r_8{:}\, a \to aa, r_9{:}\, a \to \varepsilon\} \\
P_4 &= \{r_{10}{:}\, a \to aa, r_{11}{:}\, a \to \varepsilon\}
\end{aligned}$$

Let $\Xi = \{r_6 r_8 r_{10}\}^* \{r_7 r_1\}^* \{r_2\} \{r_9 r_3\}^* \{r_4\} \{r_{11} r_5\}^*$ be a control language. Observe that every successful derivation in Γ with Ξ is of the form

$$\begin{array}{llll}(c,a,a,a) & \Rightarrow_{\Gamma}^{3(k-1)} & (c,a^k,a^k,a^k) & [(r_6r_8r_{10})^{k-1}]\\ & \Rightarrow_{\Gamma}^{2k} & (c^{k+1},\varepsilon,a^k,a^k) & [(r_7r_1)^k]\\ & \Rightarrow_{\Gamma} & (bc^{k+1},\varepsilon,a^k,a^k) & [r_2]\\ & \Rightarrow_{\Gamma}^{2k} & (b^{k+1}c^{k+1},\varepsilon,\varepsilon,a^k) & [(r_9r_3)^k]\\ & \Rightarrow_{\Gamma} & (ab^{k+1}c^{k+1},\varepsilon,\varepsilon,a^k) & [r_4]\\ & \Rightarrow_{\Gamma}^{2k} & (a^{k+1}b^{k+1}c^{k+1},\varepsilon,\varepsilon,\varepsilon) & [(r_{11}r_5)^k]\end{array}$$

for some $k \geq 1$. Clearly, $L(\Gamma, \Xi) = \{a^n b^n c^n \mid n \geq 2\}$. □

From Example 14.1.6, we see that regular-controlled pGSs can generate non-context-free languages. Moreover, notice that Ξ in Example 14.1.6 is, in fact, a union-free regular language (see [21]).

For every $n \geq 1$, let ${}_n\mathbf{pGS}$ denote the language family generated by n-pGSs. Furthermore, set

$$\mathbf{pGS} = \bigcup_{n \geq 1} {}_n\mathbf{pGS}$$

14.2 Generative Power

In this section, we prove results (I) through (III), given next.

(I) pGSs without control languages characterize only a proper subset of the family of context-free languages (Theorem 14.2.4).

(II) Every recursively enumerable language can be generated by a regular-controlled 2-pGS (Theorems 14.2.7 and 14.2.8).

(III) pGSs over unary alphabets controlled by languages from **rC** generate only the family of regular languages (Theorem 14.2.9).

Power of Pure Grammar Systems

First, we show that the language family generated by pGSs without control languages is properly included in the family of context-free languages.

Lemma 14.2.1. *Let Γ be an n-pGS satisfying $L(\Gamma) \neq \emptyset$, for some $n \geq 1$. Then, there is a 1-pGS Ω such that $L(\Omega) = L(\Gamma)$.*

Proof. Let $\Gamma = (T, \Psi, P_1, w_1, P_2, w_2, \ldots, P_n, w_n)$ be an n-pGS satisfying $L(\Gamma) \neq \emptyset$, for some $n \geq 1$. Let $z \in L(\Gamma)$. By the definition of $L(\Gamma)$, there exists

$$(w_1, w_2, \ldots, w_n) \Rightarrow_{\Gamma}^{*} (z, \varepsilon, \ldots, \varepsilon)$$

Since all components are independent of each other, there is also

$$(w_1, w_2, \dots, w_n) \Rightarrow^*_\Gamma (w_1, \varepsilon, \dots, \varepsilon) \Rightarrow^*_\Gamma (z, \varepsilon, \dots, \varepsilon)$$

Therefore, the 1-pGS $\Omega = (T, \Psi, P_1, w_1)$ clearly satisfies $L(\Omega) = L(\Gamma)$. Hence, the lemma holds. □

Lemma 14.2.2. *Let Γ be a 1-pGS. Then, $L(\Gamma)$ is context-free.*

Proof. Let $\Gamma = (T, \Psi, P, w)$ be a 1-pGS. We next construct an extended pushdown automaton M such that $L_{ef}(M) = L(\Gamma)$. Construct

$$M = \big(W, T, \Omega, R, s, \#, F\big)$$

as follows. Initially, set $W = \{s, t, f\}$, $\Omega = T \cup \{\#\}$, $R = \emptyset$, and $F = \{f\}$. Without any loss of generality, assume that $\# \notin T$. Perform (1) through (4), given next.

(1) Add $\#s \to \mathrm{rev}(w)t$ to R.
(2) For each $a \to y \in P$, add $at \to \mathrm{rev}(y)t$ to R.
(3) Add $t \to f$ to R.
(4) For each $a \in T$, add $afa \to f$ to R.

M works in the following way. It starts from $\#sz$, where $z \in T^*$. By the rule from (1), it generates the reversed version of the start string of Γ on the pushdown, ending up in $\mathrm{rev}(w)tz$. Then, by rules from (2), it rewrites $\mathrm{rev}(w)$ to a string over T. During both of these generations, no input symbols are read. To accept z, M has to end up in $\mathrm{rev}(z)tz$. After that, it moves to f by the rule from (3). Then, by using rules introduced in (4), it compares the contents of the pushdown with the input string. M accepts z if and only if the contents of the pushdown match the input string, meaning that $z \in L(\Gamma)$.

Clearly, $L_{ef}(M) = L(\Gamma)$, so the lemma holds. □

Lemma 14.2.3. *There is no n-pGS that generates $\{a, aa\}$, for any $n \geq 1$.*

Proof. By contradiction. Without any loss of generality, making use of Lemma 14.2.1, we can only consider 1-pGSs. For the sake of contradiction, assume that there exists a 1-pGS, $\Omega = (\{a\}, P, w)$, such that $L(\Omega) = \{a, aa\}$. Observe that either (i) $w = a$ or (ii) $w = aa$. These two cases are discussed next.

(i) Assume that $w = a$. Then, $a \to aa \in P$. However, this implies that $L(\Omega)$ is infinite—a contradiction.
(ii) Assume that $w = aa$. Then, $a \to \varepsilon \in P$. However, this implies that $\varepsilon \in L(\Omega)$—a contradiction.

Hence, no 1-pGS generates $\{a, aa\}$, so the lemma holds. □

Theorem 14.2.4. $\mathbf{pGS} \subset \mathbf{CF}$

Proof. Let Γ be an n-pGS, for some $n \geq 1$. If $L(\Gamma) = \emptyset$, then $L(\Gamma)$ is clearly context-free. Therefore, assume that $L(\Gamma) \neq \emptyset$. Then, by Lemma 14.2.1, there is a 1-pGS Ω such that $L(\Omega) = L(\Gamma)$. By Lemma 14.2.2, $L(\Omega)$ is context-free. Hence, $\mathbf{pGS} \subseteq \mathbf{CF}$. By Lemma 14.2.3, $\mathbf{CF} - \mathbf{pGS} \neq \emptyset$, so the theorem holds. □

Power of Controlled Pure Grammar Systems

In this section, we prove that every recursively enumerable language can be generated by a regular-controlled 2-pGS.

To do this, we need the following normal form of left-extended queue grammars (see Definition 3.3.18).

Lemma 14.2.5 (see [18]). *For every recursively enumerable language K, there exists a left-extended queue grammar, $Q = (V, T, W, F, R, g)$, such that $L(Q) = K$, $T = \mathrm{alph}(K)$, $F = \{f\}$, $W = X \cup Y \cup \{\S\}$, where $X, Y, \{\S\}$ are pairwise disjoint, and every $(a, p, y, q) \in R$ satisfies either $a \in V - T$, $p \in X$, $y \in (V - T)^*$, $q \in X \cup \{\S\}$ or $a \in V - T$, $p \in Y \cup \{\S\}$, $y \in T^*$, $q \in Y$.*

Furthermore, Q generates every $h \in L(Q)$ in this way

$$
\begin{array}{lll}
 & \#a_0 p_0 & \\
\Rightarrow_Q & a_0\#x_0 p_1 & [(a_0, p_0, z_0, p_1)] \\
\Rightarrow_Q & a_0 a_1\#x_1 p_2 & [(a_1, p_1, z_1, p_2)] \\
 & \vdots & \\
\Rightarrow_Q & a_0 a_1 \cdots a_k\#x_k p_{k+1} & [(a_k, p_k, z_k, p_{k+1})] \\
\Rightarrow_Q & a_0 a_1 \cdots a_k a_{k+1}\#x_{k+1} y_1 p_{k+2} & [(a_{k+1}, p_{k+1}, y_1, p_{k+2})] \\
 & \vdots & \\
\Rightarrow_Q & a_0 a_1 \cdots a_k a_{k+1} \cdots a_{k+m-1}\#x_{k+m-1} y_1 \cdots y_{m-1} p_{k+m} & [(a_{k+m-1}, p_{k+m-1}, y_{m-1}, p_{k+m})] \\
\Rightarrow_Q & a_0 a_1 \cdots a_k a_{k+1} \cdots a_{k+m}\#y_1 \cdots y_m p_{k+m+1} & [(a_{k+m}, p_{k+m}, y_m, p_{k+m+1})]
\end{array}
$$

where $k, m \geq 1$, $a_i \in V - T$ for $i = 0, \dots, k + m$, $x_j \in (V - T)^*$ for $j = 1, \dots, k + m$, $g = a_0 p_0$, $a_j x_j = x_{j-1} z_j$ for $j = 1, \dots, k$, $a_1 \cdots a_k x_{k+1} = z_0 \cdots z_k$, $a_{k+1} \cdots a_{k+m} = x_k$, $p_0, p_1, \dots, p_{k+m} \in W - F$ and $p_{k+m+1} = f$, $z_i \in (V - T)^*$ for $i = 1, \dots, k$, $y_j \in T^*$ for $j = 1, \dots, m$, and $h = y_1 y_2 \cdots y_{m-1} y_m$. □

Informally, the queue grammar Q in Lemma 14.2.5 generates every string in $L(Q)$ so that it passes through state §. Before it enters §, it generates only strings over $V - T$; after entering §, it generates only strings over T.

Lemma 14.2.6. *Let Q be a left-extended queue grammar satisfying*

$$\operatorname{card}\Big(\operatorname{alph}\big(L(Q)\big)\Big) \geq 2$$

and the properties given in Lemma 14.2.5. Then, there is a 2-pGS Γ and a regular language Ξ such that $L(\Gamma, \Xi) = K$.

Proof. Let $Q = (V, T, W, F, R, g)$ be a left-extended queue grammar satisfying

$$\operatorname{card}\Big(\operatorname{alph}\big(L(Q)\big)\Big) \geq 2$$

and the properties given in Lemma 14.2.5. Let $g = a_0 p_0$, $W = X \cup Y \cup \{\S\}$, and $F = \{f\}$. Assume that

$$\{0, 1\} \subseteq \operatorname{alph}\big(L(Q)\big)$$

Observe that there exist a positive integer n and an injection ι from VW to $\{0, 1\}^n - \{1^n\}$ so that ι remains an injection when its domain is extended to $(VW)^*$ in the standard way (after this extension, ι thus represents an injective homomorphism from $(VW)^*$ to $(\{0, 1\}^n - \{1^n\})^*$); a proof of this observation is simple and left to the reader. Based on ι, define the substitution ν from V to $(\{0, 1\}^n - \{1^n\})$ as

$$\nu(a) = \big\{\iota(aq) \mid q \in W\big\}$$

for every $a \in V$. Extend the domain of ν to V^*. Furthermore, define the substitution μ from W to $(\{0, 1\}^n - \{1^n\})$ as

$$\mu(q) = \big\{\iota(aq) \mid a \in V\big\}$$

for every $q \in W$. Extend the domain of μ to W^*.
Construct the 2-pGS

$$\Gamma = \big(T, \Psi, P_1, w_1, P_2, w_2\big)$$

where

$$\begin{aligned}
\Psi &= \{{}^{1}_{1y}1 \mid (a, p, y, q) \in R\} \\
&\quad \cup \{{}^{1}_{1w}1 \mid w \in \nu(y), (a, p, y, q) \in R\} \\
&\quad \cup \{{}^{2}_{1z}1 \mid z \in \mu(q), (a, p, y, q) \in R\} \\
&\quad \cup \{{}^{i}_{\varepsilon}0, {}^{i}_{\varepsilon}1 \mid i = 1, 2\} \\
&\quad \cup \{{}^{1}_{1^{n+1}}1\} \\
P_1 &= \{{}^{1}_{1y}1\colon 1 \to 1y \mid (a, p, y, q) \in R\} \\
&\quad \cup \{{}^{1}_{1w}1\colon 1 \to 1w \mid w \in \nu(y), (a, p, y, q) \in R\} \\
&\quad \cup \{{}^{1}_{\varepsilon}0\colon 0 \to \varepsilon, {}^{1}_{\varepsilon}1\colon 1 \to \varepsilon\} \\
&\quad \cup \{{}^{1}_{1^{n+1}}1\colon 1 \to 1^{n+1}\} \\
w_1 &= 1 \\
P_2 &= \{{}^{2}_{1z}1\colon 1 \to 1z \mid z \in \mu(q), (a, p, y, q) \in R\} \\
&\quad \cup \{{}^{2}_{\varepsilon}0\colon 0 \to \varepsilon, {}^{2}_{\varepsilon}1\colon 1 \to \varepsilon\} \\
w_2 &= 1^{n+1}
\end{aligned}$$

Intuitively, ${}^{i}_{y}a$ means that a is rewritten to y in the ith component. Construct the right-linear grammar

$$G = \big(N, \Psi, P, \langle f, 2\rangle\big)$$

as follows. Initially, set $P = \emptyset$ and $N = \{\$\} \cup \{\langle p, i\rangle \mid p \in W, i = 1, 2\}$, where \$ is a new symbol. Perform (1) through (5), given next.

(1) If $(a, p, y, q) \in R$, where $a \in V - T$, $p \in W - F$, $q \in W$, and $y \in T^*$, add $\langle q, 2\rangle \rightarrow {}^{1}_{1y}1\ {}^{1}_{1z}1\ \langle p, 2\rangle$ to P for each $z \in \mu(p)$.
(2) Add $\langle \S, 2\rangle \rightarrow {}^{1}_{1^{n+1}}1\ \langle \S, 1\rangle$ to P.
(3) If $(a, p, y, q) \in R$, where $a \in V - T$, $p \in W - F$, $q \in W$, and $y \in (V - T)^*$, add $\langle q, 1\rangle \rightarrow {}^{1}_{1w}1\ {}^{1}_{1z}1\ \langle p, 1\rangle$ to P for each $w \in \nu(y)$ and $z \in \mu(p)$.
(4) Add $\langle p_0, 1\rangle \rightarrow {}^{1}_{1w}1\ \$$ to P for each $w \in \nu(a_0)$.
(5) Add $\$ \rightarrow {}^{1}_{\varepsilon}0\ {}^{2}_{\varepsilon}0\ \$$, $\$ \rightarrow {}^{1}_{\varepsilon}1\ {}^{2}_{\varepsilon}1\ \$$, and $\$ \rightarrow \varepsilon$ to P.

Let $\mathscr{G} = (\Gamma, L(G))$. Before we establish the identity $L(\mathscr{G}) = L(Q)$, we explain how $\mathscr{G}$ works. In what follows, $(x, y)\ p$ means that the current configuration of Γ is (x, y) and that p is the nonterminal in the current sentential form of G. Consider the form of the derivations of Q in Lemma 14.2.5. The regular-controlled 2-pGS $\mathscr{G}$ simulates these derivations in reverse as follows. The start configuration of $\mathscr{G}$ is

$$\big(1, 1^{n+1}\big) \quad \langle f, 2\rangle$$

Rules from (1) generate $h \in L(Q)$ in the first component and encoded states p_{k+m}, $p_{k+m-1}, \dots, p_k$ in the second component

$$\big(1h, 1z1^n\big) \quad \langle \S, 2\rangle$$

where $z \in \mu(p_k p_{k+1} \cdots p_{k+m})$. Rules from (2) appends 1^n (a delimiter) to the first component

$$\big(1^{n+1}h, 1z1^n\big) \quad \langle \S, 1\rangle$$

Rules from (3) generate encoded symbols a_{k+m}, a_{k+m-1}, $\dots$, a_1 in the first component and encoded states $p_{k-1}, p_{k-2}, \dots, p_0$ in the second component

$$\big(1w1^nh, 1z'z1^n\big) \quad \langle p_0, 1\rangle$$

where $w \in \nu(a_1a_2 \cdots a_{k+m})$ and $z' \in \mu(p_0p_1 \cdots p_{k-1})$. A rule from (4) generates

$$\big(1w'w1^nh, 1z'z1^n\big) \quad \$$$

where $w' \in \nu(a_0)$ and a_0 is the start symbol of Q. Notice that

$$w'w \in \nu(a_0a_1a_2 \cdots a_{k+m})$$

and

$$z'z \in \mu(p_0 p_1 p_2 \cdots p_{k+m})$$

Finally, rules from (5) check that $1w'w1^n = 1z'z1^n$ by erasing these two strings in a symbol-by-symbol way, resulting in (h, ε).

For brevity, the following proof omits some obvious details, which the reader can easily fill in. The next claim proves the above explanation rigorously—that is, it shows how $\mathscr{G}$ generates each string of $L(\mathscr{G})$.

Claim 1. The regular-controlled 2-pGS $\mathscr{G}$ generates every $h \in L(\mathscr{G})$ in this way

$$\begin{aligned}
& (1, 1^{n+1}) \\
\Rightarrow_\Gamma\ & (1y_m, 1g_{k+m}1^n) \\
\Rightarrow_\Gamma\ & (1y_{m-1}y_m, 1g_{k+m-1}g_{k+m}1^n) \\
& \vdots \\
\Rightarrow_\Gamma\ & (1y_1 \cdots y_{m-1}y_m, 1g_k \cdots g_{k+m-1}g_{k+m}1^n) \\
\Rightarrow_\Gamma\ & (1^{n+1}h, 1g_k \cdots g_{k+m-1}g_{k+m}1^n) \\
\Rightarrow_\Gamma\ & (1t_{k+m}1^n h, 1g_{k-1}g_k \cdots g_{k+m-1}g_{k+m}1^n) \\
\Rightarrow_\Gamma\ & (1t_{k+m-1}t_{k+m}1^n h, 1g_{k-2}g_{k-1}g_k \cdots g_{k+m-1}g_{k+m}1^n) \\
& \vdots \\
\Rightarrow_\Gamma\ & (1t_1 \cdots t_{k+m-1}t_{k+m}1^n h, 1g_0 \cdots g_{k-2}g_{k-1}g_k \cdots g_{k+m-1}g_{k+m}1^n) \\
\Rightarrow_\Gamma\ & (1t_0t_1 \cdots t_{k+m-1}t_{k+m}1^n h, 1g_0 \cdots g_{k-2}g_{k-1}g_k \cdots g_{k+m-1}g_{k+m}1^n) \\
\Rightarrow_\Gamma\ & (v_1h, v_1) \\
& \qquad \Rightarrow_\Gamma (v_2h, v_2) \\
& \qquad \vdots \\
& \qquad \Rightarrow_\Gamma (v_\ell h, v_\ell) \\
& \qquad \Rightarrow_\Gamma (h, \varepsilon)
\end{aligned}$$

where $k, m \geq 1$; $h = y_1 \cdots y_{m-1}y_m$, where $y_i \in T^*$ for $i = 1, 2, \ldots, m$; $t_i \in \nu(a_i)$ for $i = 0, 1, \ldots, k+m$, where $a_i \in V - T$; $g_i \in \mu(p_i)$ for $i = 0, 1, \ldots, k+m$, where $p_i \in W - F$; $v_i \in \{0, 1\}^*$ for $i = 1, 2, \ldots, \ell$, where $\ell = |t_0t_1 \cdots t_{k+m-1}t_{k+m}1^n|$; $|v_{i+1}| = |v_i| - 1$ for $i = 0, 1, \ldots, \ell - 1$.

Proof. Examine the construction of $\mathscr{G}$. Notice that in every successful computation, $\mathscr{G}$ uses rules from step (i) before it uses rules from step (i+1), for $i = 1, 2, 3, 4$. Thus, in a greater detail, every successful computation

$$(1, 1^{n+1}) \Rightarrow_\Gamma^* (h, \varepsilon)\ [\rho]$$

where $\rho \in L(G)$, can be expressed as

$$
\begin{aligned}
& (1, 1^{n+1}) \\
\Rightarrow_{\Gamma}\ & (1y_m, 1g_{k+m}1^n) \\
\Rightarrow_{\Gamma}\ & (1y_{m-1}y_m, 1g_{k+m-1}g_{k+m}1^n) \\
& \vdots \\
\Rightarrow_{\Gamma}\ & (1y_1 \cdots y_{m-1}y_m, 1g_k \cdots g_{k+m-1}g_{k+m}1^n) \\
\Rightarrow_{\Gamma}\ & (1^{n+1}h, 1g_k \cdots g_{k+m-1}g_{k+m}1^n) \\
\Rightarrow_{\Gamma}\ & (1t_{k+m}1^n h, 1g_{k-1}g_k \cdots g_{k+m-1}g_{k+m}1^n) \\
\Rightarrow_{\Gamma}\ & (1t_{k+m-1}t_{k+m}1^n h, 1g_{k-2}g_{k-1}g_k \cdots g_{k+m-1}g_{k+m}1^n) \\
& \vdots \\
\Rightarrow_{\Gamma}\ & (1t_1 \cdots t_{k+m-1}t_{k+m}1^n h, 1g_0 \cdots g_{k-2}g_{k-1}g_k \cdots g_{k+m-1}g_{k+m}1^n) \\
\Rightarrow_{\Gamma}\ & (1t_0t_1 \cdots t_{k+m-1}t_{k+m}1^n h, 1g_0 \cdots g_{k-2}g_{k-1}g_k \cdots g_{k+m-1}g_{k+m}1^n) \\
\Rightarrow_{\Gamma}^{*}\ & (h, \varepsilon)
\end{aligned}
$$

where $k, m \geq 1$; $h = y_1 \cdots y_{m-1}y_m$, where $y_i \in T^*$ for $i = 1, 2, \ldots, m$; $t_i \in \nu(a_i)$ for $i = 0, 1, \ldots, k + m$, where $a_i \in V - T$; $g_i \in \mu(p_i)$ for $i = 0, 1, \ldots, k + m$, where $p_i \in W - F$. Furthermore, during

$$(1t_0t_1 \cdots t_{k+m-1}t_{k+m}1^n h, 1g_0 \cdots g_{k-2}g_{k-1}g_k \cdots g_{k+m-1}g_{k+m}1^n) \Rightarrow_{\Gamma}^{*} (h, \varepsilon)$$

only rules from (5) are used. Therefore,

$$1t_0t_1 \cdots t_{k+m-1}t_{k+m}1^n h \;=\; 1g_0 \cdots g_{k-2}g_{k-1}g_k \cdots g_{k+m-1}g_{k+m}1^n$$

Let $v = 1t_0t_1 \cdots t_{k+m-1}t_{k+m}1^n h$. By rules from (5), $\mathscr{G}$ makes $|v|$ steps to erase v. Consequently, $(vh, v) \Rightarrow_{\Gamma}^{*} (h, \varepsilon)$ can be expressed as

$$
\begin{aligned}
& (vh, v) \\
\Rightarrow_{\Gamma}\ & (v_1h, v_1) \\
\Rightarrow_{\Gamma}\ & (v_2h, v_2) \\
& \vdots \\
\Rightarrow_{\Gamma}\ & (v_\ell h, v_\ell) \\
\Rightarrow_{\Gamma}\ & (h, \varepsilon)
\end{aligned}
$$

where $v_i \in \{0, 1\}^*$ for $i = 1, 2, \ldots, \ell$, where $\ell = |v| - 1$, and $|v_{i+1}| = |v_i| - 1$ for $i = 0, 1, \ldots, \ell - 1$. As a result, the claim holds. □

Let $\mathscr{G}$ generate $h \in L(\mathscr{G})$ in the way described in Claim 1. Examine the construction of G to see that at this point, R contains $(a_0, p_0, z_0, p_1), \ldots, (a_k, p_k, z_k, p_{k+1}), (a_{k+1}, p_{k+1}, y_1, p_{k+2}), \ldots, (a_{k+m-1}, p_{k+m-1}, y_{m-1}, p_{k+m}), (a_{k+m}, p_{k+m}, y_m, p_{k+m+1})$, where $p_{k+m+1} = f$ and $z_i \in (V - T)^*$ for $i = 1, 2, \ldots, k$, so Q makes the generation of h in the way described in Lemma 14.2.5. Thus, $h \in L(Q)$. Consequently, $L(\mathscr{G}) \subseteq L(Q)$.

Let Q generate $g \in L(Q)$ in the way described in Lemma 14.2.5. Then, $\mathscr{G}$ generates h in the way described in Claim 1, so $L(Q) \subseteq L(\mathscr{G})$; a detailed proof of this inclusion is left to the reader.

As $L(\mathscr{G}) \subseteq L(Q)$ and $L(Q) \subseteq L(\mathscr{G})$, $L(\mathscr{G}) = L(Q)$. Hence, the lemma holds. □

Theorem 14.2.7. *Let K be a recursively enumerable language satisfying*

$$\text{card}\big(\text{alph}(K)\big) \geq 2$$

Then, there is a 2-pGS Γ and a regular language Ξ such that $L(\Gamma, \Xi) = K$.

Proof. This theorem follows from Lemmas 14.2.5 and 14.2.6. □

Theorem 14.2.8. *Let K be a unary recursively enumerable language, and let $c \notin \text{alph}(K)$ be a new symbol. Then, there is a 2-pGS Γ and a regular language Ξ such that $L(\Gamma, \Xi) = K$.*

Proof. This theorem can be proved by analogy with the proof of Theorem 14.2.7 (we use c as the second symbol in the proof of Lemma 14.2.6). □

Power of Controlled Pure Grammar Systems Over Unary Alphabets

In this section, we prove that pGSs over unary alphabets controlled by languages from **rC** generate only regular languages.

Theorem 14.2.9. *Let $\Gamma = (T, \Psi, P_1, w_1, P_2, w_2, \ldots, P_n, w_n)$ be an n-pGS satisfying* card$(T) = 1$, *for some $n \geq 1$, and let $\Xi \in$* **rC**. *Then, $L(\Gamma, \Xi)$ is regular.*

Proof. Let $\Gamma = (T, \Psi, P_1, w_1, P_2, w_2, \ldots, P_n, w_n)$ be an n-pGS satisfying card$(T) = 1$, for some $n \geq 1$, and let $\Xi \in$ **rC**. We show how to convert Γ and Ξ into an equivalent regular-controlled grammar (G, Π). Then, since card$(T) = 1$, Theorem 5.1.6 implies that $L(\Gamma, \Xi)$ is regular.

Let $\bar{G} = (\bar{N}, \Psi, \bar{\Phi}, \bar{P}, \bar{S})$ be a context-free grammar and $\bar{M}$ be a finite automaton such that $L(\bar{G}, L(\bar{M})) = \Xi$. Let $T = \{c\}$. To distinguish between the components of Γ in G, we encode c for each component. Set

$$N_{\#} = \big\{c_i \mid 1 \leq i \leq n\big\}$$

For each $i \in \{1, \ldots, n\}$, define the homomorphism τ_i from T^* to $N_{\#}^*$ as $\tau_i(c) = c_i$. For each $i \in \{1, \ldots, n\}$, set

$$R_i = \big\{r\colon \tau_i(c) \to \tau_i(y) \mid r\colon c \to y \in P_i\big\}$$

Define G as

$$G = \big(N, \{c\}, \Phi, R, S\big)$$

where

$$\begin{aligned}
N &= \{S\} \cup \bar{N} \cup \Psi \cup N_{\#} \cup \textstyle\bigcup_{1 \le i \le n} R_i, \\
\Phi &= \bar{\Phi} \cup \{s, c_1\} \cup \{r_\varepsilon \mid r \in \bar{\Phi}\} \\
R &= \bar{P} \cup \{s\colon S \to \bar{S}\tau_1(w_1)\tau_2(w_2)\cdots\tau_n(w_n)\} \\
&\quad \cup \{c_1\colon c_1 \to c\} \\
&\quad \cup \{r_\varepsilon\colon r \to \varepsilon \mid r \in \bar{\Phi}\} \\
\Lambda &= \{r_\varepsilon r \mid r\colon a \to y \in \textstyle\bigcup_{1 \le i \le n} R_i\}^* \\
\Pi &= \{s\}L(\bar{M})\Lambda\{c_1\}^* \text{ (the control language of } G)
\end{aligned}$$

Without any loss of generality, we assume that $\{S\}$, $\bar{N}$, Ψ, $N_{\#}$, and $\bigcup_{1 \le i \le n} R_i$ are pairwise disjoint. We also assume that $\bar{\Phi}$, $\{s, c_1\}$, and $\{r_\varepsilon \mid r \in \bar{\Phi}\}$ are pairwise disjoint.

In every successful derivation of every $z \in L(G, \Pi)$ in G, s is applied to S. It generates the start symbol of $\bar{G}$ and encoded start strings of each component of Γ. Indeed, instead of c, we generate c_i, where $1 \le i \le n$. Then, rules from $\bar{P}$ are used to rewrite $\bar{S}$ to a control string from $L(\bar{G}, L(\bar{M}))$. By using pairs of rules $r_\varepsilon r \in \Lambda$, G erases an occurrence of r in the current sentential form and applies r to a symbol in a proper substring of the current sentential corresponding to the component which would use r. This process is repeated until the control string is completely erased. Since $\mathrm{card}(T) = 1$, the order of used rules and the occurrence of the rewritten cs are not important. Finally, G uses $c_1\colon c_1 \to c$ to decode each occurrence of c_1 back to c, thus obtaining z. If G applies its rules in an improper way—that is, if there remain some symbols from $\bigcup_{1 \le i \le n} R_i$ or from $\{c_i \mid 2 \le i \le n\}$ after the last pair from Λ is applied—the derivation is blocked.

Based on these observations, we see that every successful derivation of every $z \in L(G, \Pi)$ in G with Π is of the form

$$\begin{aligned}
S &\Rightarrow_G \bar{S}\tau_1(w_1)\tau_2(w_2)\cdots\tau_n(w_n) && [s] \\
&\Rightarrow_G^* \rho\tau_1(w_1)\tau_2(w_2)\cdots\tau_n(w_n) && [\upsilon] \\
&\Rightarrow_G^* \tau_1(z) && [\lambda] \\
&\Rightarrow_G^* z && [\gamma]
\end{aligned}$$

where $\rho \in L(\bar{G}, L(\bar{M}))$, $\upsilon \in L(\bar{M})$, $\lambda \in \Lambda$, and $\gamma \in \{c_1\}^*$. In Γ, there is

$$(w_1, w_2, \ldots, w_n) \Rightarrow_\Gamma^* (z, \varepsilon, \ldots, \varepsilon)\ [\rho]$$

Hence, $L(G, \Pi) \subseteq L(\Gamma)$. Conversely, for every $z \in L(\Gamma)$, there is a derivation of z in G with Π of the above form, so $L(\Gamma) \subseteq L(G, \Pi)$. Therefore, $L(\Gamma) = L(G, \Pi)$, and the theorem holds. A rigorous proof of the identity $L(\Gamma) = L(G, \Pi)$ is left to the reader. □

From Theorem 14.2.9, we obtain the following corollary.

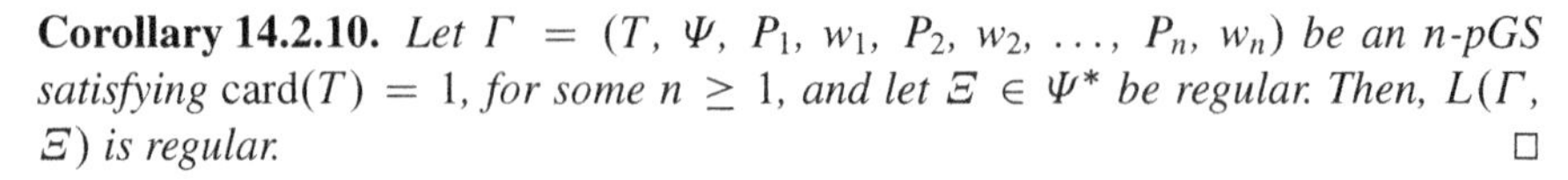

Corollary 14.2.10. *Let* $\Gamma = (T, \Psi, P_1, w_1, P_2, w_2, \ldots, P_n, w_n)$ *be an* n*-pGS satisfying* $\mathrm{card}(T) = 1$, *for some* $n \geq 1$, *and let* $\Xi \in \Psi^*$ *be regular. Then,* $L(\Gamma, \Xi)$ *is regular.* □

Notice that this result is surprising in the light of Theorems 14.2.7 and 14.2.8, which say that 2-pGSs having at least two symbols are computationally complete.

The next four open problem areas are related to the achieved results.

Open Problem 14.2.11. Let Γ be an n-pGS, for some $n \geq 1$. By Lemma 14.2.2, $L(\Gamma)$ is context-free. Is $L(\Gamma)$, in fact, regular? □

Open Problem 14.2.12. Consider proper subfamilies of the family of regular languages (see [6, 11, 21, 23]). Can we obtain Theorems 14.2.7 and 14.2.8 when the control languages are from these subfamilies? □

Open Problem 14.2.13. By Theorems 14.2.7 and 14.2.8, two components suffice to generate any recursively enumerable language by regular-controlled pGSs. What is the power of controlled pGSs with a single component? □

Open Problem 14.2.14. Let Γ be an n-pGS, for some $n \geq 1$. If no rule of Γ has ε on its right-hand side, then Γ is said to be *propagating*. What is the power of controlled propagating pGSs? □

For some preliminary solutions to the above four open problems, see [20].

References

1. Aho, A.V., Lam, M.S., Sethi, R., Ullman, J.D.: Compilers: Principles, Techniques, and Tools, 2nd edn. Addison-Wesley, Boston (2006)
2. Aydin, S., Bordihn, H.: Sequential versus parallel grammar formalisms with respect to measures of descriptional complexity. Fundam. Inform. **55**(3–4), 243–254 (2003)
3. Beek, M., Kleijn, J.: Petri net control for grammar systems. In: Formal and Natural Computing, pp. 220–243. Springer, New York (2002)
4. Bensch, S., Bordihn, H.: Active symbols in pure systems. Fundam. Inform. **76**(3), 239–254 (2007)
5. Bordihn, H., Csuhaj-Varjú, E., Dassow, J.: CD grammar systems versus L systems. In: Grammatical Models of Multi-Agent Systems. Topics in Computer Mathematics, vol. 8, pp. 18–32. Gordon and Breach Science Publishers, Amsterdam (1999)
6. Bordihn, H., Holzer, M., Kutrib, M.: Determination of finite automata accepting subregular languages. Theor. Comput. Sci. **410**(35), 3209–3222 (2009)
7. Csuhaj-Varjú, E., Vaszil, G.: On context-free parallel communicating grammar systems: synchronization, communication, and normal forms. Theor. Comput. Sci. **255**(1–2), 511–538 (2001)
8. Csuhaj-Varjú, E., Dassow, J., Păun, G.: Dynamically controlled cooperating/distributed grammar systems. Inf. Sci. **69**(1–2), 1–25 (1993)
9. Csuhaj-Varjú, E., Dassow, J., Kelemen, J., Păun, G.: Grammar Systems: A Grammatical Approach to Distribution and Cooperation. Gordon and Breach, Yverdon (1994)

10. Dassow, J., Păun, G.: Regulated Rewriting in Formal Language Theory. Springer, New York (1989)
11. Dassow, J., Truthe, B.: Subregularly tree controlled grammars and languages. In: Automata and Formal Languages, pp. 158–169. Computer and Automation Research Institute, Hungarian Academy of Sciences, Balatonfured (2008)
12. Fernau, H., Holzer, M.: Graph-controlled cooperating distributed grammar systems with singleton components. J. Autom. Lang. Comb. **7**(4), 487–503 (2002)
13. Goldefus, F.: Cooperating distributed grammar systems and graph controlled grammar systems with infinite number of components. In: Proceedings of the 15th Conference STUDENT EEICT 2009, vol. 4, pp. 400–404. Brno University of Technology, Brno, CZ (2009)
14. Mäkinen, E.: A note on pure grammars. Inf. Process. Lett. **23**(5), 271–274 (1986)
15. Martinek, P.: Limits of pure grammars with monotone productions. Fundamenta Informaticae **33**(3), 265–280 (1998)
16. Maurer, H.A., Salomaa, A., Wood, D.: Pure grammars. Inf. Control **44**(1), 47–72 (1980)
17. Meduna, A.: Automata and Languages: Theory and Applications. Springer, London (2000)
18. Meduna, A.: Two-way metalinear PC grammar systems and their descriptional complexity. Acta Cybernetica **2004**(16), 385–397 (2004)
19. Meduna, A.: Elements of Compiler Design. Auerbach Publications, Boston (2007)
20. Meduna, A., Vrábel, L., Zemek, P.: Solutions to four open problems concerning controlled pure grammar systems. Int. J. Comput. Math. (to appear)
21. Nagy, B.: Union-free regular languages and 1-cycle-free-path automata. Publicationes Mathematicae Debrecen **98**, 183–197 (2006)
22. Păun, G.: On the synchronization in parallel communicating grammar systems. Acta Inform. **30**(4), 351–367 (1993)
23. Rozenberg, G., Salomaa, A. (eds.): Handbook of Formal Languages, vol. 1: Word, Language, Grammar. Springer, New York (1997)
24. Salomaa, A.: Formal Languages. Academic, London (1973)

Part VI
Regulated Automata

This part presents the fundamentals of regulated automata. It consists of two comprehensive chapters—Chaps. 15 and 16. The former deals with self-regulating automata while the latter covers the essentials concerning automata regulated by control languages.

Chapter 15 defines and investigates *self-regulating automata* in which the selection of a rule according to which the current move is made follows from the rule applied during the previous move. Both finite and pushdown versions of these automata are covered therein.

Chapter 16 discusses automata in which the application of rules is regulated by control languages by analogy with context-free grammars regulated by control languages. This chapter considers finite and pushdown automata regulated by various control languages.

Chapter 15
Self-Regulating Automata

Abstract This chapter defines and investigates *self-regulating automata*. They regulate the selection of a rule according to which the current move is made by a rule according to which a previous move was made. Both finite and pushdown versions of these automata are investigated. The chapter is divided into two sections. Section 15.1 discusses *self-regulating finite automata*. It establishes two infinite hierarchies of language families resulting from them. Both hierarchies lie between the family of regular languages and the family of context-sensitive languages. Section 15.2 studies *self-regulating pushdown automata*. Based upon them, this section characterizes the families of context-free and recursively enumerable languages. However, as opposed to the results about self-regulating finite automata, many questions concerning their pushdown versions remain open; indeed, Sect. 15.2 formulates several specific open problem areas, including questions concerning infinite language-family hierarchies resulting from them.

Keywords Self-regulating automata • Finite and pushdown versions • Infinite hierarchies of language families • Characterizations of language families • Computational completeness • Open problems

In this chapter, as the first type of regulated automata discussed in Part IV, we define and investigate *self-regulating finite automata*. In essence, self-regulating finite automata regulate the selection of a rule according to which the current move is made by a rule according to which a previous move was made.

To give a more precise insight into self-regulating automata, consider a finite automaton M with a finite binary relation R over the set of rules in M. Furthermore, suppose that M makes a sequence of moves ρ that leads to the acceptance of a string, so ρ can be expressed as a concatenation of $n+1$ consecutive subsequences, $\rho = \rho_0\rho_1 \cdots \rho_n$, where $|\rho_i| = |\rho_j|$, $0 \leq i, j \leq n$, in which r_i^j denotes the rule according to which the ith move in ρ_j is made, for all $0 \leq j \leq n$ and $1 \leq i \leq |\rho_j|$ (as usual, $|\rho_j|$ denotes the length of ρ_j). If for all $0 \leq j < n$, $(r_1^j, r_1^{j+1}) \in R$,

A. Meduna and P. Zemek, *Regulated Grammars and Automata*,
DOI 10.1007/978-1-4939-0369-6_15,

then M represents an *n-turn first-move self-regulating finite automaton with respect to R*. If for all $0 \leq j < n$ and all $1 \leq i \leq |\rho_i|$, $(r_i^j, r_i^{j+1}) \in R$, then M represents an *n-turn all-move self-regulating finite automaton with respect to R*.

The chapter is divided into two sections. In Sect. 15.1, based on the number of turns, we establish two infinite hierarchies of language families that lie between the families of regular and context-sensitive languages. First, we demonstrate that n-turn first-move self-regulating finite automata give rise to an infinite hierarchy of language families coinciding with the hierarchy resulting from $(n + 1)$-parallel right-linear grammars (see [5, 6, 9, 10]). Recall that n-parallel right-linear grammars generate a proper language subfamily of the language family generated by $(n + 1)$-parallel right-linear grammars (see Theorem 5 in [6]). As a result, n-turn first-move self-regulating finite automata accept a proper language subfamily of the language family accepted by $(n + 1)$-turn first-move self-regulating finite automata, for all $n \geq 0$. Similarly, we prove that n-turn all-move self-regulating finite automata give rise to an infinite hierarchy of language families coinciding with the hierarchy resulting from $(n + 1)$-right-linear simple matrix grammars (see [1, 3, 10]). As n-right-linear simple matrix grammars generate a proper subfamily of the language family generated by $(n + 1)$-right-linear simple matrix grammars (see Theorem 1.5.4 in [1]), n-turn all-move self-regulating finite automata accept a proper language subfamily of the language family accepted by $(n + 1)$-turn all-move self-regulating finite automata. Furthermore, since the families of right-linear simple matrix languages coincide with the language families accepted by multi-tape non-writing automata (see [2]) and by finite-turn checking automata (see [8]), all-move self-regulating finite automata characterize these families, too. Finally, we summarize the results about both infinite hierarchies.

In Sect. 15.2, by analogy with self-regulating finite automata, we introduce and discuss *self-regulating pushdown automata*. Regarding self-regulating all-move pushdown automata, we prove that they do not give rise to any infinite hierarchy analogical to the achieved hierarchies resulting from the self-regulating finite automata. Indeed, zero-turn all-move self-regulating pushdown automata define the family of context-free languages while one-turn all-move self-regulating pushdown automata define the family of recursively enumerable languages. On the other hand, as far as self-regulating first-move pushdown automata are concerned, the question whether they define an infinite hierarchy is open.

15.1 Self-Regulating Finite Automata

The present section consists of two subsections. Section 15.1.1 defines n-turn first-move self-regulating finite automata and n-turn all-move self-regulating finite automata. Section 15.1.2 determines the accepting power of these automata.

15.1.1 Definitions and Examples

In this section, we define and illustrate n-turn first-move self-regulating finite automata and n-turn all-move self-regulating finite automata. Recall the formalization of rule labels from Definition 3.4.3 because we make use of it in this section frequently.

Definition 15.1.1. A *self-regulating finite automaton* (an *SFA* for short) is a septuple

$$M = \big(Q, \Sigma, \delta, q_0, q_t, F, R\big)$$

where

(1) $(Q, \Sigma, \delta, q_0, F)$ is a finite automaton,
(2) $q_t \in Q$ is a *turn state*, and
(3) $R \subseteq \Psi \times \Psi$ is a finite relation on the alphabet of rule labels. □

In this chapter, we consider two ways of self-regulation—first-move and all-move. According to these two types of self-regulation, two types of n-turn self-regulating finite automata are defined.

Definition 15.1.2. Let $n \geq 0$ and $M = (Q, \Sigma, \delta, q_0, q_t, F, R)$ be a self-regulating finite automaton. M is said to be an *n-turn first-move self-regulating finite automaton* (an *n-first-SFA* for short) if every $w \in L(M)$ is accepted by M in the following way

$$q_0 w \vdash_M^* f\ [\mu]$$

such that

$$\mu = r_1^0 \cdots r_k^0 r_1^1 \cdots r_k^1 \cdots r_1^n \cdots r_k^n$$

where $k \geq 1$, r_k^0 is the first rule of the form $qx \to q_t$, for some $q \in Q$, $x \in \Sigma^*$, and

$$(r_1^j, r_1^{j+1}) \in R$$

for all $j = 0, 1, \ldots, n$. □

The family of languages accepted by n-first-SFAs is denoted by $\mathbf{FSFA}_n$.

Example 15.1.3. Consider a 1-first-SFA

$$M = \big(\{s, t, f\}, \{a, b\}, \delta, s, t, \{f\}, \{(1, 3)\}\big)$$

with δ containing rules (see Fig. 15.1)

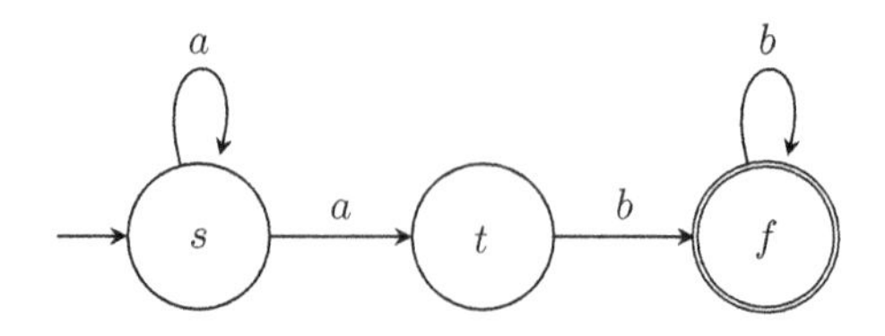

Fig. 15.1 1-turn first-move self-regulating finite automaton M

$$
\begin{aligned}
&1\colon sa \to s\\
&2\colon sa \to t\\
&3\colon tb \to f\\
&4\colon fb \to f
\end{aligned}
$$

With $aabb$, M makes

$$saabb \vdash_M sabb\,[1] \vdash_M tbb\,[2] \vdash_M fb\,[3] \vdash_M f\,[4]$$

In brief, $saabb \vdash_M^* f\,[1234]$. Observe that $L(M) = \{a^n b^n \mid n \geq 1\}$, which belongs to **CF** − **REG**. □

Definition 15.1.4. Let $n \geq 0$ and $M = (Q, \Sigma, \delta, q_0, q_t, F, R)$ be a self-regulating finite automaton. M is said to be an *n-turn all-move self-regulating finite automaton* (an *n-all-SFA* for short) if every $w \in L(M)$ is accepted by M in the following way

$$q_0 w \vdash_M^* f\,[\mu]$$

such that

$$\mu = r_1^0 \cdots r_k^0 r_1^1 \cdots r_k^1 \cdots r_1^n \cdots r_k^n$$

where $k \geq 1$, r_k^0 is the first rule of the form $qx \to q_t$, for some $q \in Q$, $x \in \Sigma^*$, and

$$(r_i^j, r_i^{j+1}) \in R$$

for all $i = 1, 2, \ldots, k$ and $j = 0, 1, \ldots, n-1$. □

The family of languages accepted by n-all-SFAs is denoted by $\mathbf{ASFA}_n$.

Example 15.1.5. Consider a 1-all-SFA

$$M = \big(\{s, t, f\}, \{a, b\}, \delta, s, t, \{f\}, \{(1,4), (2,5), (3,6)\}\big)$$

with δ containing the following rules (see Fig. 15.2)

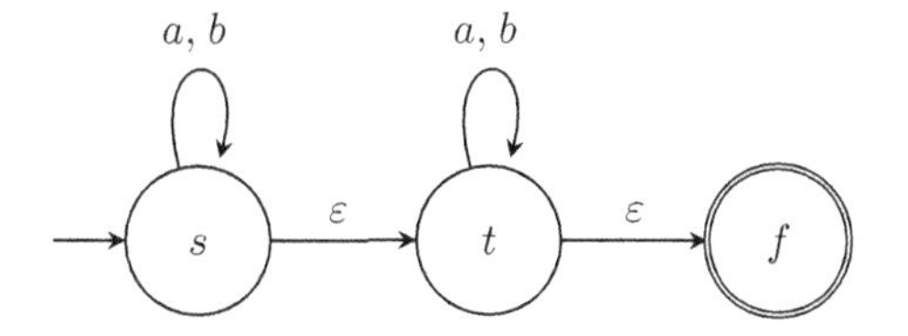

Fig. 15.2 1-turn all-move self-regulating finite automaton M

$$\begin{aligned}
&1\colon sa \to s\\
&2\colon sb \to s\\
&3\colon s \to t\\
&4\colon ta \to t\\
&5\colon tb \to t\\
&6\colon t \to f
\end{aligned}$$

With $abab$, M makes

$$sabab \vdash_M sbab\,[1] \vdash_M sab\,[2] \vdash_M tab\,[3] \vdash_M tb\,[4] \vdash_M t\,[5] \vdash_M f\,[6]$$

In brief, $sabab \vdash_M^* f\,[123456]$. Observe that $L(M) = \{ww \mid w \in \{a,b\}^*\}$, which belongs to $\mathbf{CS} - \mathbf{CF}$. □

15.1.2 Accepting Power

In this section, we discuss the accepting power of n-first-SFAs and n-all-SFAs.

n-Turn First-Move Self-Regulating Finite Automata

We prove that the family of languages accepted by n-first-SFAs coincides with the family of languages generated by so-called $(n+1)$-parallel right-linear grammars (see [5, 6, 9, 10]). First, however, we define these grammars formally.

Definition 15.1.6. For $n \geq 1$, an *n-parallel right-linear grammar* (see [5, 6, 9, 10]) (an *n-PRLG* for short) is an $(n+3)$-tuple

$$G = \big(N_1, \dots, N_n, T, S, P\big)$$

where N_i, $1 \leq i \leq n$, are pairwise disjoint *nonterminal alphabets*, T is a *terminal alphabet*, $S \notin N$ is an *initial symbol*, where $N = N_1 \cup \dots \cup N_n$, and P is a finite set of *rules* that contains these three kinds of rules

1. $S \to X_1 \cdots X_n$, $X_i \in N_i$, $1 \leq i \leq n$;
2. $X \to wY$, $X, Y \in N_i$ for some i, $1 \leq i \leq n$, $w \in T^*$;
3. $X \to w$, $X \in N$, $w \in T^*$.

For $x, y \in (N \cup T \cup \{S\})^*$,

$$x \Rightarrow_G y$$

if and only if

(1) either $x = S$ and $S \to y \in P$,
(2) or $x = y_1 X_1 \cdots y_n X_n$, $y = y_1 x_1 \cdots y_n x_n$, where $y_i \in T^*$, $x_i \in T^* N \cup T^*$, $X_i \in N_i$, and $X_i \to x_i \in P$, $1 \leq i \leq n$.

Let $x, y \in (N \cup T \cup \{S\})^*$ and $\ell > 0$. Then, $x \Rightarrow_G^\ell y$ if and only if there exists a sequence

$$x_0 \Rightarrow_G x_1 \Rightarrow_G \cdots \Rightarrow_G x_\ell$$

where $x_0 = x$, $x_\ell = y$. As usual, $x \Rightarrow_G^+ y$ if and only if there exists $\ell > 0$ such that $x \Rightarrow_G^\ell y$, and $x \Rightarrow_G^* y$ if and only if $x = y$ or $x \Rightarrow_G^+ y$.

The *language* of G is defined as

$$L(G) = \{w \in T^* \mid S \Rightarrow_G^+ w\}$$

A language $K \subseteq T^*$ is an *n-parallel right-linear language* (n-PRLL for short) if there is an n-PRLG G such that $K = L(G)$. □

The family of n-PRLLs is denoted by $\mathbf{PRL}_n$.

Definition 15.1.7. Let $G = (N_1, \ldots, N_n, T, S, P)$ be an n-PRLG, for some $n \geq 1$, and $1 \leq i \leq n$. By the *ith component* of G, we understand the 1-PRLG

$$G = (N_i, T, S', P')$$

where P' contains rules of the following forms:

1. $S' \to X_i$ if $S \to X_1 \cdots X_n \in P$, $X_i \in N_i$;
2. $X \to wY$ if $X \to wY \in P$ and $X, Y \in N_i$;
3. $X \to w$ if $X \to w \in P$ and $X \in N_i$. □

To prove that the family of languages accepted by n-first-SFAs coincides with the family of languages generated by $(n+1)$-PRLGs, we need the following normal form of PRLGs.

Lemma 15.1.8. *For every n-PRLG $G = (N_1, \ldots, N_n, T, S, P)$, there is an equivalent n-PRLG $G' = (N'_1, \ldots, N'_n, T, S, P')$ that satisfies:*

(i) if $S \to X_1 \cdots X_n \in P'$, then X_i does not occur on the right-hand side of any rule, for $i = 1, 2, \ldots, n$;
(ii) if $S \to \alpha$, $S \to \beta \in P'$ and $\alpha \neq \beta$, then $\mathrm{alph}(\alpha) \cap \mathrm{alph}(\beta) = \emptyset$.

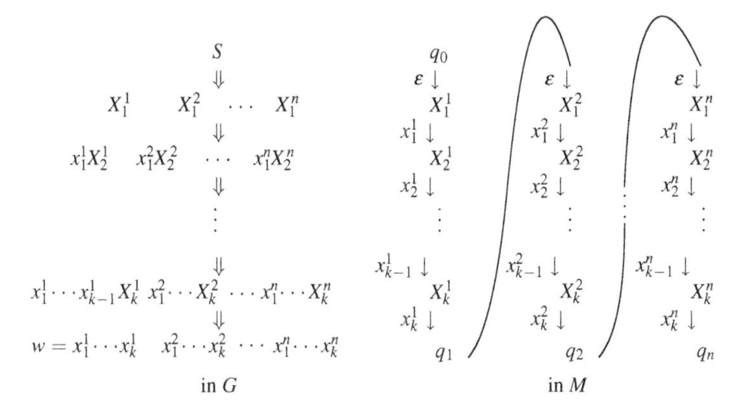

Fig. 15.3 A derivation of w in G and the corresponding acceptance of w in M

Proof. If G does not satisfy the conditions from the lemma, then we construct a new n-PRLG

$$G' = \left(N'_1, \ldots, N'_n, T, S, P'\right)$$

where P' contains all rules of the form $X \to \beta \in P$, $X \neq S$, and $N_j \subseteq N'_j$, $1 \leq j \leq n$. For each rule $S \to X_1 \cdots X_n \in P$, we add new nonterminals $Y_j \notin N'_j$ into N'_j, and rules include $S \to Y_1 \cdots Y_n$ and $Y_j \to X_j$ in P', $1 \leq j \leq n$. Clearly,

$$S \Rightarrow_G X_1 \cdots X_n \text{ if and only if } S \Rightarrow_{G'} Y_1 \cdots Y_n \Rightarrow_{G'} X_1 \cdots X_n$$

Thus, $L(G) = L(G')$. □

Lemma 15.1.9. *Let G be an n-PRLG. Then, there is an $(n-1)$-first-SFA M such that $L(G) = L(M)$.*

Proof. Informally, M is divided into n parts (see Fig. 15.3). The ith part represents a finite automaton accepting the language of the ith component of G, and R also connects the ith part to the $(i+1)$st part as depicted in Fig. 15.3.

Formally, without loss of generality, we assume $G = (N_1, \ldots, N_n, T, S, P)$ to be in the form from Lemma 15.1.8. We construct an $(n-1)$-first-SFA

$$M = \left(Q, T, \delta, q_0, q_t, F, R\right)$$

where

$$
\begin{aligned}
Q &= \{q_0, \ldots, q_n\} \cup N, N = N_1 \cup \cdots \cup N_n, \{q_0, q_1, \ldots, q_n\} \cap N = \emptyset, F = \{q_n\} \\
\delta &= \{q_i \to X_{i+1} \mid S \to X_1 \cdots X_n \in P,\ 0 \leq i < n\} \cup \\
&\quad\ \{Xw \to Y \mid X \to wY \in P\} \cup \\
&\quad\ \{Xw \to q_i \mid X \to w \in P,\ w \in T^*,\ X \in N_i,\ i \in \{1, \ldots, n\}\} \\
q_t &= q_1 \\
\Psi &= \delta \\
R &= \{(q_i \to X_{i+1}, q_{i+1} \to X_{i+2}) \mid S \to X_1 \cdots X_n \in P,\ 0 \leq i \leq n-2\}
\end{aligned}
$$

Next, we prove that $L(G) = L(M)$. To prove that $L(G) \subseteq L(M)$, consider any derivation of w in G and construct an acceptance of w in M depicted in Fig. 15.3.

This figure clearly demonstrates the fundamental idea behind this part of the proof; its complete and rigorous version is left to the reader. Thus, M accepts every $w \in T^*$ such that $S \Rightarrow_G^* w$.

To prove that $L(M) \subseteq L(G)$, consider any $w \in L(M)$ and any acceptance of w in M. Observe that the acceptance is of the form depicted on the right-hand side of Fig. 15.3. It means that the number of steps M made from q_{i-1} to q_i is the same as from q_i to q_{i+1} since the only rule in the relation with $q_{i-1} \to X_1^i$ is the rule $q_i \to X_1^{i+1}$. Moreover, M can never come back to a state corresponding to a previous component. (By a component of M, we mean the finite automaton

$$M_i = \left(Q, \Sigma, \delta, q_{i-1}, \{q_i\}\right)$$

for $1 \le i \le n$.) Next, construct a derivation of w in G. By Lemma 15.1.8, we have

$$\text{card}\left(\{X \mid (q_i \to X_1^{i+1}, q_{i+1} \to X) \in R\}\right) = 1$$

for all $0 \le i < n - 1$. Thus, $S \to X_1^1 X_1^2 \cdots X_1^n \in P$. Moreover, if $X_j^i x_j^i \to X_{j+1}^i$, we apply $X_j^i \to x_j^i X_{j+1}^i \in P$, and if $X_k^i x_k^i \to q_i$, we apply $X_k^i \to x_k^i \in P$, $1 \le i \le n$, $1 \le j < k$.

Hence, Lemma 15.1.9 holds. □

Lemma 15.1.10. *Let M be an n-first-SFA. Then, there is an $(n+1)$-PRLG G such that $L(G) = L(M)$.*

Proof. Let $M = (Q, \Sigma, \delta, q_0, q_t, F, R)$. Consider

$$G = \left(N_0, \dots, N_n, \Sigma, S, P\right)$$

where

$$\begin{aligned}
N_i &= (Q\Sigma^l \times Q \times \{i\} \times Q) \cup (Q \times \{i\} \times Q)\\
l &= \max(\{|w| \mid qw \to p \in \delta\}), 0 \le i \le n\\
P &= \{S \to [q_0x_0, q^0, 0, q_t][q_tx_1, q^1, 1, q_{i_1}][q_{i_1}x_2, q^2, 2, q_{i_2}] \cdots [q_{i_{n-1}}x_n, q^n, n, q_{i_n}] \mid\\
&\quad r_0\colon q_0x_0 \to q^0, r_1\colon q_tx_1 \to q^1, r_2\colon q_{i_1}x_2 \to q^2, \dots, r_n\colon q_{i_{n-1}}x_n \to q^n \in \delta,\\
&\quad (r_0, r_1), (r_1, r_2), \dots, (r_{n-1}, r_n) \in R,\ q_{i_n} \in F\} \cup\\
&\quad \{[px, q, i, r] \to x[q, i, r]\} \cup\\
&\quad \{[q, i, q] \to \varepsilon \mid q \in Q\} \cup\\
&\quad \{[q, i, p] \to w[q', i, p] \mid qw \to q' \in \delta\}
\end{aligned}$$

Next, we prove that $L(G) = L(M)$. To prove that $L(G) \subseteq L(M)$, observe that we make $n + 1$ copies of M and go through them similarly to Fig. 15.3. Consider a derivation of w in G. Then, in a greater detail, this derivation is of the form

$$\begin{aligned}
S &\Rightarrow_G [q_0 x_0^0, q_1^0, 0, q_t][q_t x_0^1, q_1^1, 1, q_{i_1}] \cdots [q_{i_{n-1}} x_0^n, q_1^n, n, q_{i_n}] \\
&\Rightarrow_G x_0^0 [q_1^0, 0, q_t] x_0^1 [q_1^1, 1, q_{i_1}] \cdots x_0^n [q_1^n, n, q_{i_n}] \\
&\Rightarrow_G x_0^0 x_1^0 [q_2^0, 0, q_t] x_0^1 x_1^1 [q_2^1, 1, q_{i_1}] \cdots x_0^n x_1^n [q_2^n, n, q_{i_n}] \qquad (15.1) \\
&\ \ \vdots \\
&\Rightarrow_G x_0^0 x_1^0 \cdots x_k^0 [q_t, 0, q_t] x_0^1 x_1^1 \cdots x_k^1 [q_{i_1}, 1, q_{i_1}] \cdots x_0^n x_1^n \cdots x_k^n [q_{i_n}, n, q_{i_n}] \\
&\Rightarrow_G x_0^0 x_1^0 \cdots x_k^0 x_0^1 x_1^1 \cdots x_k^1 \cdots x_0^n x_1^n \cdots x_k^n
\end{aligned}$$

and

$$r_0 \colon q_0 x_0^0 \to q_1^0, r_1 \colon q_t x_0^1 \to q_1^1, r_2 \colon q_{i_1} x_0^2 \to q_1^2, \ldots, r_n \colon q_{i_{n-1}} x_0^n \to q_1^n \in \delta$$

$$(r_0, r_1), (r_1, r_2), \ldots, (r_{n-1}, r_n) \in R$$

and $q_{i_n} \in F$.

Thus, the sequence of rules used in the acceptance of w in M is

$$\begin{aligned}
\mu = {} & (q_0 x_0^0 \to q_1^0)(q_1^0 x_1^0 \to q_2^0) \cdots (q_k^0 x_k^0 \to q_t) \\
& (q_t x_0^1 \to q_1^1)(q_1^1 x_1^1 \to q_2^1) \cdots (q_k^1 x_k^1 \to q_{i_1}) \\
& (q_{i_1} x_0^2 \to q_1^2)(q_1^2 x_1^2 \to q_2^2) \cdots (q_k^2 x_k^2 \to q_{i_2}) \qquad (15.2) \\
& \ \ \vdots \\
& (q_{i_{n-1}} x_0^n \to q_1^n)(q_1^n x_1^n \to q_2^n) \cdots (q_k^n x_k^n \to q_{i_n})
\end{aligned}$$

Next, we prove that $L(M) \subseteq L(G)$. Informally, the acceptance is divided into $n+1$ parts of the same length. Grammar G generates the ith part by the ith component and records the state from which the next component starts.

Let μ be a sequence of rules used in an acceptance of

$$w = x_0^0 x_1^0 \cdots x_k^0 x_0^1 x_1^1 \cdots x_k^1 \cdots x_0^n x_1^n \cdots x_k^n$$

in M of the form (15.2). Then, the derivation of the form (15.1) is the corresponding derivation of w in G since $[q_j^i, i, p] \to x_j^i [q_{j+1}^i, i, p] \in P$ and $[q, i, q] \to \varepsilon$, for all $0 \leq i \leq n$, $1 \leq j < k$.

Hence, Lemma 15.1.10 holds. □

The first main result of this chapter follows next.

Theorem 15.1.11. *For all $n \geq 0$,* $\mathbf{FSFA}_n = \mathbf{PRL}_{n+1}$.

Proof. This proof follows from Lemmas 15.1.9 and 15.1.10. □

Corollary 15.1.12. *The following statements hold true.*

(i) $\mathbf{REG} = \mathbf{FSFA}_0 \subset \mathbf{FSFA}_1 \subset \mathbf{FSFA}_2 \subset \cdots \subset \mathbf{CS}$
(ii) $\mathbf{FSFA}_1 \subset \mathbf{CF}$
(iii) $\mathbf{FSFA}_2 \not\subseteq \mathbf{CF}$
(iv) $\mathbf{CF} \not\subseteq \mathbf{FSFA}_n$ *for any* $n \geq 0$.
(v) *For all* $n \geq 0$, $\mathbf{FSFA}_n$ *is closed under union, finite substitution, homomorphism, intersection with a regular language, and right quotient with a regular language.*
(vi) *For all* $n \geq 1$, $\mathbf{FSFA}_n$ *is not closed under intersection and complement.*

Proof. Recall the following statements that are proved in [6].

- $\mathbf{REG} = \mathbf{PRL}_1 \subset \mathbf{PRL}_2 \subset \mathbf{PRL}_3 \subset \cdots \subset \mathbf{CS}$
- $\mathbf{PRL}_2 \subset \mathbf{CF}$
- $\mathbf{CF} \not\subseteq \mathbf{PRL}_n$, $n \geq 1$.
- For all $n \geq 1$, $\mathbf{PRL}_n$ is closed under union, finite substitution, homomorphism, intersection with a regular language, and right quotient with a regular language.
- For all $n \geq 2$, $\mathbf{PRL}_n$ is not closed under intersection and complement.

These statements and Theorem 15.1.11 imply statements (i), (ii), (iv), (v), and (vi) in Corollary 15.1.12. Moreover, observe that

$$\{a^n b^n c^{2n} \mid n \geq 0\} \in \mathbf{FSFA}_2 - \mathbf{CF}$$

which proves (iii). □

Theorem 15.1.13. *For all* $n \geq 1$, $\mathbf{FSFA}_n$ *is not closed under inverse homomorphism.*

Proof. For $n = 1$, let $L = \{a^k b^k \mid k \geq 1\}$, and let the homomorphism $h : \{a, b, c\}^* \rightarrow \{a, b\}^*$ be defined as $h(a) = a$, $h(b) = b$, and $h(c) = \varepsilon$. Then, $L \in \mathbf{FSFA}_1$, but

$$L' = h^{-1}(L) \cap c^* a^* b^* = \{c^* a^k b^k \mid k \geq 1\} \notin \mathbf{FSFA}_1$$

Assume that L' is in $\mathbf{FSFA}_1$. Then, by Theorem 15.1.11, there is a 2-PRLG

$$G = (N_1, N_2, T, S, P)$$

such that $L(G) = L'$. Let

$$k > \mathrm{card}(P) \cdot \max(\{|w| \mid X \rightarrow wY \in P\})$$

Consider a derivation of $c^k a^k b^k \in L'$. The second component can generate only finitely many as; otherwise, it derives $\{a^k b^n \mid k < n\}$, which is not regular. Analogously, the first component generates only finitely many bs. Therefore, the first component generates any number of as, and the second component generates any number of bs. Moreover, there is a derivation of the form $X \Rightarrow_G^m X$, for some $X \in N_2$, and $m \geq 1$, used in the derivation in the second component. In the first component, there is a derivation $A \Rightarrow_G^l a^s A$, for some $A \in N_1$, and $s, l \geq 1$. Then, we can modify the derivation of $c^k a^k b^k$ so that in the first component, we repeat the cycle $A \Rightarrow_G^l a^s A$ $(m+1)$-times, and in the second component, we repeat the cycle $X \Rightarrow_G^m X$ $(l+1)$-times. The derivations of both components have the same length—the added cycles are of length ml, and the rest is of the same length as in the derivation of $c^k a^k b^k$. Therefore, we have derived $c^k a^r b^k$, where $r > k$, which is not in L'—a contradiction.

For $n > 1$, the proof is analogous and left to the reader. □

Corollary 15.1.14. *For all $n \geq 1$, $\mathbf{FSFA}_n$ is not closed under concatenation. Therefore, it is not closed under Kleene closure either.*

Proof. For $n = 1$, let $L_1 = \{c\}^*$ and $L_2 = \{a^k b^k \mid k \geq 1\}$. Then,

$$L_1 L_2 = \{c^j a^k b^k \mid k \geq 1, j \geq 0\}$$

Analogously, prove this corollary for $n > 1$. □

n-Turn All-Move Self-Regulating Finite Automata

We next turn our attention to n-all-SFAs. We prove that the family of languages accepted by n-all-SFAs coincides with the family of languages generated by so-called n-right-linear simple matrix grammars (see [1, 3, 10]). First, however, we define these grammars formally.

Definition 15.1.15. For $n \geq 1$, an *n-right-linear simple matrix grammar* (see [1, 3, 10]), an *n-RLSMG* for short, is an $(n+3)$-tuple

$$G = (N_1, \dots, N_n, T, S, P)$$

where N_i, $1 \leq i \leq n$, are pairwise disjoint *nonterminal alphabets*, T is *a terminal alphabet*, $S \notin N$ is an *initial symbol*, where $N = N_1 \cup \dots \cup N_n$, and P is a finite set of *matrix rules*. A matrix rule can be in one of the following three forms

1. $[S \rightarrow X_1 \cdots X_n]$, $\quad X_i \in N_i$, $1 \leq i \leq n$;
2. $[X_1 \rightarrow w_1 Y_1, \cdots, X_n \rightarrow w_n Y_n]$, $\quad w_i \in T^*$, $X_i, Y_i \in N_i$, $1 \leq i \leq n$;
3. $[X_1 \rightarrow w_1, \cdots, X_n \rightarrow w_n]$, $\quad X_i \in N_i$, $w_i \in T^*$, $1 \leq i \leq n$.

Let m be a matrix. Then, $m[i]$ denotes the ith rule of m. For $x, y \in (N \cup T \cup \{S\})^*$,

$$x \Rightarrow_G y$$

if and only if

(1) either $x = S$ and $[S \rightarrow y] \in P$,
(2) or $x = y_1X_1 \cdots y_nX_n$, $y = y_1x_1 \cdots y_nx_n$, where $y_i \in T^*$, $x_i \in T^*N \cup T^*$, $X_i \in N_i$, $1 \leq i \leq n$, and $[X_1 \rightarrow x_1, \cdots, X_n \rightarrow x_n] \in P$.

We define $x \Rightarrow_G^+ y$ and $x \Rightarrow_G^* y$ as in Definition 15.1.6.

The *language* of G is defined as

$$L(G) = \{w \in T^* \mid S \Rightarrow_G^* w\}$$

A language $K \subseteq T^*$ is an *n-right linear simple matrix language* (an n-RLSML for short) if there is an n-RLSMG G such that $K = L(G)$. □

The family of n-RLSMLs is denoted by $\mathbf{RLSM}_n$. Furthermore, the ith component of an n-RLSMG is defined analogously to the ith component of an n-PRLG (see Definition 15.1.7).

To prove that the family of languages accepted by n-all-SFAs coincides with the family of languages generated by n-RLSMGs, the following lemma is needed.

Lemma 15.1.16. *For every n-RLSMG, $G = (N_1, \ldots, N_n, T, S, P)$, there is an equivalent n-RLSMG G' that satisfies (i) through (iii), given next.*

(i) If $[S \rightarrow X_1 \cdots X_n]$, then X_i does not occur on the right-hand side of any rule, $1 \leq i \leq n$.
(ii) If $[S \rightarrow \alpha]$, $[S \rightarrow \beta] \in P$ and $\alpha \neq \beta$, then $\mathrm{alph}(\alpha) \cap \mathrm{alph}(\beta) = \emptyset$.
(iii) For any two matrices $m_1, m_2 \in P$, if $m_1[i] = m_2[i]$, for some $1 \leq i \leq n$, then $m_1 = m_2$.

Proof. The first two conditions can be proved analogously to Lemma 15.1.8. Suppose that there are matrices m and m' such that $m[i] = m'[i]$, for some $1 \leq i \leq n$. Let

$$\begin{aligned} m &= [X_1 \rightarrow x_1, \ldots, X_n \rightarrow x_n] \\ m' &= [Y_1 \rightarrow y_1, \ldots, Y_n \rightarrow y_n] \end{aligned}$$

Replace these matrices with matrices

$$\begin{aligned} m_1 &= [X_1 \rightarrow X_1', \ldots, X_n \rightarrow X_n'] \\ m_2 &= [X_1' \rightarrow x_1, \ldots, X_n' \rightarrow x_n] \\ m_1' &= [Y_1 \rightarrow Y_1'', \ldots, Y_n \rightarrow Y_n''] \\ m_2' &= [Y_1'' \rightarrow y_1, \ldots, Y_n'' \rightarrow y_n] \end{aligned}$$

where X_i', Y_i'' are new nonterminals for all i. These new matrices satisfy condition (iii). Repeat this replacement until the resulting grammar satisfies the properties of G' given in this lemma. □

Lemma 15.1.17. *Let G be an n-RLSMG. There is an $(n-1)$-all-SFA M such that $L(G) = L(M)$.*

Proof. Without loss of generality, we assume that $G = (N_1, \ldots, N_n, T, S, P)$ is in the form described in Lemma 15.1.16. We construct an $(n-1)$-all-SFA

$$M = \big(Q, T, \delta, q_0, q_t, F, R\big)$$

where

$$\begin{aligned}
Q &= \{q_0, \ldots, q_n\} \cup N, N = N_1 \cup \cdots \cup N_n, \{q_0, q_1, \ldots, q_n\} \cap N = \emptyset \\
F &= \{q_n\} \\
\delta &= \{q_i \to X_{i+1} \mid [S \to X_1 \cdots X_n] \in P,\ 0 \le i < n\} \cup \\
&\quad \{X_i w_i \to Y_i \mid [X_1 \to w_1 Y_1, \ldots, X_n \to w_n Y_n] \in P,\ 1 \le i \le n\} \cup \\
&\quad \{X_i w_i \to q_i \mid [X_1 \to w_1, \ldots, X_n \to w_n] \in P,\ w_i \in T^*,\ 1 \le i \le n\} \\
q_t &= q_1 \\
\Psi &= \delta \\
R &= \{(q_i \to X_{i+1}, q_{i+1} \to X_{i+2}) \mid \\
&\qquad [S \to X_1 \cdots X_n] \in P,\ 0 \le i \le n-2\} \cup \\
&\quad \{(X_i w_i \to Y_i, X_{i+1} w_{i+1} \to Y_{i+1}) \mid \\
&\qquad [X_1 \to w_1 Y_1, \ldots, X_n \to w_n Y_n] \in P,\ 1 \le i < n\} \cup \\
&\quad \{(X_i w_i \to q_i, X_{i+1} w_{i+1} \to q_{i+1}) \mid \\
&\qquad [X_1 \to w_1, \ldots, X_n \to w_n] \in P,\ w_i \in T^*,\ 1 \le i < n\}
\end{aligned}$$

Next, we prove that $L(G) = L(M)$. A proof of $L(G) \subseteq L(M)$ can be made by analogy with the proof of the same inclusion of Lemma 15.1.9, which is left to the reader.

To prove that $L(M) \subseteq L(G)$, consider $w \in L(M)$ and an acceptance of w in M. As in Lemma 15.1.9, the derivation looks like the one depicted on the right-hand side of Fig. 15.3. Next, we describe how G generates w. By Lemma 15.1.16, there is matrix

$$[S \to X_1^1 X_1^2 \cdots X_1^n] \in P$$

Moreover, if $X_j^i x_j^i \to X_{j+1}^i$, $1 \le i \le n$, then

$$(X_j^i \to x_j^i X_{j+1}^i, X_j^{i+1} \to x_j^{i+1} X_{j+1}^{i+1}) \in R$$

for $1 \le i < n$, $1 \le j < k$. We apply

$$[X_j^1 \to x_j^1 X_{j+1}^1, \ldots, X_j^n \to x_j^n X_{j+1}^n] \in P$$

If $X_k^i x_k^i \to q_i$, $1 \le i \le n$, then

$$(X_k^i \to x_k^i, X_k^{i+1} \to x_k^{i+1}) \in R$$

for $1 \le i < n$, and we apply

$$[X_k^1 \to x_k^1, \ldots, X_k^n \to x_k^n] \in P$$

Thus, $w \in L(G)$.

Hence, Lemma 15.1.17 holds. □

Lemma 15.1.18. *Let M be an n-all-SFA. There is an $(n+1)$-RLSMG G such that $L(G) = L(M)$.*

Proof. Let $M = (Q, \Sigma, \delta, q_0, q_t, F, R)$. Consider

$$G = \big(N_0, \ldots, N_n, \Sigma, S, P\big)$$

where

$$\begin{aligned}
N_i &= (Q\Sigma^l \times Q \times \{i\} \times Q) \cup (Q \times \{i\} \times Q)\\
l &= \max(\{|w| \mid qw \to p \in \delta\}), 0 \le i \le n\\
P &= \{[S \to [q_0x_0, q^0, 0, q_t][q_tx_1, q^1, 1, q_{i_1}] \cdots [q_{i_{n-1}}x_n, q^n, n, q_{i_n}]] \mid\\
&\qquad r_0\colon q_0x_0 \to q^0, r_1\colon q_tx_1 \to q^1, \ldots, r_n\colon q_{i_{n-1}}x_n \to q^n \in \delta\\
&\qquad (r_0, r_1), \ldots, (r_{n-1}, r_n) \in R, q_{i_n} \in F\} \cup\\
&\quad \{[[p_0x_0, q_0, 0, r_0] \to x_0[q_0, 0, r_0], \ldots, [p_nx_n, q_n, n, r_n] \to x_n[q_n, n, r_n]]\} \cup\\
&\quad \{[[q_0, 0, q_0] \to \varepsilon, \ldots, [q_n, n, q_n] \to \varepsilon] : q_i \in Q,\ 0 \le i \le n\} \cup\\
&\quad \{[[q_0, 0, p_0] \to w_0[q_0', 0, p_0], \ldots, [q_n, n, p_n] \to w_n[q_n', n, p_n]] \mid\\
&\qquad r_j\colon q_jw_j \to q_j' \in \delta,\ 0 \le j \le n,\ (r_i, r_{i+1}) \in R,\ 0 \le i < n\}
\end{aligned}$$

Next, we prove that $L(G) = L(M)$. To prove that $L(G) \subseteq L(M)$, consider a derivation of w in G. Then, the derivation is of the form (15.1) and there are rules

$$r_0\colon q_0x_0^0 \to q_1^0, r_1\colon q_tx_0^1 \to q_1^1, \ldots, r_n\colon q_{i_{n-1}}x_0^n \to q_1^n \in \delta$$

such that $(r_0, r_1), \ldots, (r_{n-1}, r_n) \in R$. Moreover, $(r_j^l, r_j^{l+1}) \in R$, where $r_j^l\colon q_j^lx_j^l \to q_{j+1}^l \in \delta$, and $(r_k^l, r_k^{l+1}) \in R$, where $r_k^l\colon q_k^lx_k^l \to q_{i_l} \in \delta$, $0 \le l < n$, $1 \le j < k$, q_{i_0} denotes q_t, and $q_{i_n} \in F$. Thus, M accepts w with the sequence of rules μ of the form (15.2).

To prove that $L(M) \subseteq L(G)$, let μ be a sequence of rules used in an acceptance of

$$w = x_0^0x_1^0 \cdots x_k^0x_0^1x_1^1 \cdots x_k^1 \cdots x_0^nx_1^n \cdots x_k^n$$

in M of the form (15.2). Then, the derivation is of the form (15.1) because

$$[[q_j^0, 0, q_t] \to x_j^0[q_{j+1}^0, 0, q_t], \ldots, [q_j^n, n, q_{i_n}] \to x_j^n[q_{j+1}^n, n, q_{i_n}]] \in P$$

for all $q_j^i \in Q$, $1 \le i \le n$, $1 \le j < k$, and $[[q_t, 0, q_t] \to \varepsilon, \ldots, [q_{i_n}, n, q_{i_n}] \to \varepsilon] \in P$.

Hence, Lemma 15.1.18 holds. □

Next, we establish another important result of this chapter.

Theorem 15.1.19. *For all $n \ge 0$,* $\mathbf{ASFA}_n = \mathbf{RLSM}_{n+1}$.

Proof. This proof follows from Lemmas 15.1.17 and 15.1.18. □

Corollary 15.1.20. *The following statements hold true.*

(i) $\mathbf{REG} = \mathbf{ASFA}_0 \subset \mathbf{ASFA}_1 \subset \mathbf{ASFA}_2 \subset \cdots \subset \mathbf{CS}$
(ii) $\mathbf{ASFA}_1 \nsubseteq \mathbf{CF}$
(iii) $\mathbf{CF} \nsubseteq \mathbf{ASFA}_n$*, for every* $n \geq 0$.
(iv) *For all* $n \geq 0$, $\mathbf{ASFA}_n$ *is closed under union, concatenation, finite substitution, homomorphism, intersection with a regular language, and right quotient with a regular language.*
(v) *For all* $n \geq 1$, $\mathbf{ASFA}_n$ *is not closed under intersection, complement, and Kleene closure.*

Proof. Recall the following statements that are proved in [10].

- $\mathbf{REG} = \mathbf{RLSM}_1 \subset \mathbf{RLSM}_2 \subset \mathbf{RLSM}_3 \subset \cdots \subset \mathbf{CS}$
- For all $n \geq 1$, $\mathbf{RLSM}_n$ is closed under union, finite substitution, homomorphism, intersection with a regular language, and right quotient with a regular language.
- For all $n \geq 2$, $\mathbf{RLSM}_n$ is not closed under intersection and complement.

Furthermore, recall these statements proved in [7, 8].

- For all $n \geq 1$, $\mathbf{RLSM}_n$ is closed under concatenation.
- For all $n \geq 2$, $\mathbf{RLSM}_n$ is not closed under Kleene closure.

These statements and Theorem 15.1.19 imply statements (i), (iv), and (v) of Corollary 15.1.20. Moreover, observe that

$$\{ww \mid w \in \{a, b\}^*\} \in \mathbf{ASFA}_1 - \mathbf{CF}$$

(see Example 15.1.5), which proves (ii). Finally, let

$$L = \{wcw^R \mid w \in \{a, b\}^*\}$$

By Theorem 1.5.2 in [1], $L \notin \mathbf{RLSM}_n$, for any $n \geq 1$. Thus, (iii) follows from Theorem 15.1.19. □

Theorem 15.1.21, given next, follows from Theorem 15.1.19 and from Corollary 3.3.3 in [8]. However, Corollary 3.3.3 in [8] is not proved effectively. We next prove Theorem 15.1.21 effectively.

Theorem 15.1.21. $\mathbf{ASFA}_n$ *is closed under inverse homomorphism, for all* $n \geq 0$.

Proof. For $n = 1$, let $M = (Q, \Sigma, \delta, q_0, q_t, F, R)$ be a 1-all-SFA, and let $h : \Delta^* \to \Sigma^*$ be a homomorphism. Next, we construct a 1-all-SFA

$$M' = (Q', \Delta, \delta', q_0', q_t', \{q_f'\}, R')$$

accepting $h^{-1}(L(M))$ as follows. Set

$$k = \max(\{|w| \mid qw \to p \in \delta\}) + \max(\{|h(a)| \mid a \in \Delta\})$$

and

$$Q' = \{q_0'\} \cup \{[x, q, y] \mid x, y \in \Sigma^*, |x|, |y| \leq k, q \in Q\}$$

Initially, set δ' and R' to $\emptyset$. Then, extend δ' and R' by performing (1) through (5), given next.

(1) For $y \in \Sigma^*$, $|y| \leq k$, add
 $(q_0' \to [\varepsilon, q_0, y], q_t' \to [y, q_t, \varepsilon])$ to R'.
(2) For $A \in Q'$, $q \neq q_t$, add
 $([x, q, y]a \to [xh(a), q, y], A \to A)$ to R'.
(3) For $A \in Q'$, add
 $(A \to A, [x, q, \varepsilon]a \to [xh(a), q, \varepsilon])$ to R'.
(4) For $(qx \to p, q'x' \to p') \in R$, $q \neq q_t$, add
 $([xw, q, y] \to [w, p, y], [x'w', q', \varepsilon] \to [w', p', \varepsilon])$ to R'.
(5) For $q_f \in F$, add
 $([y, q_t, y] \to q_t', [\varepsilon, q_f, \varepsilon] \to q_f')$ to R'.

In essence, M' simulates M in the following way. In a state of the form $[x, q, y]$, the three components have the following meaning

- $x = h(a_1 \cdots a_n)$, where $a_1 \cdots a_n$ is the input string that M' has already read;
- q is the current state of M;
- y is the suffix remaining as the first component of the state that M' enters during a turn; y is thus obtained when M' reads the last symbol right before the turn occurs in M; M reads y after the turn.

More precisely, $h(w) = w_1 y w_2$, where w is an input string, w_1 is accepted by M before making the turn—that is, from q_0 to q_t, and yw_2 is accepted by M after making the turn—that is, from q_t to $q_f \in F$. A rigorous version of this proof is left to the reader.

For $n > 1$, the proof is analogous and left to the reader. □

Language Families Accepted by *n*-First-SFAs and *n*-All-SFAs

Next, we compare the family of languages accepted by n-first-SFAs with the family of languages accepted by n-all-SFAs.

Theorem 15.1.22. *For all* $n \geq 1$, $\mathbf{FSFA}_n \subset \mathbf{ASFA}_n$.

Proof. In [6, 10], it is proved that for all $n > 1$, $\mathbf{PRL}_n \subset \mathbf{RLSM}_n$. The proof of Theorem 15.1.22 thus follows from Theorems 15.1.11 and 15.1.19. □

Theorem 15.1.23. $\mathbf{FSFA}_n \not\subseteq \mathbf{ASFA}_{n-1}$, $n \geq 1$.

Proof. Recall that $\mathbf{FSFA}_n = \mathbf{PRL}_{n+1}$ (see Theorem 15.1.11) and $\mathbf{ASFA}_{n-1} = \mathbf{RLSM}_n$ (see Theorem 15.1.19). It is easy to see that

$$L = \{a_1^k a_2^k \cdots a_{n+1}^k \mid k \geq 1\} \in \mathbf{PRL}_{n+1}$$

However, Lemma 1.5.6 in [1] implies that

$$L \notin \mathbf{RLSM}_n$$

Hence, the theorem holds. □

Lemma 15.1.24. *For each regular language L,* $\{w^n \mid w \in L\} \in \mathbf{ASFA}_{n-1}$.

Proof. Let $L = L(M)$, where M is a finite automaton. Make n copies of M. Rename their states so all the sets of states are pairwise disjoint. In this way, also rename the states in the rules of each of these n automata; however, keep the labels of the rules unchanged. For each rule label r, include (r, r) into R. As a result, we obtain an n-all-SFA that accepts $\{w^n \mid w \in L\}$. A rigorous version of this proof is left to the reader. □

Theorem 15.1.25. $\mathbf{ASFA}_n - \mathbf{FSFA} \neq \emptyset$, *for all* $n \geq 1$, *where* $\mathbf{FSFA} = \bigcup_{m=1}^{\infty} \mathbf{FSFA}_m$.

Proof. By induction on $n \geq 1$, we prove that

$$L = \left\{(cw)^{n+1} \mid w \in \{a, b\}^*\right\} \notin \mathbf{FSFA}$$

From Lemma 15.1.24, it follows that $L \in \mathbf{ASFA}_n$.

Basis. For $n = 1$, let G be an m-PRLG generating L, for some positive integer m. Consider a sufficiently large string $cw_1cw_2 \in L$ such that $w_1 = w_2 = a^{n_1}b^{n_2}$, $n_2 > n_1 > 1$. Then, there is a derivation of the form

$$\begin{aligned} S &\Rightarrow_G^p x_1A_1x_2A_2 \cdots x_mA_m \\ &\Rightarrow_G^k x_1y_1A_1x_2y_2A_2 \cdots x_my_mA_m \end{aligned} \qquad (15.3)$$

in G, where cycle (15.3) generates more than one a in w_1. The derivation continues as

$$\begin{aligned} x_1y_1A_1 \cdots x_my_mA_m &\Rightarrow_G^r \\ x_1y_1z_1B_1 \cdots x_my_mz_mB_m &\Rightarrow_G^l x_1y_1z_1u_1B_1 \cdots x_my_mz_mu_mB_m \qquad (15.4) \\ \text{(cycle (15.4) generates no } a\text{s)} &\Rightarrow_G^s cw_1cw_2 \end{aligned}$$

Next, modify the left derivation, the derivation in components generating cw_1, so that the a-generating cycle (15.3) is repeated $(l + 1)$-times. Similarly, modify the right derivation, the derivation in the other components, so that the no-a-generating cycle (15.4) is repeated $(k + 1)$-times. Thus, the modified left derivation is of length

$$p + k(l + 1) + r + l + s = p + k + r + l(k + 1) + s$$

which is the length of the modified right derivation. Moreover, the modified left derivation generates more as in w_1 than the right derivation in w_2—a contradiction.

Induction Hypothesis. Suppose that the theorem holds for all $k \leq n$, for some $n \geq 1$.

Induction Step. Consider $n + 1$ and let

$$\{(cw)^{n+1} \mid w \in \{a,b\}^*\} \in \mathbf{FSFA}_l$$

for some $l \geq 1$. As $\mathbf{FSFA}_l$ is closed under the right quotient with a regular language, and language $\{cw \mid w \in \{a,b\}^*\}$ is regular, we obtain

$$\{(cw)^n \mid w \in \{a,b\}^*\} \in \mathbf{FSFA}_l \subseteq \mathbf{FSFA}$$

which is a contradiction. □

15.2 Self-Regulating Pushdown Automata

The present section consists of two subsections. Section 15.2.1 defines n-turn first-move self-regulating pushdown automata and n-turn all-move self-regulating pushdown automata. Section 15.2.2 determines the accepting power of n-turn all-move self-regulating pushdown automata.

15.2.1 Definitions

Before defining self-regulating pushdown automata, recall the formalization of rule labels from Definition 3.4.10 because this formalization is often used throughout this section.

Definition 15.2.1. A *self-regulating pushdown automaton* (an *SPDA* for short) M is a 9-tuple

$$M = (Q, \Sigma, \Gamma, \delta, q_0, q_t, Z_0, F, R)$$

where

(1) $(Q, \Sigma, \Gamma, \delta, q_0, Z_0, F)$ is a pushdown automaton entering a final state and emptying its pushdown,
(2) $q_t \in Q$ is a *turn state*, and
(3) $R \subseteq \Psi \times \Psi$ is a finite relation, where Ψ is an alphabet of rule labels. □

Definition 15.2.2. Let $n \geq 0$ and

$$M = (Q, \Sigma, \Gamma, \delta, q_0, q_t, Z_0, F, R)$$

be a self-regulating pushdown automaton. M is said to be an *n-turn first-move self-regulating pushdown automaton*, *n-first-SPDA*, if every $w \in L(M)$ is accepted by M in the following way

$$Z_0 q_0 w \vdash_M^* f\ [\mu]$$

such that

$$\mu = r_1^0 \cdots r_k^0 r_1^1 \cdots r_k^1 \cdots r_1^n \cdots r_k^n$$

where $k \geq 1$, r_k^0 is the first rule of the form $Zqx \to \gamma q_t$, for some $Z \in \Gamma$, $q \in Q$, $x \in \Sigma^*$, $\gamma \in \Gamma^*$, and

$$(r_1^j, r_1^{j+1}) \in R$$

for all $0 \leq j < n$. □

The family of languages accepted by n-first-SPDAs is denoted by $\mathbf{FSPDA}_n$.

Definition 15.2.3. Let $n \geq 0$ and

$$M = \big(Q, \Sigma, \Gamma, \delta, q_0, q_t, Z_0, F, R\big)$$

be a self-regulating pushdown automaton. M is said to be an *n-turn all-move self-regulating pushdown automaton* (an *n-all-SPDA* for short) if every $w \in L(M)$ is accepted by M in the following way

$$Z_0 q_0 w \vdash_M^* f\ [\mu]$$

such that

$$\mu = r_1^0 \cdots r_k^0 r_1^1 \cdots r_k^1 \cdots r_1^n \cdots r_k^n$$

where $k \geq 1$, r_k^0 is the first rule of the form $Zqx \to \gamma q_t$, for some $Z \in \Gamma$, $q \in Q$, $x \in \Sigma^*$, $\gamma \in \Gamma^*$, and

$$(r_i^j, r_i^{j+1}) \in R$$

for all $1 \leq i \leq k$, $0 \leq j < n$. □

The family of languages accepted by n-all-SPDAs is denoted by $\mathbf{ASPDA}_n$.

15.2.2 Accepting Power

In this section, we investigate the accepting power of self-regulating pushdown automata.

As every n-all-SPDA without any turn state represents, in effect, an ordinary pushdown automaton, we obtain the following theorem.

Theorem 15.2.4. $\mathbf{ASPDA}_0 = \mathbf{CF}$ □

However, if we consider 1-all-SPDAs, their power is that of phrase-structure grammars.

Theorem 15.2.5. $\mathbf{ASPDA}_1 = \mathbf{RE}$

Proof. For any $L \in RE$, $L \subseteq \Delta^*$, there are context-free languages $L(G)$ and $L(H)$ and a homomorphism $h : \Sigma^* \rightarrow \Delta^*$ such that

$$L = h\big(L(G) \cap L(H)\big)$$

(see Theorem 3.3.14). Suppose that $G = (N_G, \Sigma, P_G, S_G)$ and $H = (N_H, \Sigma, P_H, S_H)$ are in the Greibach normal form (see Definition 4.1.17)—that is, all rules are of the form $A \rightarrow a\alpha$, where A is a nonterminal, a is a terminal, and α is a (possibly empty) string of nonterminals. Let us construct an 1-all-SPDA

$$M = \big(\{q_0, q, q_t, p, f\}, \Delta, \Sigma \cup N_G \cup N_H \cup \{Z\}, \delta, q_0, Z, \{f\}, R\big)$$

where $Z \notin \Sigma \cup N_G \cup N_H$, with R constructed by performing (1) through (4), stated next.

(1) Add $(Zq_0 \rightarrow ZS_Gq, Zq_t \rightarrow ZS_Hp)$ to R.
(2) Add $(Aq \rightarrow B_n \cdots B_1aq, Cp \rightarrow D_m \cdots D_1ap)$ to R if
 $A \rightarrow aB_1 \cdots B_n \in P_G$ and
 $C \rightarrow aD_1 \cdots D_m \in P_H$.
(3) Add $(aqh(a) \rightarrow q, ap \rightarrow p)$ to R.
(4) Add $(Zq \rightarrow Zq_t, Zp \rightarrow f)$ to R.

Moreover, δ contains only the rules from the definition of R.

Next, we prove that $w \in h(L(G) \cap L(H))$ if and only if $w \in L(M)$.

Only If. Let $w \in h(L(G) \cap L(H))$. There are $a_1, a_2, \ldots, a_n \in \Sigma$ such that

$$a_1a_2 \cdots a_n \in L(G) \cap L(H)$$

and $w = h(a_1a_2 \cdots a_n)$, for some $n \geq 0$. There are leftmost derivations

$$S_G \Rightarrow_G^n a_1a_2 \cdots a_n$$

and

$$S_H \Rightarrow_H^n a_1a_2 \cdots a_n$$

of length n in G and H, respectively, because in every derivation step exactly one terminal element is derived. Thus, M accepts $h(a_1)h(a_2) \cdots h(a_n)$ as

$$
\begin{aligned}
& Zq_0h(a_1)h(a_2)\cdots h(a_n) \\
\vdash_M\ & ZS_Gqh(a_1)h(a_2)\cdots h(a_n) \\
& \vdots \\
\vdash_M\ & Za_nqh(a_n) \\
\vdash_M\ & Zq \\
\vdash_M\ & Zq_t \\
\vdash_M\ & ZS_Hp \\
& \vdots \\
\vdash_M\ & Za_np \\
\vdash_M\ & Zp \\
\vdash_M\ & f
\end{aligned}
$$

In state q, by using its pushdown, M simulates a derivation of $a_1 \cdots a_n$ in G but reads $h(a_1)\cdots\ h(a_n)$ as the input. In p, M simulates a derivation of $a_1a_2\cdots a_n$ in H but reads no input. As $a_1a_2\cdots a_n$ can be derived in both G and H by making the same number of steps, the automaton can successfully complete the acceptance of w.

If. Notice that in one step, M can read only $h(a) \in \Delta^*$, for some $a \in \Sigma$. Let $w \in L(M)$, then $w = h(a_1)h(a_2)\cdots h(a_n)$, for some $a_1, a_2, \ldots, a_n \in \Sigma$. Consider the following acceptance of w in M

$$
\begin{aligned}
& Zq_0h(a_1)h(a_2)\cdots h(a_n) \\
\vdash_M\ & ZS_Gqh(a_1)h(a_2)\cdots h(a_n) \\
& \vdots \\
\vdash_M\ & Za_nqh(a_n) \\
\vdash_M\ & Zq \\
\vdash_M\ & Zq_t \\
\vdash_M\ & ZS_Hp \\
& \vdots \\
\vdash_M\ & Za_np \\
\vdash_M\ & Zp \\
\vdash_M\ & f
\end{aligned}
$$

As stated above, in q, M simulates a derivation of $a_1a_2\cdots a_n$ in G, and then in p, M simulates a derivation of $a_1a_2\cdots a_n$ in H. It successfully completes the acceptance of w only if $a_1a_2\cdots a_n$ can be derived in both G and H. Hence, the if part holds, too.

□

Open Problems

Although the fundamental results about self-regulating automata have been achieved in this chapter, there still remain several open problems concerning them.

Open Problem 15.2.6. What is the language family accepted by n-turn first-move self-regulating pushdown automata, when $n \geq 1$ (see Definition 15.2.2)? □

Open Problem 15.2.7. By analogy with the standard deterministic finite and pushdown automata (see pp. 145 and 437 in [4]), introduce the deterministic versions of self-regulating automata. What is their power? □

Open Problem 15.2.8. Discuss the closure properties of other language operations, such as the reversal. □

References

1. Dassow, J., Păun, G.: Regulated Rewriting in Formal Language Theory. Springer, New York (1989)
2. Fischer, P.C., Rosenberg, A.L.: Multitape one-way nonwriting automata. J. Comput. Syst. Sci. **2**, 38–101 (1968)
3. Ibarra, O.H.: Simple matrix languages. Inf. Control **17**, 359–394 (1970)
4. Meduna, A.: Automata and Languages: Theory and Applications. Springer, London (2000)
5. Rosebrugh, R.D., Wood, D.: A characterization theorem for n-parallel right linear languages. J. Comput. Syst. Sci. **7**, 579–582 (1973)
6. Rosebrugh, R.D., Wood, D.: Restricted parallelism and right linear grammars. Utilitas Mathematica **7**, 151–186 (1975)
7. Siromoney, R.: Studies in the mathematical theory of grammars and its applications. Ph.D. thesis, University of Madras, Madras, India (1969)
8. Siromoney, R.: Finite-turn checking automata. J. Comput. Syst. Sci. **5**, 549–559 (1971)
9. Wood, D.: Properties of n-parallel finite state languages. Technical report, McMaster University (1973)
10. Wood, D.: m-parallel n-right linear simple matrix languages. Utilitas Mathematica **8**, 3–28 (1975)

Chapter 16
Automata Regulated by Control Languages

Abstract The present two-section chapter discusses automata in which the application of rules is regulated by control languages by analogy with context-free grammars regulated control languages (see Sect. 5.1). Section 16.1 studies this topic in terms of finite automata while Sect. 16.2 investigates pushdown automata regulated in this way. More precisely, Sect. 16.1 discusses finite automata working under two kinds of regulation—*state-controlled regulation* and *transition-controlled regulation*. It establishes conditions under which any state-controlled finite automaton can be turned to an equivalent transition-controlled finite automaton and vice versa. Then, it proves that under either of the two regulations, finite automata controlled by regular languages characterize the family of regular languages, and an analogical result is then reformulated in terms of context-free languages. However, Sect. 16.1 also demonstrates that finite automata controlled by languages generated by propagating programmed grammars with appearance checking increase their power significantly; in fact, they are computationally complete. Section 16.2 first shows that pushdown automata regulated by regular languages are as powerful as ordinary pushdown automata. Then, however, it proves that pushdown automata regulated by linear languages characterize the family of recursively enumerable languages; in fact, this characterization holds even in terms of one-turn pushdown automata.

Keywords Automata regulated by controlled languages • Finite and pushdown versions • Regular and linear control languages • Computational completeness

The present two-section chapter discusses automata in which the application of rules is regulated by control languages just like in context-free grammars regulated control languages (see Sect. 5.1). Section 16.1 studies this topic in terms of finite automata while Sect. 16.2 investigates pushdown automata regulated in this way.

A. Meduna and P. Zemek, *Regulated Grammars and Automata*,
DOI 10.1007/978-1-4939-0369-6_16, © Springer Science+Business Media New York 2014

More precisely, Sect. 16.1 studies finite automata that work under (1) *state-controlled regulation* and (2) *transition-controlled regulation*. It proves that under either of these two regulations, the family of regular languages results from finite automata controlled by regular languages. Similarly, under either of these regulations, the family of context-free languages results from finite automata controlled by context-free languages. On the other hand, finite automata controlled by languages generated by propagating programmed grammars with appearance checking (see Sect. 5.3) are much stronger. As a matter of fact, they are computationally complete as also demonstrated in Sect. 16.1.

Section 16.2 considers pushdown automata and regulates the application of their rules by control languages similarly to the same type of regulation in terms of context-free grammars (see Chap. 5). It demonstrates that this regulation has no effect on the power of pushdown automata if the control languages are regular. On the other hand, pushdown automata increase their power remarkably if they are regulated by linear languages. In fact, Sect. 16.2 proves they are computationally complete.

16.1 Finite Automata Regulated by Control Languages

The present section studies finite automata regulated by control languages. In fact, it studies two kinds of this regulation—*state-controlled regulation* and *transition-controlled regulation*. To give an insight into these two types of regulation, consider a finite automaton M controlled by a language C, and a sequence $\tau \in C$ that resulted into the acceptance of an input word w. Working under the former regulation, M has C defined over the set of states, and it accepts w by going through all the states in τ and ending up in a final state. Working under the latter regulation, M has C defined over the set of transitions, and it accepts w by using all the transitions in τ and ending up in a final state.

First, Sect. 16.1.1 defines these two types of controlled finite automata formally. After that, Sect. 16.1.2 establishes conditions under which it is possible to convert any state-controlled finite automaton to an equivalent transition-controlled finite automaton and vice versa (Theorem 16.1.5). Then, Sect. 16.1.3 proves that under both regulations, finite automata controlled by regular languages characterize the family of regular languages (Theorem 16.1.7 and Corollary 16.1.8). Section 16.1.4 shows that finite automata controlled by context-free languages characterize the family of context-free languages (Theorem 16.1.10 and Corollary 16.1.11).

After that, Sect. 16.1.5 demonstrates a close relation of controlled finite automata to programmed grammars with appearance checking (see Sect. 5.3). Recall that programmed grammars with appearance checking are computationally complete—that is, they are as powerful as phrase-structure grammars (see Theorem 5.3.4). Furthermore, the language family generated by propagating programmed grammars with appearance checking is properly included in the family of context-sensitive languages (see Theorem 5.3.4). This section proves that finite automata that are controlled by languages generated by propagating programmed grammars with appearance checking are computationally complete (Theorem 16.1.15 and Corollary 16.1.16). More precisely, state-controlled finite automata are computationally complete with $n + 1$ states, where n is the number of symbols in the accepted language (Corollary 16.1.17). Transition-controlled finite automata are computationally complete with a single state (Theorem 16.1.18).

16.1.1 Definitions

We begin by defining state-controlled and transition-controlled finite automata formally.

Definition 16.1.1. Let $M = (Q, \Sigma, R, s, F)$ be a finite automaton. Based on $\vdash_M$, we define a relation $\triangleright_M$ over $Q\Sigma^* \times Q^*$ as follows: if $\alpha \in Q^*$ and $pax \vdash_M qx$, where $p, q \in Q, x \in \Sigma^*$, and $a \in \Sigma \cup \{\varepsilon\}$, then

$$(pax, \alpha) \triangleright_M (qx, \alpha p)$$

Let $\triangleright_M^n$, $\triangleright_M^*$, and $\triangleright_M^+$ denote the nth power of $\triangleright_M$, for some $n \geq 0$, the reflexive-transitive closure of $\triangleright_M$, and the transitive closure of $\triangleright_M$, respectively.

Let $C \subseteq Q^*$ be a *control language*. The *state-controlled language of M with respect to C* is denoted by ${}_{\triangleright}L(M, C)$ and defined as

$${}_{\triangleright}L(M, C) = \left\{w \in \Sigma^* \mid (sw, \varepsilon) \triangleright_M^* (f, \alpha), f \in F, \alpha \in C\right\}$$

The pair (M, C) is called a *state-controlled finite automaton.* □

Before defining transition-controlled finite automata, recall the formalization of rule labels from Definition 3.4.3.

Definition 16.1.2. Let $M = (Q, \Sigma, \Psi, R, s, F)$ be a finite automaton. Based on $\vdash_M$, we define a relation $\blacktriangleright_M$ over $Q\Sigma^* \times \Psi^*$ as follows: if $\beta \in \Psi^*$ and $pax \vdash_M qx\ [r]$, where $r\colon pa \to q \in R$ and $x \in \Sigma^*$, then

$$(pax, \beta) \blacktriangleright_M (qx, \beta r)$$

Let $\blacktriangleright_M^n$, $\blacktriangleright_M^*$, and $\blacktriangleright_M^+$ denote the nth power of $\blacktriangleright_M$, for some $n \geq 0$, the reflexive-transitive closure of $\blacktriangleright_M$, and the transitive closure of $\blacktriangleright_M$, respectively.

Let $C \subseteq \Psi^*$ be a *control language*. The *transition-controlled language of M with respect to C* is denoted by ${}_{\blacktriangleright}L(M, C)$ and defined as

$${}_{\blacktriangleright}L(M, C) = \{w \in \Sigma^* \mid (sw, \varepsilon) \blacktriangleright_M^* (f, \beta), f \in F, \beta \in C\}$$

The pair (M, C) is called a *transition-controlled finite automaton.* □

For any family of languages $\mathscr{L}$, **SCFA**($\mathscr{L}$) and **TCFA**($\mathscr{L}$) denote the language families defined by state-controlled finite automata controlled by languages from $\mathscr{L}$ and transition-controlled finite automata controlled by languages from $\mathscr{L}$, respectively.

16.1.2 Conversions

First, we show that under certain circumstances, it is possible to convert any state-controlled finite automaton to an equivalent transition-controlled finite automaton and vice versa. These conversions will be helpful because to prove that **SCFA**($\mathscr{L}$) = **TCFA**($\mathscr{L}$) = $\mathscr{J}$, where $\mathscr{L}$ satisfies the required conditions, we only have to prove that either **SCFA**($\mathscr{L}$) = $\mathscr{J}$ or **TCFA**($\mathscr{L}$) = $\mathscr{J}$.

Lemma 16.1.3. *Let $\mathscr{L}$ be a language family that is closed under finite ε-free substitution. Then,* **SCFA**($\mathscr{L}$) $\subseteq$ **TCFA**($\mathscr{L}$).

Proof. Let $\mathscr{L}$ be a language family that is closed under finite ε-free substitution, $M = (Q, \Sigma, R, s, F)$ be a finite automaton, and $C \in \mathscr{L}$ be a control language. Without any loss of generality, assume that $C \subseteq Q^*$. We next construct a finite automaton M' and a language $C' \in \mathscr{L}$ such that ${}_{\triangleright}L(M, C) = {}_{\blacktriangleright}L(M', C')$. Define

$$M' = \big(Q, \Sigma, \Psi, R', s, F\big)$$

where

$$\begin{aligned} \Psi &= \{\langle p, a, q\rangle \mid pa \to q \in R\} \\ R' &= \{\langle p, a, q\rangle \colon pa \to q \mid pa \to q \in R\} \end{aligned}$$

Define the finite ε-free substitution π from Q^* to Ψ^* as

$$\pi(p) = \big\{\langle p, a, q\rangle \mid pa \to q \in R\big\}$$

Let $C' = \pi(C)$. Since $\mathscr{L}$ is closed under finite ε-free substitution, $C' \in \mathscr{L}$. Observe that $(sw, \varepsilon) \triangleright_M^n (f, \alpha)$, where $w \in \Sigma^*, f \in F, \alpha \in C$, and $n \geq 0$, if and only if $(sw, \varepsilon) \blacktriangleright_{M'}^n (f, \beta)$, where $\beta \in \pi(\alpha)$. Hence, ${}_{\triangleright}L(M, C) = {}_{\blacktriangleright}L(M', C')$, so the lemma holds. □

Lemma 16.1.4. *Let $\mathscr{L}$ be a language family that contains all finite languages and is closed under concatenation. Then,* **TCFA**($\mathscr{L}$) $\subseteq$ **SCFA**($\mathscr{L}$).

Proof. Let $\mathscr{L}$ be a language family that contains all finite languages and is closed under concatenation, $M = (Q, \Sigma, \Psi, R, s, F)$ be a finite automaton, and $C \in \mathscr{L}$ be a control language. Without any loss of generality, assume that $C \subseteq \Psi^*$. We next construct a finite automaton M' and a language $C' \in \mathscr{L}$ such that ${}_{\blacktriangleright}L(M, C) = {}_{\triangleright}L(M', C')$. Define

$$M' = \big(Q', \Sigma, R', s', F'\big)$$

where

$$
\begin{aligned}
Q' &= \Psi \cup \{s', \ell\} \quad (s', \ell \notin \Psi) \\
R' &= \{s' \rightarrow r \mid r{:}\, sa \rightarrow q \in R\} \cup \\
&\quad\ \{ra \rightarrow t \mid r{:}\, pa \rightarrow q, t{:}\, qb \rightarrow m \in R\} \cup \\
&\quad\ \{ra \rightarrow \ell \mid r{:}\, pa \rightarrow q \in R, q \in F\} \\
F' &= \{r \mid r{:}\, pa \rightarrow q \in R, q \in F\} \cup \{\ell\}
\end{aligned}
$$

Finally, if $s \in F$, then add s' to F'. Set $C' = \{s', \varepsilon\}C$. Since $\mathscr{L}$ is closed under concatenation and contains all finite languages, $C' \in \mathscr{L}$. Next, we argue that $_\blacktriangleright L(M, C) = {}_\triangleright L(M', C')$. First, notice that $s \in F$ if and only if $s' \in F$. Hence, by the definition of C', it is sufficient to consider nonempty sequences of moves of both M and M'. Indeed, $(s, \varepsilon) \blacktriangleright_M^0 (s, \varepsilon)$ with $s \in F$ and $\varepsilon \in C$ if and only if $(s', \varepsilon) \triangleright_{M'}^0 (s', \varepsilon)$ with $s' \in F$ and $\varepsilon \in C'$. Observe that

$$
(sw, \varepsilon) \blacktriangleright_M (p_1 w_1, r_1) \blacktriangleright_M (p_2 w_2, r_1 r_2) \blacktriangleright_M \cdots \blacktriangleright_M (p_n w_n, r_1 r_2 \cdots r_n)
$$

by

$$
\begin{aligned}
r_1{:} &\quad p_0 a_1 \rightarrow p_1 \\
r_2{:} &\quad p_1 a_2 \rightarrow p_2 \\
&\qquad \vdots \\
r_n{:} &\ p_{n-1} a_n \rightarrow p_n
\end{aligned}
$$

where $w \in \Sigma^*$, $p_i \in Q$ for $i = 1, 2, \ldots, n$, $p_n \in F$, $w_i \in \Sigma^*$ for $i = 1, 2, \ldots, n$, $a_i \in \Sigma \cup \{\varepsilon\}$ for $i = 1, 2, \ldots n$, and $n \geq 1$ if and only if

$$
(s'w, \varepsilon) \triangleright_{M'} (r_1 w, s') \triangleright_{M'} (r_2 w_1, s' r_1) \triangleright_{M'} \cdots \triangleright_{M'} (r_{n+1} w_n, s' r_1 r_2 \cdots r_n)
$$

by

$$
\begin{aligned}
s' &\rightarrow r_1 \\
r_1 a_1 &\rightarrow r_2 \\
r_2 a_2 &\rightarrow r_3 \\
&\ \vdots \\
r_n a_n &\rightarrow r_{n+1}
\end{aligned}
$$

with $r_{n+1} \in F'$ (recall that $p_n \in F$). Hence, $_\blacktriangleright L(M, C) = {}_\triangleright L(M', C')$ and the lemma holds. □

Theorem 16.1.5. *Let $\mathscr{L}$ be a language family that is closed under finite ε-free substitution, contains all finite languages, and is closed under concatenation. Then,* $\mathbf{SCFA}(\mathscr{L}) = \mathbf{TCFA}(\mathscr{L})$.

Proof. This theorem follows directly from Lemmas 16.1.3 and 16.1.4. □

16.1.3 Regular-Controlled Finite Automata

Initially, we consider finite automata controlled by regular control languages.

Lemma 16.1.6. SCFA(REG) $\subseteq$ REG

Proof. Let $M = (Q, \Sigma, R, s, F)$ be a finite automaton and $C \subseteq Q^*$ be a regular control language. Since C is regular, there is a complete finite automaton $H = (\hat{Q}, Q, \hat{R}, \hat{s}, \hat{F})$ such that $L(H) = C$. We next construct a finite automaton M' such that $L(M') = {}_{\triangleright}L(M', L(H))$. Define

$$M' = \left(Q', \Sigma, R', s', F'\right)$$

where

$$\begin{aligned} Q' &= \{\langle p, q\rangle \mid p \in Q, q \in \hat{Q}\} \\ R' &= \{\langle p, r\rangle a \to \langle q, t\rangle \mid pa \to q \in R, rp \to t \in \hat{R}\} \\ s' &= \langle s, \hat{s}\rangle \\ F' &= \{\langle p, q\rangle \mid p \in F, q \in \hat{F}\} \end{aligned}$$

Observe that a move in M' by $\langle p, r\rangle a \to \langle q, t\rangle \in R'$ simultaneously simulates a move in M by $pa \to q \in R$ and a move in H by $rp \to t \in \hat{R}$. Based on this observation, it is rather easy to see that M' accepts an input string $w \in \Sigma^*$ if and only if M reads w and enters a final state after going through a sequence of states from $L(H)$. Therefore, $L(M') = {}_{\triangleright}L(M, L(H))$. A rigorous proof of the identity $L(M') = {}_{\triangleright}L(M, L(H))$ is left to the reader. □

The following theorem shows that finite automata controlled by regular languages are of little or no interest because they are as powerful as ordinary finite automata.

Theorem 16.1.7. SCFA(REG) = REG

Proof. The inclusion **REG** $\subseteq$ **SCFA(REG)** is obvious. The converse inclusion follows from Lemma 16.1.6. □

Combining Theorems 16.1.5 and 16.1.7, we obtain the following corollary (recall that **REG** satisfies all the conditions from Theorem 16.1.5).

Corollary 16.1.8. TCFA(REG) = REG □

16.1.4 Context-Free-Controlled Finite Automata

Next, we consider finite automata controlled by context-free control languages.

Lemma 16.1.9. $\mathbf{SCFA(CF)} \subseteq \mathbf{CF}$

Proof. Let $M = (Q, \Sigma, R, s, F)$ be a finite automaton and $C \subseteq Q^*$ be a context-free control language. Since C is context-free, there is a pushdown automaton $H = (\hat{Q}, Q, \Gamma, \hat{R}, \hat{s}, \hat{Z}, \hat{F})$ such that $L(H) = C$. Without any loss of generality, we assume that $bpa \rightarrow wq \in \hat{R}$ implies that $a \neq \varepsilon$ (see Lemma 5.2.1 in [4]). We next construct a pushdown automaton M' such that $L(M') = {}_{\triangleright}L(M, L(H))$. Define

$$M' = \left(Q', \Sigma, \Gamma, R', s', Z, F'\right)$$

where

$$\begin{aligned} Q' &= \{\langle p, q\rangle \mid p \in Q, q \in \hat{Q}\} \\ R' &= \{b\langle p, r\rangle a \rightarrow w\langle q, t\rangle \mid pa \rightarrow q \in R, bpr \rightarrow wt \in \hat{R}\} \\ s' &= \langle s, \hat{s}\rangle \\ F' &= \{\langle p, q\rangle \mid p \in F, q \in \hat{F}\} \end{aligned}$$

By a similar reasoning as in Lemma 16.1.6, we can prove that $L(M') = {}_{\triangleright}L(M, L(H))$. A rigorous proof of the identity $L(M') = {}_{\triangleright}L(M, L(H))$ is left to the reader. □

The following theorem says that even though finite automata controlled by context-free languages are more powerful than finite automata, they cannot accept any non-context-free language.

Theorem 16.1.10. $\mathbf{SCFA(CF)} = \mathbf{CF}$

Proof. The inclusion $\mathbf{CF} \subseteq \mathbf{SCFA(CF)}$ is obvious. The converse inclusion follows from Lemma 16.1.9. □

Combining Theorems 16.1.5 and 16.1.10, we obtain the following corollary (recall that **CF** satisfies all the conditions from Theorem 16.1.5).

Corollary 16.1.11. $\mathbf{TCFA(CF)} = \mathbf{CF}$ □

16.1.5 Program-Controlled Finite Automata

In this section, we show that there is a language family, strictly included in the family of context-sensitive languages, which significantly increases the power of finite automata. Indeed, finite automata controlled by languages generated by propagating programmed grammars with appearance checking have the same power as phrase-structure grammars. This result is of some interest because $\mathbf{P}_{ac}^{-\varepsilon} \subset \mathbf{CS}$ (see Theorem 5.3.4).

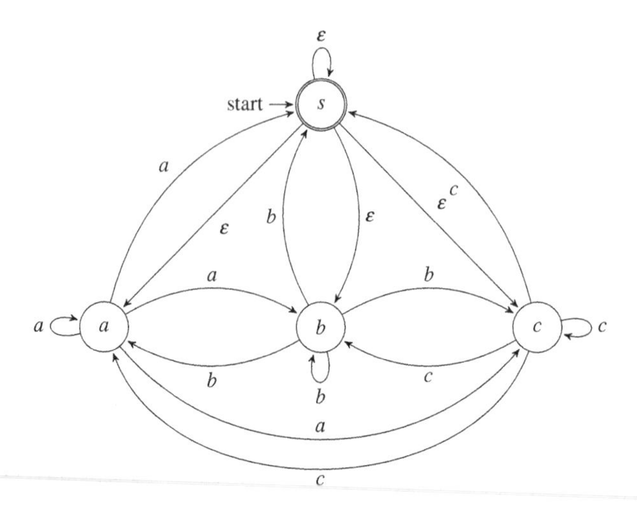

Fig. 16.1 Example of a finite automaton constructed by Algorithm 16.1.12

More specifically, we show how to algorithmically convert any programmed grammar with appearance checking G to a finite automaton M and a propagating programmed grammar with appearance checking G' such that ${}_{\triangleright}L(M, L(G')) = L(G)$. First, we give an insight into the algorithm. Then, we describe it formally and after that, we verify its correctness.

Let $T = \text{alph}(L(G))$ and let $s \notin T$ be a new symbol. From G, we construct the propagating programmed grammar with appearance checking G' such that $w \in L(G)$ if and only if $sws^k \in L(G')$, where $k \geq 1$. Then, the set of states of M will be $T \cup \{s\}$, where s is the starting and also the only final state. For every $a, b \in T$, we introduce $aa \rightarrow b$ to M. For every $a \in T$, we introduce $s \rightarrow a$ and $aa \rightarrow s$. Finally, we add $s \rightarrow s$. An example of such a finite automaton when $T = \{a, b, c\}$ can be seen in Fig. 16.1. The key idea is that when M is in a state $a \in T$, in the next move, it has to read a. Hence, with $sws^k \in L(G')$, M moves from s to a state in T, then reads every symbol in w, ends up in s, and uses k times the rule $s \rightarrow s$.

G' works in the following way. Every intermediate sentential form is of the form xvZ, where x is a string of symbols that are not erased in the rest of the derivation, v is a string of nonterminals that are erased in the rest of the derivation, and Z is a nonterminal. When simulating a rule of G, G' non-deterministically selects symbols that are erased, appends them using Z to the end of the currently generated string, and replaces an occurrence of the left-hand side of the original rule with the not-to-be-erased symbols from the right-hand side. To differentiate the symbols in x and v, v contains barred versions of the nonterminals. If G' makes an improper non-deterministic selection—that is, the selection does not correspond to a derivation in G—then G' is not able to generate a terminal string as explained in the notes following the algorithm.

Next, we describe the algorithm formally.

Algorithm 16.1.12.

Input: A programmed grammar with appearance checking $G = (N, T, \Psi, P, S)$.

Output: A finite automaton $M = (Q, T, R, s, F)$ and a propagating programmed grammar with appearance checking $G' = (N', Q, \Psi', P', S')$ such that ${}_{\triangleright}L(M, L(G')) = L(G)$.

Note: Without any loss of generality, we assume that $s, S', Z, \# \notin (Q \cup N \cup T)$ and $\ell_0, \bar{\ell}_0, \ell_s \notin \Psi$.

Method: Set $V = N \cup T$ and $\bar{N} = \{\bar{A} \mid A \in N\}$. Define the function τ from N^* to $\bar{N}^* \cup \{\ell_s\}$ as $\tau(\varepsilon) = \ell_s$ and $\tau(A_1A_2\cdots A_m) = \bar{A}_1\bar{A}_2\cdots\bar{A}_m$, where $A_i \in N$ for $i = 1, 2, \dots, m$, for some $m \geq 1$. Initially, set

$$
\begin{aligned}
Q &= T \cup \{s\}\\
R &= \{s \to s\} \cup \{s \to a \mid a \in T\} \cup \{aa \to b \mid a \in T, b \in T \cup \{s\}\}\\
F &= \{s\}\\
N' &= N \cup \bar{N} \cup \{S', Z\}\\
\Psi' &= \{\ell_0, \bar{\ell}_0, \ell_s\}\\
P' &= \{(\ell_0\colon S' \to sSZ, \{r \mid (r\colon S \to x, \sigma_r, \rho_r) \in P, x \in V^*\}, \emptyset)\} \cup\\
&\quad\ \{(\bar{\ell}_0\colon S' \to s\bar{S}Z, \{r \mid (r\colon S \to x, \sigma_r, \rho_r) \in P, x \in V^*\}, \emptyset)\} \cup\\
&\quad\ \{(\ell_s\colon Z \to s, \{\ell_s\}, \emptyset)\}
\end{aligned}
$$

Repeat (1) through (3), given next, until none of the sets Ψ' and P' can be extended in this way.

(1) **If** $(r\colon A \to y_0Y_1y_1Y_2y_2\cdots Y_my_m, \sigma_r, \rho_r) \in P$, where $y_i \in V^*, Y_j \in N$, for $i = 0, 1, \dots, m$ and $j = 1, 2, \dots, m$, for some $m \geq 1$
then

(1.1) add $\ell = \langle r, y_0, Y_1y_1, Y_2y_2, \dots, Y_my_m\rangle$ to Ψ';
(1.2) add ℓ' to Ψ', where ℓ' is a new unique label;
(1.3) add $(\ell\colon A \to y_0y_1\cdots y_m, \{\ell'\}, \emptyset)$ to P';
(1.4) add $(\ell'\colon Z \to \bar{y}Z, \sigma_r, \emptyset)$, where $\bar{y} = \tau(Y_1Y_2\cdots Y_m)$ to P'.

(2) **If** $(r\colon A \to y, \sigma_r, \rho_r) \in P$, where $y \in N^*$
then

(2.1) add $\bar{r}$ to Ψ', where $\bar{r}$ is a new unique label;
(2.2) add $(\bar{r}\colon \bar{A} \to \bar{y}, \sigma_r, \emptyset)$ to P', where $\bar{y} = \tau(y)$.

(3) **If** $(r\colon A \to y, \sigma_r, \rho_r) \in P$, where $y \in V^*$,
then

(3.1) add $\hat{r}$ to Ψ', where $\hat{r}$ is a new unique label;
(3.2) add $\hat{r}'$ to Ψ', where $\hat{r}'$ is a new unique label;
(3.3) add $(\hat{r}\colon A \to \#y, \emptyset, \{\hat{r}'\})$ to P';
(3.4) add $(\hat{r}'\colon \bar{A} \to \#y, \emptyset, \rho_r)$ to P'.

Finally, for every $t \in \Psi$, let Ψ'_t denote the set of rule labels introduced in steps (1.1), (2.1), and (3.1) from a rule labeled with t.
Replace every $(\ell' \colon Z \to \bar{y}Z, \sigma_r, \emptyset)$ from (1.4) satisfying $\sigma_r \neq \emptyset$ with

$$(\ell' \colon Z \to \bar{y}Z, \sigma'_r \cup \{\ell_s\}, \emptyset) \text{ where } \sigma'_r = \bigcup_{t \in \sigma_r} \Psi'_t$$

Replace every $(\bar{r} \colon \bar{A} \to \bar{y}, \sigma_r, \emptyset)$ from (2.2) satisfying $\sigma_r \neq \emptyset$ with

$$(\bar{r} \colon \bar{A} \to \bar{y}, \sigma'_r \cup \{\ell_s\}, \emptyset) \text{ where } \sigma'_r = \bigcup_{t \in \sigma_r} \Psi'_t$$

Replace every $(\hat{r}' \colon \bar{A} \to \#y, \emptyset, \rho_r)$ from (3.4) satisfying $\rho_r \neq \emptyset$ with

$$(\hat{r}' \colon \bar{A} \to \#y, \emptyset, \rho'_r \cup \{\ell_s\}) \text{ where } \rho'_r = \bigcup_{t \in \rho_r} \Psi'_t$$

□

Before proving that this algorithm is correct, we make some informal comments concerning the purpose of the rules of G'. The rules introduced in (1) are used to simulate an application of a rule where some of the symbols on its right-hand side are erased in the rest of the derivation while some others are not erased. The rules introduced in (2) are used to simulate an application of a rule which erases a nonterminal by making one or more derivation steps. The rules introduced in (3) are used to simulate an application of a rule in the appearance checking mode. Observe that when simulating $(r \colon A \to y, \sigma_r, \rho_r) \in P$ in the appearance checking mode, we have to check the absence of both A and $\bar{A}$. If some of these two nonterminals appear in the current configuration, the derivation is blocked because rules from (3) have empty success fields. Finally, notice that the final part of the algorithm ensures that after a rule $t \in P$ is applied, ℓ_s or any of the rules introduced in (1.3), (2.2), or (3.3) from rules in σ_t or ρ_t can be applied.

Reconsider (1). Notice that the algorithm works correctly although it makes no predetermination of nonterminals from which ε can be derived. Indeed, if the output grammar improperly selects a nonterminal that is not erased throughout the rest of the derivation, then this occurrence of the nonterminal never disappears so a terminal string cannot be generated under this improper selection.

We next prove that Algorithm 16.1.12 is correct.

Lemma 16.1.13. *Algorithm 16.1.12 converts any programmed grammar with appearance checking* $G = (N, T, \Psi, P, S)$ *to a finite automaton* $M = (Q, T, R, s, F)$ *and a propagating programmed grammar with appearance checking* $G' = (N', Q, \Psi', P', S')$ *such that* ${}_{\triangleright}L(M, L(G')) = L(G)$.

Proof. Clearly, the algorithm always halts. Consider the construction of G'. Observe that every string in $L(G')$ is of the form sws^k, where $k \geq 1$. From the construction of M, it is easy to see that $w \in {}_{\triangleright}L(M, L(G'))$ for $w \in T^*$ if and only if $sws^k \in L(G')$ for some $k \geq 1$. Therefore, to prove that ${}_{\triangleright}L(M, L(G')) = L(G)$, it is sufficient to prove that $w \in L(G)$ for $w \in T^*$ if and only if $sws^k \in L(G')$ for some $k \geq 1$.

We establish this equivalence by proving two claims. First, we prove that $w \in L(G)$ for $w \in T^+$ if and only if $sws^k \in L(G')$ for some $k \geq 1$. Then, we show that $\varepsilon \in L(G)$ if and only if $\varepsilon \in L(G')$.

The first claim, stated next, shows how G' simulates G. Recall the meaning of the ${}^{\varepsilon}$ and ${}^{\not\varepsilon}$notation from Definition 3.3.26 because this notation is frequently used in the rest of the proof.

Claim 1. If $(S, t_1) \Rightarrow_G^n ({}^{\not\varepsilon}x_0{}^{\varepsilon}X_1{}^{\not\varepsilon}x_1{}^{\varepsilon}X_2{}^{\not\varepsilon}x_2 \cdots {}^{\varepsilon}X_h{}^{\not\varepsilon}x_h, t_2) \Rightarrow_G^* (z, t_3)$, *where* $z \in T^+, t_1, t_2, t_3 \in \Psi, x_i \in V^*$ *for* $i = 0, 1, \dots, h$, $X_j \in N$ *for* $j = 1, 2, \dots, h$, *for some* $h \geq 0$ *and* $n \geq 0$, *then* $(S', \ell_0) \Rightarrow_{G'}^* (sx_0x_1x_2 \cdots x_h vZ, t_2')$, *where* $v \in \mathrm{perm}(\tau(X_1X_2 \cdots X_h)\xi), \xi \in \{s\}^*$, *and* t_2' *can be* ℓ_s *or any rule constructed from* t_2 *in (1.3), (2.2), or (3.3).*

Proof. This claim is established by induction on $n \geq 0$.

Basis. Let $n = 0$. Then, for $(S, t_1) \Rightarrow_G^0 (S, t_1) \Rightarrow_G^* (z, t_2)$, there is $(S', \ell_0) \Rightarrow_{G'} (sSZ, t_1')$, where t_1' can be ℓ_s or any rule constructed from t_1 in (1.3), (2.2), or (3.3). Hence, the basis holds.

Induction Hypothesis. Suppose that there exists $n \geq 0$ such that the claim holds for all derivations of length m, where $0 \leq m \leq n$.

Induction Step. Consider any derivation of the form

$$(S, t_1) \Rightarrow_G^{n+1} (w, t_3) \Rightarrow_G^* (z, t_4)$$

where $w \in V^+$ and $z \in T^+$. Since $n + 1 \geq 1$, this derivation can be expressed as

$$(S, t_1) \Rightarrow_G^n ({}^{\not\varepsilon}x_0{}^{\varepsilon}X_1{}^{\not\varepsilon}x_1{}^{\varepsilon}X_2{}^{\not\varepsilon}x_2 \cdots {}^{\varepsilon}X_h{}^{\not\varepsilon}x_h, t_2) \Rightarrow_G (w, t_3) \Rightarrow_G^* (z, t_4)$$

where $x_i \in V^*$ for $i = 0, 1, \dots, h$, $X_j \in V$ for $j = 1, 2, \dots, h$, for some $h \geq 0$. By the induction hypothesis,

$$(S', \ell_0) \Rightarrow_{G'}^* (sx_0x_1x_2 \cdots x_h vZ, t_2')$$

where $v \in \mathrm{perm}(\tau(X_1X_2 \cdots X_h)\xi), \xi \in \{s\}^*$, and t_2' can be ℓ_s or any rule constructed from t_2 in (1.3), (2.2), or (3.3).

Let $x = {}^{\not\varepsilon}x_0{}^{\varepsilon}X_1{}^{\not\varepsilon}x_1{}^{\varepsilon}X_2{}^{\not\varepsilon}x_2 \cdots {}^{\varepsilon}X_h{}^{\not\varepsilon}x_h$. Next, we consider all possible forms of the derivation $(x, t_2) \Rightarrow_G (w, t_3)$, covered by the following three cases—(i) through (iii).

(i) *Application of a rule that rewrites a symbol in some* x_j. Let $x_j = x_j'Ax_j''$ and $(t_2: A \to y_0Y_1y_1Y_2y_2 \cdots Y_my_m, \sigma_{t_2}, \rho_{t_2}) \in P$ for some $j \in \{0, 1, \dots, h\}$ and $m \geq 0$, where $y_i \in V^*$ for $i = 0, 1, \dots, m$, $Y_i \in N$ for $i = 1, \dots, m$, and $t_3 \in \sigma_{t_2}$ so that

$$({}^{\not\varepsilon}x_0{}^{\varepsilon}X_1{}^{\not\varepsilon}x_1{}^{\varepsilon}X_2{}^{\not\varepsilon}x_2\cdots{}^{\varepsilon}X_j{}^{\not\varepsilon}x'_j{}^{\not\varepsilon}A{}^{\not\varepsilon}x''_j\cdots{}^{\varepsilon}X_h{}^{\not\varepsilon}x_h, t_2) \Rightarrow_G$$
$$({}^{\not\varepsilon}x_0{}^{\varepsilon}X_1{}^{\not\varepsilon}x_1{}^{\varepsilon}X_2{}^{\not\varepsilon}x_2\cdots{}^{\varepsilon}X_j{}^{\not\varepsilon}y_0{}^{\varepsilon}Y_1{}^{\not\varepsilon}y_1{}^{\varepsilon}Y_2{}^{\not\varepsilon}y_2\cdots{}^{\varepsilon}Y_m{}^{\not\varepsilon}y_m\cdots{}^{\varepsilon}X_h{}^{\not\varepsilon}x_h, t_3)$$

By (1) and by the final step of the algorithm, P' contains

$$(\ell\colon A \to y_0y_1\cdots y_m, \{\ell'\}, \emptyset)$$
$$(\ell'\colon Z \to \bar{y}Z, \sigma_{\ell'}, \emptyset), \text{ where } \bar{y} = \tau(Y_1Y_2\cdots Y_m)$$

By the induction hypothesis, we assume that $t'_2 = \ell$. Then,

$$(sx_0x_1x_2\cdots x'_jAx''_j\cdots x_hvZ, \ell) \Rightarrow_{G'}$$
$$(sx_0x_1x_2\cdots x'_jy_0y_1y_2\cdots y_mx''_j\cdots x_hvZ, \ell') \Rightarrow_{G'}$$
$$(sx_0x_1x_2\cdots x'_jy_0y_1y_2\cdots y_mx''_j\cdots x_hv\bar{y}Z, t'_3)$$

with $t'_3 \in \sigma_{\ell'}$. By the final step of the algorithm, t'_3 can be ℓ_s or any rule constructed from t_3 in (1.3), (2.2), or (3.3). As $x_0x_1x_2\cdots x'_jy_0y_1y_2\cdots y_mx''_j\cdots x_hv\bar{y}Z$ is of the required form, the induction step for (i) is completed.

(ii) *Application of a rule that rewrites some* X_j. Let $(t_2\colon X_j \to y, \sigma_{t_2}, \rho_{t_2}) \in P$ for some $j \in \{1, 2, \ldots, h\}$, where $y \in N^*$ and $t_3 \in \sigma_{t_2}$ so that

$$({}^{\not\varepsilon}x_0{}^{\varepsilon}X_1{}^{\not\varepsilon}x_1{}^{\varepsilon}X_2{}^{\not\varepsilon}x_2\cdots{}^{\varepsilon}X_j{}^{\not\varepsilon}x_j\cdots{}^{\varepsilon}X_h{}^{\not\varepsilon}x_h, t_2) \Rightarrow_G$$
$$({}^{\not\varepsilon}x_0{}^{\varepsilon}X_1{}^{\not\varepsilon}x_1{}^{\varepsilon}X_2{}^{\not\varepsilon}x_2\cdots{}^{\varepsilon}y{}^{\not\varepsilon}x_j\cdots{}^{\varepsilon}X_h{}^{\not\varepsilon}x_h, t_3)$$

By the induction hypothesis, $v = v_1\bar{X}_jv_2$ for some $v_1, v_2 \in \bar{N}^*$. By (2) and by the final step of the algorithm,

$$(\bar{t}_2\colon \bar{X}_j \to \bar{y}, \sigma_{\bar{t}_2}, \emptyset) \in P' \text{ where } \bar{y} = \tau(y)$$

By the induction hypothesis, we assume that $t'_2 = \bar{t}_2$. Then,

$$(sx_0x_1x_2\cdots x_hv_1\bar{X}_jv_2Z, \bar{t}_2) \Rightarrow_{G'} (sx_0x_1x_2\cdots x_hv_1\bar{y}v_2Z, t'_3)$$

with $t'_3 \in \sigma_{\bar{t}_2}$. By the final step of the algorithm, t'_3 can be ℓ_s or any rule that was constructed from t_3 in (1.3), (2.2), or (3.3). Since $x_0x_1x_2\cdots x_hv_1\bar{y}_jv_2Z$ is of the required form, the induction step for (ii) is completed.

(iii) *Application of a rule in the appearance checking mode.* Let $(t_2\colon A \to y, \sigma_{t_2}, \rho_{t_2}) \in P$, where $A \notin \mathrm{alph}(x)$, $y \in V^*$ and $t_3 \in \rho_{t_2}$ so that

$$(x, t_2) \Rightarrow_G (x, t_3)$$

By the induction hypothesis, we assume that t'_2 was constructed from t_2 in (3.3). By (3) and by the final step of the algorithm, P' contains

$$(\hat{r}: A \rightarrow \#y, \emptyset, \{\hat{r}'\})$$
$$(\hat{r}': \bar{A} \rightarrow \#y, \emptyset, \rho_{\hat{r}'})$$

Since $A \notin \text{alph}(x)$, $\bar{A} \notin \text{alph}(x_0x_1x_2 \cdots x_h v Z)$. Then,

$$\begin{aligned}&(sx_0x_1x_2 \cdots x_h v Z, \hat{r}) \Rightarrow_{G'}\\&(sx_0x_1x_2 \cdots x_h v Z, \hat{r}') \Rightarrow_{G'}\\&(sx_0x_1x_2 \cdots x_h v Z, t_3')\end{aligned}$$

where t_3' can be ℓ_s or any rule that was constructed from t_3 in (1.3), (2.2), or (3.3). Since $x_0x_1x_2 \cdots x_h v Z$ is of the required form, the induction step for (iii) is completed.

Observe that cases (i) through (iii) cover all possible forms of $(x, t_2) \Rightarrow_G (w, t_3)$. Thus, the claim holds. □

To simplify the second claim and its proof, we define a generalization of $\Rrightarrow_{G'}$. In this generalization, we use the property that whenever a rule introduced in (1.3) or (3.3) is applied during a successful derivation, it has to be followed by its corresponding rule from (1.4) or (3.4), respectively. Let $V' = N' \cup T$. Define the binary relation $\Rrightarrow_{G'}$ over $V'^* \times \Psi'$ as

$$(x, r) \Rrightarrow_{G'} (w, t)$$

if and only if either

$$(x, r) \Rightarrow_{G'} (w, t)$$

where $r, t \in \Psi'$ such that t is not introduced in (1.4) and (3.4), or

$$(x, r) \Rightarrow_{G'} (y, r') \Rightarrow_{G'} (w, t)$$

where r is a rule introduced in (1.3) or (3.3) and r' is its corresponding second rule introduced in (1.4) or (3.4), respectively. Define $\Rrightarrow_{G'}^n$ for $n \geq 0$ and $\Rrightarrow_{G'}^*$ in the usual way.

The next claim shows how G simulates G'. Define the homomorphism ι from V'^* to V^* as $\iota(X) = X$ for $X \in V, \iota(\bar{X}) = X$ for $X \in N$, and $\iota(s) = \iota(S') = \iota(Z) = \varepsilon$.

Claim 2. If $(S', \ell_0) \Rrightarrow_{G'}^n (sxu, t) \Rrightarrow_{G'}^* (z, g)$, *where* $x \in V^+, u \in (\bar{N} \cup \{s\})^*\{Z, \varepsilon\}, z \in Q^+, t, g \in \Psi'$, *and* $n \geq 1$, *then* $(S, t_1) \Rightarrow_G^* (x_0X_1x_1X_2x_2 \cdots X_hx_h, t')$, *where* $x = x_0x_1 \cdots x_h, X_1X_2 \cdots X_h \in \text{perm}(\iota(u)), h \geq 0$, *and* t' *is the rule from which* t *was constructed or any rule in* P *if* t *was not constructed from any rule in* P.

Proof. This claim is established by induction on $n \geq 1$.

Basis. Let $n = 1$. Then, for $(S', \ell_0) \Rightarrow^{n}_{G'} (sSZ, t) \Rightarrow^{*}_{G'} (z, g)$, where $t, g \in \Psi'$ and $z \in Q^+$, there is $(S, t') \Rightarrow^{0}_{G} (S, t')$, where t' is the rule from which t was constructed. Hence, the basis holds.

Induction Hypothesis. Suppose that there exists $n \geq 1$ such that the claim holds for all derivations of length m, where $0 \leq m \leq n$.

Induction Step. Consider any derivation of the form

$$(S', \ell_0) \Rightarrow^{n+1}_{G'} (swv, p) \Rightarrow^{*}_{G'} (z, g)$$

where $n \geq 1, w \in V^+, v \in (\bar{N} \cup \{s\})^*\{Z, \varepsilon\}, z \in Q^+$, and $p, g \in \Psi'$. Since $n + 1 \geq 1$, this derivation can be expressed as

$$(S', \ell_0) \Rightarrow^{n}_{G'} (swu, t) \Rightarrow_{G'} (swv, p) \Rightarrow^{*}_{G'} (z, g)$$

where $x \in V^+, u \in (\bar{N} \cup \{s\})^*\{Z, \varepsilon\}$, and $t \in \Psi'$. By the induction hypothesis,

$$(S, t_1) \Rightarrow^{*}_{G} (x_0 X_1 x_1 X_2 x_2 \cdots X_h x_h, t')$$

where $x = x_0 x_1 \cdots x_h, X_1 X_2 \cdots X_h \in \mathrm{perm}(\iota(u)), h \geq 0$, and t' is the rule from which t was constructed or any rule in P if t was not constructed from any rule in P.

Next, we consider all possible forms of $(sxu, t) \Rightarrow_{G'} (swv, p)$, covered by the following four cases—(i) through (iv).

(i) *Application of* $(\ell_s \colon Z \to s, \{\ell_s\}, \emptyset)$. Let $t = \ell_s$, so

$$(swu'Z, \ell_s) \Rightarrow_{G'} (swu's, \ell_s)$$

where $u = u'Z$. Then, the induction step for (i) follows directly from the induction hypothesis (recall that $\iota(Z) = \iota(s) = \varepsilon$).

(ii) *Application of two rules introduced in (1).* Let $x = x'Ax''$ and $(\ell \colon A \to y_0 y_1 \cdots y_m, \{\ell'\}, \emptyset), (\ell' \colon Z \to \bar{y}Z, \sigma_{\ell'}, \emptyset) \in P'$ be two rules introduced in (1) from $(r \colon A \to y_0 Y_1 y_1 Y_2 y_2 \cdots Y_m y_m, \sigma_r, \rho_r) \in P$, where $y_i \in V^*, Y_j \in N$, for $i = 0, 1, \dots, m$ and $j = 1, 2, \dots, m$, for some $m \geq 1$, and $\bar{y} = \tau(Y_1 Y_2 \cdots Y_m)$. Then,

$$(sx'Ax''uZ, \ell) \Rightarrow_{G'} (sx'y_0 y_1 \cdots y_m x'' u \bar{y} Z, p)$$

by applying ℓ and ℓ' ($p \in \sigma_{\ell'}$). By the induction hypothesis, $t' = r$ and $x_i = x_i' A x_i''$ for some $i \in \{0, 1, \dots, h\}$. Then,

$$(x_0 X_1 x_1 X_2 x_2 \cdots X_i x_i' A x_i'' \cdots X_h x_h, r) \Rightarrow_G$$
$$(x_0 X_1 x_1 X_2 x_2 \cdots X_i x_i' y_0 Y_1 y_1 Y_2 y_2 \cdots Y_m y_m x_i'' \cdots X_h x_h, t'')$$

Clearly, both configurations are of the required forms, so the induction step is completed for (ii).

(iii) *Application of a rule introduced in (2).* Let $u = u'\bar{A}u''$ and $(\bar{r}: \bar{A} \to \bar{y}, \sigma_{\bar{r}}), \emptyset \in P'$ be a rule introduced in (2) from $(r: A \to y, \sigma_r, \rho_r) \in P$, where $y \in N^*$ and $\bar{y} = \tau(y)$. Then,

$$(sxu'\bar{A}u''Z, \bar{r}) \Rightarrow_{G'} (sxu'\bar{y}u''Z, p)$$

where $p \in \sigma_{\bar{r}}$. By the induction hypothesis, $t' = r$ and $X_i = A$ for some $i \in \{1, \dots, h\}$. Then,

$$(x_0X_1x_1X_2x_2 \cdots X_ix_i \cdots X_hx_h, r) \Rightarrow_G$$
$$(x_0X_1x_1X_2x_2 \cdots yx_i \cdots X_hx_h, t'')$$

Clearly, both configurations are of the required forms, so the induction step is completed for (iii).

(iv) *Application of two rules introduced in (3).* Let $(\hat{r}: A \to \#y, \emptyset, \{\hat{r}'\}), (\hat{r}': \bar{A} \to \#y, \emptyset, \rho_{\hat{r}'}) \in P'$ be two rules introduced in (3) from $(r: A \to y, \sigma_r, \rho_r) \in P$, where $y \in V^*$, such that $\{A, \bar{A}\} \cap \mathrm{alph}(sxuZ) = \emptyset$. Then,

$$(sxu, \hat{r}) \Rightarrow_{G'} (sxu, p)$$

by applying $\hat{r}$ and $\hat{r}'$ in the appearance checking mode ($p \in \rho_{\hat{r}'}$). By the induction hypothesis, $t' = r$ and $A \notin \mathrm{alph}(x_0X_1x_1X_2x_2 \cdots X_hx_h)$, so

$$(x_0X_1x_1X_2x_2 \cdots X_hx_h, r) \Rightarrow_G (x_0X_1x_1X_2x_2 \cdots X_hx_h, t'')$$

Clearly, both configurations are of the required forms, so the induction step is completed for (iv).

Observe that cases (i) through (iv) cover all possible forms of $(sxu, t) \Rightarrow_{G'} (swv, p)$. Thus, the claim holds. □

Consider Claim 1 with $h = 0$. Then,

$$(S, t_1) \Rightarrow_G^* (z, r)$$

implies that

$$(S', \ell_0) \Rightarrow_{G'} (szs^k, r')$$

where $k \geq 1, t_1, r \in \Psi$, and $r' \in \Psi'$. Consider Claim 2 with $x \in T^+$ and $u \in \{s\}^+$. Then,

$$(S', \ell_0) \Rightarrow_{G'}^* (sxu, t)$$

implies that

$$(S, t_1) \Rightarrow_G^* (x, t')$$

Hence, we have $w \in L(G)$ for $w \in T^+$ if and only if $sws^k \in L(G')$ for some $k \geq 1$.

It remains to be shown that $\varepsilon \in L(G)$ if and only if $\varepsilon \in L(G')$. This can be proved by analogy with proving Claims 1 and 2, where G' uses $\bar{\ell}_0$ instead of ℓ_0 (see the initialization part of the algorithm). We leave this part of the proof to the reader. Hence, ${}_{\triangleright}L(M, L(G')) = L(G)$, and the lemma holds. □

Lemma 16.1.14. $\mathbf{RE} \subseteq \mathbf{SCFA}(\mathbf{P}_{ac}^{-\varepsilon})$

Proof. Let $I \in \mathbf{RE}$ and $T = \text{alph}(I)$. By Theorem 5.3.4, there is a programmed grammar with appearance checking $G = (N, T, \Psi, P, S)$ such that $L(G) = I$. Let $M = (Q, T, R, s, F)$ and $G' = (N', Q, \Psi', P', S')$ be the finite automaton and the propagating programmed grammar with appearance checking, respectively, constructed by Algorithm 16.1.12 from G. By Lemma 16.1.13, ${}_{\triangleright}L(M, L(G')) = L(G) = I$, so the lemma holds. □

Theorem 16.1.15. $\mathbf{SCFA}(\mathbf{P}_{ac}^{-\varepsilon}) = \mathbf{RE}$

Proof. The inclusion $\mathbf{SCFA}(\mathbf{P}_{ac}^{-\varepsilon}) \subseteq \mathbf{RE}$ follows from Church's thesis. The converse inclusion $\mathbf{RE} \subseteq \mathbf{SCFA}(\mathbf{P}_{ac}^{-\varepsilon})$ follows from Lemma 16.1.14. □

Combining Theorems 16.1.5 and 16.1.15, we obtain the following corollary (recall that $\mathbf{P}_{ac}^{-\varepsilon}$ satisfies all the conditions from Theorem 16.1.5, see [1]).

Corollary 16.1.16. $\mathbf{TCFA}(\mathbf{P}_{ac}^{-\varepsilon}) = \mathbf{RE}$ □

Finally, we briefly investigate a reduction of the number of states in controlled finite automata. First, observe that the finite automaton $M = (Q, T, R, s, F)$ from Algorithm 16.1.12 has $\text{card}(T) + 1$ states. Therefore, we obtain the following corollary.

Corollary 16.1.17. *Let I be a recursively enumerable language, and let $T = \text{alph}(I)$. Then, there is a finite automaton $M = (Q, T, R, s, F)$ such that $\text{card}(Q) = \text{card}(T) + 1$, and a propagating programmed grammar with appearance checking G such that ${}_{\triangleright}L(M, L(G)) = I$.* □

In a comparison to Corollary 16.1.17, a more powerful result holds in terms of transition-controlled finite automata. In this case, the number of states can be decreased to a single state as stated in the following theorem.

Theorem 16.1.18. *Let I be a recursively enumerable language. Then, there is a finite automaton $M = (Q, T, R, s, F)$ such that $\text{card}(Q) = 1$, and a propagating programmed grammar with appearance checking G such that ${}_{\blacktriangleright}L(M, L(G)) = I$.*

Proof. Reconsider Algorithm 16.1.12. We modify the construction of $M = (Q, T, R, s, F)$ in the following way. Let $G = (N, T, \Psi, P, S)$ be the input programmed grammar with appearance checking. Construct the finite automaton

$$M = \big(Q, \Phi, R, s, F\big)$$

where

$$
\begin{aligned}
Q &= \{s\} \\
\Phi &= \{s\} \cup T \\
R &= \{s{:}\, s \to s\} \cup \{a{:}\, sa \to s \mid a \in T\} \\
F &= \{s\}
\end{aligned}
$$

Observe that $\text{card}(Q) = 1$ and that this modified algorithm always halts. The correctness of the modified algorithm—that is, the identity $_{\blacktriangleright}L(M, L(G')) = L(G)$, where G' is the propagating programmed grammar constructed by Algorithm 16.1.12—can be established by analogy with the proof of Lemma 16.1.13, so we leave the proof to the reader. The rest of the proof of this theorem parallels the proof of Lemma 16.1.14, so we omit it. □

We close this section by presenting three open problem areas that are related to the achieved results.

Open Problem 16.1.19. In general, the state-controlled and transition-controlled finite automata in Theorem 16.1.15 and Corollary 16.1.16 are non-deterministic. Do these results hold in terms of deterministic versions of these automata? □

Open Problem 16.1.20. By using control languages from **CF**, we characterize **CF**. By using control languages from $\mathbf{P}_{ac}^{-\varepsilon}$, we characterize **RE**. Is there a language family $\mathscr{L}$ such that $\mathbf{CF} \subset \mathscr{L} \subset \mathbf{P}_{ac}^{-\varepsilon}$ by which we can characterize **CS**? □

Open Problem 16.1.21. Theorem 16.1.5 requires $\mathscr{L}$ to contain all finite languages and to be closed under finite ε-free substitution and concatenation. Does the same result hold if there are fewer requirements placed on $\mathscr{L}$? □

16.2 Pushdown Automata Regulated by Control Languages

Section 16.2 consists of four subsections. In Sect. 16.2.1, we define pushdown automata that regulate the application of their rules by control languages by analogy with context-free grammars regulated in this way (see Chap. 5). In Sect. 16.2.2, we demonstrate that this regulation has no effect on the power of pushdown automata if the control languages are regular. Considering this result, we point out that pushdown automata regulated by regular languages are of little interest because their power coincides with the power of ordinary pushdown automata. In Sect. 16.2.3, however, we prove that pushdown automata increase their power remarkably if they are regulated by linear languages; indeed, they characterize the family of recursively enumerable languages. In Sect. 16.2.4, we continue with the discussion of regulated pushdown automata, but we narrow our attention to their special cases, such as one-turn pushdown automata.

16.2.1 Definitions

Without further ado, we next define pushdown automata regulated by control languages. Recall the formalization of rule labels from Definition 3.4.10 because this formalization is often used throughout the present section.

Definition 16.2.1. Let $M = (Q, \Sigma, \Gamma, R, s, S, F)$ be a pushdown automaton, and let Ψ be an alphabet of its rule labels. Let Ξ be a *control language* over Ψ; that is, $\Xi \subseteq \Psi^*$. With Ξ, M defines the following three types of accepted languages

- $L(M, \Xi, 1)$—the *language accepted by final state*
- $L(M, \Xi, 2)$—the *language accepted by empty pushdown*
- $L(M, \Xi, 3)$—the *language accepted by final state and empty pushdown*

defined as follows. Let $\chi \in \Gamma^* Q \Sigma^*$. If $\chi \in \Gamma^* F, \chi \in Q, \chi \in F$, then χ is a *1-final configuration*, *2-final configuration*, *3-final configuration*, respectively. For $i = 1, 2, 3$, we define $L(M, \Xi, i)$ as

$$L(M, \Xi, i) = \{w \mid w \in \Sigma^* \text{ and } Ssw \vdash_M^* \chi\ [\sigma] \text{ for an } i\text{-final configuration } \chi \text{ and } \sigma \in \Xi\}$$

The pair (M, Ξ) is called a *controlled pushdown automaton*. □

For any family of languages $\mathscr{L}$ and $i \in \{1, 2, 3\}$, define

$$\mathbf{RPDA}(\mathscr{L}, i) = \{L \mid L = L(M, \Xi, i), \text{ where } M \text{ is a pushdown automaton and } \Xi \in \mathscr{L}\}$$

We demonstrate that

$$\mathbf{CF} = \mathbf{RPDA}(\mathbf{REG}, 1) = \mathbf{RPDA}(\mathbf{REG}, 2) = \mathbf{RPDA}(\mathbf{REG}, 3)$$

and

$$\mathbf{RE} = \mathbf{RPDA}(\mathbf{LIN}, 1) = \mathbf{RPDA}(\mathbf{LIN}, 2) = \mathbf{RPDA}(\mathbf{LIN}, 3)$$

Some of the following proofs involve several grammars and automata. To avoid any confusion, these proofs sometimes specify a regular grammar G as $G = (N_G, T_G, P_G, S_G)$ because this specification clearly expresses that N_G, T_G, P_G, and S_G represent the components of G. Other grammars and automata are specified analogously whenever any confusion may exist.

16.2.2 Regular-Controlled Pushdown Automata

This section proves that if the control languages are regular, then the regulation of pushdown automata has no effect on their power. The proof of the following lemma presents a transformation that converts any regular grammar G and any pushdown automaton K to an ordinary pushdown automaton M such that $L(M) = L(K, L(G), 1)$.

Lemma 16.2.2. *For every regular grammar G and every pushdown automaton K, there exists a pushdown automaton M such that $L(M) = L(K, L(G), 1)$.*

Proof. Let $G = (N_G, T_G, P_G, S_G)$ be any regular grammar, and let $K = (Q_K, \Sigma_K, \Gamma_K, R_K, s_K, S_K, F_K)$ be any pushdown automaton. Next, we construct a pushdown automaton M that simultaneously simulates G and K so that $L(M) = L(K, L(G), 1)$.

Let f be a new symbol. Define M as

$$M = \big(Q_M, \Sigma_M, \Gamma_M, R_M, s_M, S_M, F_M\big)$$

where

$$\begin{aligned}
Q_M &= \{\langle qB\rangle \mid q \in Q_K, B \in N_G \cup \{f\}\}\\
\Sigma_M &= \Sigma_K\\
\Gamma_M &= \Gamma_K\\
s_M &= \langle s_K S_G\rangle\\
S_M &= S_K\\
F_M &= \{\langle qf\rangle \mid q \in F_K\}\\
R_M &= \{C\langle qA\rangle b \to x\langle pB\rangle \mid a{:}\, Cqb \to xp \in R_K, A \to aB \in P_G\}\\
&\quad \cup \{C\langle qA\rangle b \to x\langle pf\rangle \mid a{:}\, Cqb \to xp \in R_K, A \to a \in P_G\}
\end{aligned}$$

Observe that a move in M according to $C\langle qA\rangle b \to x\langle pB\rangle \in R_M$ simulates a move in K according $a{:}\, Cqb \to xp \in R_K$, where a is generated in G by using $A \to aB \in P_G$. Based on this observation, it is rather easy to see that M accepts an input string w if and only if K reads w and enters a final state after using a complete string of $L(G)$; therefore, $L(M) = L(K, L(G), 1)$. A rigorous proof that $L(M) = L(K, L(G), 1)$ is left to the reader. □

Theorem 16.2.3. *For $i \in \{1, 2, 3\}$, $\mathbf{CF} = \mathbf{RPDA}(\mathbf{REG}, i)$.*

Proof. To prove that $\mathbf{CF} = \mathbf{RPDA}(\mathbf{REG}, 1)$, notice that $\mathbf{RPDA}(\mathbf{REG}, 1) \subseteq \mathbf{CF}$ follows from Lemma 16.2.2. Clearly, $\mathbf{CF} \subseteq \mathbf{RPDA}(\mathbf{REG}, 1)$, so $\mathbf{RPDA}(\mathbf{REG}, 1) = \mathbf{CF}$. By analogy with the demonstration of $\mathbf{RPDA}(\mathbf{REG}, 1) = \mathbf{CF}$, we can prove that $\mathbf{CF} = \mathbf{RPDA}(\mathbf{REG}, 2)$ and $\mathbf{CF} = \mathbf{RPDA}(\mathbf{REG}, 3)$. □

Let us point out that most fundamental regulated grammars use control mechanisms that can be expressed in terms of regular control languages (see

Sect. 5.1). However, pushdown automata introduced by analogy with these grammars are of little or no interest because they are as powerful as ordinary pushdown automata (see Theorem 16.2.3 above).

16.2.3 Linear-Controlled Pushdown Automata

This section demonstrates that pushdown automata regulated by linear control languages are more powerful than ordinary pushdown automata. In fact, it proves that

$$\mathbf{RE} = \mathbf{RPDA}(\mathbf{LIN}, 1) = \mathbf{RPDA}(\mathbf{LIN}, 2) = \mathbf{RPDA}(\mathbf{LIN}, 3)$$

First, we prove a normal form for left-extended queue grammars (see Definition 3.3.18), which is needed later in this section.

Lemma 16.2.4. *Let K be a left-extended queue grammar. Then, there exists a left-extended queue grammar, $Q = (V, T, W, F, R, g)$, satisfying $L(K) = L(Q)$, ! is a distinguished member of $W - F$, $V = U \cup Z \cup T$ such that U, Z, T are pairwise disjoint, and Q derives every $z \in L(Q)$ in this way*

$$\begin{aligned}
\#g &\Rightarrow_Q^+ x\#b_1b_2\cdots b_n!\\
&\Rightarrow_Q xb_1\#b_2\cdots b_n y_1 p_2\\
&\Rightarrow_Q xb_1b_2\#b_3\cdots b_n y_1 y_2 p_3\\
&\quad\vdots\\
&\Rightarrow_Q xb_1b_2\cdots b_{n-1}\#b_n y_1 y_2\cdots y_{n-1} p_n\\
&\Rightarrow_Q xb_1b_2\cdots b_{n-1}b_n\# y_1 y_2\cdots y_n p_{n+1}
\end{aligned}$$

where $n \in N, x \in U^*, b_i \in Z$ for $i = 1, \dots, n$, $y_i \in T^*$ for $i = 1, \dots, n$, $z = y_1y_2\cdots y_n$, $p_i \in W - \{!\}$ for $i = 1, \dots, n-1$, $p_n \in F$, and in this derivation, $x\#b_1b_2\cdots b_n!$ is the only string containing !.

Proof. Let K be any left-extended queue grammar. Convert K to a left-extended queue grammar

$$H = \left(V_H, T_H, W_H, F_H, R_H, g_H\right)$$

such that $L(K) = L(H)$ and H generates every $x \in L(H)$ by making two or more derivation steps (this conversion is trivial and left to the reader).

Define the bijection α from W to W', where $W' = \{q' \mid q \in W\}$, as $\alpha(q) = q'$ for every $q \in W$. Analogously, define the bijection β from W to W'', where $W'' = \{q'' \mid q \in W\}$, as $\beta(q) = q''$ for every $q \in W$. Without any loss of generality, assume that $\{1, 2\} \cap (V \cup W) = \emptyset$. Set

$$\Xi = \{\langle a, q, u1v, p\rangle \mid (a, q, uv, p) \in R_H \text{ for some } a \in V, q \in W - F, v \in T^*, u \in V^*, \text{ and } p \in W\}$$
$$\Omega = \{\langle a, q, z2w, p\rangle \mid (a, q, zw, p) \in R_H \text{ for some } a \in V, q \in W - F, w \in T^*, z \in V^*, \text{ and } p \in W\}$$

Define the relation χ from V_H to $\Xi\Omega$ so for every $a \in V$

$$\chi(a) = \{\langle a, q, y1x, p\rangle\langle a, q, y2x, p\rangle \mid \langle a, q, y1x, p\rangle \in \Xi, \langle a, q, y2x, p\rangle \in \Omega, q \in W - F, x \in T^*, y \in V^*, p \in W\}$$

Define the bijection δ from V_H to V', where $V' = \{a' \mid a \in V\}$, as $\delta(a) = a'$. In the standard manner, extend δ so it is defined from V_H^* to V'^*. Finally, define the bijection φ from V_H to V'', where $V'' = \{a'' \mid a \in V\}$, as $\varphi(a) = a''$. In the standard manner, extend φ so it is defined from V_H^* to V''^*.

Define the left-extended queue grammar

$$Q = (V_Q, T_Q, W_Q, F_Q, R_Q, g_Q)$$

so that

$$\begin{aligned} V_Q &= V_H \cup \delta(V_H) \cup \varphi(V_H) \cup \Xi \cup \Omega \\ T_Q &= T_H \\ W_Q &= W_H \cup \alpha(W_H) \cup \beta(W_H) \cup \{!\} \\ F_Q &= \beta(F_H) \\ g_Q &= \delta(g_H) \end{aligned}$$

and R_Q is constructed in this way

(1) if $(a, q, x, p) \in R_H$, where $a \in V, q \in W - F, x \in V^*$, and $p \in W$, add $(\delta(a), q, \delta(x), p)$ and $(\delta(a), \alpha(q), \delta(x), \alpha(p))$ to R_Q;
(2) if $(a, q, xAy, p) \in R_H$, where $a \in V, q \in W - F, x, y \in V^*, A \in V$, and $p \in W$, add $(\delta(a), q, \delta(x)\chi(A)\varphi(y), \alpha(p))$ to R_Q;
(3) if $(a, q, yx, p) \in R_H$, where $a \in V, q \in W - F, y \in V^*, x \in T^*$, and $p \in W$, add $(\langle a, q, y1x, p\rangle, \alpha(q), \varphi(y), !)$ and $(\langle a, q, y2x, p\rangle, !, x, \beta(p))$ to R_Q;
(4) if $(a, q, y, p) \in R_H$, where $a \in V, q \in W - F, y \in T^*$, and $p \in W$, add $(\varphi(a), \beta(q), y, \beta(p))$ to R_Q.

Set $U = \delta(V_H) \cup \Xi$ and $Z = \varphi(V_H) \cup \Omega$. Notice that Q satisfies the second and the third property of Lemma 16.2.4. To demonstrate that the other two properties hold as well, observe that H generates every $z \in L(H)$ in this way

$$
\begin{aligned}
\#g_H &\Rightarrow_H^+ x\#b_1b_2\cdots b_i p_1 \\
&\Rightarrow_H xb_1\#b_2\cdots b_i b_{i+1}\cdots b_n y_1 p_2 \\
&\Rightarrow_H xb_1b_2\#b_3\cdots b_i b_{i+1}\cdots b_n y_1 y_2 p_3 \\
&\ \ \vdots \\
&\Rightarrow_H xb_1b_2\cdots b_{i-1}\#b_i b_{i+1}\cdots b_n y_1 y_2\cdots y_{i-1} p_i \\
&\Rightarrow_H xb_1b_2\cdots b_i\#b_{i+1}\cdots b_n y_1 y_2\cdots y_{i-1} y_i p_{i+1} \\
&\ \ \vdots \\
&\Rightarrow_H xb_1b_2\cdots b_{n-1}\#b_n y_1 y_2\cdots y_{n-1} p_n \\
&\Rightarrow_H xb_1b_2\cdots b_{n-1}b_n\#y_1 y_2\cdots y_n p_{n+1}
\end{aligned}
$$

where $n \geq 1, x \in V^+, b_i \in V$ for $i = 1, \ldots, n$, $y_i \in T^*$ for $i = 1, \ldots, n$, $z = y_1y_2\cdots y_n$, $p_i \in W$ for $i = 1, \ldots, n$, and $p_{n+1} \in F$. Q simulates this generation of z as follows

$$
\begin{aligned}
\#g_Q &\Rightarrow_Q^+ \delta(x)\#\chi(b_1)\varphi(b_2\cdots b_i)\alpha(p_1) \\
&\Rightarrow_Q \delta(x)\langle b_1, p_1, b_{i+1}\cdots b_n 1y_1, p_2\rangle\#\langle b_1, p_1, b_{i+1}\cdots b_n 2y_1, p2\rangle \\
&\qquad \varphi(b_2\cdots b_i b_{i+1}\cdots b_n)! \\
&\Rightarrow_Q \delta(x)\chi(b_1)\#\varphi(b_2\cdots b_n)y_1 p_2 \\
&\Rightarrow_Q \delta(x)\chi(b_1)\varphi(b_2)\#\varphi(b_3\cdots b_n)y_1 y_2 p_3 \\
&\ \ \vdots \\
&\Rightarrow_Q \delta(x)\chi(b_1)\varphi(b_2\cdots b_{n-1})\#\varphi(b_n)y_1 y_2\cdots y_{n-1} p_n \\
&\Rightarrow_Q \delta(x)\chi(b_1)\varphi(b_2\cdots b_n)\#y_1 y_2\cdots y_n p_{n+1}
\end{aligned}
$$

Q makes the first $|x| - 1$ steps of $\#g_Q \Rightarrow_Q^+ \delta(x)\#\chi(b_1)\varphi(b_2\cdots b_i)\alpha(p_1)$ according to rules introduced in (1); in addition, during this derivation, Q makes one step by using a rule introduced in (2). By using rules introduced in (3), Q makes these two steps

$$
\begin{aligned}
&\delta(x)\#\chi(b_1)\varphi(b_2\cdots b_i)\alpha(p_0) && \Rightarrow_Q \\
&\delta(x)\langle b_1, p_1, b_{i+1}\cdots b_n 1y_1, p_2\rangle\#\langle b_1, p_1, b_{i+1}\cdots b_n 2y_1, p_2\rangle \\
&\qquad \varphi(b_2\cdots b_i b_{i+1}\cdots b_n)! && \Rightarrow_Q \\
&\delta(x)\chi(b_1)\#\varphi(b_2\cdots b_n)y_1 p_2
\end{aligned}
$$

with

$$
\chi(b_1) = \langle b_1, p_0, b_{i+1}\cdots b_n 1y_1, p_1\rangle\langle b_1, p_0, b_{i+1}\cdots b_n 2y_1, p2\rangle
$$

Q makes the rest of the derivation by using rules introduced in (4). Based on the previous observation, it easy to see that Q satisfies all the four properties stated in Lemma 16.2.4, whose rigorous proof is left to the reader. □

Lemma 16.2.5. *Let Q be a left-extended queue grammar that satisfies the properties of Lemma 16.2.4. Then, there exist a linear grammar G and a pushdown automaton M such that $L(Q) = L(M, L(G), 3)$.*

Proof. Let $Q = (V_Q, T_Q, W_Q, F_Q, R_Q, g_Q)$ be a left-extended queue grammar satisfying the properties of Lemma 16.2.4. Without any loss of generality, assume that $\{@, £, ¶\} \cap (V \cup W) = \emptyset$. Define the coding ζ from V_Q^* to $\{\langle £as \rangle \mid a \in V_Q\}^*$ as $\zeta(a) = \langle £as \rangle$ (s is used as the start state of the pushdown automaton M defined later in this proof).

Construct the linear grammar $G = (N_G, T_G, P_G, S_G)$ in the following way. Initially, set

$$\begin{aligned} N_G &= \{S_G, \langle ! \rangle, \langle !, 1 \rangle\} \cup \{\langle f \rangle \mid f \in F_Q\} \\ T_G &= \zeta(V_Q) \cup \{\langle £\S s \rangle, \langle £@ \rangle\} \cup \{\langle £\S f \rangle \mid f \in F_Q\} \\ P_G &= \{S_G \to \langle £\S s \rangle \langle f \rangle \mid f \in F_Q\} \cup \{\langle ! \rangle \to \langle !, 1 \rangle \langle £@ \rangle\} \end{aligned}$$

Extend N_G, T_G, and P_G by performing (1) through (3), given next.

(1) For every $(a, p, x, q) \in R_Q$ where $p, q \in W_Q, a \in Z$, and $x \in T^*$,

$$\begin{aligned} N_G &= N_G \\ N_G &= N_G \cup \{\langle apxqk \rangle \mid k = 0, \dots, |x|\} \cup \{\langle p \rangle, \langle q \rangle\} \\ T_G &= T_G \cup \{\langle £\,\mathrm{sym}(y, k) \rangle \mid k = 1, \dots, |y|\} \cup \{\langle £apxq \rangle\} \\ P_G &= P_G \cup \{\langle q \rangle \to \langle apxq|x| \rangle \langle £apxq \rangle, \langle apxq0 \rangle \to \langle p \rangle\} \\ &\quad \cup \{\langle apxqk \rangle \to \langle apxq(k-1) \rangle \langle £\,\mathrm{sym}(x, k) \rangle \mid k = 1, \dots, |x|\} \end{aligned}$$

(2) For every $(a, p, x, q) \in R_Q$ with $p, q \in W_Q, a \in U$, and $x \in V_Q^*$,

$$\begin{aligned} N_G &= N_G \\ N_G &= N_G \cup \{\langle p, 1 \rangle, \langle q, 1 \rangle\} \\ P_G &= P_G \cup \{\langle q, 1 \rangle \to \mathrm{rev}(\zeta(x)) \langle p, 1 \rangle \zeta(a)\} \end{aligned}$$

(3) For every $(a, p, x, q) \in R_Q$ with $ap = q_Q, p, q \in W_Q$, and $x \in V_Q^*$,

$$\begin{aligned} N_G &= N_G \\ N_G &= N_G \cup \{\langle q, 1 \rangle\} \\ P_G &= P_G \cup \{\langle q, 1 \rangle \to \mathrm{rev}(x) \langle £\$s \rangle\} \end{aligned}$$

The construction of G is completed. Set $\Psi = T_G$. Ψ represents the alphabet of rule labels corresponding to the rules of the pushdown automaton M, defined as

$$M = (Q_M, \Sigma_M, \Gamma_M, R_M, s_M, S_M, \{\rceil\})$$

Throughout the rest of this proof, s_M is abbreviated to s. Initially, set

$$\begin{aligned} Q_M &= \{s, \langle\P!\rangle, \lfloor, \rceil\} \\ \Sigma_M &= T_Q \\ \Gamma_M &= \{S_M, \S\} \cup V_Q \\ R_M &= \{\langle\pounds\S s\rangle\colon S_M s \to \S s\} \cup \{\langle\pounds\S f\rangle.\S\langle\P f\rangle \to \rceil \mid f \in F_M\} \end{aligned}$$

Extend Q_M and R_M by performing (A) through (D), given next.

(A) Set $R_M = R_M \cup \{\langle\pounds bs\rangle\colon as \to abs \mid a \in \Gamma_M - \{S_M\}, b \in \Gamma_M - \{\$\}\}$.
(B) Set $R_M = R_M \cup \{\langle\pounds\$s\rangle\colon as \to a\lfloor \mid a \in V_Q\} \cup \{\langle\pounds a\rangle.a\lfloor \to \lfloor \mid a \in V_Q\}$.
(C) Set $R_M = R_M \cup \{\langle\pounds @\rangle.a\lfloor \to a\langle\P!\rangle \mid a \in Z\}$.
(D) For every $(a, p, x, q) \in R_Q$, where $p, q \in W_Q, a \in Z, x \in T_Q^*$, set

$$\begin{aligned} Q_M &= Q_M \cup \{\langle\P p\rangle\} \cup \{\langle\P qu\rangle \mid u \in \text{prefix}(x)\} \\ R_M &= R_M \cup \{\langle\pounds b\rangle.a\langle\P qy\rangle b \to a\langle\P qyb\rangle \mid b \in T_Q, y \in T_Q^*, yb \in \text{prefix}(x)\} \\ &\quad \cup \{\langle\pounds apxq\rangle.a\langle\P qx\rangle \to \langle\P p\rangle\} \end{aligned}$$

The construction of M is completed. Notice that several components of G and M have this form: $\langle x\rangle$. Intuitively, if x begins with £, then $\langle x\rangle \in T_G$. If x begins with ¶, then $\langle x\rangle \in Q_M$. Finally, if x begins with a symbol different from £ or ¶, then $\langle x\rangle \in N_G$.

First, we only sketch the reason why $L(Q)$ contains $L(M, L(G), 3)$. According to a string from $L(G)$, M accepts every string w as

$$\begin{aligned} \S w_1 \cdots w_{m-1} w_m &\vdash_M^+ \S b_m \cdots b_1 a_n \cdots a_1 s w_1 \cdots w_{m-1} w_m \\ &\vdash_M \S b_m \cdots b_1 a_n \cdots a_1 \lfloor w_1 \cdots w_{m-1} w_m \\ &\vdash_M^n \S b_m \cdots b_1 \lfloor w_1 \cdots w_{m-1} w_m \\ &\vdash_M \S b_m \cdots b_1 \langle\P q_1\rangle w_1 \cdots w_{m-1} w_m \\ &\vdash_M^{|w_1|} \S b_m \cdots b_1 \langle\P q_1 w_1\rangle w_2 \cdots w_{m-1} w_m \\ &\vdash_M \S b_m \cdots b_2 \langle\P q_2\rangle w_2 \cdots w_{m-1} w_m \\ &\vdash_M^{|w_2|} \S b_m \cdots b_2 \langle\P q_2 w_2\rangle w_3 \cdots w_{m-1} w_m \\ &\vdash_M \S b_m \cdots b_3 \langle\P q_3\rangle w_3 \cdots w_{m-1} w_m \\ &\vdots \\ &\vdash_M \S b_m \langle\P q_m\rangle w_m \\ &\vdash_M^{|w_m|} \S b_m \langle\P q_m w_m\rangle \\ &\vdash_M \S\langle\P q_{m+1}\rangle \\ &\vdash_M \rceil \end{aligned}$$

where $w = w_1 \cdots w_{m-1} w_m, a_1 \cdots a_n b_1 \cdots b_m = x_1 \cdots x_{n+1}$, and R_Q contains $(a_0, p_0, x_1, p_1), (a_1, p_1, x_2, p_2), \ldots, (a_n, p_n, x_{n+1}, q_1), (b_1, q_1, w_1, q_2), (b_2, q_2, w_2, q_3), \ldots, (b_m, q_m, w_m, q_{m+1})$. According to these members of R_Q, Q makes

$$
\begin{array}{lll}
\#a_0 p_0 & \Rightarrow_Q a_0\#y_0x_1p_1 & [(a_0, p_0, x_1, p_1)] \\
& \Rightarrow_Q a_0a_1\#y_1x_2p_2 & [(a_1, p_1, x_2, p_2)] \\
& \Rightarrow_Q a_0a_1a_2\#y_2x_3p_3 & [(a_2, p_2, x_3, p_3)] \\
& \vdots & \\
& \Rightarrow_Q a_0a_1a_2\cdots a_{n-1}\#y_{n-1}x_np_n & [(a_{n-1}, p_{n-1}, x_n, p_n)] \\
& \Rightarrow_Q a_0a_1a_2\cdots a_n\#y_nx_{n+1}q_1 & [(a_n, p_n, x_{n+1}, q_1)] \\
& \Rightarrow_Q a_0\cdots a_nb_1\#b_2\cdots b_mw_1q_2 & [(b_1, q_1, w_1, q_2)] \\
& \Rightarrow_Q a_0\cdots a_nb_1b_2\#b_3\cdots b_mw_1w_2q_3 & [(b_2, q_2, w_2, q_3)] \\
& \vdots & \\
& \Rightarrow_Q a_0\cdots a_nb_1\cdots b_{m-1}\#b_mw_1w_2\cdots w_{m-1}q_m & [(b_{m-1}, q_{m-1}, w_{m-1}, q_m)] \\
& \Rightarrow_Q a_0\cdots a_nb_1\cdots b_m\#w_1w_2\cdots w_mq_{m+1} & [(b_m, q_m, w_m, q_{m+1})]
\end{array}
$$

Therefore, $L(M, L(G), 3) \subseteq L(Q)$.

More formally, to demonstrate that $L(Q)$ contains $L(M, L(G), 3)$, consider any $h \in L(G)$. G generates h as

$$
\begin{array}{ll}
S_G \Rightarrow_G & \langle £\S s\rangle\langle q_{m+1}\rangle \\
\Rightarrow_G^{|w_m|+1} & \langle £\S s\rangle\langle q_m\rangle t_m\langle £b_mq_mw_mq_{m+1}\rangle \\
\Rightarrow_G^{|w_{m-1}|+1} & \langle £\S s\rangle\langle q_{m-1}\rangle t_{m-1}\langle £b_{m-1}q_{m-1}w_{m-1}q_m\rangle t_m\langle £b_mq_mw_mq_{m+1}\rangle \\
\vdots & \\
\Rightarrow_G^{|w_1|+1} & \langle £\S s\rangle\langle q_1\rangle o \\
\Rightarrow_G^{|w_1|+1} & \langle £\S s\rangle\langle q_1, 1\rangle\langle £@\rangle o \\
& [\langle q_1\rangle \to \langle q_1, 1\rangle\langle £@\rangle] \\
\Rightarrow_G & \langle £\S s\rangle\zeta(\mathrm{rev}(x_{n+1}))\langle p_n, 1\rangle\langle £a_n\rangle\langle £@\rangle o \\
& [\langle q_1, 1\rangle \to \mathrm{rev}(\zeta(x_{n+1}))\langle p_n, 1\rangle\langle £a_n\rangle\langle £@\rangle] \\
\Rightarrow_G & \langle £\S s\rangle\zeta(\mathrm{rev}(x_nx_{n+1}))\langle p_{n-1}, 1\rangle\langle £a_{n-1}\rangle\langle £a_n\rangle\langle £@\rangle o \\
& [\langle p_n, 1\rangle \to \mathrm{rev}(\zeta(x_n))\langle p_{n-1}, 1\rangle\langle £a_{n-1}\rangle] \\
\vdots & \\
\Rightarrow_G & \langle £\S s\rangle\zeta(\mathrm{rev}(x_2\cdots x_nx_{n+1}))\langle p1, 1\rangle\langle £a_1\rangle\langle £a_2\rangle\cdots\langle £a_n\rangle\langle £@\rangle o \\
& [\langle p_2, 1\rangle \to \mathrm{rev}(\zeta(x_2))\langle p_1, 1\rangle\langle £a_1\rangle] \\
\Rightarrow_G & \langle £\S s\rangle\zeta(\mathrm{rev}(x_1\cdots x_nx_{n+1}))\langle £\$ s\rangle\langle £a_1\rangle\langle £a_2\rangle\cdots\langle £a_n\rangle\langle £@\rangle o \\
& [\langle p_1, 1\rangle \to \mathrm{rev}(\zeta(x_1))\langle £\$ s\rangle]
\end{array}
$$

where

$$
\begin{aligned}
&n, m \geq 1 \\
&a_i \in U \text{ for } i = 1, \dots, n \\
&b_k \in Z \text{ for } k = 1, \dots, m \\
&x_l \in V^* \text{ for } l = 1, \dots, n+1 \\
&p_i \in W \text{ for } i = 1, \dots, n \\
&q_l \in W \text{ for } l = 1, \dots, m+1 \text{ with } q_1 = \,! \text{ and } q_{m+1} \in F
\end{aligned}
$$

and

$$
\begin{aligned}
t_k &= \langle £\,\mathrm{sym}(w_k, 1)\rangle \cdots \langle £\,\mathrm{sym}(w_k, |w_k| - 1)\rangle \langle £\,\mathrm{sym}(w_k, |w_k|)\rangle \\
&\quad \text{for } k = 1, \dots, m; \\
o &= t_1 \langle £ b_1 q_1 w_1 q_2\rangle \cdots \langle £§s\rangle \langle q_{m-1}\rangle t_{m-1} \langle £ b_{m-1} q_{m-1} w_{m-1} q_m\rangle t_m \\
&\quad \langle £ b_m q_m w_m q_{m+1}\rangle; \\
h &= \langle £§s\rangle \zeta(\mathrm{rev}(x_1 \cdots x_n x_{n+1}))\langle £\$\rangle \langle £a_1\rangle \langle £a_2\rangle \cdots \langle £a_n\rangle \langle £@\rangle o
\end{aligned}
$$

We describe this derivation in a greater detail. Initially, G makes $S_G \Rightarrow_G \langle £§s\rangle\langle q_{m+1}\rangle$ according to $S_G \to \langle £§s\rangle\langle q_{m+1}\rangle$. Then, G makes

$$
\begin{aligned}
& \langle £§s\rangle \langle q_{m+1}\rangle \\
\Rightarrow_G^{|w_m|+1} & \langle £§s\rangle \langle q_m\rangle t_m \langle £ b_m q_m w_m q_{m+1}\rangle \\
\Rightarrow_G^{|w_{m-1}|+1} & \langle £§s\rangle \langle q_{m-1}\rangle t_{m-1} \langle £ b_{m-1} q_{m-1} w_{m-1} q_m\rangle t_m \langle £ b_m q_m w_m q_{m+1}\rangle \\
& \vdots \\
\Rightarrow_G^{|w_1|+1} & \langle £§s\rangle \langle q_1\rangle o
\end{aligned}
$$

according to rules introduced in (1). Then, G makes

$$
\langle £§s\rangle \langle q_1\rangle o \Rightarrow_G \langle £§s\rangle \langle q_1, 1\rangle \langle £@\rangle o
$$

according to $\langle !\rangle \to \langle !, 1\rangle\langle £@\rangle$ (recall that $q_1 = !$). After this step, G makes

$$
\begin{aligned}
& \langle £§s\rangle \langle q_1, 1\rangle \langle £@\rangle o \\
\Rightarrow_G\ & \langle £§s\rangle \zeta(\mathrm{rev}(x_{n+1}))\langle p_n, 1\rangle \langle £a_n\rangle \langle £@\rangle o \\
\Rightarrow_G\ & \langle £§s\rangle \zeta(\mathrm{rev}(x_n x_{n+1}))\langle p_{n-1}, 1\rangle \langle £a_{n-1}\rangle \langle £a_n\rangle \langle £@\rangle o \\
& \vdots \\
\Rightarrow_G\ & \langle £§s\rangle \zeta(\mathrm{rev}(x_2 \cdots x_n x_{n+1}))\langle p_1, 1\rangle \langle £a_1\rangle \langle £a_2\rangle \cdots \langle £a_n\rangle \langle £@\rangle o
\end{aligned}
$$

according to rules introduced in (2). Finally, according to $\langle p_1, 1\rangle \to \mathrm{rev}(\zeta(x_1))\langle £\$\rangle$, which is introduced in (3), G makes

$$
\begin{aligned}
& \langle £§s\rangle \zeta(\mathrm{rev}(x_2 \cdots x_n x_{n+1}))\langle p_1, 1\rangle \langle £a_1\rangle \langle £a_2\rangle \cdots \langle £a_n\rangle \langle £@\rangle o \\
\Rightarrow_G\ & \langle £§s\rangle \zeta(\mathrm{rev}(x_1 \cdots x_n x_{n+1}))\langle £\$\rangle \langle £a_1\rangle \langle £a_2\rangle \cdots \langle £a_n\rangle \langle £@\rangle o
\end{aligned}
$$

If $a_1 \cdots a_n b_1 \cdots b_m$ differs from $x_1 \cdots x_{n+1}$, then M does not accept according to h. Assume that $a_1 \cdots a_n b_1 \cdots b_m = x_1 \cdots x_{n+1}$. At this point, according to h, M makes this sequence of moves

$$
\begin{aligned}
\S w_1 \cdots w_{m-1} w_m & \vdash_M^+ \S b_m \cdots b_1 a_n \cdots a_1 s w_1 \cdots w_{m-1} w_m \\
& \vdash_M \S b_m \cdots b_1 a_n \cdots a_1 \lfloor w_1 \cdots w_{m-1} w_m \\
& \vdash_M^n \S b_m \cdots b_1 \lfloor w_1 \cdots w_{m-1} w_m \\
& \vdash_M \S b_m \cdots b_1 \langle \P q_1 \rangle w_1 \cdots w_{m-1} w_m \\
& \vdash_M^{|w_1|} \S b_m \cdots b_1 \langle \P q_1 w_1 \rangle w_2 \cdots w_{m-1} w_m \\
& \vdash_M \S b_m \cdots b_2 \langle \P q_2 \rangle w_2 \cdots w_{m-1} w_m \\
& \vdash_M^{|w_2|} \S b_m \cdots b_2 \langle \P q_2 w_2 \rangle w_3 \cdots w_{m-1} w_m \\
& \vdash_M \S b_m \cdots b_3 \langle \P q_3 \rangle w_3 \cdots w_{m-1} w_m \\
& \vdots \\
& \vdash_M \S b_m \langle \P q_m \rangle w_m \\
& \vdash_M^{|w_m|} \S b_m \langle \P q_m w_m \rangle \\
& \vdash_M \S \langle \P q_{m+1} \rangle \\
& \vdash_M \urcorner
\end{aligned}
$$

In other words, according to h, M accepts $w_1 \cdots w_{m-1} w_m$. Return to the generation of h in G. By the construction of P_G, this generation implies that R_Q contains (a_0, p_0, x_1, p_1), (a_1, p_1, x_2, p_2), $\ldots, (a_{j-1}, p_{j-1}, x_j, p_j), \ldots,$ (a_n, p_n, x_{n+1}, q_1), $(b_1, q_1, w_1, q_2), (b_2, q_2, w_2, q_3), \ldots, (b_m, q_m, w_m, q_{m+1})$.

Thus, in Q,

$$
\begin{aligned}
\#a_0 p_0 & \Rightarrow_Q a_0 \# y_0 x_1 p_1 && [(a_0, p_0, x_1, p_1)] \\
& \Rightarrow_Q a_0 a_1 \# y_1 x_2 p_2 && [(a_1, p_1, x_2, p_2)] \\
& \Rightarrow_Q a_0 a_1 a_2 \# y_2 x_3 p_3 && [(a_2, p_2, x_3, p_3)] \\
& \vdots \\
& \Rightarrow_Q a_0 a_1 a_2 \cdots a_{n-1} \# y_{n-1} x_n p_n && [(a_{n-1}, p_{n-1}, x_n, p_n)] \\
& \Rightarrow_Q a_0 a_1 a_2 \cdots a_n \# y_n x_{n+1} q_1 && [(a_n, p_n, x_{n+1}, q_1)] \\
& \Rightarrow_Q a_0 \cdots a_n b_1 \# b_2 \cdots b_m w_1 q_2 && [(b_1, q_1, w_1, q_2)] \\
& \Rightarrow_Q a_0 \cdots a_n b_1 b_2 \# b_3 \cdots b_m w_1 w_2 q_3 && [(b_2, q_2, w_2, q_3)] \\
& \vdots \\
& \Rightarrow_Q a_0 \cdots a_n b_1 \cdots b_{m-1} \# b_m w_1 w_2 \cdots w_{m-1} q_m && [(b_{m-1}, q_{m-1}, w_{m-1}, q_m)] \\
& \Rightarrow_Q a_0 \cdots a_n b_1 \cdots b_m \# w_1 w_2 \cdots w_m q_{m+1} && [(b_m, q_m, w_m, q_{m+1})]
\end{aligned}
$$

Therefore, $w_1 w_2 \cdots w_m \in L(Q)$. Consequently, $L(M, L(G), 3) \subseteq L(Q)$. A proof that $L(Q) \subseteq L(M, L(G), 3)$ is left to the reader. As $L(Q) \subseteq L(M, L(G), 3)$ and $L(M, L(G), 3) \subseteq L(Q)$, $L(Q) = L(M, L(G), 3)$. Therefore, Lemma 16.2.5 holds. □

Theorem 16.2.6. *For* $i \in \{1, 2, 3\}$, $\mathbf{RE} = \mathbf{RPDA}(\mathbf{LIN}, i)$.

Proof. Obviously, $\mathbf{RPDA}(\mathbf{LIN}, 3) \subseteq \mathbf{RE}$. To prove that $\mathbf{RE} \subseteq \mathbf{RPDA}(\mathbf{LIN}, 3)$, consider any recursively enumerable language $L \in \mathbf{RE}$. By Theorem 3.3.17, $L(Q) = L$, for a queue grammar Q. Clearly, there exists a left-extended queue grammar Q' so that $L(Q) = L(Q')$. Furthermore, by Lemmas 16.2.4 and 16.2.5, $L(Q') = L(M, L(G), 3)$, for a linear grammar G and a pushdown automaton M. Thus, $L = L(M, L(G), 3)$. Hence, $\mathbf{RE} \subseteq \mathbf{RPDA}(\mathbf{LIN}, 3)$. As $\mathbf{RPDA}(\mathbf{LIN}, 3) \subseteq \mathbf{RE}$ and $\mathbf{RE} \subseteq \mathbf{RPDA}(\mathbf{LIN}, 3)$, $\mathbf{RE} = \mathbf{RPDA}(\mathbf{LIN}, 3)$.

By analogy with the demonstration of $\mathbf{RE} = \mathbf{RPDA}(\mathbf{LIN}, 3)$, we can prove that $\mathbf{RE} = \mathbf{RPDA}(\mathbf{LIN}, i)$ for $i = 1, 2$. □

16.2.4 One-Turn Linear-Controlled Pushdown Automata

In the present section, we continue with the discussion of regulated pushdown automata, but we narrow our attention to their special cases–*one-turn regulated pushdown automata*. To give an insight into one-turn pushdown automata, consider two consecutive moves made by an ordinary pushdown automaton M. If during the first move M does not shorten its pushdown and during the second move it does, then M makes a *turn* during the second move. A pushdown automaton is *one-turn* if it makes no more than one turn with its pushdown during any computation starting from a start configuration. Recall that one-turn pushdown automata characterize the family of linear languages (see [2]) while their unrestricted versions characterize the family of context-free languages (see Theorem 3.4.12). As a result, one-turn pushdown automata are less powerful than the pushdown automata.

As the most surprising result, we demonstrate that linear-regulated versions of one-turn pushdown automata characterize the family of recursively enumerable languages. Thus, as opposed to the ordinary one-turn pushdown automata, one-turn linear-regulated pushdown automata are as powerful as linear-regulated pushdown automata that can make any number of turns.

In fact, this characterization holds even for some restricted versions of one-turn regulated pushdown automata, including their atomic and reduced versions, which are sketched next.

(I) During a move, an *atomic* one-turn regulated pushdown automaton changes a state and, in addition, performs exactly one of the following three actions:

- it pushes a symbol onto the pushdown;
- it pops a symbol from the pushdown;
- it reads an input symbol.

(II) A *reduced* one-turn regulated pushdown automaton has a limited number of some components, such as the number of states, pushdown symbols, or transition rules.

We establish the above-mentioned characterization in terms of acceptance by final state and empty pushdown, acceptance by final state, and acceptance by empty pushdown.

Definition 16.2.7. An *atomic pushdown automaton* is a septuple

$$M = \left(Q, \Sigma, \Gamma, R, s, \$, F\right)$$

where

- $Q, \Sigma, \Gamma, s \in Q$, and $F \subseteq Q$ are defined as in a pushdown automaton;
- $\$$ is the *pushdown-bottom marker*, $\$ \notin Q \cup \Sigma \cup \Gamma$;
- R is a finite set of *rules* of the form $Apa \rightarrow wq$, where $p, q \in Q, A, w \in \Gamma \cup \{\varepsilon\}, a \in \Sigma \cup \{\varepsilon\}$, such that $|Aaw| = 1$. That is, R is a finite set of rules such that each of them has one of these forms

(1) $Ap \rightarrow q$ (*popping rule*)
(2) $p \rightarrow wq$ (*pushing rule*)
(3) $pa \rightarrow q$ (*reading rule*)

Let Ψ be an alphabet of *rule labels* such that $\mathrm{card}(\Psi) = \mathrm{card}(R)$, and ψ be a bijection from R to Ψ. For simplicity, to express that ψ maps a rule, $Apa \rightarrow wq \in R$, to r, where $r \in \Psi$, we write $r\colon Apa \rightarrow wq \in R$; in other words, $r\colon Apa \rightarrow wq$ means $\psi(Apa \rightarrow wq) = r$. A *configuration* of M, χ, is any string from $\{\$\}\Gamma^* Q \Sigma^*$; χ is a *start configuration* if $\chi = \$sw$, where $w \in \Sigma^*$. For every $x \in \Gamma^*, y \in \Sigma^*$, and $r\colon Apa \rightarrow wq \in R, M$ makes a *move* from configuration $\$xApay$ to configuration $\$xwqy$ according to r, written as

$$\$xApay \vdash_M \$xwqy \; [r]$$

or, simply, $\$xApay \vdash_M \$xwqy$.

Let χ be any configuration of M. M makes *zero moves* from χ to χ according to ε, symbolically written as

$$\chi \vdash_M^0 \chi \; [\varepsilon]$$

Let there exist a sequence of configurations $\chi_0, \chi_1, \dots, \chi_n$ for some $n \geq 1$ such that $\chi_{i-1} \vdash_M \chi_i \; [r_i]$, where $r_i \in \Psi$, for $i = 1, \dots, n$, then M makes n *moves* from χ_0 to χ_n according to $r_1 \cdots r_n$, symbolically written as

$$\chi_0 \vdash_M^n \chi_n \; [r_1 \cdots r_n]$$

or, simply, $\chi_0 \vdash_M^n \chi_n$. Define $\vdash_M^*$ and $\vdash_M^+$ in the standard manner.

Let $x, x', x'' \in \Gamma^*, y, y', y'' \in \Sigma^*, q, q', q'' \in Q$, and $\$xqy \vdash_M \$x'q'y' \vdash_M \$x''q''y''$. If $|x| \leq |x'|$ and $|x'| > |x''|$, then $\$x'q'y' \vdash_M \$x''q''y''$ is a *turn*. If M makes no more than one turn during any sequence of moves starting from a start configuration, then M is said to be *one-turn*.

Let Ξ be a *control language* over Ψ; that is, $\Xi \subseteq \Psi^*$. With Ξ, M defines the following three types of accepted languages:

- $L(M, \Xi, 1)$—the *language accepted by final state*
- $L(M, \Xi, 2)$—the *language accepted by empty pushdown*
- $L(M, \Xi, 3)$—the *language accepted by final state and empty pushdown*

defined as follows. Let $\chi \in \{\$\}\Gamma^* Q \Sigma^*$. If $\chi \in \{\$\}\Gamma^* F$, $\chi \in \{\$\}Q$, $\chi \in \{\$\}F$, then χ is a 1-*final configuration*, 2-*final configuration*, 3-*final configuration*, respectively. For $i = 1, 2, 3$, define $L(M, \Xi, i)$ as

$$L(M, \Xi, i) = \big\{w \mid w \in \Sigma^*, \text{ and } sw \vdash_M^* \chi\ [\sigma] \\ \text{for an } i\text{-final configuration } \chi \text{ and } \sigma \in \Xi\big\}$$

The pair (M, Ξ) is called a *controlled pushdown automaton*. □

For any family of language $\mathscr{L}$ and $i \in \{1, 2, 3\}$, define

$$\textbf{OA-RPDA}(\mathscr{L}, i) = \big\{L \mid L = L(M, \Xi, i), \text{ where } M \text{ is a one-turn} \\ \text{atomic pushdown automaton and } \Xi \in \mathscr{L}\big\}$$

We next prove that one-turn atomic pushdown automata regulated by linear languages characterize the family of recursively enumerable languages. In fact, these automata need no more than one state and two pushdown symbols to achieve this characterization.

Lemma 16.2.8. *Let Q be a left-extended queue grammar satisfying the properties of Lemma 16.2.4. Then, there is a linear grammar G and a one-turn atomic pushdown automaton $M = (\{\llcorner\}, \tau, \{0, 1\}, H, \llcorner, \$, \{\llcorner\})$ such that* $\mathrm{card}(H) = \mathrm{card}(\tau) + 4$ *and* $L(Q) = L(M, L(G), 3)$.

Proof. Let $Q = (V, \tau, W, F, R, g)$ be a queue grammar satisfying the properties of Lemma 16.2.4. For some $n \geq 1$, introduce a homomorphism f from R to X, where

$$X = \{1\}^*\{0\}\{1\}^*\{1\}^n \cap \{0, 1\}^{2n}$$

Extend f so it is defined from R^* to X^*. Define the substitution h from V^* to X^* as

$$h(a) = \big\{f(r) \mid r = (a, p, x, q) \in R \text{ for some } p, q \in W, x \in V^*\big\}$$

Define the coding d from $\{0, 1\}^*$ to $\{2, 3\}^*$ as $d(0) = 2, d(1) = 3$. Construct the linear grammar

$$G = \big(N, T, P, S\big)$$

as follows. Initially, set

$$\begin{aligned} T &= \{0, 1, 2, 3\} \cup \tau \\ N &= \{S\} \cup \{\tilde{q} \mid q \in W\} \cup \{\hat{q} \mid q \in W\} \\ P &= \{S \to \tilde{f} \mid f \in F\} \cup \{\tilde{!} \to \hat{!}\} \end{aligned}$$

Extend P by performing (1) through (3), given next.

(1) For every $r = (a, p, x, q) \in R, p, q \in w, x \in T^*$

$$P = P \cup \{\tilde{q} \to \tilde{p}d(f(r))x\}$$

(2) For every $(a, p, x, q) \in R$,

$$P = P \cup \{\hat{q} \to y\hat{p}b \mid y \in \text{rev}(h(x)), b \in h(a)\}$$

(3) For every $(a, p, x, q) \in R, ap = S, p, q \in W, x \in V^*$,

$$P = P \cup \{\hat{q} \to y \mid y \in \text{rev}(h(x))\}$$

Define the atomic pushdown automaton

$$M = (\{\lfloor\}, \tau, \{0, 1\}, H, \lfloor, \$, \{\lfloor\})$$

where H contains the following transition rules

$$\begin{aligned} &0\colon \lfloor \to 0\lfloor \\ &1\colon \lfloor \to 1\lfloor \\ &2\colon 0\lfloor \to \lfloor \\ &3\colon 1\lfloor \to \lfloor \\ &a\colon \lfloor a \to \lfloor \text{ for every } a \in \tau \end{aligned}$$

We next demonstrate that $L(M, L(G), 3) = L(Q)$. Observe that M accepts every string $w = w_1 \cdots w_{m-1}w_m$ as

$$\begin{aligned} \$w_1 \cdots w_{m-1}w_m &\vdash_M^+ \$\bar{b}_m \cdots \bar{b}_1\bar{a}_n \cdots \bar{a}_1\lfloor w_1 \cdots w_{m-1}w_m \\ &\vdash_M \$\bar{b}_m \cdots \bar{b}_1\bar{a}_n \cdots \bar{a}_1\lfloor w_1 \cdots w_{m-1}w_m \\ &\vdash_M^n \$\bar{b}_m \cdots \bar{b}_1\lfloor w_1 \cdots w_{m-1}w_m \\ &\vdash_M \$\bar{b}_m \cdots \bar{b}_1\lfloor w_1 \cdots w_{m-1}w_m \\ &\vdash_M^{|w_1|} \$\bar{b}_m \cdots \bar{b}_1\lfloor w_2 \cdots w_{m-1}w_m \\ &\vdash_M \$\bar{b}_m \cdots \bar{b}_2\lfloor w_2 \cdots w_{m-1}w_m \\ &\vdash_M^{|w_2|} \$\bar{b}_m \cdots \bar{b}_2\lfloor w_3 \cdots w_{m-1}w_m \\ &\vdash_M \$\bar{b}_m \cdots \bar{b}_3\lfloor w_3 \cdots w_{m-1}w_m \\ &\ \ \vdots \\ &\vdash_M \$\bar{b}_m\lfloor w_m \\ &\vdash_M^{|w_m|} \$\bar{b}_m\lfloor \\ &\vdash_M \$\lfloor \end{aligned}$$

according to a string of the form $\beta\alpha\alpha'\beta' \in L(G)$ where

$$\begin{aligned}
\beta &= \operatorname{rev}(f(r_m))\operatorname{rev}(f(r_{m-1}))\cdots\operatorname{rev}(f(r_1))\\
\alpha &= \operatorname{rev}(f(t_n))\operatorname{rev}(f(t_{n-1}))\cdots\operatorname{rev}(f(t_1))\\
\alpha' &= f(t_0)f(t_1)\cdots f(t_n)\\
\beta' &= d(f(r_1))w_1d(f(r_2))w_2\cdots d(f(r_m))w_m
\end{aligned}$$

for some $m, n \geq 1$ so that for $i = 1, \ldots, m$,

$$t_i = (b_i, q_i, w_i, q_{i+1}) \in R, b_i \in V - \tau, q_i, q_{i+1} \in Q, \bar{b}_i = f(t_i)$$

and for $j = 1, \ldots, n+1, r_j = (a_{j-1}, p_{j-1}, x_j, p_j), a_{j-1} \in V - \tau, p_{j-1}, p_j \in Q - F, x_j \in (V-\tau)^*, \bar{a}_j = f(r_j), q_{m+1} \in F, \bar{a}_0 p_0 = g$. Thus, in Q,

$$\begin{aligned}
\#a_0p_0 &\Rightarrow_Q a_0\#y_0x_1p_1 && [(a_0, p_0, x_1, p_1)]\\
&\Rightarrow_Q a_0a_1\#y_1x_2p_2 && [(a_1, p_1, x_2, p_2)]\\
&\Rightarrow_Q a_0a_1a_2\#y_2x_3p_3 && [(a_2, p_2, x_3, p_3)]\\
&\ \ \vdots\\
&\Rightarrow_Q a_0a_1a_2\cdots a_{n-1}\#y_{n-1}x_np_n && [(a_{n-1}, p_{n-1}, x_n, p_n)]\\
&\Rightarrow_Q a_0a_1a_2\cdots a_n\#y_nx_{n+1}q_1 && [(a_n, p_n, x_{n+1}, q_1)]\\
&\Rightarrow_Q a_0\cdots a_nb_1\#b_2\cdots b_mw_1q_2 && [(b_1, q_1, w_1, q_2)]\\
&\Rightarrow_Q a_0\cdots a_nb_1b_2\#b_3\cdots b_mw_1w_2q_3 && [(b_2, q_2, w_2, q_3)]\\
&\ \ \vdots\\
&\Rightarrow_Q a_0\cdots a_nb_1\cdots b_{m-1}\#b_mw_1\cdots w_{m-1}q_m && [(b_{m-1}, q_{m-1}, w_{m-1}, q_m)]\\
&\Rightarrow_Q a_0\cdots a_nb_1\cdots b_m\#w_1\cdots w_mq_{m+1} && [(b_m, q_m, w_m, q_{m+1})]
\end{aligned}$$

Therefore, $w_1w_2\cdots w_m \in L(Q)$. Consequently, $L(M, L(G), 3) \subseteq L(Q)$. A proof of $L(Q) \subseteq L(M, L(G), 3)$ is left to the reader.

As $L(Q) \subseteq L(M, L(G), 3)$ and $L(M, L(G), 3) \subseteq L(Q), L(Q) = L(M, L(G), 3)$. Observe that M is one-turn. Furthermore, $\operatorname{card}(H) = \operatorname{card}(\tau) + 4$. Thus, Lemma 16.2.8 holds. □

Theorem 16.2.9. *For every $L \in$ **RE**, there is a linear language Ξ and a one-turn atomic pushdown automaton $M = (Q, \Sigma, \Gamma, R, s, \$, F)$ such that $\operatorname{card}(Q) \leq 1, \operatorname{card}(\Gamma) \leq 2, \operatorname{card}(R) \leq \operatorname{card}(\Sigma) + 4$, and $L(M, \Xi, 3) = L$.*

Proof. By Theorem 3.3.17, for every $L \in$ **RE**, there is a queue grammar Q such that $L = L(Q)$. Clearly, there is a left-extended queue grammar Q' such that $L(Q) = L(Q')$. Thus, this theorem follows from Lemmas 16.2.4 and 16.2.8. □

Theorem 16.2.10. *For every $L \in$ **RE**, there is a linear language Ξ and a one-turn atomic pushdown automaton $M = (Q, \Sigma, \Gamma, R, s, \$, F)$ such that $\operatorname{card}(Q) \leq 1, \operatorname{card}(\Gamma) \leq 2, \operatorname{card}(R) \leq \operatorname{card}(\Sigma) + 4$, and $L(M, \Xi, 1) = L$.*

Proof. This theorem can be proved by analogy with the proof of Theorem 16.2.9. □

Theorem 16.2.11. *For every* $L \in \mathbf{RE}$*, there is a linear language* Ξ *and a one-turn atomic pushdown automaton* $M = (Q, \Sigma, \Gamma, R, s, \$, F)$ *such that* $\mathrm{card}(Q) \leq 1, \mathrm{card}(\Gamma) \leq 2, \mathrm{card}(R) \leq \mathrm{card}(\Sigma) + 4$*, and* $L(M, \Xi, 2) = L$.

Proof. This theorem can be proved by analogy with the proof of Theorem 16.2.9. □

From the previous three theorems, we obtain the following corollary.

Corollary 16.2.12. *For* $i \in \{1, 2, 3\}$, $\mathbf{RE} = \mathbf{OA}$ - $\mathbf{RPDA}(\mathbf{LIN}, i)$. □

We close this section by suggesting some open problem areas concerning regulated automata.

Open Problem 16.2.13. For $i = 1, \ldots, 3$, consider $\mathbf{RPDA}(\mathscr{L}, i)$, where $\mathscr{L}$ is a language family satisfying $\mathbf{REG} \subset \mathscr{L} \subset \mathbf{LIN}$. For instance, consider $\mathscr{L}$ as the family of *minimal linear languages* (see p. 76 in [3]). Compare **RE** with $\mathbf{RPDA}(\mathscr{L}, i)$. □

Open Problem 16.2.14. Investigate special cases of regulated pushdown automata, such as their deterministic versions. □

Open Problem 16.2.15. By analogy with regulated pushdown automata, introduce and study some other types of regulated automata. □

References

1. Dassow, J., Păun, G.: Regulated Rewriting in Formal Language Theory. Springer, New York (1989)
2. Harrison, M.: Introduction to Formal Language Theory. Addison-Wesley, Boston (1978)
3. Salomaa, A.: Formal Languages. Academic, London (1973)
4. Wood, D.: Theory of Computation: A Primer. Addison-Wesley, Boston (1987)

Part VII
Related Unregulated Automata

This part modifies classical automata, such as finite and pushdown automata. Since the automata resulting from this modification are closely related to regulated automata, they are included in this book, too. Part VII consists of Chaps. 17 and 18.

Chapter 17 introduces and studies *jumping finite automata* that work just like classical finite automata except that they do not read their input strings in a symbol-by-symbol left-to-right way. Instead, after reading a symbol, they can jump in either direction within their input tapes and continue making moves from there. The chapter demonstrates their fundamental properties, including results concerning closure properties and decidability. In addition, it establishes an infinite hierarchy of language families resulting from them.

Chapter 18 deals with *deep pushdown automata* that can make expansions deeper in the pushdown; otherwise, they work just like ordinary pushdown automata. It establishes an infinite hierarchy of language families resulting from these automata and points out its coincidence with a hierarchy resulting from state grammars.

Chapter 17
Jumping Finite Automata

Abstract The present chapter introduces and studies *jumping finite automata*. In essence, these automata work just like classical finite automata except that they do not read their input strings in a symbol-by-symbol left-to-right way. Instead, after reading a symbol, they can jump in either direction within their input tapes and continue making moves from there. Once an occurrence of a symbol is read, it cannot be re-read again later on. Otherwise, their definition coincides with the definition of standard finite automata. Organized into eight sections, this chapter gives a systematic body of knowledge concerning jumping finite automata. First, it formalizes them (Sect. 17.1). Then, it demonstrates their fundamental properties (Sect. 17.2), after which it compares their power with the power of well-known language-defining formal devices (Sect. 17.3). Naturally, this chapter also establishes several results concerning jumping finite automata with respect to commonly studied areas of formal language theory, such as closure properties (Sect. 17.4) and decidability (Sect. 17.5). In addition, it establishes an infinite hierarchy of language families resulting from these automata (Sect. 17.6). Finally, it studies some special topics and features, such as one-directional jumps (Sect. 17.7) and various start configurations (Sect. 17.8). Throughout its discussion, this chapter points out several open questions regarding these automata, which may represent a new investigation area of automata theory in the future.

Keywords Modification of finite automata • Accepting power • Comparison with other language-defining devises • Closure properties • Decidability • Infinite hierarchy of language families • One-directional jumps • Start configurations • Open problems • New investigation areas

Recall that the well-known notion of a classical finite automaton, M, which consists of an input tape, a read head, and a finite state control. The input tape is divided into squares. Each square contains one symbol of an input string. The symbol under the read head, a, is the current input symbol. The finite control is represented by a finite set of states together with a control relation, which is usually specified as

A. Meduna and P. Zemek, *Regulated Grammars and Automata*,
DOI 10.1007/978-1-4939-0369-6_17,

a set of computational rules. M computes by making a sequence of moves. Each move is made according to a computational rule that describes how the current state is changed and whether the current input symbol is read. If the symbol is read, the read head is shifted precisely one square to the right. M has one state defined as the start state and some states designated as final states. If M can read w by making a sequence of moves from the start state to a final state, M accepts w; otherwise, M rejects w.

As obvious, the finite state control can be viewed as a regulated mechanism in finite automata, which thus deserve our specific attention with respect to the subject of the book. However, most results concerning their classical versions are so notoriously known that their repetition might be unnecessarily. Even more importantly, these classical versions work so they often fail to reflect the real needs of current computer science. Perhaps most significantly, they fail to formalize discontinuous information processing, which is central to today's computation while it was virtually unneeded and, therefore, unknown in the past. Indeed, in the previous century, most classical computer science methods were developed for continuous information processing. Accordingly, their formal models, including finite automata, work on strings, representing information, in a strictly continuous left-to-right symbol-by-symbol way. Modern information methods, however, frequently process information in a discontinuous way [1–5, 7]. Within a particular running process, a typical computational step may be performed somewhere in the middle of information while the very next computational step is executed far away from it; therefore, before the next step is carried out, the process has to jump over a large portion of the information to the desired position of execution. Of course, classical formal models, which work on strings strictly continuously, inadequately and inappropriately reflect discontinuous information processing of this kind.

Formalizing discontinuous information processing adequately gives rise to the idea of adapting classical formal models in a discontinuous way. In this way, the present chapter introduces and studies the notion of a *jumping finite automaton*, H. In essence, H works just like a classical finite automaton except it does not read the input string in a symbol-by-symbol left-to-right way. That is, after reading a symbol, H can jump over a portion of the tape in either direction and continue making moves from there. Once an occurrence of a symbol is read on the tape, it cannot be re-read again later during computation of H. Otherwise, it coincides with the standard notion of a finite automaton, and as such, it is based upon a regulated mechanism consisting in its finite state control. Therefore, we study them in detail in this book.

More precisely, concerning jumping finite automata, this chapter considers commonly studied areas of this theory, such as decidability and closure properties, and establishes several results concerning jumping finite automata regarding these areas. It concentrates its attentions on results that demonstrate differences between jumping finite automata and their classical versions. As a whole, this chapter gives a systematic body of knowledge concerning jumping finite automata. At the same time, however, it points out several open questions regarding these automata, which may represent a new, attractive, significant investigation area of automata theory in the future.

The present chapter is organized as follows. First, Sect. 17.1 formalizes jumping finite automata. Then, Sect. 17.2 demonstrates their fundamental properties. Section 17.3 compares their power with the power of well-known language-defining formal devices. After that, this chapter establishes several results concerning jumping finite automata with respect to commonly studied areas of formal language theory, such as closure properties (Sect. 17.4) and decidability (Sect. 17.5). Section 17.6 establishes an infinite hierarchy of language families resulting from these automata. Finally, this chapter studies some special topics and features, such as one-directional jumps (Sect. 17.7) and various start configurations (Sect. 17.8).

17.1 Definitions and Examples

In this section, we define a variety of jumping finite automata discussed in this chapter and illustrate them by examples.

Definition 17.1.1. A *general jumping finite automaton* (a *GJFA* for short) is a quintuple

$$M = \big(Q, \Sigma, R, s, F\big)$$

where Q is a finite set of *states*, Σ is the *input alphabet*, $\Sigma \cap Q = \emptyset$, $R \subseteq Q \times \Sigma^* \times Q$ is a finite relation, $s \in Q$ is the *start state*, and F is a set of *final states*. Members of R are referred to as *rules* of M and instead of $(p, y, q) \in R$, we write $py \to q \in R$. A *configuration* of M is any string in $\Sigma^* Q \Sigma^*$. The binary *jumping relation*, symbolically denoted by $\curvearrowright_M$, over $\Sigma^* Q \Sigma^*$, is defined as follows. Let $x, z, x', z' \in \Sigma^*$ such that $xz = x'z'$ and $py \to q \in R$; then, M makes a *jump* from $xpyz$ to $x'qz'$, written as

$$xpyz \curvearrowright_M x'qz'$$

When there is no danger of confusion, we simply write $\curvearrowright$ instead of $\curvearrowright_M$. In the standard manner, we extend $\curvearrowright$ to $\curvearrowright^m$, where $m \geq 0$. Let $\curvearrowright^+$ and $\curvearrowright^*$ denote the transitive closure of $\curvearrowright$ and the reflexive-transitive closure of $\curvearrowright$, respectively. The *language* accepted by M, denoted by $L(M)$, is defined as

$$L(M) = \big\{uv \mid u, v \in \Sigma^*, usv \curvearrowright^* f, f \in F\big\}$$

Let $w \in \Sigma^*$. We say that M *accepts* w if and only if $w \in L(M)$. M *rejects* w if and only if $w \in \Sigma^* - L(M)$. Two GJFAs M and M' are said to be *equivalent* if and only if $L(M) = L(M')$. $\square$

Definition 17.1.2. Let $M = (Q, \Sigma, R, s, F)$ be a GJFA. M is an *ε-free GJFA* if $py \to q \in R$ implies that $|y| \geq 1$. M is of *degree n*, where $n \geq 0$, if $py \to q \in R$ implies that $|y| \leq n$. M is a *jumping finite automaton* (a *JFA* for short) if its degree is 1. □

Definition 17.1.3. Let $M = (Q, \Sigma, R, s, F)$ be a JFA. Analogously to a GJFA, M is an *ε-free JFA* if $py \to q \in R$ implies that $|y| = 1$. M is a *deterministic JFA* (a *DJFA* for short) if (1) it is an ε-free JFA and (2) for each $p \in Q$ and each $a \in \Sigma$, there is no more than one $q \in Q$ such that $pa \to q \in R$. M is a *complete JFA* (a *CJFA* for short) if (1) it is a DJFA and (2) for each $p \in Q$ and each $a \in \Sigma$, there is precisely one $q \in Q$ such that $pa \to q \in R$. □

Definition 17.1.4. Let $M = (Q, \Sigma, R, s, F)$ be a GJFA. The *transition graph* of M, denoted by $\Delta(M)$, is a multigraph, where nodes are states from Q, and there is an edge from p to q labeled with y if and only if $py \to q \in R$. A state $q \in Q$ is *reachable* if there is a walk from s to q in $\Delta(M)$; q is *terminating* if there is a walk from q to some $f \in F$. If there is a walk from p to q, $p = q_1, q_2, \dots, q_n = q$, for some $n \geq 2$, where $q_i y_i \to q_{i+1} \in R$ for all $i = 1, \dots, n-1$, then we write

$$py_1 y_2 \cdots y_n \rightsquigarrow q$$

□

Next, we illustrate the previous definitions by two examples.

Example 17.1.5. Consider the DJFA

$$M = \big(\{s, r, t\}, \Sigma, R, s, \{s\}\big)$$

where $\Sigma = \{a, b, c\}$ and

$$R = \big\{sa \to r, rb \to t, tc \to s\big\}$$

Starting from s, M has to read some a, some b, and some c, entering again the start (and also the final) state s. All these occurrences of a, b, and c can appear anywhere in the input string. Therefore, the accepted language is clearly

$$L(M) = \big\{w \in \Sigma^* \mid \text{occur}(w, a) = \text{occur}(w, b) = \text{occur}(w, c)\big\}$$

□

Recall that $L(M)$ in Example 17.1.5 is a well-known non-context-free context-sensitive language.

Example 17.1.6. Consider the GJFA

$$M = \big(\{s, t, f\}, \{a, b\}, R, s, \{f\}\big)$$

where

$$R = \big\{sba \to f, fa \to f, fb \to f\big\}$$

Starting from s, M has to read string ba, which can appear anywhere in the input string. Then, it can read an arbitrary number of symbols a and b, including no symbols. Therefore, the accepted language is $L(M) = \{a, b\}^*\{ba\}\{a, b\}^*$. □

Denotation of Language Families

Throughout the rest of this chapter, **GJFA**, $\mathbf{GJFA}^{-\varepsilon}$, **JFA**, $\mathbf{JFA}^{-\varepsilon}$, and **DJFA** denote the families of languages accepted by GJFAs, ε-free GJFAs, JFAs, ε-free JFAs, and DJFAs, respectively.

17.2 Basic Properties

In this section, we discuss the generative power of GJFAs and JFAs and some other basic properties of these automata.

Theorem 17.2.1. *For every DJFA M, there is a CJFA M' such that $L(M) = L(M')$.*

Proof. Let $M = (Q, \Sigma, R, s, F)$ be a DJFA. We next construct a CJFA M' such that $L(M) = L(M')$. Without any loss of generality, we assume that $\bot \notin Q$. Initially, set

$$M' = \big(Q \cup \{\bot\}, \Sigma, R', s, F\big)$$

where $R' = R$. Next, for each $a \in \Sigma$ and each $p \in Q$ such that $pa \to q \notin R$ for all $q \in Q$, add $pa \to \bot$ to R'. For each $a \in \Sigma$, add $\bot a \to \bot$ to R'. Clearly, M' is a CJFA and $L(M) = L(M')$. □

Lemma 17.2.2. *For every GJFA M of degree $n \geq 0$, there is an ε-free GJFA M' of degree n such that $L(M') = L(M)$.*

Proof. This lemma can be demonstrated by using the standard conversion of finite automata to ε-free finite automata (see Algorithm 3.2.2.3 in [6]). □

Theorem 17.2.3. $\mathbf{GJFA} = \mathbf{GJFA}^{-\varepsilon}$

Proof. $\mathbf{GJFA}^{-\varepsilon} \subseteq \mathbf{GJFA}$ follows from the definition of a GJFA. $\mathbf{GJFA} \subseteq \mathbf{GJFA}^{-\varepsilon}$ follows from Lemma 17.2.2. □

Theorem 17.2.4. $\mathbf{JFA} = \mathbf{JFA}^{-\varepsilon} = \mathbf{DJFA}$

Proof. $\mathbf{JFA} = \mathbf{JFA}^{-\varepsilon}$ can be proved by analogy with the proof of Theorem 17.2.3, so we only prove that $\mathbf{JFA}^{-\varepsilon} = \mathbf{DJFA}$. $\mathbf{DJFA} \subseteq \mathbf{JFA}^{-\varepsilon}$ follows from the definition of a DJFA. The converse inclusion can be proved by using the standard technique of converting ε-free finite automata to deterministic finite automata (see Algorithm 3.2.3.1 in [6]). □

The next theorem shows a property of languages accepted by GJFAs with unary input alphabets.

Theorem 17.2.5. *Let $M = (Q, \Sigma, R, s, F)$ be a GJFA such that* $\mathrm{card}(\Sigma) = 1$. *Then, $L(M)$ is regular.*

Proof. Let $M = (Q, \Sigma, R, s, F)$ be a GJFA such that $\text{card}(\Sigma) = 1$. Since $\text{card}(\Sigma) = 1$, without any loss of generality, we can assume that the acceptance process for $w \in \Sigma^*$ starts from the configuration sw and M does not jump over any symbols. Therefore, we can threat M as an equivalent general finite automaton (see Definition 3.4.1). As general finite automata accept only regular languages (see Theorems 3.4.4 and 3.4.5), $L(M)$ is regular. □

As a consequence of Theorem 17.2.5, we obtain the following corollary (recall that K below is not regular).

Corollary 17.2.6. *The language* $K = \{a^p \mid p \text{ is a prime number}\}$ *cannot be accepted by any GJFA.* □

The following theorem gives a necessary condition for a language to be in **JFA**.

Theorem 17.2.7. *Let K be an arbitrary language. Then,* $K \in \mathbf{JFA}$ *only if* $K = \text{perm}(K)$.

Proof. Let $M = (Q, \Sigma, R, s, F)$ be a JFA. Without any loss of generality, we assume that M is a DJFA (recall that $\mathbf{JFA} = \mathbf{DJFA}$ by Theorem 17.2.4). Let $w \in L(M)$. We next prove that $\text{perm}(w) \subseteq L(M)$. If $w = \varepsilon$, then $\text{perm}(\varepsilon) = \varepsilon \in L(M)$, so we assume that $w \neq \varepsilon$. Then, $w = a_1 a_2 \cdots a_n$, where $a_i \in \Sigma$ for all $i = 1, \dots, n$, for some $n \geq 1$. Since $w \in L(M)$, R contains

$$\begin{aligned} s a_{i_1} &\to s_{i_1} \\ s_{i_1} a_{i_2} &\to s_{i_2} \\ &\vdots \\ s_{i_{n-1}} a_{i_n} &\to s_{i_n} \end{aligned}$$

where $s_j \in Q$ for all $j \in \{i_1, i_2, \dots, i_n\}$, $(i_1, i_2, \dots, i_n)$ is a permutation of $(1, 2, \dots, n)$, and $s_{i_n} \in F$. However, this implies that $a_{k_1} a_{k_2} \cdots a_{k_n} \in L(M)$, where $(k_1, k_2, \dots, k_n)$ is a permutation of $(1, 2, \dots, n)$, so $\text{perm}(w) \subseteq L(M)$. □

From Theorem 17.2.7, we obtain the following two corollaries, which are used in subsequent proofs.

Corollary 17.2.8. *There is no JFA that accepts* $\{ab\}^*$. □

Corollary 17.2.9. *There is no JFA that accepts* $\{a, b\}^*\{ba\}\{a, b\}^*$. □

Consider the language of primes K from Corollary 17.2.6. Since $K = \text{perm}(K)$, the condition from Theorem 17.2.7 is not sufficient for a language to be in **JFA**. This is stated in the following corollary.

Corollary 17.2.10. *There is a language K satisfying* $K = \text{perm}(K)$ *that cannot be accepted by any JFA.* □

The next theorem gives both a necessary and sufficient condition for a language to be accepted by a JFA.

Theorem 17.2.11. *Let L be an arbitrary language. $L \in$ **JFA** if and only if $L = \text{perm}(K)$, where K is a regular language.*

Proof. The proof is divided into the only-if part and the if part.

Only If. Let M be a JFA. Consider M as a finite automaton M'. Set $K = L(M')$. K is regular, and $L(M) = \text{perm}(K)$. Hence, the only-if part holds.

If. Take $\text{perm}(K)$, where K is any regular language. Let $K = L(M)$, where M is a finite automaton. Consider M as a JFA M'. Observe that $L(M') = \text{perm}(K)$, which proves the if part of the proof. □

Finally, we show that GJFAs are stronger than JFAs.

Theorem 17.2.12. JFA ⊂ GJFA

Proof. **JFA** ⊆ **GJFA** follows from the definition of a JFA. From Corollary 17.2.9, **GJFA** − **JFA** ≠ ∅, because $\{a, b\}^*\{ba\}\{a, b\}^*$ is accepted by the GJFA from Example 17.1.6. □

Open Problem 17.2.13. Is there a necessary and sufficient condition for a language to be in **GJFA**? □

17.3 Relations with Well-Known Language Families

In this section, we establish relations between **GJFA**, **JFA**, and some well-known language families, including **FIN**, **REG**, **CF**, and **CS**.

Theorem 17.3.1. FIN ⊂ GJFA

Proof. Let $K \in$ **FIN**. Since K is a finite, there exists $n \geq 0$ such that $\text{card}(K) = n$. Therefore, we can express K as $K = \{w_1, w_2, \dots, w_n\}$. Define the GJFA

$$M = \left(\{s, f\}, \Sigma, R, s, \{f\}\right)$$

where $\Sigma = \text{alph}(K)$ and $R = \{sw_1 \rightarrow f, sw_2 \rightarrow f, \dots, sw_n \rightarrow f\}$. Clearly, $L(M) = K$. Therefore, **FIN** ⊆ **GJFA**. From Example 17.1.5, **GJFA** − **FIN** ≠ ∅, which proves the theorem. □

Lemma 17.3.2. *There is no GJFA that accepts $\{a\}^*\{b\}^*$.*

Proof. By contradiction. Let $K = \{a\}^*\{b\}^*$. Assume that there is a GJFA, $M = (Q, \Sigma, R, s, F)$, such that $L(M) = K$. Let $w = a^n b$, where n is the degree of M. Since $w \in K$, during an acceptance of w, a rule, $pa^i b \rightarrow q \in R$, where $p, q \in Q$ and $0 \leq i < n$, has to be used. However, then M also accepts from the configuration $a^i bsa^{n-i}$. Indeed, as $a^i b$ is read in a single step and all the other symbols in w are just as, $a^i ba^{n-i}$ may be accepted by using the same rules as during an acceptance of w. This implies that $a^i ba^{n-i} \in K$—a contradiction with the assumption that $L(M) = K$. Therefore, there is no GJFA that accepts $\{a\}^*\{b\}^*$. □

Theorem 17.3.3. **REG** *and* **GJFA** *are incomparable.*

Proof. **GJFA** $\not\subseteq$ **REG** follows from Example 17.1.5. **REG** $\not\subseteq$ **GJFA** follows from Lemma 17.3.2. □

Theorem 17.3.4. **CF** *and* **GJFA** *are incomparable.*

Proof. **GJFA** $\not\subseteq$ **CF** follows from Example 17.1.5, and **CF** $\not\subseteq$ **GJFA** follows from Lemma 17.3.2. □

Theorem 17.3.5. **GJFA** $\subset$ **CS**

Proof. Clearly, jumps of GJFAs can be simulated by context-sensitive grammars, so **GJFA** $\subseteq$ **CS**. From Lemma 17.3.2, it follows that **CS** $-$ **GJFA** $\neq \emptyset$. □

Theorem 17.3.6. **FIN** *and* **JFA** *are incomparable.*

Proof. **JFA** $\not\subseteq$ **FIN** follows from Example 17.1.5. Consider the finite language $K = \{ab\}$. By Theorem 17.2.7, $K \notin$ **JFA**, so **FIN** $\not\subseteq$ **JFA**. □

17.4 Closure Properties

In this section, we show the closure properties of the families **GJFA** and **JFA** under various operations.

Theorem 17.4.1. *Both* **GJFA** *and* **JFA** *are not closed under endmarking.*

Proof. Consider the language $K = \{a\}^*$. Clearly, $K \in$ **JFA**. A proof that no GJFA accepts $K\{\#\}$, where # is a symbol such that $\# \neq a$, can be made by analogy with the proof of Lemma 17.3.2. □

Theorem 17.4.1 implies that both families are not closed under concatenation. Indeed, observe that the JFA

$$M = \big(\{s, f\}, \{\#\}, \{s\# \rightarrow f\}, s, \{f\}\big)$$

accepts $\{\#\}$.

Corollary 17.4.2. *Both* **GJFA** *and* **JFA** *are not closed under concatenation.* □

Theorem 17.4.3. **JFA** *is closed under shuffle.*

Proof. Let $M_1 = (Q_1, \Sigma_1, R_1, s_1, F_1)$ and $M_2 = (Q_2, \Sigma_2, R_2, s_2, F_2)$ be two JFAs. Without any loss of generality, we assume that $Q_1 \cap Q_2 = \emptyset$. Define the JFA

$$H = \big(Q_1 \cup Q_2, \Sigma_1 \cup \Sigma_2, R_1 \cup R_2 \cup \{f \rightarrow s_2 \mid f \in F_1\}, s_1, F_2\big)$$

To see that $L(H) = \text{shuffle}(L(M_1), L(M_2))$, observe how H works. On an input string, $w \in (\Sigma_1 \cup \Sigma_2)^*$, H first runs M_1 on w, and if it ends in a final state, then it runs M_2 on the rest of the input. If M_2 ends in a final state, H accepts w. Otherwise, it rejects w. By Theorem 17.2.7, $L(M_i) = \text{perm}(L(M_i))$ for all $i \in \{1, 2\}$. Based on these observations, since H can jump anywhere after a symbol is read, we see that $L(H) = \text{shuffle}(L(M_1), L(M_2))$. □

Notice that the construction used in the previous proof coincides with the standard construction of a concatenation of two finite automata (see [6]).

Theorem 17.4.4. *Both* **GJFA** *and* **JFA** *are closed under union.*

Proof. Let $M_1 = (Q_1, \Sigma_1, R_1, s_1, F_1)$ and $M_2 = (Q_2, \Sigma_2, R_2, s_2, F_2)$ be two GJFAs. Without any loss of generality, we assume that $Q_1 \cap Q_2 = \emptyset$ and $s \notin (Q_1 \cup Q_2)$. Define the GJFA

$$H = \big(Q_1 \cup Q_2 \cup \{s\}, \Sigma_1 \cup \Sigma_2, R_1 \cup R_2 \cup \{s \rightarrow s_1, s \rightarrow s_2\}, s, F_1 \cup F_2\big)$$

Clearly, $L(H) = L(M_1) \cup L(M_2)$, and if both M_1 and M_2 are JFAs, then H is also a JFA. □

Theorem 17.4.5. GJFA *is not closed under complement.*

Proof. Consider the GJFA M from Example 17.1.6. Observe that the complement of $L(M)$ (with respect to $\{a, b\}^*$) is $\{a\}^*\{b\}^*$, which cannot be accepted by any GJFA (see Lemma 17.3.2). □

Theorem 17.4.6. JFA *is closed under complement.*

Proof. Let $M = (Q, \Sigma, R, s, F)$ be a JFA. Without any loss of generality, we assume that M is a CJFA (**JFA** = **DJFA** by Theorem 17.2.4 and every DJFA can be converted to an equivalent CJFA by Theorem 17.2.1). Then, the JFA

$$M' = \big(Q, \Sigma, R, s, Q - F\big)$$

accepts $\overline{L(M)}$. □

By using De Morgan's laws, we obtain the following two corollaries of Theorems 17.4.4–17.4.6.

Corollary 17.4.7. GJFA *is not closed under intersection.* □

Corollary 17.4.8. JFA *is closed under intersection.* □

Theorem 17.4.9. *Both* **GJFA** *and* **JFA** *are not closed under intersection with regular languages.*

Proof. Consider the language $J = \{a, b\}^*$, which can be accepted by both GJFAs and JFAs. Consider the regular language $K = \{a\}^*\{b\}^*$. Since $J \cap K = K$, this theorem follows from Lemma 17.3.2. □

Theorem 17.4.10. **JFA** *is closed under reversal.*

Proof. Let $K \in \mathbf{JFA}$. Since $\text{perm}(w) \subseteq K$ by Theorem 17.2.7 for all $w \in K$, also $\text{rev}(w) \in K$ for all $w \in K$, so the theorem holds. □

Theorem 17.4.11. **JFA** *is not closed under Kleene star or under Kleene plus.*

Proof. Consider the language $K = \{ab, ba\}$, which is accepted by the JFA

$$M = \left(\{s, r, f\}, \{a, b\}, \{sa \to r, rb \to f\}, s, \{f\}\right)$$

However, by Theorem 17.2.7, there is no JFA that accepts K^* or K^+ (notice that, for example, $abab \in K^+$, but $aabb \notin K^+$). □

Lemma 17.4.12. *There is no GJFA that accepts* $\{a\}^*\{b\}^* \cup \{b\}^*\{a\}^*$.

Proof. This lemma can be proved by analogy with the proof of Lemma 17.3.2. □

Theorem 17.4.13. *Both* **GJFA** *and* **JFA** *are not closed under substitution.*

Proof. Consider the language $K = \{ab, ba\}$, which is accepted by the JFA M from the proof of Theorem 17.4.11. Define the substitution σ from $\{a, b\}^*$ to $2^{\{a,b\}^*}$ as $\sigma(a) = \{a\}^*$ and $\sigma(b) = \{b\}^*$. Clearly, both $\sigma(a)$ and $\sigma(b)$ can be accepted by JFAs. However, $\sigma(K)$ cannot be accepted by any GJFA (see Lemma 17.4.12). □

Since the substitution σ in the proof of Theorem 17.4.13 is regular, we obtain the following corollary.

Corollary 17.4.14. *Both* **GJFA** *and* **JFA** *are not closed under regular substitution.* □

Theorem 17.4.15. **GJFA** *is closed under finite substitution.*

Proof. Let $M = (Q, \Sigma, R, s, F)$ be a GJFA, Γ be an alphabet, and σ be a finite substitution from Σ^* to 2^{Γ^*}. The language $\sigma(L(M))$ is accepted by the GJFA

$$M' = \left(Q, \Gamma, R', s, F\right)$$

where $R' = \{py' \to q \mid y' \in \sigma(y), py \to q \in R\}$. □

Since homomorphism is a special case of finite substitution, we obtain the following corollary of Theorem 17.4.15.

Corollary 17.4.16. **GJFA** *is closed under homomorphism.* □

Theorem 17.4.17. **JFA** *is not closed under ε-free homomorphism.*

Proof. Define the ε-free homomorphism φ from $\{a\}$ to $\{a, b\}^+$ as $\varphi(a) = ab$, and consider the language $\{a\}^*$, which is accepted by the JFA

$$M = \left(\{s\}, \{a\}, \{sa \to s\}, \{s\}\right)$$

Notice that $\varphi(L(M)) = \{ab\}^*$, which cannot be accepted by any JFA (see Corollary 17.2.8). □

Table 17.1 Summary of closure properties

	GJFA	JFA
Endmarking	–	–
Concatenation	–	–
Shuffle	?	+
Union	+	+
Complement	–	+
Intersection	–	+
Int. with regular languages	–	–
Kleene star	?	–
Kleene plus	?	–
Reversal	?	+
Substitution	–	–
Regular substitution	–	–
Finite substitution	+	–
Homomorphism	+	–
ε-free homomorphism	+	–
Inverse homomorphism	+	+

Since ε-free homomorphism is a special case of homomorphism and since homomorphism is a special case of finite substitution, we obtain the following corollary of Theorem 17.4.17.

Corollary 17.4.18. **JFA** *is not closed under homomorphism or under finite substitution.* □

Theorem 17.4.19. *Both* **GJFA** *and* **JFA** *are closed under inverse homomorphism.*

Proof. Let $M = (Q, \Gamma, R, s, F)$ be a GJFA, Σ be an alphabet, and φ be a homomorphism from Σ^* to Γ^*. We next construct a JFA M' such that $L(M') = \varphi^{-1}(L(M))$. Define

$$M' = \big(Q, \Sigma, R', s, F\big)$$

where

$$R' = \big\{pa \to q \mid a \in \Sigma, p\varphi(a) \rightsquigarrow q\text{in } \Delta(M)\big\}$$

Observe that $w_1sw_2 \curvearrowright^* q$ in M if and only if $w_1'sw_2' \curvearrowright^* q$ in M', where $w_1w_2 = \varphi(w_1'w_2')$ and $q \in Q$, so $L(M') = \varphi^{-1}(L(M))$. A fully rigorous proof is left to the reader. □

The summary of closure properties of the families **GJFA** and **JFA** is given in Table 17.1, where + marks closure, – marks non-closure, and ? means that the closure property represents an open problem. It is worth noting that **REG**, characterized by finite automata, is closed under all of these operations.

Open Problem 17.4.20. Is **GJFA** closed under shuffle, Kleene star, Kleene plus, and under reversal? □

17.5 Decidability

In this section, we prove the decidability of some decision problems with regard to **GJFA** and **JFA**.

Lemma 17.5.1. *Let $M = (Q, \Sigma, R, s, F)$ be a GJFA. Then, $L(M)$ is infinite if and only if $py \rightsquigarrow p$ in $\Delta(M)$, for some $y \in \Sigma^+$ and $p \in Q$ such that p is both reachable and terminating in $\Delta(M)$.*

Proof. If. Let $M = (Q, \Sigma, R, s, F)$ be a GJFA such that $py \rightsquigarrow p$ in $\Delta(M)$, for some $y \in \Sigma^+$ and $p \in Q$ such that p is both reachable and terminating in $\Delta(M)$. Then,

$$w_1 s w_2 \curvearrowright^* upv \curvearrowright^+ xpz \curvearrowright^* f$$

where $w_1 w_2 \in L(M)$, $u, v, x, z \in \Sigma^*$, $p \in Q$, and $f \in F$. Consequently,

$$w_1 s w_2 \curvearrowright^* upvy' \curvearrowright^+ xpz \curvearrowright^* f$$

where $y' = y^n$ for all $n \geq 0$. Therefore, $L(M)$ is infinite, so the if part holds.
Only If. Let $M = (Q, \Sigma, R, s, F)$ be a GJFA such that $L(M)$ is infinite. Without any loss of generality, we assume that M is ε-free (see Lemma 17.2.2). Then,

$$w_1 s w_2 \curvearrowright^* upv \curvearrowright^+ xpz \curvearrowright^* f$$

for some $w_1 w_2 \in L(M)$, $u, v, x, z \in \Sigma^*$, $p \in Q$, and $f \in F$. This implies that p is both terminating and reachable in $\Delta(M)$. Let $y \in \Sigma^+$ be a string read by M during $upv \curvearrowright^+ xpz$. Then, $py \rightsquigarrow p$ in $\Delta(M)$, so the only-if part holds. □

Theorem 17.5.2. *Both finiteness and infiniteness are decidable for* **GJFA**.

Proof. Let $M = (Q, \Sigma, R, s, F)$ be a GJFA. By Lemma 17.5.1, $L(M)$ is infinite if and only if $py \rightsquigarrow p$ in $\Delta(M)$, for some $y \in \Sigma^+$ and $p \in Q$ such that p is both reachable and terminating in $\Delta(M)$. This condition can be checked by any graph searching algorithm, such as breadth-first search (see p. 73 in [8]). Therefore, the theorem holds. □

Corollary 17.5.3. *Both finiteness and infiniteness are decidable for* **JFA**. □

Observe that since there is no deterministic version of a GJFA, the following proof of Theorem 17.5.4 is not as straightforward as in terms of regular languages and classical deterministic finite automata.

Theorem 17.5.4. *The membership problem is decidable for* **GJFA**.

Proof. Let $M = (Q, \Sigma, R, s, F)$ be a GJFA, and let $x \in \Sigma^*$. Without any loss of generality, we assume that M is ε-free (see Theorem 17.2.3). If $x = \varepsilon$, then $x \in L(M)$ if and only if $s \in F$, so assume that $x \neq \varepsilon$. Set

Table 17.2 Summary of decidability properties

	GJFA	JFA
Membership	+	+
Emptiness	+	+
Finiteness	+	+
Infiniteness	+	+

$$\Gamma = \{(x_1, x_2, \ldots, x_n) \mid x_i \in \Sigma^+, 1 \leq i \leq n, x_1 x_2 \cdots x_n = x, n \geq 1\}$$

and

$$\Gamma_p = \{(y_1, y_2, \ldots, y_n) \mid (x_1, x_2, \ldots, x_n) \in \Gamma, n \geq 1, (y_1, y_2, \ldots, y_n) \text{ is a permutation of } (x_1, x_2, \ldots, x_n)\}$$

If there exist $(y_1, y_2, \ldots, y_n) \in \Gamma_p$ and $q_1, q_2, \ldots, q_{n+1} \in Q$, for some n, $1 \leq n \leq |x|$, such that $s = q_1$, $q_{n+1} \in F$, and $q_i y_i \rightarrow q_{i+1} \in R$ for all $i = 1, 2, \ldots, n$, then $x \in L(M)$; otherwise, $x \notin L(M)$. Since both Q and Γ_p are finite, this check can be performed in finite time. □

Corollary 17.5.5. *The membership problem is decidable for* **JFA**. □

Theorem 17.5.6. *The emptiness problem is decidable for* **GJFA**.

Proof. Let $M = (Q, \Sigma, R, s, F)$ be a GJFA. Then, $L(M)$ is empty if and only if no $f \in F$ is reachable in $\Delta(M)$. This check can be done by any graph searching algorithm, such as breadth-first search (see p. 73 in [8]). □

Corollary 17.5.7. *The emptiness problem is decidable for* **JFA**. □

The summary of decidability properties of the families **GJFA** and **JFA** is given in Table 17.2, where + marks decidability.

17.6 An Infinite Hierarchy of Language Families

In this section, we establish an infinite hierarchy of language families resulting from GJFAs of degree n, where $n \geq 0$. Let $\mathbf{GJFA}_n$ and $\mathbf{GJFA}_n^{-\varepsilon}$ denote the families of languages accepted by GJFAs of degree n and by ε-free GJFAs of degree n, respectively. Observe that $\mathbf{GJFA}_n = \mathbf{GJFA}_n^{-\varepsilon}$ by the definition of a GJFA and by Lemma 17.2.2, for all $n \geq 0$.

Lemma 17.6.1. *Let Σ be an alphabet such that* $\text{card}(\Sigma) \geq 2$. *Then, for any $n \geq 1$, there is a GJFA of degree n, $M_n = (Q, \Sigma, R, s, F)$, such that $L(M_n)$ cannot be accepted by any GJFA of degree $n - 1$.*

Proof. Let Σ be an alphabet such that $\text{card}(\Sigma) \geq 2$, and let $a, b \in \Sigma$ such that $a \neq b$. The case when $n = 1$ follows immediately from the definition of a JFA, so we assume that $n \geq 2$. Define the GJFA of degree n

$$M_n = (\{s, f\}, \Sigma, \{sw \to r\}, s, \{r\})$$

where $w = ab(a)^{n-2}$. Clearly, $L(M_n) = \{w\}$. We next prove that $L(M_n)$ cannot be accepted by any GJFA of degree $n - 1$.

Suppose, for the sake of contradiction, that there is a GJFA of degree $n - 1$, $H = (Q, \Sigma, R, s', F)$, such that $L(H) = L(M_n)$. Without any loss of generality, we assume that H is ε-free (see Lemma 17.2.2). Since $L(H) = L(M_n) = \{w\}$ and $|w| > n - 1$, there has to be

$$us'xv \curvearrowright^m f$$

in H, where $w = uxv$, $u, v \in \Sigma^*$, $x \in \Sigma^+$, $f \in F$, and $m \geq 2$. Thus,

$$s'xuv \curvearrowright^m f$$

and

$$uvs'x \curvearrowright^m f$$

in H, which contradicts the assumption that $L(H) = \{w\}$. Therefore, $L(M_n)$ cannot be accepted by any GJFA of degree $n - 1$. □

Theorem 17.6.2. $\mathbf{GJFA}_n \subset \mathbf{GJFA}_{n+1}$ *for all* $n \geq 0$.

Proof. $\mathbf{GJFA}_n \subseteq \mathbf{GJFA}_{n+1}$ follows from the definition of a GJFA of degree n, for all $n \geq 0$. From Lemma 17.6.1, $\mathbf{GJFA}_{n+1} - \mathbf{GJFA}_n \neq \emptyset$, which proves the theorem. □

Taking Lemma 17.2.2 into account, we obtain the following corollary of Theorem 17.6.2.

Corollary 17.6.3. $\mathbf{GJFA}_n^{-\varepsilon} \subset \mathbf{GJFA}_{n+1}^{-\varepsilon}$ *for all* $n \geq 0$. □

17.7 Left and Right Jumps

We define two special cases of the jumping relation.

Definition 17.7.1. Let $M = (Q, \Sigma, R, s, F)$ be a GJFA. Let $w, x, y, z \in \Sigma^*$, and $py \to q \in R$; then, (1) M makes a *left jump* from $wxpyz$ to $wqxz$, symbolically written as

$$wxpyz \; {}_l\!\curvearrowright \; wqxz$$

and (2) M makes a *right jump* from $wpyxz$ to $wxqz$, written as

$$wpyxz \ {}_{r}\curvearrowright \ wxqz$$

Let $u, v \in \Sigma^* Q \Sigma^*$; then, $u \curvearrowright v$ if and only if $u \ {}_{l}\curvearrowright \ v$ or $u \ {}_{r}\curvearrowright \ v$. Extend ${}_{l}\curvearrowright$ and ${}_{r}\curvearrowright$ to ${}_{l}\curvearrowright^m$, ${}_{l}\curvearrowright^*$, ${}_{l}\curvearrowright^+$, ${}_{r}\curvearrowright^m$, ${}_{r}\curvearrowright^*$, and ${}_{r}\curvearrowright^+$, where $m \geq 0$, by analogy with extending $\curvearrowright$. Set

$${}_{l}L(M) = \{uv \mid u, v \in \Sigma^*, usv \ {}_{l}\curvearrowright^* \ f \text{ with } f \in F\}$$

and

$${}_{r}L(M) = \{uv \mid u, v \in \Sigma^*, usv \ {}_{r}\curvearrowright^* \ f \text{ with } f \in F\}$$ □

Let ${}_{l}$**GJFA**, ${}_{l}$**JFA**, ${}_{r}$**GJFA**, and ${}_{r}$**JFA** denote the families of languages accepted by GJFAs using only left jumps, JFAs using only left jumps, GJFAs using only right jumps, and JFAs using only right jumps, respectively.

Theorem 17.7.2. ${}_{r}$**GJFA** = ${}_{r}$**JFA** = **REG**

Proof. We first prove that ${}_{r}$**JFA** = **REG**. Consider any JFA, $M = (Q, \Sigma, R, s, F)$. Observe that if M occurs in a configuration of the form xpy, where $x \in \Sigma^*$, $p \in Q$, and $y \in \Sigma^*$, then it cannot read the symbols in x anymore because M can make only right jumps. Also, observe that this covers the situation when M starts to accept $w \in \Sigma^*$ from a different configuration than sw. Therefore, to read the whole input, M has to start in configuration sw, and it cannot jump to skips some symbols. Consequently, M behaves like an ordinary finite automaton, reading the input from the left to the right, so $L(M)$ is regular and, therefore, ${}_{r}$**JFA** $\subseteq$ **REG**. Conversely, any finite automaton can be viewed as a JFA that starts from configuration sw and does not jump to skip some symbols. Therefore, **REG** $\subseteq$ ${}_{r}$**JFA**, which proves that ${}_{r}$**JFA** = **REG**. ${}_{r}$**GJFA** = **REG** can be proved by the same reasoning using general finite automata instead of finite automata. □

Next, we show that JFAs using only left jumps accept some non-regular languages.

Theorem 17.7.3. ${}_{l}$**JFA** − **REG** $\neq \emptyset$

Proof. Consider the JFA

$$M = (\{s, p, q\}, \{a, b\}, R, s, \{s\})$$

where

$$R = \{sa \rightarrow p, pb \rightarrow s, sb \rightarrow q, qa \rightarrow s\}$$

We argue that

$${}_{l}L(M) = \{w \mid \text{occur}(w, a) = \text{occur}(w, b)\}$$

With $w \in \{a, b\}^*$ on its input, M starts over the last symbol. M reads this symbol by using $sa \rightarrow p$ or $sb \rightarrow q$, and jumps to the left in front of the rightmost occurrence of b or a, respectively. Then, it consumes it by using $pb \rightarrow s$ or $qa \rightarrow s$, respectively. If this read symbol was the rightmost one, it jumps one letter to the left and repeats the process. Otherwise, it makes no jumps at all. Observe that in this way, every configuration is of the form urv, where $r \in \{s, p, q\}$, $u \in \{a, b\}^*$, and either $v \in \{a, \varepsilon\}\{b\}^*$ or $v \in \{b, \varepsilon\}\{a\}^*$.

Based on the previous observations, we see that

$$_l L(M) = \{w \mid \text{occur}(w, a) = \text{occur}(w, b)\}$$

Since $L(M)$ is not regular, $_l\mathbf{JFA} - \mathbf{REG} \neq \emptyset$, so the theorem holds. □

Open Problem 17.7.4. Study the effect of left jumps to the acceptance power of JFAs and GJFAs. □

17.8 A Variety of Start Configurations

In general, a GJFA can start its computation anywhere in the input string (see Definition 17.1.1). In this section, we consider the impact of various start configurations on the acceptance power of GJFAs and JFAs.

Definition 17.8.1. Let $M = (Q, \Sigma, R, s, F)$ be a GJFA. Set

$$^bL(M) = \{w \in \Sigma^* \mid sw \curvearrowright^* f \text{ with } f \in F\},$$
$$^aL(M) = \{uv \mid u, v \in \Sigma^*, usv \curvearrowright^* f \text{ with } f \in F\},$$
$$^eL(M) = \{w \in \Sigma^* \mid ws \curvearrowright^* f \text{ with } f \in F\}.$$
□

Intuitively, b, a, and e stand for *beginning*, *anywhere*, and *end*, respectively; in this way, we express where the acceptance process starts. Observe that we simplify $^aL(M)$ to $L(M)$ because we pay a principal attention to the languages accepted in this way in this chapter. Let $^b\mathbf{GJFA}$, $^a\mathbf{GJFA}$, $^e\mathbf{GJFA}$, $^b\mathbf{JFA}$, $^a\mathbf{JFA}$, and $^e\mathbf{JFA}$ denote the families of languages accepted by GJFAs starting at the beginning, GJFAs starting anywhere, GJFAs starting at the end, JFAs starting at the beginning, JFAs starting anywhere, and JFAs starting at the end, respectively.

We show that

(1) starting at the beginning increases the acceptance power of GJFAs and JFAs, and
(2) starting at the end does not increase the acceptance power of GJFAs and JFAs.

Theorem 17.8.2. $^a\mathbf{JFA} \subset {^b\mathbf{JFA}}$

Proof. Let $M = (Q, \Sigma, R, s, F)$ be a JFA. The JFA

$$M' = \left(Q, \Sigma, R \cup \{s \rightarrow s\}, s, F\right)$$

clearly satisfies ${}^{a}L(M) = {}^{b}L(M')$, so ${}^{a}\mathbf{JFA} \subseteq {}^{b}\mathbf{JFA}$. We prove that this inclusion is, in fact, proper. Consider the language $K = \{a\}\{b\}^*$. The JFA

$$H = \big(\{s, f\}, \{a, b\}, \{sa \to f, fb \to f\}, s, \{f\}\big)$$

satisfies ${}^{b}L(H) = K$. However, observe that ${}^{a}L(H) = \{b\}^*\{a\}\{b\}^*$, which differs from K. By Theorem 17.2.7, for every JFA N, it holds that ${}^{a}L(N) \neq K$. Hence, ${}^{a}\mathbf{JFA} \subset {}^{b}\mathbf{JFA}$. □

Theorem 17.8.3. ${}^{a}\mathbf{GJFA} \subset {}^{b}\mathbf{GJFA}$

Proof. This theorem can be proved by analogy with the proof of Theorem 17.8.2. □

Lemma 17.8.4. *Let M be a GJFA of degree $n \geq 0$. Then, there is a GJFA M' of degree n such that ${}^{a}L(M) = {}^{e}L(M')$.*

Proof. Let $M = (Q, \Sigma, R, s, F)$ be a GJFA of degree n. Then, the GJFA

$$M' = \big(Q, \Sigma, R \cup \{s \to s\}, s, F\big)$$

is of degree n and satisfies ${}^{a}L(M) = {}^{e}L(M')$. □

Lemma 17.8.5. *Let M be a GJFA of degree $n \geq 0$. Then, there is a GJFA $\hat{M}$ of degree n such that ${}^{e}L(M) = {}^{a}L(\hat{M})$.*

Proof. Let $M = (Q, \Sigma, R, s, F)$ be a GJFA of degree n. If ${}^{e}L(M) = \emptyset$, then the GJFA

$$M' = \big(\{s\}, \Sigma, \emptyset, s, \emptyset\big)$$

is of degree n and satisfies ${}^{a}L(M') = \emptyset$. If ${}^{e}L(M) = \{\varepsilon\}$, then the GJFA

$$M'' = \big(\{s\}, \Sigma, \emptyset, s, \{s\}\big)$$

is of degree n and satisfies ${}^{a}L(M'') = \{\varepsilon\}$. Therefore, assume that $w \in {}^{e}L(M)$, where $w \in \Sigma^+$. Then, $s \to p \in R$, for some $p \in Q$. Indeed, observe that either ${}^{e}L(M) = \emptyset$ or ${}^{e}L(M) = \{\varepsilon\}$, which follows from the observation that if M starts at the end of an input string, then it first has to jump to the left to be able to read some symbols.

Define the GJFA $\hat{M} = (Q, \Sigma, \hat{R}, s, F)$, where

$$\hat{R} = R - \big\{su \to q \mid u \in \Sigma^+, q \in Q, \text{and there is no } x \in \Sigma^+ \text{ such that } sx \rightsquigarrow s \text{ in } \Delta(M)\big\}$$

The reason for excluding such $su \to q$ from $\hat{R}$ is that M first has to use a rule of the form $s \to p$, where $p \in Q$ (see the argumentation above). However, since $\hat{M}$

starts anywhere in the input string, we need to force it to use $s \rightarrow p$ as the first rule, thus changing the state from s to p, just like M does.

Clearly, $\hat{M}$ is of degree n and satisfies ${}^{e}L(M) = {}^{a}L(\hat{M})$, so the lemma holds. □

Theorem 17.8.6. ${}^{e}\mathbf{GJFA} = {}^{a}\mathbf{GJFA}$ *and* ${}^{e}\mathbf{JFA} = {}^{a}\mathbf{JFA}$

Proof. This theorem follows from Lemmas 17.8.4 and 17.8.5. □

We also consider combinations of left jumps, right jumps, and various start configurations. For this purpose, by analogy with the previous denotations, we define ${}^{b}_{l}\mathbf{GJFA}$, ${}^{a}_{l}\mathbf{GJFA}$, ${}^{e}_{l}\mathbf{GJFA}$, ${}^{b}_{r}\mathbf{GJFA}$, ${}^{a}_{r}\mathbf{GJFA}$, ${}^{e}_{r}\mathbf{GJFA}$, ${}^{b}_{l}\mathbf{JFA}$, ${}^{a}_{l}\mathbf{JFA}$, ${}^{e}_{l}\mathbf{JFA}$, ${}^{b}_{r}\mathbf{JFA}$, ${}^{a}_{r}\mathbf{JFA}$, and ${}^{e}_{r}\mathbf{JFA}$. For example, ${}^{b}_{r}\mathbf{GJFA}$ denotes the family of languages accepted by GJFAs that perform only right jumps and starts at the beginning.

Theorem 17.8.7. ${}^{a}_{r}\mathbf{GJFA} = {}^{a}_{r}\mathbf{JFA} = {}^{b}_{r}\mathbf{GJFA} = {}^{b}_{r}\mathbf{JFA} = {}^{b}_{l}\mathbf{GJFA} = {}^{b}_{l}\mathbf{JFA} = \mathbf{REG}$

Proof. Theorem 17.7.2, in fact, states that ${}^{a}_{r}\mathbf{GJFA} = {}^{a}_{r}\mathbf{JFA} = \mathbf{REG}$. Furthermore, ${}^{b}_{r}\mathbf{GJFA} = {}^{b}_{r}\mathbf{JFA} = \mathbf{REG}$ follows from the proof of Theorem 17.7.2 because M has to start the acceptance process of a string w from the configuration sw—that is, it starts at the beginning of w. ${}^{b}_{l}\mathbf{GJFA} = {}^{b}_{l}\mathbf{JFA} = \mathbf{REG}$ can be proved analogously. □

Theorem 17.8.8. ${}^{e}_{r}\mathbf{GJFA} = {}^{e}_{r}\mathbf{JFA} = \{\emptyset, \{\varepsilon\}\}$

Proof. Consider JFAs $M = (\{s\}, \{a\}, \emptyset, s, \emptyset)$ and $M' = (\{s\}, \{a\}, \emptyset, s, \{s\})$ to see that $\{\emptyset, \{\varepsilon\}\} \subseteq {}^{e}_{r}\mathbf{GJFA}$ and $\{\emptyset, \{\varepsilon\}\} \subseteq {}^{e}_{r}\mathbf{JFA}$. The converse inclusion also holds. Indeed, any GJFA that starts the acceptance process of a string w from ws and that can make only right jumps accepts either $\emptyset$ or $\{\varepsilon\}$. □

Open Problem 17.8.9. What are the properties of ${}^{e}_{l}\mathbf{GJFA}$ and ${}^{e}_{l}\mathbf{JFA}$? □

Notice that Open Problem 17.7.4, in fact, suggests an investigation of the properties of ${}^{a}_{l}\mathbf{GJFA}$ and ${}^{a}_{l}\mathbf{JFA}$.

A Summary of Open Problems

Within the previous sections, we have already pointed out several specific open problems concerning them. We close the present chapter by pointing out some crucially important open problem areas as suggested topics of future investigations.

(I) Concerning closure properties, study the closure of **GJFA** under shuffle, Kleene star, Kleene plus, and under reversal.
(II) Regarding decision problems, investigate other decision properties of **GJFA** and **JFA**, like equivalence, universality, inclusion, or regularity. Furthermore, study their computational complexity. Do there exist undecidable problems for **GJFA** or **JFA**?

(III) Section 17.7 has demonstrated that GJFAs and JFAs using only right jumps define the family of regular languages. How precisely do left jumps affect the acceptance power of JFAs and GJFAs?

(IV) Broaden the results of Sect. 17.8 concerning various start configurations by investigating the properties of ${}_{l}^{e}\mathbf{GJFA}$ and ${}_{l}^{e}\mathbf{JFA}$.

(V) Determinism represents a crucially important investigation area in terms of all types of automata. In essence, the non-deterministic versions of automata can make several different moves from the same configuration while their deterministic counterparts cannot—that is, they make no more than one move from any configuration. More specifically, the deterministic version of classical finite automata require that for any state q and any input symbol a, there exists no more than one rule with qa on its left-hand side; in this way, they make no more than one move from any configuration. As a result, with any input string w, they make a unique sequence of moves. As should be obvious, in terms of jumping finite automata, this requirement does not guarantee their determinism in the above sense. Modify the requirement so it guarantees the determinism.

References

1. Baeza-Yates, R., Ribeiro-Neto, B.: Modern Information Retrieval: The Concepts and Technology behind Search, 2nd edn. Addison-Wesley Professional, Boston (2011)
2. Bouchon-Meunier, B., Coletti, G., Yager, R.R. (eds.): Modern Information Processing: From Theory to Applications. Elsevier Science, New York (2006)
3. Buettcher, S., Clarke, C.L.A., Cormack, G.V.: Information Retrieval: Implementing and Evaluating Search Engines. The MIT Press, Cambridge (2010)
4. Grossman, D.A., Frieder, O.: Information Retrieval: Algorithms and Heuristics, 2nd edn. Springer, Berlin (2004)
5. Manning, C.D., Raghavan, P., Schütze, H.: Introduction to Information Retrieval. Cambridge University Press, New York (2008)
6. Meduna, A.: Automata and Languages: Theory and Applications. Springer, London (2000)
7. Nisan, N., Schocken, S.: The Elements of Computing Systems: Building a Modern Computer from First Principles. The MIT Press, Cambridge (2005)
8. Russell, S., Norvig, P.: Artificial Intelligence: A Modern Approach, 2nd edn. Prentice-Hall, New Jersey (2002)

Chapter 18
Deep Pushdown Automata

Abstract The present chapter defines *deep pushdown automata*, which represent a very natural modification of ordinary pushdown automata. While the ordinary versions can expand only the pushdown top, deep pushdown automata can make expansions deeper in the pushdown; otherwise, they both work identically. This chapter proves that the power of deep pushdown automata is similar to the generative power of regulated context-free grammars without erasing rules. Indeed, just like these grammars, deep pushdown automata are stronger than ordinary pushdown automata but less powerful than context-sensitive grammars. More precisely, they give rise to an infinite hierarchy of language families coinciding with the hierarchy resulting from n-limited state grammars. The present chapter is divided into two sections—Sects. 18.1 and 18.2. The former defines and illustrates deep pushdown automata. The latter establishes their accepting power, formulates some open problem areas concerning them, and suggests introducing new deterministic and generalized versions of these automata.

Keywords Modified pushdown automata • Deep pushdown expansions • Accepting power • Infinite hierarchy of language families • State grammars

From a slightly broader perspective, regulated grammars—that is, the central topic of this book—can be seen as naturally modified versions of ordinary context-free grammars, and by now, based upon the previous passages of this book, it should be obvious that formal language theory has introduced a great number of them. Many of these modified context-free grammars define language families lying between the families of context-free and context-sensitive languages. To give a quite specific example, an infinite hierarchy of language families between the families of context-free and context-sensitive languages was established based on n-limited state grammars (see Sect. 5.4). As a matter of fact, most regulated context-free grammar without erasing rules are stronger than context-free grammars but no more powerful than context-sensitive grammars. Compared to the number of grammatical modifications, there exist significantly fewer modifications of pushdown automata

A. Meduna and P. Zemek, *Regulated Grammars and Automata*,

DOI 10.1007/978-1-4939-0369-6_18,

although automata theory has constantly paid some attention to their investigation (see Chap. 3 of [2–5, 7, 10–13]). Some of these modifications, such as *finite-turn pushdown automata* (see [3, 13]), define a proper subfamily of the family of context-free languages. On the other hand, other modifications, such as *two-pushdown automata* (see [9, 10]), are as powerful as phrase-structure grammars. As opposed to the language families generated by regulated context-free grammars without erasing rules, there are hardly any modifications of pushdown automata that define a language family between the families of context-free and context-sensitive languages. It thus comes as no surprise that most of these modified context-free grammars, including n-limited state grammars, lack any automaton counterpart.

To fill this gap, the present chapter defines *deep pushdown automata*, which represent a very natural modification of ordinary pushdown automata, and proves that their power is similar to the generative power of regulated context-free grammars without erasing rules. Indeed, these automata are stronger than ordinary pushdown automata but less powerful than context-sensitive grammars. More precisely, they give rise to an infinite hierarchy of language families coinciding with the hierarchy resulting from n-limited state grammars. In this sense, deep pushdown automata represent the automaton counterpart to these grammars, so the principal objective of this chapter is achieved.

From a conceptual viewpoint, the introduction of deep pushdown automata is inspired by the well-known conversion of a context-free grammar to an equivalent pushdown automaton M frequently referred to as the *general top-down parser* for the grammar (see, for instance, p. 176 in [11], p. 148 in [6], p. 113 in [8], and p. 444 in [9]). Recall that during every move, M either *pops* or *expands* its pushdown depending on the symbol occurring on the pushdown top. If an input symbol a occurs on the pushdown top, M compares the pushdown top symbol with the current input symbol, and if they coincide, M pops the topmost symbol from the pushdown and proceeds to the next input symbol on the input tape. If a nonterminal occurs on the pushdown top, the parser expands its pushdown so it replaces the top nonterminal with a string. M accepts an input string x if it makes a sequence of moves so it completely reads x, empties its pushdown, and enters a final state; the latter requirement of entering a final state is dropped in some books (see, for instance, Algorithm 5.3.1.1.1 in [9] or Theorem 5.1 in [1]).

A deep pushdown automaton, ${}_{deep}M$, represents a slight generalization of M. Indeed, ${}_{deep}M$ works exactly as M except that it can make expansions of depth m so ${}_{deep}M$ replaces the mth topmost pushdown symbol with a string, for some $m \geq 1$. We demonstrate that the deep pushdown automata that make expansions of depth m or less, where $m \geq 1$, are equivalent to m-limited state grammars, so these automata accept a proper language subfamily of the language family accepted by deep pushdown automata that make expansions of depth $m + 1$ or less. The resulting infinite hierarchy of language families obtained in this way occurs between the families of context-free and context-sensitive languages. For every positive integer n, however, there exist some context-sensitive languages that cannot be accepted by any deep pushdown automata that make expansions of depth n or less.

Apart from all these theoretical results, deep pushdown automata fulfill a useful role in pragmatically oriented computer science as well. Indeed, based on them, we build up a parser in Sect. 20.3

The present chapter is divided into two sections—Sects. 18.1 and 18.2. The former defines and illustrates deep pushdown automata. The latter establishes their accepting power, formulates some open problem areas concerning them, and suggests introducing new deterministic and generalized versions of these automata.

18.1 Definitions and Examples

Without further ado, we define the notion of a deep pushdown automata, after which we illustrate it by an example. Let $\mathbb{N}$ denote the set of all positive integers.

Definition 18.1.1. A *deep pushdown automaton* is a septuple

$$M = \left(Q, \Sigma, \Gamma, R, s, S, F\right)$$

where

- Q is a finite set of *states*;
- Σ is an *input alphabet*;
- Γ is a *pushdown alphabet*, $\mathbb{N}$, Q, and Γ are pairwise disjoint, $\Sigma \subseteq \Gamma$, and $\Gamma - \Sigma$ contains a special *bottom symbol*, denoted by #;
- $R \subseteq (\mathbb{N} \times Q \times (\Gamma - (\Sigma \cup \{\#\})) \times Q \times (\Gamma - \{\#\})^+)$ $\cup (\mathbb{N} \times Q \times \{\#\} \times Q \times (\Gamma - \{\#\})^* \{\#\})$ is a finite relation;
- $s \in Q$ is the *start state*;
- $S \in \Gamma$ is the *start pushdown symbol*;
- $F \subseteq Q$ is the set of *final states*.

Instead of $(m, q, A, p, v) \in R$, we write $mqA \rightarrow pv \in R$ and call $mqA \rightarrow pv$ a *rule*; accordingly, R is referred to as the *set of rules* of M. A *configuration* of M is a triple in $Q \times T^* \times (\Gamma - \{\#\})^* \{\#\}$. Let χ denote the set of all configurations of M. Let $x, y \in \chi$ be two configurations. M *pops* its pushdown from x to y, symbolically written as

$$x \ {}_p\!\vdash y$$

if $x = (q, au, az)$, $y = (q, u, z)$, where $a \in \Sigma$, $u \in \Sigma^*$, $z \in \Gamma^*$. M *expands* its pushdown from x to y, symbolically written as

$$x \ {}_e\!\vdash y$$

if $x = (q, w, uAz)$, $y = (p, w, uvz)$, $mqA \rightarrow pv \in R$, where $q, p \in Q$, $w \in \Sigma^*$, $A \in \Gamma$, $u, v, z \in \Gamma^*$, and $\mathrm{occur}(u, \Gamma - \Sigma) = m - 1$. To express that M makes $x \ {}_e\!\vdash y$ according to $mqA \rightarrow pv$, we write

$$x \ {}_e\!\vdash y \ [mqA \rightarrow pv]$$

We say that $mqA \rightarrow pv$ is a *rule of depth* m; accordingly, $x\ {}_e\vdash y\ [mqA \rightarrow pv]$ is an *expansion of depth* m. M makes a *move* from x to y, symbolically written as

$$x \vdash y$$

if M makes either $x\ {}_e\vdash y$ or $x\ {}_p\vdash y$. If $n \in \mathbb{N}$ is the minimal positive integer such that each rule of M is of depth n or less, we say that M *is of depth* n, symbolically written as ${}_nM$. In the standard manner, we extend ${}_p\vdash$, ${}_e\vdash$, and $\vdash$ to ${}_p\vdash^m$, ${}_e\vdash^m$, and $\vdash^m$, respectively, for $m \geq 0$; then, based on ${}_p\vdash^m$, ${}_e\vdash^m$, and $\vdash^m$, we define ${}_p\vdash^+$, ${}_p\vdash^*$, ${}_e\vdash^+$, ${}_e\vdash^*$, $\vdash^+$, and $\vdash^*$.

Let M be of depth n, for some $n \in \mathbb{N}$. We define the *language accepted by* ${}_nM$, $L({}_nM)$, as

$$L({}_nM) = \left\{w \in \Sigma^* \mid (s, w, S\#) \vdash^* (f, \varepsilon, \#) \text{ in } {}_nM \text{ with } f \in F\right\}$$

In addition, we define the *language that* ${}_nM$ *accepts by empty pushdown*, $E({}_nM)$, as

$$E({}_nM) = \left\{w \in \Sigma^* \mid (s, w, S\#) \vdash^* (q, \varepsilon, \#) \text{ in } {}_nM \text{ with } q \in Q\right\}$$ □

For every $k \geq 1$, ${}_{deep}\mathbf{PDA}_k$ denotes the family of languages defined by deep pushdown automata of depth i, where $1 \leq i \leq k$. Analogously, ${}^{empty}_{deep}\mathbf{PDA}_k$ denotes the family of languages defined by deep pushdown automata of depth i by empty pushdown, where $1 \leq i \leq k$.

The following example gives a deep pushdown automaton accepting a language from

$$\left({}_{deep}\mathbf{PDA}_2 \cap {}^{empty}_{deep}\mathbf{PDA}_2 \cap \mathbf{CS}\right) - \mathbf{CF}$$

Example 18.1.2. Consider the deep pushdown automaton

$${}_2M = \left(\{s, q, p\}, \{a, b, c\}, \{A, S, \#\}, R, s, S, \{f\}\right)$$

with R containing the following five rules

$$\begin{array}{lll} 1sS \rightarrow qAA & 1qA \rightarrow fab & 1fA \rightarrow fc \\ 1qA \rightarrow paAb & 2pA \rightarrow qAc & \end{array}$$

On $aabbcc$, M makes

$$
\begin{aligned}
(s, aabbcc, S\#) \ {}_e\vdash\ & (q, aabbcc, AA\#) && [1sS \rightarrow qAA] \\
{}_e\vdash\ & (p, aabbcc, aAbA\#) && [1qA \rightarrow paAb] \\
{}_p\vdash\ & (p, abbcc, AbA\#) && \\
{}_e\vdash\ & (q, abbcc, AbAc\#) && [2pA \rightarrow qAc] \\
{}_e\vdash\ & (q, abbcc, abbAc\#) && [1qA \rightarrow fab] \\
{}_p\vdash\ & (f, bcc, bAc\#) && \\
{}_p\vdash\ & (f, cc, Ac\#) && \\
{}_e\vdash\ & (f, cc, Ac\#) && [1fA \rightarrow fc] \\
{}_p\vdash\ & (f, cc, cc\#) && \\
{}_p\vdash\ & (f, c, c\#) && \\
{}_p\vdash\ & (f, \varepsilon, \#) &&
\end{aligned}
$$

In brief, $(s, aabbcc, S\#) \vdash^* (f, \varepsilon, \#)$. Observe that $L({}_2M) = E({}_2M) = \{a^n b^n c^n \mid n \geq 1\}$, which belongs to **CS** – **CF**. □

18.2 Accepting Power

In the present section, we establish the main results of this chapter. That is, we demonstrate that deep pushdown automata that make expansions of depth m or less, where $m \geq 1$, are equivalent to m-limited state grammars, so these automata accept a proper subfamily of the language family accepted by deep pushdown automata that make expansions of depth $m + 1$ or less. Then, we point out that the resulting infinite hierarchy of language families obtained in this way occurs between the families of context-free and context-sensitive languages. However, we also show that there always exist some context-sensitive languages that cannot be accepted by any deep pushdown automata that make expansions of depth n or less, for every positive integer n.

To rephrase these results briefly and formally, we prove that

$$
{}_{deep}\mathbf{PDA}_1 = {}^{empty}_{deep}\mathbf{PDA}_1 = \mathbf{CF}
$$

and for every $n \geq 1$,

$$
{}^{empty}_{deep}\mathbf{PDA}_n = {}_{deep}\mathbf{PDA}_n \subset {}^{empty}_{deep}\mathbf{PDA}_{n+1} = {}_{deep}\mathbf{PDA}_{n+1} \subset \mathbf{CS}
$$

After proving all these results, we formulate several open problem areas, including some suggestions concerning new deterministic and generalized versions of deep pushdown automata.

Lemma 18.2.1. *For every state grammar G and for every $n \geq 1$, there exists a deep pushdown automaton of depth n, ${}_nM$, such that $L(G, n) = L({}_nM)$.*

Proof. Let $G = (V, W, T, P, S)$ be a state grammar and let $n \geq 1$. Set $N = V - T$. Define the homomorphism f over $(\{\#\} \cup V)^*$ as $f(A) = A$, for every $A \in \{\#\} \cup N$, and $f(a) = \varepsilon$, for every $a \in T$. Introduce the deep pushdown automaton of depth n

$$_nM = \big(Q, T, \{\#\} \cup V, R, s, S, \{\$\}\big)$$

where

$$Q = \{S, \$\} \cup \big\{\langle p, u\rangle \mid p \in W, u \in N^*\{\#\}^n, |u| \leq n\big\}$$

and R is constructed by performing the following four steps

(1) for each $(p, S) \to (q, x) \in P$, $p, q \in W$, $x \in V^+$, add

$1sS \to \langle p, S\rangle S$ to R;

(2) if $(p, A) \to (q, x) \in P$, $\langle p, uAv\rangle \in Q$, $p, q \in W$, $A \in N$, $x \in V^+$, $u \in N^*$, $v \in N^*\{\#\}^*$, $|uAv| = n$, $p \notin \text{states}_G(u)$, add

$|uA|\langle p, uAv\rangle A \to \langle q, \text{prefix}(uf(x)v, n)\rangle x$ to R;

(3) if $A \in N$, $p \in W$, $u \in N^*$, $v \in \{\#\}^*$, $|uv| \leq n - 1$, $p \notin \text{states}_G(u)$, add

$|uA|\langle p, uv\rangle A \to \langle p, uAv\rangle A$ and
$|uA|\langle p, uv\rangle\# \to \langle p, uv\#\rangle\#$ to R;

(4) for each $q \in W$, add

$1\langle q, \#^n\rangle\# \to \$\#$ to R.

$_nM$ simulates n-limited derivations of G so it always records the first n nonterminals occurring in the current sentential form in its state (if there appear fewer than n nonterminals in the sentential form, it completes them to n in the state by #s from behind). $_nM$ simulates a derivation step in the pushdown and, simultaneously, records the newly generated nonterminals in the state. When G successfully completes the generation of a terminal string, $_nM$ completes reading the string, empties its pushdown, and enters the final state \$.

To establish $L(G, n) = L(_nM)$, we first prove two claims.

Claim 1. Let $(p, S)\ {}_n\!\Rightarrow^m (q, dy)$ *in* G*, where* $d \in T^*, y \in (NT^*)^*, p, q \in W, m \geq 0$*. Then,* $(\langle p, S\rangle, d, S\#) \vdash^* (\langle q, \text{prefix}(f(y\#^n), n)\rangle, \varepsilon, y\#)$ *in* $_nM$.

Proof. This claim is proved by induction on $m \geq 0$.

Basis. Let $i = 0$, so $(p, S)\ {}_n\!\Rightarrow^0 (p, S)$ in $G, d = \varepsilon$ and $y = S$. By using rules introduced in steps (1) and (4),

$$(\langle p, S\rangle, \varepsilon, S\#) \vdash^* (\langle p, \text{prefix}(f(S\#^n), n)\rangle, \varepsilon, S\#) \text{ in } _nM$$

so the basis holds.

Induction Hypothesis. Assume that the claim holds for all m, $0 \leq m \leq k$, where k is a non-negative integer.

Induction Step. Let $(p, S)\ {}_n{\Rightarrow}^{k+1}\ (q, dy)$ in G, where $d \in T^*$, $y \in (NT^*)^*$, $p, q \in W$. Since $k + 1 \geq 1$, express $(p, S)\ {}_n{\Rightarrow}^{k+1}\ (q, dy)$ as

$$(p, S)\ {}_n{\Rightarrow}^{k}\ (h, buAo)\ {}_n{\Rightarrow}\ (q, buxo)\ [(h, A) \rightarrow (q, x)]$$

where $b \in T^*$, $u \in (NT^*)^*$, $A \in N$, $h, q \in W$, $(h, A) \rightarrow (q, x) \in P$, max-suffix$(buxo, (NT^*)^*) = y$, and max-prefix$(buxo, T^*) = d$. By the induction hypothesis,

$$(\langle p, S\rangle, w, S\#) \vdash^* (\langle h, \text{prefix}(f(uAo\#^n), n)\rangle, \varepsilon, uAo\#) \text{ in } M$$

where w = max-prefix$(buAo, T^*)$. As $(p, A) \rightarrow (q, x) \in P$, step (2) of the construction introduces rule

$$|uA|\langle h, \text{prefix}(f(uAo\#^n), n)\rangle A \rightarrow \langle q, \text{prefix}(f(uxo\#^n), n)\rangle x \text{ to } R$$

By using this rule, ${}_nM$ simulates $(buAo, h)\ {}_n{\Rightarrow}\ (buxo, q)$ by making

$$(\langle h, \text{prefix}(f(uAo\#^n), n)\rangle, \varepsilon, uAo\#) \vdash (\langle q, z\rangle, \varepsilon, uxo\#)$$

where $z = \text{prefix}(f(uxo\#^n), n)$ if $x \in V^+ - T^+$ and $z = \text{prefix}(f(uxo\#^n), n-1) = \text{prefix}(f(uo\#^n), n-1)$ if $x \in T^+$. In the latter case ($z = \text{prefix}(f(uo\#^n), n-1)$, so $|z| = n - 1$), ${}_nM$ makes

$$(\langle q, \text{prefix}(f(uo\#^n), n-1)\rangle, \varepsilon, uxo\#) \vdash (\langle q, \text{prefix}(f(uo\#^n), n)\rangle, \varepsilon, uxo\#)$$

by a rule introduced in (3). If $uxo \in (NT^*)^*$, $uxo = y$ and the induction step is completed. Therefore, assume that $uxo \neq y$, so $uxo = ty$ and $d = wt$, for some $t \in T^+$. Observe that $\text{prefix}(f(uxo\#^n), n) = \text{prefix}(f(y\#^n), n)$ at this point. Then, ${}_nM$ removes t by making $|t|$ popping moves so that

$$(\langle q, \text{prefix}(f(uxo\#^n), n)\rangle, t, ty\#)\ {}_p{\vdash}^{|t|}\ (\langle q, \text{prefix}(f(y\#^n), n)\rangle, \varepsilon, y\#^n)$$

Thus, putting the previous sequences of moves together, we obtain

$$\begin{aligned}(p, wt, S\#^n)\ &\vdash^*\ (\langle q, \text{prefix}(f(uxo\#^n), n)\rangle, t, ty\#)\ [1sS \rightarrow qAA]\\ &{}_p{\vdash}^{|t|}\ (\langle q, \text{prefix}(f(y\#^n), n)\rangle, \varepsilon, y\#)\end{aligned}$$

which completes the induction step. □

By the previous claim for $y = \varepsilon$, if $(p, S)\ {}_n{\Rightarrow}^*\ (q, d)$ in G, where $d \in T^*$, $p, q \in W$, then

$$(\langle p, S\rangle, d, S\#) \vdash^* (\langle q, \text{prefix}(f(\#^n), n)\rangle, \varepsilon, \#) \text{ in } {}_nM$$

As prefix$(f(\#^n), n) = \#$ and R contains rules introduced in (1) and (4), we also have

$$\begin{aligned}(s, d, S\#) &\vdash (\langle p, S\rangle, d, S\#)\\ &\vdash^* (\langle q, \#^n, n\rangle\rangle, \varepsilon, \#)\\ &\vdash^* (\$, \varepsilon, \#) \text{ in } {}_nM\end{aligned}$$

Thus, $d \in L(G)$ implies that $d \in L({}_nM)$, so $L(G, n) \subseteq L({}_nM)$.

Claim 2. Let $(\langle p, S\#^{n-1}\rangle, c, S\#) \vdash^m (\langle q, \text{prefix}(f(y\#^n), n)\rangle, \varepsilon, by\#)$ *in* ${}_nM$ *with* $c, b \in T^*$, $y \in (NT^*)^*$, $p, q \in W$, *and* $m \geq 0$. *Then,* $(p, S)\ {}_n\Rightarrow^* (q, cby)$ *in* G.

Proof. This claim is proved by induction on $m \geq 0$.

Basis. Let $m = 0$. Then, $c = b = \varepsilon$, $y = S$, and

$$(\langle p, S\#^{n-1}\rangle, \varepsilon, S\#) \vdash^0 (\langle q, \text{prefix}(f(S\#^n), n)\rangle, \varepsilon, S\#) \text{ in } {}_nM$$

As $(p, S)\ {}_n\Rightarrow^0 (p, S)$ in G, the basis holds.

Induction Hypothesis. Assume that the claim holds for all m, $0 \leq m \leq k$, where k is a non-negative integer.

Induction Step. Let

$$(\langle p, S\#^{n-1}\rangle, c, S\#) \vdash^{k+1} (\langle q, \text{prefix}(f(y\#^n), n)\rangle, \varepsilon, by\#) \text{ in } {}_nM$$

where $c, b \in T^*$, $y \in (NT^*)^*$, $p, q \in W$ in ${}_nM$. Since $k + 1 \geq 1$, we can express

$$(\langle p, S\#^{n-1}\rangle, c, S\#) \vdash^{k+1} (\langle q, \text{prefix}(f(y\#^n), n)\rangle, \varepsilon, by\#)$$

as

$$\begin{aligned}(\langle p, S\#^{n-1}\rangle, c, S\#) &\vdash^k \alpha\\ &\vdash (\langle q, \text{prefix}(f(y\#^n), n)\rangle, \varepsilon, by\#) \text{ in } {}_nM\end{aligned}$$

where α is a configuration of ${}_nM$ whose form depend on whether the last move is (i) a popping move or (ii) an expansion, described next.

(i) Assume that $\alpha\ {}_p\vdash (\langle q, \text{prefix}(f(y\#^n), n)\rangle, \varepsilon, by\#)$ in ${}_nM$. In a greater detail, let $\alpha = (\langle q, \text{prefix}(f(y\#^n), n)\rangle, a, aby\#)$ with $a \in T$ such that $c = \text{prefix}(c, |c| - 1)a$. Thus,

$$\begin{aligned}(\langle p, S\#^{n-1}\rangle, c, S\#) &\vdash^k (\langle q, \text{prefix}(f(y\#^n), n)\rangle, a, aby\#)\\ &{}_p\vdash (\langle q, \text{prefix}(f(y\#^n), n)\rangle, \varepsilon, by\#)\end{aligned}$$

Since $(\langle p, S\#^{n-1}\rangle, c, S\#) \vdash^k (\langle q, \text{prefix}(f(y\#^n), n)\rangle, a, aby\#)$, we have

$$(\langle p, S\#^{n-1}\rangle, \text{prefix}(c, |c| - 1), S\#) \vdash^k (\langle q, \text{prefix}(f(y\#^n), n)\rangle, \varepsilon, aby\#)$$

By the induction hypothesis, $(p, S)\ {}_n\Rightarrow^* (q, \text{prefix}(c, |c| - 1)aby)$ in G. As $c = \text{prefix}(c, |c| - 1)a$, $(p, S)\ {}_n\Rightarrow^* (q, cby)$ in G.

(ii) Assume that $\alpha\ {}_e\vdash (\langle q, \text{prefix}(f(y\#^n), n)\rangle, \varepsilon, by\#)$ in ${}_nM$. Observe that this expansion cannot be made by rules introduced in steps (1) or (4). If this expansion is made by a rule introduced in (3), which does not change the pushdown contents at all, the induction step follows from the induction hypothesis. Finally, suppose that this expansion is made by a rule introduced in step (2). In a greater detail, suppose that

$$\alpha = (\langle o, \text{prefix}(f(uAv\#^n), n)\rangle, \varepsilon, uAv\#)$$

and ${}_nM$ makes

$$(\langle o, \text{prefix}(f(uAv\#^n), n)\rangle, \varepsilon, uAv\#)\ {}_e\vdash (\langle q, \text{prefix}(f(uxv\#^n), n)\rangle, \varepsilon, uxv\#)$$

by using

$$|f(uA)|\langle o, \text{prefix}(f(uAv\#^n), n)\rangle A \rightarrow \langle q, \text{prefix}(f(uxv\#^n), n)\rangle x \in R$$

introduced in step (2) of the construction, where $A \in N$, $u \in (NT^*)^*$, $v \in (N \cup T)^*$, $o \in W$, $|f(uA)| \leq n$, $by\# = uxv\#$. By the induction hypothesis,

$$(\langle p, S\#^{n-1}\rangle, c, S\#) \vdash^k (\langle o, \text{prefix}(f(uAv\#^n), n)\rangle, \varepsilon, uAv\#) \text{ in } {}_nM$$

implies that $(p, S)\ {}_n\Rightarrow^* (o, cuAv)$ in G. From

$$|f(uA)|\langle o, \text{prefix}(f(uAv\#^n), n)\rangle A \rightarrow \langle q, \text{prefix}(f(uxv\#^n), n)\rangle x \in R$$

it follows that $(o, A) \rightarrow (q, x) \in P$ and $A \notin \text{states}_G(f(u))$. Thus,

$$\begin{aligned}(p, S)\ &{}_n\Rightarrow^* (o, cuAv)\\ &{}_n\Rightarrow\ (q, cuxv) \text{ in } G\end{aligned}$$

Therefore, $(p, S)\ {}_n\Rightarrow^* (q, cby)$ in G because $by\# = uxv\#$. □

Consider the previous claim for $b = y = \varepsilon$ to see that

$$(\langle p, S\#^{n-1}\rangle, c, S\#) \vdash^* (\langle q, \text{prefix}(f(\#), n)\rangle, \varepsilon, \#^n) \text{ in } {}_nM$$

implies that $(p, S)\ {}_n\Rightarrow^* (q, c)$ in G. Let $c \in L({}_nM)$. Then,

$$(s, c, S\#) \vdash^* (\$, \varepsilon, \#) \text{ in } {}_nM$$

Examine the construction of ${}_nM$ to see that $(s, c, S) \vdash^* (\$, \varepsilon, \#)$ starts by using a rule introduced in (1), so $(s, c, S) \vdash^* (\langle p, S\#^{n-1}\rangle, c, S\#)$. Furthermore, notice that this sequence of moves ends $(s, c, S) \vdash^* (\$, \varepsilon, \varepsilon)$ by using a rule introduced in step (4). Thus, we can express

$$(s, c, \#) \vdash^* (\$, \varepsilon, \#)$$

as

$$\begin{aligned}(s, c, \#) &\vdash^* (\langle p, S\#^{n-1}\rangle, c, S\#)\\ &\vdash^* (\langle q, \text{prefix}(f(\#^n), n)\rangle, \varepsilon, \#)\\ &\vdash \ \ (\$, \varepsilon, \#) \text{ in } {}_nM\end{aligned}$$

Therefore, $c \in L({}_nM)$ implies that $c \in L(G, n)$, so $L({}_nM) \subseteq L(G, n)$.

As $L({}_nM) \subseteq L(G, n)$ and $L(G, n) \subseteq L({}_nM)$, $L(G, n) = L({}_nM)$. Thus, Lemma 18.2.1 holds. □

Lemma 18.2.2. *For every $n \geq 1$ and every deep pushdown automaton ${}_nM$, there exists a state grammar G such that $L(G, n) = L({}_nM)$.*

Proof. Let $n \geq 1$ and ${}_nM = (Q, T, V, R, s, S, F)$ be a deep pushdown automaton. Let Z and $\$$ be two new symbols that occur in no component of ${}_nM$. Set $N = V - T$. Introduce sets

$$C = \{\langle q, i, \triangleright\rangle \mid q \in Q, 1 \leq i \leq n - 1\}$$

and

$$D = \{\langle q, i, \triangleleft\rangle \mid q \in Q, 0 \leq i \leq n - 1\}$$

Moreover, introduce an alphabet W such that $\text{card}(V) = \text{card}(W)$, and for all i, $1 \leq i \leq n$, an alphabet U_i such that $\text{card}(U_i) = \text{card}(N)$. Without any loss of generality, assume that V, Q, and all these newly introduced sets and alphabets are pairwise disjoint. Set $U = \bigcup_{i=1}^{n} U_i$. For each i, $1 \leq i \leq n-1$, set $C_i = \{\langle q, i, \triangleright\rangle \mid q \in Q\}$ and for each i, $0 \leq i \leq n - 1$, set $D_i = \{\langle q, i, \triangleleft\rangle \mid q \in Q\}$. Introduce a bijection h from V to W. For each i, $1 \leq i \leq n$, introduce a bijection ${}_ig$ from N to U_i. Define the state grammar

$$G = \big(V \cup W \cup U \cup \{Z\}, Q \cup C \cup D \cup \{\$\}, T, P, Z\big)$$

where P is constructed by performing the following steps

(1) add $(s, Z) \to (\langle s, 1, \triangleright\rangle, h(S))$ to P;
(2) for each $q \in Q$, $A \in N$, $1 \leq i \leq n - 1$, $x \in V^+$, add

(2.1) $(\langle q, i, \triangleright\rangle, A) \to (\langle q, i + 1, \triangleright\rangle, {}_ig(A))$ and
(2.2) $(\langle q, i, \triangleleft\rangle, {}_ig(A)) \to (\langle p, i - 1, \triangleleft\rangle, A)$ to P;

(3) if $ipA \rightarrow qxY \in R$, for some $p, q \in Q$, $A \in N$, $x \in V^*$, $Y \in V$, $i = 1, \ldots, n$, add

$$(\langle p, i, \triangleright\rangle, A) \rightarrow (\langle q, i-1, \triangleleft\rangle, xY) \text{ and}$$
$$(\langle p, i, \triangleright\rangle, h(A)) \rightarrow (\langle q, i-1, \triangleleft\rangle, xh(Y)) \text{ to } P;$$

(4) for each $q \in Q$, $A \in N$, add

$$(\langle q, 0, \triangleleft\rangle, A) \rightarrow (\langle q, 1, \triangleright\rangle, A) \text{ and}$$
$$(\langle q, 0, \triangleleft\rangle, h(Y)) \rightarrow (\langle q, 1, \triangleright\rangle, h(Y)) \text{ to } P;$$

(5) for each $q \in F$, $a \in T$, add

$$(\langle q, 0, \triangleleft\rangle, h(a)) \rightarrow (\$, a) \text{ to } P.$$

G simulates the application of $ipA \rightarrow qy \in R$ so it makes a left-to-right scan of the sentential form, counting the occurrences of nonterminals until it reaches the ith occurrence of a nonterminal. If this occurrence equals A, it replaces this A with y and returns to the beginning of the sentential form in order to analogously simulate a move from q. Throughout the simulation of moves of ${}_nM$ by G, the rightmost symbol of every sentential form is from W. G completes the simulation of an acceptance of a string x by ${}_nM$ so it uses a rule introduced in step (5) of the construction of P to change the rightmost symbol of x, $h(a)$, to a and, thereby, to generate x.

We next establish $L(G, n) = L({}_nM)$. To keep the rest of the proof as readable as possible, we omit some details in what follows. The reader can easily fill them in.

Claim 1. $L(G, n) \subseteq L({}_nM)$

Proof. Consider any $w \in L(G, n)$. Observe that G generates w as

$$\begin{aligned} (p, Z) \;{}_n\!\Rightarrow\; & (\langle s, 1, \triangleright\rangle, h(S)) \; [(s, Z) \rightarrow (\langle s, 1, \triangleright\rangle, h(S))] \\ {}_n\!\Rightarrow^*\; & (f, yh(a)) \\ {}_n\!\Rightarrow\; & (\$, w) \end{aligned}$$

where $f \in F$, $a \in T$, $y \in T^*$, $ya = w$, $(s, Z) \rightarrow (\langle s, 1, \triangleright\rangle, h(S))$ in step (1) of the construction of P, $(\langle q, 0, \triangleleft\rangle, h(a)) \rightarrow (\$, a)$ in (5), every

$$u \in \text{strings}\big((\langle s, 1, \triangleright\rangle, h(S)) \;{}_n\!\vdash^* (q, yh(a))\big)$$

satisfies $u \in (V \cup U)^* W$, and every step in

$$(\langle s, 1, \triangleright\rangle, h(S)) \;{}_n\!\vdash^* (f, yh(S))$$

is made by a rule introduced in (2) through (4). Indeed, the rule constructed in (1) is always used in the first step and a rule constructed in (5) is always used in the very last step of any successful generation in G; during any other step, neither of them can be applied. Notice that the rule of (1) generates $h(S)$. Furthermore, examine the rules of (2) through (4) to see that by their use, G always produces a string that has exactly one occurrence of a symbol from W in any string and this occurrence appears as the rightmost symbol of the string; formally,

$$u \in \text{strings}\big((\langle s, 1, \triangleright\rangle, h(S)) \;{}_n\!\Rightarrow^* (f, yh(a))\big)$$

implies that $u \in (V \cup U)^* W$. In a greater detail,

$$(\langle s, 1, \triangleright\rangle, h(S)) \;{}_n\!\Rightarrow^* (f, yh(a))$$

can be expressed as

$$\begin{array}{llllll}
(q_0, z_0) & {}_n\!\Rightarrow^* (c_0, y_0) & {}_n\!\Rightarrow (d_0, u_0) & {}_n\!\Rightarrow^* (p_0, v_0) & {}_n\!\Rightarrow \\
(q_1, z_1) & {}_n\!\Rightarrow^* (c_1, y_1) & {}_n\!\Rightarrow (d_1, u_1) & {}_n\!\Rightarrow^* (p_1, v_1) & {}_n\!\Rightarrow \\
 & \vdots & \vdots & \vdots & \vdots \\
(q_m, z_m) & {}_n\!\Rightarrow^* (c_m, y_m) & {}_n\!\Rightarrow (d_m, u_m) & {}_n\!\Rightarrow^* (p_m, v_m) & {}_n\!\Rightarrow \\
(q_{m+1}, z_{m+1}) & & & &
\end{array}$$

for some $m \geq 1$, where $z_0 = h(S)$, $z_{m+1} = yh(a)$, $f = q_{m+1}$, and for each j, $0 \leq j \leq m$, $q_j \in C_1$, $p_j \in D_0$, $z_j \in V^* W$, and there exists $i_j \in \{1, \dots, n\}$ so $c_j \in C_{i_j}$, $y_j \in T^* C_1 T^* C_2 \cdots T^* C_{i_j - 1} V^* W$, $d_j \in D_{i_j - 1}$, $u_j \in T^* C_1 T^* C_2 \cdots T^* D_{i_j - 1} V^* W$, and

$$(q_j, z_j) \;{}_n\!\Rightarrow^* (c_j, y_j) \;{}_n\!\Rightarrow (d_j, u_j) \;{}_n\!\Rightarrow^* (p_j, v_j) \;{}_n\!\Rightarrow (q_{j+1}, z_{j+1})$$

satisfies (i)–(iv), given next.

For brevity, we first introduce the following notation. Let w be any string. For $i = 1, \dots, |w|$, $\lfloor w, i, N \rfloor$ denotes the ith occurrence of a nonterminal from N in w, and if such a nonterminal does not exist, $\lfloor w, i, W \rfloor = 0$; for instance, $\lfloor ABABC, 2, \{A, C\} \rfloor$ denotes the underlined $\underline{A}$ in $AB\underline{A}BC$.

(i) $(q_j, z_j) \;{}_n\!\Rightarrow^* (c_i, y_i)$ consists of $i_j - 1$ steps during which G changes $\lfloor z_j, 1, N \rfloor, \dots, \lfloor z_j, i_j - 1, N \rfloor$ to ${}_1 g(\langle \lfloor z_j, 1, N \rfloor, 2 \rangle), \dots, {}_{i_j} g(\langle \lfloor z_j, i_j - 1, N \rfloor, i_j - 1 \rangle)$, respectively, by using rules of (2.1) in the construction of P;

(ii) if $i_j \leq \text{occur}(z_j, N)$, then $(c_j, y_j) \;{}_n\!\Rightarrow (d_j, u_j)$ have to consist of a step according to $(\langle q, i, \triangleright\rangle, A_j) \to (\langle q, i - 1, \triangleleft\rangle, x_j X_j)$, where $\lfloor z_j, i_j, N \rfloor$ is an occurrence of $A_j, x_j \in V^*, X_j \in V$, and if $i_j = \text{occur}(z_j, N \cup W)$, then $(c_j, y_j) \;{}_n\!\Rightarrow (d_j, u_j)$ consists of a step according to $(\langle p, i, \triangleright\rangle, h(A_j)) \to (\langle q, i - 1, \triangleleft\rangle, x_j h(X_j))$ constructed in (3), where $\lfloor z_j, i_j, N \cup W \rfloor$ is an occurrence of $h(A_j), x_j \in V^*, X_j \in V$;

(iii) $(d_j, u_j) \;{}_n\!\Rightarrow^* (p_j, v_j)$ consists of $i_j - 1$ steps during which G changes ${}_{i_j} g(\langle \lfloor z_j, i_j - 1, N \rfloor, i_j - 1 \rangle), \dots, {}_1 g(\langle \lfloor z_j, 1, N \rfloor, 1 \rangle)$ back to $\lfloor z_j, i_j - 1, N \rfloor, \dots, \lfloor z_j, 1, N \rfloor$, respectively, in a right-to-left way by using rules constructed in (2.2);

(iv) $(p_j, v_j) \; _n\Rightarrow (q_{j+1}, z_{j+1})$ is made by a rule constructed in (4).

For every

$$\begin{aligned}(q_j, z_j) \; _n\Rightarrow^* & \; (c_j, y_j) \\ _n\Rightarrow & \; (d_j, u_j) \\ _n\Rightarrow^* & \; (p_j, v_j) \\ _n\Rightarrow & \; (q_{j+1}, z_{j+1}) \text{ in } G\end{aligned}$$

where $0 \leq j \leq m$, $_nM$ makes

$$(q_j, o_j, \text{suffix}(z_j, t_j)) \vdash^* (q_{j+1}, o_{j+1}, \text{suffix}(z_{j+1}, t_{j+1}))$$

with $o_0 = w$, $z_0 = S\#$, $t_{j+1} = |\,\text{max-prefix}(z_{j+1}, T^*)|$, $o_{j+1} = \text{suffix}(o_j, |o_j| + t_{j+1})$, where $o_0 = w$, $z_0 = S\#$, and $t_0 = |z_0|$. In this sequence of moves, the first move is an expansion made according to $i_j q_j A_j \to q_{j+1} x_j X_j$ (see steps (2) and (3) of the construction) followed by t_{j+1} popping moves (notice that $i_j \geq 2$ implies that $t_{j+1} = 0$). As $f \in F$ and $ya = w$, $w \in L(_nM)$. Therefore, $L(G, n) \subseteq L(_nM)$.

□

Claim 2. $L(_nM) \subseteq L(G, n)$

Proof. This proof is simple and left to the reader. □

As $L(_nM) \subseteq L(G, n)$ and $L(G, n) \subseteq L(_nM)$, we have $L(G, n) = L(_nM)$, so this lemma holds true. □

Theorem 18.2.3. *For every $n \geq 1$ and for every language L, $L = L(G, n)$ for a state grammar G if and only if $L = L(_nM)$ for a deep pushdown automaton $_nM$.*

Proof. This theorem follows from Lemmas 18.2.1 and 18.2.2. □

By analogy with the demonstration of Theorem 18.2.3, we can establish the next theorem.

Theorem 18.2.4. *For every $n \geq 1$ and for every language L, $L = L(G, n)$ for a state grammar G if and only if $L = E(_nM)$ for a deep pushdown automaton $_nM$.*

□

The main result of this chapter follows next.

Corollary 18.2.5. *For every $n \geq 1$,*

$${}_{deep}^{empty}\mathbf{PDA}_n = {}_{deep}\mathbf{PDA}_n \subset {}_{deep}\mathbf{PDA}_{n+1} = {}_{deep}^{empty}\mathbf{PDA}_{n+1}$$

Proof. This corollary follows from Theorems 18.2.3 and 18.2.4 above and from Theorem 5.4.3, which says that the m-limited state grammars generate a proper subfamily of the family generated by $(m + 1)$-limited state grammars, for every $m \geq 1$. □

Finally, we state two results concerning **CF** and **CS**.

Corollary 18.2.6. ${}_{deep}\mathbf{PDA}_1 = {}^{empty}_{deep}\mathbf{PDA}_1 = \mathbf{CF}$

Proof. This corollary follows from Lemmas 18.2.1 and 18.2.2 for $n = 1$, and from Theorem 5.4.3, which says that one-limited state grammars characterize **CF**. □

Corollary 18.2.7. *For every* $n \geq 1$, ${}_{deep}\mathbf{PDA}_n = {}^{empty}_{deep}\mathbf{PDA}_n \subset \mathbf{CS}$.

Proof. This corollary follows from Lemmas 18.2.1 and 18.2.2, Theorems 18.2.3 and 18.2.4, and from Theorem 5.4.3, which says that $\mathbf{ST}_m$, for every $m \geq 1$, is properly included in **CS**. □

Open Problems

Finally, we suggest two open problem areas concerning deep pushdown automata.

Determinism

This chapter has discussed a general version of deep pushdown automata, which work non-deterministically. Undoubtedly, the future investigation of these automata should pay a special attention to their deterministic versions, which fulfill a crucial role in practice. In fact, we can introduce a variety of deterministic versions, including the following two types. First, we consider the fundamental strict form of determinism.

Definition 18.2.8. Let $M = (Q, \Sigma, \Gamma, R, s, S, F)$ be a deep pushdown automaton. We say that M is *deterministic* if for every $mqA \to pv \in R$,

$$\mathrm{card}\big(\{mqA \to ow \mid mqA \to ow \in R, o \in Q, w \in \Gamma^+\} - \{mqA \to pv\}\big) = 0$$

□

As a weaker form of determinism, we obtain the following definition.

Definition 18.2.9. Let $M = (Q, \Sigma, \Gamma, R, s, S, F)$ be a deep pushdown automaton. We say that M is *deterministic with respect to the depth of its expansions* if for every $q \in Q$

$$\mathrm{card}\big(\{m \mid mqA \to pv \in R, A \in \Gamma, p \in Q, v \in \Gamma^+\}\big) \leq 1$$

because at this point from the same state, all expansions that M can make are of the same depth. □

To illustrate, consider, for instance, the deep pushdown automaton ${}_2M$ from Example 18.1.2. This automaton is deterministic with respect to the depth of its expansions; however, it does not satisfy the strict determinism. Notice that ${}_nM$ constructed in the proof of Lemma 18.2.1 is deterministic with respect to the depth of its expansions, so we obtain this corollary.

Corollary 18.2.10. *For every state grammar G and for every $n \geq 1$, there exists a deep pushdown automato ${}_nM$ such that $L(G, n) = L({}_nM)$ and ${}_nM$ is deterministic with respect to the depth of its expansions.* □

Open Problem 18.2.11. Can an analogical statement to Corollary 18.2.10 be established in terms of the strict determinism? □

Generalization

Let us note that throughout this chapter, we have considered only true pushdown expansions in the sense that the pushdown symbol is replaced with a nonempty string rather than with the empty string; at this point, no pushdown expansion can result in shortening the pushdown length. Nevertheless, the discussion of moves that allow deep pushdown automata to replace a pushdown symbol with ε and, thereby, shorten its pushdown represent a natural generalization of deep pushdown automata discussed in this chapter.

Open Problem 18.2.12. What is the language family defined by deep pushdown automata generalized in this way? □

References

1. Aho, A.V., Ullman, J.D.: The Theory of Parsing, Translation and Compiling, Volume I: Parsing. Prentice-Hall, New Jersey (1972)
2. Courcelle, B.: On jump deterministic pushdown automata. Math. Syst. Theory **11**, 87–109 (1977)
3. Ginsburg, S., Spanier, E.H.: Control sets on grammars. Theory Comput. Syst. **2**(2), 159–177 (1968)
4. Ginsburg, S., Greibach, S.A., Harrison, M.: One-way stack automata. J. ACM **14**(2), 389–418 (1967)
5. Greibach, S.A.: Checking automata and one-way stack languages. J. Comput. Syst. Sci. **3**, 196–217 (1969)
6. Harrison, M.: Introduction to Formal Language Theory. Addison-Wesley, Boston (1978)
7. Kolář, D., Meduna, A.: Regulated pushdown automata. Acta Cybernetica **2000**(4), 653–664 (2000)
8. Lewis, H.R., Papadimitriou, C.H.: Elements of the Theory of Computation. Prentice-Hall, New Jersey (1981)
9. Meduna, A.: Automata and Languages: Theory and Applications. Springer, London (2000)
10. Meduna, A.: Simultaneously one-turn two-pushdown automata. Int. J. Comput. Math. **2003**(80), 679–687 (2003)
11. Rozenberg, G., Salomaa, A. (eds.): Handbook of Formal Languages, vol. 1: Word, Language, Grammar. Springer, New York (1997)
12. Sakarovitch, J.: Pushdown automata with terminating languages. In: Languages and Automata Symposium, RIMS 421, Kyoto University, pp. 15–29 (1981)
13. Valiant, L.: The equivalence problem for deterministic finite turn pushdown automata. Inf. Control **81**, 265–279 (1989)

Part VIII
Applications

This part demonstrates applications of regulated language models discussed earlier in this book. First, it describes these applications and their perspectives from a general viewpoint. Then, it adds several case studies to show quite specific real-world applications. It narrows its attention to regulated grammars rather than regulated automata, and it selects only three application areas—biology, compilers, and linguistics. Part VIII consists of Chaps. 19 and 20.

Chapter 19 comments applications of regulated grammars and their perspectives in general. It describes their potential applications in terms of the specification and translation of languages, including both programming and natural languages. This chapter also explains how the grammatical regulation of information processing fulfills a crucially important role in many areas of biology.

To illustrate the general comments quite realistically, Chap. 20 presents several specific case studies concerning the three scientific areas selected in Chap. 19. Specifically, these case studies deal with compiler writing, computational linguistics, and molecular biology.

Chapter 19
Applications: Overview

Abstract This chapter makes several general remarks about applications of regulated language models discussed earlier in this book. It concentrates its attention to three application areas—biology, compilers, and linguistics. The chapter consists of two sections. Section 19.1 describes applications of regulated grammars at present. Section 19.2 suggests their applications in the near future.

Keywords Regulated grammars • Applications • Perspectives • Linguistics • Compilers • Biology

As already stated in Chap. 1, this book is primarily and principally meant as a theoretical treatment of regulated grammars and automata. Nevertheless, to demonstrate their practical importance, we make some general remarks regarding their applications in the present two-section chapter. However, since applications of regulated grammars and automata overflow so many scientific areas, we only cover a very restricted selection of these applications. Indeed, in Sect. 19.1, we concentrate our attention primarily to applications related to grammar-based regulation in three application areas concerning them—linguistics, compilers, and biology. We give an insight into the current applications of regulated grammars in these three areas, which are further illustrated by quite specific case studies in Chap. 20. In Sect. 19.2, we make several general remarks concerning application-oriented perspectives of regulated grammars and automata in the near future, and we illustrate them by examples.

19.1 Current Applications

In the present section, we demonstrate applications of regulated grammars in linguistics, compilers, and biology. In a greater detail, we further illustrate them by case studies covered in Chap. 20.

A. Meduna and P. Zemek, *Regulated Grammars and Automata*,

DOI 10.1007/978-1-4939-0369-6_19,

Linguistics

In terms of English syntax, grammatical regulation can specify a number of relations between individual syntax-related elements of sentences in natural languages. For instance, relative clauses are introduced by *who* or *which* depending on the subject of the main clause. If the subject in the main clause is a person, the relative clause is introduced by *who*; otherwise, it starts by *which*. We encourage the reader to design a regulated grammar that describes this dependency (consult [8]).

In other natural languages, there exist syntax relations that can be elegantly handled by regulated grammars, too. To illustrate, in Spanish, all adjectives inflect according to gender of the noun they characterize. Both the noun and the adjective may appear at different parts of a sentence, which makes their syntactical dependency difficult to capture by classical grammars; obviously, regulated grammars can describe this dependency in a more elegant and simple way.

Apart from description, specification, and transformation of language syntax, regulated grammars can be applied to other linguistically oriented fields, such as *morphology* (see [2, 4]).

Compilers

Rather than cover a complete process of computer compilation related to grammar-based regulation, we restrict our attention primarily to *parsing* (see [1, 3, 5, 9, 11, 16]), already discussed in Sects. 6.7 and 8.1 in a rather theoretical way. In addition, we briefly sketch applications in compilation phases controlled by parsing, such as syntax-directed translation.

Ordinary parsers represent crucially important components of compilers, and they are traditionally underlined by ordinary context-free grammars. As their name indicates, regulated parsers are based upon regulated context-free grammars. Considering their advantages, including properties (I) through (IV) listed next, it comes as no surprise that they become increasingly popular in modern compilers design.

(I) Regulated parsers work in a faster way than classical parsers do. Indeed, ordinary parsers control their parsing process so they consult their parsing tables during every single step. As opposed to this exhaustively busy approach, in regulated parsers, regulated grammatical mechanisms take control over the parsing process to a large extent; only during very few pre-determined steps, they consult their parsing tables to decide how to continue the parsing process under the guidance of regulating mechanism. Such a reduction of communication with the parsing tables obviously results into a significant acceleration of the parsing process as a whole.

(II) Regulated context-free grammars are much stronger than ordinary context-free grammars. Accordingly, parsers based upon regulated grammars are more powerful than their ordinary versions. As an important practical consequence, they can parse syntactical structures that cannot be parsed by ordinary parsers.
(III) Regulated parsers make use of their regulated mechanisms to perform their parsing process in a deterministic way.
(IV) Compared to ordinary parsers, regulated parsers are often written more succinctly and, therefore, readably as follows from reduction-related results concerning the number of their components, such as nonterminals and rules, achieved earlier in this book (see Sects. 4.4.2, 4.6.2, 4.7.4, 6.4, 10.2.2, 10.3.2, and 10.4.2).

From a more general point of view, some fundamental parts of compilers, such as syntax-directed translators, run within the compilation process under the parser-based regulation. Furthermore, through their symbol tables, parsers also regulate exchanging various pieces of information between their components, further divided into several subcomponents. Indeed, some parts of modern compilers may be further divided into various subparts, which are run in a regulated way, and within these subparts, a similar regulation can be applied again, and so on. As a matter of fact, syntax-directed translation is frequently divided into two parts, which work concurrently. One part is guided by a precedence parser that works with expressions and conditions while the other part is guided by a predictive parser that processes the general program flow. In addition, both parts are sometimes further divided into several subprocesses or threads. Of course, this two-parser design of syntax-directed translation requires an appropriate regulation of compilation as a whole. Indeed, prior to this syntax-directed translation, a pre-parsing decomposition of the tokenized source program separates the syntax constructs for both parsers. On the other hand, after the syntax-directed translation based upon the two parsers is successfully completed, all the produced fragments of the intermediate code are carefully composed together so the resulting intermediate code is functionally equivalent to the source program. Of course, handling compilation like this requires a proper regulation of all these compilation subphases.

To give one more example in terms of modern compiler design, various optimization methods are frequently applied to the generation of the resulting target code to speed the code up as much as possible. This way of code generation may result from an explicit requirement in the source program. More often, however, modern compilers themselves recognize that a generation like this is appropriate within the given computer framework, so they generate the effective target code to speed up its subsequent execution. Whatever they do, however, they always have to guarantee that the generated target code is functionally equivalent to the source program. Clearly, this design of compilers necessitates an extremely careful control over all the optimization routines involved, and this complicated control has to be based upon a well developed theory of computational regulation. Within formal language theory, which has always provided compilation techniques with their formal models, this control can be accomplished by regulated grammars, which naturally and elegantly formalize computational regulation.

Biology

As the grammatical regulation of information processing fulfills a crucially important role in biology as a whole, it is literally impossible to cover all these applications in this scientific filed. Therefore, we restrict our attention only to microbiology, which also makes use of the systematically developed knowledge concerning these grammars significantly. Even more specifically, we narrow our attention to *molecular genetics* (see [14, 17, 18]). A regulation of information processing is central to this scientific field although it approaches this processing in a specific way. Indeed, genetically oriented studies usually investigate how to regulate the modification of several symbols within strings that represent a molecular organism. To illustrate a modification like this, consider a typical molecular organism consisting of several groups of molecules; for instance, take any organism consisting of several parts that slightly differ in behavior of DNA molecules made by specific sets of enzymes. During their development, these groups of molecules communicate with each other, and this communication usually influences the future behavior of the whole organism. A simulation of such an organism might be formally based upon regulated grammars, which can control these changes at various places. Consequently, genetic dependencies of this kind represent another challenging application area of regulated grammars in the future.

To sketch the applicability of regulated grammars in this scientific area in a greater detail, consider one-sided forbidding grammars, studied earlier in Sect. 6.2. These grammars can formally and elegantly simulate processing information in molecular genetics, including information concerning macromolecules, such as DNA, RNA, and polypeptides. For instance, consider an organism consisting of DNA molecules made by enzymes. It is a common phenomenon that a molecule m made by a specific enzyme can be modified unless molecules made by some other enzymes occur either to the left or to the right of m in the organism. Consider a string w that formalizes this organism so every molecule is represented by a symbol. As obvious, to simulate a change of the symbol a that represents m requires forbidding occurrences of some symbols that either precede or follow a in w. As obvious, one-sided forbidding grammars can provide a string-changing formalism that can capture this forbidding requirement in a very succinct and elegant way. To put it more generally, one-sided forbidding grammars can simulate the behavior of molecular organisms in a rigorous and uniform way. Application-oriented topics like this obviously represent a future investigation area concerning one-sided forbidding grammars.

As already stated, we give several in-depth case studies concerning linguistics, compilers, and biology in Chap. 20.

19.2 Perspectives

In the near future, highly regulated information processing is expected to intensify rapidly and significantly. Indeed, to take advantage of highly effective parallel and mutually connected computers as much as possible, a modern software product simultaneously run several processes, each of which gather, analyze and modify various elements occurring within information of an enormous size, largely spread and constantly growing across the virtually endless and limitless computer environment. During a single computational step, a particular running process selects a finite set of mutually related information elements, from which it produces new information as a whole and, thereby, completes the step. In many respects, the newly created information affects the way the process performs the next computational step, and from a more broadly perspective, it may also significantly change the way by which the other processes work as well. Clearly, a product conceptualized in this modern way requires a very sophisticated regulation of its computation performed within a single process as well as across all the processes involved.

As already explained in Chap. 1, computer science urgently needs to express regulated computation by appropriate mathematical models in order to express its fundamentals rigorously. Traditionally, formal language theory provides computer science with various automata and grammars as formal models of this kind. However, classical automata and grammars, such as ordinary finite automata or context-free grammars, represent unregulated formal models because they were introduced several decades ago when hardly any highly regulated computation based upon parallelism and distribution occurred in computer science. As an inescapable consequence, these automata and grammars fail to adequately formalize highly regulated computation. Consequently, so far, most theoretically oriented computer science areas whose investigation involve this computation simplify their investigation so they reduce their study to quite specific areas in which they work with various ad-hoc simplified models without any attempt to formally describe highly regulated computation generally and systematically. In this sense, theoretical computer science based upon unregulated formal models is endangered by approaching computation in an improper way, which does not reflect the expected regulated computation in the future at all. Simply put, rather than shed some light on fundamental ideas of this processing, this approach produces little or no relevant results concerning future computation.

Taking into account this unsatisfactory and dangerous situation occurring in the very heart of computational theory, the present book has paid an explicit attention to modifying automata and grammars so they work in a regulated way. As a result of this modification, the resulting regulated versions of grammars and automata can properly and adequately underlie a systematized theory concerning general ideas behind future regulated information processing. Out of all these regulated grammars and automata, we next select three types and demonstrate the way they can appropriately act as formal models of regulated computation. Namely, we choose

(1) Scattered context grammars (see Sect. 4.7);
(2) Regulated grammar systems (see Chap. 13);
(3) Regulated pushdown automata (see Sect. 16.2).

(1) In general, the heart of every grammar consists of a finite set of rules, according to which the grammar derives sentences. The collection of all sentences derived by these rules forms the language generated by the grammar. Most classical grammars perform their derivation steps in a strictly sequential way. To illustrate, context-free grammars work in this way because they rewrite a single symbol of the sentential form during every single derivation step (see [7, 10, 12, 15, 19]).

As opposed to strictly sequential grammars, the notion of a scattered context grammar is based upon finitely many sequences of context-free rules that are simultaneously applied during a single derivation step. Beginning from its start symbol, the derivation process, consisting of a sequence of derivation steps, successfully ends when the derived strings contain only terminal symbols. A terminal word derived in this successful way is included into the language of this grammar, which contains all strings derived in this way. As obvious, this way of rewriting makes scattered context grammars relevant to regulated information processing as illustrated next in terms of computational linguistics.

Consider several texts such that (a) they all are written in different natural languages, but (b) they correspond to the same syntactical structure, such as the structure of basic clauses. With respect to (b), these texts are obviously closely related, yet we do not tend to compose them into a single piece of information because of (a). Suppose that a multilingual processor simultaneously modifies all these texts in their own languages so all the modified texts again correspond to the same syntactical structure, such as a modification of basic clauses to the corresponding interrogative clauses; for instance, *I said that* would be changed to *Did I say that?* in English. At this point, a processor like this needs to regulate its computation across all these modified texts in mutually different languages. As obvious, taking advantage of their simultaneous way of rewriting, scattered context grammars can handle changes of this kind while ordinary unregulated context-free grammars cannot.

(2) Classical grammar systems combine several grammars (see [6]). All the involved grammars cooperate according to some protocol during their derivations. Admittedly, compared to isolated grammars, these grammar systems show several significant advantages, including an increase of the generative power and, simultaneously, a decrease of their descriptional complexity. In essence, the classical grammar systems can be classified into cooperating distributed (CD) and parallel communicating (PC) grammar systems (see [6]). CD grammar systems work in a sequential way. Indeed, all the grammars that form components of these systems have a common sentential form, and every derivation step is performed by one of these grammars. A cooperation protocol dictates the way by which the grammars cooperate. For instance, one grammar performs precisely k derivation steps, then another grammar works in this way, and so on, for a positive integer k. In addition, some stop conditions are given to determine when the grammar systems become

inactive and produce their sentences. For example, a stop condition of this kind says that no grammar of the system can make another derivation step. Many other cooperating protocols and stop conditions are considered in the literature (see [6] and Chap. 4 in [13] for an overview). As opposed to a CD grammar system, a PC grammar system works in parallel. The PC grammatical components have their own sentential forms, and every derivation step is performed by each of the components with its sentential form. A cooperation protocol is based on a communication between the components through query symbols. More precisely, by generating these query symbols, a component specifies where to insert the sentential form produced by another component. Nevertheless, even PC grammar systems cannot control their computation across all their grammatical components simultaneously and globally.

Regulated grammar systems, discussed in Chap. 13, are based upon classical grammar systems, sketched above, because they also involve several grammatical components. However, these regulated versions can regulate their computation across all these components by finitely many sequences of nonterminals or rules while their unregulated counterparts cannot. As illustrated next, since the unregulated grammar systems fail to capture regulated information processing across all the grammatical components, they may be inapplicable under some circumstances while regulated grammar systems are applicable.

Consider regulated information processing concerning digital images. Suppose that the processor composes and transforms several fragments of these images into a single image according to its translation rules. For instance, from several digital images that specify various parts of a face, the processor produces a complete digital image of the face. Alternatively, from a huge collection of files containing various image data, the translator selects a set of images satisfying some prescribed criteria and composes them into a single image-data file. Of course, the processor makes a multi-composition like this according to some compositional rules. As obvious, a proper composition-producing process like this necessities a careful regulation of all the simultaneously applied rules, which can be elegantly accomplished by regulated grammar systems that control their computation by sequences of rules. On the other hand, a regulation like this is hardly realizable based upon unregulated grammar systems, which lack any rule-controlling mechanism.

(3) Classical pushdown automata work by making moves during which they change states (see [7,10,12,15,19]). As a result, this state mechanism is the only way by which they can control their computation. In practice, however, their applications may require a more sophisticated regulation, which cannot be accomplished by state control. Frequently, however, the regulated versions of pushdown automata (see Chap. 16) can handle computational tasks like this by their control languages, so under these circumstance, they can act as appropriate computational models while their unregulated counterparts cannot as illustrated next in terms of parsing.

Consider a collection of files, each of which contains a portion of a source program that should by parsed as a whole by a syntax analyzer, underlain by a pushdown automaton. By using a simple control language, we can prescribe the order in which the syntax analyzer should properly compose all these fragmented pieces of code stored in several different files, after which the entire code composed in this way is parsed. As obvious, we cannot prescribe any global composition like this over the collection of files by using any classical pushdown automata, which does not regulate its computation by any control language.

To summarize this section, regulated grammars and automata represent appropriate formal models of highly regulated computation, which is likely to fulfill a central role in computer science as a whole in the near future. As such, from a theoretical perspective, they will allow us to express the theoretical fundamentals of this computation rigorously and systematically. From a more pragmatic perspective, based upon them, computer science can create a well-designed methodology concerning regulated information processing. Simply put, as their main perspective in near future, regulated grammars and automata allow us to create (a) a systematized body of knowledge representing an in-depth theory of highly regulated computation and (b) a sophisticated methodology concerning regulated information processing, based upon this computation.

References

1. Aho, A.V., Lam, M.S., Sethi, R., Ullman, J.D.: Compilers: Principles, Techniques, and Tools, 2nd edn. Addison-Wesley, Boston (2006)
2. Aronoff, M., Fudeman, K.: What is Morphology (Fundamentals of Linguistics). Wiley-Blackwell, New Jersey (2004)
3. Bal, H., Grune, D., Jacobs, C., Langendoen, K.: Modern Compiler Design. John Wiley & Sons, Hoboken (2000)
4. Bauer, L.: Introducing Linguistic Morphology, 2nd edn. Georgetown University Press, Washington, DC (2003)
5. Cooper, K.D., Torczon, L.: Engineering a Compiler. Morgan Kaufmann Publishers, San Francisco (2004)
6. Csuhaj-Varjú, E., Dassow, J., Kelemen, J., Păun, G.: Grammar Systems: A Grammatical Approach to Distribution and Cooperation. Gordon and Breach, Yverdon (1994)
7. Harrison, M.: Introduction to Formal Language Theory. Addison-Wesley, Boston (1978)
8. Huddleston, R., Pullum, G.: A Student's Introduction to English Grammar. Cambridge University Press, New York (2005)
9. Kennedy, K., Allen, J.R.: Optimizing Compilers for Modern Architectures: A Dependence-Based Approach. Morgan Kaufmann Publishers, San Francisco (2002)
10. Meduna, A.: Automata and Languages: Theory and Applications. Springer, London (2000)
11. Meduna, A.: Elements of Compiler Design. Auerbach Publications, Boston (2007)
12. Rozenberg, G., Salomaa, A. (eds.): Handbook of Formal Languages, Vol. 1: Word, Language, Grammar. Springer, New York (1997)
13. Rozenberg, G., Salomaa, A. (eds.): Handbook of Formal Languages, Vol. 2: Linear Modeling: Background and Application. Springer, New York (1997)
14. Russel, P.J.: iGenetics: A Molecular Approach, 3rd edn. Benjamin Cummings Publishing, San Francisco (2009)
15. Salomaa, A.: Formal Languages. Academic Press, London (1973)
16. Srikant, Y.N., Shankar, P.: The Compiler Design Handbook. CRC Press, London (2002)

17. Strachan, T., Read, A.: Human Molecular Genetics, 4th edn. Garland Science, New York (2010)
18. Watson, J.D., Baker, T.A., Bell, S.P., Gann, A., Levine, M., Losick, R.: Molecular Biology of the Gene, 6th edn. Benjamin Cummings Publishing, San Francisco (2007)
19. Wood, D.: Theory of Computation: A Primer. Addison-Wesley, Boston (1987)

Chapter 20
Case Studies

Abstract The present chapter covers several specific case studies concerning three scientific areas of regulated grammars—linguistics, biology, and parsing, which are discussed in the previous chapter from a general viewpoint. It consists of three sections, each of which is devoted to one of these areas. Section 20.1 demonstrates applications of scattered context grammars in linguistics. It concentrates its attention to many complicated English syntactical structures and demonstrates how scattered context grammars allow us to explore them clearly, elegantly, and precisely. Section 20.2 considers grammars with context conditions and applies them in microbiology, which appears of great interest at present. Section 20.3 applies scattered context grammars to parsing, which fulfills a crucially important role in compiler writing.

Keywords Applied regulated grammars • Case studies • Linguistics • English syntax • Microbiology • Compilers • Parsing

The previous chapter has discussed applications of regulated grammars and automata from a rather general standpoint. The present chapter gives several quite specific case studies concerning three scientific areas—linguistics, biology, and parsing. It consists of three sections, each of which is devoted to one of these areas. The first two case studies are based upon regulated grammars rather than regulated automata while the third section utilizes both regulated automata and grammars.

In Sect. 20.1, we demonstrate applications of scattered context grammars (see Sect. 4.7) in linguistics. As obvious, these grammars are useful to every linguistic field that formalizes its results by strings in which there exist some scattered context dependencies spread over the strings. Since numerous linguistic areas, ranging from discourse analysis, through psycholinguistics up to neurolinguistics, formalize and study their results by using strings involving dependencies of this kind, describing applications of scattered context grammars in all these areas would be unbearably sketchy and, therefore, didactically inappropriate. Instead of an encyclopedic approach like this, we narrow our attention to the investigation

A. Meduna and P. Zemek, *Regulated Grammars and Automata*,
DOI 10.1007/978-1-4939-0369-6_20,

of *English syntax* (see [3, 10]), which describes the rules concerning how words relate to each other in order to form well-formed grammatical English sentences. We have selected syntax of this language because the reader is surely familiar with English very well. Nevertheless, analogical ideas can be applied to members of other language families, including Indo-European, Sino-Tibetan, Niger-Congo, Afro-Asiatic, Altaic, and Japonic families of languages. We explore several common linguistic phenomena involving scattered context in English syntax and explain how to express these phenomena by scattered context grammars.

However, even within the linguistics concerning English syntax, we cannot be exhaustive in any way. Rather, we consider only selected topics concerning English syntax and demonstrate how scattered context grammars allow us to explore them clearly, elegantly, and precisely. Compared to the previous parts of this book, which are written in a strictly mathematical way, we discuss and describe scattered context grammars less formally here because we are interested in demonstrating real applications rather than theoretical properties. Specifically, we primarily use scattered context grammars to transform and, simultaneously, verify that the English sentences under discussion are grammatical.

In Sect. 20.2, we consider parallel grammars with context conditions (see Chap. 10) and apply them in biology. These grammars are useful to every biological field that formalizes its results by some strings and studies how these strings are produced from one another under some permitting or, in contrast, forbidding conditions. As numerous areas of biology formalize and study their results in this way, any description of applications that cover more than one of these areas would be unbearably sketchy, if not impossible. Therefore, we concentrate our attention on a single application area—*microbiology*, which appears of great interest at present (see [2, 8, 17]).

Finally, in Sect. 20.3, we apply LL scattered context grammars (see Sect. 4.7.5) to *parsing*, which fulfills a crucially important role in compiler writing (see [1, 4, 6, 11, 12, 16]).

20.1 Linguistics

The present section consists of Sects. 20.1.1, 20.1.2, and 20.1.3. Section 20.1.1 connects the theoretically oriented discussion of scattered context grammars given earlier in this book and the pragmatically oriented discussion of these grammars applied to English syntax in the present section. Then, Sect. 20.1.2 modifies scattered context grammars to their transformational versions, which are easy to apply to syntax-related modifications of sentences. Most importantly, Sect. 20.1.3 describes English syntax and its transformations by methods based upon the transformational versions of scattered context grammars.

20.1.1 Syntax and Related Linguistic Terminology

In the linguistic study concerning English syntax, we discuss and describe the principles and rules according to which we correctly construct and transform grammatical English sentences. To give an insight into the discussion of English syntax, we open this section by some simple examples that illustrate how we connect the theoretically oriented discussion of scattered context grammars with the application-oriented discussion of English syntax. Then, we introduce the basic terminology used in syntax-oriented linguistics.

Introduction Through Examples

Observe that many common English sentences contain expressions and words that mutually depend on each other although they are not adjacent to each other in the sentences. For example, consider this sentence

He usually goes to work early.

The subject (*he*) and the predicator (*goes*) are related; sentences

**He usually go to work early.*

and

**I usually goes to work early.*

are ungrammatical because the form of the predicator depends on the form of the subject, according to which the combinations **he... go* and **I... goes* are illegal (throughout this section, * denotes ungrammatical sentences or their parts). Clearly, any change of the subject implies the corresponding change of the predicator as well. Linguistic dependencies of this kind can be easily and elegantly captured by scattered context grammars. Let us construct a scattered context grammar that contains this rule

$$(\text{He}, \text{goes}) \rightarrow (\text{We}, \text{go})$$

This rule checks whether the subject is the pronoun *he* and whether the verb *go* is in third person singular. If the sentence satisfies this property, it can be transformed to the grammatically correct sentence

We usually go to work early.

Observe that the related words may occur far away from each other in the sentence in question. In the above example, the word *usually* occurs between the subject and the predicator. While it is fairly easy to use context-sensitive grammars to model context dependencies where only one word occurs between the related words, note that the number of the words appearing between the subject and the predicator can be virtually unlimited. We can say

He almost regularly goes to work early.

but also

He usually, but not always, goes to work early.

and many more grammatical sentences like this. To model these context dependencies by ordinary context-sensitive grammars, many auxiliary rules have to be introduced to send the information concerning the form of a word to another word, which may occur at the opposite end of the sentence. As opposed to this awkward and tedious description, the single scattered context rule above is needed to perform the same job regardless of the number of the words appearing between the subject and the predicator.

We next give another example that illustrates the advantage of scattered context grammars over classical context-sensitive grammars under some circumstances. Consider these two sentences

John recommended it.

and

Did John recommend it?

There exists a relation between the basic clause and its interrogative counterpart. Indeed, we obtain the second, interrogative clause by adding *did* in front of *John* and by changing *recommended* to *recommend* while keeping the rest of the sentence unchanged. In terms of scattered context grammars, this transformation can be described by the scattered context rule

$$(\text{John}, \text{recommended}) \rightarrow (\text{Did John}, \text{recommend})$$

Clearly, when applied to the first sentence, this rule performs exactly the same transformation as we have just described. Although this transformation is possible by using an ordinary context rule, the inverse transformation is much more difficult to achieve. The inverse transformation can be performed by a scattered context rule

$$(\text{Did}, \text{recommend}) \rightarrow (\varepsilon, \text{recommended})$$

Obviously, by erasing *did* and changing *recommend* to *recommended*, we obtain the first sentence from the second one. Again, instead of *John* the subject may consist of a noun phrase containing several words, which makes it difficult to capture this context dependency by ordinary context-sensitive grammars.

Considering the examples above, the advantage of scattered context grammars is more than obvious: scattered context grammars allow us to change only some words during the transformation while keeping the others unchanged. On the other hand, context-sensitive grammars are inconvenient to perform transformations of this kind. A typical context-sensitive grammar that performs this job usually needs many more context-sensitive rules by which it repeatedly traverses the transformed sentence in question just to change very few context dependent words broadly spread across the sentence.

Terminology

Taking into account the intuitive insight given above, we see that there are structural rules and regularities underlying syntactically well-formed English sentences and their transformations. Although we have already used some common linguistic notions, such as subject or predicator, we now introduce this elementary linguistic terminology more systematically so we can express these English sentences in terms of their syntactic structure in a more exact and general way. However, we restrict this introduction only to the very basic linguistic notions, most of which are taken from [9, 10].

Throughout the rest of this section, we narrow our discussion primarily to verbs and personal pronouns, whose proper use depends on the context in which they occur. For instance, *is, are, was*, and *been* are different forms of the same verb *be*, and their proper use depends on the context in which they appear. We say that words in these categories *inflect* and call this property *inflection*. Verbs and personal pronouns often represent the key elements of a clause—the *subject* and the *predicate*. In simple clauses like

She loves him.

we can understand the notion of the subject and the predicate so that some information is "predicated of" the subject (*she*) by the predicate (*loves him*). In more complicated clauses, the best way to determine the subject and the predicate is the examination of their syntactic properties (see [9] for more details). The predicate is formed by a *verb phrase*—the most important word of this phrase is the verb, also known as the *predicator*. In some verb phrases, there occur several verbs. For example, in the sentence

He has been working for hours.

Table 20.1 Paradigms of English verbs

Form	Paradigm	Person	Example
Primary	Present	3rd sg	*She* walks *home.*
		Other	*They* walk *home.*
	Preterite		*She* walked *home.*
Secondary	Plain form		*They should* walk *home.*
	Gerund-participle		*She is* walking *home.*
	Past participle		*She has* walked *home.*

the verb phrase contains three verbs—*has, been*, and *working*. The predicator is, however, always the first verb of a verb phrase (*has* in the above example). In this study, we focus on the most elementary clauses—*canonical clauses*. In these clauses, the subject always precedes the predicate, and these clauses are positive, declarative, and without subordinate or coordinate clauses.

Next, we describe the basic categorization of verbs and personal pronouns, and further characterize their inflectional forms in a greater detail.

Verbs

We distinguish several kinds of verbs based upon their grammatical properties. The set of all verbs is divided into two subsets—the set of *auxiliary verbs*, and the set of *lexical verbs*. Further, the set of auxiliary verbs consists of *modal verbs* and *non-modal verbs*. The set of modal verbs includes the following verbs—*can*, *may*, *must*, *will*, *shall*, *ought*, *need*, *dare*; the verbs *be*, *have*, and *do* are non-modal. All the remaining verbs are lexical. In reality, the above defined classes overlap in certain situations; for example, there are sentences, where *do* appears as an auxiliary verb, and in different situations, *do* behaves as a lexical verb. For simplicity, we do not take into account these special cases in what follows.

Inflectional forms of verbs are called *paradigms*. In English, every verb, except for the verb *be*, may appear in each of the six paradigms described in Table 20.1 (see [9]). Verbs in *primary form* may occur as the only verb in a clause and form the head of its verb phrase (predicator); on the other hand, verbs in *secondary form* have to be accompanied by a verb in primary form.

The verb *be* has nine paradigms in its neutral form. All primary forms have, in addition, their negative contracted counterparts. Compared to other verbs, there is one more verb paradigm called *irrealis*. The irrealis form *were* (and *weren't*) is used in sentences of an unrealistic nature, such as

I wish I were rich.

All these paradigms are presented in Table 20.2.

Table 20.2 Paradigms of the verb *be*

Form	Paradigm	Person	Neutral	Negative
Primary	Present	1st sg	*am*	*Aren't*
		3rd sg	*Is*	*Isn't*
		other	*Are*	*Aren't*
	Preterite	1st sg, 3rd sg	*Was*	*Wasn't*
		other	*Were*	*Weren't*
	Irrealis	1st sg, 3rd sg	*Were*	*Weren't*
Secondary	Plain form		*Be*	–
	Gerund-participle		*Being*	–
	Past participle		*Been*	–

Table 20.3 Personal pronouns

Non-reflexive				
Nominative	Accusative	Genitive		
Plain		Dependent	Independent	Reflexive
I	*Me*	*My*	*Mine*	*Myself*
You	*You*	*Your*	*Yours*	*Yourself*
He	*Him*	*His*	*His*	*Himself*
She	*Her*	*Her*	*Hers*	*Herself*
It	*It*	*Its*	*Its*	*Itself*
We	*Us*	*Our*	*Ours*	*Ourselves*
You	*You*	*Your*	*Yours*	*Yourselves*
They	*them*	*their*	*theirs*	*Themselves*

Personal Pronouns

Personal pronouns exhibit a great amount of inflectional variation as well. Table 20.3 summarizes all their inflectional forms. The most important for us is the class of pronouns in *nominative* because these pronouns often appear as the subject of a clause.

20.1.2 Transformational Scattered Context Grammars

As we have already mentioned, we primarily apply scattered context grammars to transform grammatical English sentences to other grammatical English sentences. To do so, we next slightly modify scattered context grammars so they start their derivations from a language rather than a single start symbol. Even more importantly, these grammars define transformations of languages, not just their generation.

Definition 20.1.1. A *transformational scattered context grammar* is a quadruple

$$G = (V, T, P, I)$$

where

- V is the *total vocabulary*;
- $T \subset V$ is the set of terminals (or the *output vocabulary*);
- P is a finite set of scattered context rules;
- $I \subset V$ is the *input vocabulary*.

The derivation step is defined as in scattered context grammars (see Definition 4.7.1). The *transformation T that G defines from $K \subseteq I^*$* is denoted by $T(G, K)$ and defined as

$$T(G, K) = \{(x, y) \mid x \Rightarrow_G^* y, x \in K, y \in T^*\}$$

If $(x, y) \in T(G, K)$, we say that *x is transformed to y* by G; x and y are called the *input* and the *output sentence*, respectively. □

As already pointed out, while scattered context grammars generate strings, transformational scattered context grammars translate them. In a sense, however, the language generated by any scattered context grammar $G = (V, T, P, S)$ can be expressed by using a transformational scattered context grammar $H = (V, T, P, \{S\})$ as well. Observe that

$$L(G) = \{y \mid (S, y) \in T(H, \{S\})\}$$

Before we make use of transformational scattered context grammars in terms of English syntax in the next section, we give two examples to demonstrate a close relation of these grammars to the theoretically oriented studies given previously in this book. To link the theoretical discussions given earlier in this book to the present section, the first example presents a transformational scattered context grammar that works with purely abstract languages. In the second example, we discuss a transformational scattered context grammar that is somewhat more linguistically oriented.

Example 20.1.2. Define the transformational scattered context grammar

$$G = (V, T, P, I)$$

where $V = \{A, B, C, a, b, c\}$, $T = \{a, b, c\}$, $I = \{A, B, C\}$, and

$$P = \{(A, B, C) \to (a, bb, c)\}$$

For example, for the input sentence $AABBCC$,

$$AABBCC \Rightarrow_G aABbbcC \Rightarrow_G aabbbbcc$$

Therefore, the input sentence $AABBCC \in I^*$ is transformed to the output sentence $aabbbbcc \in T^*$, and

$$(AABBCC, aabbbbcc) \in T(G, I^*)$$

If we restrict the input sentences to the language $L = \{A^n B^n C^n \mid n \geq 1\}$, we get

$$T(G, L) = \{(A^n B^n C^n, a^n b^{2n} c^n) \mid n \geq 1\}$$

so every $A^n B^n C^n$, where $n \geq 1$, is transformed to $a^n b^{2n} c^n$. □

In the following example, we modify strings consisting of English letters by a transformational scattered context grammar, and in this way, we relate these grammars to lexically oriented linguistics—that is, the area of linguistics that concentrates its study on vocabulary analysis and dictionary design.

Example 20.1.3. We demonstrate how to lexicographically order alphabetic strings and, simultaneously, convert them from their uppercase versions to lowercase versions. More specifically, we describe a transformational scattered context grammar G that takes any alphabetic strings that consist of English uppercase letters enclosed in angle brackets, lexicographically orders the letters, and converts them to the corresponding lowercases. For instance, G transforms $\langle XXUY \rangle$ to $uxxy$.

More precisely, let J and T be alphabets of English uppercases and English lowercases, respectively. Let $\prec$ denote the *lexical order* over J; that is, $A \prec B \prec \cdots \prec Z$. Furthermore, let h be the function that maps the uppercases to the corresponding lowercases; that is, $h(A) = a, h(B) = b, \ldots, h(Z) = z$. Let i denote the inverse of h, so $i(a) = A, i(b) = B, \ldots, i(z) = Z$. Let $N = \{\hat{a} \mid a \in T\}$. We define the transformational scattered context grammar

$$G = (V, T, P, I)$$

where T is defined as above, $I = J \cup \{\langle, \rangle\}$, $V = I \cup N \cup T$, and P is constructed as follows:

(1) For each $A, B \in I$, where $A \prec B$, add $(B, A) \to (A, B)$ to P;
(2) For each $a \in T$, add $(\langle) \to (\hat{a})$ to P;
(3) For each $a \in T$ and $A \in J$, where $i(a) = A$, add $(\hat{a}, A) \to (a, \hat{a})$ to P;
(4) For each $a, b \in T$, where $i(a) \prec i(b)$, add $(\hat{a}) \to (\hat{b})$ to P;
(5) For each $a \in T$, add $(\hat{a}, \rangle) \to (\varepsilon, \varepsilon)$ to P.

Set $K = \{\langle\} J^* \{\rangle\}$. For instance, G transforms $\langle ORDER \rangle \in K$ to $deorr \in T^*$ as

$$\begin{aligned} \langle ORDER \rangle &\Rightarrow_G \langle OEDRR \rangle \Rightarrow_G \langle DEORR \rangle \\ &\Rightarrow_G \hat{d} DEORR \rangle \Rightarrow_G d\hat{d} EORR \rangle \Rightarrow_G d\hat{e} EORR \rangle \Rightarrow_G de\hat{e} ORR \rangle \\ &\Rightarrow_G de\hat{o} ORR \rangle \Rightarrow_G deo\hat{o} RR \rangle \Rightarrow_G deo\hat{r} RR \rangle \Rightarrow_G deor\hat{r} R \rangle \\ &\Rightarrow_G deorr\hat{r} \rangle \Rightarrow_G deorr \end{aligned}$$

so $(\langle ORDER\rangle, deorr) \in T(G, K)$. Clearly, G can make the same transformation in many more ways; on the other hand, notice that the set of all transformations of $\langle ORDER\rangle$ to $deorr$ is finite.

More formally, we claim that G transforms every $\langle A_1 \dots A_n\rangle \in K$ to $b_1 \dots b_n \in T^*$, for some $n \geq 0$, so that $i(b_1) \dots i(b_n)$ represents a permutation of $A_1 \dots A_n$, and for all $1 \leq j \leq n-1$, $i(b_j) \prec i(b_{j+1})$ (the case when $n = 0$ means that $A_1 \dots A_n = b_1 \dots b_n = \varepsilon$). To see why this claim holds, notice that $T \cap I = \emptyset$, so every successful transformation of a string from K to a string from T^* is performed so that all symbols are rewritten during the computation. By rules introduced in (1), G lexicographically orders the input uppercases. By a rule of the form $(\langle) \rightarrow (\hat{a})$ introduced in (2), G changes the leftmost symbol $\langle$ to $\hat{a}$. By rules introduced in (3) and (4), G verifies that the alphabetic string is properly ordered and, simultaneously, converts its uppercase symbols into the corresponding lowercases in a strictly left-to-right one-by-one way. Observe that a rule introduced in (2) is applied precisely once during every successful transformation because the left-to-right conversion necessities its application, and on the other hand, no rule can produce $\langle$. By a rule from (5), G completes the transformation; notice that if this completion is performed prematurely with some uppercases left, the transformation is necessary unsuccessful because the uppercases cannot be turned to the corresponding lowercases. Based upon these observations, it should be obvious that G performs the desired transformation. □

Having illustrated the lexically oriented application, we devote the next section solely to the applications of transformational scattered context grammars in English syntax.

20.1.3 Scattered Context in English Syntax

In this section, we apply transformational scattered context grammars to English syntax. Before opening this topic, let us make an assumption regarding the set of all English words. We assume that this set, denoted by T, is finite and fixed. From a practical point of view, this is obviously a reasonable assumption because we all commonly use a finite and fixed vocabulary of words in everyday English (purely hypothetically, however, this may not be the case as illustrated by the study that closes this section). Next, we subdivide this set into subsets with respect to the classification of verbs and pronouns described in Sect. 20.1.1:

- T is the set of all words including all their inflectional forms;
- $T_{\mathrm{V}} \subset T$ is the set of all verbs including all their inflectional forms;
- $T_{\mathrm{VA}} \subset T_{\mathrm{V}}$ is the set of all auxiliary verbs including all their inflectional forms;
- $T_{\mathrm{V}pl} \subset T_{\mathrm{V}}$ is the set of all verbs in plain form;
- $T_{\mathrm{PPn}} \subset T$ is the set of all personal pronouns in nominative.

To describe all possible paradigms of a verb $v \in T_{Vpl}$, we use the following notation

- $\pi_{3rd}(v)$ is the verb v in third person singular present;
- $\pi_{pres}(v)$ is the verb v in present (other than third person singular);
- $\pi_{pret}(v)$ is the verb v in preterite.

There are several conventions we use throughout this section in order to simplify the presented case studies, given next.

- We do not take into account capitalization and punctuation. Therefore, according to this convention,

  He is your best friend.

  and

  he is your best friend

  are equivalent.
- To make the following studies as simple and readable as possible, we expect every input sentence to be a canonical clause. In some examples, however, we make slight exceptions to this rule; for instance, sometimes we permit the input sentence to be negative. The first example and the last example also demonstrate a simple type of coordinated canonical clauses.
- The input vocabulary is the set $I = \{\langle x \rangle \mid x \in T\}$, where T is the set of all English words as stated above. As a result, every transformational scattered context grammar in this section takes an input sentence over I and transforms it to an output sentence over T. For instance, in the case of the declarative-to-interrogative transformation,

  $$\langle \text{he} \rangle \; \langle \text{is} \rangle \; \langle \text{your} \rangle \; \langle \text{best} \rangle \; \langle \text{friend} \rangle$$

  is transformed to

  is he your best friend

  As we have already mentioned, we omit punctuation and capitalization, so the above sentence corresponds to

  Is he your best friend?

Next, we give several studies that describe how to transform various kinds of grammatical sentences to other grammatical sentences by using transformational scattered context grammars.

Clauses with **Neither** *and* **nor**

The first example shows how to use transformational scattered context grammars to negate clauses that contain the pair of the words *neither* and *nor*, such as

Neither Thomas nor his wife went to the party.

Clearly, the words *neither* and *nor* are related, but there is no explicit limit of the number of the words appearing between them. The following transformational scattered context grammar G converts the above sentence to

Both Thomas and his wife went to the party.

In fact, the constructed grammar G is general enough to negate every grammatical clause that contains the pair of the words *neither* and *nor*.

Set $G = (V, T, P, I)$, where $V = T \cup I$, and P is defined as follows:

$$\begin{aligned} P = & \big\{\big(\langle\text{neither}\rangle, \ \langle\text{nor}\rangle\big) \rightarrow (\text{both}, \ \text{and})\big\} \\ & \cup \big\{\big(\langle x\rangle\big) \rightarrow (x) \mid x \in T - \{\text{neither}, \text{nor}\}\big\} \end{aligned}$$

For example, for the above sentence, the transformation can proceed in this way

$$\begin{aligned} & \langle\text{neither}\rangle \ \langle\text{thomas}\rangle \ \langle\text{nor}\rangle \ \langle\text{his}\rangle \ \langle\text{wife}\rangle \ \langle\text{went}\rangle \ \langle\text{to}\rangle \ \langle\text{the}\rangle \ \langle\text{party}\rangle \\ & \quad \Rightarrow_G \text{both} \ \langle\text{thomas}\rangle \ \text{and} \ \langle\text{his}\rangle \ \langle\text{wife}\rangle \ \langle\text{went}\rangle \ \langle\text{to}\rangle \ \langle\text{the}\rangle \ \langle\text{party}\rangle \\ & \quad \Rightarrow_G \text{both thomas and} \ \langle\text{his}\rangle \ \langle\text{wife}\rangle \ \langle\text{went}\rangle \ \langle\text{to}\rangle \ \langle\text{the}\rangle \ \langle\text{party}\rangle \\ & \quad \Rightarrow_G \text{both thomas and his} \ \langle\text{wife}\rangle \ \langle\text{went}\rangle \ \langle\text{to}\rangle \ \langle\text{the}\rangle \ \langle\text{party}\rangle \\ & \quad \Rightarrow_G^5 \text{both thomas and his wife went to the party} \end{aligned}$$

The rule

$$\big(\langle\text{neither}\rangle, \ \langle\text{nor}\rangle\big) \rightarrow (\text{both}, \ \text{and})$$

replaces *neither* and *nor* with *both* and *and*, respectively. Every other word $\langle w\rangle \in I$ is changed to $w \in T$. Therefore, if we denote all possible input sentences, described in the introduction of this example, by K, $T(G, K)$ represents the set of all negated sentences from K, and

$$\begin{aligned} & \big(\langle\text{neither}\rangle \ \langle\text{thomas}\rangle \ \langle\text{nor}\rangle \ \langle\text{his}\rangle \ \langle\text{wife}\rangle \ \langle\text{went}\rangle \ \langle\text{to}\rangle \ \langle\text{the}\rangle \ \langle\text{party}\rangle, \\ & \quad \text{both thomas and his wife went to the party}\big) \in T(G, K) \end{aligned}$$

Existential Clauses

In English, clauses that indicate an existence are called *existential*. These clauses are usually formed by the dummy subject *there*; for example,

There was a nurse present.

However, this dummy subject is not mandatory in all situations. For instance, the above example can be rephrased as

A nurse was present.

We construct a transformational scattered context grammar G that converts any canonical existential clause without the dummy subject *there* to an equivalent existential clause with *there*.

Set $G = (V, T, P, I)$, where $V = T \cup I \cup \{X\}$ (X is a new symbol such that $X \notin T \cup I$), and P is defined as follows:

$$\begin{aligned} P = \big\{ & \big(\langle x\rangle, \langle\text{is}\rangle\big) \rightarrow (\text{there is } xX, \varepsilon), \\ & \big(\langle x\rangle, \langle\text{are}\rangle\big) \rightarrow (\text{there are } xX, \varepsilon), \\ & \big(\langle x\rangle, \langle\text{was}\rangle\big) \rightarrow (\text{there was } xX, \varepsilon), \\ & \big(\langle x\rangle, \langle\text{were}\rangle\big) \rightarrow (\text{there were } xX, \varepsilon) \mid x \in T\big\} \\ \cup & \big\{\big(X, \langle x\rangle\big) \rightarrow (X, x) \mid x \in T\big\} \\ \cup & \big\{(X) \rightarrow (\varepsilon)\big\} \end{aligned}$$

For the above sample sentence, we get the following derivation

$$\begin{aligned} & \langle\text{a}\rangle\langle\text{nurse}\rangle\langle\text{was}\rangle\langle\text{present}\rangle \\ & \Rightarrow_G \text{there was a } X \langle\text{nurse}\rangle\langle\text{present}\rangle \\ & \Rightarrow_G \text{there was a } X \text{ nurse } \langle\text{present}\rangle \\ & \Rightarrow_G \text{there was a } X \text{ nurse present} \\ & \Rightarrow_G \text{there was a nurse present} \end{aligned}$$

A rule from the first set has to be applied first because initially there is no symbol X in the sentential form and all other rules require X to be present in the sentential form. In our case, the rule

$$\big(\langle\text{a}\rangle, \langle\text{was}\rangle\big) \rightarrow (\text{there was a } X, \varepsilon)$$

is applied; the use of other rules from this set depends on what tense is used in the input sentence and whether the subject is in singular or plural. The rule non-deterministically selects the first word of the sentence, puts *there was* in front of it, and the symbol X behind it; in addition, it erases *was* in the middle of the sentence. Next, all words $\langle w\rangle \in I$ are replaced with $w \in T$ by rules from the second set. These rules also verify that the previous non-deterministic selection was made at the beginning of the sentence; if not, there remains a word $\langle w\rangle \in I$ in front of X that cannot be rewritten. Finally, the derivation ends by erasing X from the sentential form.

This form of the derivation implies that if we denote the input existential clauses described in the introduction of this example by K, $T(G, K)$ represents the set of these clauses with the dummy subject *there*. As a result,

$$\big(\langle\text{a}\rangle\langle\text{nurse}\rangle\langle\text{was}\rangle\langle\text{present}\rangle, \text{there was a nurse present}\big) \in T(G, K)$$

Interrogative Clauses

In English, there are two ways of transforming declarative clauses into interrogative clauses depending on the predicator. If the predicator is an auxiliary verb, the interrogative clause is formed simply by swapping the subject and the predicator. For example, we get the interrogative clause

Is he mowing the lawn?

by swapping *he*, which is the subject, and *is*, which is the predicator, in

He is mowing the lawn.

On the other hand, if the predicator is a lexical verb, the interrogative clause is formed by adding the dummy *do* to the beginning of the declarative clause. The dummy *do* has to be of the same paradigm as the predicator in the declarative clause and the predicator itself is converted to its plain form. For instance,

She usually gets up early.

is a declarative clause with the predicator *gets*, which is in third person singular, and the subject *she*. By inserting *do* in third person singular to the beginning of the sentence and converting *gets* to its plain form, we obtain

Does she usually get up early?

To simplify the following transformational scattered context grammar G, which performs this conversion, we assume that the subject is a personal pronoun in nominative.

Set $G = (V, T, P, I)$, where $V = T \cup I \cup \{X\}$ (X is a new symbol such that $X \notin T \cup I$), and P is defined as follows:

$$
\begin{aligned}
P = {} & \big\{\big(\langle p\rangle, \langle v\rangle\big) \rightarrow (vp, X) \mid v \in T_{\mathrm{VA}}, p \in T_{\mathrm{PPn}}\big\} \\
\cup {} & \big\{\big(\langle p\rangle, \langle \pi_{\mathrm{pret}}(v)\rangle\big) \rightarrow (\text{did } p, vX), \\
& \quad \big(\langle p\rangle, \langle \pi_{\mathrm{3rd}}(v)\rangle\big) \rightarrow (\text{does } p, vX), \\
& \quad \big(\langle p\rangle, \langle \pi_{\mathrm{pres}}(v)\rangle\big) \rightarrow (\text{do } p, vX) \mid v \in T_{\mathrm{Vpl}} - T_{\mathrm{VA}}, p \in T_{\mathrm{PPn}}\big\} \\
\cup {} & \big\{\big(\langle x\rangle, X\big) \rightarrow (x, X), \\
& \quad \big(X, \langle y\rangle\big) \rightarrow (X, y) \mid x \in T - T_{\mathrm{V}}, y \in T\big\} \\
\cup {} & \big\{(X) \rightarrow (\varepsilon)\big\}
\end{aligned}
$$

For sentences whose predicator is an auxiliary verb, the transformation made by G proceeds as follows:

$$
\begin{aligned}
&\langle\text{he}\rangle\ \langle\text{is}\rangle\ \langle\text{mowing}\rangle\ \langle\text{the}\rangle\ \langle\text{lawn}\rangle\\
&\quad\Rightarrow_G \text{ is he } X\,\langle\text{mowing}\rangle\ \langle\text{the}\rangle\ \langle\text{lawn}\rangle\\
&\quad\Rightarrow_G \text{ is he } X \text{ mowing } \langle\text{the}\rangle\ \langle\text{lawn}\rangle\\
&\quad\Rightarrow_G \text{ is he } X \text{ mowing the } \langle\text{lawn}\rangle\\
&\quad\Rightarrow_G \text{ is he } X \text{ mowing the lawn}\\
&\quad\Rightarrow_G \text{ is he mowing the lawn}
\end{aligned}
$$

The derivation starts by the application of a rule from the first set, which swaps the subject and the predicator, and puts X behind them. Next, rules from the third set rewrite every word $\langle w\rangle \in I$ to $w \in T$. Finally, X is removed from the sentential form.

The transformation of the sentences in which the predicator is a lexical verb is more complicated:

$$
\begin{aligned}
&\langle\text{she}\rangle\ \langle\text{usually}\rangle\ \langle\text{gets}\rangle\ \langle\text{up}\rangle\ \langle\text{early}\rangle\\
&\quad\Rightarrow_G \text{ does she } \langle\text{usually}\rangle \text{ get } X\,\langle\text{up}\rangle\ \langle\text{early}\rangle\\
&\quad\Rightarrow_G \text{ does she usually get } X\,\langle\text{up}\rangle\ \langle\text{early}\rangle\\
&\quad\Rightarrow_G \text{ does she usually get } X \text{ up } \langle\text{early}\rangle\\
&\quad\Rightarrow_G \text{ does she usually get } X \text{ up early}\\
&\quad\Rightarrow_G \text{ does she usually get up early}
\end{aligned}
$$

As the predicator is in third person singular, a rule from

$$\left\{(\langle p\rangle, \langle\pi_{3rd}(v)\rangle) \rightarrow (\text{does } p, vX) \mid v \in T_{\mathit{Vpl}} - T_{\mathit{VA}}, p \in T_{\mathit{PPn}}\right\}$$

is applied at the beginning of the derivation. It inserts *does* to the beginning of the sentence, converts the predicator *gets* to its plain form *get*, and puts X behind it. Next, rules from

$$\left\{(\langle x\rangle, X) \rightarrow (x, X) \mid x \in T - T_V\right\}$$

rewrite every word $\langle w\rangle \in I$ appearing in front of the predicator to $w \in T$. Notice that they do not rewrite verbs—in this way, the grammar verifies that the first verb in the sentence was previously selected as the predicator. For instance, in the sentence

He has been working for hours.

has must be selected as the predicator; otherwise, the derivation is unsuccessful. Finally, the grammar rewrites all words behind X, and erases X in the last step as in the previous case.

Based on this intuitive explanation, we can see that the set of all input sentences K described in the introduction of this example is transformed by G to $T(G, K)$, which is the set of all interrogative sentences constructed from K. Therefore,

$$\big(\langle\text{he}\rangle\ \langle\text{is}\rangle\ \langle\text{mowing}\rangle\ \langle\text{the}\rangle\ \langle\text{lawn}\rangle,\ \text{is he mowing the lawn}\big) \in T(G, K),$$
$$\big(\langle\text{she}\rangle\ \langle\text{usually}\rangle\ \langle\text{gets}\rangle\ \langle\text{up}\rangle\ \langle\text{early}\rangle,\ \text{does she usually get up early}\big) \in T(G, K)$$

Question Tags

Question tags are special constructs that are primarily used in spoken language. They are used at the end of declarative clauses, and we customarily use them to ask for agreement or confirmation. For instance, in

Your sister is married, isn't she?

isn't she is a question tag, and we expect an answer stating that she is married. The polarity of question tags is always opposite to the polarity of the main clause—if the main clause is positive, the question tag is negative, and vice versa. If the predicator is an auxiliary verb, the question tag is formed by the same auxiliary verb. For lexical verbs, the question tag is made by using *do* as

He plays the violin, doesn't he?

There are some special cases that have to be taken into account. First, the verb *be* has to be treated separately because it has more paradigms than other verbs and the question tag for first person singular is irregular:

I am always right, aren't I?

Second, for the verb *have*, the question tag depends on whether it is used as an auxiliary verb, or a lexical verb. In the first case, *have* is used in the question tag as

He has been working hard, hasn't he?

in the latter case, the auxiliary *do* is used as

They have a dog, don't they?

To explain the basic concepts as simply as possible, we omit the special cases of the verb *have* in the following transformational scattered context grammar G, which supplements a canonical clause with a question tag. For the same reason, we only sketch its construction and do not mention all the created rules explicitly. In addition, we suppose that the subject is represented by a personal pronoun.

Set $G = (V, T, P, I)$, where $V = T \cup I \cup \{X, Y\}$ (X, Y are new symbols such that $X, Y \notin T \cup I$), and P is defined as follows:

$$\begin{aligned}
P = &\big\{\big(\langle p\rangle, \langle \text{will}\rangle, \langle x\rangle\big) \to (p, \text{will } X, Yx \text{ won't } p),\\
&\quad \big(\langle p\rangle, \langle \text{won't}\rangle, \langle x\rangle\big) \to (p, \text{won't } X, Yx \text{ will } p),\\
&\quad \cdots \mid p \in T_{\text{PPn}}, x \in T\big\}\\
&\cup \big\{\big(\langle \text{I}\rangle, \langle \text{am}\rangle, \langle x\rangle\big) \to (\text{I}, \text{am } X, Yx \text{ aren't I}),\\
&\quad \big(\langle \text{you}\rangle, \langle \text{are}\rangle, \langle x\rangle\big) \to (\text{you}, \text{are } X, Yx \text{ aren't you}),\\
&\quad \cdots \mid x \in T\big\}\\
&\cup \big\{\big(\langle p\rangle, \langle v\rangle, \langle x\rangle\big) \to (p, vX, Yx \text{ doesn't } p),\\
&\quad \big(\langle q\rangle, \langle v\rangle, \langle x\rangle\big) \to (q, vX, Yx \text{ don't } q) \mid\\
&\quad p \in \{\text{he}, \text{she}, \text{it}\}, q \in T_{\text{PPn}} - \{\text{he}, \text{she}, \text{it}\}, v \in T_{\text{V}} - T_{\text{VA}}, x \in T\big\}\\
&\quad \vdots\\
&\cup \big\{\big(\langle x\rangle, X\big) \to (x, X),\\
&\quad \big(X, \langle y\rangle, Y\big) \to (X, y, Y) \mid x \in T - T_{\text{V}}, y \in T\big\}\\
&\cup \big\{(X, Y) \to (\varepsilon, \varepsilon)\big\}
\end{aligned}$$

First, we describe the generation of question tags for clauses whose predicator is an auxiliary verb:

$$\begin{aligned}
&\langle \text{I}\rangle\ \langle \text{am}\rangle\ \langle \text{always}\rangle\ \langle \text{right}\rangle\\
&\quad \Rightarrow_G \text{I am } X \langle \text{always}\rangle Y \text{ right aren't I}\\
&\quad \Rightarrow_G \text{I am } X \text{ always } Y \text{ right aren't I}\\
&\quad \Rightarrow_G \text{I am always right aren't I}
\end{aligned}$$

Here, the rule

$$\big(\langle \text{I}\rangle, \langle \text{am}\rangle, \langle \text{right}\rangle\big) \to (\text{I}, \text{am } X, Y \text{ right aren't I})$$

initiates the derivation. When it finds *I am* at the beginning of the sentence, it generates the question tag *aren't I* at its end. In addition, it adds X behind *I am* and Y in front of *right aren't I*. Next, it rewrites all words from $\langle w\rangle \in I$ to $w \in T$. It makes sure that the predicator was chosen properly by rules from

$$\big\{\big(\langle x\rangle, X\big) \to (x, X) \mid x \in T - T_{\text{V}}\big\}$$

similarly to the previous example. In addition, rules from

$$\big\{\big(X, \langle y\rangle, Y\big) \to (X, y, Y) \mid x \in T - T_{\text{V}}, y \in T\big\}$$

check whether the question tag was placed at the very end of the sentence. If not, there remains some symbol from the input vocabulary behind Y that cannot be rewritten. Finally, the last rule removes X and Y from the sentential form.

When the predicator is a lexical verb in present, the question tag is formed by *does* or *do* depending on person in which the predicator occurs:

$$\begin{aligned}&\langle\text{he}\rangle\ \langle\text{plays}\rangle\ \langle\text{the}\rangle\ \langle\text{violin}\rangle\\&\quad\Rightarrow_G \text{he plays } X\langle\text{the}\rangle Y \text{ violin doesn't he}\\&\quad\Rightarrow_G \text{he plays } X \text{ the violin } Y \text{ doesn't he}\\&\quad\Rightarrow_G \text{he plays the violin doesn't he}\end{aligned}$$

The rest of the derivation is analogous to the first case.

Based on these derivations, we can see that the set of all input sentences K described in the introduction of this example is transformed by G to $T(G,K)$, which is the set of all sentences constructed from K that are supplemented with question tags. Therefore,

$$\big(\langle\text{I}\rangle\ \langle\text{am}\rangle\ \langle\text{always}\rangle\ \langle\text{right}\rangle,\ \text{I am always right aren't I}\big) \in T(G,K),$$
$$\big(\langle\text{he}\rangle\ \langle\text{plays}\rangle\ \langle\text{the}\rangle\ \langle\text{violin}\rangle,\ \text{he plays the violin doesn't he}\big) \in T(G,K)$$

Generation of Grammatical Sentences

The purpose of the next discussion, which closes this section, is sixfold—(1) through (6), stated below.

(1) We want to demonstrate that ordinary scattered context grammars, discussed earlier in this book, can be seen as a special case of transformational scattered context grammars, whose applications are discussed in the present section.

(2) As pointed out in the notes following the general definition of a transformational scattered context grammar (see Definition 20.1.1), there exists a close relation between ordinary scattered context grammars and transformational scattered context grammars. That is, for every scattered context grammar $G = (V, T, P, S)$, there is a transformational scattered context grammar $H = \big(V, T, P, \{S\}\big)$ satisfying

$$L(G) = \big\{y \mid (S, y) \in T\big(H, \{S\}\big)\big\}$$

and in this way, $L(G)$ is defined by H. Next, we illustrate this relation by a specific example.

(3) From a syntactical point of view, we want to show that scattered context grammars can generate an infinite non-context-free grammatical subset of English language in a very succinct way.

(4) In terms of morphology—that is, the area of linguistics that studies the structure of words and their generation—we demonstrate how to use transformational scattered context grammars to create complicated English words within English sentences so that the resulting words and sentences are grammatically correct.
(5) As stated in the beginning of the present section, so far we have assumed that the set of common English words is finite. Next, we want to demonstrate that from a strictly theoretical point of view, the set of all possible well-formed English words, including extremely rare words in everyday English, is infinite. Indeed, L, given next, includes infinitely many words of the form

$$\begin{gathered} \textit{(great-)}^i \textit{grandparents} \\ \textit{(great-)}^i \textit{grandfathers} \\ \textit{(great-)}^i \textit{grandmothers} \end{gathered}$$

for all $i \geq 0$, and purely theoretically speaking, they all represent well-formed English words. Of course, most of them, such as

great-great-great-great-great-great-great-great-great-grandfathers

cannot be considered as common English words because most people never use them during their lifetime.
(6) We illustrate that the language generation based upon scattered context grammars may have significant advantages over the generation based upon classical grammars, such as context-sensitive grammars.

Without further ado, consider the language L consisting of these grammatical English sentences:

Your grandparents are all your grandfathers and all your grandmothers.

Your great-grandparents are all your great-grandfathers and all your great-grandmothers.

Your great-great-grandparents are all your great-great-grandfathers and all your great-great-grandmothers.

$\vdots$

In brief,

$$\begin{aligned} L = \{ & \text{your } \{\text{great-}\}^i \text{grandparents are all your } \{\text{great-}\}^i \text{grandfathers} \\ & \text{and all your } \{\text{great-}\}^i \text{grandmothers} \mid i \geq 0 \} \end{aligned}$$

Introduce the scattered context grammar $G = (V, T, P, S)$, where

$$T = \{\text{all, and, are, grandfathers, grandmothers, grandparents, great-, your}\}$$

$V = T \cup \{S, \#\}$, and P consists of these three rules

$$
\begin{aligned}
(S) &\rightarrow (\text{your \#grandparents are all your \#grandfathers} \\
&\qquad \text{and all your \#grandmothers}) \\
(\#, \#, \#) &\rightarrow (\text{\#great-}, \text{\#great-}, \text{\#great-}) \\
(\#, \#, \#) &\rightarrow (\varepsilon, \varepsilon, \varepsilon)
\end{aligned}
$$

Obviously, this scattered context grammar generates L; formally, $L = L(G)$. Consider the transformational scattered context grammar $H = \big(V, T, P, \{S\}\big)$. Notice that

$$L(G) = \Big\{y \mid (S, y) \in T\big(H, \{S\}\big)\Big\}$$

Clearly, L is not context-free, so its generation is beyond the power of context-free grammars. It would be possible to construct a context-sensitive grammar that generates L. However, a context-sensitive grammar like this would have to keep traversing across its sentential forms to guarantee the same number of occurrences of *great-* in the generated sentences. Compared to this awkward way of generating L, the scattered context grammar G generates L in a more elegant, economical, and effective way.

In this section, we have illustrated how to transform and generate grammatical sentences in English by using transformational scattered context grammars, which represent a very natural linguistic apparatus straightforwardly based on scattered context grammars. However, from a more general perspective, we can apply these grammars basically in any area of science that formalizes its results by strings containing some scattered context dependencies.

20.2 Biology

This section consists of Sects. 20.2.1 and 20.2.2. Section 20.2.1 presents two case studies of biological organisms whose development is affected by some abnormal conditions, such as a virus infection. From a more practical point of view, Sect. 20.2.2 discusses parametric 0L grammars (see [14]), which represent a powerful and elegant implementation tool in the area of biological simulation and modeling today. More specifically, we extend parametric 0L grammars by context conditions and demonstrate their use in models of growing plants.

20.2.1 Simulation of Biological Organisms

Case Study 1. Consider a cellular organism in which every cell divides itself into two cells during every single step of healthy development. However, when a virus

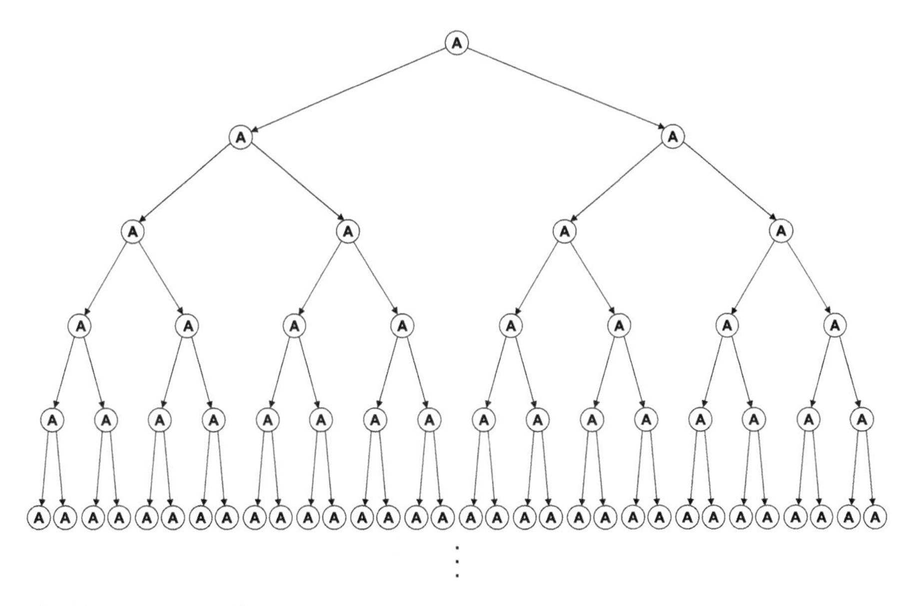

Fig. 20.1 Healthy development

infects some cells, all of the organism stagnates until it is cured again. During the stagnation period, all of the cells just reproduce themselves without producing any new cells. To formalize this development by a suitable simple semi-conditional L grammar (see Sect. 10.3), we denote a healthy cell and a virus-infected cell by A and B, respectively, and introduce the simple semi-conditional 0L grammar

$$G = \big(\{A, B\}, P, A\big)$$

where P contains the following rules

$$\begin{array}{ll} (A \rightarrow AA, 0, B) & (B \rightarrow B, 0, 0) \\ (A \rightarrow A, B, 0) & (B \rightarrow A, 0, 0) \\ (A \rightarrow B, 0, 0) & \end{array}$$

Figure 20.1 describes G simulating a healthy development while Fig. 20.2 gives a development with a stagnation period caused by the virus. □

In the next case study, we discuss an 0L grammar that simulates the developmental stages of a red alga (see [15]). Using context conditions, we can modify this grammar so that it describes some unhealthy development of this alga that leads to its partial death or degeneration.

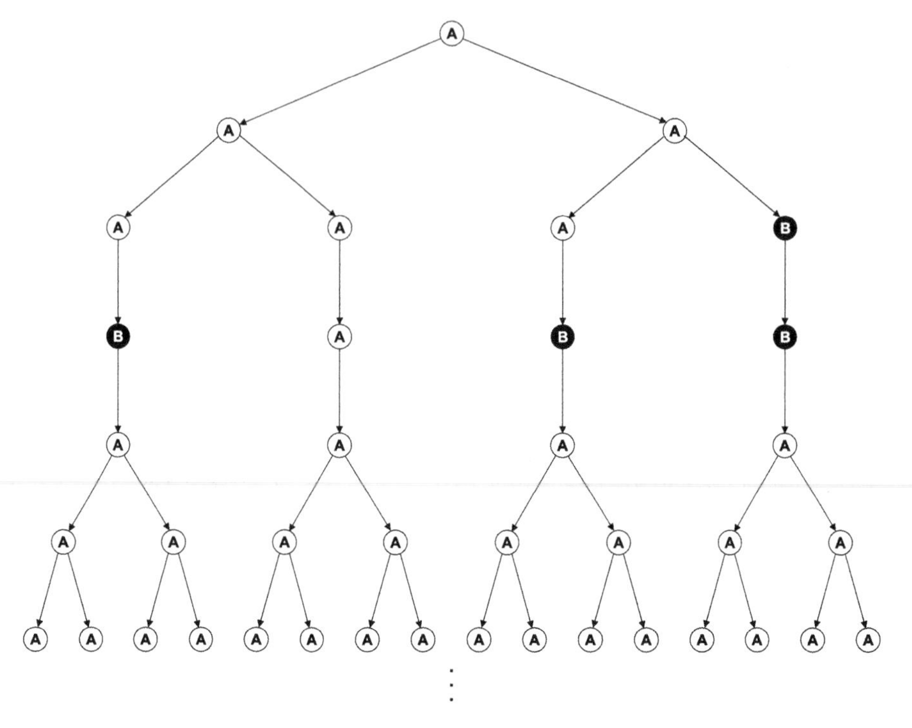

Fig. 20.2 Development with a stagnating period

Case Study 2. Consider an 0L grammar

$$G = (V, P, 1)$$

where

$$V = \{1, 2, 3, 4, 5, 6, 7, 8, [,]\}$$

and the set of rules P contains

$$\begin{array}{lllll} 1 \to 23 & 2 \to 2 & 3 \to 24 & 4 \to 54 & [\, \to [\\ 5 \to 6 & 6 \to 7 & 7 \to 8[1] & 8 \to 8 &]\, \to] \end{array}$$

From a biological viewpoint, expressions in fences represent branches whose position is indicated by 8s. These branches are shown as attached at alternate sides of the branch on which they are born. Figure 20.3 gives a biological interpretation of the developmental stages formally specified by the next derivation, which contains 13 strings corresponding to stages (a) through (m) in this figure.

$$
\begin{aligned}
1 &\Rightarrow_G 23 \\
&\Rightarrow_G 224 \\
&\Rightarrow_G 2254 \\
&\Rightarrow_G 22654 \\
&\Rightarrow_G 227654 \\
&\Rightarrow_G 228[1]7654 \\
&\Rightarrow_G 228[23]8[1]7654 \\
&\Rightarrow_G 228[224]8[23]8[1]7654 \\
&\Rightarrow_G 228[2254]8[224]8[23]8[1]7654 \\
&\Rightarrow_G 228[22654]8[2254]8[224]8[23]8[1]7654 \\
&\Rightarrow_G 228[227654]8[22654]8[2254]8[224]8[23]8[1]7654 \\
&\Rightarrow_G 228[228[1]7654]8[227654]8[22654]8[2254]8[224]8[23]8[1]7654
\end{aligned}
$$

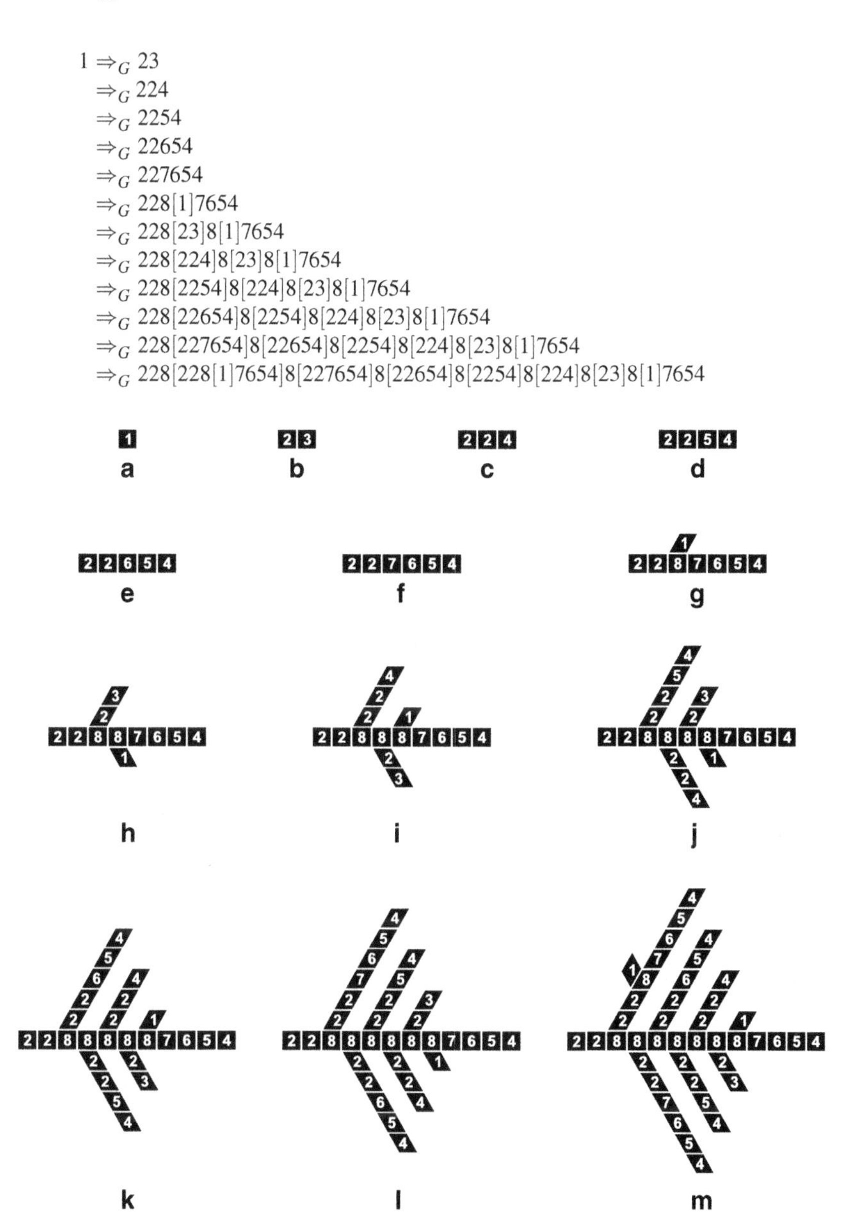

Fig. 20.3 Healthy development

$$
\begin{aligned}
1 &\Rightarrow_G 23 \\
&\Rightarrow_G 224 \\
&\Rightarrow_G 2254 \\
&\Rightarrow_G 22654 \\
&\Rightarrow_G 227654 \\
&\Rightarrow_G 228[1]7654 \\
&\Rightarrow_G 228[23]8[1]7654 \\
&\Rightarrow_G 228[224]8[23]8[1]7654 \\
&\Rightarrow_G 228[2254]8[224]8[23]8[1]7654 \\
&\Rightarrow_G 228[22654]8[2254]8[224]8[23]8[1]7654 \\
&\Rightarrow_G 228[227654]8[22654]8[2254]8[224]8[23]8[1]7654 \\
&\Rightarrow_G 228[228[1]7654]8[227654]8[22654]8[2254]8[224]8[23]8[1]7654
\end{aligned}
$$

Death. Let us assume that the red alga occurs in some unhealthy conditions under which only some of its parts survive while the rest dies. This dying process starts from the newly born, marginal parts of branches, which are too young and weak to survive, and proceeds toward the older parts, which are strong enough to live under these conditions. To be quite specific, all the red alga parts become gradually dead except for the parts denoted by 2s and 8s. This process is specified by the following 0L grammar G with forbidding conditions. Let $W = \{a' \mid a \in V\}$. Then,

$$G = \big(V \cup W, P, 1\big)$$

where the set of rules P contains

$$
\begin{array}{ll}
(1 \to 23, W) & (1' \to 2', \{3', 4', 5', 6', 7'\}) \\
(2 \to 2, W) & (2' \to 2', \emptyset) \\
(3 \to 24, W) & (3' \to \varepsilon, \{4', 5', 6', 7'\}) \\
(4 \to 54, W) & (4' \to \varepsilon, \emptyset) \\
(5 \to 6, W) & (5' \to \varepsilon, \{4'\}) \\
(6 \to 7, W) & (6' \to \varepsilon, \{4', 5'\}) \\
(7 \to 8[1], W) & (7' \to \varepsilon, \{4', 5', 6'\}) \\
(8 \to 8, W) & \\
([\to [, \emptyset) & \\
(] \to], \emptyset) &
\end{array}
$$

and for every $a \in V$,

$$(a \to a', \emptyset) \qquad (a' \to a', \emptyset)$$

Figure 20.4 pictures the dying process corresponding to the next derivation, whose last eight strings correspond to stages (a) through (h) in the figure.

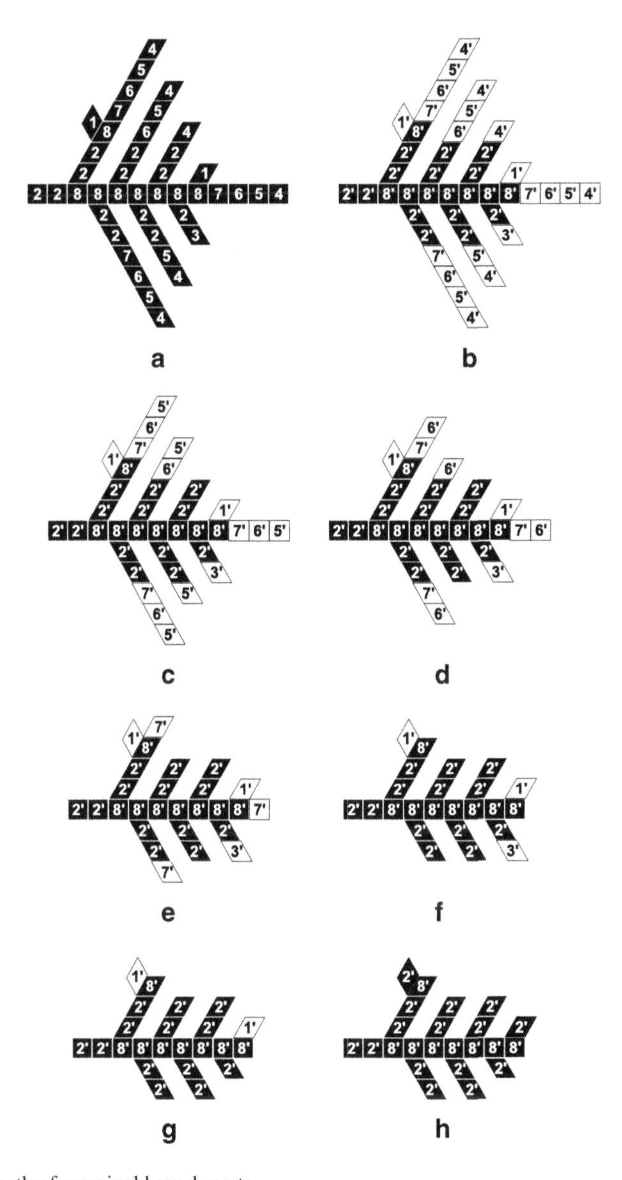

Fig. 20.4 Death of marginal branch parts

$$
\begin{aligned}
1 &\Rightarrow_G^* 228[228[1]7654]8[227654]8[22654]8[2254]8[224]8[23]8[1]7654 \\
&\Rightarrow_G 2'2'8'[2'2'8'[1']7'6'5'4']8'[2'2'7'6'5'4']8'[2'2'6'5'4']8'[2'2'5'4']8'[2'2'4'] \\
&\quad 8'[2'3']8'[1']7'6'5'4' \\
&\Rightarrow_G 2'2'8'[2'2'8'[1']7'6'5']8'[2'2'7'6'5']8'[2'2'6'5']8'[2'2'5']8'[2'2']8'[2'3'] \\
&\quad 8'[1']7'6'5' \\
&\Rightarrow_G 2'2'8'[2'2'8'[1']7'6']8'[2'2'7'6']8'[2'2'6']8'[2'2']8'[2'2']8'[2'3']8'[1']7'6' \\
&\Rightarrow_G 2'2'8'[2'2'8'[1']7']8'[2'2'7']8'[2'2']8'[2'2']8'[2'2']8'[2'3']8'[1']7' \\
&\Rightarrow_G 2'2'8'[2'2'8'[1']]8'[2'2']8'[2'2']8'[2'2']8'[2'2']8'[2'3']8'[1'] \\
&\Rightarrow_G 2'2'8'[2'2'8'[1']]8'[2'2']8'[2'2']8'[2'2']8'[2'2']8'[2']8'[1'] \\
&\Rightarrow_G 2'2'8'[2'2'8'[2']]8'[2'2']8'[2'2']8'[2'2']8'[2'2']8'[2']8'[2']
\end{aligned}
$$

Degeneration. Consider circumstances under which the red alga has degenerated. During this degeneration, only the main stem was able to give a birth to new branches while all the other branches lengthened themselves without any branching out. This degeneration is specified by the forbidding 0L grammar $G = (V \cup \{D, E\}, P, 1)$, with P containing

$(1 \rightarrow 23, \emptyset)$	$(2 \rightarrow 2, \emptyset)$	$(3 \rightarrow 24, \emptyset)$	$(4 \rightarrow 54, \emptyset)$
$(5 \rightarrow 6, \emptyset)$	$(6 \rightarrow 7, \emptyset)$	$(7 \rightarrow 8[1], \{D\})$	$(8 \rightarrow 8, \emptyset)$
$([\rightarrow [, \emptyset)$	$(] \rightarrow], \emptyset)$	$(7 \rightarrow 8[D], \emptyset)$	
$(D \rightarrow ED, \emptyset)$	$(E \rightarrow E, \emptyset)$		

Figure 20.5 pictures the degeneration specified by the following derivation, in which the last 10 strings correspond to stages (a) through (j) in the figure:

$$
\begin{aligned}
1 &\Rightarrow_G^* 227654 \\
&\Rightarrow_G 228[D]7654 \\
&\Rightarrow_G 228[ED]8[D]7654 \\
&\Rightarrow_G 228[E^2D]8[ED]8[D]7654 \\
&\Rightarrow_G 228[E^3D]8[E^2D]8[ED]8[D]7654 \\
&\Rightarrow_G 228[E^4D]8[E^3D]8[E^2D]8[ED]8[D]7654 \\
&\Rightarrow_G 228[E^5D]8[E^4D]8[E^3D]8[E^2D]8[ED]8[D]7654 \\
&\Rightarrow_G 228[E^6D]8[E^5D]8[E^4D]8[E^3D]8[E^2D]8[ED]8[D]7654 \\
&\Rightarrow_G 228[E^7D]8[E^6D]8[E^5D]8[E^4D]8[E^3D]8[E^2D]8[ED]8[D]7654 \\
&\Rightarrow_G 228[E^8D]8[E^7D]8[E^6D]8[E^5D]8[E^4D]8[E^3D]8[E^2D]8[ED]8[D]7654
\end{aligned}
$$

□

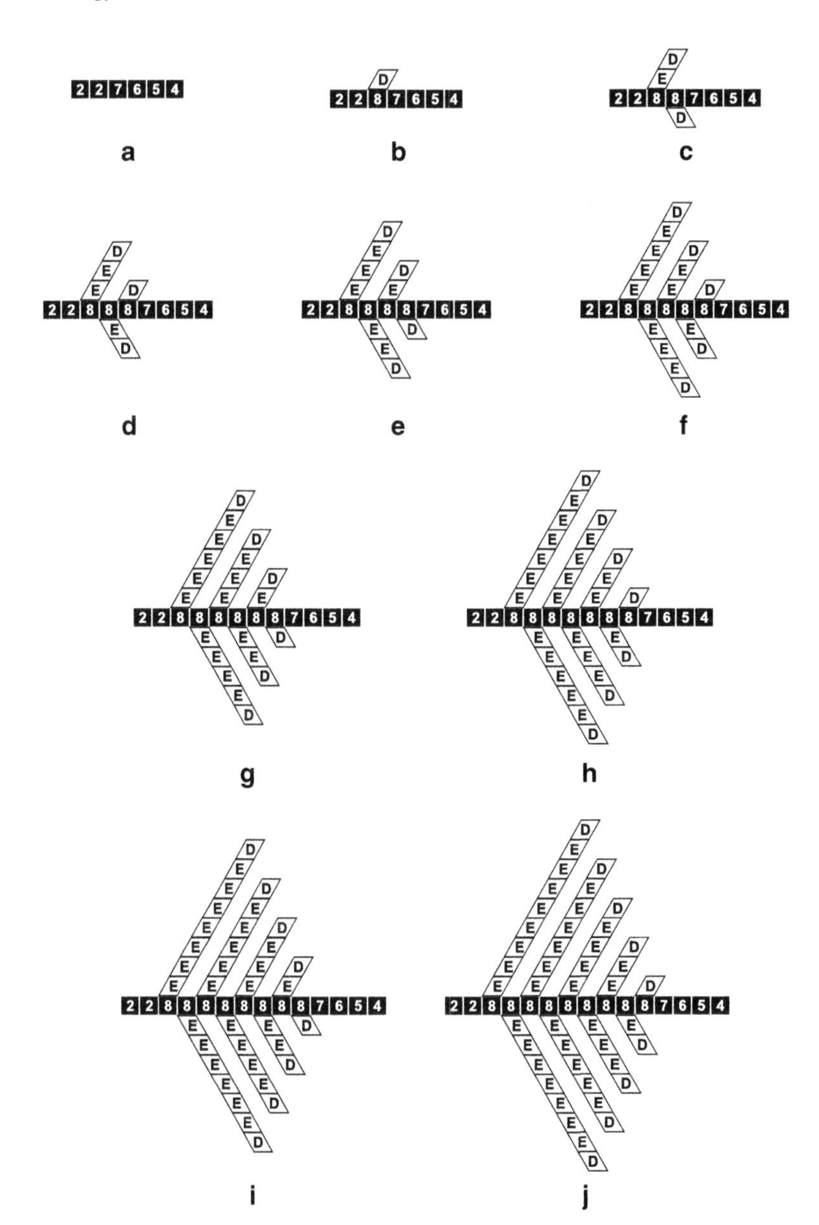

Fig. 20.5 Degeneration

20.2.2 Implementation

In this section, we describe *parametric 0L grammars* (see [14]) and their extension by context conditions. We make this description from a purely practical point of view to clearly demonstrate how these grammars are implemented and used.

Case Study 3. *Parametric 0L grammars* (see [13, 14]) operate on strings of modules called *parametric words*. A *module* is a symbol from an alphabet with an associated sequence of *parameters* belonging to the set of real numbers. Rules of parametric 0L grammars are of the form

$$predecessor\ [\,:\ logical\ expression\,]\ \rightarrow\ successor$$

The *predecessor* is a module having a sequence of formal parameters instead of real numbers. The *logical expression* is any expression over predecessor's parameters and real numbers. If the logical expression is missing, the logical truth is assumed. The *successor* is a string of modules containing expressions as parameters; for example,

$$A(x)\ :\ x < 7\ \rightarrow\ A(x+1)D(1)B(3-x)$$

Such a rule *matches* a module in a parametric word provided that the symbol of the rewritten module is the same as the symbol of the predecessor module, both modules have the same number of parameters, and the value for the logical expression is true. Then, the module can be rewritten by the given rule. For instance, consider $A(4)$. This module matches the above rule since A is the symbol of rule's predecessor, there is one actual parameter, 4, in $A(4)$ that corresponds to the formal parameter x in $A(x)$, and the value for the logical expression $x < 7$ with $x = 4$ is true. Thus, $A(4)$ can be rewritten to $A(5)D(1)B(-1)$.

As usual, a parametric 0L grammar can rewrite a parametric word provided that there exists a matching rule for every module that occurs in it. Then, all modules are simultaneously rewritten, and we obtain a new parametric word.

Parametric 0L grammars with context conditions. Next, we extend the parametric 0L grammars by permitting context conditions. Each rule of a *parametric 0L grammar with permitting conditions* has the form

$$predecessor\ [\ ?\ context\ conditions]\ [\,:\ logical\ expression]\ \rightarrow\ successor$$

where *predecessor*, *logical expression*, and *successor* have the same meaning as in parametric 0L grammars, and *context conditions* are some permitting context conditions separated by commas. Each condition is a string of modules with formal parameters. For example, consider

$$A(x)\ ?\ B(y),\ C(r,z)\ :\ x < y + r\ \rightarrow\ D(x)E(y+r)$$

This rule matches a module in a parametric word w provided that the predecessor $A(x)$ matches the rewritten module with respect to the symbol and the number of parameters and there exist modules matching to $B(y)$ and $C(r,z)$ in w such that the value for logical expression $x < y+r$ is true. For example, this rule matches $A(1)$ in $C(3,8)D(-1)B(5)H(0,0)A(1)F(3)$ because there are $C(3,8)$ and $B(5)$ such that $1 < 5+3$ is true. If there are more substrings matching the context condition, any of them can be used.

Having described the parametric 0L grammars with permitting conditions, we next show how to use them to simulate the development of some plants.

In nature, developmental processes of multicellular structures are controlled by the quantity of substances exchanged between modules. In the case of plants, growth depends on the amount of water and minerals absorbed by the roots and carried upward to the branches. The model of branching structures making use of the resource flow was proposed by Borchert and Honda in [5]. The model is controlled by a *flux* of resources that starts at the base of the plant and propagates the substances toward the apexes. An apex accepts the substances, and when the quantity of accumulated resources exceeds a predefined threshold value, the apex bifurcates and initiates a new lateral branch. The distribution of the flux depends on the number of apexes that the given branch supports and on the type of the branch—plants usually carry greater amount of resources to straight branches than to lateral branches (see [5] and [13]).

The following two examples (I) and (II) illustrate the idea of plants simulated by parametric 0L grammars with permitting conditions.

(I) Consider the model

$$\begin{aligned}
\text{start}: \quad & I(1,1,e_{root})\,A(1) \\
p_1: \quad & A(id)\;?\;I(id_p,c,e)\;:\;id == id_p \wedge e \geq e_{th} \\
& \rightarrow [+(\alpha)\,I(2*id+1,\gamma,0)\,A(2*id+1)]/(\pi)\,I(2*id,1-\gamma,0) \\
& \quad A(2*id) \\
p_2: \quad & I(id,c,e)\;?\;I(id_p,c_p,e_p)\;:\;id_p == \lfloor id/2 \rfloor \\
& \rightarrow I(id,c,c*e_p)
\end{aligned}$$

This L grammar describes a simple plant with a constant resource flow from its roots and with a fixed distribution of the stream between lateral and straight branches. It operates on the following types of modules.

- $I(id,c,e)$ represents an internode with a unique identification number id, a distribution coefficient c, and a flux value e.
- $A(id)$ is an apex growing from the internode with identification number equal to id.
- $+(\varphi)$ and $/(\varphi)$ rotate the segment orientation by angle φ (for more information, consult [13]).
- [and] enclose the sequence of modules describing a lateral branch.

We assume that if no rule matches a given module $X(x_1,\ldots,x_n)$, the module is rewritten by an implicit rule of the form

$$X(x_1,\ldots,x_n) \rightarrow X(x_1,\ldots,x_n)$$

That is, it remains unchanged.

At the beginning, the plant consists of one internode $I(1, 1, e_{root})$ with apex $A(1)$, where e_{root} is a constant flux value provided by the root. The first rule, p_1, simulates the bifurcation of an apex. If an internode preceding the apex $A(id)$ reaches a sufficient flux $e \geq e_{th}$, the apex creates two new internodes I terminated by apexes A. The lateral internode is of the form $I(2*id+1, \gamma, 0)$ and the straight internode is of the form $I(2*id, 1-\gamma, 0)$. Clearly, the identification numbers of these internodes are unique. Moreover, every child internode can easily calculate the identification number of its parent internode; the parent internode has $id_p = \lfloor id/2 \rfloor$. The coefficient γ is a fraction of the parent flux to be directed to the lateral internode. The second rule, p_2, controls the resource flow of a given internode. Observe that the permitting condition $I(id_p, c_p, e_p)$ with $id_p = \lfloor id/2 \rfloor$ matches only the parent internode. Thus, p_2 changes the flux value e of $I(id, c, e)$ to $c * e_p$, where e_p is the flux of the parent internode, and c is either γ for lateral internodes or $1 - \gamma$ for straight internodes. Therefore, p_2 simulates the transfer of a given amount of parent's flux into the internode. Figure 20.6 pictures 12 developmental stages of this plant with e_{root}, e_{th}, and γ set to 12, 0.9, and 0.4, respectively. The numbers indicate the flow values of internodes.

It is easy to see that this model is unrealistically simple. Since the model ignores the number of apexes, its flow distribution does not depend on the size of branches, and the basal flow is set to a constant value. However, it sufficiently illustrates the technique of communication between adjacent internodes. Thus, it can serve as a template for more sophisticated models of plants, such as the following model.

(II) We discuss a plant development with a resource flow controlled by the number of apexes. This example is based on Example 17 in [13].

$$
\begin{aligned}
\text{start}: \;& N(1)\, I(1, straight, 0, 1)\, A(1) \\
p_1: \;& N(k) \;\rightarrow\; N(k+1) \\
p_2: \;& I(id, t, e, c) \;?\; N(k),\; A(id) \\
& : id == 1 \\
& \rightarrow\; I(id, t, \sigma_0 2^{(k-1)\eta^k}, 1) \\
p_3: \;& I(id, t, e, c) \;?\; N(k),\; I(id_s, t_s, e_s, c_s),\; I(id_l, t_l, e_l, c_l) \\
& : id == 1 \;\wedge\; id_s == 2*id \;\wedge\; id_l == 2*id+1 \\
& \rightarrow\; I(id, t, \sigma_0 2^{(k-1)\eta^k}, c_s + c_l) \\
p_4: \;& I(id, t, e, c) \;?\; I(id_p, t_p, e_p, c_p),\; I(id_s, t_s, e_s, c_s),\; I(id_l, t_l, e_l, c_l) \\
& : id_p == \lfloor id/2 \rfloor \;\wedge\; id_s == 2*id \;\wedge\; id_l == 2*id+1 \\
& \rightarrow\; I(id, t, \delta(t, e_p, c_p, c), c_s + c_l) \\
p_5: \;& Id(id, t, e, c) \;?\; I(id_p, t_p, e_p, c_p),\; A(id_a) \\
& : id_p == \lfloor id/2 \rfloor \;\wedge\; id_a == id \\
& \rightarrow\; I(id, t, \delta(t, e_p, c_p, c), 1) \\
p_6: \;& A(id) \;?\; I(id_p, t_p, e_p, c_p) \\
& : id == id_p \;\wedge\; e_p \geq e_{th} \\
& \rightarrow\; [+(\alpha)\, I(2*id+1, lateral, e_p*(1-\lambda), 1)\, A(2*id+1)] \\
& \qquad /(\pi)\, I(2*id, straight, e_p*\lambda, 1)\, A(2*id)
\end{aligned}
$$

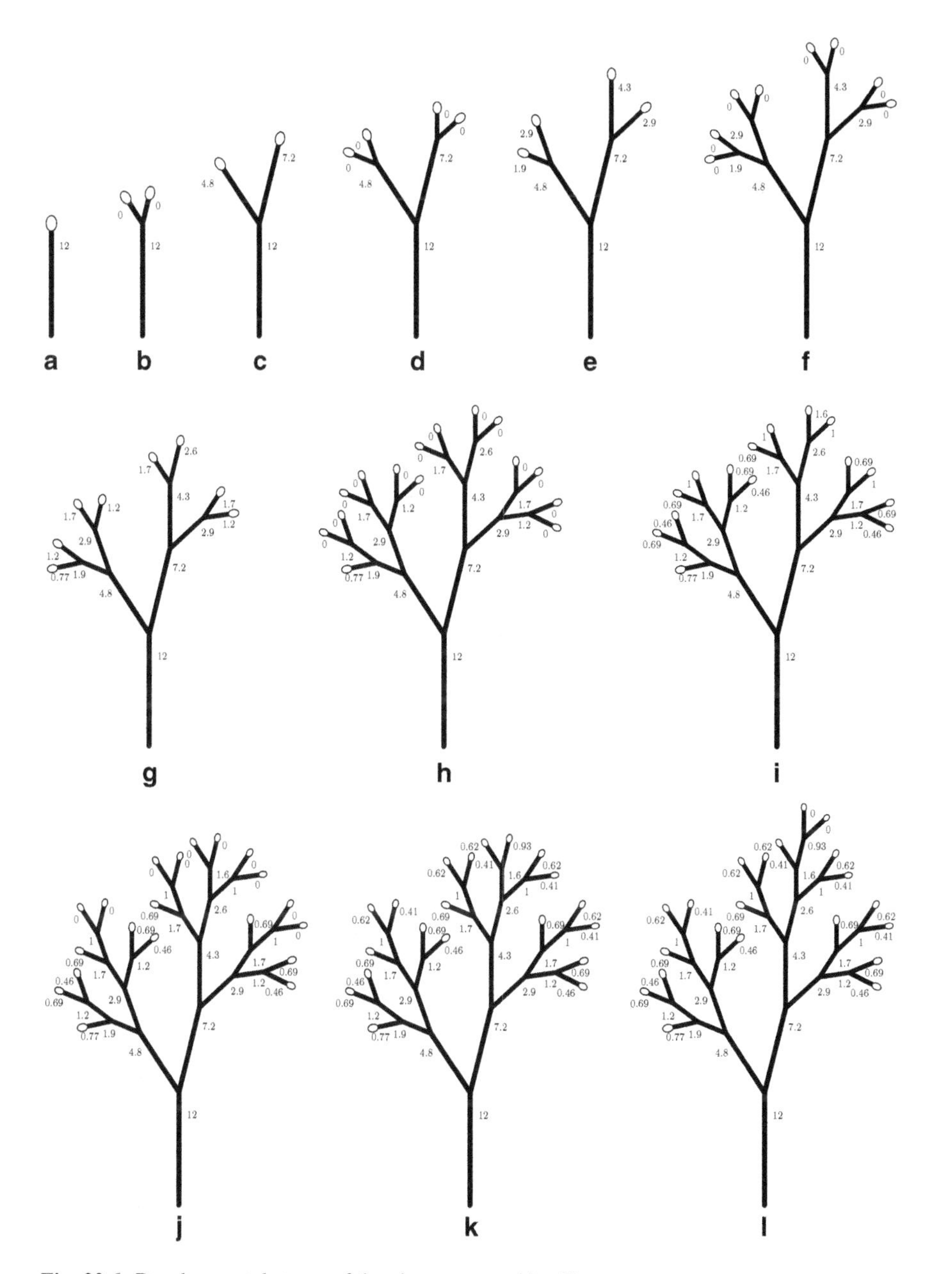

Fig. 20.6 Developmental stages of the plant generated by (I)

This L grammar uses the following types of modules.

- $I(id, t, e, c)$ is an internode with a unique identification number id, where t is a type of this internode, $t \in \{straight, lateral\}$, e is a flux value, and c is a number of apexes the internode supports.
- $A(id)$ is an apex terminating the internode id.
- $N(k)$ is an auxiliary module, where k is the number of a developmental cycle to be done by the next derivation.
- $+(\varphi)$, $/(\varphi)$, [and] have the same meaning as in the previous example.

The flux distribution function, δ, is defined as

$$\delta(t, e_p, c_p, c) = \begin{cases} e_p - e_p(1-\lambda)((c_p - c)/c) & \text{if } t = straight, \\ e_p(1-\lambda)(c/(c_p - c)) & \text{if } t = lateral \end{cases}$$

The development starts from $N(1)\, I(1, straight, 0, 1)\, A(1)$ containing one straight internode with one apex. In each derivation step, by application of p_4, every inner internode $I(id, t, e, c)$ gets the number of apexes of its straight $(I(id_s, t_s, e_s, c_s))$ and lateral $(I(id_l, t_l, e_l, c_l))$ descendant. Then, this number is stored in c. Simultaneously, it accepts a given part of the flux e_p provided by its parent internode $I(id_p, t_p, e_p, c_p)$. The distribution function δ depends on the number of apexes in the given branch and in the sibling branch, and on the type of this branch (straight or lateral). The distribution factor λ determines the amount of the flux that reaches the straight branch in case that both branches support the same number of apexes. Otherwise, the fraction is also affected by the ratio of apex counts. Rules p_2 and p_3 rewrite the basal internode, calculating its input flux value. The expression used for this purpose, $\sigma_0 2^{(k-1)\eta^k}$, was introduced by Borchert and Honda to simulate a sigmoid increase of the input flux; σ_0 is an initial flux, k is a developmental cycle, and η is a constant value scaling the flux change. Rule p_5 rewrites internodes terminated by apexes. It keeps the number of apexes set to 1, and by analogy with p_4, it loads a fraction of parent's flux by using the δ function. The last rule, p_6, controls the addition of new segments. By analogy with p_1 in the previous example, it erases the apex and generates two new internodes terminated by apexes. Figure 20.7 shows 15 developmental stages of a plant simulation based on this model.

Obviously, there are two concurrent streams of information in this model. The bottom-up (acropetal) stream carries and distributes the substances required for the growth. The top-down (basipetal) flow propagates the number of apexes that is used for the flux distribution. A remarkable feature of this model is the response of a plant to a pruning. Indeed, after a branch removal, the model redirects the flux to the remaining branches and accelerates their growth.

Let us note that this model is a simplified version of the model described in [13], which is very complex. Under this simplification, however, $c_p - c$ may be equal to zero as the denominator in the distribution function δ. If this happens, we change this zero value to the proper non-zero value so that the number of apexes supported by

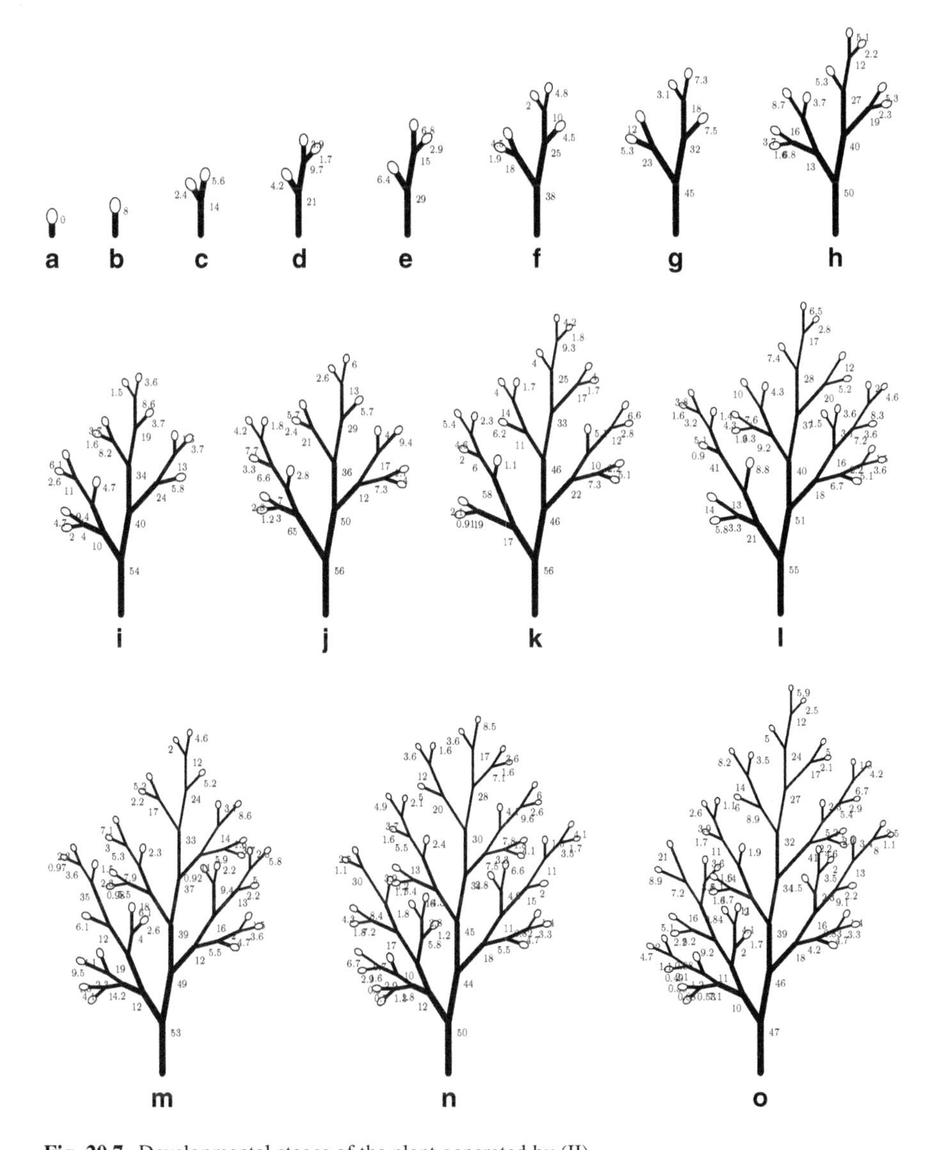

Fig. 20.7 Developmental stages of the plant generated by (II)

the parent internode corresponds to the number of apexes on the straight and lateral branches growing from the parent internode. Consult [13] for a more appropriate, but also more complicated solution of this problem.

From the presented examples, we see that with permitting conditions, parametric 0L grammars can describe sophisticated models of plants in a very natural way. Particularly, compared to the context-sensitive L grammars, they allow one to refer to modules that are not adjacent to the rewritten module, and this property makes them more adequate, succinct, and elegant. □

20.3 Compilers

In this section, we describe parsing based upon LL k-linear scattered context grammars (see Sect. 4.7.5). First, we present the underlying formal model for our parser. Then, we discuss implementation details and parsing complexity.

20.3.1 Underlying Formal Model

As the underlying formal model, we use deep pushdown automata, discussed in Chap. 18. Recall that these automata represent a generalization of the classical pushdown automata consisting of allowing expansions deeper in the pushdown. Informally, during every move, a deep pushdown automaton M either pops or expands its pushdown. If an input symbol a occurs on the pushdown top, M compares the pushdown top symbol with the current input symbol, and if they coincide, M pops the topmost symbol from the pushdown and proceeds to the next input symbol on the input tape. If a non-input symbol (a nonterminal in our case) occurs on the pushdown top, M expands its pushdown. However, as opposed to an ordinary pushdown automaton, M can perform this expansion deeper in the pushdown. More precisely, M can make an expansion of depth n so it replaces the nth topmost non-input pushdown symbol with a string, for some $n \geq 1$. M accepts an input string w if M can make a sequence of moves so it reads w and empties its pushdown (we do not consider final states as they are not needed in our parser).

Let $G = (N, T, P, S)$ be an LL k-linear scattered context grammar, for some $k \geq 0$. An equivalent deep pushdown automaton of degree k, M, can be constructed as follows. For every scattered context rule

$$(A_1, A_2, \dots, A_m) \rightarrow (x_1, x_2, \dots, x_m)$$

we introduce the following m deep pushdown automaton rules

$$\begin{array}{ll} m s A_m & \rightarrow s_{m-1} x_m \\ (m-1) s_{m-1} A_{m-1} & \rightarrow s_{m-2} x_{m-1} \\ & \vdots \\ 1 s_1 A_1 & \rightarrow s x_1 \end{array}$$

where s is the start state of M, s_1 through s_{m-1} are new, unique states, and the meaning of $npA \rightarrow qx$ is that M expands the nth non-input symbol in the pushdown—which has to be A—to x, and moves to state q. Notice that we have to start by rewriting A_m first; otherwise, if some A_i, where $i < m$, is rewritten to a string of terminals, this construction may fail to work properly. Furthermore, observe that we can use such an approach because the number of occurrences of nonterminals that can simultaneously appear in the pushdown is limited by k.

Fig. 20.8 The LL table corresponding to G from Example 4.7.30

	a	b	c	d
S	1			1
A	2			3

By using standard techniques (see, for instance, [12]), we can create an LL table from fcore(G). For example, the LL table corresponding to fcore(G) from Example 4.7.30 is shown in Fig. 20.8. Observe that we only have to consider the leftmost components of rules of G. Therefore, the resulting LL table does not need to necessarily include all nonterminals if they do not appear as the leftmost nonterminals in the rules of G. Otherwise, the created parser works like any ordinary top-down parser except that expansions might be done deeper in the pushdown. Furthermore, notice that the created parser is deterministic.

20.3.2 Implementation

Next, we discuss implementation details and parsing complexity. Let $G = (N, T, P, S)$ be an LL k-linear scattered context grammar, for some $k \geq 0$, and M be an equivalent deep pushdown automaton created from G in the way described above. Since the expansions of M can be performed deeper in the pushdown, we use doubly linked lists (see [7]) to effectively implement the pushdown storage and operations over it. For convenience, the first item of the list always points to the top of the pushdown. Furthermore, since the number of occurrences of nonterminals that can appear in the pushdown is limited by k, we use an additional auxiliary array of k elements, where each element i contains a pointer to the ith nonterminal in the pushdown. Since k does not depend on the length of the input string, all operations with this array can be done in $O(1)$—that is, constant to the length of the input in both time and space.

Let w be a string over T and let n be the length of w. For convenience, we assume that the input string is always ended with some end marker, like \$ (\$ $\notin T$). The parsing algorithm is implemented as follows. Let a denote the current input symbol and let S be the only symbol in the pushdown. We repeat the next two steps, (i) and (ii), until either (1) $a =$ \$ (the input is read) and the pushdown is empty, in which case we output *yes*, or (2) none of the steps can be done, in which case we output *no*.

(i) If the topmost symbol on the pushdown is a nonterminal, A_1, and the intersection of A_1 and a in the LL table constructed from fcore(G) contains a rule

$$(A_1, A_2, \ldots, A_m) \to (x_1, x_2, \ldots, x_m)$$

we successively expand $A_m, A_{m-1}, \ldots, A_1$ to $x_m, x_{m-1}, \ldots, x_1$, respectively, by using standard list operations and the auxiliary array in $O(1)$.

(ii) If a appears on the top of the pushdown, we pop it by removing the first item of the list and replace it with the second item in $O(1)$ (if there is no second item, we set it to some special value, like *null*, indicating that the pushdown is empty). Furthermore, let a be the next input symbol.

Based on this description, we see that the overall parsing complexity of our parser is $O(n)$—that is, linear to the length of the input in both time and space.

References

1. Aho, A.V., Lam, M.S., Sethi, R., Ullman, J.D.: Compilers: Principles, Techniques, and Tools, 2nd edn. Addison-Wesley, Boston (2006)
2. Alberts, B., Bray, D., Hopkin, K., Johnson, A., Lewis, J., Raff, M., Roberts, K., Walter, P.: Essential Cell Biology, 3rd edn. Garland Science, New York (2009)
3. Baker, C.L.: English Syntax, 2nd edn. The MIT Press, Cambridge (1995)
4. Bal, H., Grune, D., Jacobs, C., Langendoen, K.: Modern Compiler Design. John Wiley & Sons, Hoboken (2000)
5. Borchert, R., Honda, H.: Control of development in the bifurcating branch system of *Tabebuia Rosea*: A computer simulation. Bot. Gaz. **145**(2), 184–195 (1984)
6. Cooper, K.D., Torczon, L.: Engineering a Compiler. Morgan Kaufmann Publishers, San Francisco (2004)
7. Cormen, T.H., Stein, C., Rivest, R.L., Leiserson, C.E.: Introduction to Algorithms, 2nd edn. McGraw-Hill Higher Education, New York (2001)
8. Cowan, M.K., Talaro, K.P.: Microbiology: A Systems Approach. McGraw-Hill, New York (2008)
9. Huddleston, R., Pullum, G.: The Cambridge Grammar of the English Language. Cambridge University Press, New York (2002)
10. Huddleston, R., Pullum, G.: A Student's Introduction to English Grammar. Cambridge University Press, New York (2005)
11. Kennedy, K., Allen, J.R.: Optimizing Compilers for Modern Architectures: A Dependence-Based Approach. Morgan Kaufmann Publishers, San Francisco (2002)
12. Meduna, A.: Elements of Compiler Design. Auerbach Publications, Boston (2007)
13. Prusinkiewicz, P., Hammel, M., Hanan, J., Měch, R.: L-systems: From the theory to visual models of plants. In: Proceedings of the 2nd CSIRO Symposium on Computational Challenges in Life Sciences. CSIRO Publishing, Collingwood, Victoria, Australia (1996)
14. Prusinkiewicz, P., Lindenmayer, A.: The Algorithmic Beauty of Plants. Springer, New York (1990)
15. Salomaa, A.: Formal Languages. Academic Press, London (1973)
16. Srikant, Y.N., Shankar, P.: The Compiler Design Handbook. CRC Press, London (2002)
17. Talaro, K.P., Chess, B.: Foundations in Microbiology. McGraw-Hill, New York (2011)

Part IX
Conclusion

This final part of the book closes its discussion by adding remarks regarding its coverage. Most of these remarks concern new investigation trends and open problem areas. The historical development of regulated grammars and automata is described as well, and this description includes many relevant bibliographic comments and references. A summary of all the material ends this part. Part IX consists of Chaps. 21 and 22.

Chapter 21 describes the entire development of regulated grammars and automata while supporting this description by many bibliographic and historical comments. It also sketches new investigation trends and formulates several open problem areas. Chapter 22 summarizes all the material covered in this book.

Chapter 21
Concluding Remarks

Abstract This three-section chapter makes several final remarks concerning the material covered in this book. First, Sect. 21.1 describes new investigation trends closely related to regulated grammars and automata. Second, Sect. 21.2 suggests several open problem areas to the future study concerning the subject of this book. Finally, Sect. 21.3 gives an overview of significant studies published on the subject of this book from a historical perspective. This chapter contains an unusually great number of references in order to demonstrate the significance of all its coverage in a specific and detailed way.

Keywords New trends • Open problems • Bibliography • History

Supported by more than two hundreds of references, this chapter makes several final remarks concerning the material covered in this book with a special focus on its current and future developments. First, it suggests several new investigation areas concerning regulated grammars and automata (Sect. 21.1). Then, it lists the most important open problems resulting from the study of this book and explains why they are significant (Sect. 21.2). Finally, this chapter chronologically summarizes the concepts and results achieved in most significant studies on the subject of this book (Sect. 21.3). As a result, this summary also represents a theoretical background of the entire book, so the references included in this summary are highly recommended for a further study concerning regulated grammars and automata.

As a whole, this book represents a theoretically oriented study, which contributes to the knowledge about formal languages and their models. Accordingly, all the three sections of this summarizing chapter treat their topics from a primarily theoretical viewpoint rather than a practical standpoint.

A. Meduna and P. Zemek, *Regulated Grammars and Automata*,
DOI 10.1007/978-1-4939-0369-6_21,

21.1 New Trends and Their Expected Investigation

In this section, we point out three new directions in the investigation of regulated grammars and automata. In addition, we make several suggestions regarding their future investigation.

An Algebraic Approach to Regulated Grammars and Automata

From an algebraic viewpoint, various kinds of regulation can be viewed as restrictions placed upon relations by which the regulated models define their languages. Indeed, regulated grammars are based on restrictions placed upon derivations while regulated automata restrict the way they make moves. From this point of view, the investigation of regulated grammars is closely related to many algebraically oriented studies in formal language theory. Investigate how to replace some of the previous regulating mechanisms by suitable relation-domain restrictions and vice versa. Furthermore, study how some well-known special cases of these relations affect the resulting language-defining power. Specifically, perform this study under the assumptions that these relations represent functions, injections, or surjections. The algebraic theory of formal languages and their automata is discussed in a great number of articles and books, some of which are summarized in Chap. 6 through 11 of [170]. Furthermore, [45,47,48,61,184] represent a systematic introduction to this area of formal language theory.

Combining Regulated Grammars and Automata

In formal language theory, the overwhelming majority of language-defining devices is based on rewriting systems that represent either grammars or automata. Although it is obviously quite natural to design language-defining devices based on a combination of both grammars and automata and, thereby, make the scale of language-defining models much richer and broader, only a tiny minority of these models is designed in this combined way.

In terms of regulated language-defining models, state grammars and so-called #-rewriting systems represent regulated models that have features of both regulated grammars and regulated automata. Introduced several decades ago, state grammars (see Sect. 5.4) represent a classical grammatical model of regulation, which has been covered in this monograph in detail. On the other hand, *#-rewriting systems* have been introduced relatively recently (see [80,81,83,85,87]). These systems generate languages just like any grammars. On the other hand, like regulated automata, they use simple state-based regulation during their language-generation process. These systems characterize an infinite hierarchy of language families resulting from programmed grammars of finite index (see [85]). As a result, to put it from a broader perspective, regulated systems of this combined kind are naturally related to some classical results about formal languages, on which they can shed light in an alternative way. Therefore, it is highly expectable that formal language theory will introduce and investigate many more rewriting systems based upon a combination of regulated grammars and automata.

Regulated Translation-Defining Models

Regulated grammars and automata discussed in this book generate languages. As obvious, they can be easily and naturally modified to regulated translation-defining models by analogy with the modification of context-free grammars and pushdown automata to context-free translation grammars and pushdown transducers, respectively (see [3]). Most probably, formal language theory will open their investigation of regulated translation-defining models by studying their properties from a theoretical point of view by analogy with other well-known studies of formal translation, including [2,3,14,16,54,90,161,176]. Simultaneously, however, we can expect a struggle to apply them to the translation of programming as well as natural languages. As a matter of fact, to some extent, [15, 57–59] have already sketched applications concerning the specification and translation of natural languages in this way.

21.2 Open Problem Areas

Throughout this book, we have already formulated many open problems. Out of them, we next select and repeat the most important questions, which deserve our special attention. To see their significance completely, however, we suggest that the reader returns to the referenced parts of the book in order to view these questions in the full context of their formulation and discussion in detail.

1. Over the last four decades, formal language theory has struggled to determine the precise impact of erasing rules to the power of regulated grammars. Indeed, it is still an open question whether regular-controlled, matrix, programmed, and forbidding grammars are equivalent to their propagating versions (see Chaps. 4 and 5). Section 7.2 presents a partial solution to this problem in terms of regular-controlled grammars. However, in general, this important question has not been answered yet. For some very recent results regarding this topic, see [200–203].
2. By Theorem 4.7.6, we can convert any propagating scattered context grammar into an equivalent context-sensitive grammar. However, it is a long-standing open problem whether these two types of grammars are, in fact, equivalent.

3. Consider the results in Sect. 4.7.4 concerning the reduction of scattered context grammars. While one-nonterminal versions of scattered context grammars do not generate the entire family of recursively enumerable languages (see Theorem 4.7.13), their two-nonterminal versions do (see Theorem 4.7.18). Therefore, regarding the number of nonterminals, this open problem area has been completely solved. By Theorem 4.7.19, the two-context-sensitive rule versions of scattered context grammars characterize the family of recursively enumerable languages. On the other hand, the generative power of their one-context-sensitive rule versions has not been determined yet.
4. Chapter 6 discusses one-sided random context grammars and their special versions. Specifically, it shows that one-sided random context grammars with erasing rules have the same power as ordinary random context grammars (see Corollary 6.2.23). What is the generative power of left random context grammars? What is the role of erasing rules in this left variant? What is the generative power of one-sided forbidding grammars? Furthermore, in Sect. 6.4.3, we have proved that any recursively enumerable language is generated by a one-sided random context grammar having no more than two right random context rules. Does this result hold with one or even zero right random context rules?
5. All the uniform rewriting discussed in Chap. 11 is obtained for grammars with erasing rules. In the proof techniques by which we have achieved this rewriting, these rules fulfill a crucial role. Indeed, these techniques cannot be straightforwardly adapted for grammars without erasing rules. Can we achieve some uniform rewriting for grammars without erasing rules in a different way?
6. Return to LRC-ET0L grammars and their variants, discussed in Sect. 10.4. Recall that ET0L and EPT0L grammars have the same generative power (see Theorem 3.3.23). Do LF-E0L and LF-EP0L grammars have the same power? Are LP-E0L and LP-EP0L grammars equally powerful? What is the relation between the language families generated by ET0L grammars and by LP-E0L grammars? What is the generative power of LF-E0L grammars?
7. Chapter 13 gives the basics of multigenerative grammar systems. Recall that they are based upon a simultaneous generation of several strings, which are composed together by some basic operation, such as concatenation, after their generation is completed. Consider other operations, like intersection, and study languages generated in this way by multigenerative grammars systems. Furthermore, study multigenerative grammars systems based on special cases of context-free grammars. Specifically, what is the generative power of multigenerative grammar systems based upon regular or linear grammars?
8. In Sect. 16.1, we have proved that state-controlled and transition-controlled finite automata regulated by languages generated by propagating programmed grammars with appearance checking characterize the family of recursively enumerable languages (see Theorem 16.1.15 and Corollary 16.1.16). Let us point out, however, that these automata are, in a general case, non-deterministic. Does this characterization hold in terms of their deterministic versions, too? Furthermore, try to achieve an analogical characterization of the family of context-sensitive languages.

9. Consider jumping finite automata and their general versions, discussed in Chap. 17. Theorem 17.2.11 gives a necessary and sufficient condition for a language to belong to the family defined by jumping finite automata. Does there exist a similar necessary and sufficient condition for general jumping finite automata as well? Furthermore, how precisely do left jumps affect the power of these automata? Is the family of languages defined by general jumping finite automata closed under shuffle, Kleene star, Kleene plus, or under reversal? Are there any undecidable problems concerning the family of languages accepted by these automata?
10. Reconsider deep pushdown automata, discussed in Sect. 18. In its conclusion, this section discusses two special types of these automata—deterministic deep pushdown automata and deep pushdown automata whose expansions can erase symbols inside of their pushdowns. Determine the language families defined by these two variants.

21.3 Bibliographical and Historical Remarks

This section gives an overview of the crucially important studies published on the subject of this book from a historical perspective. As this book represents primarily a theoretically oriented treatment, we concentrate our attention primarily on theoretical studies.

Although the present treatment of regulated grammars and automata is self-contained, some background in formal language theory is definitely helpful to grasp the material of this book easily. As an introduction to formal language theory, we recommend [43,48,55,56,69,89,96,122,158,175,177]. The three-volume *Handbook of Formal Languages* (see [170–172]) gives an overview of the recent important trends in formal language theory.

For a summary of the fundamental knowledge about regulated rewriting published by 1989, consult [26]. Furthermore, [97] and Chap. 3 of [171] give a brief overview of recent results concerning regulated grammars. [148] summarizes recent results concerning various transformations of regulated grammars. More specifically, it concentrates its attention on algorithms that transform these grammars and some related regulated language-defining models so the resulting transformed models are equivalent and, in addition, satisfy some prescribed properties.

Context-Based Grammatical Regulation

The classical normal forms from Sect. 4.1.1 were established in [46, 79, 162]. The two new normal forms appearing in this book are completely new. Consult page 180 in [170] for a summary of normal forms of phrase-structure grammars.

The uniform generation of sentences by phrase-structure grammars, discussed in Sect. 4.1.2, has been investigated in [121].

Conditional grammars were introduced in [44]. Several variants of these grammars were discussed in [18, 27, 30, 31, 67, 68, 70, 78, 113, 159, 164, 166, 169, 173, 185, 186]. The crucial concepts of these grammars and results concerning them are summarized in [130].

Random context grammars were introduced in [191]. Strictly speaking, in [191], their definition coincides with the definition of permitting grammars in this book. Forbidding grammars, also known as *N-grammars* (see [163]), together with other variants of random context grammars were originally studied by Lomkovskaya in [91–93]. After these studies, many more papers discussed these grammars, including [6, 32–35, 100, 146, 201]. In [28, 103, 107], simplified versions of random context grammars, called *restricted context-free grammars*, were studied. Moreover, [20, 51, 82, 101] studied grammar systems with their components represented by random context grammars.

Generalized forbidding grammars were introduced in [112] and further investigated in [104, 144, 146].

Semi-conditional and simple semi-conditional grammars were introduced and investigated in [166] and [131], respectively. Their descriptional complexity was studied in [98, 105, 143, 146, 160, 186, 187].

Originally, scattered context grammars were defined in [53]. Their original version disallowed erasing rules, however. Four years later, [188] generalized them to scattered context grammars with erasing rules (see also [117]). The following studies represent the most important studies that have discussed these grammars: [17, 19, 29, 36, 38, 39, 52, 53, 76, 77, 99, 100, 102, 106, 108–110, 114, 116, 117, 119, 120, 123–128, 133–138, 146, 147, 157, 165, 182, 183, 187, 188]. Generalized restricted erasing from Sect. 7.3 was studied in [137]. The coincidental extension operation—the topic of Sect. 8.2—was investigated in [127, 128]. LL variants of scattered context grammars, discussed in Sect. 4.7.5, were studied in [63–65, 71, 140]). Uniform generation of languages by scattered context grammars was investigated in [125]. For an in-depth overview of scattered context grammars and their applications, consult [139] and the references given therein.

One-sided random context grammars were introduced in [149]. Their special variants, left forbidding and left permitting grammars, were originally introduced in [51] and [20], respectively. The nonterminal complexity of one-sided random context grammars was investigated in [152]. Several normal forms of these grammars were established in [198]. Leftmost derivations were studied in [153]. The generalized version of one-sided forbidding grammars was introduced and investigated in [154]. A list of open problems concerning these grammars appears in [199]. Finally, a reduction of the number of right random context rules and LL versions of one-sided random context grammars appear in this book for the first time.

Sequential rewriting over word monoids has been studied in [111, 118]. Moreover, [10–12] investigate sequential rewriting over free groups.

Rule-Based Grammatical Regulation

Grammars regulated by regular control languages over the set of rules were introduced in [50]. The workspace conditions given in Sect. 7.2 were established in [150] (see also [196]). Generation of sentences with their parses by these grammars, discussed in Sect. 8.1, was investigated in [156].

Matrix grammars were first defined and studied in [1]. For some very recent results regarding the elimination of erasing rules from these grammars, see [200–203].

Programmed grammars were introduced in [168]. Their non-determinism has been investigated in [7, 13, 141, 189, 190]. Some other recent papers include [37, 40, 42].

State grammars were defined by Kasai in [66]. A generalized version of these grammars with erasing rules was originally studied in [60].

Regulated Grammars: Parallelism

In general, context-regulated ET0L grammars have been studied in [8, 24, 24, 26, 26, 145, 146, 173, 179–181]). Context-conditional ET0L grammars were studied in Sect. 4.2.1 in [146]. Forbidding ET0L grammars were introduced and investigated in [145]. Simple semi-conditional ET0L grammars were introduced in [181] and further studied in [75]. Left versions of ET0L grammars were introduced and studied in [155]. Their nonterminal complexity was investigated in [197]. The uniform generation of sentences by EIL grammars, discussed in Sect. 4.1.2 has been investigated in [121]. Parallel rewriting over word monoids was studied in [76, 115].

Let us finally add that there also exist regulated versions of *(uniformly) limited ET0L grammars* (see [41, 192–195]) and ET0L grammars regulated by other mechanisms, such as mechanisms based upon control languages (see [5, 25, 49] and Chap. 8 in [26]).

Regulated Grammar Systems

Multigenerative grammar systems based upon leftmost derivations were introduced in [94]. Their general versions were studied in [95]. Controlled pure grammar systems were introduced and investigated in [153]. Moreover, [142] gives a preliminary solution to four open problems raised in [153].

Regulated Automata

Self-regulated finite and pushdown automata were introduced in [132]. Finite automata regulated by control languages appear in this book for the very first time.

For a study of finite automata controlled by Petri nets, see [62]. Regulated pushdown automata were introduced in [72]. Their special versions, referred to as one-turn linear-regulated pushdown automata, were studied in [73] (see also [74, 174]). *Blackhole pushdown automata*, which are closely related to regulated pushdown automata, were introduced and investigated in [21, 22].

Related Unregulated Automata

Jumping finite automata were introduced in [151]. Other related models involving discontinuity include *nested word automata* [4], *bag automata* [23], and *input-revolving finite automata* [9]. Deep pushdown automata were proposed and studied in [129]. For more results related to these automata, consult [84, 86, 88, 167, 178].

References

1. Abraham, S.: Some questions of language theory. In: Proceedings of the 1965 conference on Computational linguistics, pp. 1–11. Association for Computational Linguistics, New York (1965)
2. Aho, A.V., Lam, M.S., Sethi, R., Ullman, J.D.: Compilers: Principles, Techniques, and Tools, 2nd edn. Addison-Wesley, Boston (2006)
3. Aho, A.V., Ullman, J.D.: The Theory of Parsing, Translation and Compiling, Volume I: Parsing. Prentice-Hall, New Jersey (1972)
4. Alur, R., Madhusudan, P.: Adding nesting structure to words. J. ACM **56**(3), 16:1–16:43 (2009)
5. Asveld, P.R.J.: Controlled iteration grammars and full hyper-AFL's. Inform. Contr. **34**(3), 248–269 (1977)
6. Atcheson, B., Ewert, S., Shell, D.: A note on the generative capacity of random context. South African Comput. J. **36**, 95–98 (2006)
7. Barbaiani, M., Bibire, C., Dassow, J., Delaney, A., Fazekas, S., Ionescu, M., Liu, G., Lodhi, A., Nagy, B.: The power of programmed grammars with graphs from various classes. J. Appl. Mathe. Comput. **22**(1–2), 21–38 (2006)
8. Beek, M., Csuhaj-Varjú, E., Holzer, M., Vaszil, G.: On competence in CD grammar systems. In: Developments in Language Theory, Lecture Notes in Computer Science, vol. 3340, pp. 3–14. Springer, Berlin (2005)
9. Bensch, S., Bordihn, H., Holzer, M., Kutrib, M.: On input-revolving deterministic and nondeterministic finite automata. Informat. Comput. **207**(11), 1140–1155 (2009)
10. Bidlo, R., Blatný, P., Meduna, A.: Formal models over free groups. In: 1st Doctoral Workshop on Mathematical and Engineering Methods in Computer Science, pp. 193–199. Faculty of Information Technology BUT, Brno, CZ (2005)
11. Bidlo, R., Blatný, P., Meduna, A.: Automata with two-sided pushdowns defined over free groups generated by reduced alphabets. Kybernetika **2007**(1), 21–35 (2007)
12. Bidlo, R., Blatný, P., Meduna, A.: Context-free and E0L derivations over free groups. Schedae Informaticae **2007**(16), 14–24 (2007)
13. Bordihn, H., Holzer, M.: Programmed grammars and their relation to the LBA problem. Acta Informatica **43**(4), 223–242 (2006)
14. Brookshear, J.G.: Theory of Computation. Benjamin Cummings Publishing, San Francisco (1989)

15. Čermák, M., Horáček, P., Meduna, A.: Rule-restricted automaton-grammar tranduscers: Power and linguistic applications. Math. Appl. **1**(1), 13–35 (2012)
16. Chomsky, N.: Syntactic Structures. Mouton, New York (2002)
17. Cremers, A.B.: Normal forms for context-sensitive grammars. Acta Informatica **3**, 59–73 (1973)
18. Csuhaj-Varjú, E.: On grammars with local and global context conditions. Int. J. Comput. Math. **47**, 17–27 (1992)
19. Csuhaj-Varjú, E., Vaszil, G.: Scattered context grammars generate any recursively enumerable language with two nonterminals. Inf. Process. Lett. **110**(20), 902–907 (2010)
20. Csuhaj-Varjú, E., Masopust, T., Vaszil, G.: Cooperating distributed grammar systems with permitting grammars as components. Rom. J. Infor. Sci. Tech. **12**(2), 175–189 (2009)
21. Csuhaj-Varjú, E., Masopust, T., Vaszil, G.: Blackhole state-controlled regulated pushdown automata. In: Second Workshop on Non-Classical Models for Automata and Applications (NCMA 2010), pp. 45–56 (2010)
22. Csuhaj-Varjú, E., Masopust, T., Vaszil, G.: Blackhole pushdown automata. Fundamenta Informaticae **112**(2–3), 137–156 (2011)
23. Daley, M., Eramian, M., McQuillan, I.: Bag automata and stochastic retrieval of biomolecules in solution. In: Implementation and Application of Automata, Eighth International Conference CIAA 2003, Santa Barbara, CA, 2003, no. 2759 in Lecture Notes in Computer Science, pp. 239–250. Springer, New York (2003)
24. Dassow, J.: On cooperating distributed grammar systems with competence based start and stop conditions. Fundamenta Informaticae **76**, 293–304 (2007)
25. Dassow, J., Fest, U.: On regulated L systems. Rostocker Mathematisches Kolloquium **25**, 99–118 (1984)
26. Dassow, J., Păun, G.: Regulated Rewriting in Formal Language Theory. Springer, New York (1989)
27. Dassow, J., Păun, G., Salomaa, A.: Grammars based on patterns. Int. J. Found. Comput. Sci. **4**(1), 1–14 (1993)
28. Dassow, J., Masopust, T.: On restricted context-free grammars. J. Comput. Syst. Sci. **78**(1), 293–304 (2012)
29. Ehrenfeucht, A., Rozenberg, G.: An observation on scattered grammars. Inf. Process. Lett. **9**(2), 84–85 (1979)
30. Ehrenfeucht, A., Kleijn, H.C.M., Rozenberg, G.: Adding global forbidding context to context-free grammars. Theor. Comput. Sci. **37**, 337–360 (1985)
31. Ehrenfeucht, A., Pasten, P., Rozenberg, G.: Context-free text grammars. Acta Informatica **31**, 161–206 (1994)
32. Ewert, S., Walt, A.: A shrinking lemma for random forbidding context languages. Theor. Comput. Sci. **237**(1–2), 149–158 (2000)
33. Ewert, S., Walt, A.: A pumping lemma for random permitting context languages. Theor. Comput. Sci. **270**(1–2), 959–967 (2002)
34. Ewert, S., Walt, A.: The power and limitations of random context. In: Grammars and Automata for String Processing: from Mathematics and Computer Science to Biology, pp. 33–43. Taylor and Francis, London (2003)
35. Ewert, S., Walt, A.: Necessary conditions for subclasses of random context languages. Theor. Comput. Sci. **475**, 66–72 (2013)
36. Fernau, H.: Scattered context grammars with regulation. Ann. Bucharest Univ. Math-Inform. Ser. **45**(1), 41–49 (1996)
37. Fernau, H.: Nonterminal complexity of programmed grammars. Theor. Comput. Sci. **296**(2), 225–251 (2003)
38. Fernau, H., Meduna, A.: A simultaneous reduction of several measures of descriptional complexity in scattered context grammars. Inf. Process. Lett. **86**(5), 235–240 (2003)
39. Fernau, H., Meduna, A.: On the degree of scattered context-sensitivity. Theor. Comput. Sci. **290**(3), 2121–2124 (2003)

40. Fernau, H., Stephan, F.: How powerful is unconditional transfer? — When UT meets AC. In: Developments in Language Theory, pp. 249–260 (1997)
41. Fernau, H., Wätjen, D.: Remarks on regulated limited ET0L systems and regulated context-free grammars. Theor. Comput. Sci. **194**(1–2), 35–55 (1998)
42. Fernau, H., Freund, R., Oswald, M., Reinhardt, K.: Refining the nonterminal complexity of graph-controlled, programmed, and matrix grammars. J. Autom. Lang. Combin. **12**(1–2), 117–138 (2007)
43. Floyd, R.W., Beigel, R.: The Language of Machines: An Introduction to Computability and Formal Languages. Computer Science Press, New York (1994)
44. Fris, I.: Grammars with partial ordering of the rules. Inf. Control **12**, 415–425 (1968)
45. Gathen, J., Gerhard, J.: Modern Computer Algebra, 2nd edn. Cambridge University Press, New York (2003)
46. Geffert, V.: Normal forms for phrase-structure grammars. Theor. Inf. Appl. **25**(5), 473–496 (1991)
47. Gilbert, W.J., Gilbert, W.J.: Modern Algebra with Applications (Pure and Applied Mathematics: A Wiley Series of Texts, Monographs and Tracts), 2nd edn. Wiley-Blackwell, New Jersey (2003)
48. Ginsburg, S.: Algebraic and Automata-Theoretic Properties of Formal Languages. Elsevier Science, New York (1975)
49. Ginsburg, S., Rozenberg, G.: T0L schemes and control sets. Inf. Control **27**, 109–125 (1974)
50. Ginsburg, S., Spanier, E.: Finite-turn pushdown automata. SIAM J. Control **4**, 429–453 (1968)
51. Goldefus, F., Masopust, T., Meduna, A.: Left-forbidding cooperating distributed grammar systems. Theor. Comput. Sci. **20**(3), 1–11 (2010)
52. Gonczarowski, J., Warmuth, M.K.: Scattered versus context-sensitive rewriting. Acta Informatica **27**, 81–95 (1989)
53. Greibach, S.A., Hopcroft, J.E.: Scattered context grammars. J. Comput. Syst. Sci. **3**(3), 233–247 (1969)
54. Gries, D.: Compiler Construction for Digital Computers. Wiley, New York (1971)
55. Harrison, M.: Introduction to Formal Language Theory. Addison-Wesley, Boston (1978)
56. Hopcroft, J.E., Ullman, J.D.: Formal Languages and Their Relation to Automata. Addison-Wesley, Boston (1969)
57. Horáček, P.: On generative power of synchronous grammars with linked rules. In: Proceedings of the 18th Conference STUDENT EEICT 2012, vol. 3, pp. 376–380. Brno University of Technology, Brno, CZ (2012)
58. Horáček, P., Meduna, A.: Regulated rewriting in natural language translation. In: 7th Doctoral Workshop on Mathematical and Engineering Methods in Computer Science, pp. 35–42. Faculty of Information Technology BUT, Brno, CZ (2011)
59. Horáček, P., Meduna, A.: Synchronous versions of regulated grammars: Generative power and linguistic applications. Theor. Appl. Inform. **24**(3), 175–190 (2012)
60. Horváth, G., Meduna, A.: On state grammars. Acta Cybernetica **1988**(8), 237–245 (1988)
61. Ito, M.: Algebraic Theory of Automata and Languages, 2nd edn. World Scientific Publishing Company, Singapore (2003)
62. Jantzen, M., Kudlek, M., Zetzsche, G.: Finite automata controlled by Petri nets. In: Proceedings of the 14th Workshop; Algorithmen und Werkzeuge für Petrinetze, Technical Report Nr. 25/2007, pp. 57–62. Universität Koblenz-Landau (2007)
63. Jirák, O.: Delayed execution of scattered context grammar rules. In: Proceedings of the 15th Conference and Competition EEICT 2009, pp. 405–409. Brno University of Technology, Brno, CZ (2009)
64. Jirák, O.: Table-driven parsing of scattered context grammar. In: Proceedings of the 16th Conference and Competition EEICT 2010, pp. 171–175. Brno University of Technology, Brno, CZ (2010)
65. Jirák, O., Kolář, D.: Derivation in scattered context grammar via lazy function evaluation. In: 5th Doctoral Workshop on Mathematical and Engineering Methods in Computer Science, pp. 118–125. Masaryk University (2009)

66. Kasai, T.: An hierarchy between context-free and context-sensitive languages. J. Comput. Syst. Sci. **4**, 492–508 (1970)
67. Kelemen, J.: Conditional grammars: Motivations, definition, and some properties. In: Proceedings on Automata, Languages and Mathematical Systems, pp. 110–123. K. Marx University of Economics, Budapest (1984)
68. Kelemen, J.: Measuring cognitive resources use (a grammatical approach). Comput. Artif. Intell. **8**(1), 29–42 (1989)
69. Kelley, D.: Automata and Formal Languages. Prentice-Hall, New Jersey (1995)
70. Kleijn, H.C.M., Rozenberg, G.: Context-free-like restrictions on selective rewriting. Theor. Comput. Sci. **16**, 237–239 (1981)
71. Kolář, D.: Scattered context grammars parsers. In: Proceedings of the 14th International Congress of Cybernetics and Systems of WOCS, pp. 491–500. Wroclaw University of Technology (2008)
72. Kolář, D., Meduna, A.: Regulated pushdown automata. Acta Cybernetica **2000**(4), 653–664 (2000)
73. Kolář, D., Meduna, A.: One-turn regulated pushdown automata and their reduction. Fundamenta Informaticae **2001**(21), 1001–1007 (2001)
74. Kolář, D., Meduna, A.: Regulated automata: From theory towards applications. In: Proceeding of 8th International Conference on Information Systems Implementation and Modelling (ISIM'05), pp. 33–48 (2005)
75. Kopeček, T., Meduna, A.: Simple-semi-conditional versions of matrix grammars with a reduced regulating mechanism. Comput. Inform. **2004**(23), 287–302 (2004)
76. Kopeček, T., Meduna, A., Švec, M.: Simulation of scattered context grammars and phrase-structured grammars by symbiotic E0L grammars. In: Proceeding of 8th International Conference on Information Systems, Implementation and Modelling (ISIM'05), pp. 59–66. Brno, CZ (2005)
77. Král, J.: On multiple grammars. Kybernetika **1**, 60–85 (1969)
78. Kral, J.: A note on grammars with regular restrictions. Kybernetika **9(3)**, 159–161 (1973)
79. Kuroda, S.Y.: Classes of languages and linear-bounded automata. Inf. Control **7**(2), 207–223 (1964)
80. Křivka, Z.: Deterministic #-rewriting systems. In: Proceedings of the 13th Conference and Competition EEICT 2007, vol. 4, pp. 386–390. Brno University of Technology, Brno, CZ (2007)
81. Křivka, Z.: Rewriting Systems with Restricted Configurations. Faculty of Information Technology, Brno University of Technology, Brno, CZ (2008)
82. Křivka, Z., Masopust, T.: Cooperating distributed grammar systems with random context grammars as components. Acta Cybernetica **20**(2), 269–283 (2011)
83. Křivka, Z., Meduna, A.: Generalized #-rewriting systems of finite index. In: Proceedings of 2nd International Workshop on Formal Models (WFM'07), pp. 197–204. Silesian University, Opava, CZ (2007)
84. Křivka, Z., Meduna, A., Schönecker, R.: General top-down parsers based on deep pushdown expansions. In: Proceedings of 1st International Workshop on Formal Models (WFM'06), pp. 11–18. Ostrava, CZ (2006)
85. Křivka, Z., Meduna, A., Schönecker, R.: Generation of languages by rewriting systems that resemble automata. Int. J. Found. Comput. Sci. **17**(5), 1223–1229 (2006)
86. Křivka, Z., Meduna, A., Schönecker, R.: Reducing deep pushdown automata and infinite hierarchy. In: 2nd Doctoral Workshop on Mathematical and Engineering Methods in Computer Science, pp. 214–221. Brno University of Technology, Brno, CZ (2006)
87. Křivka, Z., Meduna, A., Smrček, J.: n-right-linear #-rewriting systems. In: 3rd Doctoral Workshop on Mathematical and Engineering Methods in Computer Science, pp. 105–112. Brno University of Technology, Brno, CZ (2007)
88. Leupold, P., Meduna, A.: Finitely expandable deep PDAs. In: Automata, Formal Languages and Algebraic Systems: Proceedings of AFLAS 2008, pp. 113–123. Hong Kong University of Scinece and Technology (2010)

89. Lewis, H.R., Papadimitriou, C.H.: Elements of the Theory of Computation. Prentice-Hall, New Jersey (1981)
90. Lewis, P.M., Rosenkrantz, D.J., Stearns, R.E.: Compiler Design Theory. Addison-Wesley, Boston (1976)
91. Lomkovskaya, M.V.: Conditional grammars and intermediate classes of languages (in Russian). Soviet Mathematics – Doklady **207**, 781–784 (1972)
92. Lomkovskaya, M.V.: On c-conditional and other commutative grammars (in Russian). Nauchno-Tekhnicheskaya Informatsiya **2**(2), 28–31 (1972)
93. Lomkovskaya, M.V.: On some properties of c-conditional grammars (in Russian). Nauchno-Tekhnicheskaya Informatsiya **2**(1), 16–21 (1972)
94. Lukáš, R., Meduna, A.: Multigenerative grammar systems. Schedae Informaticae **2006**(15), 175–188 (2006)
95. Lukáš, R., Meduna, A.: Multigenerative grammar systems and matrix grammars. Kybernetika **46**(1), 68–82 (2010)
96. Martin, J.C.: Introduction to Languages and the Theory of Computation, 3rd edn. McGraw-Hill, New York (2002)
97. Martín-Vide, C., Mitrana, V., Păun, G. (eds.): Formal Languages and Applications, chapter 13, pp. 249–274. Springer, Berlin (2004)
98. Masopust, T.: An improvement of the descriptional complexity of grammars regulated by context conditions. In: 2nd Doctoral Workshop on Mathematical and Engineering Methods in Computer Science, pp. 105–112. Faculty of Information Technology BUT, Brno, CZ (2006)
99. Masopust, T.: Scattered context grammars can generate the powers of 2. In: Proceedings of the 13th Conference and Competition EEICT 2007, vol. 4, pp. 401–404. Brno University of Technology, Brno, CZ (2007)
100. Masopust, T.: On the descriptional complexity of scattered context grammars. Theor. Comput. Sci. **410**(1), 108–112 (2009)
101. Masopust, T.: On the terminating derivation mode in cooperating distributed grammar systems with forbidding components. Int. J. Found. Comput. Sci. **20**(2), 331–340 (2009)
102. Masopust, T.: Bounded number of parallel productions in scattered context grammars with three nonterminals. Fundamenta Informaticae **99**(4), 473–480 (2010)
103. Masopust, T.: Simple restriction in context-free rewriting. J. Comput. Syst. Sci. **76**(8), 837–846 (2010)
104. Masopust, T., Meduna, A.: Descriptional complexity of grammars regulated by context conditions. In: LATA '07 Pre-proceedings. Reports of the Research Group on Mathematical Linguistics 35/07, Universitat Rovira i Virgili, pp. 403–411 (2007)
105. Masopust, T., Meduna, A.: Descriptional complexity of semi-conditional grammars. Inf. Process. Lett. **104**(1), 29–31 (2007)
106. Masopust, T., Meduna, A.: On descriptional complexity of partially parallel grammars. Fundamenta Informaticae **87**(3), 407–415 (2008)
107. Masopust, T., Meduna, A.: On context-free rewriting with a simple restriction and its computational completeness. RAIRO-Theor. Inform. Appl.-Informatique Théorique et Appl. **43**(2), 365–378 (2009)
108. Masopust, T., Meduna, A., Šimáček, J.: Two power-decreasing derivation restrictions in generalized scattered context grammars. Acta Cybernetica **18**(4), 783–793 (2008)
109. Masopust, T., Techet, J.: Leftmost derivations of propagating scattered context grammars: A new proof. Discrete Math. Theor. Comput. Sci. **10**(2), 39–46 (2008)
110. Mayer, O.: Some restrictive devices for context-free grammars. Inf. Control **20**, 69–92 (1972)
111. Meduna, A.: Context-free derivations on word monoids. Acta Informatica **1990**(27), 781–786 (1990)
112. Meduna, A.: Generalized forbidding grammars. Int. J. Comput. Math. **36**(1–2), 31–38 (1990)
113. Meduna, A.: Global context conditional grammars. J. Autom. Lang. Combin. **1991**(27), 159–165 (1991)
114. Meduna, A.: Scattered rewriting in the formal language theory. In: Missourian Annual Conference on Computing, pp. 26–36. Columbia, US (1991)

115. Meduna, A.: Symbiotic E0L systems. Acta Cybernetica **10**, 165–172 (1992)
116. Meduna, A.: Canonical scattered rewriting. Int. J. Comput. Math. **51**, 122–129 (1993)
117. Meduna, A.: Syntactic complexity of scattered context grammars. Acta Informatica **1995**(32), 285–298 (1995)
118. Meduna, A.: Syntactic complexity of context-free grammars over word monoids. Acta Informatica **1996**(33), 457–462 (1996)
119. Meduna, A.: Four-nonterminal scattered context grammars characterize the family of recursively enumerable languages. Int. J. Comput. Math. **63**, 67–83 (1997)
120. Meduna, A.: Economical transformations of phrase-structure grammars to scattered context grammars. Acta Cybernetica **13**, 225–242 (1998)
121. Meduna, A.: Uniform rewriting based on permutations. Int. J. Comput. Math. **69**(1–2), 57–74 (1998)
122. Meduna, A.: Automata and Languages: Theory and Applications. Springer, London (2000)
123. Meduna, A.: Generative power of three-nonterminal scattered context grammars. Theor. Comput. Sci. **2000**(246), 279–284 (2000)
124. Meduna, A.: Terminating left-hand sides of scattered context grammars. Theor. Comput. Sci. **2000**(237), 424–427 (2000)
125. Meduna, A.: Uniform generation of languages by scattered context grammars. Fundamenta Informaticae **44**, 231–235 (2001)
126. Meduna, A.: Descriptional complexity of scattered rewriting and multirewriting: An overview. J. Autom. Lang. Combin. **2002**(7), 571–577 (2002)
127. Meduna, A.: Coincidental extension of scattered context languages. Acta Informatica **39**(5), 307–314 (2003)
128. Meduna, A.: Erratum: Coincidental extension of scattered context languages. Acta Informatica **39**(9), 699 (2003)
129. Meduna, A.: Deep pushdown automata. Acta Informatica **2006**(98), 114–124 (2006)
130. Meduna, A., Csuhaj-Varjú, E.: Grammars with context conditions. EATCS Bulletin **32**, 112–124 (1993)
131. Meduna, A., Gopalaratnam, A.: On semi-conditional grammars with productions having either forbidding or permitting conditions. Acta Cybernetica **11**, 307–323 (1994)
132. Meduna, A., Masopust, T.: Self-regulating finite automata. Acta Cybernetica **18**(1), 135–153 (2007)
133. Meduna, A., Techet, J.: Generation of sentences with their parses: the case of propagating scattered context grammars. Acta Cybernetica **17**, 11–20 (2005)
134. Meduna, A., Techet, J.: Canonical scattered context generators of sentences with their parses. Theor. Comput. Sci. **2007**(389), 73–81 (2007)
135. Meduna, A., Techet, J.: Maximal and minimal scattered context rewriting. In: FCT 2007 Proceedings, pp. 412–423. Budapest (2007)
136. Meduna, A., Techet, J.: Reduction of scattered context generators of sentences preceded by their leftmost parses. In: DCFS 2007 Proceedings, pp. 178–185. High Tatras, SK (2007)
137. Meduna, A., Techet, J.: Scattered context grammars that erase nonterminals in a generalized k-limited way. Acta Informatica **45**(7), 593–608 (2008)
138. Meduna, A., Techet, J.: An infinite hierarchy of language families generated by scattered context grammars with n-limited derivations. Theor. Comput. Sci. **410**(21), 1961–1969 (2009)
139. Meduna, A., Techet, J.: Scattered Context Grammars and Their Applications. WIT Press, Southampton (2010)
140. Meduna, A., Vrábel, L., Zemek, P.: LL leftmost k-linear scattered context grammars. In: AIP Conference Proceedings, vol. 1389, pp. 833–836. American Institute of Physics, Kassandra, Halkidiki, GR (2011)
141. Meduna, A., Vrábel, L., Zemek, P.: On nondeterminism in programmed grammars. In: 13th International Conference on Automata and Formal Languages, pp. 316–328. Computer and Automation Research Institute, Hungarian Academy of Sciences, Debrecen, HU (2011)

142. Meduna, A., Vrábel, L., Zemek, P.: Solutions to four open problems concerning controlled pure grammar systems. Int. J. Comput. Math. (to appear)
143. Meduna, A., Švec, M.: Reduction of simple semi-conditional grammars with respect to the number of conditional productions. Acta Cybernetica **15**, 353–360 (2002)
144. Meduna, A., Švec, M.: Descriptional complexity of generalized forbidding grammars. Int. J. Comput. Math. **80**(1), 11–17 (2003)
145. Meduna, A., Švec, M.: Forbidding ET0L grammars. Theor. Comput. Sci. **2003**(306), 449–469 (2003)
146. Meduna, A., Švec, M.: Grammars with Context Conditions and Their Applications. Wiley, New Jersey (2005)
147. Meduna, A., Židek, S.: Scattered context grammars generating sentences followed by derivation trees. Theor. Appl. Inform. **2011**(2), 97–106 (2011)
148. Meduna, A., Zemek, P.: Regulated Grammars and Their Transformations. Faculty of Information Technology, Brno University of Technology, Brno, CZ (2010)
149. Meduna, A., Zemek, P.: One-sided random context grammars. Acta Informatica **48**(3), 149–163 (2011)
150. Meduna, A., Zemek, P.: Workspace theorems for regular-controlled grammars. Theor. Comput. Sci. **412**(35), 4604–4612 (2011)
151. Meduna, A., Zemek, P.: Jumping finite automata. Int. J. Found. Comput. Sci. **23**(7), 1555–1578 (2012)
152. Meduna, A., Zemek, P.: Nonterminal complexity of one-sided random context grammars. Acta Informatica **49**(2), 55–68 (2012)
153. Meduna, A., Zemek, P.: One-sided random context grammars with leftmost derivations. In: LNCS Festschrift Series: Languages Alive, vol. 7300, pp. 160–173. Springer, New York (2012)
154. Meduna, A., Zemek, P.: Generalized one-sided forbidding grammars. Int. J. Comput. Math. **90**(2), 127–182 (2013)
155. Meduna, A., Zemek, P.: Left random context ET0L grammars. Fundamenta Informaticae **123**(3), 289–304 (2013)
156. Meduna, A., Zemek, P.: On the generation of sentences with their parses by propagating regular-controlled grammars. Theor. Comput. Sci. **477**(1), 67–75 (2013)
157. Milgram, D., Rosenfeld, A.: A note on scattered context grammars. Inf. Process. Lett. **1**, 47–50 (1971)
158. Moll, R.N., Arbib, M.A., Kfoury, A.J.: An Introduction to Formal Language Theory. Springer, New York (1988)
159. Navrátil, E.: Context-free grammars with regular conditions. Kybernetika **6**(2), 118–125 (1970)
160. Okubo, F.: A note on the descriptional complexity of semi-conditional grammars. Inf. Process. Lett. **110**(1), 36–40 (2009)
161. Pagen, F.G.: Formal Specifications of Programming Language: A Panoramic Primer. Prentice-Hall, New Jersey (1981)
162. Penttonen, M.: One-sided and two-sided context in formal grammars. Inf. Control **25**(4), 371–392 (1974)
163. Penttonen, M.: ET0L-grammars and N-grammars. Inf. Process. Lett. **4**(1), 11–13 (1975)
164. Păun, G.: On the generative capacity of conditional grammars. Inf. Control **43**, 178–186 (1979)
165. Păun, G.: On simple matrix languages versus scattered context languages. Informatique Théorique et Appl. **16**(3), 245–253 (1982)
166. Păun, G.: A variant of random context grammars: semi-conditional grammars. Theor. Comput. Sci. **41**(1), 1–17 (1985)
167. Quesada, A.A., Stewart, I.A.: On the power of deep pushdown stacks. Acta Informatica **46**(7), 509–531 (2009)
168. Rosenkrantz, D.J.: Programmed grammars and classes of formal languages. J. ACM **16**(1), 107–131 (1969)

169. Rozenberg, G.: Selective substitution grammars (towards a framework for rewriting systems). Part 1: Definitions and examples. Elektronische Informationsverarbeitung und Kybernetik **13**(9), 455–463 (1977)
170. Rozenberg, G., Salomaa, A. (eds.): Handbook of Formal Languages, Vol. 1: Word, Language, Grammar. Springer, New York (1997)
171. Rozenberg, G., Salomaa, A. (eds.): Handbook of Formal Languages, Vol. 2: Linear Modeling: Background and Application. Springer, New York (1997)
172. Rozenberg, G., Salomaa, A. (eds.): Handbook of Formal Languages, Volume 3: Beyond Words. Springer, Berlin (1997)
173. Rozenberg, G., Solms, S.H.: Priorities on context conditions in rewriting systems. Inf. Sci. **14**(1), 15–50 (1978)
174. Rychnovský, L.: Regulated pushdown automata revisited. In: Proceedings of the 15th Conference STUDENT EEICT 2009, pp. 440–444. Brno University of Technology, Brno, CZ (2009)
175. Salomaa, A.: Formal Languages. Academic Press, London (1973)
176. Sippu, S., Soisalon-Soininen, E.: Parsing Theory. Springer, New York (1987)
177. Sipser, M.: Introduction to the Theory of Computation, 2nd edn. PWS Publishing Company, Boston (2006)
178. Solár, P.: Parallel deep pushdown automata. In: Proceedings of the 18th Conference STUDENT EEICT 2012, vol. 3, pp. 410–414. Brno University of Technology, Brno, CZ (2012)
179. Solms, S.H.: Some notes on ET0L languages. Int. J. Comput. Math. **5**, 285–296 (1976)
180. Sosík, P.: The power of catalysts and priorities in membrane systems. Grammars **6**(1), 13–24 (2003)
181. Švec, M.: Simple semi-conditional ET0L grammars. In: Proceedings of the International Conference and Competition Student EEICT 2003, pp. 283–287. Brno University of Technology, Brno, CZ (2003)
182. Techet, J.: A note on scattered context grammars with non-context-free components. In: 3rd Doctoral Workshop on Mathematical and Engineering Methods in Computer Science, pp. 225–232. Brno University of Technology, Brno, CZ (2007)
183. Techet, J.: Scattered context in formal languages. Ph.D. thesis, Faculty of Information Technology, Brno University of Technology, Brno (2008)
184. Truss, J.: Discrete Mathematics for Computer Scientists (International Computer Science Series), 2nd edn. Addison-Wesley, Boston (1998)
185. Urbanek, F.J.: A note on conditional grammars. Revue Roumaine de Mathématiques Pures at Appliquées **28**, 341–342 (1983)
186. Vaszil, G.: On the number of conditional rules in simple semi-conditional grammars. In: Descriptional Complexity of Formal Systems, pp. 210–220. MTA SZTAKI, Budapest, HU (2003)
187. Vaszil, G.: On the descriptional complexity of some rewriting mechanisms regulated by context conditions. Theor. Comput. Sci. **330**(2), 361–373 (2005)
188. Virkkunen, V.: On scattered context grammars. Acta Universitatis Ouluensis **20**(6), 75–82 (1973)
189. Vrábel, L.: A new normal form for programmed grammars. In: Proceedings of the 17th Conference STUDENT EEICT 2011, vol. 3. Brno University of Technology, Brno, CZ (2011)
190. Vrábel, L.: A new normal form for programmed grammars with appearance checking. In: Proceedings of the 18th Conference STUDENT EEICT 2012, vol. 3, pp. 420–425. Brno University of Technology, Brno, CZ (2012)
191. Walt, A.: Random context grammars. In: Proceedings of Symposium on Formal Languages, pp. 163–165 (1970)
192. Wätjen, D.: Regulation of k-limited ET0L systems. Int. J. Comput. Math. **47**, 29–41 (1993)
193. Wätjen, D.: Regulation of uniformly k-limited T0L systems. J. Inf. Process. Cybern. **30**(3), 169–187 (1994)

194. Wätjen, D.: On regularly controlled k-limited T0L systems. Int. J. Comput. Math. **55**(1–2), 57–66 (1995)
195. Wätjen, D.: Regulations of uniformly k-limited ET0L systems and their relations to controlled context-free grammars. J. Autom. Lang. Combin. **1**(1), 55–74 (1996)
196. Zemek, P.: k-limited erasing performed by regular-controlled context-free grammars. In: Proceedings of the 16th Conference STUDENT EEICT 2011, vol. 3, pp. 42–44. Brno University of Technology, Brno, CZ (2010)
197. Zemek, P.: On the nonterminal complexity of left random context E0L grammars. In: Proceedings of the 17th Conference STUDENT EEICT 2011, vol. 3, pp. 510–514. Brno University of Technology, Brno, CZ (2011)
198. Zemek, P.: Normal forms of one-sided random context grammars. In: Proceedings of the 18th Conference STUDENT EEICT 2012, vol. 3, pp. 430–434. Brno University of Technology, Brno, CZ (2012)
199. Zemek, P.: One-sided random context grammars: Established results and open problems. In: Proceedings of the 19th Conference STUDENT EEICT 2013, vol. 3, pp. 222–226. Brno University of Technology, Brno, CZ (2013)
200. Zetzsche, G.: Erasing in Petri net languages and matrix grammars. In: DLT '09: Proceedings of the 13th International Conference on Developments in Language Theory, pp. 490–501. Springer, New York (2009)
201. Zetzsche, G.: On erasing productions in random context grammars. In: ICALP'10: Proceedings of the 37th International Colloquium on Automata, Languages and Programming, pp. 175–186. Springer, New York (2010)
202. Zetzsche, G.: A sufficient condition for erasing productions to be avoidable. In: DLT'11: Developments in Language Theory, Lecture Notes in Computer Science, vol. 6795, pp. 452–463. Springer, Berlin (2011)
203. Zetzsche, G.: Toward understanding the generative capacity of erasing rules in matrix grammars. Int. J. Comput. Math. **22**(2), 411–426 (2011)

Chapter 22
Summary

Abstract This chapter sums up all the material covered in this book.

Keywords Summary

Subject and Purpose

This monograph deals with formal language theory, which represents a branch of mathematics that formalizes languages and devices that define them (see [1]). In other words, this theory represents a mathematically systematized body of knowledge concerning languages in general. It defines languages as sets of finite sequences consisting of symbols. As a result, this general definition encompasses almost all languages, including natural languages as well as artificial languages, such as programming languages.

The strictly mathematical approach to languages necessitates introducing language-defining models. Formal language theory has classified these models into two basic categories—grammars and automata. *Grammars* define strings of their language so their rewriting process generates them from a special start symbol. *Automata* define strings of their language by rewriting process that starts from these strings and ends in a special final string.

Regulated grammars and automata—that is, the central subject of this book—are based upon these language-defining models extended by an additional mathematical mechanism that prescribes the use of rules during the generation of their languages. From a practical viewpoint, an important advantage of these models consists in controlling their language-defining process and, therefore, operating in a more deterministic way than general models, which perform their derivations in a completely unregulated way. More significantly, the regulated versions of language models are stronger than their unregulated versions. Considering these significant advantages and properties, regulated grammars and automata fulfill a highly beneficial role in many kinds of language-related work conducted by a broad

A. Meduna and P. Zemek, *Regulated Grammars and Automata*,

DOI 10.1007/978-1-4939-0369-6_22,

variety of scientists, ranging from mathematicians through computer scientists up to linguists and geneticists. The principle purpose of the present monograph is to summarize key results about them.

Topics Under Investigation

This monograph restricts its attention to four important topics concerning regulated versions of grammars and automata—their power, properties, reduction, and convertibility.

As obvious, the *power* of the regulated language models under consideration represents perhaps the most important information about them. Indeed, we always want to know the language family defined by these models.

A special attention is paid to algorithms that arrange regulated grammars and automata so they satisfy some prescribed *properties* while the generated languages remain unchanged because many language processors strictly require their satisfaction. From a theoretical viewpoint, these properties frequently simplify proofs demonstrating results about these grammars and automata.

The *reduction* of regulated grammars and automata also represents an important investigation area of this book because their reduced versions define languages in a succinct and easy-to-follow way. As obvious, this reduction simplifies the development of language processing technologies, which then work economically and effectively.

Of course, the same languages can be defined by different language models. We obviously tend to define them by the most appropriate models under given circumstances. Therefore, whenever discussing different types of equally powerful language models, we also study their mutual *convertibility*. More specifically, given a language model of one type, we explain how to convert it to a language model of another equally powerful type so both the original model and the model produced by this conversion define the same language.

Organization and Coverage

All the text is divided into nine parts, each of which consists of several chapters. Every part starts with an abstract that summarizes the contents of its chapters. Altogether, the book contains twenty-two chapters.

Part I

Part I, consisting of Chap. 1 through 3, gives an introduction to this monograph in order to express all its discussion clearly and make the book completely self-contained. It places all the coverage of the book into scientific context and reviews important mathematical concepts with a focus on formal language theory.

Chapter 1 gives an introduction to the subject of this monograph—regulated grammars and automata. It explains the reason why this subject fulfills an important role in science. The chapter conceptualizes regulated grammars and automata and places them into general scientific context. It demonstrates that the study of these grammars and automata represents a vivid investigation area of today's computer science. Specifically, this investigation is central to formal languages theory. Chapter 1 also gives a step-by-step insight into all the topics covered in this book. Its conclusion sketches important applications of regulated grammars and automata in many scientific areas.

Chapter 2 gives the mathematical background of this monograph. It reviews all the necessary mathematical concepts to grasp the topics covered in the book. These concepts primarily include fundamental areas of discrete mathematics. First, this chapter reviews basic concepts from set theory. Then, it gives the essentials concerning relations and their crucially important special cases, namely, functions. Finally, this chapter reviews fundamental concepts from graph theory.

Chapter 3 covers selected areas of formal language theory needed to follow the rest of this book. It introduces the basic terminology concerning strings, languages, operations, and closure properties. Furthermore, it overviews a large variety of grammars, automata and language families resulting from them. Apart from the classical rudiments of formal language theory, Chap. 3 covers several less known areas of this theory, such as fundamentals concerning parallel grammars, because these areas are also needed to grasp some topics of this monograph.

Part II

Part II, consisting of Chaps. 4 and 5, gives the fundamentals of regulated grammars. It distinguishes between context-based regulated grammars and rule-based regulated grammars. First, it gives an extensive and thorough coverage of regulated grammars that generate languages under various context-related restrictions. Then, it studies grammatical regulation underlain by restrictions placed on the use of rules.

Chapter 4 gives an extensive and thorough coverage of regulated grammars that generate their languages under various contextual restrictions. First, it considers classical grammars, including phrase-structure and context-sensitive grammars, as contextually regulated grammars while paying a special attention to their normal forms and uniform rewriting. Then, this chapter deals with *context-conditional grammars*. More specifically, it introduces their general version and establishes key results about them. Furthermore, it studies special cases of context-conditional grammars—namely, *random context grammars*, *generalized forbidding grammars*, *semi-conditional grammars*, and *simple semi-conditional grammars*. Chapter 4 closes its coverage by discussing *scattered context grammars*, which regulate their language generation so they simultaneously rewrite several prescribed nonterminals scattered throughout sentential forms during derivation steps.

Chapter 5 explores regulated grammars underlain by restrictions placed on the use of rules. Four types of regulated grammars of this kind are covered–namely, *regular-controlled*, *matrix*, *programmed*, and *state grammars*. Regular-control grammars control the use of rules by regular languages over rule labels. Matrix grammars represent, in fact, special cases of regular-control grammars whose control languages have the form of the iteration of finite languages. Programmed grammars regulate the use of their rules by relations over rule labels. Finally, state grammars regulate the use of rules by states in a way that strongly resembles the finite-state control of finite automata.

Part III

Part III, consisting of Chap. 6 through 9, covers several special topics concerning regulated grammars. First, it studies special cases of context-based regulated grammars. Then, it discusses erasing of symbols by regulated grammars. Finally, this part investigates an algebraic way of grammatical regulation.

Chapter 6 discusses *one-sided random context grammars* as special cases of random context grammars. In every one-sided random context grammar, the set of rules is divided into the set of left random context rules and the set of right random context rules. When applying a left random context rule, the grammar checks the existence and absence of its permitting and forbidding symbols, respectively, only in the prefix to the left of the rewritten nonterminal. Analogously, when applying a right random context rule, it checks the existence and absence of its permitting and forbidding symbols, respectively, only in the suffix to the right of the rewritten nonterminal. Otherwise, it works just like any ordinary random context grammar. This chapter demonstrates that propagating versions of one-sided random context grammars characterize the family of context-sensitive languages, and with erasing rules, they characterize the family of recursively enumerable languages. Furthermore, it discusses the generative power of several special cases of one-sided random context grammars. Specifically, it proves that one-sided permitting grammars, which have only permitting rules, are more powerful than context-free grammars; on the other hand, they are no more powerful than scattered context grammars. One-sided forbidding grammars, which have only forbidding rules, are equivalent to selective substitution grammars. Finally, left forbidding grammars, which have only left-sided forbidding rules, are only as powerful as context-free grammars. Chapter 6 also establishes four normal forms of one-sided random context grammars and studies their reduction with respect to the number of nonterminals and rules. It also places various leftmost restrictions on derivations in one-sided random context grammars and investigates how they affect the generative power of these grammars. In addition, it discusses generalized versions of one-sided forbidding grammars. Chapter 6 closes its discussion by investigating parsing-related variants of one-sided random context grammars in order to demonstrate their future application-related perspectives.

Chapter 7 studies how to eliminate erasing rules, having the empty string on their right-hand sides, from context-free grammars and their regulated versions. The chapter points out that this important topic still represents a largely open problem area in the theory of regulated grammars. It describes two methods of eliminating erasing rules from ordinary context-free grammars. One method is based upon a well-known technique, but the other represents a completely new algorithm that performs this elimination. Chapter 7 also establishes workspace theorems for regular-controlled grammars. In essence, these theorems give derivation conditions under which erasing rules can be removed from these grammars. Furthermore, the chapter discusses the elimination of erasing rules in terms of scattered context grammars. First, it points out that scattered context grammars with erasing rules characterize the family of recursively enumerable languages while their propagating versions do not. In fact, propagating scattered context grammars cannot generate any non-context-sensitive language, so some scattered context grammars with erasing rules are necessarily unconvertible to equivalent propagating scattered context grammars. Chapter 7 establishes a sufficient condition under which this conversion is always possible.

Chapter 8 studies regulated grammars that are modified so they generate their languages extended by some extra symbols that represent useful information related to the generated languages. It explains how to transform any regular-controlled grammar with appearance checking G to a propagating regular-controlled with appearance checking H whose language $L(H)$ has every sentence of the form $w\rho$, where w is a string of terminals in G and ρ is a sequence of rules in H, so that (i) $w\rho \in L(H)$ if and only if $w \in L(G)$ and (ii) ρ is a parse of w in H. Consequently, for every recursively enumerable language K, there exists a propagating regular-controlled grammar with appearance checking H with $L(H)$ of the above-mentioned form so K results from $L(H)$ by erasing all rules in $L(H)$. Analogical results are established for regular-controlled grammars without appearance checking and for these grammars that make only leftmost derivations. Chapter 8 also studies a language operation referred to as coincidental extension, which extend strings by inserting some symbols into the languages generated by propagating scattered context grammars.

Chapter 9 defines the relation of a direct derivation in grammars over free monoids generated by finitely many strings. It explains that this modification can be seen as a very natural context-based grammatical regulation; indeed, a derivation step is performed on the condition that the rewritten sentential form occurs in the free monoids generated in this modified way. First, it defines the above modification rigorously and explains its relation to the subject of this book. Then, it demonstrates that this modification results into a large increase of the generative power of context-free grammars. In fact, even if the free monoids are generated by strings consisting of no more than two symbols, the resulting context-free grammars are as powerful as phrase-structure grammars.

Part IV

Part IV, consisting Chaps. 10 through 12, studies parallel versions of regulated grammars. First, it studies generalized parallel versions of context-free grammars, generally referred to as *regulated ET0L grammars*. Then, it studies how to perform the parallel generation of languages in a uniform way. Finally, it studies algebraically regulated parallel grammars.

Chapter 10 deals with ET0L grammars, which can be viewed as generalized parallel versions of context-free grammars. More precisely, there exist three main conceptual differences between them and context-free grammars. First, instead of a single set of rules, they have finitely many sets of rules. Second, the left-hand side of a rule may be formed by any grammatical symbol, including a terminal. Third, all symbols of a string are simultaneously rewritten during a single derivation step. The present chapter studies ET0L grammars regulated in a context-conditional way. Specifically, by analogy with sequential context-conditional grammars, this chapter discusses *context-conditional ET0L grammars* that capture this dependency so each of their rules may be associated with finitely many strings representing permitting conditions and, in addition, finitely many strings representing forbidding conditions. A rule like this can rewrite a symbol if all its permitting conditions occur in the rewritten current sentential form and, simultaneously, all its forbidding conditions do not. Otherwise, these grammars work just like ordinary ET0L grammars. Apart from the basic version of context-conditional ET0L grammars, this chapter investigates three variants of the basic version—*forbidding ET0L grammars*, *simple semi-conditional ET0L grammars*, and *left random context ET0L grammars*. It concentrates its attention on establishing the generative power of all these variants of context-conditional ET0L grammars.

Chapter 11 discusses how to perform the parallel generation of languages in a uniform way with respect to the rewritten strings. More precisely, it transforms grammars that work in parallel so they produce only strings that have a uniform permutation-based form. In fact, this chapter makes the regulated rewriting uniform in terms of both partially and totally parallel grammars. Indeed, it represents the semi-parallel language generation by scattered context grammars, which belong to the most important types of regulated grammars, and demonstrates how to transform scattered context grammars so they produce only strings that have a uniform permutation-based form. In addition, Chap. 11 represents the totally parallel generation of languages by EIL grammars and presents an analogical transformation for them.

Chapter 12 studies the regulation of grammatical parallelism so it defines parallel derivations over free monoids generated by finitely many strings. The grammatical parallelism is represented by E0L grammars (an E0L grammar is an ET0L grammar with a single set of rules). The chapter demonstrates that this regulation results into a large increase of the generative power of ordinary E0L grammars, even if the strings that generate free monoids consist of no more than two symbols. In fact, the E0L grammars regulated in this way are computationally complete.

Part V

Part V, consisting of Chaps. 13 and 14, studies sets of mutually communicating grammars working under regulating restrictions. First, it studies their regulation based upon a simultaneous generation of several strings composed together by some basic operation after the generation is completed. Then, it studies their regulated pure versions, which have only one type of symbols.

Chapter 13 discusses regulated versions of grammar systems, referred to as *multigenerative grammar systems*, which consist of several components represented by context-free grammars. Their regulation is based upon a simultaneous generation of several strings, which are composed together by some basic operation, such as concatenation, after their generation is completed. The chapter first defines the basic versions of multigenerative grammar systems. During one generation step, each of their grammatical components rewrites a nonterminal in its sentential form. After this simultaneous generation is completed, all the generated strings are composed into a single string by some common string operation, such as union and concatenation. It shows that these systems characterize the family of matrix languages. In addition, it demonstrates that multigenerative grammar systems with any number of grammatical components can be transformed to equivalent two-component versions of these systems. Then, Chap. 13 discusses leftmost versions of multigenerative grammar systems in which each generation step is performed in a leftmost manner. That is, all the grammatical components of these versions rewrite the leftmost nonterminal occurrence in their sentential forms; otherwise, they work as the basic versions. It proves that leftmost multigenerative grammar systems are more powerful than their basic versions because they are computational complete. It demonstrates that leftmost multigenerative grammar systems with any number of grammatical components can be transformed to equivalent two-component versions of these systems.

Chapter 14 studies pure grammar systems, which have only terminals. They generate their languages in the leftmost way, and in addition, this generative process is regulated by control languages over rule labels. The chapter concentrates its attention on investigating the generative power of these systems. It establishes three major results. First, without any control languages, these systems do not even generate some context-free languages. Second, with regular control languages, these systems characterize the family of recursively enumerable languages, and this result holds even if these systems have no more than two components. Finally, this chapter considers control languages as languages that are themselves generated by regular-controlled context-free grammars; surprisingly enough, with control languages of this kind, these systems over unary alphabets define nothing but regular languages.

Part VI

Part VI, consisting of Chaps. 15 and 16, presents the fundamentals of regulated automata. First, it studies self-regulating automata. Then, it covers the essentials concerning automata regulated by control languages.

Chapter 15 investigates *self-regulating automata*, which regulate the selection of a rule according to which the current move is made by a rule according to which a previous move was made. Both finite and pushdown versions of these automata are investigated. More specifically, first, it discusses self-regulating finite automata. It establishes two infinite hierarchies of language families resulting from them. Both hierarchies lie between the family of regular languages and the family of context-sensitive languages. Then, the chapter studies self-regulating pushdown automata. Based upon them, the chapter characterizes the families of context-free and recursively enumerable languages. However, as opposed to the results about self-regulating finite automata, many questions concerning their pushdown versions remain open; indeed, Chap. 15 closes its discussion by formulating several specific open problem areas, including questions concerning infinite language-family hierarchies resulting from them.

Chapter 16 studies finite and pushdown automata in which the application of rules is regulated by control languages. More specifically, first, it discusses finite automata working under two kinds of regulation—*state-controlled regulation* and *transition-controlled regulation*. It establishes conditions under which any state-controlled finite automaton can be turned to an equivalent transition-controlled finite automaton and vice versa. Then, it proves that under either of the two regulations, finite automata controlled by regular languages characterize the family of regular languages, and an analogical result is then reformulated in terms of context-free languages. However, the chapter also demonstrates that finite automata controlled by languages generated by propagating programmed grammars with appearance checking increase their power significantly; in fact, they are computationally complete. Then, Chap. 16 turns its attention to pushdown automata regulated by control languages. It shows that these automata regulated by regular languages are as powerful as ordinary pushdown automata. On the other hand, it also proves that pushdown automata regulated by linear languages characterize the family of recursively enumerable languages.

Part VII

Part VII, consisting of Chaps. 17 and 18, studies modified versions of classical automata closely related to regulated automata—namely, jumping finite automata and deep pushdown automata.

Chapter 17 gives a systematic body of knowledge concerning *jumping finite automata*, which work just like classical finite automata except that they do not read their input strings in a symbol-by-symbol left-to-right way. Instead, after reading a symbol, they can jump in either direction within their input tapes and continue making moves from there. Once an occurrence of a symbol is read, it cannot be re-read again later on. Otherwise, their definition coincides with the definition of standard finite automata. First, this chapter formalizes jumping finite automata and illustrates them by examples. Then, it demonstrates their fundamental properties, after which it compares their power with the power of well-known language-defining formal devices. Naturally, this chapter also establishes several results concerning jumping finite automata with respect to commonly studied areas of formal language theory, such as closure properties and decidability. In addition, it establishes an infinite hierarchy of language families resulting from these automata. Finally, it studies some special topics and features, such as one-directional jumps and various start configurations. Throughout its discussion, this chapter points out several open questions regarding these automata, which may represent a new investigation area of automata theory in the future.

Chapter 18 defines *deep pushdown automata*, which represent a very natural modification of ordinary pushdown automata. While the ordinary versions can expand only the pushdown top, deep pushdown automata can make expansions deeper in the pushdown; otherwise, they both work identically. This chapter first defines and illustrates deep pushdown automata. Then, it proves that the power of deep pushdown automata is similar to the generative power of regulated context-free grammars without erasing rules. Indeed, just like these grammars, deep pushdown automata are stronger than ordinary pushdown automata but less powerful than context-sensitive grammars. More precisely, they give rise to an infinite hierarchy of language families coinciding with the hierarchy resulting from n-limited state grammars. The chapter closes its discussion by pointing out some open problem areas concerning deep pushdown automata, including a variety of new deterministic and generalized versions of these automata.

Part VIII

Part VIII, consisting of Chaps. 19 and 20, demonstrates applications of regulated language models. It narrows its attention to regulated grammars rather than automata. First, it describes these applications and their perspectives from a general viewpoint. Then, it adds several case studies to show quite specific real-world applications concerning computational linguistics, molecular biology, and compiler writing.

Chapter 19 covers important applications of regulated language models discussed earlier in this book. It narrows its attention to three application areas—biology, compilers, and linguistics. Within their framework, it describes current applications and, in addition, suggests several application-related perspectives in the near future.

Chapter 20 gives several specific case studies concerning three scientific areas of regulated grammars—linguistics, biology, and parsing—in order to illustrate the discussion of Chap. 19 by some real-world examples. First, it demonstrates applications of scattered context grammars in linguistics. It concentrates its attention to many complicated English syntactical structures and demonstrates how scattered context grammars allow us to explore them clearly, elegantly, and precisely. Second, it considers grammars with context conditions and applies them in microbiology, which appears of great interest at present. Finally, it applies scattered context grammars to parsing, which fulfills a crucially important role in compiler writing.

Part IX

Part IX, consisting of Chaps. 21 and 22, closes the entire book by adding several remarks concerning its coverage. First, it sketches the entire development of regulated grammars and automata. Then, it points out many new investigation trends and long-time open problems. Finally, it briefly summarizes all the material covered in the text.

Chapter 21 makes several remarks concerning the material covered in this book. First, it describes new investigation trends closely related to regulated grammars and automata. Second, it suggests many open problem areas to the future study concerning the subject of this book. Finally, it gives an overview of significant studies published on the subject of this book from a historical perspective. This chapter contains a great number of references in order to demonstrate the significance of all its coverage in a specific and detailed way.

Chapter 22 is the present chapter, which has summed up the coverage of this monograph as a whole.

References

1. Rozenberg, G., Salomaa, A. (eds.): Handbook of Formal Languages, Volumes 1 through 3, Springer, New York (1997)

Language Family Index

Family	Page	Formal model
FIN	19	-
RE	22	Phrase-structure grammar
CS	24	Context-sensitive grammar
CF	24	Context-free grammar
LIN	24	Linear grammar
REG	24	Regular grammar
RLIN	25	Right-linear grammar
S	28	Selective substitution grammar
$\mathbf{S}^{-\varepsilon}$	28	Propagating selective substitution grammar
0L	29	0L grammar
E0L	29	E0L grammar
EP0L	29	EP0L grammar
ET0L	29	ET0L grammar
EPT0L	29	EPT0L grammar
$\mathbf{EPDA}_f$	34	Extended pushdown automaton accepting by final state
$\mathbf{EPDA}_e$	34	Extended pushdown automaton accepting by empty pushdown
$\mathbf{EPDA}_{ef}$	34	Extended pushdown automaton accepting by final state and empty pushdown
$\mathbf{PDA}_f$	35	Pushdown automaton accepting by final state
$\mathbf{PDA}_e$	35	Pushdown automaton accepting by empty pushdown
$\mathbf{PDA}_{ef}$	35	Pushdown automaton accepting by final state and empty pushdown
$\mathbf{PS}[.j]$	48	Phrase-structure grammar with l-uniform rewriting
$\mathbf{PS}[j.]$	48	Phrase-structure grammar with r-uniform rewriting

(continued)

A. Meduna and P. Zemek, *Regulated Grammars and Automata*,
DOI 10.1007/978-1-4939-0369-6,

(continued)

Family	Page	Formal model
CG	57	Conditional grammar
$\mathbf{CG}^{-\varepsilon}$	57	Propagating conditional grammar
RC	64	Random context grammar
$\mathbf{RC}^{-\varepsilon}$	64	Propagating random context grammar
For	64	Forbidding grammar
$\mathbf{For}^{-\varepsilon}$	64	Propagating forbidding grammar
Per	64	Permitting grammar
$\mathbf{Per}^{-\varepsilon}$	64	Propagating permitting grammar
GF	69	Generalized forbidding grammar
$\mathbf{GF}^{-\varepsilon}$	69	Propagating generalized forbidding grammar
SC	84	Semi-conditional grammar
$\mathbf{SC}^{-\varepsilon}$	84	Propagating semi-conditional grammar
SSC	88	Simple semi-conditional grammar
$\mathbf{SSC}^{-\varepsilon}$	88	Propagating simple semi-conditional grammar
SCAT	123	Scattered context grammar
$\mathbf{SCAT}^{-\varepsilon}$	123	Propagating scattered context grammar
$\mathbf{SCAT}_k$	328	Scattered context grammar erasing its nonterminals in a generalized k-restricted way
$\mathbf{LL}\text{-}_k\mathbf{SCAT}$	152	LL k-linear scattered context grammar
$\mathbf{SCAT}[.i/j]$	432	Scattered context grammar with l-uniform rewriting
$\mathbf{SCAT}[i/j.]$	432	Scattered context grammar with r-uniform rewriting
rC	160	Regular-controlled grammar
$\mathbf{rC}^{-\varepsilon}$	160	Propagating regular-controlled grammar
$\mathbf{rC}_{ac}$	160	Regular-controlled grammar with appearance checking
$\mathbf{rC}_{ac}^{-\varepsilon}$	160	Propagating regular-controlled grammar with appearance checking
$\mathbf{rC}^{ws}$	309	Regular-controlled grammar satisfying the workspace condition
$\mathbf{rC}_{ac}^{ws}$	309	Regular-controlled grammar with appearance checking satisfying the workspace condition
M	162	Matrix grammar
$\mathbf{M}^{-\varepsilon}$	162	Propagating matrix grammar
$\mathbf{M}_{ac}$	162	Matrix grammar with appearance checking
$\mathbf{M}_{ac}^{-\varepsilon}$	162	Propagating matrix grammar with appearance checking
P	165	Programmed grammar
$\mathbf{P}^{-\varepsilon}$	165	Propagating programmed grammar
$\mathbf{P}_{ac}$	165	Programmed grammar with appearance checking

(continued)

(continued)

Family	Page	Formal model
$\mathbf{P}_{ac}^{-\varepsilon}$	165	Propagating programmed grammar with appearance checking
${}_1\mathbf{P}$	175	Programmed grammar in the one-ND rule normal form
ST	186	State grammar
$\mathbf{ST}_n$	186	n-limited state grammar
ORC	198	One-sided random context grammar
$\mathbf{ORC}^{-\varepsilon}$	198	Propagating one-sided random context grammar
OFor	198	One-sided forbidding grammar
$\mathbf{OFor}^{-\varepsilon}$	198	Propagating one-sided forbidding grammar
OPer	198	One-sided permitting grammar
$\mathbf{OPer}^{-\varepsilon}$	198	Propagating one-sided permitting grammar
LRC	198	Left random context grammar
$\mathbf{LRC}^{-\varepsilon}$	198	Propagating left random context grammar
LFor	198	Left forbidding grammar
$\mathbf{LFor}^{-\varepsilon}$	198	Propagating left forbidding grammar
LPer	198	Left permitting grammar
$\mathbf{LPer}^{-\varepsilon}$	198	Propagating left permitting grammar
$\mathbf{ORC}({}_{\mathrm{lm}}\overset{1}{\Rightarrow})$	246	One-sided random context grammar using type-1 leftmost derivations
$\mathbf{ORC}({}_{\mathrm{lm}}\overset{2}{\Rightarrow})$	248	One-sided random context grammar using type-2 leftmost derivations
$\mathbf{ORC}({}_{\mathrm{lm}}\overset{3}{\Rightarrow})$	253	One-sided random context grammar using type-3 leftmost derivations
$\mathbf{ORC}^{-\varepsilon}({}_{\mathrm{lm}}\overset{1}{\Rightarrow})$	246	Propagating one-sided random context grammar using type-1 leftmost derivations
$\mathbf{ORC}^{-\varepsilon}({}_{\mathrm{lm}}\overset{2}{\Rightarrow})$	248	Propagating one-sided random context grammar using type-2 leftmost derivations
$\mathbf{ORC}^{-\varepsilon}({}_{\mathrm{lm}}\overset{3}{\Rightarrow})$	253	Propagating one-sided random context grammar using type-3 leftmost derivations
$\mathbf{RC}({}_{\mathrm{lm}}\overset{1}{\Rightarrow})$	255	Random context grammar using type-1 leftmost derivations
$\mathbf{RC}({}_{\mathrm{lm}}\overset{2}{\Rightarrow})$	255	Random context grammar using type-2 leftmost derivations
$\mathbf{RC}({}_{\mathrm{lm}}\overset{3}{\Rightarrow})$	255	Random context grammar using type-3 leftmost derivations
$\mathbf{RC}^{-\varepsilon}({}_{\mathrm{lm}}\overset{1}{\Rightarrow})$	255	Propagating random context grammar using type-1 leftmost derivations
$\mathbf{RC}^{-\varepsilon}({}_{\mathrm{lm}}\overset{2}{\Rightarrow})$	255	Propagating random context grammar using type-2 leftmost derivations

(continued)

(continued)

Family	Page	Formal model
RC$^{-\varepsilon}({}_{\mathrm{lm}}^{3}\Rightarrow)$	255	Propagating random context grammar using type-3 leftmost derivations
LL - CF	276	LL context-free grammar
LL - ORC	276	LL one-sided random context grammar
WM	352	Context-free grammar over word monoids
WM$^{-\varepsilon}$	352	Propagating context-free grammar over word monoid
C - E0L	368	Conditional E0L grammar
C - EP0L	368	Conditional EP0L grammar
C - ET0L	368	Conditional ET0L grammar
C - EPT0L	368	Conditional EPT0L grammar
F - E0L	375	Forbidding E0L grammar
F - EP0L	375	Forbidding EP0L grammar
F - ET0L	375	Forbidding ET0L grammar
F - EPT0L	375	Forbidding EPT0L grammar
SSC - E0L	397	Simple semi-conditional E0L grammar
SSC - EP0L	397	Simple semi-conditional EP0L grammar
SSC - ET0L	397	Simple semi-conditional ET0L grammar
SSC - EPT0L	397	Simple semi-conditional EPT0L grammar
LRC - E0L	412	Left random context E0L grammar
LRC - EP0L	412	Left random context EP0L grammar
LRC - ET0L	412	Left random context ET0L grammar
LRC - EPT0L	412	Left random context EPT0L grammar
LF - E0L	412	Left forbidding E0L grammar
LF - EP0L	412	Left forbidding EP0L grammar
LF - ET0L	412	Left forbidding ET0L grammar
LF - EPT0L	412	Left forbidding EPT0L grammar
LP - E0L	412	Left permitting E0L grammar
LP - EP0L	412	Left permitting EP0L grammar
LP - ET0L	412	Left permitting ET0L grammar
LP - EPT0L	412	Left permitting EPT0L grammar
RC - ET0L	426	Random context ET0L grammar
RC - EPT0L	426	Random context EPT0L grammar
EIL$[.j]$	439	EIL grammar with l-uniform rewriting
EIL$[j.]$	439	EIL grammar with r-uniform rewriting
WME0L	446	E0L grammar over word monoids
WMEP0L	446	Propagating E0L grammar over word monoids
SE0L	446	Symbiotic E0L grammar
SEP0L	446	Symbiotic EP0L grammar

(continued)

(continued)

Family	Page	Formal model
$\mathbf{MGR}_{n,X}$	475	n-generative rule-synchronized grammar system in the X mode, where $X \in \{union, conc, first\}$
pGS	495	Pure grammar system
${}_n\mathbf{pGS}$	495	n-component pure grammar system
FSFA	511	First-move self-regulated finite automaton
ASFA	512	All-move self-regulated finite automaton
PRL	514	Parallel right-linear grammar
RLSM	520	Right-linear simple matrix grammar
FSPDA	527	First-move self-regulated pushdown automaton
ASPDA	527	All-move self-regulated pushdown automaton
$\mathbf{SCFA}(\mathscr{L})$	534	State-controlled finite automaton controlled by languages from $\mathscr{L}$
$\mathbf{TCFA}(\mathscr{L})$	534	Transition-controlled finite automaton controlled by languages from $\mathscr{L}$
RPDA	548	Regulated pushdown automaton
OA - RPDA	560	One-turn atomic regulated pushdown automaton
JFA	571	Jumping finite automaton
$\mathbf{JFA}^{-\varepsilon}$	571	ε-free jumping finite automaton
DJFA	571	Deterministic jumping finite automaton
GJFA	571	Jumping finite automaton
$\mathbf{GJFA}^{-\varepsilon}$	571	ε-free jumping finite automaton
${}_{deep}\mathbf{PDA}_k$	590	Deep pushdown automaton of depth k
${}^{empty}_{deep}\mathbf{PDA}_k$	590	Deep pushdown automaton of depth k accepting by empty pushdown

Subject Index

Symbols
1-final configuration, 548, 560
2-final configuration, 548, 560
2-limited propagating scattered context grammar, 124
3-final configuration, 548, 560

A
acceptance
  by empty pushdown, 548, 560
  by final state, 548, 560
  by final state and empty pushdown, 548, 560
activated symbol, 28
acyclic graph, 13
all-move self-regulating
  finite automaton, 512
  pushdown automaton, 527
almost identity, 19
alph(), 16
alphabet, 16
  of language, 17
  of string, 16
  unary, 16
appearance checking
  mode, 158
  set, 158, 161
atomic pushdown automaton, 559
automaton, 4
  finite, 32
    complete, 32
    controlled, 533
    deterministic, 32
    general, 31
    jumping, 569, 570
  pushdown, 34
    atomic, 559
    blackhole, 660
    controlled, 548, 559
    deep, 589, 600
    extended, 33
  self-regulating
    finite, 511, 512
    pushdown, 526, 527
auxiliary verb, 620

B
bag automaton, 660
bijection, 12
binary
  normal form, 169
  operation, 12
  relation, 11
blackhole pushdown automaton, 660
bottom symbol, 589
#-rewriting system, 654

C
C-E0L grammar, 368
C-EP0L grammar, 368
C-EPT0L grammar, 367
C-ET0L grammar, 367
canonical clause, 620
card(), 10
cardinality, 10
Cartesian product, 11
cf-rules(), 58
cf-sim(), 311
characterization, 20
Chomsky
  hierarchy, 25
  normal form, 47

A. Meduna and P. Zemek, *Regulated Grammars and Automata*,
DOI 10.1007/978-1-4939-0369-6, © Springer Science+Business Media New York 2014

Church's thesis, 23
CJFA, *see* complete jumping finite automaton
closure
 of language, 18
 property, 12
 under binary operation, 12
 under linear erasing, 20
 under restricted homomorphism, 20
 under unary operation, 12
coding, 19
coincident language families, 20
coincidental extension, 346
complement
 of language, 17
 of set, 10
complete
 finite automaton, 32
 jumping finite automaton, 570
computational
 completeness, 23
 incompleteness, 23
concatenation
 of languages, 17
 of strings, 16
conditional rule, 56
configuration, 32, 33, 164, 493, 559, 569, 589
 start, 493, 559
context condition, 642
context-based regulation, 5
context-conditional
 E0L grammar, 368
 ET0L grammar, 367
 grammar, 56
context-free
 grammar, 24
  over word monoid, 352
 language, 24
 rule, 120
 simulation, 312
  partial, 311
context-sensitive
 grammar, 24
 language, 24
 rule, 120
control language, 494
 of atomic pushdown automaton, 560
 of regular-controlled grammar, 156
 of regulated pushdown automaton, 548
 of state-controlled finite automaton, 533
 of transition-controlled finite automaton, 533
control word, *see* parse
controlled
 finite automaton, 533
 pushdown automaton, 548
  atomic, 560
conversion, 4
core grammar
 of matrix grammar, 161
 of regular-controlled grammar, 156
 of scattered context grammar, 311
core(), 311
cycle, 13

D

dcs(), 126, 142
deep pushdown automaton, 589
 of depth n, 590
degree
 of C-ET0L grammar, 367
 of context sensitivity
  of phrase-structure grammar, 142
  of scattered context grammar, 126
 of context-conditional grammar, 56
 of context-free grammar over word monoid, 352
 of general jumping finite automaton, 570
 of generalized one-sided random context grammar, 258
 of WME0L grammar, 446
derivation
 leftmost, 25
  of type-1, 245
  of type-2, 248
  of type-3, 253
 rightmost, 25
 tree, 30
 word, *see* parse
descriptional complexity, 224
deterministic
 deep pushdown automaton, 600
  with respect to depth of expansions, 600
 finite automaton, 32
 jumping finite automaton, 570
 rule, 181
difference
 of languages, 17
 of sets, 10
direct
 derivation, 21, 27–30, 56, 120, 161, 164, 185, 194, 257, 352, 367, 412, 446, 493
  leftmost, 25
  rightmost, 25
 descendant, 13
 move, 32, 34, 590
 predecessor, 13

directed graph, 13
disjoint sets, 11
DJFA, *see* deterministic jumping finite automaton
domain, 12
domain(), 12

E
E(m, n)L grammar, 29
E0L grammar
  over word monoid, 446
  symbiotic, 446
edge, 13
effective proof, ix
EIL grammar, 30
empty
  language, 17
  set, 10
  string, 16
English syntax, 616
enter, 13
E0L grammar, 29
EP0L grammar, 29
ε-free
  family of languages, 19
  general finite automaton, 32
  general jumping finite automaton, 570
  grammar, *see* propagating grammar
  homomorphism, 19
  jumping finite automaton, 570
  substitution, 19
ε-nonterminal, 282
EPT0L grammar, 29
equality
  of language families, 19
  of languages, 17
equally powerful formal models, 20
equivalent formal models, 20
erasing rule, 21
erasure, 31
ET0L grammar, 29
exhaustive
  left quotient, 18
  right quotient, 18
existential clause, 626
expansion
  of depth m, 590
  of pushdown, 588, 589
extended
  pushdown automaton, 33
  Szilard language, 331

F
F-E0L grammar, 375
F-EP0L grammar, 375
F-EPT0L grammar, 375
F-ET0L grammar, 375
failure field, 164
family
  of languages, 19
  of sets, 11
fcore(), 150
fin(), 17
final
  state, 32, 569, 589
finite
  automaton, 32
  language, 17
  relation, 11
  sequence, 11
  set, 10
  substitution, 19
first Geffert normal form, 42
first-component core grammar, 150
first-move self-regulating
  finite automaton, 511
  pushdown automaton, 527
floor(), 294
flux, 643
forbidding
  ET0L grammar, 375
  grammar, 64
frontier, 13
function, 12
  partial, 12
  total, 12

G
Geffert normal form, 42
general
  finite automaton, 31
  jumping finite automaton, 569
  top-down parser, 588
generalized
  forbidding grammar, 69
  one-sided forbidding grammar, 257
  restricted erasing, 312
generated language, 195, 258, 352, 367, 446, 494, 514, 520
  in the concatenation mode, 459, 476
  in the first mode, 459, 476
  in the union mode, 459, 476
generator, 352
gerund-participle, 620

gf-grammar, *see* generalized forbidding grammar
GJFA, *see* general jumping finite automaton
grammar, 4
  C-E0L, 368
  C-EP0L, 368
  C-ET0L, 367
  context-conditional, 56
    E0L, 368
    ET0L, 367
  context-free, 24
    over word monoid, 352
  context-sensitive, 24
  controlled by a bicoloured digraph, 310
  E(m,n)L, 29
  E0L
    over word monoid, 446
    symbiotic, 446
  EIL, 30
  E0L, 29
  EPT0L, 29
  ET0L, 29
  F-E0L, 375
  F-EP0L, 375
  F-EPT0L, 375
  F-ET0L, 375
  forbidding, 64
    ET0L, 375
  generalized
    forbidding, 69
    one sided forbidding, 257
  left
    forbidding, 195
    permitting, 195
    random context, 195
  LF-E0L, 412
  LF-EP0L, 412
  LF-EPT0L, 412
  LF-ET0L, 412
  linear, 24
  LL
    context-free, 269
    k-linear scattered context, 150
    one-sided random context, 270
  LL(k)
    context-free, 269
    one-sided, 270
  LP-E0L, 412
  LP-EP0L, 412
  LP-EPT0L, 412
  LP-ET0L, 412
  LRC-E0L, 412
  LRC-EP0L, 412
  LRC-EPT0L, 412
  LRC-ET0L, 412
  matrix, 161
    with appearance checking, 161
  0L, 29
    parametric, 642
  one-sided
    forbidding, 195
    permitting, 195
    random context, 194
  permitting, 64
  phrase-structure, 20
  programmed, 164
    with appearance checking, 163
  queue, 27
  random context, 64
  regular, 24
  regular-controlled, 156
    with appearance checking, 157
  right-linear, 25
  scattered context, 119
    transformational, 622
  selective substitution, 28
  semi-conditional, 84
  simple semi-conditional, 88
  simple semi-conditional ET0L, 396
  SSC-E0L, 396
  SSC-EP0L, 396
  SSC-EPT0L, 396
  SSC-ET0L, 396
  state, 184
  symmetric s-grammar, 212
grammar system
  multigenerative
    nonterminal-synchronized, 477
    rule-synchronized, 459, 476
  pure, 493
    controlled, 494
graph, 13
  transition, 570
Greibach normal form, 47

H

homomorphism, 19
  ε-free, 19
  inverse, 19

I

i-th component of an n-component parallel right-linear grammar, 514
identity of sets, 11

infinite
  language, 17
  relation, 11
  sequence, 11
  set, 10
inflection, 619
initial
  pushdown symbol, 33
  symbol, 513, 519
injection, 12
input
  alphabet, 32, 569, 589
  sentence, 622
  vocabulary, 622
input-revolving finite automaton, 660
insert(), 315
intersection
  of languages, 17
  of sets, 10
inverse
  homomorphism, 19
  relation, 11
irrealis form, 620

J
JFA, *see* jumping finite automaton
join(), 315
jump, 569
jumping
  finite automaton, 570
  relation, 569

K
k-limited workspace, 294
k-linear condition, 150
Kleene star, 18
Kuroda normal form, 41

L
labeled tree, 13
language, 17, 24
  accepted, 32, 34, 569, 590
  context-sensitive, 24
  empty, 17
  finite, 17
  generated, 21, 27–30, 57, 120, 156, 158, 161, 164, 185, 412
    using leftmost derivations, 25
    using rightmost derivations, 26
  linear, 24
  recursively enumerable, 22
  regular, 24
  right-linear, 25
  unary, 17
leaf, 13
leave, 13
left
  forbidding
    context, 194, 257, 412
    ET0L grammar, 412
    grammar, 195
    rule, 257
  jump, 580
  parse, 25
  permitting
    context, 194, 412
    ET0L grammar, 412
    grammar, 195
  random context
    ET0L grammar, 412
    grammar, 195
    nonterminal, 231
    rule, 194
left(), 380
left-extended queue grammar, 27
left-hand side, 21
leftmost
  derivation, 25
    of type-1, 245
    of type-2, 248
    of type-3, 253
  grammar system
    nonterminal-synchronized, 477
    rule-synchronized, 476
  symbol, 16
leftmost-applicable rule, 269
len(), 120
length of string, 16
lexical
  order, 623
  verb, 620
LF-E0L grammar, 412
LF-EP0L grammar, 412
LF-EPT0L grammar, 412
LF-ET0L grammar, 412
lhs(), 21, 120
lhs-replace(), 315
linear
  erasing, 19
  grammar, 24
  rule, 149
LL
  condition, 150
  context-free grammar, 269
  k-linear scattered context grammar, 150
  one-sided random context grammar, 270

LL(k)
  context-free grammar, 269
  one-sided random context grammar, 270
LMGN, *see* leftmost nonterminal-synchronized grammar system
LMGR, *see* leftmost rule-synchronized grammar system
lms(), 16
logical expression, 642
LP-E0L grammar, 412
LP-EP0L grammar, 412
LP-EPT0L grammar, 412
LP-ET0L grammar, 412
LRC-E0L grammar, 412
LRC-EP0L grammar, 412
LRC-EPT0L grammar, 412
LRC-ET0L grammar, 412

M
match, 642
matrix, 161
  grammar, 161
    with appearance checking, 161
  rule, 519
max(), 11
max-len(), 17
max-prefix(), 17
max-suffix(), 17
maximum context sensitivity, 127
mcs(), 127
member, 10
MGR, *see* rule-synchronized grammar system
microbiology, 616
min(), 11
modal verb, 620
module, 642
molecular genetics, 608
morphism, *see* homomorphism
morphology, 606
move, 32, 33, 590
multigenerative grammar system, 459, 476, 477
multigraph, 13

N
$\mathbb{N}$, 589
$\mathbb{N}_0$, 10
n-all-SFA, *see* n-turn all-move self-regulating finite automaton
n-all-SPA, *see* n-turn first-move self-regulating pushdown automaton
n-component pure grammar system, 493
n-first-SFA, *see* n-turn first-move self-regulating finite automaton
n-first-SPA, *see* n-turn first-move self-regulating pushdown automaton
n-generative rule-synchronized grammar system, 459
N-grammar, 658
n-language, 459, 476, 477
n-limited direct derivation, 185
n-LMGN, *see* leftmost n-generative nonterminal-synchronized grammar system
n-LMGR, *see* leftmost n-generative rule-synchronized grammar system
n-MGR, *see* n-generative rule-synchronized grammar system
n-parallel right-linear grammar, 513
n-pGS, *see* n-component pure grammar system
n-PRLG, *see* n-parallel right-linear grammar
n-right-linear simple matrix grammar, 519
n-RLSMG, *see* n-right-linear simple matrix grammar
n-turn all-move self-regulating finite automaton, 512
n-turn all-move self-regulating pushdown automaton, 527
n-turn first-move self-regulating finite automaton, 511
n-turn first-move self-regulating pushdown automaton, 527
nested word automaton, 660
nlrcn(), 231
node, 13
nominative, 621
non-identical sets, 11
non-modal verb, 620
nonterminal
  alphabet, 20, 194, 257, 412, 513, 519
  complexity, 224
nonterminal-synchronized grammar system, 477
nrrcn(), 231
null, 650
number
  of left random context nonterminals, 231
  of right random context nonterminals, 231

O
occur(), 16
ocs(), 127
0L grammar, 29
  parametric, 642

with context conditions, 642
with permitting conditions, 642
ond(), 175
one-ND rule normal form, 175
one-sided
forbidding grammar, 195
permitting grammar, 195
random context grammar, 194
one-turn atomic pushdown automaton, 559
operation
binary, 12
unary, 12
ordered tree, 13
output
sentence, 622
vocabulary, 622
overall
context sensitivity, 127
non-determinism, 175

P

pair, 11
pairwise disjoint, 11
paradigm, 620
parallel
right-linear grammar, 513
uniform rewriting, 439
parameter, 642
parametric
0L grammar, 642
with context conditions, 642
with permitting conditions, 642
word, 642
parse, 22, 331
parsing, 606, 616
partial function, 12
past participle, 620
Penttonen normal form, 41
perm(), 16
permitting grammar, 64
permutation
of language, 17
of string, 16
pGS, *see* pure grammar system
phrase-structure grammar, 20
plain form, 620
pop of pushdown, 588, 589
popping rule, 559
positive closure, 18
power, 4
of language, 17
of relation, 12
of string, 16
set, 11
predecessor, 642
predicate, 619
predicator, 619
Predict(), 269, 270
prefix
of string, 16
prefix(), 16
present form, 620
preterite, 620
primary form, 620
procedure, 22
production, *see* rule
programmed grammar, 164
with appearance checking, 163
propagating
context-conditional grammar, 56
context-free grammar over word monoid, 352
ET0L grammar, 29
generalized one-sided forbidding grammar, 267
left
forbidding grammar, 195
permitting grammar, 195
random context grammar, 195
matrix grammar, 162
with appearance checking, 162
one-sided
forbidding grammar, 195
permitting grammar, 195
random context grammar, 195
phrase-structure grammar, 21
programmed grammar, 164
with appearance checking, 164
pure grammar system, 504
regular-controlled grammar, 160
with appearance checking, 160
scattered context
grammar, 120
language, 120
selective substitution grammar, 28
proper
prefix, 16
subset, 10
suffix, 16
superset, 10
property, 4
PSCAT = CS problem, 123
pure grammar system, 493
pushdown
alphabet, 33, 589
automaton, 34
pushdown-bottom marker, 559
pushing rule, 559

Q
quadruple, 11
question tag, 630
queue grammar, 27
quintuple, 11
quotient
  left, 18
    exhaustive, 18
  right, 18
    exhaustive, 18

R
random context grammar, 64
range, 12
range(), 12
reachable state, 570
reading rule, 559
recursively enumerable language, 22
red alga, 635
reduction, 4
reflexive-transitive closure, 12
regular
  grammar, 24
  language, 24
regular-controlled
  grammar, 156
    with appearance checking, 157
  pure grammar system, 494
regulated
  automaton, 6
  grammar, 4
relation
  binary, 11
  finite, 11
  infinite, 11
  inverse, 11
  on, 11
  over, 11
restricted
  context-free grammar, 658
  erasing
    generalized, 312
  homomorphism, 19
reversal
  of language, 17
  of string, 16
rev(), 16
rewriting rule, *see* rule
rhs(), 21, 120
right
  forbidding
    context, 194, 257
    rule, 257
  jump, 581
  linear simple matrix grammar, 519
  parse, 26
  permitting context, 194
  random context
    nonterminal, 231
    rule, 194
right-hand side, 21
right-linear
  grammar, 25
  language, 25
rightmost derivation, 25
rms(), 16
root, 13
rule, 21, 28, 29, 32, 33, 56, 119, 164, 412, 493, 513, 559, 589
  conditional, 56
  erasing, 21
  label, 21, 32, 34, 493, 559
  of depth m, 590
  popping, 559
  pushing, 559
  reading, 559
rule-based regulation, 5
rule-synchronized grammar system, 459, 476

S
s-grammar, *see* selective substitution grammar
sc-grammar, *see* semi-conditional grammar
scattered context
  grammar, 119
  language, 120
second Geffert normal form, 42
secondary form, 620
selective substitution grammar, 28
selector, 28
self-regulating
  finite automaton, 511
  pushdown automaton, 526
semi-conditional grammar, 84
semi-parallel uniform rewriting, 432
sentence, 21
sentential
  form, 21
  n-form, 459, 476, 477
septuple, 11
sequence, 11
  of rule labels, 22
  of rules, 22
sequential uniform rewriting, 47
set, 10
  empty, 10
  finite, 10
  infinite, 10

sextuple, 11
SFA, *see* self-regulating finite automaton
shuffle, 18
shuffle(), 18
simple semi-conditional
  ET0L grammar, 396
  grammar, 88
SPDA, *see* self-regulating pushdown automaton
split(), 316
SSC-E0L grammar, 396
SSC-EP0L grammar, 396
SSC-EPT0L grammar, 396
SSC-ET0L grammar, 396
ssc-grammar, *see* simple semi-conditional grammar
start
  configuration, 493, 559
  from anywhere, 582
  from the beginning, 582
  from the end, 582
  pushdown symbol, 589
  state, 32, 569, 589
  string, 29, 30, 493
  symbol, 20, 28, 56, 120, 184, 194, 257, 367
state, 32, 184, 569, 589
  grammar, 184
  reachable, 570
  terminating, 570
state-controlled
  finite automaton, 533
  language, 533
string, 16
strings(), 185
sub(), 16
subject, 619
subset, 10
  of language families, 20
substitution, 19
  finite, 19
substring, 16
subtree, 13
success field, 164
successful
  derivation, 21
  n-limited generation, 185
successor, 642
suffix of string, 16
suffix(), 16
superset, 10
surjection, 12
sym(), 16
symbiotic E0L grammar, 446
symbol, 16
  activated, 28
symbol-exhaustive
  left quotient, 18
  right quotient, 18
symmetric s-grammar, 212
syntax, 617
Szilard word, *see* parse, *see* parse

T

terminal
  alphabet, 20, 28–30, 56, 119, 184, 194, 257, 367, 513, 519
  derivation, 21
terminating state, 570
total
  alphabet, 21, 28–30, 56, 119, 164, 184, 194, 257, 367
  function, 12
  vocabulary, 622
transformation, 4, 622
transformational scattered context grammar, 622
transition, 32, 33
  graph, 570
transition-controlled
  finite automaton, 533
  language, 533
transitive closure, 12
tree, 13
triple, 11
tuple, 11
turn, 559
  state, 511, 526
type-1 leftmost derivation, 245
type-2 leftmost derivation, 248
type-3 leftmost derivation, 253

U

unary
  alphabet, 16
  language, 17
  operation, 12
uniform rewriting
  parallel, 439
  semi-parallel, 432
  sequential, 47
uniformly limited ET0L grammars, 659
union
  of languages, 17
  of sets, 10
universal language, 17
universe, 10

V
verb phrase, 619

W
walk, 13
wm-grammar, 352
WME0L grammar, 446
word, *see* string
workspace
    of derivation, 294
    of string, 294
    theorem
        for phrase-structure grammars, 26
        for regular-controlled grammars, 309

Z
zero moves, 32, 35

Printed in the United States
By Bookmasters